WORLD RAILWAY ACCIDENTS 1991-2005

Francis Voisey

ISBN: 978-1-9162011-3-2

This book is dedicated to signalman Brian Stone, who first awakened my lifelong interest in railways way back in 1966.

Thanks to the staff of the Exeter Reference Library and the University of Exeter Library for their assistance while I was researching this book, to my friend June Nightingale for encouraging me to go ahead and get it finished when my enthusiasm was waning a little, and to my publishing team at KINDLE BOOK PUBLISHING, without whose guidance and assistance it is doubtful if I would have been able to navigate the mysteries of getting my first book into print.

CONTENTS

INTRODUCTION

This book is the ninth in a series of volumes detailing railway accidents through the years, worldwide. A good many books have been written by various authors about railway accidents, some concentrating on single disasters, others covering accidents in a given country or geographical area. Accidents in countries such as Great Britain, Germany, the Netherlands, the United States, Australia, New Zealand and so on have been documented comprehensively in one or more books but no attempt has been made previously to detail every significant accident that has occurred over two centuries, worldwide. This is clearly an ambitious project, but eight volumes dealing with the earlier years having been completed. This book will attempt to detail the accidents that occurred between 1991 and 2005. Previous volumes have covered the periods 1811-80, 1881-1900, 1901-15, 1916-30, 1931-45, 1946-60, 1961-75 and 1976-90. One more volume is in preparation to bring the story up to the present day. As before, an attempt will be made to include all accidents involving five or more fatalities to train passengers. A good number of less serious accidents, which have been considered to be of interest, have also been included. The accidents have been grouped into chapters according to their primary causes and are dealt with, in general, chronologically within each chapter.

The gathering of information about accidents in distant parts of the world has not been easy, though it is much simpler nowadays than when this project began in the days before the internet. Outside of Great Britain and Ireland few countries seem to have had a formal body charged with investigating all serious railway accidents until recent years, though one or two major disasters, such as the Bussey Bridge failure in the United States and the Baptiste Creek collision in Canada, were the subject of full formal investigations even in the early days. Otherwise reliance has largely had to be placed on contemporary newspaper reports and previously published works on the subject, though the author has tried as far as possible to keep to primary sources. If the book appears to be biased towards accidents in certain countries such as Great Britain, the United States and India, this is probably partly because these are the countries about which information on accidents has been easiest to obtain, though these were also in any case the countries where many of the most serious accidents occurred.

Contemporary spellings of place names have been used where appropriate and for the many American and Canadian accidents, terms such as railroad, engineer, caboose and so on have been used instead of railway, driver, brake van etc. By the period covered in this volume the 24-hour clock was mostly in use, except in the United States. The 12-hour and 24-hour clocks will both be used depending upon the custom in the country concerned. Distances likewise will be given in miles and yards/feet or kilometres and metres depending on the practice in the country concerned.

By 1991 railways had spread to almost all parts of the world, though some small countries such as Malta and Trinidad had by that date lost theirs and in many countries unprofitable lines had been closed.

The names and the boundaries of countries have undergone many changes over the years. The names of various places have also changed. The names of locations in use at the time an accident occurred have been given, together with present names where appropriate, as for example Bombay (now Mumbai). Local spellings of place names have also been used where it has been felt appropriate, for example Roma for Rome.

In order for the record to be as complete as possible, included at the start of several chapters of this volume are accounts of some accidents from earlier days, information upon which has only come to light since the publication of the volume in question.

CHAPTER 1

DERAILMENTS: TRACK DEFECTS

The tracks over which trains run are obviously a vital part of the railway and any problems with
The tracks over which trains run are obviously a vital part of the railway and any problems with the track are likely to lead to the train being derailed. By 1991 the technology involved in laying railway track had much advanced since the early days, but train speeds and weights had also much increased, while accidents occurred from time to time due to failures to keep the track in a good state of maintenance, to insufficient allowance being made for hot weather expanding the rails, to damage caused to the track from floodwaters and so on. Also included in this chapter, though they were not strictly due to defects in the track, are accidents due to trains having become derailed on open sets of points. Broken rails were another hazard that caused mishaps and worse from time to time.

Train No. 2311 of the Northern Railway, the Up Howrah – Kalka Mail, travelling at 106 km/h, was derailed at 08:55 on Friday 15th February 1991 at Karchana station, Uttar Pradesh, India. Fifteen people lost their lives in the accident and 76 were injured, 28 of them seriously.

The investigation into the accident noted that Karchana was a B Class station on the double line with Standard III interlocking and multiple-aspect colourlight signalling, controlled from East and West cabins, each working Outer and Inner Distant, Home, Starting and Advanced Starting signals. The points and locked bars were operated by rodding from the cabins and there was track circuiting between the Starters and Advanced Starters. There was Absolute Block Working with Lock and Block instruments, the cabin instruments being slotted from the stationmaster's office.

Driver Raziul Rahman of train No. 2311 told the inquiry into the accident that his train had left Mughalsarai at 06:58. The signals at Karchana had all been Green and he had been preparing to exchange signals with the East cabin when his engine had jumped at a set of facing points, he had applied the brakes in emergency and the train had become derailed. Sending the assistant driver to the assistant stationmaster's office to protect the other running lines he had gone to the East cabin but the switchman had not been there. He had then hidden in a field, fearing violence from a crowd that had collected at the station.

Diesel Assistant Rajendra Singh corroborated his driver's evidence and stated that at the assistant stationmaster's office he had heard a permanent way inspector asking why the train had been allowed to run through, when he had been called to check the points. No one had been at the points when their train reached them, though a permanent way gang had been nearby.

Driver Mewalal of train No. 5114 stated that his train had been stopped at the East cabin Home signal from 07:11 to 07:16, before passing through safely, with no jerk at the facing points leading to the Up loop.

Driver Bajrang Kumar of train DBSI 975 likewise testified that he had passed through Karchana at 08:34 with Green signals and no jerk at the facing points.

Guard S.L. Sonker of train No. 2311 stated that his train had stopped with a snatch at 08:57. His seating-cum-luggage rake had been one of three coaches that had stopped outside of the facing points. He had gone first to the assistant stationmaster's office, then to the East cabin. Like his driver he had found no one there; levers 9, 11 and 12 had been pulled.

Karchana Assistant Stationmaster (ASM) Asha Ram Shukla stated that he had given permission for train No. 5114 Up to run through the station. Then at 07:12 the East cabin switchman had informed him that the Normal indication of his facing points was not lit. The electrical signal maintainer had been at the station, had been told of the problem, had taken the relay room key and had gone away. The East cabin switchman subsequently informed him (the ASM) that the facing points indication had returned and trains Nos. 5114, 4083 and DBSI had passed through the station safely. Permission had then been granted for train No. 2311 to pass through but instead it had become derailed at the East cabin facing points leading to the Up Loop.

In answer to questions the ASM stated that normally the giving out of the relay keys was entered in the register but a mistake had been made. It was not correct to run trains through when the relay room key was out but he had thought that the facing points had been properly set and locked. There had been no failure of the signalling during the night.

Switchman Ramji stated that he had been on duty in the East cabin. There had been no difficulties up until his failing to get a points indication before the arrival of train No. 5114, when he had informed the ASM, Electrical Signal Maintainer (ESM) Nageshwar Prasad had arrived and had gone to the relay room. Some time later the points indication had returned and trains Nos. 4114, 4083 and DBSI had passed through safely. The ESM had still been in the relay room when train No. 2311 had arrived and become derailed. After the accident he and the ESM had left the scene to avoid angry passengers.

In answer to questions Switchman Ramji testified that it had been after receiving the Up Badrapur train on the Up Loop and resetting the points for the Up Main that the points indication had failed. He had then cycled to the points and checked that they were correctly set. On returning he had found the ESM doing something in the relay room and told him that he ought not to be, but the ESM had stated that he had the authority of the ASM, which the latter had confirmed. He repeated that the ESM had been in the relay room at the moment of the derailment.

Switchmen Dhani Ram and Shyam Krishna Malviya from Karchan West cabin, Section Controller Bhagwal Giri at Allahabad and Chief Traction Loco Controller A.P. Chakrabarty gave evidence.

Electrical Signal Maintainer Prasad stated that he had been with his khalasis (workmen) Indramani Kushawaha and Om Prakash Patel and had not been to the East cabin before the accident.

Khalashi Kushawaha (who was stated to have appeared terrified at the inquiry) agreed that he had been with the ESM in the tools room before the accident. He refused to say any more. At the inquiry he had at first tendered what he maintained was his written statement, a document that had clearly been written by someone else. Khalasi Patel gave similar evidence.

Mechanical Signal Maintainer Rajman at Meja Road, Electrical Signal Maintainer Ram Chander at Meja Road, Mechanical Signal Maintainer (Night) Harbanah Singh at Meja Road, and Signal Inspector A.K. Chakrabarty gave evidence.

Signal Inspector Mohammad Nasiruddin described examining the facing points after the accident and seeing, through the window of the East cabin, which had then been locked up, that levers Nos. 9, 11 and 12 were pulled over. He described the damage to the points, which he believed had caused the derailment. Assistant Signal and Telecommunications Engineer Atul Kumar gave similar evidence.

Assistant Engineer A. Pandey stated that he had been riding on the engine of train No. 2311. When the derailment occurred he had left the engine along with the driver, diesel assistant and Permanent Way Inspector (PWI) V.K. Singh of Mirzapur, who had also been riding in the cab. When asked how the derailment had occurred Driver Ramnan and PWI Singh said that the points had not been properly set, causing the train to take two roads, derailing 11 coaches. There had been no one in the East cabin and communications had been down; a railwayman in the ASM's office had been trying to contact Control. He had heard the Karchana PWI say that ESM Prasad had been talking to his two khalasis near the East cabin and saying that there had been a gap in the facing points. The relay door had been open at this time. The ESM had gone towards the ASM's office, telling his khalasis not to say anything about the matter.

Permanent Way Inspector Singh stated that approaching Karchana he had seen a gap between the left-hand tongue rail and the stock rail and wondered how the signals could be Clear with the points standing open, when the derailment had occurred. After the accident he had seen that the left-hand tongue rail had been in the set position. The fishplates linking the tongue rail and lead rail had separated and the outer fishplate had been missing. All the bolts had been broken. He had not noticed the position of the facing point lock.

Permanent Way Inspector A.K. Kohli at Karchana testified that coming towards the East cabin he had heard ESM Prasad talking loudly with his khalasis, who had been coming from No. 8 points (the facing points involved), saying that they were not set properly and there was a gap in them. He (Kohli) had said that the signals should not be lowered but the switchman had replied that they were already lowered and a train was approaching. He (Kohli) had then witnessed the derailment and hurrying to inform the ASM, he had seen the ESM locking the door to the relay room. The ESM had then also gone to the ASM's office with his khalasis, warning them not to say anything about the matter.

In answer to questions the PWI added that work on the Up Loop involving deep screening and sleeper renewal had been in progress since 4th February. The work usually began at 09:00 after a 'block' for 6-7 hours had been obtained. He had not checked the levers in the East cabin after the derailment. No one had complained to him about the points.

Permanent Way Inspector A. Chakravorty at Meja Road stated that he had been deputed to preserve evidence and had prepared sketches, made track readings and so on.

Trolleyman Chhote Lal at Karchana spoke of seeing ESM Prasad and his khalasis near the East cabin, the upper and lower doors of which had been open. Later he had seen the khalasis running towards the ESM, saying that the "JHIRI," the Kalka Mail, had derailed.

Trolleyman Hari Shanker gave evidence that he had also seen the block fitter (ESM Prasad) running towards the East cabin, probably because there was something wrong with points.

Gangman Paras Nath spoke of seeing PWI Kohli and ESM Prasad talking to one another near the East cabin before the derailment. The switchman had been in his cabin and the relay room door had been open.

Gangman Mithai Lal, Permanent Way Mistry (foreman or supervisor) Dinesh Kumar, Mate Jagdish Bahadur, Gangman Baharat Lal, Gangman Bajrangi Lal, Gangman Mitthu Passi, Gateman Ram Karan, Government Railway Police Inspector Babu Lal Ojha, Railway Protection Force Inspector S.N. Singh and Railway Medical Superintendent Dr. P.D. Srivastava all tendered evidence.

Commissioner of Railway Safety K.J.N. Kutty detailed tests and observations made at the site.

Discussing the accident, the commissioner fixed the time that it had happened at 08:54 and stated that the speed of the train at the moment of derailment, from the Hasler recorder, had been 106 km/h.

Driver Rahman, Diesel Assistant Singh and Assistant Engineer Pandy had, the commissioner continued, no particular reason to look at the facing points and thus had not noticed that they were gaping but PWI Singh had seen that they were standing partly open. PWI Kohli had seen the khalasis running from the facing points shouting to ESM Prasad that they were open and while he (the PWI) had been remonstrating with Switchman Singh the derailment had occurred. The switchman denied speaking to the PWI but admitted clearing his signals for train No. 2311 while the ESM had been in the relay room, while the ASM admitted that the ESM had taken the keys to the relay room when there had been a problem, before the arrival of train No. 5114, and that the keys had only been returned after the accident. An engineering department keyman had seen the ESM doing something near the points and detector box.

The ESM and his khalasis vehemently denied having been near the points before the derailment but could not produce any witnesses to say that they had been elsewhere.

Three mechanisms could be propounded to account for the derailment:

(a) That there had been an adequate gap between the left switch and stock rails for the locomotive flange to enter it and thus lead to a "no road" derailment.

(b) That the derailment had occurred due to the absence of the heel block, distance block and one fishplate, allegedly removed by the engineering gangs without a proper 'block.'

(c) That there had been an obstruction on the rail, perhaps some tool left by a workman. A crowbar subsequently recovered from a Gangman of Permanent Way Gang No. 18 was suspected to have been involved.

From the evidence it was clear that the engine had derailed first.

Given the time of day and the generally good standard of the track, causes such as sabotage and track defects could be excluded. There was no evidence of any component having fallen from the train, of a defect in the rolling stock, or of a broken rail.

After a detailed examination of the possibilities the commissioner decided that the missing track components and obstruction (crowbar) theories could be dismissed and it was reasonably clear that there must have been a "no road" derailment of the left wheels of the locomotive. If this was the case it should not have been possible to clear the signals for the train, the rear coaches of the train should not have been able to take the correct route (as they had) and the points should not have been correctly set, locked and undamaged after the accident.

Everything pointed, Commissioner Kutty stated, to ESM Prasad and his khalasis having been in the relay room of the East cabin at the time of the derailment and that the points had been disconnected and unlocked at the time that train No. 2311 was approaching, due to unauthorised inference by the signalling staff, in collusion with the train passing staff. It was easy to get the signals to show false Clear aspects by manipulating the relays.

A Northern Railway letter dated 6th October 1988 stated that when a signal failure was observed in the face of an approaching train, trains should only be dealt with by piloting. A joint circular of 1st February 1990 ordered that when the ASM's relay room key had been removed, the ASM must be extremely vigilant, and take all possible precautions in the reception/dispatch of trains to prevent any accidents occurring. In all unusual cases of signal or points failure the points must be clamped and padlocked. In the current case the ASM had allowed the ESM to take the relay room keys while train No. 2311 was approaching, in violation of the instructions, so as to avoid delay, an apparently not uncommon practice. The signalling staff had also violated instructions by beginning work without the issue of a

Disconnection Memo, which would have forced the operating staff to take the necessary precautions.

The commissioner concluded that the derailment had occurred through the locomotive taking "no road" at the first facing points, which had not been set and locked as required, the signals being taken off through unauthorised procedures.

Primarily responsible for the accident were ESM Prasad and Khalasi Kushwaha. As he had only been appointed three weeks earlier Khalasi Patel was responsible to a lesser degree. ASM Shukla and Switchman Ramji had contributed to the accident.

Under remarks and recommendations Commissioner Kutty noted that there had been at least seven accidents on the Northern Railway due to unauthorised interference by signalling staff between April 1988 and December 1990. Instructions issued by the Railway Administration to eradicate such practices needed to be implemented forcefully.

Failures of the signalling and interlocking systems were apparently much too frequent, leading to shortcuts and/or unauthorised procedures to avoid delays. There needed to be an all-out effort to upgrade the signalling and interlocking to reduce the number of such failures.

Arrangements facilitating the quick and easy delinking of mechanical interlocking systems should be reviewed to prevent such facile delinking. [1]

At about 03:30 Cuba Standard Time a passenger train from Habana to Guantanamo was derailed just after passing through Santo Domingo, Cuba, in Villa Clara province, on Saturday 6th April 1991. Six or seven coaches were derailed, according to different reports, and at least 56 people lost their lives and 240 were injured.[2] Preliminary reports stated that the track was in poor condition, implying that track defects had caused the derailment. [3]

On Wednesday 31st July 1991, Amtrak passenger train No. 82, the *Silver Star*, bound from Tampa and Miami, Florida, to New York, New York, was derailed at Lugoff, South Carolina, United States, its rear six coaches leaving the track. Of the 407 passengers on board eight were killed, 12 were seriously hurt and 53 were slightly hurt. Twelve of the 16 on-board service crew members were also slightly hurt but none of the six train crew.

The train consisted of two diesel electric locomotives, three baggage cars and 15 passenger cars. The derailment occurred at 5:01 a.m. at a set of facing points (the Orlon crossover) leading to Du Pont siding, serving the E.I. Du Pont plant, the derailed cars colliding with nine hopper cars stabled on the siding.[4] The accident was dealt with in a handful of papers, some of which stated that 'at least 13 people were killed and 70 seriously injured.' Passenger Stephen Clark from Philadelphia, Pennsylvania, was interviewed by a reporter. He related that the person sat next to him had died and added, "It was devastating. It was awful," and, "I'm breathing, praise the Lord." Lugoff High School, the paper informed its readers, had been set up as a shelter for the injured. [5]

The National Transportation Safety Board (NTSB) investigated the accident. Its report stated that three minutes after the derailment the engineer attempted to radio the CSX Transportation

---

[1] Government of India Ministry of Civil Aviation (Commission of Railway Safety) Railway Accident Investigation Report on Derailment of 2311 Up Kalka Mail at Karchana Station on Northern Railway on 15th February 1991.

[2] *Port Arthur News*, Port Arthur, Texas, United States, 7th April 1991

[3] *The Hawk Eye*, Burlington, Iowa, United States, 7th April 1991

[4] National Transportation Safety Board report

[5] *New Castle News*, New Castle, Pennsylvania, United States, 31st July 1991

dispatcher in Jacksonville, Florida, but although the dispatcher could hear the engineer, the latter could not hear the dispatcher, because of a faulty transmitter board at Lugoff. Eventually, at 5:08 a.m., the engineer was able to speak to staff at Cayce Yard instead. This call was overheard by the Jacksonville dispatcher, as was the engineer giving the location of the accident as "the Lugoff crossover," and the assistant dispatcher called the emergency services at 5:12 a.m. The Kershaw County Sheriff's Department located the accident at 5:24 a.m.

Train No. 82, the NTSB report noted, consisted of two trains: No. 82, from Tampa, formed of engine No. 403 and eight cars, plus train No. 92, from Miami, formed of engine No. 257 and ten cars; these trains had been combined at Auburndale, Florida. The combined train was in charge of an unnamed conductor with Engineer T.A. Hill, Qualifying Engineer C.P. Peterson and Fireman J.Q. Palmer.

The track at the site of the accident consisted of 115·25 lb continuous welded rail with a road bed of crushed granite. The Orlon crossover consisted of a number 10 turnout with two 16-foot, 6-inch switch points and a number 10 frog, and had been installed in 1962. After the accident, in the words of the report, 'the main track crossover was found to have the connecting rod disconnected from the switch stand crank. The switch point was not secured to the stock rail, the cross pin that attached the switch stand crank to its spindle was not in place, and the crank had dropped onto the safety plate. The cross pin was found near the switch stand.'

Track inspection records indicated that in the three months leading up to the accident the switch had been examined 55 times, four times by Roadmaster L.E. Bowers, 44 times by Track Inspector J.R. McLain and Trackman Q.A. Martin and seven times by other employees. The last annual inspection of the switch had been on 26th October 1990, though the switch had not then been taken apart, as per the standing instructions for the annual inspection. Inspection records for the three months preceding the accident did not record any defects. Track Inspector McLain said that every now and then there were complaints from train crews that the switch was hard to operate, in response to which the points were cleaned and lubricated with graphite. Roadmaster Bowers stated that he had not noticed a broken cross pin, found after the accident by NTSB investigators, when he had made his last inspection on 8th July and that at that time the switch had worked normally. Trackman Martin was the last person to examine the switch at about 10:15 a.m. on the day before the accident, when he said he had operated it without difficulty.

Analysing the accident, the NTSB noted that the operation of the train had not contributed to it and the signalling had functioned correctly, while the switch stand had functioned as designed. Describing the course of the derailment, the NTSB surmised that the train, travelling north, had passed over Lachicotte Road Crossing and arrived at the switch, which the two engines and first 12 cars had passed over safely, before the connecting rod of the switch had disengaged, allowing the points to open and derailing the last six cars. Trains had passed through the switch, with its cross pin apparently missing, for some time before it had finally failed completely and caused the accident.

In the weeks before the accident the switch had been examined a number of times by the roadmaster, a track inspector, a signal supervisor and others, none of whom had noticed the broken cross pin, cross level deficiencies or a nail used instead of a cotter pin. 'The inspectors could and should have seen the switch deficiencies during a normal inspection and, with appropriate action, could have prevented the accident.'

The NTSB report continued, 'Although the track inspector, the trackman and the roadmaster were sufficiently trained and experienced to do the inspection and maintenance, the condition of the switch indicated that the CSXT and FRA {Federal Railroad Administration} procedures had not been followed. If the switch had been inspected and maintained to comply with the CSXT and the FRA requirements, such as the requirement that the connecting rod must be securely fastened, the worn, broken, fouled and missing parts would have been noted and corrected, and the accident probably would not have occurred. These deficiencies indicate that the maintenance and inspection procedures of the CSXT on this roadmaster's territory before the accident were inadequate and ineffective. The roadmaster and the track inspectors stated that they were not doing the track and switch inspections as required because the track surface repairs and other maintenance needs that they found during the day did not allow them enough time to do that work and to complete the required inspection schedule.'

Detailing the lack of time available for proper track inspection, the NTSB wrote, 'A CSXT inspector needs 8 hours 4 minutes each day to inspect switches and track (5 hours 44 minutes and 2 hours 20 minutes respectively). This time expenditure allows no time for other activities, such as travelling to and from inspection areas, clearing the track for trains, or normal maintenance and repair. Because an inspector must engage in these other activities, the time he can spend on inspections would be depleted accordingly. The roadmaster has an additional and time consuming responsibility, checking the quality of the inspections done by the track inspector and the trackman.

'Initially, the track inspector stated that he had adequate time to fulfil his inspection requirements and that he could meet the requirements if he did not have anything else to do. He noted, however, that often he had something else to do, such as tamping, track surface smoothing, or other duties that were unknown when he began an inspection trip. Thus, he qualified his earlier statement and said that taking care of maintenance and other duties did not leave him enough time to get his inspections done and that this had been the situation for about 3 or 4 years. He attributed the situation to the elimination of some positions and the reassignment of personnel. He also reported that he did not do a detailed inspection each time he inspected a switch, but that the monthly inspections were more detailed and thorough. He added that as long as the switch threw well and the points fitted up, he did not generally check everything. The track inspector's statements that he did not have sufficient time to adequately do his work, and that during his inspections he checked only whether the switch points fitted properly, indicate a definite need for an evaluation of his work schedule.

'The roadmaster reported that he worked 10 or 11 hour days to complete his duties. Even working these hours, he could not complete the 50 per cent of inspections for which he was responsible. The reports for the 3 months before the accident showed that he had done only 4 (7 per cent) of the 55 recorded track inspections. His inability to do half of the inspections increased the number that the track inspector and the trackman had to do, and they already had more work than they could manage. The roadmaster stated that although the assigned manpower for his territory had been reduced by half, the time for inspection remained adequate. According to the CSXT, automation and mechanisation of the work have compensated for the reduction of the work force.'

The report noted that after the accident the public address system had not been used to locate any passengers who had medical training, while the on-board train staff had not been required to have first-aid training, as recommended by the board in its report on a train fire at Gibson, California, in 1982.

The emergency response had been delayed by seven minutes due to inadequate information as to the location of the train, the dispatcher's telephone list of emergency responders being out of date, the dispatcher giving an incorrect phone number to the sheriff, the non-functioning transmitter at Lugoff and the use by the dispatcher of railroad jargon, which had confused the emergency personnel.

Although drug or alcohol use by the train crew had not been suspected as having been a factor in the accident, timely post-accident toxicological testing had not been done.

The NTSB detailed nine conclusions regarding the accident:

(a) The train operation, the signal system, the weather, the design of the switch stand and the passenger train equipment had not been factors in the accident.

(b) The switch stand cross pin had fractured and been missing from the spindle and crank for some time before the accident.

(c) The CSX Transportation Inc. track maintenance and inspection process had not been adequate to detect and correct in a timely manner the problem that existed in the Orlon crossover switch.

(d) The Orlon crossover switch had not been properly maintained for some time before the accident.

(e) The on-board service crew members had failed to follow appropriate established emergency procedures, such as using the public address system to inform passengers about the emergency and give related instructions.

(f) The on-board service crew members had not been required to attend periodic training in first aid or emergency procedures, such as the use of the public address system to locate passengers who had medical expertise and who might have been able to render assistance.

(g) Had new instructions on emergency radio procedures been published in the CSX Transportation Inc. timetable, the locomotive crew might have broadcast a more detailed description of the train location and eliminated the confusion and delay.

(h) The inability of the dispatcher to respond over the Lugoff base radio station because of the transmitter board failure had slowed the emergency response effort.

(i) The post-accident alcohol and drug testing had not been conducted soon enough after the accident to provide meaningful test results.

The NTSB found that the probable cause of the accident had been the opening of the switch points under the Amtrak train because of a poorly maintained switch, as a result of inadequate track inspections, switch maintenance and management oversight.

The NTSB made six recommendations:

To CSX Transportation Inc.:

(a) It should review and revise, as necessary, existing practices to ensure that track supervisors reviewed their subordinates' track inspections and that switch inspections were adequately documented.

(b) It should review and revise, as necessary, manpower schedules for track and switch inspections to ensure that the track and switch standards of the Federal Railroad Administration and the CSX Transportation Inc. could be met.

(c) It should maintain an up-to-date emergency response telephone list.

(d) It should instruct dispatchers on the use of terminology that could be readily understood by local emergency personnel, when advising them of train locations after an accident.

(e) It should revise post-accident alcohol and drug testing procedures to ensure timely specimen collection.

To the National Railroad Passenger Corporation {Amtrak}:

(a) It should require that all on-board service personnel periodically took training in the emergency operating rules and in first-aid, cardiopulmonary resuscitation, and the use of the public address system during train emergencies. [6]

The *Tirumala Express* of the South Central Railway, train No. 7488 Up from Visakapatnam to Tirupati, was derailed at 02:55 on Saturday 21st March 1992 at Km. 230/23 between Tettu and Kavali stations, Andhra Pradesh, India, on the Vijayawada – Gudur broad-gauge electrified section of the railway. Five people were either killed outright or died of their injuries and another 68 were hurt, 12 seriously. [7]

Driver T. Vijayakumar told a newspaper reporter that his train had left Vijayawada at 23:20 on Friday and had been running at between 90 and 100 km/h at the moment of the derailment, when he had felt a sudden jerk and the brakes had come on automatically as he had seen some coaches behind him overturn. He said that, "Everything happened in seconds." He had not seen any damage to the track or fishplates having been removed and did not think the derailment had been due to sabotage.

Thirteen of the 21 coaches of the train were derailed and the S8, S9 and S10 coaches fared the worst. The accident was fortunate inasmuch as the coaches overturned on the western side of the track. Had they overturned on the eastern side, they would have fallen down a 30ft ditch.

Trains were cancelled or diverted after the accident while the uninjured passengers on the express were forwarded to Tirupati on buses. [8]

An investigation into the accident determined that it had been caused by a broken rail. The Commission of Railway Safety report on it contained three major recommendations:

(a) The instructions laid down in the Long Welded Rails manual should be strictly observed.

(b) The standard of inspections by officers at different levels should be upgraded. The supply of magnifying glasses and thermometers to permanent way gangs should be ensured.

(c) All flash butt welded joints done in welding plants and all thermit welds carried out in the field should be tested.[9]

Negligence was apparently the cause of what at the time was probably the most serious railway accident in the history of South Korea, which occurred at Gupo station, near Pusan, on Sunday 28th March 1993 (ten years later a considerably greater catastrophe would occur in Daegu Tunnel – see Tunnel Accidents chapter).

The 1993 accident had its origins in work undertaken by the Korea Electric Power Corporation in connection with the laying of underground electric cables. The work of laying the cables had been in progress for two years. A short time before the arrival of a Mugunghwa-ho passenger train,[10] dynamite was used in the construction of a tunnel 15 metres below the railway tracks. No notice had been given to the railway of the intended work. Shortly after the explosion had been set off, a crowded express arrived at the location, ten minutes away from the end of its five-hour journey from Seoul. Driver Loh Jin Hwan, 32, who escaped with slight injuries, said afterwards, "I saw the ground crumbling away in front of me and immediately applied the brakes but it was too late." A hole about 5 metres deep and with a

[6] National Transportation Safety Board report

[7] Annual Report for 1991-92 by Chief Commissioner of Railway Safety, Lucknow

[8] *The Indian Express*, Madras, India, 22nd March 1992

[9] Government of India Ministry of Civil Aviation and Tourism Commission of Railway Safety Annual Report for 1991-92

[10] Wikipedia, Mugunghwa-ho

diameter of about 15 metres had opened up in front of the train, into which its engine, power car and leading carriage fell, while the next three carriages overturned.

The train had a seating capacity of 500 but was carrying about 700 passengers, so many were standing, and the death toll was correspondingly high. By Tuesday the number of known dead had risen to 79 and it was feared that it could rise further, as many of the 105 injured were in a serious condition. About 1,500 rescue workers laboured amongst the twisted wreckage in driving rain and some of the injured reportedly had to lie about on the muddy ground for several hours before they could be taken to hospital. The rain-sodden earth had possibly also contributed to the track subsidence. [11] [12]

Northern Railway train No. 3049, the Up Howrah – Amritsar express, was derailed at Km. 1441/3 between Mewa Nawada and Seohara stations, Uttar Pradesh, India, on the Moradabad – Saharanpur broad-gauge double-line non-electrified section of the Moradabad Division of the railway at 00:20 on Wednesday 20th July 1994. The driver of the diesel locomotive pulling the 13-coach train felt a heavy jerk and the train came to a halt between Km. 1441/3 and Km. 1441/12. The un-derailed engine and derailed two leading coaches stopped about 250 metres ahead of the third to seventh coaches, which were capsized, the eighth to tenth coaches, which were derailed, and the rear three coaches, which remained on the track. Seven people died and 15 were injured, 11 seriously. The accident was apparently due to a broken rail. [13]

A passenger train from Świnoujście to Kraków was derailed at Reptowo, northern Poland, at about 12:25 on Monday 5th May 1997.[14] The rear four or five coaches of the train left the rails as it was travelling at 120 km/h and collided violently with a freight train standing on an adjoining line. Police spokesman Krzysztof Rusinowicz stated that, "The impact was so powerful that one carriage was torn in half and another was completely destroyed." Twelve people lost their lives and 30 were injured, half of them seriously. An air ambulance took the most seriously injured to hospital. Police Inspector Stanislaw Bukowski blamed the derailment provisionally on a defective set of points, while a commission of inquiry was appointed to investigate the disaster. Deputy Prime Minister Jarosław Kalinowski went to the scene of the accident. [15]

The danger of multiple tracks is of course that if a train on one line suffers an accident, which then results in part of it fouling another line, there is a chance that the initial accident will be followed by a collision with a train on that line, leading to more serious consequences. Such was the case on the Northern Railway in India on Thursday 26th November 1998. Train No. 2903, the Up *Golden Temple Mail* (formerly the *Frontier Mail*), had passed through the station at Khanna in the Punjab at 03:10. A minute earlier train No. 3152, the Down Jammu Tawi – Sealdah express, had passed through the station at Chawapail, Punjab, almost 11 kilometres away, running 18 minutes late. At about 03:13 the tenth to 18th coaches from the engine of train No. 2903 were derailed, due, it was later found, to a broken rail, and they fouled the opposite running line almost directly in front of train No. 3152, the engine and seven coaches of which were derailed in turn in the resulting collision. Driver Ashwini Kumar and Assistant Driver Satbir brought their engine and the leading coaches of train No. 2903 to a

---

[11] *European Stars and Stripes*, Darmstadt, Germany 29th and 30th March 1993
[12] *Pacific Stars and Stripes*, Tokyo, Japan, 30th and 31st March 1993
[13] Annual Report for 1994-95 by Chief Commissioner of Railway Safety, Lucknow
[14] *The Medicine Hat News*, Medicine Hat, Alberta, Canada, 6th May 1997
[15] *The Times*, London, England, 6th May 1997

stand three kilometres down the line.

There had been an estimated 2,500 passengers on the two trains and early on, it was known that the results of the accident were very serious, an initial report speaking of 150 dead and another 250 injured. This proved not to be an exaggeration; the final figures were 212 dead and 138 injured. Driver Subhash Chander and Assistant Driver Sewa Ram Murti of train No. 3152 were among those who lost their lives. Medical Relief vans and Accident Relief trains were sent from Ambala and Ludhiana and the injured were taken to a number of hospitals: the Post Graduate Institute in Chandigarh, Rajendra Medical College Hospital in Patiala, Dayanand Medical College and the Christian Medical College hospitals in Ludhiana and the civil hospitals in Ludhiana, Samrala, Khanna, Doraha, Mandi Gobindgarh and Sirhind. As might be expected, Railway Minister Nitish Kumar went to the scene. [16]

'Ram Parkash, 14, one of the survivors in the Sealdah Express, said many of his co-passengers were pilgrims returning from the Vaishnodevi shrine in Jammu. Railway officials said four soldiers of the Indian Army and a jawan {soldier} of the Indo-Tibetan Border Police were among those killed.

'A soldier trapped in a mangled coach was rescued. Heavy-duty cranes, which had reached the site were not pressed into service to untangle and remove the mass of coaches so as to save those who may be still trapped inside alive, an official said. Instead, men armed with gas cutters were trying to cut through the mangled mass of metal.

'The rescue workers were finding it difficult to lift the diesel engine weighing nearly 110 tonnes despite the fact that four heavy duty cranes — two belonging to the Railways, one from the Pong Dam and one from the Army — had been pressed into service to clear the wreckage. The death toll is expected to mount further once that bogie was reached. "It is a massive job", Mr R S Grover, Divisional Railway Manager, Ambala, told TNS {Tribune News Service} at the accident site.

'"We will have to cut open the roof of the compartment once we remove the engine, which in turn is weighed down by another smashed bogie lying on its side." About 100 bodies had been identified by the kith and kin of the victims and claimed by this afternoon. An accident information centre has been set up at the college grounds manned by policemen, railway officials and doctors, several local voluntary bodies have also opened counters for supplying food, tea and refreshments to the visitors.

'A steady stream of distraught kith and kin of the accident victims poured into the college compound where the bodies had been kept on ice. Many of them were seen crying as they lifted shrouds to identify their relatives. Most of the bodies were badly mutilated. While police officials prepared "panchnamas" {investigation records} the railway officials made on-the-spot payment of the ex gratia of Rs 25,000 to the relatives of each of the victims. Wooden crates were supplied to those wanting to carry the bodies to distant places.

'According to Dr G.P. Chander, Civil Surgeon, Ludhiana, all the bodies which had not been claimed so far were being shifted to Patiala. Mr A Venuparsad, an IAS {Indian Administrative Service} officer, who along with Mr Arun Goel, Deputy Commissioner, Ludhiana, was supervising rescue operations, said that he did not expect any survivors in the wreckage. Patient handling of the operations by the rescue workers saved many lives. One accident victim, trapped in the wreckage, asked them repeatedly to amputate both his legs in order to extricate him. But rescue workers worked steadily with gas cutters to reach him through the wreckage and bring him out with only minor injuries. An Army officer was similarly rescued, losing only three fingers of his hand.

---

[16] https://indiankanoon.org/doc/235936/

'Mr SK Mehta, General Manager, Northern Railways, said that one railway track had been cleared and repaired. Efforts were being made to clear and repair the second track also as soon as possible. Traffic on it was expected to be restored before long.' [17]

A gateman witnessed the derailment and attempted to have the Sealdah express warned with a telephone call but it was already in the section and there was no way that its crew could be contacted. Divisional Manager R.S. Grover at Ambala stated, "We do not have wireless sets in railway engines. Thereafter we are thinking of providing walkie-talkie sets to the driver and the guard."

Local residents were quick to come to the assistance of the victims of the accident, as another report made clear. 'Residents of four villages around the site of the rail disaster yesterday rose like one man to rescue the victims of the ill-fated trains and provide succour to them. Village of Kauri, Daudpur, Libra and Buha Majra were the first to come to the aid of the victims. "I was fast asleep when I heard a loud noise shortly after three in the morning," recalls Gurmail Singh (30), a resident of village Kauri, whose pucca {solid, permanent} house is nearest to the site of the accident." I got up immediately and came out of my house but could not see anything. I was badly frightened. I saw that many other residents of the village had also woken up and were peering into the darkness curiously. I was still debating what to do next when some Army jawans came running and summoned me and others for help." After some hesitation, he and other residents rushed to the site of the accident.

'Announcements were made on the loudspeaker of the village gurdwara {Sikh community centre} and soon the entire village was up and about. "The scene at the accident site was indescribable. Coaches had been smashed. There was hue and cry all around. It was pitch dark. Some of us broke open the panes and cut the iron grills with axes and saws so as to gain access to the blood-drenched victims."

'Gurmail Singh says that he could not say how many victims were rescued and how many were dead and how many alive. There was no time. We just tried to get everybody out." In the meantime, more residents came with tractor trailers and trucks and worked in the light provided by their head lamps.

'Village women set up langar {a free community meal in a gurdwara} and prepared tea, "we collected a lot of rice straw and set it on fire to keep the victims warm," recalls Mrs Pritam Kaur. "We also provided them with tea and blankets. Official help came much later."

'A good Samaritan from Khanna town showed a four-year-old boy to the TNS team. "I found him crying inconsolably near the accident site yesterday," said Mr Yash Pal Gomi who lives near police station (city), Khanna. "I brought him home and comforted him. He cried non-stop almost throughout the day. Today, he is better and has given his name as Shubam. He says that his father and mother have died in the accident. They were residents of Rohini, New Delhi, and were returning home after a visit to the Vaishno Devi," Mr Gomi can be contacted on telephone numbers: 101, and 22602. Among those dead in the accident was Mr Ramesh Chand (36), an assistant officer in the LIC {Local Indian Council} posted at Amritsar. He had gone to Delhi to cast his vote in Wednesday's election. He died in the train crash while on his way back home to Amritsar.' [18]

The Commission of Railway Safety determined that the initial derailment had resulted from multiple fractures in a length of continuous welded rail due to fatigue as a result of a railhead flaw. It made a series of recommendations:

(a) The flasher lights on locomotives should be improved so that it was easier to warn the drivers of trains on adjoining lines in cases where they were likely to be obstructed.

[17] https://m.rediff.com/news/1998/nov/26train.htm

[18] https://www.tribuneindia.com/1998/98nov28/head1.htm

(b) Communication facilities between the crews of passenger trains needed improvement.

(c) The Indian Railways' infrastructure facilities for the ultrasonic testing of rails needed augmenting.

(d) All manually operated ultrasonic equipment should be capable of automatically recording defects found.

(e) More self-propelled rail testing cars were required on the Indian Railways system.

(f) A need-based ultrasonic testing system related to traffic density should be adopted.

(g) Rail steel suppliers should ensure adequate quality control by the use of on-line equipment to achieve better rail quality. [19]

Train No. 7007 Up of the South Central Railway, the *Godavari Express*, was derailed on Friday 4th June 1999 between Ghanapur and Ippaguda stations, Andhra Pradesh, India (since 2014 both places have been in Telangana state). The accident occurred on the Kazipet – Secunderabad broad gauge section of the Secunderabad Division of the railway. Fourteen coaches of the train left the track at 03:15 and five of them overturned. A newspaper reported that, 'The first-class air conditioned coach was the worst hit. Out of the 17 passengers in this coach, five died. Six others managed to escape with minor injuries. According to reports, the trapped passengers in this coach had been crying for help until 06:30 hours. Subsequently, no cries or shrieks were heard. "It will take some more time to break open the coach and take out the passengers. We do not know whether they are dead or alive," said an official. An Indian Administrative Service officer, Priyadarshi Dash, was among those rescued from the ill-fated coach.'

The surviving passengers were not happy at the time that it took for the relief services to reach them. One of them told a reporter, "It was around 3 or 3:15 a.m., when most of us were fast asleep, when we felt a big jolt and heard a loud noise. We were thrown out of our seats with the impact. As we woke up, we found that our bogies had derailed and there was darkness all around." However, "The relief teams and doctors came only around 7 a.m. or so. We were taken to Ghanpur from where we were transported to the city in State Transport Corporation buses." The survivors complained that had the relief operation begun a couple of hours earlier some of those who died might have been saved.

An initial report spoke of six dead and 60 injured[20] and a second report of 12 dead and 80 injured. The official figures were that there were 9 dead and 35 injured, six of them seriously. Those who died included the Andhra Pradesh Minister for School Education, Devineni Venkatramana.

The train was said to have been travelling at 90 km/h when its driver applied the brakes in emergency upon hearing a loud bang from the air conditioned coach. [21]

The Commission of Railway Safety determined that the derailment had been caused by a fatigued alumino-thermic weld in the continuous welded rail in the left-hand rail of the Up line.

The Commission made a couple or recommendations in relation to the accident:

(a) The flange of conventional alumino-thermic welds should be tested by 80 degree hand, probably at the frequency prescribed in paragraph 7.8.2.1 of the ultrasonic flaw detection manual 1998.

---

[19] www.itsasafety.org/ITSA/deadly_indian_train_accident_pro.htm

[20] https://m.rediff.com/news/1999/jun/04ap.htm

[21] https://m.rediff.com/news/1999/jun/04ap1.htm

(b) The criteria for the replacement of resistance element welds needed to be reviewed as the existing criteria was too stringent to be followed in the field. [22]

As southbound Teito Rapid Transit Authority (TRTA) Hibiya line commuter train A861S, from Kitasenju to Kikuna, consisting of an eight-car electric multiple unit (EMU) emerged from a tunnel just short of the elevated station at Nakameguro, Japan, in suburban Tokyo, travelling at 12 to 13 km/h, the rearmost coach of the train left the rails. A trackside lead rail for setting off track maintenance vehicles had caused the derailed car to slew further to the right until it collided with the fifth and sixth coaches of an oncoming Tobu Railway train B801T from Nakameguro to Takenotsuka, another eight-car EMU, at 09:01 on Wednesday 8th March 2000. The side of a coach of the Hibiya line train, in which there were 240 passengers, was described as having been 'torn out from end to end.' Neither train was moving at more than 40 km/h. The southbound train quickly stopped after its motorman had applied the brakes and the northbound train stopped when the brake pipe broke between the fifth and sixth cars.

Passenger Motoshi Yamabe, 30, told a reporter, "My body was lifted up from the seat with a 'boom' and an impact and I wondered what had happened."

An unidentified 21-year-old commuter said, "I saw a huge lump of metal penetrating my car, and everybody was panicking. Many passengers were collapsing."

Two passengers were killed more or less outright and over 60 were hurt, three of whom later succumbed to their injuries.

Kiyoshi Terashema, the president of the TRTA, apologised for the accident and stated, "We will conduct a thorough investigation without delay to ensure a similar accident doesn't happen again." Meanwhile Prime Minister Keizo Obuchi instructed the Transport Ministry to open its own thorough and speedy investigation. [23] [24] [25] [26]

Occurring in the centre of Tokyo, the accident caused considerable public concern. It was investigated by the Railway Accident Investigation Working Group, whose report ran to 107 pages and listed seven factors that had contributed to the derailment:

(1) Unbalanced static wheel load, which had caused extremely decreased wheel load and increased lateral thrust to the outer wheels while running on the curve.

(2) Increased coefficient of rail-wheel friction due to the setting of a rail lubricator, causing undesirable increased lateral thrust.

(3) Track irregularities, especially twist and diminishing cant, which had caused decreased outer wheel load.

(4) The performance of the axle suspension had possibly contributed to undesirably decreased outer wheel load.

(5) The combination of rail and wheel profiles had led to increased later thrust.

(6) Low speed running at 12 to 13 km/h had prevented an increase in outer wheel load due to centrifugal force.

(7) High air suspension rigidity against bogie rotation had led to increased lateral thrust.

---

[22] Government of India Ministry of Civil Aviation and Tourism Commission of Railway Safety Annual Report for 1999-2000

[23] *The Brownsville Herald*, Brownsville, Texas, United States, 8th March 2000

[24] *The Chronicle*, Centralia/Chehalis, Washington, United States, 8th March 2000

[25] *The Lima News*, Lima, Ohio, United States, 9th March 2000

[26] *Hutchinson News*, Hutchinson, Kansas, United States, 9th March 2000

A combination of these factors, the report concluded, had caused the flanges of the outer wheels to ride up over the rail head until derailment had occurred.

In order to prevent a recurrence of the accident the report recommended stricter control of static wheel load, proper management of track regularity – especially flatness, appropriate management of rail profile after grinding, increase of flange angle from 60° to 65° or 70°, and installation of anti-derailment guard rails on sharp curves. Also further studies of improved methods of measuring static wheel load, improved rolling stock performance negotiating curves, better track design on sharp curves and improved rolling stock design to minimise passenger injuries in accidents.

The many sharp curves of the railway arose from the fact that as far as possible it had been built beneath public roads, to minimise the need for the purchase of expensive land. [27]

In response to the report the TRTA adjusted wheel weight ratios and introduced stricter guidelines for them, installed five new control systems costing 40 to 50 million yen each, changed the flange angles of railcars to 70°, undertook periodic shaving operations to remove scratches and dents and restore the original surface and shape of track on sharp curves, commissioned the Technical Research Institute to determine optimal track shape and lengthened the transition curve at the point of derailment from 30 to 40 metres. [28]

A derailment that occurred at Hatfield, England, at 12:23 on Tuesday 17th October 2000 was historically hardly a major disaster, resulting as it did in four passenger fatalities and some 70 or more injured, but it provoked a reaction, which, to the author, was out of all proportion to its magnitude.

By this date British Railways had been privatised and as a result, although the track and infrastructure were under the control of the Railtrack London North Eastern company, the train involved in the accident was being operated by a different so-called train-operating company, Great North Eastern Railway (GNER).

The express involved in the accident had left London King's Cross for Leeds at 12:10. As it was travelling at 185 km/h it and was being operated by a trainee driver, under the supervision of an experienced colleague, it became derailed at a spot south of Hatfield station. The front power car and two leading coaches remained on the rails as it travelled another 900 metres before stopping; the remainder of the train left the track. At the same time the train separated into three portions and the restaurant car, the eighth vehicle, overturned onto its side, collided with a pole supporting the overhead line equipment and was badly damaged. Passengers Robert Alcorn, Steve Arther, Leslie Gray and Peter Monkhouse, all of whom had been in the restaurant car, were killed. The other Mark 4 coaches of the train stood up well to the accident.

Passenger Stephen Morgan told a reporter: "I just heard a bang. I was walking towards the bar and heard a bang, then the train started swaying and it went over. I just held on for dear life. The carriage I was in was a bit busted up but nobody was really badly injured where I was."

Another passenger, Justin Rowlett, who chanced to be a Channel Four News television reporter himself, said: "The roof has been completely ripped off the restaurant car. There is a guy with a broken leg still inside the train and two people with severe gashes. The first-class carriages have completely come off the rails. The carriages at the back of the train looked like they were whipping around and have gone into some of the supports for the electric rails. I can see mangled seats, newspapers and bags in there. I'm told you can see dead bodies lying by

---

[27] https://www.ejrcf.or.jp/jrtr/jrtr33/pdf/f26_edi.pdf

[28] https://www.shippai.org/fkd/en/cfen/CA1000608.html

the rails." [29]

A preliminary investigation into the derailment determined that a rail had fragmented as trains had passed over it and that the likely cause of this was "rolling contact fatigue" (defined as multiple surface-breaking cracks). Such cracks were caused by high loads where the wheels contacted the rail. Repeated high loading caused fatigue cracks to grow. When these cracks reached a critical size, the rail failed. Portions of the failed track at Hatfield were reassembled and numerous such fatigue cracks were identified. These had contributed to the spalling of the running surface to around five millimetres (0.2 in) deep and 100 millimetres (3.9 in) long.

The problem had been known about before the accident; a letter from the infrastructure company Railtrack in December 1999 had warned that the existing Railtrack Line Specification was insufficient to guard against this type of fatigue. Replacement rails were made available, but they were never delivered to the correct location for installation.

Since privatisation, Railtrack had divested the engineering knowledge of British Rail to contractors. While it had comprehensive maintenance procedures that might have prevented the accident if followed appropriately, later investigation showed there was a serious problem with the experience and working knowledge of the staff involved. In a subsequent interview, the Zone Quality Standards Manager said, "I do not have knowledge of railway engineering nor railway safety," completely contrary to the written requirements for the role. In May 1999, the Head of Track had said that insufficiently skilled work was causing more rails to break. Railtrack did not know how many other cases of rail fatigue around the network could lead to a similar accident and after the derailment it consequently imposed over 1,800 emergency speed restrictions and instigated a costly nationwide track replacement programme. The author has a clear memory of a journey from Exeter to Plymouth post-Hatfield during which the train crawled the whole length of Exminster straight, a piece of track suitable for fast running, because of the emergency speed restrictions, adding a considerable time to the journey.

The speed restrictions and track replacement work continued to cause significant disruption on a majority of the national network for over a year, while the disruption and Railtrack's spiralling costs eventually caused the company to enter administration at the insistence of Transport Secretary Stephen Byers and its replacement by the not-for-dividend company Network Rail under Byers's successor Alistair Darling.

The train operating companies were likewise adversely affected by the disruption, losing an estimated 19% of their revenue in the year following the crash. Freight operator EWS was cancelling up to 400 trains per week as a result, whilst estimates put the Freightliner company's resultant losses at £1 million per month. The cost to the entire UK economy of the disruption was estimated at £6 million per day.

The Institute of Rail Welding (IoRW) was set up in 2002 by The Welding Institute (TWI) and Network Rail as a consequence of the recommendations in the investigation report to provide a focus for individuals and organisations involved in rail welding and facilitate the adoption of best practice.

In 2003, five managers and two companies – Network Rail (as successors of Railtrack) and the division of contractors Balfour Beatty that maintained the track – were charged with manslaughter and a breach of health and safety charges in connection with the accident. The managers, Anthony Walker (Balfour Beatty's rail maintenance director), Nicholas Jeffries (its civil engineer), Alistair Cook and Sean Fugill of Railtrack (asset managers for the London North-East zone), and track engineer Keith Lee, all denied the charges. The corporate manslaughter charges against Railtrack/Network Rail and some of its executives were dropped

[29] http://news.bbc.../976653.st

in September 2004, but the other charges stood. The trial began in January 2005. On 14th July, the judge, Mr. Justice Mackay, instructed the jury to acquit all defendants on charges of manslaughter. A few days later, Balfour Beatty changed its plea to guilty on the health and safety charges, and on 6th September, Network Rail was found guilty of breaching health and safety law. Network Rail were fined £3.5 million while Balfour Beatty were fined £10 million. [30] [31]

On Friday 10th May 2002 a West Anglia Great Northern train left London King's Cross station at 12:45 heading for King's Lynn via Cambridge. At 12:55, travelling at 156 km/h, the four-coach Class 365 Electric multiple unit (EMU) passed over a set of points, No 2182A, just south of the station at Potters Bar, England. These points moved under the train, causing the rear bogie of the third coach and the entire fourth carriage to derail. This caused the fourth coach to become detached and cross onto the adjacent line where it flipped into the air. The momentum threw the carriage into the station, where one end of the carriage struck Darkes Lane bridge parapet, destroying the masonry and sending debris onto the road below. It then mounted the platform and slid along before coming to rest under the platform canopy at 45 degrees. The leading three coaches remained on the tracks and came to a stop approximately 400 metres north of the station due to an automatic application of the brakes. Six passengers travelling on board the train were killed, including the broadcaster Austen Kark (whose wife Nina Bawden was seriously injured) and the philosopher Jonael Schickler. A seventh person, 80-year-old pedestrian Agnes Quinlivan, was killed by the masonry falling from the bridge over Darkes Lane as she walked along the street.

One person involved in the accident had already been in a bigger disaster three years earlier. A survivor of the Hillsborough football disaster has revealed how he cheated death by seconds in Friday's rail crash. Karl Brodrick, 44, was pushing a food trolley on the train as it hurled towards Potters Bar. He recalled how he left the fourth carriage to begin serving passengers in the next carriage just 10 seconds before the accident happened.

'Mr. Brodrick, of King's Lynn, spoke of his "incredible luck" after surviving the crash and the Hillsborough disaster which cost 96 lives in 1989. At that time, he was a supporter in the Leppings Lane stand where many of the fans died. He was discharged from hospital today and went straight to the scene of the crash. He said he earlier told a vicar: "I was at Hillsborough. This is the second time I have walked away from one of these things. There has got to be a reason why I have walked away."

'Mr. Brodrick, who works for Cambridgeshire-based firm Express Catering Services, told how he was rushing through his shift in the hope of being able to relax on the train for the latter part of the journey to King's Lynn. He said: "I had only just entered the third carriage when I felt the whole train shake. The driver really slammed on the brakes. The trolley fell over me and then on top of me. I crashed down on the corner of a seat and then the trolley landed on top of me. I was lying on my back in the aisle." He went on: "I can remember the people who I served on the train...I can recall one young man on the rear carriage. I served him a Stella beer and moved on. But when I was in the third carriage he came through the door and asked me for a packet of crisps. He was in the third carriage with me but was making his way back to the fourth carriage when the train crashed. I really hope that a 45p packet of crisps saved his life, but I don't know what happened to him."' [32]

The Rail Safety and Standards Board report on the derailment found that the points had

---

[30] Report of Independent Investigation Board on Hatfield derailment

[31] Wikipedia – Hatfield rail crash

[32] .../sm_58601/07/2002Ananova – Crash survivor was also caught up in Hillsborough tragedy

been poorly maintained and that this was the principal cause of the accident. The bolts that held the stretcher bars, maintaining the distance between the pair of point blades, had come loose or gone missing, causing the point blades to move apart when the train had passed over them. The points had been fully inspected on 1st May by a team working for the private railway maintenance firm Jarvis plc and there had been a further visual inspection of them on 9th May, the day before the crash, with no problems reported.

However, that evening, a West Anglia Great Northern station announcer who had been travelling on the down fast line had reported a "rough ride" at Potters Bar whilst going over that same place on the track, points No. 2182A. A Railtrack London North Eastern engineering supervisor had been sent to make an inspection, but due to an apparent misunderstanding by the King's Cross signal box staff, he had been sent in error to examine the Up Fast line to check the track and points there and thus had not found the "loose nuts", which had subsequently led to the accident.

Initially, after the accident, Jarvis had claimed that the points' poor condition must be due to sabotage of some sort, and that its maintenance could not be to blame. No solid evidence of any sabotage ever came to light, and the Rail Safety and Standards Board (RSSB) report found that other sets of points in the Potters Bar area showed similar, less serious maintenance deficiencies and concluded that the poor state of maintenance "probably arose from a failure to understand fully the design and safety requirements".

Further investigations by the RSSB found that heavy and constant vibrations on the stretcher bars and their bolts had caused them in turn to vibrate and oscillate until their nuts had fallen off the bolts. These had since been replaced by two-part locking nuts instead of the main nuts having half-size locking nuts to hold them in place.

The tragedy sparked a debate about whether private maintenance firms were paying too little attention to training and safety and in 2003, Network Rail announced that it was taking all track maintenance in-house, ending the use of private contractors, except for large-scale renewal or development projects.

On 28th April 2004, Jarvis sent a letter to the victims' families, admitting liability for the accident. The company said that it would formally accept "legally justified claims" after making a financial provision of £3,000,000.

In November 2010, the Office of Rail Regulation said Network Rail and Jarvis Rail would be charged under the Health and Safety at Work Act. The case was heard at Watford Magistrates' Court in February 2011 and on 13th May Network Rail was fined £3 million for safety failings related to the crash. [33] [34]

Train No. 1932, the *Freccia della Laguna* express, which had started from Palermo at 16:00, was derailed just before 19:00 on the evening of Saturday 20th July 2002, near the station at Rometta Marea, Sicily, Italy, while travelling at about 105 km/h. E656 Class engine No. 32 left the rails at a bridge over the Saponara stream and the following four coaches swerved across the line and demolished a railway owned building in which two families of railwaymen lived, neither of whom was fortunately at home at the time. The fifth coach was slightly derailed, the rear two coaches remained on the track, and the bridge was partly demolished.

A large number of locals from the town were quickly on the scene, accompanied by the town mayor, but found that there was little they could do without proper equipment. Fortunately, this soon arrived in the form of firemen under the command of Baldassare

---

33 Rail Safety and Standards Board report on the Potters Bar derailment

34 Wikipedia – Potters Bar rail accidents

Battaglia – trained in alpine, cave and river rescue – carabinieri, ambulances, soldiers from the Aosta Mechanised Infantry Brigade under Brigadier General Adriano Santini and two helicopters from Palermo.

It was soon learnt that six people had lost their lives in the accident, one of them being Driver Saverio Nania, 45, while about 30 others were injured. The passengers who died included four from the same Moroccan family: Miloudi Abdhelakim, 75; his wife Fatima Fauhreddine, 59; their son Ali, 33; and their daughter Hanja, 42. Hanja's children Ola Defi, 8, and her brother Abdelaziz, 6, survived because they were shielded from the main impact of the crash by their maternal grandparents. The children's father, Ali Defi, was away working in Saudi Arabia. Injured, and with the children in hospital, were their uncle Arbi Abdhelakim, 35, and his Italian wife Carmela, 28. Arbi lamented, "Our family is destroyed. In an instant I lost my parents, my sister and my brother."

Also among the injured was Driver Marcello Raneri, who was trapped in his engine with his dead fellow driver and was seriously hurt. The injured were taken to two hospitals in Messina and some to the hospitals in Milazzo and Barcellona Pozzo di Gotto.

Several surviving passengers were interviewed by reporters. Giovanni Andronico, 29, from Messina, who escaped with slight injuries, related, "We were calmly sitting in the leading carriage when a sudden skidding of the train alerted us that something was wrong. The next moment suitcases and the bricks of a house rained down upon us. There were injured all around me and I had a very lucky escape."

Giovanni Russo, 46, from the Palermo area, said, "I heard a loud bang and then there was chaos. I got up and saw that our train had ended up against a house. I managed to climb out through a window and not being hurt was able to help other passengers to get out. There were injured and also dead all around me."

Antonio Romeo, 27, from the Calabria region, stated, "It was a mess. I was in the third coach, which went straight into a house. We managed to get off and saw the bodies of two men in an adjoining carriage."

Getano Gianni, his wife and two children, who were one of the two families who had lived in the demolished house, had gone to the seaside after having lunch and came back to find their home in ruins. Equally lucky was retired railwayman Filippo Currò, 56. Normally from about 18:00 he rested in his deckchair in the garden, watching the trains go by, and would have still been there when the express arrived. On this evening, as chance would have it, he had reluctantly agreed a few minutes before 19:00 to accompany his wife on a shopping trip.

An initial theory that a failure of the bridge had caused the derailment was soon disproved, as was a theory that some defect in the train itself had been responsible. Possibilities such as a landslide or an obstruction on the line were also ruled out. Focus then shifted to a problem with the track itself.

Antonio Nania, 19, the son of the dead driver, declared, "It's not my father's fault. He was a scrupulous, attentive, person, who would never have made a mistake that would have jeopardised people's lives." His mother, widow Concetta Crescenti, agreed, telling investigators that her late husband had spoken a number of times about rough riding in the Rometta area, claiming that there was something wrong with the road. Other trains drivers concurred. In response to these reports some maintenance work had been done on the track and a temporary speed restriction to 60 km/h had been imposed, which had been lifted and the permitted speed had been raised to the usual 120 km/h only three days before the accident.

Three inquiries were launched into the accident, one on the part of the judiciary by Luigi Croce, the Messina public prosecutor; one on behalf of the Ferrovie dello Stato; and a third on the part of Esposito Construction, the contractors who had done the repairs to the track, which repairs reportedly had included the removal of 80% of the sleepers, which had been worn out.

The accident received extensive coverage in at least one Italian newspaper, including allegations that the accident would not have occurred had the government proceeded sooner with modernising the line between Palermo and Messina.

At the close of the investigation the derailment was conclusively attributed to track defects. [35] [36]

A passenger train from Ye to Tavoy, Myanmar, had only travelled about three kilometres from its starting point on Saturday 27th December 2003 when it was derailed, causing the reported death of a least five people and serious injuries to another 20, who were taken to a local hospital. There was no official statement concerning the accident but locals claimed that the military government had forced them to construct that railway with cheap materials and using unpaid and unskilled labour, implying that the derailment was likely due to track defects. [37]

Seven passengers, four of them women, reportedly lost their lives in a derailment on Sunday 1st August 2004 near Aunglan, Myanmar. As in the derailment the previous December, no information regarding the accident was released by the military authorities and it was again blamed on their building lines in haste, using substandard old tracks and forced labour from the local populace rather than modern machinery and equipment. [38]

The rear four of the six carriages of Philippines National Railway (PNR) train No.580 from Legaspi City to Manila, which was carrying about 350 passengers, were derailed and fell about 12 metres into a ravine near Padre Burgos, the Philippines, at about 02:30 on Friday 12th November 2004. The engine and leading two carriages remained on the track, though the carriages were overturned. Varying figures were given for the number of casualties in the accident but it would seem that there were 13 dead and about 160 injured.

Police, soldiers and local residents worked to rescue the victims of the accident, using any tools available. Vehicles were commandeered by the police to carry the injured to Camp Nakar Hospital and the MMG Clinic in Lucena, the Jane County Hospital in Pagbilao and the Quezon Provincial Medical Centre, while the railway arranged for a breakdown train to be sent to the scene. [39]

Conductor Melquiades del Pilar related that the economy carriage in which he had been riding when the train derailed had shuddered as the train negotiated a bend. "It was very quick. It felt like we were being pulled over," he said.

Passenger Rolando Cruz alleged that the train had been going too fast, saying, "It was speeding. Suddenly the lights went out and we felt the train coming off the tracks. Our coach overturned three times." [40]

Passenger Donito Briones said, "I suddenly felt the wheels of the train detach. Then it {his coach} tumbled over, knocking us around." In the darkness his aunt had cried, "My God, my God." They had waited for over an hour to be rescued before Briones had crawled out through

[35] *L'Unita*, Roma, Italy, 21st, 22nd, 23rd, 24th, and 25th July 2002

[36] Wikipedia – Rometta Marea derailment

[37] www.dvb.no/english/2003/december/20031229c.html

[38] http://english.dvb.no/news.php?id=2843

[39] www.mb.com.ph/issues/2004/11/13/MAIN2004111322407_print.html

[40] www.reuters.com/printerFriendlyPopup.jhtml?type=worldNews&storyID=67925...

a partly opened door and helped the three other members of his family to escape.

Passenger Edgar Palmiano had been travelling with seven other family members and had awoken in an overturned coach. He had then helped his other relatives out through an open window. His mother had suffered bruises, his father had suspected broken bones and one of his children had sustained a serious head injury. [41]

Following the initial investigation, Transportation Secretary Leandro Mendoza stated that the derailment had been due to the shifting of a rail as the train had rounded a curve. Spikes that should have held the track in place had been missing. He declared, "Based on observations, the colour and texture of the holes left by the spikes showed that they had long been missing and could not, therefore, have been part of a deliberate act to derail the train. Hence the committee, even at this early stage, has ruled out sabotage or terrorist action as a possible cause of the accident. The rails themselves may not show any signs of damage but the rest of the railway track components, especially the wooden ties {sleepers} and metal spikes need replacement or rehabilitation due to ageing, damage or pilferage." He added that the trains were suffering from many defects such as defective gauges, including speedometers, defective airbags and imperative emergency brakes. A thorough survey of the whole PNR system was recommended.

During the investigation Driver Santos Leano Jr. and Assistant Driver Benjamin Jallogo Sago had been tested for drugs.[42] The tests had been negative. Allegations that the train had been speeding were not really confirmed, though it had been travelling at the marginally excessive speed of 27.05 km/h (the permitted speed was 25 km/h). [43]

Fast passenger train No. 211, which should have left Lahore station at 17:30 for Narowal, was derailed near Babakwal station, Pakistan, at about 18:15 on Saturday 5th March 2005. The rear carriage of the train left the rails, overturned, and was carried a kilometre and a half before the train stopped. Two passengers were killed outright, four succumbed on the way to the Mayo Hospital in Lahore and another 19 were critically hurt. At least 20 others were less badly hurt. The ill-fated coach had a capacity of 86 but because it was the weekend it had been crammed with about 150 passengers, including 20 riding on its roof.

The derailment was blamed on the track being out of gauge. Operations General Manager Abdul Wahab Awan had reportedly been warned of the poor condition of the track following an inspection a fortnight before the accident. Following the accident he suspended six railwaymen, including Divisional Engineer (Track) Akhtar Mehmood Khatak, Assistant Engineer Saeedullah, Driver Muhammad Mushtaq and Guard Muhammad Jameel.

Compensation of 100,000 rupees was awarded to the relatives of the deceased and 50,000 rupees to the seriously injured. [44]

---

[41] www.manilatimes.net/national/2004/nov/13/yehey/top_stories/20041113top1.html
[42] www.sunstar.com.ph/static/man/2004/11/16/news/mendoza.forms.team.to.prob...
[43] http://news.inq7.net/common/print.php?index=-2&tstory_id=19136&site_id=19
[44] www.hipakistan.com/en/detail.php?newsId=76589&F_catID=&f_type=source

CHAPTER 2

DERAILMENTS - BROKEN TYRES, BROKEN AXLES AND OTHER VEHICLE DEFECTS

One of the causes of derailments of trains other than problems with the track is defects in the trains themselves, such as broken tyres, wheels, and axles. Such accidents were not as common by the twentieth century as in the early days of railways, but they still happened from time to time, due, for example, to maintenance failures.

International express train D 82 from Innsbruck, Austria, to København, Denmark, was running through Northeim, Germany, travelling at 120 km/h, at 01:30 on Sunday 15th November 1992, at the same time as local goods train Ng 64 683 from Seelze to Göttingen was travelling in the other direction on the opposite running line. As the trains passed one another, 11 coaches of the express were struck by some of the waggons of the goods train and seven of its coaches were derailed. Worst affected were two sleeping cars, the sides of which were ripped away and which, in addition, fell from an underbridge onto the federal highway below. Eleven people lost their lives as a result and another 52 were injured.

As the goods train had passed through Northeim station its second vehicle, a four-axle coach for permanent way staff, had lost a buffer, which in falling onto the track had caused the derailment of the seventh waggon of that train. Another 14 of the 43 waggons of the Göttingen freight train had been thrown off the rails in turn, some of them blocking the line on which the express was running. Occupation of the track circuits had then put the signals to Danger in front of the express, the driver of which had made an emergency brake application, but it was already too close to the site of the derailment and the collision could not be averted.

An investigation by Deutsche Bundesbahn (DB) found that the derailed staff vehicle had begun life as a passenger coach in the 1940s, had been rebuilt and converted into a luggage van in 1965 and then rebuilt again and converted into a staff vehicle in 1982. It had a special form of buffer construction, whereby rather than the buffers being directly connected to its framework, they were screwed to an extension piece, which in turn was welded to the cross member at the end of the longitudinal beam. This weld had clearly failed, probably due to metal fatigue, and the buffer had then fallen off. According to the DB no similar accident of its type had occurred before. The defect was not one that could have been detected by the visual inspection of the rolling stock made before the train's departure. As an immediate measure, after the accident an internal examination of all other staff vehicles was ordered.[45]

[45] Katastrophen der deutschen Bahnen Teil II by H.J. Ritzau, pages 203-4

A Northern Railway Up DSJ goods train left the yard at Kanpur at 19:15 on Saturday 16th January 1993. As it was subsequently running between Block Hut I and Ambiapur, Uttar Pradesh, India, it became derailed, separated into two parts and some of its waggons fouled the Down line. Almost at once the obstruction was run into by train No. 2302 Down, the *Rajdhani Express* from Howrah to New Delhi, which was derailed in turn. Four passengers and four railwaymen lost their lives in the accident and 265 passengers were injured, 13 seriously.

The DRJ goods train consisted of WAG-4 electric locomotive No. 20919, 60 Container Rail Terminal (CRT) waggons loaded with cement from Satna and a brake van.

The *Rajdhani Express* consisted of WAP-1 electric locomotive No. 22012 and 19 all-steel coaches.

The double-track electrified line was signalled on the Absolute Block System with Siemens General Electric block instruments. The stations were interlocked to Standard III with multiple-aspect colourlight signals.

Assistant Stationmaster (ASM) D.P. Singh in Block Hut I was the first witness to give evidence at the Commission of Railway Safety inquiry into the accident. He stated that as the goods train had passed at 21:49 he had seen sparks coming from the wheel of the 13th or 14th waggon from the rear of the train, whereupon he had informed the traction power controller to switch off the overhead line equipment (OLE) and had replaced the Down Home signal, which had been cleared for the express. A few moments later he had heard a loud noise and informed the section controller of this. There had been no unusual noise coming from the goods train, just sparks visible; its speed had been about 70 km/h. No one at Rura station, to his rear, had warned him of any problem with the goods train. He had been unable to contact the traction power controller (TPC) despite repeated attempts and had then tried the section controller, who had not at once replied but the OLE had been switched off before he heard the noise of the collision. He had shown a Red light to the guard of the goods train but received no response from the guard, whom he could not see. The *Rajdhani Express* had entered the block section from Ambiapur at 21:49; it usually took 1½ to 2 minutes to pass through the section, while the goods train took 3 to 4 minutes.

Assistant Stationmaster A.K. Rai at Rura station stated that he had seen sparks coming from the 14 waggon of the goods train at about 21:46, whereupon he had informed Tundla Control, who had said that the train had been examined at the station in the rear. He had then informed the ASM at Block Hut I and Allahabad Control. He had not exchanged private numbers as the brake binding had been very light. He had also shouted to the guard of the goods train but received no response. The train had been travelling at about 70 km/h.

Confronted with ASM Rai's statement, ASM Singh reiterated that he had had no message from Rura station. The Tundla and Allahabad controllers also denied having been informed of any problem by Rura station. Guard K.D. Singh of the goods train was one of the fatalities.

Section Controller P.M. Tripathi at Allahabad stated that the ASM at Block Hut Z had informed him of brake blocks jamming on the goods train whereupon he had arranged for it to be stopped and examined at Bhaupur station. This had been done and the driver and guard had been unable to find anything wrong. He had then not thought it necessary to advise Rura, the last station under his jurisdiction, to look for any problem, nor had he advised Tundla Control.

Section Controller Jai Pal Sharma at Tundla stated that the ASM at Block Hut I had advised him of a dangerous vehicle on the goods train at 21:50 and to switch the OLE off, which he had instructed the TPC to do. Although Rura was outside of his jurisdiction he had been advised by Block Hut Z that the goods train had been stopped for examination at Bhaupur station but this had been after the express had left Ambiapur. The Rura ASM had not mentioned any problem.

Assistant Stationmaster Neeraj Pandey at Bhaupur testified that upon learning from Block Hut Z that there was a problem with the brake blocks of the goods train he had informed the section controller, replaced his Starter signal and sent a memo via Porter Ram Mohan Trivedi to the train crew that the brake blocks were jamming in a waggon 13th to 15th from the brake van. The driver, diesel assistant, guard and porter had examined 22 to 23 waggons from the rear and found nothing. The train had left at 21:13.

Porter Ram Sajiwan at Block Hut I spoke of seeing sparks coming from the 13th waggon, showing a Red light to the guard and informing the ASM. He had heard the noise of the derailment, the OLE had then been turned off, followed shortly by the sound of the collision.

Leverman Om Prakash Pathak and Porter Shiv Narain at Rura described exchanging signals with the driver and guard of the goods train. Porter Narain did not recall the Rura ASM telling anyone of any problem with the train.

Porter Trivedi recalled giving the memo from the Bhaupur ASM to the crew of the goods train. He and the assistant driver had checked the south side of the train and the driver and guard had looked at the north side of the train, examining the under gear with a torch and checking the axle boxes for overheating. The job had taken 30 to 35 minutes.

Crossing Keeper Kishan Gopal stated that the goods train had been coming to a halt at his gate when the express had passed, then there had been a loud noise. He had tried to inform Block Hut I but could get no response.

Switchman R.C. Shukla at Bhaupur said that the ASM had not asked him to specifically watch the DSJ goods train; he had noticed nothing wrong with it.

Assistant Stationmaster J.P. Tiwari at Block Hut Z stated that as the DSJ goods train passed him he had seen that the brakes of the 14th and 15th waggons were jammed at the Howrah end and had informed Bhaupur East cabin, exchanging private numbers.

Porter Ram Chandler at Block Hut Z described seeing minor sparks coming from the 14th waggon from the rear of the goods train but no hot box or other dangerous condition.

Assistant Stationmaster Panch Bahadur and Porter M. Salim at Block Hut FF, Assistant Stationmaster G.K. Khan and Porter Ram Pal at Maitha station and Assistant Stationmaster M. Anjan Ali and Porter Dilip Kumar at Block Hut Y all testified that they had exchanged 'all right' signals with the driver and guard and had observed nothing wrong with the train. (Block Hut Z lay between Panki and Bhaupur stations, Block Hut FF was between Bhaupur and Maitha stations, Block Hut Y lay between Maitha and Rura stations and Block Hut I was between Rura and Ambiapur stations.)

Driver A.K. Upadhyaya of the DSJ goods train testified to being stopped at Bhaupur at 20:26 and the porter giving a memo that the 15th or 16th waggon from the brake van should be checked for brakes binding. With his diesel assistant, the guard and the porter, these waggons had been carefully checked on both sides and nothing wrong had been found. After leaving Bhaupur at 21:15 the journey had been uneventful until the vacuum had dropped and his train had stopped with a jerk, at the moment that the *Rajdhani Express* passed on the Down line. He added that although the 15th and 16th waggons had been specially examined, 20 waggons had been checked in all.

Diesel Assistant Ganga Ram corroborated the statement of his driver.

Driver Satya Prakash said that he had been a passenger on the express. After the accident he had made his way to the rear of the goods train, finding neither guard nor tail lamp, then to Block Hut I, which had been deserted (because the ASM and porter had gone to the accident site) and he had informed Control. Driver N.N. Avashti stated the same.

Guard N.K. Singh of the express stated that his train had passed through Ambiapur at about 120 km/h. Immediately afterwards there had been two strong jerks and severe bumping, the lights had gone out and the train had stopped. He had struck his head as the train stopped.

Head Train Examiner D.C. Verna, Train Examiners O.P. Upadhyaya and Y. Gupta, Coaching Depot Inspector Roop Ram, Carriage and Waggon Superintendent K.M. Gupta, Train Examiner D.N. Shah, and Assistant Mechanical Engineer N.G.J. Naidu all tendered evidence regarding the examination of the DSJ goods train. A number of other witnesses also gave evidence.

Commissioner of Railway Safety V.S. Dutta detailed tests and observations made after the accident.

Discussing the accident, the commissioner determined that the initial derailment had occurred at 21:50 and the subsequent collision at 21:51.

The goods train would have been travelling at 70 km/h at the moment of derailment and the express at 130 km/h.

The initial derailment had occurred at Km 1066/15. At about Km 1067/3-5 the train had parted between the 44th and 45th waggons from the engine and other waggons had derailed, capsized and fouled the Down line. Marks on the 45th waggon, WR 81330, indicated that it had been the first to derail.

There was no indication of sabotage or track defects having caused the derailment, or of overloading of the first waggon to derail.

The ASM at Block Hut Z had observed apparently jammed brakes on what he had thought was the 15th or 16th waggon from the rear of the train and the porter there had seen sparks. The train had then been stopped and examined at Bhaupur and nothing found. The staff at Block Hut FF, Maitha station and Block Hut Y had observed nothing. The ASM at Rura stated that he had seen sparks and informed Block Hut I and Tunda and Allahabad Control, which all three denied. The ASM and porter at Block Hut I had observed sparks and shouted and signalled to the guard of the goods train just before the derailment.

After the accident the axle box of the 17th waggon from the rear had been found on the south side of the track near the point of derailment, also some rollers, the remnants of the roller cage and the bearing rolls. The roller-bearing axle cap had worked out due to three securing studs having broken. The Delhi end wheel set had not been in position and the Howrah end wheel tread and flanges had been badly battered. The axle journal had turned blue, indicating that it had overheated. The CRT axle box had two roller bearings and one spacer ring. The outer roller bearing had been completely missing; the inner race of the inner rolling bearing had been in position. The three axle cap retaining studs had broken and the broken pieces had still been available on the axle journal face. The axle guard had been bent and bore extensive battering marks. Three of the four brake blocks had been in position, that of the left trailing wheel had been missing. The brake blocks of the right trailing wheel and leading wheels had been of full thickness, in good condition, with minimum wearing and with no sign of continuous brake binding. The left leading brake block had been broken and one portion with its clip in position had uniform wear over the entire surface.

The evidence established that the south side axle box of the Howrah end wheel set of waggon No. 81330 WR CRT, 17th from the rear of the train, had worked out during the run due to the failure of its securing and locking arrangement, causing the derailment.

The commissioner considered that the defect in the waggon had probably existed before the train left Kanpur yard and could have been spotted by the fitters deputed to examine the north side of the train there, had they been vigilant. However, considering the nature of the defect and the scope of their duty they could not be considered responsible for the derailment, though their conduct had been blameworthy.

The type of defect was such that it could easily have escaped notice during the examination at Bhaupur, carried out in darkness by staff who lacked adequate knowledge of train examination, working with inadequate light from torches and hand lamps and looking for defects such as jammed brakes and hot boxes, which were not present. The guard, driver, assistant driver and porter could not therefore be held responsible for having failed to detect the defect.

Commissioner Dutta concluded that the cause of the derailment of the Up DSJ goods train had been the working out of the south-side axle box of the Howrah end wheel from the axle journal of loaded CRT waggon No. 81330 WR, marshalled 45th from the engine. Signals indicating a major defect in the form of the rubbing/hitting of the north-side wheel with the axle guard and other parts had produced sparks, which had been visible for a sufficient distance but had not been observed and acted upon by the staff concerned.

No one could be held primarily responsible for the derailment.

Assistant Stationmaster Khan at Maitha and Porters Saleem and Kumar at Block Huts FF and Y were secondarily responsible for not properly observing the north side of the train and noticing the disturbed condition of the undergear of the affected waggon, which should have been emitting sparks and making a noise.

Assistant Stationmaster Roy (sic – spelt Rai earlier in the report) at Rura was secondarily responsible for having informally told Control and the station ahead of having seen sparks but not having sent the required signal as per the Block Working Manual for the train to be stopped and examined at the next station.

Section Controller Sharma at Tundla was secondarily responsible for not attaching sufficient importance to the information received from the Rura ASM and not arranging for the train to be stopped.

Assistant Stationmaster Singh at Block Hut I and Section Controller Sharma at Tundla were blameworthy for giving false evidence that the Rura ASM had not advised them that he had observed sparks coming from the goods train.

Assistant Stationmaster Pandey at Bhaupur and Section Controller Tripathi at Allahabad were blameworthy for not having obtained a written memo from the driver and guard of the goods train giving the results of their examination of the goods train before authorising it to proceed. The section controller was also blameworthy for not having advised the ASMs at Block Hut FF, Maitha station and Block Hut Y of the need to carefully watch the train, after the reports of sparks from Block Hut Z and the driver of train No. 992 Down.

No one was responsible for the secondary collision, which had been unavoidable.

Under remarks and recommendations the commissioner stated that:

(a) the Research Design and Standards Organisation should carry out an in-depth study of the possible causes of the working out of the axle boxes of CRT and Box-type waggons and suggest remedial measures.

(b) The design of the "L"-type axle guard fitted in CRT and Box-type waggons should be reviewed for possible replacement with the "U"-type axle guards fitted on other rolling stock, which type prevented axle boxes from coming out of their seats.

(c) The prevailing practice at Satna station of piecemeal waggon examination, formation of train loads and dispatch without further carriage and waggon examination should be stopped. Whenever the composition of a train was modified by 10 or more waggons it should be given a complete examination.

(d) Control office staff should compulsorily attend Safety Camps at least every two years and be regularly counselled through talks and posters to highlight their responsibility for strict compliance with safety rules, both by themselves and by staff on line.

(e) The pattern of train examinations should be reviewed, so that at stations where engines were changed or trains reversed train examiners were deputed for axle box feeling, vacuum cylinder release and vacuum continuity, the train examination being made "Safe to Run."

(f) Drivers, guards and pointsmen should be given adequate training in train examination for the occasions when they needed to perform such tasks.

(g) At stations where facilities existed, such as Satna, the brake power of trains should be pre-tested, preferably not using the train engine. [46]

A major disaster with similarities to that described above near Ambiapur occurred less than six months later in India, this time at Samalpatti on the Southern Railway in Tamil Nadu state. Some Container Rail Terminal (CRT) waggons of an Up Katpadi Junction (KPD Jct) goods train were derailed as the train ran through the station at about 55 km/h. They fouled the down line and were run into by train No. 6635, a Down Kurla – Cochin express. The accident occurred at 23:30 on Tuesday 1st June 1993. Twenty-one passengers died in the accident and 44 were injured, 31 of them seriously.

The KPD Jct goods train consisted of Class WDM-2 diesel electric locomotive No. 17265, 50 CRT waggons and a brake van.

Train No. 6635 consisted of Class WDM-2 diesel electric locomotive No. 17818 and 16 coaches.

Samalpatti was a Class B station provided with multiple-aspect colourlight signalling worked from A and B signal cabins, situated at the Jolarpettai Junction and Erode Junction ends of the station respectively, and Standard III interlocking. The double line was worked on the Absolute Block System with tokenless block instruments.

The accident was investigated by the Commission of Railway Safety.

Driver P. Ayyavoo of the KPD Jct goods train told the inquiry into the disaster that passing through Samalpatti station, the vacuum on his engine had dropped from 49cm to 35cm. He had stopped the train, had found that the locomotive front hose pipe was "blowing" and had attended to the leakage. The vacuum had then returned to 49cm and he had restarted. Because of a curve it had not been possible to exchange signals with his guard. He had then proceeded cautiously to Kagankarai, where the switchman at the West cabin indicated that his train had parted, as had the stationmaster there. Examining his train, he found that he only had 23 waggons and the brake hose pipe at the rear of the last vehicle was dangling. He could not explain how he had been able to obtain 49cm of vacuum after the train had become divided and was sure that neither he nor his diesel assistant had dummied (sealed off) the brake pipe at the rear of the 23rd waggon. Nothing unusual had occurred on the journey between Dasampatti and Samalpatti.

Diesel Assistant P. Arjunan generally corroborated the evidence of his driver, though he mentioned another stoppage due to vacuum trouble earlier in the journey. He stated that he had looked back after coming to a straight stretch of track after leaving Samalpatti but had not seen any signal from the guard.

---

[46] Government of India Ministry of Civil Aviation (Commission of Railway Safety) Railway Accident Investigation Report on Collision of 2302 Down Rajdhani Express with Derailed Waggons of Up DSJ Goods Train at Km 1067/4-6 Between Bluck Hut I and Ambiapur Stations on Etawah – Kanpur Broad Gauge Double Line Electrified Section of Allahabad Division of Northern Railway at 21:51 Hrs on 16-1-93.

Guard M. Ramaswamy of the goods train deposed that he had checked his train at Uttukuli and found all to be correct. The train had stopped after leaving Uttukuli due to a leak of vacuum at the dummied end of the brake van, which had been dealt with. There had been no problems after leaving Dasampatti until he had seen a white light at the gate lodge at the engineering crossing before Samalpatti and had felt a jerk as the train had passed there, travelling at 75 to 80 km/h. The Samalpatti signals had been Clear and he had seen and acknowledged a Green hand signal from the switchman in Samalpatti B cabin. Just after his van had passed that cabin he had felt a heavy jerk, which had thrown him down, and the vacuum had fallen to zero. Alighting from his van he had asked the B cabin switchman what had happened and the latter replied that his colleague in A cabin had advised that the train was divided. Just then, he had heard the loud noise of the collision. He denied having heard any shouting, having seen any Danger signals, or any stones having been thrown at his van at the engineering crossing. He had attributed the jerk at that crossing to a sudden increase in the train's speed. He denied having dropped off to sleep at any time on the journey.

Driver V.K. Paramasivan of train No. 6635 stated that the Samalpatti signals had been off for his train. As he had repeated the signal aspects, there had been a heavy sound and his engine had run into an obstruction. He had been trapped in his cab for 40 to 45 minutes in great pain before being released, when he had lost consciousness.

Diesel Assistant P. Nanjappan concurred with his driver's statement. He, too, had been injured and had lost consciousness after the collision, only coming to again in hospital.

Driver M. Paul Raj of Down Mettur Dam (MTDM) goods train stated that his train had been regulated at Samalpatti for the Down express. He had seen the Up KPD Jct goods train stop suddenly, as if by an emergency brake application, then after about two minutes had heard the noise of the collision. He had then realised that a serious accident had occurred and that waggons of his own train were also derailed. He put the speed of the Up goods at 50 to 55 km/h and had noticed nothing peculiar with it before it stopped suddenly.

AC Assistant P. Sasindran of the MTDM goods train agreed with his driver's statement. He put the speed of the Up goods at 60 to 65 km/h.

Guard S. Venkataraman of the MTDM goods train said that it had come to a stand on Road No. 1 at Samalpatti at 21:58. At about 23:25 he had seen the Up KPD Jct goods train pass without a brake van or tail lamp. He had immediately alighted from his van and run towards the stationmaster's office but before he reached it, the Down express had passed and he had heard the noise of the collision. He had afterwards assisted with the rescue work. He had shown a hand signal to the engine of the KPD Jct goods but received no response. He put the speed of the Up goods at 40 to 45 km/h.

Switchman P. Kumaran in Samalpatti B cabin testified that he had given Line Clear for the Up KPD Jct goods at 23:03 and received Train Entering Section at 23:13, when he had informed the duty stationmaster and his colleague in A cabin, exchanging private numbers. As the train had been passing his Home signal the gatekeeper at the engineering crossing informed him of sparking in the middle of the train. The train had then been approaching his cabin and he had tried to inform the stationmaster and the A cabin switchman but had only been able to contact the latter, who told him that the train was divided. He had already obtained Line Clear from Dasampatti for the express and cleared his Down signals for it. He, himself had not seen any sparking from the train and he had shown a Green hand signal to its guard. He had not thought of showing the guard a Red hand signal or throwing his Up signals back. He had only tried to contact the A cabin and the stationmaster after the goods had passed him. The brake van of the goods had come to a stand just past his cabin and he had spoken to its guard, who said that he was injured. At that point he had seen the headlights of the express and been preparing to exchange hand signals with it, when he had heard the sound of the collision. He added that once or twice a year he received messages from the crossing keeper regarding a 'hot box' or broken waggon spring and in the three minutes that goods trains took from the engineering crossing there was usually time to reroute them onto the loop line for examination. On the night of the accident the message from the crossing keeper had been delayed. There was no practice of throwing the Home signal back in the face of a driver. He put the speed of the goods train at 65 km/h.

Switchman K.E. Parthasarathy in Samalpatti A cabin stated that after clearing his signals for both the goods and express trains he had been awaiting their arrival when the B cabin switchman advised him that there were sparks coming from a waggon of the goods train. He had left his cabin to see for himself, by which time the engine of the goods had passed his Advanced Starting signal. Sparks had been coming from the last five or seven vehicles, then he had seen that the train was incomplete and advised Kagankarai B cabin on the block bell and by telephone that the goods train was divided. Just then, the Down express had passed and he had exchanged signals with its crew. A moment afterwards the power of the overhead line equipment had failed and the B cabin switchman had told him on the phone of hearing a big noise, whereupon he had called Kagankarai B and told the switchman there not to clear his signals for mail train No 6319, which he (Parthasarathy) had just accepted.

Stationmaster III V.M. Sathis at Samalpatti related that as the goods train had passed the station house he had seen that it was incomplete and only its front portion had run through. Pointsman K. Raju, on the Up side, had shouted that the train was incomplete at the same moment. The down express was then running through the station; he had heard a big noise and realised that there had been an accident. He had then organised the rescue work. He stated that after giving the slot for the goods and express trains he had gone outside when his telephone had rung and the B cabin switchman advised him of seeing sparks on the goods train. The stationmaster said he had wanted to see for himself and noted the sparks but more importantly, had seen that the train was incomplete. He had then informed Control and the A cabin switchman of this; the latter was already aware of it.

Asked what action the stationmaster should take in such a case, a witness stated that if the train was a good distance away, he would advise the switchman to replace the Home signal. Were it closer, he would replace the Advanced Starter. In the present case he had not considered a mere report of sparks to be serious enough to take such action, as sparks emanating from train wheels were a common occurrence.

Pointsman Raju described seeing that the goods train was divided while he was standing on the Up side to exchange signals with it and his shouting to alert the stationmaster.

Stationmaster III V. Undaysarasad at Samalpatti stated that he had been off duty at the time of the accident but had rushed to the station upon being advised of it by the pointsman.

Stationmaster V.T. Varadarajan at Samalpatti stated that he had also been off duty at the time of the accident but had likewise rushed to the station upon being advised of it. He estimated that the running time from the crossing to the B cabin was four or five minutes and the switchman would normally come out onto the balcony to exchange signals with the train crews and only answer the telephone after the train had passed. Had he time, the best way for the switchman to react would be first to replace his signals, then inform the stationmaster.

Crossing Keeper N. Kali stated that he had been on duty at Crossing No. 98 when the goods train had passed. Standing outside his gate lodge, he had seen the wheel of one waggon entirely engulfed in sparks. There had also been a "tuk-tuk" noise similar to that found when there was a brake defect. He had immediately begun blowing his whistle and turned his hand lamp to Danger. He had seen the guard sitting in his van leaning on his elbows with his head bowed. Shouting and stone throwing had not attracted the guard's attention. As soon as the brake van had passed he had informed the B cabin switchman. Five minutes later, two gangmen had told him that they had also seen an apparent 'hot-box' on the train.

Gangmen M. Ramaswamy and S. Ramalingham confirmed that they both had seen the problem, one stating that a wheel of the goods train had been "burning furiously with heavy smoke and flames." They had shown Danger hand signals to the brake van and shouted and thrown stones at it. They had seen the guard sat in his van; he had not responded in any way to their warnings. The two gangmen disputed which of them had reached the gate lodge first.

Train Examiners P.S. Vinod and M. Unnikrishnan, Chief Train Examiner C. Radhakrishnan and Head Train Examiner T.M. Krishnan – all at Podanur Junction, Head Train Examiner K. Valsaraj at Mangalore and Train Examiner S. Devaraju at Cochin Harbour Terminus, detailed the examinations they had carried out on the waggons of the goods train.

Stationmaster III K. Sushil at Dasampatti stated that he had observed the goods train as it passed through his station at 23:14 and had seen nothing wrong with it.

Sweeper-cum-porter K. Selvan at Dasampatti stated that he had likewise observed the goods train from the opposite side of the line and all had appeared in order.

Switchman P. Palanivelu in Dasampatti A cabin stated that he had exchanged signals with the diesel assistant and the guard of the goods train. He had seen nothing wrong with it.

Switchman P. Sidda Reddy in Dasampatti B cabin gave a similar statement.

Switchman M.V. Ravi in Kagankarai B cabin stated that he had received the 6 – 3 ('Train Divided') bell signal from Samalpatti A cabin at 23:26, when he had advised the duty stationmaster and his colleague in the B cabin and the train had been received on the loop line.

Stationmaster A. Chandrasekaran at Kagankarai corroborated Switchman Ravi's statement.

Various other witnesses gave evidence relating to matters such as measures taken after the accident.

Commissioner of Railway Safety G. Ram Mohan detailed tests and observations made as part of the inquiry.

Discussing the accident, the commissioner fixed the time of the initial derailment at about 23:20. The train had then covered another 4.4 kilometres before waggon ER 30603 had wholly derailed and the train had parted when passing through Samalpatti station at 23:27. The subsequent collision the commissioner fixed as having occurred at 23:30.

The express had been travelling at 85 km/h when it had collided with the derailed waggons.

The commissioner described the events leading up to the accident. The KPD Jct goods train had been formed of 50 CRT waggons loaded with bags of cement from the ACC Cement Factory at Madukarai. It had been intensively examined at Podanur Junction. It had then left there at 15:15 and Uttukuli at 17:45. Between Perundurai and Totiyapalayam the train had stopped due to a loss of vacuum, which the diesel assistant had rectified by attending to the brake hose pipe at the long hood end of the locomotive, with the guard likewise attending to the brake hose at the end of his van. A banking engine had assisted the train over the Lokur ghats between Karuppur and Lokur, the train leaving Lokur at 22:10. It had passed Dasampatti at 23:18 and between there and Samalpatti CRT wagon ER 30603, marshalled 24th from the engine, had become derailed. The derailment had been due to an old crack, which had developed over a period of time. A total rupture of the axle had not occurred at once but the train had run with the right wheel on the sleepers, caused by the inward bend of the axle, until contact with the stretcher bars of points at Samalpatti station had caused the final breakage of the axle, total derailment of the affected waggon and following ones, the separation of the train and the capsizing of the derailed waggons foul of the opposite running line. The derailed waggons had also collided with those of the Down MTDM goods train.

The Gatekeeper (crossing keeper) Kali at Crossing No. 98 had seen that there was a problem and attempted without success to attract the attention of Guard Ramaswamy of the goods train, as had two gangmen. Gatekeeper Kali had then immediately informed Switchman Kumaran in Samalpatti B cabin of the problem but instead of at once replacing his signals the latter had decided to check the train himself. By the time he had done this it had been too late for him to stop it. He had informed the stationmaster and his mate in A cabin but showed a Green hand signal to the guard of the goods. Stationmaster Sathis and Switchman Parthasarathy in Samalpatti A cabin had similarly watched the train and had seen that it was divided, and the switchman had duly advised his colleague in Kagankarai B cabin. The crew of the express had had Clear signals, had exchanged signals with Switchman Parthasarathy in the A cabin and with the curve and darkness had had no chance of seeing the obstruction ahead of them.

The report on the accident continues with a detailed review of the metallurgical examination of the failed axle, a comprehensive analysis of the actions of the different railwaymen involved, a discussion of the safety record of CRT waggons and so on.

In his conclusions Commissioner Mohan stated that the accident had resulted from a series of events beginning with the cold breakage of the trailing axle of a loaded CRT waggon on the KPD Jct goods train. The waggon had derailed all wheels in Samalpatti station; 16 waggons behind it had also derailed, four of them fouling the Down line and being run into by the express. The derailment had not, however, happened all at once. A gatekeeper and two gangmen had seen the problem and had made every effort to alert the guard but the latter had not been keeping a good lookout. The gatekeeper had then immediately telephoned the Samalpatti A switchman but the latter had not taken any immediate action to stop the train either.

The primary cause of the accident had been an equipment failure for which no one was responsible.

Secondary responsibility for the accident rested with Guard Ramaswamy and Switchman Kumaran. The guard had failed to keep a good lookout to satisfy himself that the train was proceeding in a safe and proper manner or to keep a lookout when passing over a manned level crossing. He had violated General Rule No. 4.43 and Southern Railway Rule 4.43 (ii). The switchman must have been aware after receiving the telephone call from the gateman that safety would be in jeopardy if the train were allowed to run further and would have had ample time to throw back his Home signal and stop the train. He had violated General Rules Nos. 2.11 (1) (a) and 2.11 (2).

Driver Ayyavoo and Diesel Assistant Arjunan of the KPD Jct goods train had been negligent in failing to keep a good lookout when passing through Samalpatti station, had failed to realise that their train had parted, and had been alerted to a problem only after covering two kilometres. Even after stopping, they had only performed a perfunctory check and restarted without even waiting for their guard's signal. They had violated General Rules Nos. 4.40, 4.41 and 4.42 and Southern Railway Rule No. 4.42 (iii) (e).

The commissioner made eight recommendations:

(a) The Railway Board might review the policy of concentrating the CRT rakes in the two zones in the south and might consider ordering their dispersal over a larger area of Indian Railway's broad gauge network.

(b) The Research Design and Standards Organisation (RDSO) might undertake a review of the design of CRT waggons and suggest improvements for increasing their stability when running empty and while negotiating curves. Measures for strengthening their springs and making them less prone to breakage of plates, for strengthening their couplers and avoiding failure of the locking mechanism and for reducing the failure of their roller bearings needed to be evolved.

(c) At present there was no limit fixed for the age of axles and there were no instructions to withdraw from service any axle however old it might be, unless internal flaws were indicated in ultrasonic testing. The RDSO might undertake studies to determine whether a need existed to fix an age for axles beyond which, on consideration of preventing cold breakages, they needed to be withdrawn from service.

(d) Posts requiring creation in various carriage depots to meet the increased workload should be created expeditiously. If there were problems for the same, in view of the policy in economising in main power done, the Southern Railway Administration might consider the possibility of getting the repairs to waggon stock in sick lines carried out by contractual agencies, while train examination and certification were only done by the departmental staff.

(e) The Southern Railway Administration should strictly adhere to the provisions of paragraph 142, Chapter 1 of the Indian Railways Establishment Manual for filling certain percentages of the posts of train examiners and head train examiners Grade I by direct recruitment of Engineering Degree and Diploma holders.

(f) A drive should be launched all over the Southern Railway to educate train examination staff on the importance of making all hose pipe connections before issuing a brake power certificate.

(g) The Southern Railway should keep up the tempo and clear the arrears in the routine overhaul of CRT waggons. The railway should also desist in future from taking policy decisions to skip or to postpone the routine overhaul and the periodic overhaul of these waggons.

(h) A drive should be launched in the Southern Railway to educate the station staff on the importance of paying prompt attention to any report of an unsafe condition made by any person and the taking of all possible action to avert an accident. [47]

Northeastern Railway train No. 5609, the Up Avadh – Assam express, was passing through the station at Mairwa, Bihar, India, on the Chapra – Bhatni broad-gauge single-line section of the Varanasi Division of the railway, at 21:28 on Tuesday 3rd August 1993, when it was derailed at the facing points at the entrance to the station. The rear, and the third and fourth coaches from the rear of the train, were the ones that left the rails. Twenty-three passengers died and two were seriously injured. The bodies of the victims were found about half a kilometre in the rear of the point where the train came to a stand after the derailment and it was theorised that they must have been riding on the roof of the train. The injured were taken to the Civil hospital in Siwan. [48]

The Commission of Railway Safety determined that the cause of the accident had been the working out of the bottom-side link pin and both swing links on the left of the leading trolley of a coach.

The commission recommended that instructions to station and train staff to stop roof travelling should be reiterated. To set an example, no train should be allowed to move out of a divisional headquarters station without special staff being deployed to remove all passengers from roofs. Additionally wide publicity should be made at stations and on trains against roof

---

[47] Government of India Ministry of Civil Aviation (Commission of Railway Safety) Railway Accident Investigation Report on Collision of No. 6635 Down Kurla – Cochin Express with the Derailed Waggons of Up KPD Jct Goods Train at Km. 244/2 at Samalpatti Station on the Jolarpettai – Erode Broad Gauge Double Line Electrified Section of Palghat Division of Southern Railway at about 23-30 Hours on 01-06-1993.

[48] https://eparlib.sansad.in/bitstream/123456789/823247/1/10_VII_04-08-1993_p319_p321_PII.pdf

travelling, specifying the penalty for doing so. [49]

A mixed train from Santa Cruz to Yacuiba, Mexico, was derailed at 13:00 on Sunday 4th September 1994 in the Mora area, about 91 kilometres south of the Santa Cruz. Five flat waggons and four other waggons left the rails. Twelve people died and another 20 were injured. Most or all the victims were poor people who had been riding illegally on the flat waggons. A relief train was sent from Santa Cruz and the injured were taken to various medical centres. The cause of the derailment was unclear; possible causes suggested were the softening of an embankment and the excessive speed of the train. [50]

On Thursday 22nd September 1994 a major catastrophe occurred near Tolunda, Angola, in Namibe province, on the main line of the Empresa do Caminho de Ferro de Benguela. A freight train was involved, loaded with blocks of granite, and which was also carrying a large number of passengers, described perhaps a little unkindly in one account of the accident as 'fare dodgers,' though they probably had no other means of travel. The train was derailed and fell down a 10-metre-deep ravine in a remote area, with the result that it was a long time before help arrived. Although Angola was still in the midst of a civil war at the time of the accident there was no known fighting in the area where the derailment occurred and it was attributed not to sabotage but to a failure of the brakes, perhaps due to lack of maintenance. [51]

Early reports spoke of 70[52] and then 146 dead, while about a week later the Angop press agency stated that the official number of fatalities had risen to 300, with another 147 injured. [53] [54]

A passenger train consisting of an engine and 12 carriages was derailed and four of the carriages overturned at Tukh Crossing (also spelt Toukh), Egypt, just north of Cairo, on Monday 29th May 1995. By the following day the police were reporting that 15 people had died and 20 were injured, while more victims might still be buried in the wreckage. At least one of those who was injured was trapped in the wreckage for hours before he could be released.[55] The derailment was apparently due to an engine defect, as the driver had attempted to reduce speed when it occurred, after 'discovering a mechanical problem' in his locomotive. [56]

Intercity Express (ICE) No. 884 *Wilhelm Conrad Röntgen* departed from München Hbf at 05:47 on Wednesday 3rd June 1998 for Hamburg-Altona. At about 10:33 it left Hannover Hbf with power car No. 401 051 leading. About 24 minutes later, at 10:57, at kilometre 55.1, it was travelling at some 200 km/h and was approximately six kilometres from the station at Eschede, Germany, when a tyre on the third axle of leading passenger car broke, due to material failure. The tyre then unwound, drilled through the floor of the coach and became jammed between two seats. The train continued to Eschede station, damaging the track en route, and at a set of points there at 10:59 a wheel rim struck a check rail, lifting the car from

---

[49] Government of India Ministry of Civil Aviation and Tourism Commission of Railway Safety Annual Report for 1993-94

[50] https://www.noticiasfides.com/nacional/sociedad/de-fatal-accidente-ferroviario-en-santa-cruz-271587

[51] *Trouw*, Amsterdam, the Netherlands, 27th September 1994

[52] *The Times*, London, England, 26th September 1994

[53] *Leeuwarder courant*, Leeuwarden, the Netherlands, 30th September 1994

[54] *Amigoe*, Curaçao, 30th September 1994

[55] *The Ottumwa Courier*, Ottumwa, Iowa, United States, 30th May 1995

[56] *European Stars and Stripes*, Darmstadt, Germany, 31st May 1995

the track. A derailed wheel then struck a second set of facing points, causing them to open beneath the train, diverting the third passenger car of the train sideways. This coach then struck the pillars of a road overbridge 200 metres beyond the points, causing it to collapse. The fourth coach passed beneath the collapsing bridge, then fell down an embankment, while the bridge, weighing about 200 tonnes, fell onto the rear half of the fifth car. The sixth car was buried in the debris of the bridge and the following cars filed into one another in a zig-zag fashion, back to the 12th and last passenger car. The rear power car was derailed. The separation of the train caused the automatic brakes to apply and the front power car and two leading cars came to a stand down the line, the detached power car only stopping three kilometres further on, its driver unaware of what had happened behind him. A Volkswagen Golf automobile, which had been parked on the overbridge, fell with it and led to incorrect initial reports that it was the fall of the vehicle onto the line that had led to the derailment. The Golf had belonged to two signal technicians, who also died in the accident.

The accident resulted in the death of 101 people, including at least four railwaymen, and the injury of another 108, 70 of them seriously. The number of fatalities equalled that from the 1939 Markdorf collision and was exceeded in Germany only by the Genthin accident on the same day as Markdorf. It is also the worst accident in the world involving a high-speed train, at least up to the beginning of March 2025.

Passenger Jörg Dittman was sat close to the spot where the tyre had broken through the armrest between the seats where his wife and son were sitting. He at once moved them to another coach and then left them to go and to tell the conductor in the third car what had happened. The latter had himself noticed that the was train vibrating abnormally, nevertheless he told Herr Dittman that he would have to investigate the matter before stopping his train. It took the conductor a minute to get to the affected coach, by which time it was beginning to sway from side to side. Herr Dittman had not been able to find an emergency brake handle in the corridor and had not noticed that there was one in his own compartment. As the conductor reached the first car and Herr Dittman was about to show him the damage caused by the wheel, the final derailment occurred.

Witnesses living nearby described the noise of the crash as "startling," "terribly loud," and "like a plane crash." Some of them imagined at first that it had been an air crash. They were the first on the scene, while an official at Eschede station alerted the emergency services. The police declared an emergency at 11:02, a major emergency at 11:07 and as the magnitude of the disaster became clear it was upgraded again by the Celle local government at 12:30. Over 2,000 rescuers from the fire service, police, ambulance staff and so on attended the scene, including 37 doctors who had been at a conference in Hannover and units of the British Army stationed in Germany.

An even worse disaster was averted because ICE No. 787 *Karl Adam* from Hamburg to Hannover had passed under the bridge in the opposite direction only two minutes before the derailment. [57]

The investigation into the catastrophe naturally focussed on the failure of the wheel that had caused it. It emerged that the original ICE trains had been equipped with single-cast wheelsets (monobloc wheels). After a while in service these suffered from problems of resonance and vibration when trains were travelling at their cruising speeds, due to uneven wear and metal fatigue. In attempt to deal with the problem it was decided to improve the suspension of the cars, by the use of a rubber damping ring between the outer steel tyres and the steel body of the wheel. Such a solution had previously been used successfully on trams, running of course at much lower speeds. The re-designed Class 064 wheels were successful in

[57] https://www.ndr.de/geschichte/schauplaetze/Zugunglueck-in-Eschede-Katastrophe-mit-101-Toten,eschede18.html

eliminating the vibration and were introduced on high-speed trains forthwith. At the time there were no facilities in Germany for the fatigue testing of wheels and train manufacturers and operating companies in other countries such as France and Japan were not consulted, so as to benefit from their experience.

Back in early 1992 the Fraunhofer-Institut für Betriebsfestigkeit und Systemzuverlässigkeit (Institute for Structural Durability and System Reliability) in Darmstadt had warned Deutsche Bahn of its concerns regarding a possible wheel failure. Then in July 1997 Üstra, the company operating trams in Hannover, had discovered problems with fatigue cracks in the wheels of its trams, operating at only 24 km/h, and had warned other operators, including the railway company. Deutsche Bahn had replied that it had not noticed any problems in its trains.

The Fraunhofer Institute was itself asked to investigate the disaster. It determined that the modelling during the design phase of the wheels had failed to consider dynamic repetitive forces and that the design had thus lacked an adequate margin of safety. Overlooked during the design had been four different factors:

(a) The tyres had flattened into an ellipse with each wheel revolution (about 500,000 times during a typical day in service on an ICE train), with corresponding fatigue effects.

(b) Unlike the monobloc wheels, cracks could form on the inside as well as the outside of the tyre.

(c) The dynamic forces increased as the tyre wore thinner, resulting in crack growth.

(d) Flat spots and ridges or swellings of the tyre dramatically increased the dynamic forces and greatly accelerated wear.

The inspection of the tyres at the maintenance depot in München had been undertaken visually with the aid of a flashlight rather than with the use of ultrasonic light profile and measuring beam detection equipment. Ultrasonic testing, which would have detected fine cracks and metal fatigue not discernible by visual inspection, had been discontinued in 1994 because it resulted in too many false positive results. Other defects in the wheel had been discovered in an inspection on the day before the accident but because they had not been considered to have any safety implications, the wheelset had not been replaced. In addition there were computer records of eight reports by staff in previous weeks of rough running due to flat spots on the wheel but these had not been evaluated as a safety problem.

Had either Herr Dittman or the conductor operated the emergency alarm at once, the disaster would have been averted. The passenger had not known where the alarm handle was, while the actions of the conductor in first verifying for himself that there was indeed a problem were held to have been in accordance with normal procedures and thus could not be found fault with.

The accident had clearly been made more serious by the existence of the central bridge pier in the path of the derailed coach.

Legal proceedings against various individuals were opened after the disaster. Those against the conductor who had failed to stop the train when Herr Dittman notified him of a problem, against three employees of the München maintenance depot and against an employee of the wheel manufacturer were all dropped. Charges of negligent homicide and causing bodily harm against a departmental head and a senior technical officer of Deutsche Bahn and an operations engineer of the tyre manufacturing plant were proceeded with. After 54 days the judges discontinued the proceedings on a payment of 10,000 euros from each of the defendants, on the grounds that only further trials could have determined if they had been aware of the risk of wheels breaking and in any case they could not be held to be seriously culpable. As might be expected, survivors of the disaster were not happy that the cases had been dropped.

Various measures were adopted in the aftermath of the disaster:

(a) The limit for acceptable wear on wheel tires was increased from 854 to 890mm tread diameter (new tyres were 920mm).

(b) The use of wheelsets of the affected type was prohibited until proof of their operational safety had been provided.

(c) Rubber sprung wheels were not re-introduced.

(d) The extent to which points existed before bridges and tunnels was to be examined and they were not to be installed at such locations on new lines.

(e) Ultrasonic testing of wheels would be carried out after a set mileage rather than after a set time interval.

(f) The dual-control principle whereby two or more people confirmed or approved an item would be used for important inspections.

(g) Additional emergency exit windows would be provided on ICE trains, which could be broken from the inside using an emergency hammer and would also improve access for emergency workers.

(h) A new cantilever bridge was built at Eschede lacking pillars.

(i) Conversion to air-sprung MD-530 bogies was considered to reduce noise in the coaches, and more frequent turning of wheels and spacers between rails and sleepers to obtain smoother running. [58]

Some of the waggons of a Central Railway Up BYC/CRC goods train became derailed and overturned, fouling the opposite running line, between Farah and Baad stations, Uttar Pradesh, India, at 02:16 on Friday 16th July 1999, apparently due to a defect in one or more of the waggons. The exact details are not available but the accident apparently resulted in damage to the overhead line equipment on the Agra Cantonment – Mathura Junction section of the Jhansi Division of the railway. Unfortunately the seriousness of what had happened was, it would seem, not appreciated by either the driver of the goods train or the section controller and so train No. 2615 Down, the Chennai – New Delhi *Grand Trunk Express,* was not halted. Instead, it collided with the derailed waggons, whereby 21 people were killed and 152 were injured, 63 of them seriously. [59]

The injured were taken to the civil and military hospitals, the J.P. Krishna Nursing Home and the Swarna Jayanti Hospital in Mathura. Railway Minister Nitish Kumar, Railway Board chairman V.K. Agarwal and other officials supervised the rescue and relief operations and later the railway minister visited the injured in hospital. A special train forwarded the uninjured passengers to New Delhi. [60]

'Mr A.K. Goel, who was travelling on the ill-fated train, after arriving in the Capital said, "For a moment I thought I was dead. But a few minutes later I regained sense and realised that I was alive. As all doors were smashed in the accident, I came out of the general second-class coach through a window." The Meerut based businessman said he heard a bang around 2.20 a.m. and realised his bogie had overturned. When he came out of his compartment, he found passengers running here and there in panic and those caught in ill-fated bogies shouting for help.

[58] Wikipedia – Eisenbahnunfall von Eschede

[59] Government of India Ministry of Civil Aviation and Tourism Commission of Railway Safety Annual Report for 1999-2000

[60] https://m.rediff.com/news/1999/jul/16rail.htm

'Major Singh, another passenger, who came to the Capital by boarding a bus near the accident site, said, "I was travelling in the fourth coach from the engine when was sleeping on the upper berth when the train started rattling. Suddenly, I fell down and everything became dark. Construction workers working near the accident site were the first persons to rush to the spot and help the injured. I suffered a minor head injury and was discharged after first-aid," he said.' [61]

The Commission of Railway Safety made several recommendations in relation to the accident:

(a) Subsidiary Rule 17.03.10 should be recast to indicate the proper chronology of the precautions to be taken by the Control staff in the event of the overhead line equipment (OHE) being found faulty, before the healthy OHE was energised.

(b) The Central Railway administration should continue to organise the foundation courses for the controllers and also for their panels in each Division and this should be completed by 31st December 1999.

(c) The Central Railway administration should arrange a crash-training programme to impart proper and adequate disaster management and first-aid training to all mobile commercial staff.

(d) The Railway Board should prescribe Fit to Run Examination of the intensively examined empty rakes, which were subsequently loaded at the next train examiner examination station, and an endorsement to this effect should be made on the brake power certificate already issued by the intensive examination station. [62]

Five bogies (carriages) of train No. 20 Down, the *Khushhal Khan Khattak Express*, en route from Peshawar to Karachi, were derailed at about 09:30 on Tuesday 9th January 2001 near Kotri, Pakistan, in Sindh province, 100 kilometres north of Karachi. An early report spoke of eight dead, six of whom had been identified, and that the injured, six of whom were in a critical condition, had been taken to hospitals in Hyderabad and Jamshoro City, while relief centres had been set up in Peshawar, Karachi, and elsewhere. [63]

Another report stated that there were at least 15 dead and 200 injured and that faulty brakes, due to poor maintenance, had been the cause of the disaster. Communications Minister Javed Ashraf told a news agency that the railwaymen believed responsible had been suspended, pending an investigation. [64]

An assistant stationmaster had informed Control that there was a problem with the train and it had subsequently been stopped and examined by officials at Budhapur station, but they were unable to find anything wrong with it, so the train had been allowed to proceed. A female passenger, who escaped with slight injuries in the accident, had also noticed that the third carriage of the train was not running properly at the next station at Unarpur, according to a railway official who spoke to a reporter, but the train had not stopped and a few kilometres further on the derailment had occurred. [65]

---

[61] www.tribuneindia.com/99jul17/head2.htm

[62] Government of India Ministry of Civil Aviation and Tourism Commission of Railway Safety Annual Report for 1999-2000

[63] *Kuwait News Agency*, Kuwait City, Kuwait, 9th January 2001

[64] *The Capital*, Annapolis, Maryland, 9th January 2001

[65] *The News International*, Karachi, Pakistan, 10th January 2001

CHAPTER 3

OTHER DERAILMENTS

Other chapters in this book cover derailments caused by track defects, defects with the trains themselves and those caused deliberately by acts of vandalism or sabotage. This chapter will deal with derailments due to other causes such as excessive speed, collisions with landslides and other accidental obstructions, traction shocks and such like, together with those for which no cause has been ascertained or is known.

At least 50 people were believed to have died when a Chemin de Fer Djibouto-Éthiopien freight (mixed?) train bound for Djibouti went off the rails near Dire Dawa, Ethiopia, on Saturday 22nd June 1991. More than 100 passengers on the overcrowded train were injured and others were trapped in 13 overturned vehicles, according to a radio report. [66]

At about 00:53 on Saturday 19th October 1991 South Central Railway No. 7055 Up, a Narsapur – Hyderabad express, was derailed on a set of facing points, diverted onto a Loop line and collided with stabled goods train No. BAM 54 Special. The accident occurred at Pondugula station, Andhra Pradesh, India, on the Nadikude – Miryalaguda section of the railway. Thirteen people died in the accident and 101 were injured, 39 seriously.

The derailment was determined to have been due to a combination of three factors:

(a) Track defects with a twist exceeding 6 mm/m.

(b) The driver of the train exceeding the maximum permissible speed.

(c) The wheels of the locomotive had been approaching condemning limits.

The Commission of Railway Safety report on the accident contained four important recommendations:

(a) Proper attention should be given to the maintenance of locomotive wheel profiles to correct dimensions. Research should be conducted to elucidate the cause of biased wear on locomotive wheels.

(b) The Railway should consider withdrawing locomotives having wheel wear approaching condemnable limits from passenger service and deploying them to haul goods trains.

(c) The track maintenance had been extremely poor, due to lack of care during the construction of embankments; permanent posts for track maintenance should be created; ballast shortages should be recouped.

(d) The quality of work by the construction wing of the signal and telecommunications

[66] *The Sunday Independent*, London, England, 23rd June 1991

department needed improvement. [67]

At 20:17 on Wednesday 30th October 1991 Southern Railway train No. 2627, the *Karnataka Express* from Bangalore City to New Delhi was derailed at Km 55/9-10 between Makalidurga and Thondebhavi stations, Karnataka, India. Torrential rain over the previous two days had resulted in a large boulder rolling down from the top of a cutting onto the track minutes before the arrival of the train. Twenty-nine people lost their lives and 21 were hurt, 18 of them seriously.

The primary cause of the derailment was determined to have been a 'sudden convulsion of nature.' A secondary cause was identified as 'certain inadequacies in the system of monsoon patrolling as organised for the vulnerable location where the accident took place.'

The Commission of Railway Safety report on the disaster contained four important recommendations:

(a) The existing system of track patrolling during the monsoon season on the Makhlighat of Bangalore Dharmavaram section needed to be revised, and a new system introduced based on the posting of stationary watchmen at vulnerable locations, as per the provisions of the Indian Railways Permanent Way Manual.

(b) Flare signals (fusees) must be issued as standard equipment for monsoon patrolmen (both stationary and mobile) on the Makli ghats.

(c) The existing permanent speed restriction in the Makli ghats needed to be reviewed and should be made more restrictive over stretches with highly restricted visibility and over vulnerable locations during the monsoon season only.

(d) Vulnerable locations, particularly in ghat sections, needed to be correctly identified and should be specifically listed in the handing-over notes of engineers in the Division. [68]

Northern Railway train No. 3, a Pathankot – Baijnath Paprola passenger service, was derailed at about 15:00 on Saturday 7th December 1991 between Janawala Shehr and Harsar Dehri stations, Himachal Pradesh, India, on the Pathankot – Jogindernagar narrow-gauge section of the railway. The accident occurred on a 12-degree curve while the train had been running at high speed, with passengers riding on its roof and hanging onto its sides. Twenty-seven people died and 44 were injured, 22 seriously.

It was determined that the derailment had resulted from the overturning of heavily and unevenly loaded coaches as the train had been negotiating a curve at high speed.

The Commission of Railway Safety report on the derailment contained three important recommendations:

(a) Practical measures should be implemented to prevent narrow-gauge passenger trains from running when they were overcrowded to an unsafe degree.

(b) The couplers used on the Kangra Valley Railway stock relied on an improvised shackle to prevent mishaps. These couplers should be quickly replaced. The design should be examined for improvement.

(c) The mechanism for limiting speeds installed on Class ZDM-3 locomotives might be examined in consultation with the Research Design and Standards Organisation and

[67] Government of India Ministry of Civil Aviation and Tourism Commission of Railway Safety Annual Report for 1991-92
[68] Ibid

correct driving procedures enforced. [69]

Sixteen people reportedly died and another 47 were seriously injured when a passenger train plunged into a river at an unstated location in central Tanzania on about Friday 17th April 1992, according to a report from a railway official. [70]

In November 1992, in connection with an extension of the yard at Hoofddorp, the Netherlands, between Amsterdam and Schiphol, a temporary S bend was incorporated into what had previously been a straight stretch of track. A tamper was initially employed to do the work but then broke down, so the job had to be completed without it, and the resulting curves were sharper than had been intended. On completion of the work on Monday 23rd the line was reopened to traffic with a speed restriction of 90 km/h, as compared to the previous 140 km/h. Six trains then passed through the curves, the crews of which complained of difficulty in negotiating them at the permitted speed, so at 08:26 that morning the speed limit was reduced to 30 km/h, then after a further inspection it was raised again to 60 km/h.

All more or less went well then until the afternoon of Saturday 28th November when an international express from Paris Nord to Amsterdam Centraal, diverted due to work on the Amsterdam – Haarlem – Leiden line, was derailed. Only the rear coach of the train left the track and one passenger was injured. Sleepers were damaged over a distance of 1,200 metres and had to be replaced. The 30 km/h speed restriction was reinstated, then on Sunday 29th November it was raised to 60 km/h over part of the section, though it remained at 30 km/h for 500 metres at the Hoofddorp end. According to Nederlandse Spoorwegen (NS) this first derailment had occurred at kilometre 19.7 and was due to a broken rail (a Wikipedia account states that like the second derailment, it was due to excessive speed, but this does not agree with an account in a contemporary newspaper).

At about 07:12 on Monday 30th November Inter City train No. 2127 from Amsterdam Centraal to Vlissingen, consisting of two Koploper electric multiple sets, seven cars in all, arrived at the curve from the opposite direction to that of the previously derailed train. There was no tachograph on the train but it was later estimated that it had been travelling at about 123 km/h approaching the location, reduced only to 106 km/h on the curve. Again the train was derailed. This time the whole of it left the rails and the leading carriage overturned and slid along the ground, while two carriages further along the train also overturned. This time also the results were more serious; five people in the leading car died and 33 others were injured. [71]

A newspaper reporter interviewed passengers and others. "'It was as if we ran over something," declared A.C. Eelzak, one of the passengers on the derailed train. He felt a severe shock as the train jumped off the track. "I threw my arms over my head and felt that I was being lifted out of my seat." No one was seriously hurt in the part of the train he was in. Afterwards he was able to exit the train via the vestibule between two cars and made his way towards the overturned front car. "We shifted a tree under which a boy was trapped," he related.

'Dr J.C. Bakker was a passenger on a train coming towards Amsterdam from his home town of Leiden. His train was stopped because of a "delay" ahead, then there was a call over the intercom asking if there were any doctors on the train. He duly reported himself, along with a fellow doctor and a nurse. When they learnt what had happened, they made their way to the wrecked train, from which the cries and moans of the injured could be heard for some time after the derailment.

---

[69] Ibid

[70] *International Herald Tribune*, Paris, France, 18th – 19th April 1992

[71] *NRC Handelsblad*, Amsterdam, the Netherlands, 1st and 11th December 1992

'Meanwhile the majority of passengers from the part of the train which had suffered less in the accident had already made their way on foot back along the track to Hoofddorp station.

'René Kater, from Den Haag, who was a daily traveller on the train, related that two or three times the previous week there had been a severe jolt at the location of the derailment, and that commuters who had been standing in the crowded train had had a job to keep their feet. Even the conductor had almost fallen over and he had told his passengers that a new piece of track had been laid.

'Mevrouw {Mrs} P. Blanson-Henkemans from Amsterdam confirmed René Kater's account. She said, "After we passed through Hoofddorp the train began to shake violently. A fellow passenger told me that this had happened before, when the train had been shuddering so badly that people had been thrown about."

'Frank von Roermund from Leiden had a similar story. He had been on a train which had left Schiphol for Leiden at 21:15 on Friday evening. "At Hoofddorp," he recalled, "the train began to rock violently. I think that it was almost derailed. Luggage tumbled off the racks and people were thrown from their seats. We said to one another, 'We might not get there.'"

'Construction workers at the scene of the accident were just having their coffee when they heard a great crash and saw flashes of light. "I thought the construction crane next to our shack had fallen," related bricklayer J. Vermont, 38. "When we came outside though, we found that it was a train, which was lying on its side 50 metres from our hut, on the other side of a ditch. People were screaming for help and there was a considerable panic. Here and there people were lying on the grass, having been thrown from the train. Vermont and his fellow workers immediately rushed to help. "We had to cross the deep ditch to reach the train. It could have been deep enough for us to drown in but you don't think of that in the heat of the moment. Some of the uninjured passengers from the rear of the train joined us to assist those in the front of the train and we did what we could, shining our construction lights into the wreckage. Then we began building a temporary footbridge across the ditch."' [72]

The second, more serious derailment, had, according to NS, occurred at kilometre 19.9, 200 metres from the first, though the location of the two derailments was also stated as having been at kilometre 19.7 for the second and kilometre 19.5 for the first. They were, at any rate, not at the same point. There was no doubt that the second derailment had been due to excessive speed. The driver of the Vlissingen train agreed that he had been issued with the original notice relating to the 90 km/h restriction, TSB (Tidelijke Snelheids Beperkingen – Temporary Speed Restrictions) 49 (applicable to week 49), but maintained that he had not received the supplementary (Aanvullend) notice ATSB 49-1 advising him of the lower speed restrictions, to 60 km/h and 30 km/h, at different points. The restrictions were also indicated by lineside signs but the driver pleaded that from the high cab of the Koploper unit he had not been able to see the signs properly and had read them as 90. [73]

A Portos e Caminhos de Ferro de Moçambique mixed train travelling from the port of Nacala to landlocked Malawi was derailed in the Malema district of northern Mozambique on Monday 13th September 1993. Initially it had been stated that there were 21 deaths but a later police report indicated that 77 bodies had been recovered and another 56 people were injured. Many of those who died had been Malawians. The train was stated to have been carrying 700 tonnes of freight and an unknown number of passengers. The lower number of fatalities originally reported was blamed on the fact that the relatives of many of those who had died had removed their bodies for private burial. It was not immediately known what had caused

[72] *Leeuwarder courant*, Leeuwarden, the Netherlands, 30th November 1992

[73] Wikipedia – Hoofddorp train accident

the derailment. [74] [75]

A Central Railway Up MT BCX Special goods train collided with a buffalo in the station yard at Birlanagar Junction, Madhya Pradesh, India, in the Jhansi Division of the railway, at about 04:00 on Wednesday 24th November 1993. As a result of the collision some of its waggons were derailed and they were then run into by train No. 4067, the Down *Malwa Express* from Indore to New Delhi. Thirteen people lost their lives and 61 more were injured, 34 of them seriously.

The Commission of Railway Safety noted that '"Cattle runover" is quite frequent on the "Shabradi Route" between New Delhi and Bhopal. The extant instructions contained in "General and Subsidiary Rules" do not stipulate for a train to be stopped in such eventualities and examined for its safe onward journey. Suitable supplementary instructions are required to be issued.' [76]

A major catastrophe occurred on the Uganda Railway, near Bukedea, Uganda, at about 16:00 on Sunday 2nd January 1994. A mixed train consisting of nine or ten waggons and three passenger coaches was involved. The train offered a cheap mode of transport for farmers to take their products to markets and because of the low fares was heavily used. Thus Mr Charles Isaal, a Bukedea businessman, related, "I could board a train from Bukedea railway station to Gulu for less than 14 shillings. By then, 12 smoked leaves of tobacco cost 4 shillings, but I could come and sell at 12 shillings." Due to the cheap transport and highly profitable tobacco business he had expanded into buying and selling other agricultural products and the train made commodities such as soap, salt and paraffin easily available.

No official record exists of the accident. Mr Alfred Orikodi, who witnessed it, believed that between 50 and 80 people were killed outright, while scores had been injured. "The train fell with a very loud thud, which horrified many residents. And since it happened during the turbulent times, it took residents some time to reach the scene," he remembered. The sights he saw, he related, had been terrifying, many victims being disfigured, with missing body parts. "The waggons fell on the left-hand side, meaning it blocked the doors, which made it extremely difficult for the people to escape and rescuers to help. Many were also stuck amidst sacks of produce," he added. He said that a total of 10 waggons had skidded off the track and overturned. The train had been heavily loaded with both people and agricultural products, so the accident claimed many human and animal lives. [77]

Reports in the American Press and a Jamaican newspaper put the location of the accident as a flat stretch of line near Mbele and stated that 30 passengers had died and 44 had been injured. As well as carrying livestock and farm produce the waggons were described as having carried charcoal and firewood. [78]

Various theories were put forward to account for the derailment. Mr Stanley Okwi Etole, vice chairperson of Kakungur village, said that after the accident rumours had circulated that the train driver had been drunk. "Some of the survivors, who later camped at Bukedea Senior Secondary School, testified that the train was speeding. It had first knocked dead an old

---

[74] *The Daily Gleaner*, Kingston, Jamaica, 17th September 1993

[75] *The Lethbridge Herald*, Lethbridge, Alberta, Canada, 16th September 1993

[76] Government of India Ministry of Civil Aviation and Tourism Commission of Railway Safety Annual Report for 1993-94

[77] https://www.monitor.co.ug/uganda/news/national/ghosts-of-bukedea-train-crash-ruin-businesses-for-28-years-3910312

[78] *The Daily Gleaner*, Kingston, Jamaica, 5th January 1994

woman in Mukura, then it had also hit a cow at Ochekereni, before arriving at Bukedea station in Okunguro complex. The accident happened just two kilometres after the train left the station," he explained.

On the other hand, clerical officer Eminpasia Ekoroi told the *Daily Monitor* newspaper that the derailment had resulted from poor maintenance and interference with the track. He stated, "The railway maintenance workers had taken a long time without fixing the rails due to the insurgency in Teso. So on the fateful day, the train found that thieves had stolen some rail support metals and loosened the rail bolts at the spot of the accident."

The accident meant the end of the use of the railway for passenger traffic, reportedly leading to the failure of numerous businesses. [79]

A packed train carrying mostly black commuters from a predominantly Zulu area jumped the rails in a rugged, heavily wooded valley at Marianhill, near Pinetown, Natal, South Africa, at about 05:20 on Tuesday 8th March 1994. Eight of the ten, or nine of the eleven, coaches of the train fell down an embankment 20 miles west of Durban. Police Major Bala Naidoo stated that at least 63 people had been killed and 370 were injured, 58 of them seriously.

'The cause of the derailment was not immediately clear, though survivors said the train had been going too fast. Witnesses said that bodies were tangled with wreckage and belongings were scattered about the track. The train was carrying about 850 commuters at the time of the accident, said Mike Asefovitz, a spokesman for South Africa's railway corporation, Spoornet. "It was peak-hour traffic for us. People were on their way to work. The nine coaches careered off the track on a sharp bend near Marianhill, outside Pinetown."

'Thabo Mazibuko, a survivor, said: "There were many people trapped who could not breathe." Rescue efforts were hampered by the terrain, and the dead and injured had to be carried more than 500 yards to ambulances and buses.' [80]

Another source stated, 'A crane began lifting the crashed train cars and more bodies were found underneath, said Clive Serrurier, of the Pinetown Protection Services, which is leading the rescue effort. "It's estimated there are now between 50 and 88 dead, and it seems it's going to be toward the higher figure because they still have quite a few coaches to pick up," he said.

'An African National Congress official blamed political motives for the deaths and injuries to black commuters. Rail officials said they could not rule out the possibility that the derailment was deliberate.

'Rescue workers struggled to carry victims across steep terrain to ambulances. Helicopters and a military plane helped transport casualties to area hospitals.

'Police pulled bloody handbags, shoes, and severed limbs from the wrecked train cars as survivors and spectators watched from the surrounding hills.

'"We just got dizzy. We didn't know whether we were falling over the hills or what," said Fios Xaba, who escaped with minor injuries.

'Rail officials said the train's engineer, who was not injured, was being given blood tests. The engineer was not identified.

'At this point of time, we don't know what caused the train to derail, but we are not ruling out the possibility of a bomb or foul play," said Sylvia Mbhele, a spokeswoman from the rail company Spoornet.

---

[79] https://www.monitor.co.ug/uganda/news/national/ghosts-of-bukedea-train-crash-ruin-businesses-for-28-years-3910312

[80] *The Times*, London, England, 9th March 1994

'The train was carrying about 800 people, almost all of them black, from the Kato Ridge suburb to central Durban. Durban is about 565 kilometres south east of Johannesburg.

'A local ANC official, Sithunywa Dube, claimed the accident was politically motivated and demanded an independent inquiry. He said when he arrived at the scene an hour after the crash, little was being done to help victims. He said there were only five ambulances at the scene when he arrived and that the first helicopter did not arrive to evacuate victims until 2½ hours after the crash.

'There are fears that terrorist attacks will increase in the run-up to South Africa's first all-race national elections in April. The election is opposed by several white and black groups.

'The crash was the worst train accident in South Africa since 1965, when 91 people died in a crash outside Durban. Excessive speed was blamed in that crash.' [81]

According to the author's notes there was a major derailment in Shaba province, Zaire, on Monday 21st March 1994, in which 43 lives were lost. [82]

Late on Sunday 3rd April 1994 a passenger train bound for Rawalpindi was derailed and its engine and two coaches fell about 12 metres down an embankment near Sohawa, Pakistan, 70 kilometres south of Islamabad. Thirteen people lost their lives and at least 45 were injured. [83] [84]

Twenty-five passengers reportedly died when a train running on a new line completed only the year before was derailed and overturned at Pinlaung, Myanmar, in Kayah state, on Sunday 15th May 1994. [85]

At least 21 people lost their lives and 51 were injured on Thursday 11th August 1994 in an accident that occurred near Dvuluchnoye, Russia, 120 kilometres south-east of Belgorod. The initial event was the derailing of a freight train from Ukraine, five container waggons of which broke loose and one or more fouled the adjoining line, on which a local commuter passenger train was running. As a result of the collision with the derailed waggons at least the two rear and possibly four carriages of the passenger train were also derailed and overturned. [86] [87]

An intercity passenger train from Athens to Volos on the Athens – Thessaloniki main line was derailed when it entered the station at Tithorea, Greece, on Sunday 11th September 1994 at high speed. Two coaches of the train were derailed, five of the more than 100 passengers lost their lives and another 35 were injured. The through route that the train normally took was blocked by another train, the engine of which had failed, and the points were therefore set for a diverging route via a low-speed crossover. [88]

An Organismos Sidirodromon Ellados (OSE) (Hellenic Railways Organisation) investigation determined that the train had been travelling at 140 km/h at the moment of the

---

[81] *The Lethbridge Herald*, Lethbridge, Alberta, Canada, 8th March 1994

[82] Author's personal notes, source mislaid

[83] https://www.upi.com/Archives/1994/04/04/Thirteen-die-in-Pakistan-rail-accident/7051765432000/

[84] *Hutchinson News*, Hutchinson, Kansas, United States, 5th April 1994

[85] www.jrtr.net/jrtr03/topics.html

[86] *The Bakersfield Californian*, Bakersfield, California, United States, 12th August 1994

[87] *Winnipeg Free Press*, Winnipeg, Manitoba, Canada, 12th August 1994

[88] *European Stars and Stripes*, Darmstadt, Germany, 12th September 1994

derailment at a point where the maximum permissible speed was 50 km/h and blamed it on the driver's failure to reduce speed and obey restrictive signal indications. Driver Giorgos Grammatakis and Conductor Andreas Mitolius subsequently found themselves in Court at Lamia and were sentenced to 59 and 33½ months' imprisonment respectively. The driver admitted his responsibility for the accident at his trial and expressed his sympathy with the relatives of the dead and injured.

Mr Epaminondas Koukos, president of the Greek Transport Workers Union, later blamed the accident and others on the fact that the railway rules and regulations were outdated and were unchanged from the era of steam traction, when trains ran at an average speed of only 40 km/h. He called for an updating of them, also complaining that most of the rolling stock and track was at least 30 years old and that there was a lack of staff training. [89]

What was described as the worst railway accident in Hungary for a quarter century occurred on Friday 2nd December 1994 at Szajol station, when an express from Nyiregyhaza to Budapest, travelling at full speed, was derailed. The second coach of the train left the rails and dragged three other coaches with it. A railwayman who witnessed the accident related, "There was a tremendous explosion, like the end of the world." He added that people were still buried in the wreckage four hours after the accident.

A reporter wrote, 'The number of dead is expected to rise as rescuers search under the train's cars and in the remains of a three floor railway worker's hostel and a waiting room, which were demolished in the accident Friday at Szajol in central Hungary.

'Police said that more than 100 people were injured, some critically.

'One of the rail cars smashed through the ground floor rooms of the hostel next to the track, bringing the two upper floors and roof crashing down on top of it. Two first-class cars mounted the platform and demolished a waiting room. An electric power pylon collapsed over the cars where they lay embedded in the remains.

'"I felt the carriage jerk and I was thrown over to the other side of the compartment," said Lazlo Bakos, 22, from his hospital bed after undergoing surgery. "The lights went out and there were five, six, seven more jerks."' [90]

Another report stated, 'At least 60 people were injured, said police spokesman Laszlo Garamvoelgyi.

'Television showed pictures of battered and broken railway cars covered in bricks and rescue crews dragging broken bodies from the wreckage. Rescue workers said part of the destroyed building held sleeping quarters for railway personnel and their families. Defence Minister Gyoregy Keleti said dogs normally used to locate victims of mining disasters found three surviving children who were dug from the rubble.

'The death toll {put in this report at 24} was expected to go higher, with temperatures below freezing and some people still trapped in the wreckage, said Garamvoelgyi.'

Doctors at a hospital near Szolnok, close to the site of the accident, said they admitted 26 people. Other hospitals also took in injured. [91]

By Saturday the number of dead had risen to 29, Transport Minister Karoly Lotz stated that a switching mistake had caused the accident and that an investigation was in progress.

---

[89] *MixanikosOse*, Athens, Greece, 11th September 2019

[90] *The Medicine Hat News*, Medicine Hat, Alberta, Canada, 3rd December 1994

[91] *The Lethbridge Herald*, Lethbridge, Alberta, Canada, 3rd December 1994

Southern Railway train No. 6529, the Down *Udyan Express* from Bombay Victoria (now Mumbai CST) to Bangalore City was derailed at Km. 14/4-16 between Yelahanka Junction and Chennasandra station, Karnataka, India, on the Bangalore City – Dharmavaram Junction broad-gauge single-line non-electrified section of the Bangalore Division of the railway, at about 07:10 on Saturday 31st December 1994. The accident happened on a two-degree curve. The train separated between the second coaches with a gap of about 250 metres between the two engines and leading two coaches and the remaining 19 coaches. The first 16 coaches of the train left the track. Eight people died and 48 were hurt, 21 seriously.

The derailment was blamed on the presence of a broken coil spring at the primary suspension level in one of the coaches and the excessive speed of the train on the curve, leading to the off-loading of the broken spring and its consequent dropping out of its seat.

Several recommendations were made by the Commission of Railway Safety in relation to the accident:

(a) Indian Railways should undertake a comprehensive study of spring breakage trends in Integral Coach Factory all-coil coaches all over the system and its implications for the safe running of trains.

(b) The Railway Administration must take remedial measures to arrest the high incidence of spring breakages in coaches primarily maintained in the Coaching Depot at Bangalore City.

(c) The working of the Coaching Depot at Solapur on the Central Railway was unsatisfactory and needed improvement.

(d) The sanctioned work of deep screening and track renewal on the Yelahanka – Baiyyappanahalli section in the Bangalore Division of the Southern Railway should be completed expeditiously so that the sectional speed could be raised to 100 km/h.

(e) The implementation of recent instructions of the Railway Board regarding the foregoing of secondary maintenance for trains logging up less than 2,500 kilometres on a round trip should be kept in abeyance. [92]

A train of the Eusko Trenbideak (Basque Railways) (Ferrocarriles Vascos in Spanish) was derailed on a curve as it left the station at Éibar, Spain, at 13:02 on Monday 27th February 1995. The narrow-gauge train was described as consisting of an engine and two coaches, in charge of Conductor Saturnino Ferro, 48, with Driver Juan Saavedra. The engine and leading carriage overturned as they fell down an embankment bordering the Ego River, though they fortunately did not go into the water. The train had a capacity of 200 passengers and was carrying about 50 at the time, five of whom died and 33 were hurt. Those who lost their lives were named as Maria Esther Irureta Aguirregomezkorta, Roberto Huidrobo López (who died in hospital), Josu Muguruza, Garbiñe Odriozola and Maria Rosario BF.

A couple named Aurora and Paco, who lived closed by, were the first rescuers on the scene. Aurora said, "I heard a loud noise. I looked out of the window and saw the huge cloud of dust next to the tracks. I immediately realised that it was something big and said to myself, I have to help. There were about twelve bodies on the embankment. What surprised me most was the silence."

As to the cause of the accident, the head of the Basque Government Emergency Department announced soon afterwards that it did not appear that it had been due to excessive

[92] Government of India Ministry of Civil Aviation and Tourism Commission of Railway Safety Annual Report for 1994-95

speed.[93] Despite this statement, the subsequent investigation determined that excessive speed through the curve, at kilometre 50 of the San Sebastián – Bilbao line, had in fact caused the accident. Carlos Garcia Cañibano, the president of the railway company, stated that an analysis of the tachograph of the engine showed a speed of 79 km/h at the curve, where the maximum permitted speed was only 35 km/h.

Driver Saavedra had been involved in previous accidents at Herrera in 1990 and Loyola in 1991, the latter due to running past a signal at danger. If, at the conclusion of the inquiry into the present accident, he and the conductor were found responsible for it, as seemed likely, they risked dismissal. [94]

On Sunday 12th January 1997 Ferrovie dello Stato rapido (express) train No. P 9415 from Milano Central to Roma Termini, the *Botticelli*, an electric Pendolino high-speed tilting train, started out at 12:55 with 167 passengers. At 13:20 it was forced to stop out of course because of a problem with a faulty door but shortly afterwards resumed its journey. Six minutes later it became derailed on the entrance curve 400 metres from Piacenza station, Italy. As a result of the derailment its leading vehicle overturned and collided with the supports of the overhead line equipment, one of which cut through it. The six following coaches were derailed and only the rear two remained on the track. Drivers Lidio De Sanctis, 55, and Pasquale Sorbo, 46, two stewardesses, two railway police officers and two passengers were killed. Francesco Cossiga, a leading Italian politician and a former President of the Republic, had a lucky escape, as he had been riding in the leading carriage but had gone to the dining car just before the derailment. [95]

Another politician who had been on the train was Member of Parliament Piero Melograni, who told a reporter that he had supposed from the number of uniformed police at Milano Central station that there must be someone of importance on the train and had discovered that it was the former president of the country when he (Melograni) had entered the dining car, and he joined the president at a table. He said: "We were sitting talking and suddenly we realised that we must have hit something. I remember that I had just started eating. The train did not seem to be going very fast. We heard the train swerve; plates fell off the table and in a moment we were tipped over onto our right side. Outside the window all that I could see was the ground. There were a few moments of panic but no one in our coach was hurt. Someone shouted that we must keep still, lest the train overturn again, and we waited for assistance, which came very quickly, when we were told to get off the train." [96]

Signor Melograni stated that he was firmly of the opinion that the derailment had been an attempt on the life of ex-President Cossiga, a theory discounted by investigators into the accident, as was a theory put forward by some of the media that the two train drivers had been drunk. A third theory, advanced on the part of some of the railway unions, was that the driveshaft of the leading locomotive had failed and falling onto the track had formed an obstruction, which had thrown the train off the line. This theory was also ruled out, it being pointed out that FIAT Ferroviaria had previously modified the locomotives by constructing a steel containment grid around their driveshafts, in order to prevent just such an accident.

The Press had also supposed that the Pendolino had been travelling at excessive speed and an examination of the tachograph from the train showed that this had indeed been the case. The derailment had occurred on the sharpest curve on the Milano – Bologna main line, over which the maximum permitted speed was 105 km/h, while the tachograph indicated that the train had been travelling at 160 km/h. The accident was thus apparently the result of driver error.

---

[93] *El Pais*, Madrid, Spain, 28th February 1995

[94] Ibid, 3rd March 1995

[95] Wikipedia – Incidente ferroviario di Piacenza

[96] *L'Unità*, Roma, 13th January 1997

Twenty-five Ferrovie dello Stato executives found themselves in Court in 2001 charged with multiple counts of manslaughter, causing injury through negligence and causing a railway disaster. All were acquitted on the grounds that they had not directly caused the accident.

From 2003 the railway started installing the Sistema Controllo Marcia Treno (SCMT) train control system to prevent a repetition of such a derailment. [97]

Intercity passenger train *Miguel de Unamuno* was proceeding towards Hendaye, Spain, at 140 km/h on the evening of Easter Monday, 31st March 1997. It was almost full, with 243 passengers plus five train crew. Many of its passengers were returning home from holidays on the Catalan coast, others were just setting out on their holidays. Normally the *Miguel de Unamuno* crossed an opposing Vitoria – Pamplona – Castejón regional train at Etxarri Aranatz station but on Easter Monday it was running 12 minutes late, so the crossing of the two trains had been altered by Control at Miranda de Ebro to the previous station at Uharte Arakil, Spain (formerly spelt Huarte Araquil).

Two kilometres short of Uharte Arakil the Señal de Avanzada (Distant signal) was displaying a Yellow triangle, indicating that the speed of the train must be reduced to 30 km/h at the semaphore Señal de Entrada (Home signal), so as to negotiate the points at the entrance to the crossing loop. Driver José García Fernández (49) and Assistant Driver Miguel Ángel Marinetto (32) failed to react to this warning and the train continued at full speed. The inquiry into the resulting accident ruled out any possibility that the Señal de Avanzada had been displaying Via Libre, indicating that the train could continue at unreduced speed, but considered that the two engine crew could have been dazzled by the setting sun.

It was only when the train was close to the Señal de Entrada, 200 metres from the points, that the engine crew realised that its semaphore arm was at 45 degrees, the Anuncio de Parada position, indicating that the points were set for the diverging route, through which the speed of the train should not exceed 30 km/h. Even then, the emergency brakes were not applied, only the locomotive's direct brake, and the train was still moving at 137 km/h at the crossover. The locomotive and leading coach managed to negotiate the points and entered the crossing loop still on the rails but the remainder of the train was derailed. The third coach mounted the second and overturned, while the fourth coach came to a stand at right angles to the track at the points. It was 19:43.

Two minutes after the derailment a resident called the emergency services. He could not offer much information other than that a passenger train had derailed and that it looked serious. A medical team from Irurtzun health centre, an ambulance from Etxarri Aranatz and firemen from Alsasua were dispatched to the scene. After a few minutes the Civil Guard arrived at the site, the full magnitude of the disaster became apparent and a full emergency response was set in motion. A helicopter, which had been engaged in traffic control, landed on a nearby road and evacuated two very seriously injured victims and all available public and private ambulances were activated. Some 500 people, including doctors, police, firemen, Red Cross volunteers and so on were eventually involved.

Eighteen people lost their lives in the disaster, 103 injured were taken to hospitals in Pamplona and 12 slightly injured to Vitoria Hospital.

Passenger Isaro Beita Leibar, 14, from Gipuzkoa, said that she had been riding in a second-class coach, sat next to the window. "I was watching the film," she said, "when suddenly the train jumped." She lost consciousness briefly and, "When I came to, I was lying on the ground, and I realised that the train had overturned. I was lying near the track and a

---

[97] Wikipedia – Incidente ferroviario di Piacenza

man in his 20s, who told me that his name was Andoni, was holding my hand and telling me to stay calm." Isaro believes that it was this man who had pulled her from the train. "I'm not certain, but that's what I think, and if he reads this, I'd like to thank him," she added. Isaro suffered from a concussion and a broken ankle.

Another passenger, José de la Poza, 62, had been travelling from Barcelona to Zumárraga with his 16-year-old grandson, Garikoitz Amenabar de la Poza. The two had spent five days in the Catalan capital visiting relatives. José de la Poza, who had been blind since birth, worked at the ONCE (National Council of Education). "I thought I was going to the other side. I thought I was going to the other world. Seriously, I thought I was going to die," he told a reporter. Asked how he felt now, he replied, "I don't feel too good, could be better, but all that's left is to recover. It's like being reborn, like the Carlos Cano song. I know I have a broken rib and I'm in a lot of pain. But the important thing is that it's all over." Asked if he had spoken to his family, José replied, "My son called me yesterday and my daughter came to the clinic to be with me. I was reassuring them. I told them I was fine. When the derailment happened, I was listening to the news on the radio but not paying much attention to it, while my grandson was watching television. I could tell that the train was going too fast and then it felt like an earthquake. Thank God I am here to tell the tale; others were not so lucky. Afterwards, I could hear little children, women and men screaming. I was screaming too. I couldn't get out of where I was, even if I wanted to, because I was trapped under the seat and was being struck in the chest. My grandson was very upset and didn't know what to do but, in the end, nothing happened to him. They had to cut my trousers with a knife to pull my leg out, bit by bit, and my trousers stayed there, caught up in the wreckage. It was only about five minutes before the ambulance arrived, but the minutes felt like hours. Meanwhile, they covered me with a blanket and placed me on the side of the road. I felt cold and dizzy." [98]

Driver Fernández and Assistant Driver Marinetto later stood trial in Pamplona Criminal Court No. 1 and both were found guilty of negligent homicide and causing injury through gross negligence. The driver was sentenced to two and a half years' imprisonment and the assistant driver to a fine of 120,000 pesetas, changed on appeal to a minimum term of imprisonment. Neither railwayman was actually imprisoned as both were pardoned by the Court. [99]

The line was subsequently equipped with the Anuncío de Señales y Frenado Automático (ASFA) automatic train protection system, already in use on a number of other lines in Spain, and which should have prevented the disaster by applying the brakes of the train when its driver had failed to do so.

About five people died in the derailment of a train near Lagos, Nigeria, on about Tuesday 15th April 1997. A very brief report on the accident stated that the train had been overloaded due to a lack of buses on the fifth week of fuel scarcity. [100]

A passenger train from Nakhon Si Thammarat province to Bangkok was derailed at an unstated location in the Na Bon district of Thailand on Friday 6th February 1998, killing nine people and injuring over 100. The police conducted an investigation into the accident and there were complaints that the State Railway of Thailand (SRT) had not been as cooperative

[98] https://www.diariodenavarra.es/noticias/navarra/2022/03/31/25-aniversario-accidente-tren-uharte-arakil-522467-300.html
[99] https://www.noticiasdenavarra.com/navarra/2022/04/03/emotivo-recuerdo-uharte-arakil-victimas-2084012.html
[100] www.cohdn.ca/ndj/NIGERIA%20TODAY%20WEDNESDAY%2016%20APRIL...

with them as it might have been. The SRT was also accused of only paying 10,000 baht (about US $300 at 2025 exchange rates) compensation to the families of the victims and of paying little attention to the injured. SRT governor Saravut Thammasiri resigned three days after the accident.

The derailment was possibly due to excessive speed. [101]

A passenger train from Turku to Pietsämäki, consisting of a locomotive and 11 carriages, was derailed at 13:49 on Friday 6th March 1998 on the approach to the yard at Jyväskylä, Finland, with serious consequences; nine of the approximately 300 passengers and the driver died and 94 people were injured, eight of them seriously. Damage amounted to the equivalent of about $4,000,000 US.

Passenger Veli Jokinen related, "As I was making my way back from the toilet, the train began to shake, people screamed, and a suitcase fell on my head. There was a lot of shouting. People were in shock and they were bleeding from wounds." He was able to retreat to the next carriage as his began to overturn. [102]

As a result of the derailment the engine turned onto its side, skidded across two lanes of road, turned onto its opposite side and collided with a bridge pillar. The leading two coaches had initially followed the engine, then they became separated from it and overturned. Two of the passengers who lost their lives had been riding in the leading coach and seven in the second coach; they had fallen through coach windows when the coaches overturned and then been crushed beneath them.

The investigation into the disaster determined that it had resulted from excessive speed. The train had approached Jyväskylä travelling at a speed of 143 km/h and had still been travelling at 110 km/h when it had been derailed at a facing crossover, through which speed was limited to 35 km/h. A brake application had been made 840 metres short of the crossover, followed by a second, stronger application 320 metres from the crossover and two seconds later by a third, emergency application, which had not begun to take effect until a point 100 metres before the crossover.

The report on the accident identified a number of factors that had contributed to it:

(a) The driver had been tired. Five hours of rest, which he had had before starting the journey, had been of poor quality because of inferior sound insulation and an echo effect in the rest rooms of the railway hostel where he had been.

(b) Approaching Jyväskylä, both the driver and the assistant driver (referred to as a lookout driver in the English translation of the Finnish accident report) had been occupied in making coffee and hence had not been paying full attention to the signals.

(c) Up to a year previously, trains approaching Jyväskylä from the south had used a different crossover, located further in advance, through which speed was restricted only to 80 km/h. On most of the passenger trains the driver had worked through Jyväskylä, either as a driver or an assistant driver, the train had been routed via the higher speed connection. That the driver had assumed this train was going to be routed through this crossover was also suggested by the setting of the speed control to 60 km/h and the manner in which the train had been braked prior to the emergency application, which would have been appropriate for the higher-speed crossover.

(d) The signal indications were not as helpful as they could have been. Yellow and Green

[101] *Bangkok Post*, Bangkok, Thailand, 17th February 1998

[102] https://www.spokesman.com/stories/1998/mar/07/passenger-train-in-finland-derails-killing-at/

aspects were displayed regardless of which crossover the train was routed through. Were a train to be routed through the second, higher-speed crossover, a figure 8 was also displayed at the signal, while for the first, lower-speed crossover, no indication was shown. The assistant driver had realised the situation at the last moment and had shouted a warning to his colleague; this was what had prompted the emergency brake application.

The accident investigators were critical of what they regarded as a casual attitude to certain rules and regulations on the part of some operating staff and a failure to update rules and instructions on the part of management.

The accident report noted that it was fortunate the derailment had occurred in an urban area. The alarm had been raised within a minute of the derailment and the first emergency services had been on the scene in seven minutes. About 37 minutes after the accident the last casualty had been ready to be taken to hospital. The rescue operation was considered to have worked well.

In addition to inquiring into the accident, members of the investigation commission were noted to have visited the nine families to inform them of the progress of the inquiry, rather than the families having to rely entirely on media reports.

The accident report included a number of recommendations. The principal recommendation involved an improvement in the signal aspects to indicate more clearly through which crossover a train was routed. Other recommendations concerned the provision of optimal seat belts in new and refurbished day coaches of long-distance trains, the planning of the layouts of railway yards, the updating of train drivers' instructions, improvements in railwaymen's rest areas at depots and some matters relating to the rescue operation. [103]

A freight train consisting of an engine and 22 waggons was derailed at a point in the vicinity of Goudiri, Senegal, on about Saturday 18th July 1998. Six people lost their lives in the derailment on the Société nationale des chemins de fer du Sénégal, which occurred during a severe rainstorm. It was believed that the train had been carrying an inflammable liquid, possibly petrol. [104]

Another Société nationale des chemins de fer du Sénégal freight train, composed of an engine and six waggons, bound for Bamako, was derailed at a point stated to have been 545 kilometres east of Dakar at about 10:00 on Friday 30th October 1998. The stated location would be in the vicinity of Goudiri, Senegal, and was thus not far from the site of the July accident. Seven people died in this second accident, including two women and a child, and two others were slightly hurt. The line remained blocked two days later. [105]

Central Railway train No. 8209 Down, a Khurla – Howrah express, was derailed between Mhasawad and Shiroli stations, Maharashtra, India, on the Bilaspur Division of the railway, at 05:41 on Thursday 3rd December 1998. Eleven coaches left the rails and five of them were badly damaged. Eight people were killed outright and 42 were injured, two seriously. The injured were taken to Jalgaon civil hospital. [106]

---

103 https://turvallisuustutkinta.fi/material/attachments/otkes/tutkintaselostukset/en/raideliikenneonnettomuuksientutkinta/1998/a11998r_tutkintaselostus/a11998r_tutkintaselostus.pdf

104 *Danger Ahead!* Historic Railway Disaster, 18th July 1998

105 https://allafrica.com/stories/199811010054.html

106 www.mumbai-central.com/grapevine/msg00517.html

A Kenya Railways overnight passenger from Nairobi to the port of Mombasa, popular with foreign tourists as well as locals, was derailed at 04:15 on Wednesday 24th March 1999 on the edge of Tsavo National Park, near a location known as Man Eaters Junction, so named because a number of workmen engaged in building the railway in the 19th century had reportedly been killed by the big cats. There were stated to have been 614 people on the train, including at least 86 French tourists, 20 British nationals, 8 Germans and 6 Americans.

The three locomotives and two waggons at the front of the train remained on the track but the coach, which was the third vehicle behind the engines, left the track, followed by 11 more coaches that tumbled down a low embankment and were ripped apart. 'Blue blankets, sheets and curtains were strewn amongst the wreckage along about 120 feet of track.' Twenty-seven Kenyans and five foreigners lost their lives and 254 were injured, 115 of whom were admitted to hospital, most being flown to Nairobi and Mombasa from the remote site amongst dusty scrubland. A few were treated in the hospital at Voi, closer to the scene.[107] According to an Associated Press reporter no large-scale rescue operation had been underway to free trapped passengers even when he had arrived at the scene, eight hours after the accident.

An English newspaper wrote, 'Nine British geography students on holiday in Kenya helped to pull the dying and injured from a train crash that killed at least 32 people. Most of the British students, all in their twenties, were asleep in the second-class carriages when the train jumped the tracks near a junction called Man-eaters, about 200 miles south-east of Nairobi. One of the students was travelling first-class with his girlfriend. They escaped with minor injuries.

'None of the British students was seriously injured. After being taken to Nairobi hospital they were allowed to go to a nearby hotel. One woman was kept in hospital with back injuries. One of the group, Emma Crowfield, 20, said: "It was about 4.30 a.m. and we were all asleep in bunks. I woke up when the train started to speed up. Our carriage rolled over and ended up on its side. We wanted to get out pretty quickly. We climbed out of the windows. Some were open and some we had to smash. We managed to get on top of the carriage. It was a complete wreck. We started to try and pull people out of the train. It was really quiet – no real screaming or crying, although you could hear people who were trapped asking for help. I pulled out a few people. One had facial injuries and was covered in blood. It was carnage. There were obviously a lot of dead people. It was quite horrific. It was pitch-black and we were also worried about the animals, because we were in a national park. Some people started to light fires. The first-class carriages were fine and were still on the rails, but most of the train was a complete mess. There was metal sticking out everywhere."

'The students were heading for Kenya's Indian Ocean resorts after completing a field trip on Tuesday, with 30 other students and tutors from Sussex University. They had been in Kenya looking at rural developments and had also visited Zimbabwe and the Victoria Falls.

'Daniel Burstow, 20, another student, said he had to break windows to help other passengers to safety. "I was awake and felt the train was bouncing around quite viciously. I immediately knew something was wrong. The next thing I knew was that there was a rather large noise and a lot of commotion." He said he saw "a lot of dead and dying. There was a big problem in that all the doors were locked and nobody could get out. We had to break windows to get them all out." Windows and doors of the overnight train are always locked to prevent thieves jumping on at stops along the route.

'Miss Crowfield added, "I am flying home but a few are planning to carry on with the journey. We all just suffered cuts and bruises, although one girl is in hospital. It was meant to

[107] *The Medicine Hat News*, Medicine Hat, Alberta, Canada, 26th March 1999

be a relaxing trip. Everyone is really shaken up."' [108]

'Michael Greenwood, 39, of Ft. Lauderdale, Fla., who suffered only minor injuries, said he was riding in the last of the 13 passenger cars. "All the doors were locked and nobody had the keys to open them," he said at Nairobi Hospital. Greenwood said some survivors had to climb out windows and walk about a mile to the main Nairobi – Mombasa highway, where they were helped by Chinese road workers.

'Beatrice Perez, one of a group of 13 French tourists on the train who was brought to Nairobi by the air force, said that the train sped up just before the accident.' [109]

Another report stated, 'Survivors said many people were sleeping when the train derailed – and some claimed the train was going too fast. "It was running at an unusual speed," said Naaman Marete, a Kenya Railways passenger service inspector and a frequent traveller on the train. "All of a sudden we went around a big bend and then we were off the track," Marete said, sitting in the steps of an upturned coach near the mangled tracks.

'Rescuers – volunteers, police and soldiers – picked through the overturned cars, searching for bodies and collecting personal belongings. A bulldozer moved aside bent and broken cars, many separated from the twisted metal frames of their wheel bases. It was not immediately known whether the portion of the railway where the accident occurred had been damaged by recent heavy rains.

'At Nairobi Hospital, where 26 seriously injured people were admitted, a woman with back injuries cried out in pain as medics carried her on a stretcher from an ambulance.

'Ticket Inspector Abel Otete was in the first car off the track and was injured when the car rolled twice. He still managed to climb out of the car window. "I was saying to myself, 'I am dying,'" he said. "I was waiting for the last hit before I died. When (the car) finally landed I saw that I was still alive."' [110]

After the accident there was an allegation from members of the train crew that there had been a brake failure. The investigation into the disaster concluded, however, that in fact there had been nothing wrong with the brakes, but that they had been mishandled by the driver, so that when the train had reached the curve, where speed was limited to 50 km/h, it had still been travelling at 93 km/h. The statements of a number of surviving passengers that the train had been running fast immediately before the derailment had therefore been correct. Transport Minister William ole Ntimama announced to parliament that the driver responsible had been arrested. [111]

A passenger train from Wuchang in Hubei province to Zhanjiang in Guangdong province, carrying over 1,000 people, was derailed at 22:48 on the evening of Friday 9th July 1999 near the city of Hengyang, China, in Hunan province. Twelve of the 18 coaches of the train left the rails and four overturned. Eight people died and 61 were injured.[112] Railway Minister Liu Zhijun, who went to the accident site, stated that the train's tachograph had been recovered and a preliminary investigation indicated that excessive speed might have caused the derailment.[113] Excessive speed was quickly confirmed as the cause and a number of officials considered to have been in some way to blame received penalties ranging from "public criticism" to "demerits on their records." The two railwaymen held to have been directly

[108] *The Times*, London, England, 25th March 1999

[109] *European Stars and Stripes*, Darmstadt, Germany, 25th March 1999

[110] *Fairbanks Daily News Miner*, Fairbanks, Alaska, United States, 25th March 1999

[111] http://danger-ahead.railfan.net/reports/rep99/voi_02.html

[112] *The Orange County Register*, Santa Ana, California, United States, 11th March 1999

[113] *China News Digest*, News Global

responsible for the accident were to be "punished according to the law by the judicial departments." [114]

At 00:13 on Sunday 6th February 2000 train D 203, the international *Schweiz Express* from Amsterdam Central to Basel, was badly derailed at Brühl, Germany. Nine of its 241 passengers died in the accident and another 149 were injured.

The passenger station at Brühl was situated two kilometres on the Bonn (Basel) side of the freight station. On the night of the accident the normal right-hand track for the train (Track No.1) had been taken out of use at Brühl for engineering work and as there was no mains crossover at the freight station, trains were being diverted onto the left-hand line (Track No. 2) at the previous station at Hürth-Kalscheuren. The normal maximum permitted line speed of 160 km/h was reduced to 120 km/h within the area of an engineering occupation but as Track No. 2 was not signalled for wrong direction movements train speeds were further restricted through the section to 40 km/h. There was also no direct connection between Track No. 2 and Track No. 1 at Brühl passenger station, at the far end of the engineering occupation, trains being instead routed there first via a crossover to Track No. 3, this crossover only being suitable for traversing at speeds of up to 50 km/h.

Train D 203 received the correct Caution indication at the Hürth-Kalscheuren Vorsignal (Distant signal) informing the driver that the Ausfahrsignal (Starting signal) there would only be clear to a restricting Hp aspect. Driver Sascha B., aged 28, duly acknowledged the Punktförmigen Zugbeeinflussung (PZB) warning at the Vorsignal, braked his train and negotiated the crossover from Track No. 1 to Track No. 2 safely. His train was then accelerated to 130 km/h. Approaching Brühl freight station the Einfahrvorsignal (Distant signal for the Home signal) was also displaying Caution and the Einfahrsignal (Home signal) was at Danger. Driver B. duly braked his train again. With the train being on Track No. 2 it was not possible for the fahrdienstleiter (signalman) to clear the Home signal, only a subsidiary signal, which he then did. The train proceeded at 40 km/h past the beginning of the engineering work, but after travelling about a kilometre further, Driver B. accelerated, first to 90 km/h and then again to about 120 km/h, rather than continuing at 40 km/h to the next Stop signal, as he should have done. Arriving at the crossover from Track No. 2 and Track No. 3 at Brühl passenger station (No. 48 points), his train attempted to negotiate it at a speed of 122 km/h. The driver's reaction when he realised what was happening was recorded; he exclaimed: "My God, my God, the points! This can't be happening!" His train, travelling much too fast to negotiate the low-speed connection, was duly derailed on the crossover, the engine and leading carriages going down an embankment and the engine crashing into a lineside house, with four coaches coming to rest a few metres from other houses. The fourth and fifth coaches of the train also buckled at an angle of 45 degrees. Only the last coach of the train remained on the rails.

Residents of the street adjoining the railway, who had been awoken by the noise, were the first on the scene, and the emergency services soon arrived. In total about 850 rescuers were involved. Difficulties arose during the rescue operation due to the darkness, the restricted location, the fact that many of the passengers were foreigners from a total of 15 different countries, some of whom spoke little or no German, and so on, so the rescue operation could only proceed slowly. It was two and a half hours before the last injured could be extricated and a good deal longer before the bodies of the last dead could be removed. Driver B. was among the injured, being taken to hospital suffering from shock. Recovery of the wreckage took several days.

---

[114] *Danger Ahead!* China: Derailment caused by excessive speed

Dr Dirk Jan Scholte, a Breda neurologist, had been a passenger in one of the sleeping cars, on his way to a medical conference in Basel. He told a reporter: "First there was a jolt, then a rumbling and a shaking. I could hear people screaming. Fortunately, our coach didn't plunge down the embankment."

The family in the house struck by the locomotive related: "We were already asleep. Suddenly there was this terrible thunder. The whole house shook. It was like a massive earthquake. Everything fell over." They escaped unhurt, though their badly damaged house had subsequently to be demolished and rebuilt.

The accident was investigated by the Eisenbahn-Bundesamt, which found that Driver B. had not completed his initial training with Deutsche Bahn but had instead done so with Köln Häfen und Güterverkehr (Harbour and Goods Transport), after which he had been re-hired by Deutsche Bahn without re-examination.

The investigation noted errors in the notices regarding temporary speed restrictions: the maximum permissible speed on Track No. 2 being stated as 120 km/h, instead of 40 km/h, as it should have been, given that there were no fixed running signals for working in the wrong direction, with a restriction to 90 km/h on Track No. 1 for a period after the engineering work had been completed. This might have confused Driver B.

The signalling of the train had, the investigation considered, been carried out correctly.

It was not clear if a trackside speed restriction sign had been in place at the time of the derailment.

In addition, although train radios were provided they had not been working, so the drivers could not be notified of the intended routing of the train by the signalman by this means.

Early in 2001 the Köln public prosecutor filed nine charges of negligent homicide and 149 charges of causing bodily harm through negligence against Driver B. and three other railwaymen, though proposed charges against the signalman and an operations manager were not proceeded with. After a 23-day trial the cases against the three railwaymen were dismissed by a Köln court in October, though they were ordered to pay fines of between 7,000 and 20,000 Deutsche Marks to charity.

Multiple changes to the railway's operating procedures were made after the accident, while driver training was improved, for example with the use of simulators. Advanced systems of train control, such as the European Rail Traffic Management System (ERTMS), which would have prevented the train being driven at excessive speed, were also gradually introduced. [115] [116]

Seven people died and at least 24 were hurt when four coaches of a passenger train left the track near Thaton, Myanmar, 30 kilometres south-east of Yangon, on Sunday 16th April 2000. The train had been on its way from Mottama to Yangon. Railway Minister Pan Aung and other officials went to the scene of the accident. Those who died comprised two men, three women and two children. [117]

The Grand Magal of Touba is an annual religious pilgrimage of Sufi Muslims of the Mouride Brotherhood, held in the city of Touba, Senegal, since 1928, a year after the death of Cheikh Ahmadou Mamba, the founder of the order, who is buried there. On Monday 22nd May

---

[115] https://www.express.de/koelner-umland/tragoedie-vor-25-jahren-das-bruehler-zugunglueck-von-2000-951807

[116] Wikipedia - Eisenbahnunfall von Brühl

[117] http://theburmanetnews.editthispage.com/stories/storyReader$328

2000, a train reportedly carrying thousands of Mouride Brotherhood pilgrims from Dakar, the capital, to Touba, on the eve of the celebrations, had travelled about 40 kilometres when it was derailed near Diamniadio, Senegal. Ten or 12 people died according to different reports and at least 210 more were injured. Many passengers had been crushed when the Société nationale des chemins de fer du Sénégal coaches overturned and rescue workers were still busy some hours after the accident evacuating the injured.[118] [119] Nine carriages were derailed. Many ambulances and a French military helicopter went to the scene.[120] The victims were given first-aid in the small hospital at Mbour before being transferred to Dakar for further treatment.

Thirteen people, ten adults and three children lost their lives and about 36 were injured when a Kenya Railways passenger train from Butere to Kisumu was derailed near Lela station, Kenya, on Tuesday 15th August 2000 (the location of the accident is also variously given by different sources as Kowango, Kisumu, Masena and Yala). Nineteen of the injured passengers were admitted to hospital and three were stated to be in a serious condition. There were about 400 passengers on the train.[121] Most of the victims were crushed when the coaches overturned on them. There was heavy rain during and after the accident, impeding rescue operations, which reportedly had to be suspended at one point before all those trapped had been released.

A BBC report suggested that the train had run away before becoming derailed,[122] while railway officials at the scene suggested that the train had been overloaded and travelling at excessive speed.[123] An inquiry determined that excessive speed had indeed been the cause of the accident. [124]

A major disaster that occurred between Sarai Banjara and Sadhoogarh stations, Punjab, India, on Saturday 2nd December 2000 was precipitated by the derailment of a Down Ajitwal – New Bongaigaon food grain goods train. The accident happened on the Ambala – Ludhiana double-line section of the Ambala Division of the Northern Railway. Some of the derailed waggons fouled the opposite running line where they were run into at 05:42 by train No. 3005, the Up Howrah – Amritsar Mail. Five coaches of the express were derailed and overturned, 45 people were killed and 149 were hurt, 77 of them seriously.

A newspaper reported that, 'Many villagers carrying lanterns rushed to the site after hearing the loud bang and started pulling out people from the badly-mangled bogies and rushed them to hospitals on their tractors before the officials reached the spot. Cranes have been put into service to lift the mangled bogies and rescuers were using blowtorches to cut through the affected coaches to pull out the trapped passengers. A PTI {Press Trust of India} correspondent who visited the spot reported a ghastly site with three coaches of the passenger train placed above the engine and two others lying derailed with people trapped inside screaming for help. Many of them were badly injured. Villagers who rushed to the site and helped in the rescue operation, also provided tea and blankets to the survivors. They also made announcements over the loudspeakers of religious shrines urging people to provide relief material.'

---

[118] https://www.tvanouvelles.ca/2000/05/22/12-morts-et-20-blesses-dans-le-deraillement-dun-train-au-senegal
[119] https://www.senenews.com/actualites/notification/aujourdhui-22-mai-2000-en-partance-pour-touba-un-accident-de-train-fait-des-morts-et-des-centaines-de-blesses_310743.html
[120] www.ananova.net/news/story/sm_4922.html
[121] https://www.thenewhumanitarian.org/news/2000/08/18/20-killed-train-crash
[122] BBC Africa News, 21st August 2000
[123] https://www.emerald.com/insight/content/doi/10.1108/dpm.2001.07310cac.005/full/html
[124] allAfrica.com 30th December 2000

About 70 of the injured were taken to the Fatehgarh Sahib Hospital, about 50 to the Government Rajindra Medical College Hospital in Patila, 32 to the Postgraduate Institute of Medical Education and Research Hospital in Chandigarh.

Railway Minister Mamata Banerjee was expected to fly to the Punjab to supervise the rescue operations. Meanwhile it was announced that a payment of 15,000 rupees would be made to the next of kin of the dead, with 5,000 rupees going to the seriously injured and 500 rupees to those who were slightly injured. Several trains were cancelled and others were diverted. Special enquiry counters were opened at Howrah, Sealdah, Asansol, Dhanbad, Patna and Mughalsarai stations. [125]

Ms Banerjee offered to resign after the accident but Prime Minister Atal Bihari Vajpayee turned down her resignation.

The Commission of Railway Safety investigators were unsure as to the cause of the initial derailment, stating that it had probably resulted from either a broken rail or from a defect in one of the waggons of the food train. The following collision they blamed on the failure of the crew of the food train to warn their colleagues on the express.

Ten recommendations were made by the Commission relating to the accident:

(a) The Northern Railway Subsidiary Rules regarding the switching on of the flasher light of the locomotive of a train in distress needed to be made more specific on the lines of the instructions issued vide Railway Board letter No.2000/SAFETY(A&R)/19/44 dated 8.12.2000. General Rule 6.03 might also need some amendment.

(b) Locomotive flasher lights should be redesigned so that they automatically switched on if the brake pressure fell to low levels, as recommended by the then Commissioner of Railway Safety in his report on the similar collision near Khanna on 26.11.1998.

(c) Portable control phones/portable control emergency phones should be provided on all locomotives, irrespective of whether they were hauling passenger or goods trains, and should be checked for their proper functioning by competent telecommunications staff before being loaded onto an engine.

(d) The routine overhaul of BCN/BOXN waggons should be brought onto a sound footing and the level of the officer who could permit the running of waggons whose routine overhaul was overdue should be specified.

(e) Registers with the list of items printed on each page should be supplied to the train examiners for the intensive examination of waggons, in replacement of the current daily books.

(f) The length of continuous welded rails should be limited to a maximum of three kilometres.

(g) Further studies should be carried out by the Research Design and Standards Organisation on the gauge corner fatigue of the type found at the point of derailment.

(h) There was a need to carry out computer-based simulation studies on train derailments along with rail/wheel interaction studies.

(i) Corrective action should be taken on the various issues/irregularities that had come to notice during the investigation into the accident.

(j) A decision needed to be taken on how to deal with the centre buffer couplers of old supplies with inferior specifications. [126]

---

[125] https://m.rediff.com/news/2000/dec/02train3.htm

[126] Government of India Ministry of Civil Aviation and Tourism Commission of Railway Safety Annual Report for 2000-01

Thirteen people, unauthorised passengers on a freight train travelling from Dire Dawa to Addis Ababa, lost their lives and 10 others were slightly hurt when the train was derailed and broke into three parts, or broke into three parts and then derailed, late on Thursday 28th or early on Friday 29th December 2000, near the town of Mieso, Ethiopia. The injured were taken to hospital in nearby Asbeteferi. Police officer Haile Cekadu commented unsympathetically that freight trains were not permitted to carry passengers and those who boarded them did so at their own risk, ignoring the fact that probably due to poverty and/or a lack of alternatives the illegal riders might have had little choice. [127]

A serious derailment at Heck, England, on Railtrack London North Eastern, which occurred at 06:13 on Wednesday 28th February 2001, resulted from a train running into an obstruction, in the form of a motor vehicle that had fallen onto the line from a motorway overbridge, followed by the coincidental arrival of a second train on the scene. (The official report on the accident gives the location of the accident as Great Heck, the name of the nearby village, but the correct historical railway name of the place has simply been Heck, so that name will be used here.)

At 04:40 on the morning of the accident 36-year-old Gary Hart had left his home in Strubby, Lincolnshire, to drive the 233 kilometres to his job in Wigan. He was driving a Land Rover Defender, which was towing a trailer loaded with a Renault Savanna estate car.

Meanwhile, Great North Eastern Railway passenger train IF23 from Newcastle to London King's Cross had set off at 04:45. It consisted of a leading Driving Van Trailer (DVT), nine Mark 4 coaches, and a trailing Class 91 locomotive. The InterCity 225 was operating in a push-pull configuration; in normal operations, the DVT was the lead vehicle for southbound services such as this one, remotely controlling the pushing Class 91 locomotive at the rear. There were 99 people on board. Coming in the opposite direction on the Down line was freight train 6G34, the 05:00 Freightliner service from Immingham to Ferrybridge power station, which consisted of a Class 66 locomotive and 16 loaded coal waggons.

Just before the arrival of the passenger train, the Land Rover driven by Hart had left the M62 motorway along which it had been travelling, at a point immediately before a bridge over the railway. The vehicle had then gone down an embankment onto the southbound (Up) main line. Having had little sleep before he set off to work, Hart had fallen asleep at the wheel. Failing to manoeuvre the Land Rover off the line, Hart got out of it and called the emergency services on his mobile phone but while he was still on the phone, train IF23 arrived and hit the vehicle. The leading DVT was derailed and fouled a set of trailing points leading to some sidings, which then deflected the DVT into the path of the Down freight train. As a result of the two collisions the DVT was almost destroyed and all coaches of the passenger train were derailed, some overturning into a field. The engine and nine waggons of the freight train were also derailed, some of them ending up in the garden of a lineside house, flattening a caravan and a garage. At the moment of collision, the passenger train had been travelling at 142 km/h and the freight train at 87 km/h, giving a closing speed of no less than 229 km/h. Six passengers, Driver John Weddle of the passenger train, Driver Stephen Dunn of the freight train, chef Paul Taylor and customer operations leader Raymond Robson of the passenger train, were all killed, while 82 people were injured, including Driver Andrew Hill of the freight train, who had been riding as a route instructor to his colleague.

A BBC television report on the accident included the stories of survivors. 'Former railway worker Raymond Brindley smashed glass and escaped through the emergency door of the rear carriage before racing to the nearest lineside telephone and alerting the signalman to the

[127] *Panafrican News Agency*, 20th December 2000

accident. "The track seemed to give way under us; the lights went out and we seemed to be going everywhere in all directions. It was utter chaos. There was a lot of people shouting, with the diesel fumes they were worried it would catch alight."

'One of the injured, Virginia Shaw, is married to a survivor of a similar disaster which killed nine people. Bill Shaw, from York, was in the Lockington rail crash in July 1986, just a few miles away from the sight of the Selby accident, when a train struck a van on a level crossing. Having escaped uninjured with his two children, he had told his wife what to do in the event of a derailment. "It's a one in a million chance – both incidents involved a vehicle on the track. I've always said…get down on the floor, get hold of something and hold on tight. Every time there is a train crash, I can imagine what it is like because I have been there. I know the despair that people are feeling." His wife called him soon after the crash on her mobile telephone. "She didn't want me to go through what she went through, hearing about the crash from the media and not knowing if I was alive or dead," explained Mr. Shaw.

'Janine Edwards, 22, of York, was travelling to London for a drama college interview. She said: "I heard screaming and shouting and the lights went out. I held onto a table in front of me and then there was a huge impact. One lady, who was travelling with her daughter, had been flung into the air and was lying in the next corridor. Her leg was trapped. The man opposite me was streaming with blood. The window next to him was smashed and the frame had come out and hit him. I think two people managed to scramble out of the carriage, but I don't know how. Both ends were crushed. Another lady in the carriage said she could feel a bone coming through her leg. It was like a rollercoaster ride. It was strange, you could even feel the impact against your face as the train tried to brake. There are so many people badly injured. I feel so lucky. If I had been in the front two carriages, I think I would have been killed."

Dai Edwards told BBC News that he went to the scene after his 22-year-old daughter Janine rang him from the wreckage using her mobile. She said: "Dad, the train's crashed." He found her at the crash site and took her to hospital. She is in hospital with an injured back.' [128]

The accident formed the subject of no less than four different reports, by the Health and Safety Executive (HSE), the Health and Safety Commission (HSC), the Highways Agency (HA), and the Department of Transport (DoT).

The HSE report noted that there had been no defects in the railway infrastructure that could have contributed to the accident and that the motorway had complied with current standards. The rolling stock of the passenger train had performed well in the accident given the high speeds involved. The freight locomotive had suffered severe damage to its front, and its design had lacked energy-absorbing features, the report stating: "Its design does not appear to complement the crashworthy considerations that guide the construction of passenger vehicles that use the same rail network." Some of the freight waggons had caused heavy damage to the passenger coaches due to their angular design.

The HSE report recommended further research into the behaviour of vehicles in derailments, considering factors such as vehicle construction, obstacle deflectors and bogie retention and energy absorption. Future freight vehicle designs should avoid aggressive corners at points of possible interface with passenger trains. Tripwires, which could detect the incursion of motor vehicles, similar presumably to those used to warn trains of landslides, were also suggested.

The HSC report decided that the risk of motor vehicles entering the railway other than at a level crossing and then causing an accident of the magnitude of Heck was minimal and that there were considerably greater risks to both road and rail traffic. Risks of such rare accidents as Heck could be reduced by educating motorists, reducing vandalism and careless behaviour

[128] http://news.bbc.co.uk/.../119389.st

and by preventing incursions from reaching the railway line, or reducing the severity of such collisions with trains as did occur.

The HA report likewise found a low risk of a road vehicle reaching a rail line and that the existing standards for safety barriers alongside roads were adequate, while recommending further research.

The DoT report suggested a risk score for places where road vehicles could get onto the railway, with action to be taken at places with high scores.

Gary Hart was subsequently convicted of ten counts of causing death by dangerous driving and sentenced to five years' imprisonment. [129] [130]

Late on Sunday 17th June 2001, as most of its passengers were asleep, five coaches of the *Thall Express* were derailed near Muzaffargarh, Pakistan. Five of the 700 passengers on the train were killed outright and many were trapped in the wreckage, some very badly hurt. About 40 in all were injured and were taken to hospitals in Muzaffargarh and Multan. The cause of the derailment was not immediately known. [131]

The *Udarata Menike* (Up-Country Maiden) express, which had left Badulla at 05:55 on Saturday 18th August 2001, was grossly overcrowded when it became derailed between Bujjamuwa and Yattalgoda stations, Sri Lanka, at about 15:00. Three coaches, one of them the Observation Saloon, left the track. Fourteen or possibly 15 people were killed and about 40 were injured. Twelve injured were being treated at Alawwa Hospital, seven critically hurt were treated at Kurunegala Hospital and two had been transferred to Colombo National Hospital. One the two tracks at the point of derailment was damaged and other services were delayed while repairs were being carried out. The driver escaped unhurt and subsequently reported himself to the police station at Alawwa, accompanied by a lawyer.

Passenger Tharangani Lanka Kumarasinghe, 23, told her story to a reporter, relating that the train had been so full that passengers had been riding on the footboards and that she had had a very lucky escape. "I was riding in the narrow gangway between two carriages, at the exact point where the train separated. There was a loud cracking sound and I grabbed hold of an iron bar. At some point I lost my hold but my blouse became entangled in another steel part of the coach and that was what saved me," she said, clearly still suffering from shock. She recalled that the train had been running very fast, with its coaches swaying, and she thought that at that speed she would reach her destination a good deal earlier than expected. She escaped with a severely bruised face.

Minister of Transport Dinesh Gunawardane appointed a commission under Mr Justice G.L.M. de Silva, a retired High Court Judge, to investigate the accident, while the railway opened its own inquiry under General Manager W.A.R. Weragama. Deputy Minister of Transport Kumar Welgama visited the injured in Alawwa and Kurunegala Hospitals. Minister Gunawardane noted that there had been previous major accidents at the same location and warned that stern action would be taken following the conclusion of the two investigations into the derailment.

Several factors apparently contributed to the accident. Railway officials who examined the site of the derailment stated that the track had only recently been re-laid and it had not compacted sufficiently to bear the stress of a fast-moving train. Deputy Minister Welgama

---

[129] Health and Safety Executive report on Great Heck derailment

[130] Wikipedia – Selby rail crash

[131] *Ananova*, 18th June 2001

pointed out that in such circumstances adequate warning signs should have been erected to caution drivers. It was not stated if they had been, but the speed of the train had apparently not been reduced. The gross overloading of the express might also have affected its stability and contributed to the derailment. [132]

Nine coaches of an 18-coach passenger train from Port Elizabeth to Johannesburg were derailed and reportedly 'concertinaed' close to Greenlands station, near Koppies, South Africa, in the northern Free State province, on the morning of Sunday 16th December 2001. Four children and three adults lost their lives. Eleven other people were treated in hospital, four of whom were admitted, and over 40 were treated at the scene for cuts and bruises and for whiplash injuries. Driver Dirk Swart was unhurt but suffered from shock. [133]

An investigation into the accident determined that due to a signal failure trains were being diverted onto a loop line, through a crossover over which speed was limited to 30 km/h. The driver had failed to reduce speed and the derailment had followed. Three days after the accident the driver and his assistant were suspended by Spoornet and at the end of February 2002 Driver Swart was dismissed. At that time both a disciplinary procedure against the assistant driver and a police investigation were still in progress.

General Secretary Chris de Vos of the United Transport and Allied Trade Union (UTATU) claimed that Driver Swart was innocent, that several factors beyond his control had contributed to the accident, and that insufficient protective measures had been in place, which the disciplinary board had not considered. The general secretary added that UTATU intended to take the driver's case to arbitration. He would not elaborate on what the missing protective measures were, for fear of prejudicing the driver's appeal. [134]

Ferrocarriles Nacionales de Cuba train No. 11, from Habana to Santiago, carrying 800 passengers, including 140 members of the national university students' federation, who were bound for a gathering in the east of the country, left the track at the entrance to the station at Coliseo, Cuba, just before 19:00 on Tuesday 10th December 2002. A freight waggon and four coaches of the 12 vehicles of the train 'toppled and twisted in an area muddied by recent rains' and two other coaches were partly derailed. One of the derailed carriages came close to striking the station platform.

Local residents complained that the train often sped through the town and witnesses stated that the rails had appeared to 'scream' immediately before the derailment. One of them, Yolanda Rodriguez, added that "Everyone ran into the street to help." [135]

Fifteen people died in the accident, including one who succumbed to their injuries in hospital, while another 73 were injured, 11 of them seriously. Those who lost their lives included Ulises Regino Ochoa, 22, a medical student from El Salvador; and Driver Esteban Garcia. The injured were taken to hospitals in Cardenas, Jovellanos and Matenzas. Most of those who lost their lives had been riding in the leading passenger coach, which struck a concrete structure, which cut through the length of it, ripping out all its windows.

Orlando Olivera, 29, described as the train mechanic, declared that the train had been travelling at its regular speed of 105 km/h when it left the rails. He said, "It was terrible. We heard a noise. Then we crashed and everything was dark." He added, "This is my seventh

[132] *Daily News*, Colomo, Sri Lanka, 20th August 2001
[133] *Mail and Guardian*, Johannesburg, South Africa, 17th December 2001
[134] https://www.news24.com/news24/death-train-driver-fired-20020227
[135] *The Medicine Hat News*, Medicine Hat, Alberta, Canada, 12th December 2002

accident. I'm not working here any more. I am young and there is too much danger, too many accidents caused by negligence."

Gregorio Bencomo, the deputy director of the railways, commented, "The system is very safe and should not fail...But something has failed."

The reason for the accident was not immediately clear. Railwaymen blamed a possible failure of a set of semiautomatic points, due to a power cut, while excessive speed was possibly an alternative cause.

The main line between Cuba's two largest cities was of course blocked and two days after the derailment rescue workers with cranes were still engaged in clearing it.

On the previous Friday, 22 people had been injured in a level crossing collision with a bus in Habana. [136] [137]

At 06:24 on Friday 31st January 2003 Down State Rail Authority (SRA) passenger train No. C311, a four-car Outer Suburban Tangara electric multiple unit (EMU) set, left Sydney Central station for Port Kembla. The train stopped at 11 stations, and left the last of them, Waterfall, New South Wales, Australia, at 07:13. About a minute and a half later and slightly less than two kilometres further on the whole train left the rails on a left-hand curve. Its leading coach collided with a stanchion supporting the overhead line equipment (OLE) and the first two coaches then collided with the side of a rock cutting, while the roof of the fourth coach struck another stanchion. Upon leaving Waterfall the train had been carrying 47 passengers, a driver and a guard. Six passengers and the driver died in the accident and 41 passengers and the guard were injured. Those who died had all been ejected from the train.

At the time of the accident Up passenger train No. C412 was approaching with 800 passengers and the Up line was fouled by the wreckage but activation of the track circuits by metal parts of the train put signal No. 770 back to Danger and the Up train was stopped four kilometres short of the obstruction. [138]

Office worker Arnouska Zehalko, 21, was a passenger on train No. C311, riding in the rear car. She related afterwards, "I could see the front carriage come in front. I could see it turn over, and the next minute we were all flipping over. Dust was everywhere, people were yelling and screaming and asking for help, a big mess." She escaped with a slight leg injury. [139]

The accident was investigated by a special commission headed by the Hon. Peter McInemey and a 96-page report on it was duly published.

The report noted that at the location of the derailment the line was on a falling gradient of 1 in 80 and there was a left-hand curve with a radius of 239 metres and a speed restriction of 60 km/h. A little further on, this curve was followed by a right-hand one with a radius of 203 metres and a speed restriction of 50 km/h.

Analysis determined that the train had entered the 239 metres curve at a speed of at least 117 km/h, without either the driver or the guard intervening to apply the brakes.

The driver was aged 53 and had been operating trains for 27 years. He had not mentioned any health problems to his wife before the accident. He had a medical history of pericarditis and at a medical examination in May 2001 had been found to be overweight with a Body Mass Index 34.1 (so in the obese range). In September 2002 his cholesterol had been 7.9 (the

---

136 *The Daily Gleaner*, Kingston, Jamaica, 12th December 2002

137 https://groups.io/g/AllAboardRailDiscussion/topic/nytimes_train_crash_kills/36023942

138 https://www.parliament.nsw.gov.au/tp/files/34400/waterfall-final-report.pdf

139 *Playground Daily News*, Fort Walton Beach, Florida, United States, 1st February 2003

healthy upper level is 5.5). His post-mortem showed extensive atherosclerosis with an 85% risk of a major heart attack within the next five years. The forensic pathologist who conducted the post-mortem commented that, "The state of his heart was such that if he had been brought in without the trauma – in other words, had he collapsed in the street and been brought in without the multiple trauma brought about by the accident itself – the state of his heart would have led you to conclude that that was the likely cause of death."

The report concluded that the driver had had a stable job and family life and there was no indication that he had intended to harm himself. His two early shifts after extended leave made fatigue and a microsleep a possibility, while from the post-mortem findings there was a strong possibility of his having been incapacitated from a heart attack in the minutes before the derailment.

The guard was 39 years old with 21 years' service. He was on his third successive early shift and it was likely that he had been tired and had experienced a microsleep after leaving Waterfall station.

The report went on to discuss the deadman's pedal system on the accident train, which should have stopped the train in the event of its driver becoming incapacitated. It was noted that the weight of a driver as heavy as the one involved in the accident (118 kg) would have been sufficient to keep the deadman's pedal depressed, while there was anecdotal evidence that some drivers wedged the pedals down by wedging the sticks of red flags in them, or by placing heavy bags on them. A vigilance control device, as fitted on some other Australian trains, might have prevented the accident. Simulation tests with volunteers of a similar weight to the dead driver indicated that the deadman's pedal could have been easily accidentally depressed by his body.

The report pointed out that the SRA trainee notebook of guard's duties stated that, "The guard is responsible for the safety of all passengers on the train and must be prepared to stop a train immediately if an emergency situation arises." However, there was no evidence that the training programme for guards provided them with the critical decision-making skills to act correctly in an emergency situation. There was also a reluctance of some guards to challenge drivers and a general expectation that guards would not apply the brakes in emergency without first communicating with the driver.

The positioning of the speed restriction boards and the timetabled speeds of the train were considered.

As to the events after the accident, it was noted that a passenger had notified the police of the accident using a mobile phone at 07:20 while after several failed attempts, due to poor reception, the guard had been able to notify Waterfall station of the derailment at 07:30. There had initially been some confusion as to the location of the accident and in communications between railway control and the police.

There was a discussion of the safety culture of the railway.

Analysing the accident, the report concluded that the train had been driven normally until it departed from Waterfall station. Full power had then been applied and remained applied until the train had left the track on the curve. There was no evidence of any defect with the rolling stock or of the brakes having been applied. From the post-mortem evidence it was highly probable that the driver had been totally incapacitated after leaving Waterfall station, from a heart attack, and hence incapable of controlling the train.

The driver's medical history of hyperlipidaemia, obesity and age should have led to further examinations. Other high-risk drivers should likewise be subject to more detailed examinations.

The SRA had been aware for 14 years of the deficiencies of the deadman's pedal system. There was no evidence that the dead driver had deliberately tried to circumvent the deadman's pedal but it was likely that the weight of his incapacitated body had kept it depressed. A vigilance control would probably have prevented the accident.

The guard could have prevented the accident but a variety of factors, such as his having to realise soon enough that the train was out of control despite not having been trained to recognise such situations, tiredness and a reluctance to take the handling of the train out of the control of the driver, had prevented him from doing so.

The report listed 25 findings in total: 11 main findings (1-11) and 14 additional findings (12-25):

1. The driver and the guard of train C311 had been correctly certified.
2. The journey of train C311 had progressed uneventfully from Sydney Central to Waterfall.
3. Train C311 had departed Waterfall on time at 0712:51.
4. The departure from Waterfall had been normal, probably with notch two or three selected on the master controller and the driver probably using the deadman foot pedal.
5. Full power had been applied for a short time after the train had cleared Waterfall platform, earlier than it should have been, but had not been shut off in the area where normal driving technique would dictate that it should have been.
6. The driver had not been known as a person who regularly exceeded the speed board limits.
7. The deadman system could be circumvented.
8. The driver's weight and position at the controls had probably coincided with the conditions that could cause inadvertent circumvention of the deadman system.
9. It was highly probable the driver had suffered some form of incapacitation, possibly related to his cardiac function.
10. The guard had not been adequately trained or conditioned to intervene in an over-speed condition and had made no attempt to contact the driver or to apply the emergency brake.
11. The C311 train had continued to accelerate out of control down the straight portion of track, achieving a maximum speed of about 117 km/h.
12. The C311 train had overturned about 1,930 m south of Waterfall, after entering a left-hand curve where the overturning speed was about 110 km/h.
13. The deadman system had been susceptible to intentional circumvention through a variety of methods.
14. Some drivers had intentionally circumvented the operation of the deadman system.
15. There had been weaknesses in the SRA certification regime, which could result in an uncertified person being rostered to do rail safety work.
16. The Regulator had not required operators to specify and validate standards for professional qualifications and competence of rail safety specialists as part of the accreditation process.
17. Some individuals in safety specialist positions in the rail industry had not had the equivalent professional education, skills and experience that would be expected of a safety practitioner in other safety-critical industries.
18. The design position of speed boards had not been consistent with Network Rule NSG 604.
19. The actual position of speed boards had not always been consistent with the design position.
20. The overall timetable for C311 had been theoretically achievable but only by perfect train and driver performance. There had been certain sections over which the timetable could not be achieved under any circumstances, while complying with the speed boards.

21. Network Rule NSG 604, introduced in November 2001, had fundamentally changed the interpretation of speed boards but neither the physical position of the speed boards nor the timetable had been reviewed to account for this change.

22. Two separate databases could be accessed to provide the same training information relating to crew certification but provided different results on the certification status of the guard.

23. The emergency equipment that had been stored under the guard's seat had not been secure in a roll-over accident.

24. No record of application or approval of the driver's secondary employment had been discovered during examination of SRA documents.

25. The SRA approach to 'joiner rights' had circumvented some of the safety-related controls that were part of the train crew sign-on process.

The report concluded with no less than 66 recommendations:

1. The board of SRA should prepare a strategic plan to bridge the gap between existing safety and risk management practices and best practices in these areas, which clearly sets out:

(a) Plans to improve the organisational safety culture.

(b) Identified target dates.

(c) The provision of appropriate safety and performance measures to test the efficacy of the strategy.

(d) The embedding of skilled safety specialists in operational areas.

(e) Demonstration to the Regulator that there is individual accountability for safety at all levels.

The SRA board should report quarterly to the Regulator on the implementation and success of this plan.

2. The SRA internal safety auditor should report directly to the SRA board.

3. Pending the adoption of a national standard through the National Transport Commission (NTC) process, the Rail Safety Regulator should mandate medical standards for train crews and other railway safety workers to provide for predictive and preventive management of potentially incapacitating medical conditions, with due regard for functional-based risk assessment of each category of worker. Such standards should make provision for psychiatric conditions as well as the more traditional medical conditions.

4. The medical standards mandated by the Rail Safety Regulator should have disqualification provisions for individuals with potentially incapacitating medical or psychiatric conditions, particularly for single-driver operations.

5. The Rail Safety Regulator should prepare a set of guidelines in relation to the application of medical standards in the rail industry for use by medical examiners who conducted periodic medical examinations of safety-critical rail personnel. The guidelines should:

(a) Set out the minimum qualifications required of a medical examiner conducting examinations.

(b) Refer the examiner to the mandated standards required of personnel in the rail industry.

(c) Provide a plain English explanation of the work functions of each category of rail safety worker, so that Examiners could consider this in assessing rail personnel.

6. The medical standards mandated by the Rail Safety Regulator should prescribe a process of

regular review to ensure that the standards were maintained at a level of best practice in medical and occupational health aspects.

7. All operators should expedite the application of the revised medical standards to rail safety workers by applying risk profiling to identify and prioritise potentially high-risk individuals.

8. All accredited operators should take immediate action to ensure that no cab-based equipment could be deliberately or inadvertently used to circumvent the deadman system or any other safety device.

9. The SRA should address as a matter of urgency the latent technical deficiencies identified in the deadman system fitted to the Tangara (and other) trains, particularly in relation to deadman foot pedals and the risk associated with drivers with a body mass greater than 110kg.

10. The Rail Safety Regulator should commission an appropriately funded project to research options available to integrate contemporary technology with existing and developing deadman, vigilance and speed envelope systems. The research should include risk assessments of each option. In particular, the research should examine tamper-proof processes in which driver alertness and/or vigilance and incapacitation could be continuously monitored and managed.

11. The Rail Safety Regulator should use the results of the study into deadman systems to define guidelines against which deadman systems will be assessed and audited.

12. Rail operators should ensure that the progressive FAID {Fatigue Audit InterDyne™} scores (or the equivalent) for individuals were assessed on both master rosters and actual hours worked, as part of the fatigue management program.

13. Rail operators should immediately review and, if necessary, update rostering policies to accommodate an acclimatisation period for rail safety workers returning to shift work after an absence or leave.

14. The Rail Safety Regulator in conjunction with industry should undertake a research project to identify appropriate standards in rostering practices for shift workers returning to work from leave.

15. Rail safety workers returning to work from leave should be provided with a familiarisation period to allow for a review of safety documentation that had been issued during the leave and an update on configuration changes that might have occurred during the absence. Local managers should be accountable for verifying and recording that crews had been provided with and understood this information.

16. The Rail Safety Regulator should collaborate with SRA and other operators to develop a guideline for fatigue management across NSW, including a risk-based determination of the upper acceptable fatigue limit for train crews and other rail safety workers. A transitional period should be defined for the adoption of these guidelines.

17. All NSW rail operators should develop awareness campaigns and formal training programs to educate safety-critical workers on strategies that could assist in minimising the effects that fatigue might have on their operational performance.

18. The SRA should have a formal Safety Culture Survey independently undertaken, with particular emphasis on identifying ALL of the factors that had an adverse effect on safety culture. The survey should include but not be confined to, various managers' interpretations of on-time running imperatives, and the regulator should be consulted on the survey methodology.

19. The SRA should clearly specify crew responsibilities and expected crew behaviours in emergency situations.

20. The SRA should expedite the integration of "crew concept" values and activities into initial and recurrent training programs associated with train drivers and train guards. This integration should seek to maximise the potential for drivers and guards to operate as interdependent crewmembers who were supportive and cooperative of each other during train operations.

21. The SRA should incorporate a risk-based approach to training course design, to ensure that, where appropriate, identifiable hazards were treated by the provision of suitable training.

22. The Rail Safety Regulator should develop and apply audit criteria to assess the effectiveness of SRA's risk-based approach to training.

23. Auditors of safety-critical training should be conversant with the principles of contemporary risk management and have access to expertise in training and development.

24. The SRA should nominate a position to be accountable for developing an integrated certification regime, which would ensure rail safety workers in all categories remained competent and correctly certified and could not be rostered or signed on for rail safety work without holding correct and current certification.

25. The SRA should review and amend the train crew performance assessment process to ensure it was effectively assessed and recorded against all identifiable performance requirements and was accountable.

26. The SRA should conduct and provide to the Rail Safety Regulator a systematic risk analysis that provided a full and complete understanding of whether not providing internal Emergency Door Releases presented an increased or decreased risk to passengers.

27. The Rail Safety Regulator should investigate performance-based rolling stock evacuation procedures and equipment and mandate an appropriate standard for all rail operators.

28. The SRA should conduct testing to determine if the existing Tangara external Emergency Door Releases would function as intended under all foreseeable conditions, including when a carriage was on its side and a secondary means of activation. Corrective action should be taken if recommended by these tests.

29. Officers of the Rail Safety Regulator who were involved in accreditation and audit should be trained in Safety Management Systems and Risk Management.

30. The Rail Safety Regulator should expand the existing audit regime to include specific Safety Management Systems audits of accredited operators.

31. The Rail Safety Regulator should be provided with all appropriate resources to fulfil its accreditation, audit and investigation responsibilities.

32. The SRA should amend its Safety Management System to include a dynamic hazard register and reporting system, applied across all divisions and recorded centrally, and the Rail Safety Regulator should include this in its Safety Management System audits.

33. The SRA should initiate a program to ensure that employees at every level of the organisation understood and carried out their responsibilities under the Safety Management System.

34. The SRA should conduct a formal review of how the Tangara performed in this accident in relation to crashworthiness. The results of this assessment to be presented to the Regulator.

35. The SRA should further assess the dynamic performance of Tangara to identify any factor

that might promote overturning.

36. The Rail Safety Regulator should require all accredited rail organisations in NSW to ensure that any risks associated with configuration changes were managed by the application of a configuration change control process.
37. The Rail Safety Regulator should develop and apply an audit function specifically to the staff certification records of operators.
38. The Rail Safety Regulator should develop and implement a capacity to randomly and regularly sample that rail safety workers were carrying the correct and current certification.
39. The Rail Safety Regulator should develop a guideline for a standard form of recording the certification details of individuals.
40. The Rail Safety Regulator should have continuous access to the certification databases that were maintained by operators. Field staff conducting safety inspections should have ready access to certification information.
41. The RIC should review the design position of all speed boards, to ensure consistency with NSG 604.
42. The RIC should have a continuous program to review the actual position of all speed boards, to ensure consistency with design position, and that an accountable position certified the accuracy of speed boards.
43. The SRA should provide the Regulator with a program to appoint an accountable person to review its timetable, using a risk-based approach, and eliminate any conditions where the timetable could not be achieved while maintaining compliance with the track speed, stopping pattern and other operational safety requirements, and that the accountable position should certify the timetable as free of incompatible goals.
44. The SRA should maintain evidence of an ongoing program for an accountable person to review its timetable, using a risk-based approach, and certify that it was free of incompatible goals. This would be particularly relevant to configuration changes that might occur.
45. The Rail Safety Regulator should issue a guideline on the demonstrable competencies required of a 'rail safety specialist' that was consistent with the definition of a qualified safety professional as it appeared in the safety industry.
46. The Rail Safety Regulator should require that any safety specialist within the rail industry not in possession of adequate qualifications, commenced a professional development program to be completed within a nominated time, to achieve the skills consistent with the definition of a 'rail safety specialist' as a condition of retaining their position.
47. The SRA should implement a train operations quality assurance program, based on routine datalogger surveillance, which would support training and other programs designed to ensure safe driver behaviour.
48. The SRA should secure all cab-based features and equipment so that they could not become projectiles during a derailment or roll-over.
49. The SRA should rename the crew competency training program to accurately reflect its objective.
50. The emergency procedures within the Rail Management Centre should be reviewed and developed so as to enable the most expeditious identification, evaluation and response to emergency situations, and to ensure unambiguous and coordinated communication within the Rail Management Centre and with external contacts, including stations, signal boxes and emergency services.

51. The Rail Management Centre staff should be trained in the evaluation of information and implementation of emergency procedures, with the procedures validated through exercises.

52. Train guards should be provided with effective and reliable communications equipment for use in emergencies, with regard to the potential for one or both crew members becoming incapacitated.

53. The emergency plans of RIC/SRA should make provision for facilitating as fully as possible, the provision of information to emergency services, and their access to all railway property.

54. The SRA emergency plan should clearly state the role of the Rail Commander at major accident sites. This should include immediate actions, site control liaison, decision-making authority, site safety, site security and the training required to fulfil this role.

55. A definition and procedures should be developed for providing a form of interim electrical system isolation, that provided accident site safety and permitted rescue operations to commence, before full isolation procedures had been completed.

56. The Rail Management Centre emergency plans should require the Electrical Operations Centre to implement the procedure mentioned in No. 55, whenever information indicated that a derailment or other major incident had occurred on a main line, until it could be established that no hazard existed.

57. The emergency plans of SRA, RIC and other operators should reflect the investigative obligations of the Rail Safety Regulator, under the Rail Safety Act 2002, and the requirement to preserve evidence.

58. The Rail Safety Regulator should develop a critical response capability and accident investigation procedure applicable to its role in transport accidents.

59. A Memorandum of Understanding, or other protocol, should be developed between the Rail Safety Regulator and the NSW Police regarding rail accident investigation in NSW, with provision to extend this protocol to other transportation modes for which the Regulator was responsible.

60. The infrastructure owner should conduct regular emergency procedures exercises involving the rail industry and emergency services, to be held in the various regions.

61. The SRA should provide local managers with criteria against which secondary employment applications could be assessed. These criteria should include opportunities to assess fatigue through FAID.

62. The SRA should introduce robust controls in relation to the enforcement of the crew secondary employment policy.

63. The SRA should impose robust controls upon the provision of safety documentation to rail safety workers commencing work.

64. The Regulator and SRA should jointly investigate 'joiner rights' to identify all the risks associated with the practice and to identify any controls that may be applied to the identified risks.

65. The Rail Safety Regulator should be appropriately resourced and skilled to fulfil its role.

66. That the Rail Safety Regulator and SRA should meet so that the Regulator could provide any clarification SRA required in the interpretation of these recommendations and so that the Regulator and SRA could agree on the implementation actions required. The SRA should produce implementation plans, including timescales, for each recommendation,

which were acceptable to the Regulator, and report quarterly to the Regulator on the implementation of these plans. [140]

Train No. 904, a Karwar – Mumbai Central holiday special, travelling at 75 km/h, was derailed at about 22:30 on Sunday 22nd June 2003 between Rajapur Road and Vaibhavradi stations, Maharashtra, India, on the Konkan Railway subsidiary zone, when it ran into a landslide. Monsoon rains had lashed the area for several days before the accident. Three unreserved and one air-conditioned coach left the track. The death toll, initially quoted as 17,[141] was later corrected to 23,[142] and then to 34[143] as more bodies were discovered and the final figure was 52.[144] Darkness and continuing heavy rain hampered the rescue work, which was carried out by railwaymen assisted by nearly 150 army personnel from the Signals Training Centre at Panjim. There were about 60 injured who were taken to hospitals in Kankavil, Oros, and Rajapur and to the Goa Medical College in Panjim.

The landslide, which had blocked the line at the exit from the Narli Tunnel, included one or more large boulders. There were 46 locations on the 760-kilometre-long Konkan route identified as being 'vulnerable' to landslides and where the ground overlooking the railway was provided with protective netting, but the site of the accident was not one of them. Deputy Chief Security Commissioner Alok Bohra commented that, "Konkan Railway has the most difficult terrain in the entire country and we remain extremely vigilant all the time." The Konkan was the only zone in India to employ a geologist. A newspaper added that, 'The loose red soil, along with the basalt rock, has started sinking as the region had received about 190 mm of rainfall in the three days prior to the mishap.' [145]

Following the disaster the railway stated that it was beginning the installation of a new safety device, the Raksha Dhaga, consisting of a strong nylon cord extending along the length of a cutting, which if broken automatically triggered a warning light and hooter. The device would initially be installed at 200 cuttings. 'The company is also working on a design for electronic rain gauges which, coupled with the anti-collision device technology, will automatically control the train's speed during heavy showers in the monsoon. The railway has already conducted boulder netting, rock bolting, micropiling and geo-matting along its routes.' [146]

A major disaster occurred between Mindouli and Kingoyi, Republic of Congo, in the Pool department, on Monday 22nd March 2004. A waggon of a freight train of the Chemin de fer Congo – Océan, carrying food supplies and an unknown number of unauthorised passengers, became detached from the remainder of the train and fell into a ravine. About 30 (another source specified 31) 'stowaways' who had been riding on the waggon at the time lost their lives. It was unknown if the derailment had been an accident or due to sabotage. Government spokesman Alain Akouala stated that a team of experts from the railway and the Ministry of Transport had gone to the scene to investigate. The 512-kilometre line between Pointe-Noire and Brazzaville had previously been subject to various acts of sabotage but the track was also in a dilapidated condition, so that train speeds were limited to 25 to 30 km/h. [147]

[140] https://www.parliament.nsw.gov.au/tp/files/34400/waterfall-final-report.pdf
[141] http://sify.com/printer_friendly.php?id=13178976&ctid=2&lid=1
[142] http://sify.com/printer_friendly.php?id=13179279&ctid=2&lid=1
[143] www.rediff.com/cms/print.jsp?docpath=/news/2003/jun23rail1.htm
[144] http://web.mid-day.com/news/city/2003/august/60092.htm
[145] www.rediff.com/cms/print.jsp?docpath=/news/2003/jun24rail1.htm
[146] http://cities.expressindia.com/fullstory.php?newsid=56456
[147] https://www-congopage-com.translate.goog/Catastrophe-ferroviaire-sur-le?_x_tr_sch=http&_x_tr_sl=fr&_x_tr_tl=en&_x_tr_hl=en&_x_tr_pto=sc

It was the turn of Turkey to suffer a major disaster at 19:45 on Thursday 22nd July 2004. At this time Turkey had just (a month earlier) inaugurated a service of accelerated, if not high-speed, trains and it was one of these, named *Yakup Kadri Karaosmanoğlu*, after a 20th-century Turkish novelist, journalist, diplomat and politician, which was involved in the accident. The train was on its way from Haydarpaşa station in Istanbul to Ankara when it left the track on a curve near Mekece station. The engine and leading carriage remained on the track and afterwards continued to Osmaneli station but four coaches further back the train overturned, while two collided with one another and were badly damaged. There had been 234 passengers and nine railwaymen on the train, of whom 41 were killed and 80 were injured. [148]

Mayor Feridun Turan of the nearby town of Pamakova, who had rushed to the scene, told a reporter, "There are decapitated heads, limbs all around. It's a bloodbath."

Doctor Elcin Ozumturk, a passenger on the train who along with his wife had survived the crash, recounted, "We were flung into the air, then the coach fell sideways. People were screaming. There was smoke and shattered glass everywhere. We broke a window and crawled out."

After the accident Suleyman Karaman, head of Türkiye Cumhuriyeti Devleyt Demiryollari (TCCD) (the state railway) maintained that the train had been running at a lower speed than normal at the time of the derailment and that, "The accident is not linked to high speed."

Surviving passengers thought otherwise. Oguz Sarisin told NTV television from his hospital bed, "The train was going really fast and went out of control." Another passenger, Abdullah Cagiridiken, declared, "I saw the speedometer on the wall of the restaurant car just 15 minutes before the crash. The dial was indicating 132 kilometres/hour."

NTV television also rebroadcast a June interview with transport expert Aydin Erel, who had stated back then, "The tracks are not secure. I would never ride on that train."

Prime Minister Recep Tayyip Erdogan, however, was keen to offer reassurance, stating, "We carried out the necessary studies." (Before launching the higher speed train.) "The system is safe." [149]

Despite the statement of the TCCD manager, an investigation into the disaster found that in fact it had been due to excessive speed and that the train had attempted to negotiate the curve in question at a speed of 132 km/h instead of the permitted 80 km/h. As the theoretical overturning speed on the curve was 130 km/h, this was marginally too high a speed to avoid disaster and had resulted in one of the left-hand wheels of the second coach lifting from the rail, overbalancing the train and leading to a general derailment.

Other factors had also played a part. There had been no warning signs preceding the curve, the time allowed for the total journey of 5 hours and 15 minutes was deemed to be too little (a minimum of 6 hours was suggested) and would have encouraged train crews to exceed speed limits in order to maintain schedules, and the whole infrastructure of the railway was not really suitable for high-speed running.

The railway unions protested that it was not so much the railwaymen directly involved who had been responsible for the accident as the TCCD management and the government, who had pushed for the introduction of faster trains despite warnings from experts that the unimproved 19th-century railway was not suitable for such an innovation. Others, such as Turhan Çakar, Chairman of the Consumer Rights Association, and Mete Orer, formerly Undersecretary to the Ministry of Transport, agreed with the union's stance.

---

[148] Pamukova train derailment

[149] *Los Angeles Times*, Los Angeles, California, United States, 23rd July 2004

There were calls for the resignation of Binali Yildirim, the Minister of Transport, Maritime and Communication, but he refused to do so, stating that the train involved, which had an average speed of 106 km/h, did not fit with the scientific designation of a high-speed train, which was one of a train with an average speed of 250 km/h or more and that, as in 85 per cent of railway accidents around the world, the human element had been the principal factor in the derailment. He also denied that no preparatory work had been done before the inauguration of the faster trains, stating that 180 sets of points and rails had been replaced.

In the end, in November 2014, it was only the train crew who ended up in Sakarya Second High Criminal Court. Driver Fikret Karabulut was sentenced to three years, one month and 15 days imprisonment and a fine of 1,500 lira, while Assistant Driver Recep Sönmez was given a suspended sentence of one year and 15 days and a 152-lira fine. Conductor Köksal Coşkun was acquitted. [150]

The unions and others had filed for action against the TCCD, alleging that the accident had resulted from the "high-speed train" project, which had been intended to make busy passenger lines more profitable so that the TCCD could be broken up and prepared for privatisation. No action was eventually taken against any other officials, while the sentences on the two railwaymen were not carried out either, because the case had exceeded the seven-and-a-half-year statute of limitations by the time that it came to court. [151] [152] [153]

An accident with similarities to that at Ufton Crossing, England, only a couple of months earlier, occurred on Union Pacific Railroad tracks at Glendale, California, United States, on Wednesday 26th January 2005. Again it involved an attempt at suicide that ended up killing a number of innocent people, though in this case the intended suicide himself survived.

At 6:03 a.m. on the day of the accident southbound Metrolink passenger train No. 100 collided with a sports utility vehicle (SUV), which had been abandoned on the tracks near Chevy Chase Drive level crossing. Derailed and jack-knifed by the initial collision with the SUV, train No. 100 then collided both with northbound Metrolink train No. 901 and with a Union Pacific ballast train stabled on an adjoining siding. A fire caused by spilt diesel fuel then broke out in the wreckage. Ten passengers and Conductor Thomas Ormiston, 51, of train No. 901, lost their lives, while 177 people were injured. [154]

'The train rumbled down the tracks as it always did, and Elaine Watson sat comfortably in her seat, immersed in her novel. There were the usual pre-dawn commuters, the ones who read the newspaper and barely looked up, the ones who liked to chat and others who slept or put on headphones. They were accountants, bank workers and public service employees, commuters who preferred the train over Southern California's congested freeways. On an average weekday, 38,000 people ride Metrolink trains through six counties. Some, like Los Angeles County Sheriff's Deputy James Tutino, rode only a few times per month. Some rode every day; Watson had taken the 25-minute ride from Northridge to Glendale to her job at an insurance company for four years.

'"GLEN-DALE," the conductor's voice bellowed, as always, and she slid her book back into her bag, then rose to wait by the door. But the train didn't slow. "There's something really wrong here," Watson thought, hurrying back to her seat. As rocks banged the sides of

---

150 https://www.hurriyet.com.tr/gundem/sakaryadaki-hizlandirilmis-tren-kazasi-davasinda-karar-27642680

151 https://web.archive.org/web/20141129053125/http://www.radikal.com.tr/haber.php?haberno=123027

152 https://arsiv.sol.org.tr/index.php?yazino=28160

153 https://www.hurriyet.com.tr/gundem/hizli-tren-kazasina-zaman-asimi-19863560

154 Wikipedia – 2005 Glendale train crash

the train, she braced herself as best she could, feet dug into the floor, hands tightly grasping a metal pole in front of her. Was it an earthquake? Debris being dragged by the train? Passengers weren't sure.

'On a neighbouring track, train No. 901 was headed in the opposite direction, away from downtown Los Angeles, and passengers Eduardo Fernandez and Danny Suarez were talking about the Super Bowl. The two were train buddies and always ended up chatting during the early commute. The 901 was almost beside the northbound train now, ready to zip between the strip malls that fill this typical inner suburb just north of Los Angeles.

'The Costco warehouse is right beside the tracks, and Ruben Landa, a 44-year-old repairman, was outside rummaging through a storage cage for boards and shelving just yards away from the whir of trains. He hardly noticed the noise anymore. Then, the train roared like thunder, only louder, and with a horrific boom. Lights inside the cars flickered out as the No. 100 train screeched along, now cockeyed and wheezing as it derailed and slid into a parked freight train. No. 901 was flying by at the same time, and the southbound train jack-knifed into it.

'Aboard train 100, the crash sent Elaine Watson lurching forward. She banged her head on the pole she was clutching. In the same car Steve Toby, who had been catching a nap before he started his day as soundman for Los Angeles City Council meetings, was thrown across the aisle, landing on top of two women. The car was moving sideways now, and his left leg was entangled in a mass of metal. When the car ground to a halt, an eerie silence followed. Diesel fumes and smoke filled the air. Flames shot up into the morning drizzle.

'Landa froze at the sight of a train car split open like an aluminium can. Passengers were slammed against tables and pinned by chairs. Seat cushions and fallen ceiling tiles jumbled together. Shattered glass crunched beneath the feet of commuters who were starting to escape. But others could not, and their moans and screams grew louder as Landa ran towards the trains, trying to help anyone he could. Other Costo employees, like Jenny Doll, grabbed fire extinguishers and went, too. Voices came from every direction. "I'm back here!" "Get me out! Get me out!" "Please help!"

'Landa dove into one car and pulled out a man lodged underneath a seat. Commuters were so stunned that some weren't trying to get themselves out. As Landa tried to aid them, Doll, 30, was trying to put the fire out. Others pitched in, too. Ice cream truck driver Dean Jæschke rushed towards the flames. On his mind was the car accident he had been in years ago when hot oil had seared and scarred his body. No one should burn like that, he thought.

'It was frantic inside some cars, calm in others. One trapped man used his own blood to write a note on a seat bottom. Using the heart symbol, he wrote, "I love my kids," and "I love Leslie." He made it out alive.

'Suarez, who'd been chatting about the Super Bowl just moments before, now did what he could to help. He noticed a man lying face down, and he tapped his shoe. "You need help?" Suarez yelled. But the man was dead. Suarez helped others get out, including his friend Fernandez, who had a huge gash on his forehead.

'Toby yanked on his thigh, finally dislodging it, as helicopters now buzzed overhead. His leg was broken.

'The Costco parking lot was turning into a triage center. Shopping carts were used to move some of the injured. Green mats were for passengers with minor injuries; a red mat was for serious injuries.

'Jæschke was able to free Scott McKeown, but he died soon after. McKeown, a father of two, who was acting telecommunications manager for the city of Pasadena, lived long enough to thank rescuers for not letting him burn alive. James Turbino, a 23-year veteran of the sheriff's department, died too.

'Train pals Suarez and Fernandez ended up at the same hospital. After helping Fernandez and others escape, Suarez had stumbled out of the train, exhausted, not realizing he was bleeding internally. The next day, Fernandez called Suarez. "Thanks for looking out for me," he said. Both will be back on the train again soon.' [155]

The SUV driver, Juan Manuel Alvarez, 25, had reportedly made previous suicide attempts, was a methamphetamine user, was said to be prone to delusional behaviour and had been experiencing difficulties in his marriage. Initially it was believed that he had driven onto the tracks at the level crossing in a suicide attempt, then had 'chickened out' at the last moment and abandoned his vehicle. Later an alternative theory was put forward by the police that he had had no intention of killing himself but staged the accident in an effort to gain attention from his estranged wife. To the author the abandoned suicide attempt theory seems much more likely. At any rate Alvarez was subsequently charged with and convicted of 11 counts of murder with special circumstances, while the prosecution sought the death penalty under an 1873 law, passed in the Jesse James era, under which causing a death by train wrecking was a capital offence. Acquitted of train wrecking, Alvarez was found guilty of 11 counts of first degree murder under special circumstances and one count of arson and was sentenced to 11 consecutive life sentences, with no possibility of parole. [156]

A long-distance passenger train that had started from Hanoi on Friday 11th March 2005 was derailed at 11:50 on Saturday 12th March at kilometre 752·5 of the North – South main line, near the hamlet of Dá Bàn and about three miles west of the station at Lăng Cô, Vietnam. Eight of the 13 carriages of the train left the rails near the Hải Vân mountain pass. An early report stated that of the 500 passengers and 29 crew on the train, which was bound for Ho Chi Minh City, five had been killed outright and six fatally hurt. Another 200 were injured, a good number seriously. 'The owner of Sao Bien Restaurant at the foot of Lăng Cô Bridge witnessed a man stuck under a carriage but could do nothing to save his life. "When I reached him the man was still alive and gave his home phone number for me to call his family. We could not pull him out. When rescuers arrived and released him he died from exhaustion," said the restaurant owner.[157] There was no road access to the site, which rescuers had to reach by water, injured passengers being taken by boats and canoes to Route IA, from where they were taken to Phú Lộc District Hospital, Huế Central Hospital and the Polyclinic in Da Nang. The single track was blocked for 26 hours. [158]

A police investigation, including a study of the tachograph recovered from the train locomotive, determined that the derailment had been due to excessive speed, as the train had been travelling at 68 km/h at a point where speed was limited to 40 km/h. At least one passenger also claimed that the train had been gathering speed just before the accident, while a lineside witness told Vietnamese Television that, "The train was running too fast. Then we heard a loud noise. We ran {and found the derailed train} and began breaking carriage windows to get the injured people out."

By Monday 14th the number of dead had risen to 13 and 55 people were still being treated in hospital.

On Wednesday 16th it was stated that the police had arrested Driver Bui Thai Son and charged him with violating railway regulations and disregarding the speed limit, leading to serious results. He was remanded in prison for four months by a court in Thừa Thiên – Huế province and admitted that his train had been going too fast, saying that he had realised this

---

155 http://www.montereyherald.com/mld/montereyherald/news/10767924.htm?template=c...
156 Wikipedia – 2005 Glendale train crash
157 www.thanhnienews.com/print.php?catid=3&newsid=5542
158 Wikipedia – 2005 Phú Lộc derailment

too late and had belatedly applied the brakes. [159]

Driver Bui Thai Son was found guilty at his subsequent trial and sentenced to 13 years' imprisonment, while Assistant Driver Hà Minh Tâm received seven years.[160] Four other senior railway officials lost their jobs; one was placed on probation and five were reprimanded for their responsibility for the accident. [161]

Six unauthorised passengers, described as 'stowaways,' who were riding on top of a waggon loaded with bananas, lost their lives on Friday 25th March 2005, when a mixed train was derailed near M'Foumbou, Republic of Congo, after colliding with a tree. Jacky Trimardeau, managing director of the Chemin de fer Congo – Océan, commented that, "The accident happened on an old stretch of track where trees fall onto the line every time that there's bad weather."[162] Two other trespassers were badly hurt and two were slightly hurt. [163]

Japan suffered a major railway catastrophe at 09:19 on Monday 25th April 2005, its most serious accident for 42 years. It occurred between Tsukaguchi and Amagasaki stations on the Fukuchiyama line of the West Japan Railway (JR West). The train involved in the accident, just after the morning rush hour, was No. 5418M, a limited-stop "Rapid" commuter service from Takarazuka to Dōshishamae, which consisted of a seven-car 207 series electric multiple unit (EMU). It was carrying about 700 passengers at the time of the accident.

Like a number of the other derailments in this chapter, the accident was due to excessive speed on a curve. In this case, the right-hand curve in question had a radius of 304 metres and a speed limit of 70 km/h, while the data recorder on the accident train showed a speed of 116 km/h at the moment of derailment, 10 km/h more than the theoretical lowest speed that would result in derailment. The first four cars of the train left the track completely. The leading car struck the parking lot of a lineside apartment building, the second car collided with the external wall of the same building and was almost totally crushed by the third and fourth cars, themselves being pushed by the fifth car. The accident resulted in the death of 106 passengers and Driver Ryūjirō Takami, 23, and the injury of 562 passengers, thus most of the other people on the train.

As might be expected, the catastrophe formed the subject of a very thorough investigation by the Aircraft and Railway Accidents Investigation Commission, the English translation of whose report runs to 118 pages.

The report included the witness statements of a number of surviving train passengers, a couple of which will be derailed here.

A male passenger stated, "There was nothing unusual after we left Itami station but a few moments after leaving Tsukaguchi station there was a violent brake application. I looked ahead and saw that we were approaching a sharp bend and were still travelling at considerable speed. It seemed impossible that the train could negotiate the curve. I immediately felt myself being pushed outwards by the centrifugal force. I grabbed a hand rail and desperately held on to it. I still hoped that somehow the train would get around the curve but the right-hand side of the car rose up into the air and never entirely regained the track. Then I felt the left-hand wheels become derailed and could see that we were heading straight for the block of flats. After the accident I found myself partially buried by my lower legs under several other

[159] https://www.emerald.com/insight/content/doi/10.1108/dpm.2006.07315cac.008/full/html

[160] www.thanhniennews.com/print.php?catid=3&newsid=15014

[161] www.thanhnienews.com/print.php?catid=3&newsid=6302

[162] www.reuters.co.co/locales/c_newsArticle.jsp;:424543ef192ab1d3d13?ty...

[163] www.ami.mr/fr/Archives2005/mars/26/3.htm

passengers. I was badly hurt, with amongst other injuries a punctured lung."

A woman passenger in her 20s related, "After leaving Itami station there was an announcement that the next station would be Amagasaki, followed by an apology for the train having overrun the proper stopping point at Itami station. As we neared the curve where the train derailed I was talking to a friend about how fast we were going. Then the windows began rattling, the train braked abruptly, there was a piercing squeaky sound such as I had never heard before and the coach began to lean over to the left. Because I was standing at right-angles to the direction in which the train was tilting the other passengers fell down to the left before I did. Over the shoulder of my friend, who was standing facing me, I could see the shifting of the second car through the rear window of ours. When our car had tilted about 45 degrees the lights in it went out. About that time I could see the second car tilting over as ours was. I saw a woman holding onto a strap in the second car fall over and then closed my eyes. The moment after the lights went out the car caved in with a 'thump' and vibrations which felt five or six times more powerful than those of the Great Hanshin Earthquake {of January 1995, which killed 5,000 people}. There was chaos in the car. It felt like being in a washing machine; I was struck all over my body before I finally fell on someone. I escaped from the coach from climbing up some wire netting into the apartment car park, as did three or four other passengers. The shin of my left leg was crushed and I had other injuries."

The investigation determined that there had been no problems with the track or the brake system of the train and the focus then shifted to the actions of Driver Takami. Even before the error that cost him his life, he had not had a good day. Twenty-five minutes before the accident he had run past a Red signal, bringing the automatic train stop (ATS) into action to halt the train. Then about four minutes before the derailment he had overshot the platform at Itami station by three car lengths and had had to set back, in the process losing 90 seconds (this was the incident referred to above by a passenger, for which the conductor had broadcast an apology).

After leaving Itami he had been in a discussion with the conductor over the intercom in an attempt to prevent the latter from reporting Itami overshoot to Control, asking, "Will you overlook?"

The conductor had replied, "It was some overrun," and hung up.

Unbeknown to the driver, the conductor had at that point received a complaint from a passenger about there having been no announcement regarding the delay at Itami. The conductor had then made the announcement and reported the overshoot to Control. It was noted that the train had regained 30 seconds of the delay by the time it passed Tsukaguchi station, moving at 120 km/h. The inquiry commissioners considered that after leaving Itami, Driver Takami had been preoccupied with his two infractions, looking for some excuse to make for them and fearing being sent for retraining. He had previously had to undergo three periods of retraining extending over 18 days, as well as receiving four reprimands or warnings.

All railways worldwide are keen that their trains are punctual but on the West Japan Railway, or perhaps Japanese railways in general, the quest for punctuality and the penalties imposed on railwaymen who failed to achieve it, appear to have been extreme. In part at least this was doubtless due to the fact that at stations such as Amagasaki, rapid and local trains stopped at the same time on the opposite sides of island platforms, so that passengers could transfer between them. Because of the tight schedules imposed by the timetables, a slight delay to one train could cascade throughout the day, while commuters depended on the near-perfect timing of trains to get to work on time.

Drivers on British Railways could be issued with so-called Form 1's for perceived offences, otherwise known as 'Please Explains,' which were basically charge sheets asking them to explain their behaviour/actions. If subsequently found guilty at a disciplinary procedure they could be punished with suspension, loss of travel facilities, warnings and so on, or in very serious cases, such as being grossly drunk on duty or punching a passenger, they could be instantly dismissed, but there was nothing equivalent to the retraining programmes which, as well as financial penalties, JR West imposed on its drivers for offences such as late running, which penalties to British eyes seem harsh and humiliating. These penalties or so-called retaining programmes, known as nikkin kyōiku, included such activities as weeding and grass-cutting duties, severe verbal abuse and the forcing of employees to repent by writing extensive reports. It was not unlikely, therefore, that Driver Takami's thoughts might have been more on how to avoid having to undergo another retraining session than on driving his train. Even when he did finally apply the brakes four seconds before the derailment he had only used the service brake, presumably because use of the emergency brake would also have had to be explained.

As to the probable cause of the disaster the Accident Commission thus considered that the driver's delay in applying the brakes for the curve had been likely due to his attention having been diverted to listening to the radio conversation between the conductor and Control after he had asked the conductor to make a false report, and to the driver trying to think of an excuse to avoid being put on an 'off the train' retraining course. The Commission noted that the disciplinary action for making out a false report would have been ever harsher than that imposed for the original offence.

The Commission made three recommendations to the Ministry of Land, Infrastructure, Transport and Tourism:

(a) A non-punitive reporting system should be developed to motivate crew members to report incidents, while research and studies should be conducted on methods and procedures to make a comprehensive analysis of information on a variety of incidents, and the results analysed in an effective manner.

(b) The use of train radios by drivers should be limited to emergency situations and drivers should be prohibited from making notes of radio conversations while driving. A means of Control sending instructions to drivers by text, which could be read by drivers when the train was stationary, should be developed, while the alternative of radio messages to conductors should be utilised as much as possible.

(c) It was important that the manufacturers of railway safety equipment understood the applicable laws and regulations so that sufficient quality control was exercised over the design, production and quality of such equipment. In addition, because the maintenance of vehicle and railway facilities was increasingly being outsourced, it was important that contractors' staff involved in the actual maintenance work knew and understood the applicable laws and regulations.

The Commission recommended a series of measures to be taken by JR West:

(a) Training on driving techniques should be improved and strengthened by:

(1) Introducing training based on cases of lapses of attention noted in analyses of incidents.

(2) The use of easily understandable driving simulators and training materials.

(3) Making drivers fully aware of the dangers of exceeding speed limits.

Off-the-train retraining courses, which some drivers considered penalties, should be altered so that there was less emphasis on spiritual training and more on practical driving techniques, as outlined above, so that they could properly be called retraining.

(b) Brake equipment should be improved so that the difference in deceleration for each brake handle position was small, regardless of whether or not regenerative braking was used, so that drivers did not need to pay attention to the status of the regenerative braking.

The actual brake deceleration should not exceed the specified standard value by an amount larger than to ensure safety, so as to prevent unnecessary brake application by the ATS.

Measures should be taken to deal with the problem of non-activation of the brake when the brake handle was placed between the 8-notch position of the service brake and the emergency position.

The differences in brake equipment between different types of railway cars should be minimised to reduce the burden on train drivers and allow them to pay more attention to safety.

(d) Train operation control systems should be improved to give the highest priority to human lives, for example by a manual describing the safest actions in the event of a derailment, such as immediately cutting off the electric power supply.

(e) Signs such as those warning of curves should be improved so they were more easily and reliably recognisable.

The Commission made a series of proposals to improve the safety performances of coaches during accidents. Studies should be undertaken to modify coaches so that sufficient space was retained to ensure passenger safety even in the event of an accident and the internal fittings of coaches should be improved to minimise the risk of injury from them.

The Commission listed actions already taken since the disaster by JR West:

(a) The company had developed a Safety Improvement Plan.

(b) The company had increased the number of ATS speed-check functions on curves on the Fukuchiyama line from 105 at the time of the accident to 1,370, and on crossovers from 663 at the time of the accident to 1,696.

(c) A backup power supply for train-protection radio equipment was to be provided in all cars with driving cabs by the end of September 2006.

(d) The rules were being revised to prohibit drivers from using train radios except when stopped at stations.

(e) The accuracy of the speedometers of trains was improved.

(f) Emergency brake and one-touch emergency equipment was being installed on all trains.

(g) Before the line was reopened on 19th June a revised train operation plan was brought into use whereby the minimum stoppage times of Up trains at Nakayamadera and Itami stations was increased and the maximum operation speed through the section was reduced.

(f) A safety laboratory was established to research safety-related management systems and human factors.

(g) Safety management regulations in which the responsibilities of top management were clearly defined were established.

Finally, the report of the Commission listed measures already taken by the Ministry of Land, Infrastructure, Transport and Tourism:

(a) JR West had been directed to develop a Safety Improvement Plan.

(b) Railway operators were directed to inspect their train operation plans to check whether they contained the appropriate time margins.

(c) Railway operators were directed to improve the functions of their ATS equipment.

(d) Railway operators were directed to inspect their train protection manuals and repeatedly provide train protection training courses for drivers and conductors.

(e) Railway operators were directed to secure the accuracy of their speedometers.

(f) The ordinance for technical standards was amended to require railway operators to install equipment recording the travelling of trains.

(g) The Railway Business Act was amended to require railway operators to establish safety management regulations, report their contents and employ a chief safety management officer. [164]

A feature of the disaster that could not be addressed was the building codes, which permitted residential buildings to be erected very close to the railway in metropolitan areas on the assumption that trains would never be derailed. Doubtless the great pressure on building space was a factor.

JR West President Takeshi Kakiuchi resigned at the end of 2005, his place being taken by Vice President Masao Yamazaki. The latter stood trial in July 2009 charged with negligence in connection with the disaster because ATS speed control had not, at the time of the accident, been installed on the curve. He was acquitted, though the railway company was criticised by the court for a faulty risk assessment of the curve. [165]

---

[164] Report of the Aircraft and Railway Accidents Investigation Commission on the Train Derailment Accident between Tsukaguchi and Amagasaki Stations of the Fukuchiyama Line of the West Japan Railway Company

[165] Wikipedia – Amagasaki derailment

CHAPTER 4

REAR COLLISIONS

The risks associated with one train running into the back of another had been low in the formative years of the railways when trains were few and speeds not very high, but as the years progressed, train speeds and traffic intensity often grew faster than advances in brake power, rolling stock construction and signalling, and the number of rear collisions increased. By the last quarter of the 20th century there had been great advances in the provision of brake power, in signalling, and in the use of devices to prevent drivers from disobeying signals, but disastrous rear collisions continued to occur from time to time.

At about 22:27 on Sunday 6th January 1991 Eastern Railway SG 45 Up Budge Budge – Sealdah electric multiple unit (EMU) local passenger train ran into the back of EC 018 Jumbo Special goods train from New Alipore (Calcutta) to Naihati at Km. 6/34 of the Up Passenger line between Budge Budge and Ballygunj stations, West Bengal, India, on the quadruple-track section of the Sealdah Division of the railway. The driving cab of the EMU mounted the brake van of the goods train and both vehicles were badly damaged. Ten people lost their lives, including the motorman of the passenger train and the guard of the goods train, and three were injured, one seriously. The weather was described as 'smoggy' with poor visibility. The area had been in darkness because of a power supply due to load shedding.

The accident was blamed on the motorman of the passenger train, who had passed semi-automatic signal No. 8 at Lake Gardens station, which had been showing a Single Yellow, but then had continued without the motorman driving with the required caution and subsequently failed to stop before proceeding past the next semi-automatic, SL-4, which had been Red, with its 'A' marker lit.

The Commission of Railway Safety made three important recommendations in connection with the accident:

(a) Only qualified motormen having suitable experience should be deployed to run passenger trains.

(b) The bustee [shanty town] on either side of the track should be properly segregated to prevent bustee dwellers from crossing the line at uncontrolled points.

(c) Ballygunj station should be provided with four-aspect automatic signalling to ensure that the signalling was uniform between Sealdah South and Mile 5B cabin. [166] [167]

---

[166] Government of India Ministry of Civil Aviation and Tourism Commission of Railway Safety Annual Report for 1990-91

[167] *The Indian Express*, Madras, India, 8th January 1991

The *Tezgam Express*, one of the premier trains of Pakistan Railways, which was often used by upper- and middle-class travellers who could not obtain air tickets, was well into its 1,450-kilometre, 21-hour, journey from Karachi to Rawalpindi when it collided with the rear of a freight train in the station at Ghotki, Pakistan, in rural Sindh province, about 650 kilometres north of Karachi. The 18-coach train, carrying about 800 passengers, was reportedly travelling at 65 km/h at the time and initial reports spoke of at least 100 dead and over 200 injured, 45 or more of the injured being in a serious or critical condition. The number of fatalities was at first, it was stated, feared likely to rise to 200 or even 300, but later it was instead revised down to 55. The accident happened at 01:10 on Saturday 8th June 1991. Eight or 12 carriages of the express and four waggons at the back of the freight train were derailed. The latter train, loaded with fertiliser, had been bound for Lahore.

Passenger Mohammad Iqbal related, "It was dark and almost everybody was sleeping. Suddenly, the whole train was shaking, just like an earthquake. People panicked."

"Witnesses said it was like a bomb exploding. The whole place shook," a railwayman related. The sound of the collision was heard in adjoining villages, whose residents hurried to the station to assist. Reports spoke of hundreds of soldiers and civilians labouring in the blistering 43° Celsius (110° Fahrenheit) heat to free victims trapped in the wreckage.

As often happens in the aftermath of major catastrophes the local health services were all but overwhelmed by the number of casualties. A doctor at a local hospital told a reporter, "The dead and wounded are lying on the floor. We don't have that many beds." [168]

Relief trains were sent to the scene with supplies and to take the injured to hospitals. [169]

The driver of the express, who survived, told authorities that he had received a Green signal to pass through the station but coming around a corner, he had then seen the freight train in his path, when he had applied the brakes in emergency.

Investigators stated that they could not immediately exclude sabotage but that negligence on the part of the officials at Ghotki station was thought much more likely. Five railwaymen were suspended pending the outcome of an investigation, among them Assistant Stationmaster Nizamuddin Channa, who had absconded after the accident and for whom police launched a manhunt.[170] Meanwhile, Prime Minister Nawaz Sharif went to Ghotki and ordered the suspension of Pakistan Railways General Manager Zafarullah Zureshi, also pending the results of the inquiry. Zureshi had also been general manager when two previous disasters occurred. Assistant Stationmaster Channa was later apprehended, or gave himself up, and was charged with 55 counts of murder.[171] Even prior to the accident, lawmakers and other officials had been arguing that Pakistan's railways badly needed modernising, that many railway employees were poorly trained and that mechanical signalling needed replacing by its electronic counterpart. [172]

The government promised to pay 50,000 rupees (about $2,100) compensation to the families of the dead and 1,000 rupees (about $415) to the injured victims of the accident. [173]

As a passenger train bound for Aswan was pulling out of the station at Badrshein, Egypt, on Thursday 13th February 1992, it was run into by a freight train whose driver was overrunning

---

[168] *The Lethbridge Herald*, Lethbridge, Alberta, Canada, 8th June 1991
[169] *Pacific Stars and Stripes*, Tokyo, Japan, 9th June 1991
[170] *Brandon Sun*, Brandon, Manitoba, Canada, 10th June 1991
[171] *The Daily Gleaner*, Kingston, Jamaica, 11th June 1991
[172] *The Daily Times*, Farmington, New Mexico, United States, 9th June 1991
[173] *European Stars and Stripes*, Darmstadt, Germany, 10th June 1991

signals. Eleven passengers, ten men and a woman, were killed and 30 were injured. [174]

A serious collision occurred in the station at Tsunduru Junction, Andhra Pradesh, India, on the Gudur – Vijayawada broad-gauge, double-line electrified section of the Vijayawada Division of the South Central Railway at about 21:10 on Sunday 5th April 1992. A goods train had been stopped at the Tsunduru Home signal and had barely restarted from the signal towards the station when it was run into from behind by train No. 423, a Bitragunta – Vijayawada passenger service. As a result of the collision the engine and leading coach of the passenger train were wrecked and the second coach and four waggons of the goods train were derailed. Upon information regarding the accident being received medical relief vans with doctors and divisional officers were sent to the scene from both Bitragunta and Vijayawada.

A statement in Parliament by the Minister of Railways the following day was to the effect that nine people had died and 20 had been injured, three seriously. Those who lost their lives included the driver and assistant driver of the passenger train.[175] However, another report stated that there were 11 confirmed deaths by early Monday and that later the bodies of another 10 victims had been recovered from the wreckage, raising the death toll to 21. This report added that a waggon (passenger coach?) had been twisted into a 'U' shape by the impact, the noise of which could be heard two kilometres away. Local villagers were said to have lit bonfires to assist in the rescue work and some uninjured or slightly injured passengers were to climb out of the overturned coaches with the aid of ropes and ladders but the rescue operation only gained momentum at daybreak.

Travelling at 80 km/h, the passenger train had reportedly run past three signals at Danger prior to the collision. [176]

At least 37 lives were lost in a rear collision at Kirodimalnagar station, Madhya Pradesh, India, on the Bilaspur Division of the South Central Railway, at 00:20 on Friday 4th September 1992 (since late 2000 the station has been in Chhattisgarh state). A goods train was standing in the station when it was run into from behind by a passenger train from Nagpur, in Maharashtra state, to Tatanagar in Bihar state. Most of the casualties occurred in two coaches of the passenger train, though a third coach was derailed and the rear vehicles of the goods train were smashed. About 100 people were injured, at least 20 of whom were admitted to hospital. A signalling error was blamed for the accident. [177]

At least 14 people were killed and 44 injured, 13 of them seriously, when the *Chenab Express*, en route from Karachi to Peshawar, ran into the rear of a freight train loaded with wheat at Reti, Pakistan, about 470 kilometres north of Karachi and 110 kilometres from Sukkur in Sindh province. The accident happened on Sunday 1st November 1992. A thousand troops were sent to the scene and recovered 14 bodies from the leading three of the 15 coaches of the express. A spokesman said, "There are numerous bodies inside the twisted wreckage," and ambulance officials warned that the number of dead could rise. The army cordoned off the area, while military doctors set up a first-aid centre. The critically injured were flown by helicopters to Karachi and other injured were taken to the nearest local hospital, 80 kilometres away.

---

[174] *The Daily Review*, Towanda, Pennsylvania, United States, 14th February 1992
[175] https://eparlib.nic.in/bitstream/123456789/10534/1/10_III_06041992_p288_p288_t274.pdf
[176] https://www.irfca.org/members/archive/messages.php?index=42
[177] *The Times-Reporter*, New Philadelphia, Ohio, United States, 4th September 1992

Initial reports blamed a signalling error for the accident and the government set up an inquiry, while Assistant Stationmaster Abdul Rahim at Reti reportedly fled the scene.[178] Other reports stated that the Reti stationmaster (the same person?) had been arrested and charged with negligence, while four other railwaymen had been suspended. [179]

Wikipedia states that a disastrous rear collision occurred in China on Saturday 10th July 1993, in which 40 people died and 48 were injured in an accident involving a passenger train and a freight train, due to 'the illegal driving of a train driver.' No location is given and the quoted source of the information on the accident is no longer accessible. It has not been possible to trace any other reference to this collision. [180]

Just before midnight on Monday 27th September 1993 a crowded passenger train from Marrakesh to Tangier was standing in the station at Temara, Morocco, when it was run into from behind by a goods train carrying, reportedly, naphtha, or according to one account, propane. The engine and tank waggons of the goods train were described as having been crushed together in a heap on the roof of the passenger train, while an explosion and fire followed. The coaches of the passenger train were described as having been quickly reduced to a charred wreck and rescuers spoke of many passengers being trapped in their seats and suffocated by the fumes or burnt to death before they could be released. A railway spokesman stated that there had been about 400 passengers on the train and that at least 25 passengers in each coach had died. One report spoke of 100 dead but other accounts stated that there were only 14 dead and 70 or 80 injured. The freight train had reportedly overrun a Stop signal. [181] [182] [183]

Western Railway train No. 5063, the Up *Avadh Express*, ran into the rear of a stationary Up GIT goods train on Wednesday 16th February 1994 at Bangrod station, Madhya Pradesh, India, on the Nagda – Ratlam section of the Ratlam division of the railway. Eight people died in the accident and 32 were injured, two of them seriously.

The accident was blamed on the express being run at greater than the maximum permitted speed.

The following recommendations made by the Commission of Railway Safety are believed to be related to this accident.

(a) There was no effective system of checking the speed recorder graphs of the locomotives and of the taking of suitable action against erring drivers who ran at speeds in excess of those permissible at any location. A system should be evolved to ensure the proper use of speed recorders. Even during the engine inspection the inspectors should check the inter-section running time through the previous section to find whether or not the drivers had been overspeeding.

(b) Suitable correction slips to the General and Subsidiary Rules should be issued as under:

(1) While giving instructions to stations regarding precedence for following trains, the controller on duty should remind the assistant stationmasters (ASMs) on duty regarding the point connections for the line intended to pass the trains.

---

[178] *Daily Herald*, Arlington Heights, Illinois, United States, 2nd November 1992
[179] *New Castle News*, New Castle, Pennsylvania, United States, 2nd November 1992
[180] Wikipedia – Lits of rail accidents in China
[181] *Stavanger Aftenbladet*, Stavanger, Norway, 29th September 1993
[182] *The Guardian*, London, England, 29th September 1993
[183] *The Times*, London, England, 29th September 1993

(2) While granting Line Clear, ASMs should ascertain from Cabinmen/Switchmen on duty (who were in independent charge of panel operations) about the correct setting of points, with the exchange of private numbers.

(3) A blank signal under a "Complete power supply off" situation should be treated as a 'defective' signal and the instructions contained in General Rules Nos. 3.08, 3.69 and 3.70 should be made applicable for such situations.

(c) It appeared that the "Psycho testing" of drivers, once every two years, for those aged 45 years and over, was not being done systematically as per the instructions contained in the Railway Board's letter No. 69/Safety/RAIC/68/88 Part B, dated 22.12.79. This needed to be monitored closely.

(d) While trenching work was being carried out for the laying of Signal and Telecommunications (S&T) cables, a proper route survey should be conducted jointly by the electrical, engineering and S&T officials (both open line and construction branches) and plans should be approved by all the concerned departmental officers. Till such time, the site should not be handed over to the contractors in charge of cable laying works. On the execution of the work, proper completion plans should be made available to the maintenance officials. [184]

Twelve, 15, 22 or 24 people lost their lives according to different reports in a rear collision between Rustavi and Tbilisi, Georgia, possibly near Gurjaani, late on the night of Saturday 13th August 1994. At least 16 others were injured. A freight train consisting of tank waggons loaded with heating oil or diesel fuel apparently stalled on an incline between two stations and was stopped awaiting assistance, when due to a signalling error or a breakdown in the signalling a passenger train was admitted into the occupied section. Two coaches of the passenger train were said to have been twisted into a tangle of twisted metal in what must have been a violent collision, though there was no report of a fire. Georgian leader Eduard Shevardnadze visited the site of the accident the next day, at which time rescuers were still at work searching for victims. A report on the accident stated that: 'Television footage showed bodies being pulled from hunks of twisted metal and train seats ripped from the floor of the cabins,' [185] [186] [187] [188]

The previous year there had been a series of terrorist attacks on the railway, probably linked to the long-running war between Azerbaijan and Armenia, Georgia's neighbours to the south, but there was no evidence that the collision had been due to sabotage; rather it was blamed on human error. [189]

India suffered a major catastrophe at 02:45 on Sunday 20th August 1995. It happened on the Northern Railway near the station at Firozabad, in Uttar Pradesh state. Train No. 4023, the Up *Kalindi Express*, had passed the West cabin Up Starter signal when it collided with a nilgai (a type of sturdy antelope). As a result of the impact with the animal the vacuum brake pipe was damaged, bringing the train to a stand a few metres short of the Advanced Starting signal. Before

---

[184] Government of India Ministry of Civil Aviation and Tourism Commission of Railway Safety Annual Report for 1993-94

[185] *Port Arthur News*, Port Arthur, Texas, United States, 15th August 1994

[186] *Daily Herald*, Arlington Heights, Illinois, United States, 15th August 1994

[187] *The Independent*, London, England, 15th August 1994

[188] https://www.klik.com.my/item/story/838430/24-die-in-georgia-train-crash

[189] https://www.upi.com/Archives/1994/08/14/A-train-crash-in-the-former-Soviet-republic-of/6980776836800/

the damage could be repaired and the train could be restarted, it was run into from behind by train No. 2801, the Up *Purshotam Express* from Puri, which had passed through the station at 02:54. Travelling at 70 km/h, the colliding train destroyed three coaches of the stationary one, while the engine and two leading coaches of the colliding train were derailed. There had been 2,200 passengers aboard the two trains, most of whom had naturally been asleep at the time. Officially 298 passengers and 12 railwaymen lost their lives in the disaster and 252 were hurt, 152 of them seriously. According to some unofficial sources the true death toll exceeded 800. [190]

'Rescue workers were still pulling bodies from the tangled wreckage Monday morning and expected the death toll to rise. Cranes were being used to lift the smashed cars, which had telescoped into each other. "Every time we pick something up, there's a body underneath it," said M.N. Chopra, the railway's divisional manager.

'The accident outside Firozabad railway station, in northern India, occurred when a signalman sent the *Purshotam Express* onto a track without realising that the *Kalindi Express* had stopped ahead, the Press Trust of India news agency said, quoting unnamed railway officials. Police were looking for the signalman, who disappeared after the accident.

'"The entire area was reverberating with cries and shrieks," said Manas Patnaik, who was travelling from the eastern state of Orissa to New Delhi {the destination of both expresses}. "I stumbled several times on severed limbs and some people – I don't know whether they were sleeping or dead," Patnaik, 29, told the United News press agency.

'Among the dead were 22 teenage athletes returning from a training camp and dozens of soldiers travelling with their regiments. The injured were admitted to hospitals in Firozabad and in the nearby towns of Tundla and Agra. Bodies were stretched out in rows in loosely tied bed sheets from sleeper cars, then taken to a makeshift mortuary in a large hall in town.' [191]

'Subodh Kumar Gupta, who was standing near the door of the *Kalindi Express* when it stopped on the track, said he saw the other train approaching, but thought it was on another track. "The next thing I knew, I was crushed under a seat and the coach had overturned," he said from his bed at the Sarojini Naidu Hospital, stroking the cast on his left shoulder. [192]

By Tuesday the disposal of the bodies of the dead had become a matter of some urgency, as a newspaper reported, 'Fearful of disease, officials today began burning hundreds of bloating, decomposing, corpses recovered from the worst train wreck in India's history. Tractors carried the shrouded bodies of Hindu victims to a field behind the city's crematorium, near the tracks where two trains collided last week, killing 348 people. Although cremation is a Hindu rite, no priests were present to chant prayers at the burning of 216 unidentified victims. Wrapped in sheets, the bodies were laid side by side, 20 at a time, on wooden funeral pyres 150 yards long.

'Twelve corpses identified as Muslims were buried according to Islamic custom. Relatives identified 120 other bodies and took them away for private ceremonies.

'"We can't keep the bodies any longer. We are in the midst of a populated area, and there is a threat of disease spreading," said Ashok Tamta, the superintendent of police.

'Firozabad, a city of 300,000 people near the crash site, does not have a mortuary, and was unable to get enough ice from neighbouring towns to preserve the bodies. Before the cremation, officials left the decaying corpses piled up outside a hospital and in a park to give people a last chance of identifying them.' [193]

---

[190] Wikipedia – Firozabad rail collision
[191] *Daily Press*, Victorville, California, United States, 21st August 1995
[192] *The Transcript*, North Adams, Massachusetts, United States, 21st August 1995
[193] Ibid, 22nd August 1995

The following day an American newspaper carried a very good report detailing how the collision had occurred. 'For years, Gorelai Sharma had spent up to 12 hours a day at the switches of Firozabad station, successfully directing scores of trains. But on Sunday at 2.45 a.m., he inadvertently gave the Green light to the *Purshotam Express*, travelling at 60 mph. It slammed into the *Kalindi Express*, killing 348 people in the worst train accident in India's history. That is the account of railway officials. Sharma has said nothing about it – he fled soon after the crash.

'Sharma's colleagues said the veteran switchman was a man who kept alert and never drank alcohol. "He has taught me many things. I don't know how he could have made the mistake," said switchman Khem Chand Sharma. The two men are not related.

'Sharma, who had come on duty at 1 a.m., apparently fainted on the floor of the two storey Firozabad West signal cabin when he saw what had happened, his colleagues said. In the confusion after the crash, Sharma vanished. His family fled their two-room, railway supplied apartment, behind the station. A handwritten notice from Indian Railways posted on the door of the apartment notified Sharma that he had been suspended.

'More than 100 trains pass daily through this central Indian town on the route from New Delhi to Calcutta. About 70 usually passed during Sharma's 12 hour overnight shift. Railway switchmen are relatively well paid by Indian standards, earning an average $113 a month and getting free accommodation. But they work punishing hours – seven days in a row, then one day off. "We have only five signalmen instead of nine which are supposed to be here," said Khem Chand Sharma as he pulled two of the 28 huge iron levers which were before him. "We are often working 12 hour shifts and get little rest." [194]

Meanwhile, New Delhi Assistant Commissioner of Traffic Police Lala Ram commented that the large number of cows allowed to roam freely around the country looking for shelter from the monsoon rains were a serious problem. 'Hindus, who make up 82 per cent of India's 900 million people, consider cows to be holy creatures. Strict Hindus don't eat beef and face strict punishment for purposely killing a cow.' [195]

The two Distant, Home and Starter signals had reportedly all been Clear for the colliding express.

The Commission of Railway Safety made a number of recommendations following the catastrophe:

(a) Signalling and Telecommunications safety devices: The highest priority needed to be accorded to immediately providing, at double-line stations with end cabins, the following safety devices at a minimum:

(1) Track circuiting from fouling mark to fouling mark on the main line.

(2) 'Advance Starter located nearest to the cabin say up to 180 metres beyond the loop line trailing point, 60 metres beyond last trailing point, if any, at fouling mark of facing point, if any subject to K. Oil tail lamp at the location being visible from cabin. If Advanced Starter is continued at greater distance, track circuiting be provided from Starter to Advance Starter, During doublings and other major alterations of yard, track circuiting from fouling mark to block section limit to be ensured.' (The paragraph is reproduced more or less verbatim from the annual report. Its exact meaning is not entirely clear to the author.)

---

[194] *The Independent*, Gallup, New Mexico, United States, 23rd August 1995
[195] *Desert Dispatch*, Barstow, California, United States, 24th August 1995

(3) The circuit should be designed to ensure that the departure signals were On when the slot was given for the Home signal. For the reception of trains on the Loop line, set for the sand hump/siding, only the Starter need be proved in the On position.

(4) Sequential track circuiting should be provided beyond the Home signal.

(5) The Main line Starter and Advanced Starter should be replaced automatically on the passage of the train.

(6) The stationmaster should have a positive slide control for the Home signal, i.e. allowing it to be pulled and replaced for one train.

(7) The stationmaster should have a positive slide control over the last stop signal.

(b) On double line sections of line with permitted speeds of 110 km/h and above, the following additional facilities should be provided to improve safety standards on such high-speed routes:

(1) Track circuits from the fouling mark to the block section limit.

(2) Block proving by track circuiting or axle counters.

(3) Four-aspect colourlight signalling.

(4) The auxiliary warning system.

(c) At stations with multiple-aspect signalling, the Off aspect of Loop line Starter signals should be changed from showing Green/Through to Yellow/Caution even when the Advanced Starter was taken off.

(d) Fifty per cent of the posts of switchmen were filled by the promotion of levermen/cabinmen with a minimum Class VI education. This level was not adequate for the type of work required to be done by the switchman. The minimum educational qualification should be raised to Class X.

(e) As a special measure, there should be direct recruitment of switchmen with educational and other criteria comparable to those for assistant stationmasters.

(f) The Model Room training and examination of switchmen, in both practical and theoretical aspects, should be provided in the training module.

(g) Section controllers should be debarred from interacting directly with switchmen in cabins provided with telephone connections to control, except during emergencies. During normal work, interaction with Control should only be through the stationmaster.

(h) When the duty hours of the switchmen were currently from 01:00 to 09:00, 09:00 to 17:00 and 17:00 to 01:00, they should be changed to 06:00 to 14:00, 14:00 to 10:00 and 10:00 to 06:00 to provide better rest for more staff during the night time. The duty hours of other station staff could be staggered by half an hour.

(i) Kerosene oil type lamps for train tail lamps and at level crossings should be banned. Only battery-operated flasher-type lamps should be used for tail lamps. The changeover should be done within a phased time-bound programme.

(j) Instructions on the positioning of seating-cum-luggage-rake (SLR) coaches at the originating station of the train as per the approved marshalling order should be reiterated. Such coaches should normally be positioned at either end of the train. Only up to two slip coaches or special coaches extra to the approved marshalling order should be attached outside of the SLRs on longer trains. For short trains on branch lines with only one SLR there could be three coaches outside of the SLR as per existing instructions.

(k) The state governments might be requested to evolve procedures and rules, so that:

(1) Local officials could waive post-mortem examinations for casualties in accidents involving the public transport system.

(2) Dead bodies and death certificates should be handed over expeditiously by the district authorities to the next of kin arriving from far-off places.

(l) The Railway should take appropriate action to ensure the proper recording of the speed of all mail/express trains.

(m) The Railway should take appropriate action to ensure the availability of the Emergency Lighting Kit Box as per existing instructions. [196]

At least 75 passengers died and 76 were injured when an Egyptian State Railway Cairo – Luxor – Aswan express ran into the rear of a local train from Giza or Cairo to Beni Suef, which was carrying mostly workers bound for Fayoum (also spelt Faiyum), at Badrshein, Egypt, on the morning of Thursday 21st December 1995. The railway at this point follows the valley of the River Nile and is often, as on the day of the accident, blanketed with fog. At the point of the accident, 25 kilometres south of Cairo, wooden shops stood alongside and close to the line. Ten of them were demolished by the derailed trains but fortunately at the time of the accident, which according to one account was just before 08:00, they had not yet opened. "Thank God…not that many people were around," exclaimed Mahmoud Abdel-Shafi, who had rushed to the scene upon hearing the noise of the collision. With other townspeople he had helped pull about 25 bodies from the wreckage, including those of two children.

An Associated Press report gave a detailed account of the disaster. 'More than four hours after the crash, bloodied bodies were still being carried away on stretchers. Four bodies could be seen inside passenger cars. A blood-splattered child's school bag lay alongside twisted metal.

'Police Brig. Mohammed el-Mahdi said 75 people were dead and 76 injured. He expected the death toll to rise as rescuers found bodies inside the wreckage.

'At least three cars of one train were thrown atop the other, and other cars derailed. Lt. Gen. Mustafa Sabri, a civil defence official, said five train cars were destroyed.

'Rescue workers brought in a crane to help dismantle wreckage. Hundreds of police officers and soldiers kept away onlookers as rescuers searched for bodies and survivors. Ambulances cruised nearby streets, appealing over loudspeakers for blood donations.

'Outside the morgue of Badrshein's hospital, about 200 yards from the tracks, four men wailed after identifying the body of a relative. "Where are you, Mohammed?" one man cried, referring to the dead relative. A nurse at the hospital, Asmaa Abdel-Moneim, described the morgue as "very bloody." There is one dead woman who was found carrying her dead baby and with one of her legs cut off," she said. "There are some complete bodies and some very mutilated." Thirty-one bodies were taken to the hospital and 17 were identified. When the hospital's beds were filled up, other injured were taken to hospitals in nearby towns and to Cairo.' [197]

As usual, details of the accident varied with different accounts, for example some stated that the accident had occurred rather later, at 08:30, and that it was the local train that had run into the rear of the express and not the other way round. Another Associated Press report added that, 'Townspeople said that there had been concern about the line through Badrshein because the town does not have a guarded crossing gate. They said 10 people were killed last

---

[196] Government of India Ministry of Civil Aviation and Tourism Commission of Railway Safety Annual Report for 1995-96

[197] *The Herald Journal*, Logan, Utah, United States, 21st December 1995

year when a train hit a bus in the town. The train tracks in Badrshein are lined with shops. A day before the crash, tens of thousands of people were at the town's Wednesday market there, shopkeepers said. They said the railway recently won a court order to have the shops removed, But the shopkeepers complained they rented the stores from the city council. "We are caught in the middle," said Mohammed Kafafi, a butcher whose shop was destroyed in the crash.'

A third Associated Press report added, '"There was screaming, people jumping from the train's windows, and lots of blood," said Samia Mohammed, 13. She witnessed the wreck... Thirty-one bodies were brought to the {Badrshein} hospital and 17 were identified. Nineteen others were taken to the Hawamdiya hospital, 1½ miles away. Other dead and injured were taken to Cairo.

'El-Mahdi, the police official, said the accident occurred when a local train travelling from Cairo to Fayoum bringing labourers to work ploughed into the Cairo – Luxor express, which had slowed up in the fog. Many tourists visit Luxor for its ancient temples, but el-Mahdi said no foreigners were reported aboard the train.

'Ibrahim Tantawi, 35, an assistant engineer on the Cairo – Luxor train, said the engineer was forced to lower the train's speed. "We couldn't see a (yard) in front. The driver even stuck his head out of the window to try and see," he said from a hospital bed, his right arm bandaged and his forehead scratched."' [198] (If the Luxor express was in front, how did its assistant engineer end up among the injured? It seems more likely that it was the express that ran into the local train.)

The day after the accident, Driver Mahmoud Mahmoud, who had been in charge of the engine of the colliding train, whichever that was, and who was under guard in hospital after having had his right leg amputated, was charged with having caused the accident. It was alleged that in spite of the fog he had been speeding at 110 km/h and had thus run past the signals protecting the train in front. [199]

A long-distance passenger train from Alexandria to Aswan had reduced speed at Al-Radisiyah, Egypt, where work was being carried out on the track, when it was run into from behind by a following freight train, on the morning of Monday 3rd February 1997. Eleven passengers in the rear carriage of the passenger train were killed and eight were injured. This time, a preliminary investigation blamed a signalling error, whereby the freight train had been admitted to the occupied section, for the accident. [200]

A major catastrophe occurred in China at 10:50 on Tuesday 29th April 1997. Local train No. 818 was stopped on track No. 4 in Rongjiawan station, Hunan province, when it was run into by train No. 324 from Kunming to Zhengzhou. Thirteen coaches of one or the other train were thrown off the track, at least 90 people died and about 300 were injured. Over 1,100 medical staff and police rushed to the scene. Despite the magnitude of the accident traffic on the important south – north Beijing – Guangzhou main line was not affected. [201]

Rescue work was still in progress on Wednesday morning.

The Xinhua news agency stated that the railway remained a popular form of long-distance travel in the country and that Chinese railways had handled 942 million passengers in 1996.

---

[198] *Valley Morning Star*, Harlingen, Texas, United States, 22nd December 1995
[199] *Pacific Stars and Stripes*, Tokyo, Japan, 24th December 1995
[200] *The Southern Illinoisian*, Carbondale, Illinois, United States, 4th February 1997
[201] *The Brownsville Herald*, Brownsville, Texas, United States, 30th April 1997

The overtaxed system was prone to accidents though. [202]

A mistake in the signalling had led to train No. 324 being given a clear road into track No. 4 despite the presence of the local train. [203]

Central Railway train No. 6317, the Down *Himsagar Express* from Kanyakumari to Jamm Tawi, was just pulling out of the station at Faridabad, Haryana, India, on the Mathura – Tukhlakabad section of the Jhansi Division of the railway, when it was run into from behind at 23:06 on Sunday 27th July 1997. The colliding train was No. 2627, the Down Bangalore – New Delhi *Karnataka Express*. Three coaches of the *Himsagar Express* and both engines and seven coaches of the *Karnataka Express* were derailed. Thirteen people were killed and 67 were injured, 21 of them seriously.

According to a newspaper report the *Karnataka Express* had been travelling at 110 km/h when it hit the *Himsagar Express*, which itself was running 12 hours late. The *Karnataka Express* was not booked to stop at Faridabad. Together the trains were carrying an estimated 1,200 passengers, many of whom were asleep.

A reporter wrote, 'The track was littered today with smashed suitcases, shoes and bed linen. Two coaches on the *Karnataka Express* looked like crumpled shoe boxes. Another luggage coach was demolished and both engines derailed.

'"I woke up to find the luggage in the coach tumbling. Before I knew it the train had tumbled to the right," said Peter Paul, a 35-year-old passenger, still dazed more than six hours after the collision.

'"Hereafter I will never feel safe on trains," said S. Roberts, who makes the three-day journey from northern Punjab state where he works to his southern home state Kerala once a year.

'"Obviously the driver of the *Karnataka Express* ignored a stop signal. It's human failure," said S.A. Zaidi, a senior manager of the state-owned railway. Full details, however, would be known only after officials spoke to the driver of the *Karnataka Express*, who was hospitalised with a broken arm, Zaidi said.' [204] This statement of a railway official, made before the investigation into the accident had even begun, would seem to have been well premature, but the Commission of Railway Safety did later come to the same conclusion. It attributed the collision to three causes:

(a) Train No. 2627 Down had passed the Down Main Home signal at Danger.

(b) The assistant stationmaster in A cabin had given Line Clear for train No. 2627 Down without having first set the road for the less important third line.

(c) The Western Railway controllers at Gangapur City had booked a totally ill-qualified driver to work train No. 2627 between Gangapur City and Tughlakabad.

The Commission of Railway Safety made a series of recommendations in relation to the accident:

(a) The Central and Western Railway Administrations should issue detailed instructions encompassing all the concerned departments on the actions to be taken whenever trains were diverted from one system to another in emergencies. The current tendency to have a "step-motherly attitude" by the foster Railway should be strictly curbed.

[202] *Pacific Stars and Stripes*, Tokyo, Japan, 1st May 1997

[203] Wikipedia – Rongjiawan train disaster

[204] *The Robesonian*, Lumberton, North Carolina, United States, 28th July 1997

(b) Whenever in emergencies mail or express trains were run by goods drivers, the following precautions should be strictly observed:

(1) The Rajdhani/Shatabdi Express should not be run by them under any circumstances.

(2) They should be drawn from a panel of senior goods drivers who had been fully screened by traction and safety officers. The panelled drivers might belong to either the A or B category (Safety). None from the C category should be empanelled. Such panels should be prominently displayed at crew booking points.

(3) On occasions where such screened drivers were not available, the booked crew should be accompanied on the footplate by a locomotive inspector to monitor the crew.

(c) The Railway Board should seriously study the difficulties faced by the Zonal Railways in filling up the posts of crew controllers and loco inspectors. The extant policy of running supervisors with reference to their selection and emoluments needed to be reviewed to make it more attractive to drivers to get on promotion as supervisors. THEIR PAY SHOULD BE PROTECTED {upper case words in original}.

(d) A drive should be initiated by the Western Railway administration to set right "out of order" breathalysers, currently available at all the crew booking points. They should also be supplemented by the procurement of portable equipment.

(e) Whenever certain running staff were confirmed to be habitual drinkers, they should be ruthlessly taken out of the "running cadre" once and for all and deployed on alternate stationary duties on clemency grounds.

(f) There was a need to induct a larger percentage of running staff, while filling up the cadre of Assistant Mechanical/Electrical Engineers for better and effective crew control.

(g) The Railway Board should immediately lay down the policy of providing the Automatic Warning System on busy 'A' routes of Indian Railways, on a time-bound basis.

(h) Voice recorders should be indigenously developed and installed in all locomotive cabs to improve accident investigations.

(i) All Indian Railways divisional safety officers should be grilled on the importance of "Resetting points against the blocked line" for effective implementation in the divisions. [205]

A rear collision reminiscent of the catastrophe at Firozabad less than three years earlier (see above) occurred at 21:50 on Monday 5th January 1998 between Karna and Masit stations, Uttar Pradesh, India, on the Rosa – Balamau section of the Moradabad Division of the Northern Railway. Train No. 4258 Down, the *Kashi Vishwanath Express*, had stopped in the section after colliding with a nilgai (large antelope). While it was stationary it was run into from behind by train No. 316 Down, a Bareilly – Varanasi passenger train. An initial report stated that 48 people had been killed and 62 more injured. The rescue operation was continuing at the time of this report and the number and names of the passengers in two unreserved coaches were unknown. Nine of the injured had been admitted to the Railway Hospital in Lucknow, six to the King George Medical College and Hospital in Lucknow and one to the Civil Hospital at Hardoi. The number of fatalities would eventually be stated as 51, and of injured as 66 (41 of them being serious cases).

Railway Minister Ram Vilas Paswan and Railway Board Chairman V.K. Agarwal were

[205] Government of India Ministry of Civil Aviation and Tourism Commission of Railway Safety Annual Report for 1997-98

amongst those who went to the scene, while the Railway opened inquiry counters at Varanasi, Lucknow, New Delhi and Hazrat Nizamuddin stations. Uttar Pradesh Chief Minister Kalyan Singh announced ex gratia payments of 50,000 rupees to the next of kin of those killed and 15,000 rupees to those injured. [206]

The collision was blamed on the unauthorised clearing of the Karna Starter signal, with train No. 316 then passing the Advanced Starter signal at Danger and entering the occupied section.

The Commission of Railway Safety made several recommendations relating to the accident:

(a) Northern Railway Supplementary Rule 3.78/2 should stipulate a maximum permissible speed of 60 km/h on broad-gauge lines and 40 km/h on metre-gauge lines when visibility was restricted by fog or for any other reason.

(b) At double-line stations with tokenless block working, the Starter signals should be released through the Advanced Starters, where the latter were provided.

(c) The automatic/auxiliary warning system should be provided over at least 500 kilometres of track per year.

(d) More sanctions and funds should be provided for the Emergency Universal Communication facility between drivers, guards and stationmasters, with priority being given to double-line sections affected by fog.

(e) In new signalling installations, central panels should be provided in preference to end panels.

(f) A Supplementary Rule should be provided under General Rule 8.03 to stipulate that at B class stations the block overlap during thick fog or tempestuous weather should be extended up to the trailing point/Starter signal.

(g) A special drive should be instituted to improve the contents of the Station Working Rules where required and to ensure that the stations were worked as per the rules. [207]

Seventeen people died and 30 others were injured when the *Chittagong Mail* from Dhaka collided with some waggons between Naoti and Nangolkot, Bangladesh, on Sunday 10th May 1998. A least one carriage of the passenger train was thrown into a lineside ditch and the Dhaka – Chittagong line was blocked for many hours.[208] Upon news of the accident being received relief trains were sent to the scene from Pahartali, Chittagong; and from Akhara, the first named arriving at 11:00. Eight of the injured were admitted to Nagolkot Thana Health Complex, seven to Comilla Medical College Hospital and three to Laksam Thana Health Complex; the others were released after treatment. The line was reopened at 17:10. Communications Minister Anwar Hossain and senior railway officials, followed later by Prime Minister Sheikh Hasina, went to the accident scene.

Eleven of the 52 waggons of a goods train from Chittagong Port to Tejgaon had become detached and had come to a stand in the section, two kilometres from Naoti. The goods train had arrived at Naoti at 04:18 and stopped there to cross the mail train. The person on duty at Naoti failed to realise that the goods had arrived incomplete and the mail train was therefore allowed to proceed, the collision following at 04:52

---

[206] https://m.rediff.com/news/1998/jan/06up.htm

[207] Government of India Ministry of Civil Aviation and Tourism Commission of Railway Safety Annual Report for 1997-98

[208] https://www.bridgemanimages.com/en/noartistknown/mangled-train-coach-lays-in-ditch-in-bangladesh-1998-05-11-photo/photograph/asset/8068468

The Naoti stationmaster and the guard of the goods train were suspended pending an investigation. It was suspected that the guard, who had remained in his brake van at the rear of the detached waggons, had fallen asleep. Government and railway (internal) inquiries were opened into the accident. [209]

Seven people died and 25 were injured in a collision that occurred at 05:00 on Wednesday 13th January 1999 at Darus station, Sudan, not far from Abu Zeid in West Kordofan wilayat (province), 350 kilometres southwest of Khartoum. Two passenger trains were involved; one had broken down and the other collided with it. [210]

A serious collision occurred near Glenbrook, New South Wales, Australia, at 08:22 on Thursday 2nd December 1999. Occurring on a busy commuter line, it was the subject of much public interest and formed the subject of a public inquiry conducted by Justice Peter Aloysius McInerney, the final report of which (there had been two previous interim reports) extended to 289 pages.

The State Rail Authority line on which the accident occurred was fully track-circuited with colourlight signals. At about 08:01 on the morning of the accident the power supply failed, causing both signal No. 41.6 at Glenbrook station and signal No. 40.8 ahead of it to return to Danger. At 08:04 an *Indian Pacific* Perth-to-Sydney interstate passenger train arrived at signal No. 41.6 at Red, which Driver David Willoughby wrongly assumed must be because there was a train in front. When the signal failed to clear he alighted from his engine to contact the Penrith signalman from the signal post telephone. He then found that the box containing the telephone was locked, something he had not come across before, and had to return to his locomotive for the key to it before he could finally contact the signalman and get permission to proceed. Restarting, he proceeded at extreme caution towards signal No. 40.8, as per the Rules. His train had been stopped for seven minutes at signal No. 41.6 and took almost another eight minutes to reach signal No. 40.8, only about a kilometre further on. (The signal numbers were apparently given in miles from Sydney Central, although the inquiry report gives distances in kilometres; if the signal numbers continued to indicate their distance from Sydney in miles at the time of the accident, then the two signals were 0.8 miles or about 1.2 kilometres apart.) Finding signal No. 40.8 also at Red, Driver Willoughby had again attempted to contact the Penrith signalman but found that its press-to-ring button was missing and assumed wrongly that it would not therefore work, so after returning to his engine he had waited one minute and then proceeded cautiously again, according to the Rules (Stop and Proceed working, as per the Ballymacarrett accident of 1945 in Northern Ireland and the Winsford accident of 1962 in England).

Following the *Indian Pacific* train was double-decker interurban passenger train No. W534. Driver Kevin Sinnett of this train had already been informed by Controller Michael Browne at Sydney that signal No. 41.6 had failed and that he should "just trip past it." Instead he had used the signal post telephone to contact the Signalman Damien Mulholland at Penrith. The conversation between them went:

Driver Sinnett: "I'm right to go past it, am I mate?"

Signalman Mulholland: "Yeah mate, you certainly are."

Relying on what he had been told by the controller and the Penrith signalman, Driver Sinnett then assumed that he was right to proceed and, in addition, that there was no train in

[209] *The Daily Star*, Dhaka, Bangladesh, 11th May 1998

[210] https://web.peacelink.it/africa/scio/month_32.html

front of him. Complicating matters was the fact that Signalman Mulholland received no indication on his panel of the location of the *Indian Pacific* train, as the locations of trains in the automatic signalling area were not displayed in his box.

Confident that he had a clear line in front of him, Driver Sinnett thus set off and was travelling at a speed of 50 km/h when, coming around a curve and through a cutting, he saw the rear vehicle of the *Indian Pacific* about 100 metres in front of him. An emergency brake application was at that point too late to prevent the collision, as a result of which seven passengers at the front of the leading coach of the interurban train were killed and 51 people were badly enough hurt to be taken to hospital.[211] At the last moment Driver Sinnett had left the driving cab and run back through the train to warn the passengers; he survived with serious injuries. The rear coach of the *Indian Pacific* was a car transporter vehicle, which was devoid of passengers, hence the deaths and most, at least, of the injuries were in the colliding train.[212] A train coming in the opposite direction was stopped short of the accident site.

Passenger Michiel Irik, 46, a Leura father of three and music teacher who had been travelling to Sydney with his fiancée, Danielle Diamond, related his experiences of the accident to a reporter: "I usually catch the 6.45am from Leura but I was running a bit late so I got the later train. We were sitting in the second carriage on the top deck, on the right-hand side, about halfway down the carriage. We'd been on board 50 minutes and we were just talking and intending to do a bit of composing on the train. Everything was fine as we left Glenbrook station, then, all of a sudden, there was this almighty sound of grinding and huge forces pushing and shoving together. It took 10 seconds and then the whole thing came to a halt. There was no warning whatsoever. A lot of the seats in the second carriage slid off the floor and were thrown forward. Some passengers were thrown out of their seats and some were pinned in their seats against the wall in front of them. Luckily for me I was braced by my guitar, which was between my knees, and that's what stopped me from being thrown. Danielle was thrown onto the ground and the contents of her bag thrown all over the carriage. There was a bit of electrical flashing in the carriage and flickering of lights. Initially there was a calm in the carriage but as people became aware of what happened a bit of hysteria broke out.

"Our carriage was standing at about a 30-degree tilt to the left. The first carriage was derailed and 45 degrees across the track. The undercarriage was ripped off and the train carriage was corkscrewed a bit. I stayed in the carriage to check if anyone was hurt for about 10 minutes until Danielle became hysterical and we started to leave. We had to climb over seats and I could see other carriages had air conditioning units falling down. To get out I had to sit on the landing and lower myself down about two and a half metres. I stayed there and helped some primary school children to get off for about 15 or 20 minutes. The children were mixed in their reactions. Some thought it was exciting and an adventure, some didn't know what happened and others were hysterical. While we were on the train my main concern was getting away because it was precariously tilted. Looking back at the impact, I can understand how people talk about things happening in slow motion. As it unfolded you could see every second of it because it's so abnormal. I was very aware of every minute detail. Danielle has a phobia about train travel and was anxious before we left. She didn't want to sit in the front carriage. I think in hindsight it's lucky." [213]

The inquiry into the accident found that the training and experience of Signalman Mulholland had been deficient in several respects. He had been unaware that consecutive signals could fail safe, to display Stop, if the fault affected the overlap section and had assumed that the *Indian Pacific* freight would not be stopping at the next signal but would be

---

[211] Report of Special Commission of Inquiry into the Glenbrook Rail Accident

[212] Wikipedia – Glenbrook rail accident

[213] https://messui.polygonal-moogle.com/incident/1999_glenbrook.pdf

proceeding to Penrith and thus be clear of the area, when he had authorised the interurban train to proceed. He had failed to use other means available to him to check the position of the *Indian Pacific*, such as using two-way radio, contacting trains going in the opposite direction or telephoning the stationmaster at the next station at Lapstone, to determine if the *Indian Pacific* had passed through there. The obvious method would have been to contact the *Indian Pacific* on the two-way radio. It transpired that the radio would, if used, have enabled him to contact the driver of the *Indian Pacific*, because he had done so after the accident. In addition, Signalman Mulholland had not known the procedure for managing consecutive trains through an automatic section of track when a signal failure had occurred. This was not only a deficiency in his training but it also reflected the inadequate nature of the operational rule that was then in force.

The language used by Signalman Mulholland in his communications with the two drivers had, the inquiry considered, been colloquial and imprecise. Following the exchange with Driver Willoughby, it was obvious that Signalman Mulholland had thought the *Indian Pacific* would proceed at normal speed to Penrith and be well clear, and on that assumption he had authorised the following interurban train to proceed. From Driver Willoughby's perspective, he had thought the result of the communication with Signalman Mulholland was that he should proceed with extreme caution to the next signal and then act in accordance with the operational rules, if that signal was other than at proceed or caution. The conversations of the different railwaymen demonstrated a lack of clarity, a lack of precision and a failure to comply with the communication protocols in which these men should have been trained and should have been required to use.

Signalman Mulholland had not been trained to provide Driver Sinnett of the interurban train with all the relevant information he needed, and in particular, the critical information that he, Signalman Mulholland, did not know the location of the *Indian Pacific*. Had that information been communicated to Driver Sinnett, it was probable that the latter would have proceeded in a much more cautious manner.

The training of the drivers of both the interurban train and the *Indian Pacific* had been deficient. Driver Sinnett had not had proper training in the operation and effect of Rule No. 245. This is apparent from the fact that he had not proceeded, as the rule required, with extreme caution after passing an automatic signal at stop. Through no fault of his, Driver Willoughby had not been trained in the operation of signal post telephones, to the extent that he had not been told that the press-to-ring button was not an essential feature of their operation, so that the telephone would work perfectly well even if that button were missing. The assumption he had made, was that he would not be able to contact the signalman on that signal telephone because the button was missing, and this had led him to abandon those attempts. If he had persevered and made contact with the signalman, it might have alerted the signalman to the fact that the *Indian Pacific* was in the path of the approaching interurban train, in time for the signalman to stop the latter train and avoid the collision.

Rule No. 245, which was designed to guide the employees involved in managing trains through an automatic section of track following a signal failure, was defective in its content and language. It did not deal with the situation of two consecutive trains in a section of track with such a long overlap circuit and, in addition, it was confusing and ambiguous. If any of the railwaymen had been properly trained in the procedure, which appeared doubtful, then they had not understood it. The ignorance and confusion about the rule, which had manifested themselves at the time of the accident and during the course of the evidence in the first stage of the hearings, demonstrated fundamental defects in the rules as a primary means of managing the safety of railway operations and in the way in which those rules were expressed and taught.

The equipment provided for the communication of safety-critical information was antiquated and inadequate. The driver of the *Indian Pacific* was required to alight from his train, cross the tracks, then call on a signal post telephone by turning a handle to generate a current, which would sound a buzzer in the Penrith signal box. The delay occasioned by the need to use signal post telephones was a factor the inquiry identified as being significant in contributing to this accident. The interurban train was fitted with a Metronet radio, which permitted almost instant contact with the signal box. What was significant, however, was that one train had almost instant communications technology and the crew of the other was required to use a system of communication with the signalman that was cumbersome and time-consuming. Ironically, the *Indian Pacific* was fitted with modern communications technology, including a satellite telephone, but because of the provisions of the rule, the crew had not been permitted to use that technology to contact the Penrith signalman to seek authority to pass an automatic signal at stop, but had to use the signal post telephone. The combination of the two incompatible methods of communication meant that the headway between the two trains had necessarily been reduced.

There had been, stated Justice McInerney, a lack of consideration of the safety implications of the actions undertaken and this had probably been the single greatest defect in the safety management of the rail organisations involved. There is no evidence to suggest that employees were properly trained, or that training was regularly reinforced, therefore there had been no understanding that the degraded operational environment, arising from the signal failure at Glenbrook, had introduced additional risk to rail operations. Consequently, there had been a lack of awareness among the various railwaymen regarding the care required to be taken in managing trains through the section, to ensure that the risk of collision was properly controlled. This had been even more important given that the main focus during normal operations was to maintain on-time running. Consequently, in the absence of effective training and regular reinforcement of the importance of carefully managing a degraded system to ensure that safety was the highest priority, it was only to be expected that the operational staff would be motivated by their everyday goal of ensuring on-time running.

It was this emphasis on on-time running that Justice McInerney considered had motivated Controller Browne to contact Driver Sinnett of the interurban train and advise him that, "It's only an auto, so just trip past it." It had also been a desire to keep the interurban commuter train running in accordance with its timetable, which, it seemed, had influenced Signalman Mulholland to assume the *Indian Pacific* was clear and that he should authorise the interurban train to pass signal 41.6, even though, at the time, he did not know where the *Indian Pacific* train was. This lack of proper appreciation of the safety considerations in an inherently dangerous operation had been a very significant contributing factor to the accident. The employees involved were neither irresponsible nor reckless men. The problem was that they had not been trained and no emphasis had ever been placed upon the primary importance of safety in the conduct of rail operations.

In keeping with modern inquiries into railway accidents elsewhere, the report on the accident concluded with a large number of recommendations, no less than 95 in all:

(1) Selection processes for all safety-critical staff should include psychometric testing.

(2) The training of railway employees should include:

(i) The development of safe behaviour as the principal object of training.

(ii) Emphasis on teaching of the safety rationale for all rules and procedures.

(iii)Practical examples drawn from Australian and overseas experience to demonstrate the consequences of failure to apply operational rules and procedures correctly or in a thoughtful manner.

(iv)An appropriate balance between the practical work experience and classroom components of any training program.

(v) The use of modern, interactive simulators as a core component of training programs.

(vi)Emphasis on the importance of teamwork in rail operations, including ensuring that operational employees had a clear understanding of the duties, roles and pressures involved in the work of other operational occupational groups.

(3) Trainers of safety-critical staff should have and maintain operational experience.

(4) Trainers of safety-critical staff should develop and maintain their training skills.

(5) The performance of training organisations and individual trainers should be regularly assessed by accredited rail organisations and audited by the Rail Safety Inspectorate.

(6) The processes and techniques used for the assessment of the competency of safety-critical staff should be upgraded and strengthened to ensure that such assessments were effective and regularly performed.

(7) There should be random auditing by the Rail Safety Inspectorate of the assessments of the competence of safety-critical employees.

(8) All train drivers should have comprehensive route knowledge at all times.

(9) The Rail Safety Inspectorate should conduct random audits of drivers to determine their competency and the adequacy of their route knowledge.

(10) All trains should be fitted with data loggers to enable, among other things, train driver performance to be monitored.

(11) Train drivers with less than three years' driving experience should be classed as provisional drivers.

(12) All provisional drivers rostered on standby should travel with experienced drivers.

(13) A class of principal driver should be created to instruct provisional drivers.

(14) The position of team leader should be created to be responsible for a group of approximately 30 drivers, to act as a mentor and to instruct them individually or collectively on any safety-related matter.

(15) Each team leader should be responsible for the technical competence and safety behaviour of each driver in his team.

(16) The Rail Safety Inspectorate should conduct random audits to determine whether the team leader system was being implemented effectively.

(17) No trackside worker should be required to be solely responsible for his own protection.

(18) All trackside work supervisors should be trained to assess and control risks to trackside workers.

(19) The Rail Safety Inspectorate should conduct random audits of the safety protection of trackside workers.

(20) The project to rewrite the rules should be given the highest priority.

(21) The objectives of the project to rewrite the rules should be to:

(i) Develop rules structured around a core set of fundamental principles.

(ii) Eliminate undesirable or unnecessary material within the rules.

(iii) Eliminate undesirable or unnecessary rules.

(iv) Ensure the rules are concise and easy to read and expressed without unnecessary narrative content.

(v) Use diagrams and illustrations when appropriate.

(22) Continual and detailed input into the redevelopment and redrafting of the rules should be sought from persons with expertise and experience in:

(vi) Training, both in the development of training programs and the teaching of rules.

(vii) Operational activities including train drivers, signalmen, guards, train controllers, worksite supervisors and any other occupation within the rail environment that might have to apply the rules in their day-to-day duties.

(viii) Human factors.

(ix) Engineering expertise in each of the railway engineering disciplines.

(x) Drafting of operational procedures in other hazardous industries.

(23) Handbooks should be prepared for distribution to persons employed in specific safety-critical railway occupational groups and should contain the particular rules relevant to each group.

(24) The Rail Safety Inspectorate should be responsible for approving all redrafted rules.

(25) The Rail Safety Inspectorate should ensure there was proper testing of the rules to ensure that they were unambiguous and easily understood.

(26) The Rail Safety Inspectorate should instigate and develop a standard for railway communications within twelve months of its establishment.

(27) The Rail Safety Inspectorate should ensure that the standard for railway communications, once developed, was fully implemented.

(28) Until a uniform and integrated communications system had been implemented in accordance with the standard, all types of communications equipment should be permitted for the communication of safety-critical information.

(29) No train should be operated without being equipped with operative radio communications equipment.

(30) The existing communications protocols should be reviewed and redeveloped following consultation with other relevant organisations.

(31) The revised communications protocols should incorporate a requirement that drivers be informed of route changes.

(32) The revised communications protocols should incorporate a requirement that drivers be informed of the likely location of any trackside workers they might encounter.

(33) All communications protocols should be strictly enforced by accredited rail organisations.

(34) It should be a condition of accreditation that rail organisations strictly controlled the use of any private audio or visual device in areas where safety-critical communications occurred.

(35) The Rail Safety Inspectorate should conduct random audits of compliance by accredited rail organisations with the communications protocols.

(36) The Rail Safety Inspectorate should supervise a trial of train-to-train communications to evaluate their advantages and disadvantages.

(37) If the trial satisfied the Rail Safety Inspectorate that train-to-train communications should be introduced, then they should be implemented as soon as possible.

(38) The existing Network Control centres should be modernised by centralising the train control function, including the functions currently performed by signalmen.

(39) Train controllers should be provided with the necessary support to enable them to effectively and safely control the movement of trains from a central location, including:

(xi) Computerised train control systems, which provided a real-time display of the position of trains and computer-generated solutions to assist controllers to minimise or avoid disruptions to normal operations.

(xii) Ensuring that all support functions required by train controllers were located within the same centralised train control rooms.

(40) No train controller should be required to manage disruptions to normal operations without the immediate personal assistance of a senior supervisor.

(41) There should be random breath testing by authorised officers of the Rail Safety Inspectorate of railway employees engaged in safety-critical work.

(42) There should be drug testing of railway employees involved in an accident or incident.

(43) The Rail Safety Inspectorate should examine the advantages and disadvantages of introducing a system that enabled the immediate and reliable assessment of the fitness to commence duties of safety-critical employees.

(44) All accredited rail organisations, the Department of Transport, the Office of the Rail Regulator, the Rail Safety Inspectorate and the Rail Accident Investigation Board should each avail themselves of the information and expertise in respect of rail safety management that existed in overseas rail organisations.

(45) The second interim report recommended that there be established a Rail Safety Inspectorate.

(46) The primary function of the Rail Safety Inspectorate should be the accreditation of rail organisations in New South Wales.

(47) The Rail Safety Inspectorate should refuse accreditation to any organisation unless it had satisfied, in addition to any other matters, that:

(xiii) It had a rigorous and robust safety management system that conformed to the highest international standards of safety management and practice.

(xiv) It had an effective safety management plan for the implementation, monitoring and ongoing improvement of its safety management systems.

(xv) The members of the board, the Chief Executive Officer and all other officers holding senior managerial positions considered the safety of the organisation's activities as their first priority.

(xvi) It had an effective system for identifying safety risks in its operations and had effective mechanisms for controlling those risks, monitoring the effectiveness of the controls, and adjusting the controls accordingly.

(xvii) It had an effective system for determining the priority of activities for removing, reducing or controlling particular risks.

(xviii) It had the resources, including sufficient numbers of employees, to ensure that the safety of rail operations could be maintained under any circumstance.

(48) The Rail Safety Inspectorate should be required to make public all notices of accreditation issued by it.

(49) The Rail Safety Inspectorate should have the responsibility to ensure that each accredited rail organisation complied with its accreditation and any conditions and restrictions specified in the accreditation.

(50) The Rail Safety Inspectorate should be given the power to impose a range of sanctions, including prosecution of individual board members, chief executive officers and the accredited organisations, to enforce compliance with the accreditation and any conditions or restrictions specified in the accreditation.

(51) The Rail Safety Inspectorate should be given the power to conduct safety audits of any accredited organisation.

(52) The Rail Safety Inspectorate should be given the power to inspect any person or thing that might give rise to an unsafe activity or outcome on the rail network.

(53) All safety audit reports of the Rail Safety Inspectorate should be made public.

(54) The Minister for Transport should be given the power to direct the Rail Safety Inspectorate to conduct a safety audit or inspection of an accredited organisation.

(55) The report of any audit or inspection directed by the Minister for Transport should be made public.

(56) The Rail Safety Inspectorate should be given the power to serve any accredited organisation, or any person who appeared to be employed by or otherwise associated with an accredited organisation, with a written notice requiring specified action to be taken or stopped, which an authorised officer of the Rail Safety Inspectorate had reasonable cause to believe might give rise to an unsafe activity or outcome on the rail network.

(57) Legislation should be introduced to make it an offence, attracting substantial penalties, for failure to comply with such a notice.

(58) The Rail Safety Inspectorate should be given the power to approve any variation to an accredited organisation's safety management system, including internal structural changes, provided that the Rail Safety Inspectorate first received a disposition statement and was satisfied that a proper safety validation process had been conducted and that the variation would not reduce the level of safety of rail operations.

(59) Legislation should be introduced to make it an offence for an accredited organisation to vary the safety management system with which it had obtained accreditation without the prior written approval of the Rail Safety Inspectorate.

(60) The Rail Safety Inspectorate should be given the power to examine proposed appointments and existing appointments to the board and senior management positions, including that of the chief executive officer, of an accredited organisation to enable it to satisfy itself that any such appointee or proposed appointee had an appropriate level of understanding and commitment to the safety of the rail operations in which the organisation was, or was seeking to be, involved.

(61) The Rail Safety Inspectorate should be given the power, if not so satisfied, to provide a written report to that effect to the person or persons responsible for making the appointment.

(62) The Rail Safety Inspectorate should be given the power to reject a safety management plan of an accredited organisation if the plan was, in the opinion of the Rail Safety Inspectorate, inadequate in any respect.

(63) The Rail Safety Inspectorate should be given the power to allocate or remove the responsibility for any particular safety matter to or from an accredited organisation.

(64) Authorised officers of the Rail Safety Inspectorate should be given the powers to enter upon land, including premises and rolling stock, and to require an accredited organisation or any person who appeared to be or to have been engaged in any rail activity, to produce any document, including a document in electronic form, or anything

an authorised officer reasonably believed related to a matter that did or could affect the safety of rail operations.

(65) Authorised officers of the Rail Safety Inspectorate should be given the power to require any person to provide information orally, electronically, or in writing, which the authorised officer reasonably believed did or might affect the safety of rail operations.

(66) The legislation should make it an offence to fail to provide the document, thing or information requested.

(67) The legislation should make it an offence to provide false or misleading information to an authorised officer.

(68) The Rail Safety Inspectorate should be given the power to monitor and ensure compliance by accredited rail organisations with the recommendations made in any report of the Rail Accident Investigation Board.

(69) The legislation should provide that any accredited rail organisation that was affected by any recommendation made in a report of the Rail Accident Investigation Board, within 60 days of the release of the report, inform the Rail Safety Inspectorate in writing, as to each such recommendation, whether it accepted or rejected the recommendation in whole or in part and, if rejected in whole or in part, provide written reasons for such rejection.

(70) The Rail Safety Inspectorate should be given the power to require an accredited organisation to inform it in writing how it proposed to implement a recommendation made in a report of the Rail Accident Investigation Board and the proposed timetable for its implementation.

(71) The legislation should provide that in the event that the Rail Safety Inspectorate did not agree with the reasons for the rejection in whole or in part of any such recommendation or alternatively, if any such recommendation was accepted in whole or in part by the accredited organisation, but the Rail Safety Inspectorate considered that the proposed remedial action was either not to be carried out in a timely manner or was inadequate, then the Rail Safety Inspectorate should have the power to direct that the remedial action be concluded within such time and in such manner as the Rail Safety Inspectorate might specify in writing and the accredited organisation should be required to comply with such direction.

(72) The legislation should provide that the Minister for Transport might, by written notice to the accredited organisation, and the Rail Safety Inspectorate, extend the time for completion of the remedial action and, if such extension was granted, the Minister must provide written reasons for extending the time.

(73) The legislation should provide that the Rail Safety Inspectorate give written reasons to the Minister for Transport for any action or failure to take action against an accredited rail organisation in relation to any non-compliance by that accredited rail organisation with the terms of its accreditation or with any recommendation contained in an investigation report of the Rail Accident Investigation Board.

(74) The legislation should provide that all notices or correspondence passing between the Rail Safety Inspectorate, the Minister and an accredited rail organisation relating to any recommendation contained in a report of the Rail Accident Investigation Board be made public.

(75) The Rail Safety Inspectorate should be provided with the necessary funding to retain experts, including specialists in engineering, organisational safety, statistical analysis and human factors, and to employ or retain legal officers and to otherwise finance its activities.

(76) The Rail Safety Inspectorate should be within the Department of Transport.

(77) The legislation whereby the Rail Safety Inspectorate was created should provide for its independence from ministerial control.

(78) The legislation whereby the Rail Safety Inspectorate was created should provide for its independence from and paramountcy over the Office of the Rail Regulator created by the Transport Administration (Rail Management) Amendment Act 2000.

(79) A project team should be established within the Rail Safety Inspectorate, over and above its normal staff establishment, for the specific purpose of ensuring that the recommendations in this final report were implemented by each relevant accredited organisation and that the Rail Safety Inspectorate should report in writing to the Minister for Transport at not less than six-monthly intervals regarding the implementation of these recommendations and all such reports should be made public.

(80) The second interim report recommended the establishment of a Rail Accident Investigation Board.

(81) The Rail Accident Investigation Board should have as its primary role the independent, impartial and unbiased investigation of accidents and incidents for the purpose of identifying any matter that might have or did contribute to an incident or accident or that might contribute to an incident or accident in circumstances similar to those that occurred.

(82) The legislation should provide that the Rail Accident Investigation Board might conduct its own investigations or require an accredited rail organisation to conduct an investigation and provide it with a report.

(83) The legislation should provide that any incident or accident involving an accredited organisation be notified to the Rail Accident Investigation Board in writing as soon as practicable after its occurrence and in any event no later than 24 hours after the occurrence.

(84) The Rail Accident Investigation Board should have the power to conduct public hearings, at which witnesses could be compelled to attend and be examined.

(85) The Rail Accident Investigation Board should collect, analyse and report on data relating to rail safety matters within New South Wales.

(86) The Rail Accident Investigation Board should have as one of its functions the collection and analysis of information in relation to rail safety from interstate and overseas.

(87) The Rail Accident Investigation Board should have as one of its functions the ongoing liaison with overseas rail safety organisations, including membership of and participation in international railway organisations and conferences.

(88) The legislation should provide that the Rail Accident Investigation Board be required to provide such information to the Department of Transport, the Rail Safety Inspectorate and any accredited rail organisation.

(89) The legislation should provide that proceedings of the Rail Accident Investigation Board and communications made in the course of its investigations might not be disclosed, other than by the Board, and might not be used in any legal or other proceedings except a prosecution for perjury or a prosecution for an offence under the relevant rail legislation.

(90) The legislation should provide that, save for coronial proceedings, an investigator authorised by the Board was not compellable as a witness in any court proceedings.

(91) The legislation should provide that any statement by a member or officer of the Rail Accident Investigation Board relating to an investigation was inadmissible in any legal,

disciplinary or other proceedings.

(92) The legislation should provide that no member or officer of the Rail Accident Investigation Board might disclose any information obtained by the Board in the course of the discharge by it of its functions.

(93) The Rail Accident Investigation Board should maintain a confidential system for the reporting to it of any incident which had or might have caused an unsafe activity or outcome in the course of rail operations.

(94) The Rail Accident Investigation Board should make public each of its investigation reports.

(95) The Rail Accident Investigation Board should publish an annual report to be tabled in Parliament. [214]

Five people died and at least nine others were hurt, two critically, when a passenger train from Habana to Santiago de Cuba (another report states that it was running from Santiago de Cuba to Habana) ran into the rear of a Sugar Ministry freight train at Guayacanes, Cuba, a few kilometres from Ciego de Avila, probably on Wednesday 14th March 2001. [215] [216]

A major accident occurred in South Africa at 16:00 on Tuesday 6th February 2002 when a passenger train ran into the rear of a freight train at Charlottedale station, north of Durban. Most of the victims were schoolchildren returning home from KwaDukuza (previously named Stanger) to Durban after lessons. Six coaches of the passenger train were derailed and many passengers were trapped in them. Early on, it was known that a number of people had lost their lives and the final death toll was 24, with another 168 injured. Sixteen of those who died were children.

Local newspapers contained numerous stories of those who had survived, and some who had not.

Fifteen-year-old Njabulo Ntuli, a pupil of Groutville High School in KwaDukuza, was one of the survivors, but his best friend, Thamsanqua Gumede, was killed. Ntuli himself was trapped in the wreckage until he could be cut free by rescuers and suffered badly cut legs. He related, "We heard that there was a ghost train on the track and we were talking about it, then suddenly there was this bang. I was trapped and could not get out."

Bruce Bodozu, who was also 15 and suffered injuries to both his legs, said, "All I heard was a loud bang and the screams of the passengers." He had been on the way home from school with his sister. Their father, Sibonelo, stated, "This was an accident. It could happen at any time. I'm not angry with Metrorail. But if the cause was negligence, the company should take steps to ensure it doesn't happen again."

Cleopatra Dube, 14, who escaped with bad bruises, said that many pupils had tried to escape by jumping through the train windows and dropping several metres to the ground.

A newspaper account wrote, 'In the chaos immediately following the accident, parents were looking for their lost children among the weeping injured. A man approached reporters crying and asking them to help him find his two daughters Thuli and Nthobi Dlamini, aged 14 and 15 years. "I think they are alive, but I do not know where they were taken or how to get there. I have no car." He was referred to a volunteer who was taking down the names of the injured as they were loaded into the ambulances.

---

[214] Report of Special Commission of Inquiry into the Glenbrook Rail Accident

[215] *Herald Sun*, Melbourne, Victoria, Australia, 15th March 2001

[216] http://news.bbc.co.uk/1/hi/world/americas/1220637.stm

'On Wednesday people flocked to the KwaDukuza mortuary, where the bodies of the deceased were laid out for identification. Among the searchers was Sidney Ntuli, who had been driving around the whole night searching for his wife, Mlungile Shiza (38) a teacher. Ntuli said he had not slept and the last meal he had had was breakfast on Tuesday. Shiza's sister-in-law, also a teacher, missed the train by a minute and took a taxi home. When she arrived there she heard about the accident and phoned Ntuli. "I last ate yesterday morning. I do not feel like eating. I am worried about my children (aged three, two and 16). All I want is just to find her, whether she is dead or alive," he said.

'Mvusi Bukosini, from Groutville, lost both his 13-year-old son Bongeni and his 19-year-old brother, Mduduzi Mkhize. Mduduzi was the president of Tshelenkosi Secondary School in Shakasville. His body was damaged beyond recognition and he was identified only by the cell phone he still had on him. Mduduzi's two sisters Khanyisile (14) and Poppie, six, were still missing on Wednesday afternoon. Bukosini said he got to the scene as soon as he could. "I saw bodies being taken out and I thought I saw Bongeni. When I looked closer I saw that it was Bongeni."

'Twenty-four-year-old Thembelani Ngcobo from KwaDukuza was in the second coach from the front. Despite a broken leg he managed to crawl to safety through the windows of the wreck. "We tried to get out as soon as we could, but we could not because the doors were jammed and we had to get out through the windows. We could not save everyone, those who were okay tried to pull out those who were still inside the coaches. People were crying. Children were crying for their parents and parents were crying for their children. It was terrible! Others were praying and I am thankful to God that I survived." He was taken to the Stanger hospital, where more than a hundred injured passengers were treated in the space of a few hours.

'On Wednesday morning school principals from the schools in the area tried to ascertain which of their pupils had been on the train. At the Stanger South Secondary School at least 18 pupils were reported missing. Two of them had died. At the Nkukweni primary school a teacher died and several pupils were missing.

'Metrorail on Wednesday morning established an assistance centre at the scene of the accident to assist families with transport and information about missing loved ones. Metrorail officials will meet families on February 13 to determine further support.

'Meanwhile, the terms of reference for a commission of inquiry into the KwaDukuza train crash which claimed more than 20 lives is expected to be finalised on Thursday, Public Enterprises Minister Jeff Radebe told reporters on Wednesday.' [217]

The circumstances of the accident are not entirely clear from published reports but the events leading up to it apparently began with the stoppage of a Spoornet freight train due to brake problems. Coincidentally there was a theft of copper signalling cables, which put three signals out of action and meant that there was no communication between the controlling signal box and drivers, so that hand signalling with flags had to be resorted to. Thefts of cables were common in South Africa and elsewhere, the copper being sold on the black market.

Other factors were possibly also involved. Christo Vos, General Secretary of the United Transport and Allied Trade Union, alleged that there were major problems with insufficient periods of rest between shifts for train drivers, outdated equipment and missing safety measures. Urging President Thabo Mbeki to intervene in the struggle to secure better conditions for rail workers, he declared, "If the situation continues as it is at the moment, we will have no other option than to bring all the trains to a standstill. We'd rather inconvenience commuters than kill more of them." [218]

---

[217] *News 24*, Durban, South Africa, 6th February 2002

[218] *Mail and Guardian*, Johannesburg, South Africa, 6th February 2002

Metrorail official Sifiso Lukhele, the acting regional manager in the Durban area, stated, "There will be a thorough investigation because it's the public's right to know what happened, especially so soon after the KwaMashu train crash," referring to an accident near Durban on 27th December, when the driver had died and 51 passengers had been injured when two commuter trains collided. "If anyone is responsible, the necessary measures will be taken," he vowed, offering his condolences to bereaved families and those of the injured, and he declared that, "Metrorail will look into compensation."

Less helpfully, at the funeral of a number of the victims Education Minister Kader Asmal blamed moral decay for the accident(!). He proclaimed, "It is such moral decay that can make workers neglect their work and give their best. It is such moral decay that has contributed to the levels of crime and violence. Such moral decay undermines the very fabric of society, making it more and more difficult for those of goodwill ... to live productive and worthwhile lives," adding that the accident was one of the "most avoidable and tragic events in the history of our new democracy".

Asmal did, however, also pledge 10,000 rand to the Train Accident Disaster Fund, established by the KwaDukuza Municipality, to assist families afterwards. Transnet contributed 100,000 rand to the fund, while the provincial department of education and the KwaDukuza council also contributed large amounts. [219]

South Central Railway train No. 7064 up, a Secunderabad – Manmad express, collided with a stationary goods train in Ghatnandur station, Maharashtra, India, on the Secunderabad Division of the railway, at 01:30 on Friday 3rd January 2003. A SLR (seating cum luggage rake) coach, a general coach and a three-tier air-conditioned coach were derailed, 11 passengers were killed outright, seven were fatally hurt and 32 were injured. [220]

Railway Minister Nitish Kumar stated: "It's a total failure. The staff has failed. Whatever has happened is due to human failure." He promised 1 lakh (100,000 rupees) compensation for the families of those who had died, 15,000 rupees for victims who had suffered grievous injuries and 5,000 rupees for those with simple injuries. [221]

A preliminary report indicated that the passenger train had been admitted onto the main track on which the goods train was waiting rather than being diverted onto a loop line. Assistant Stationmaster Prashant Kumar Verma and Switchmen Vinayak Neelakantha and Shesharao Sayedu at Ghatnandur were suspended after the accident, together with Traffic Inspector Srinivas Suryavanshi and Divisional Safety Officer G. Krishnaih. [222]

Some of the surviving passengers were later interviewed by reporters: '"It was the sheer presence of mind of the driver that a disaster was averted," said an Australia-based NRI (Non-resident Indian) Indrajit Sen, who got down at the Aurangabad station from a relief train carrying other survivors from the accident site near Ghatnandur station. Sen, who hails from Raipur, his wife and two daughters were travelling in the S-7 coach of the express. The Sen family was bound from the pilgrimage town of Tirupati to Shirdi to pray at the Sai Baba shrine there. "At Ghatnandar station people fell on top of one another and there was confusion but I believe it was the driver whose presence of mind averted a major disaster," he recounted.

'Some others attributed their survival from the nearly fatal accident to the divine. "We thought we were dead but the Sai Baba saved us," said elderly couple Malkondaiyya and Uttsamma, both in their seventies. And the driver, they said, had been "a divine instrument

[219] *News 24*, Durban, South Africa, 18th February 2002

[220] *Rediff.com*, 3rd January 2003

[221] Government of India Press Release 3rd January 2003

[222] www.outlookindia.com/pti_print.asp?id=109271

that brought the train to a halt well in time."

'"A loud noise was heard and people were falling on top of each other," said passenger Rajana Deshpande, who was travelling with her two daughters in the S-8 coach. She was returning from a visit to her ailing mother in Hyderabad. "The kindness and kindly help of residents of Ghatnadar helped avoid mass casualties," she said.

'But for 72-year-old Annarao Belure the confusion at the accident site deepened as he lost his bag containing money and some articles. He was heading for Aurangabad to visit his widowed daughter. The Chairman of the Aurangabad Municipal Corporation (AMC), Nandkumar Ghodele, however, helped him and provided him with an ambulance and first aid. He was put on the relief train while local reporters helped him get a pannchanamma {a witness testimony} registered with the Railway Police.' [223]

A further report stated that Driver Rajaiah Kamble of the goods train had been suspended for refusing an instruction from Control to move his train from the main line, where it had stood for over three hours, because he said that his hours of duty had expired and he wished to be relieved. [224]

The Commission of Railway Safety concluded that: 'Rear end collision occurred due to human error, reception signals for main line, which was already occupied by Up Parli N Goods train, were most probably taken off for No. 7064 Up. The accident falls under the category of *Failure of Railway Staff.*' [225]

Another rear collision occurred in South Africa just after 17:00 on Tuesday 7th January 2003, when a Metrorail passenger train from Cape Town to Wellington collided with a Spoornet freight train. The accident occurred between Kraaifontein and Muldersvlei, in Western Cape province. The leading four coaches of the passenger train were derailed, ten passengers lost their lives and about 100 were injured, four critically. It took rescuers more than four hours to release all the trapped passengers and Driver Mervyne Matthee, aged about 56, who lost his left foot. The injured were taken to Tygerberg Hospital and Durbanville, Stellenbosch and Paarl medical clinics, mostly by road ambulances, with the most seriously hurt being conveyed by an air ambulance. The bodies of the dead were taken to a mortuary in Paarl.

Metrorail Rescue Director Dr Cleeve Robertson stated, "We were fortunate. This was still holiday time. Those trains in the evening about 17:00 are usually packed to capacity, so this could have been an absolute disaster if this had happened during peak working season."

Metrorail Manager André Harrison announced that a joint investigation into the accident would be opened with Spoornet and the police. Community Safety Member of the Executive Council Leonard Ramatlekane added that a docket on a criminal investigation had also been opened.

While the lines were blocked, trains were being diverted via Stellenbosch, though on Tuesday evening it was hoped to have one track open for single-line working in time for the Wednesday morning rush hour. This proved, however, to be too optimistic, a further announcement stating that it was hoped to reopen one line at 15:00 on Wednesday, in time for the evening rush hour, and the other not until about 18:00 on Saturday. [226]

Driver Matthee found himself on trial in Paarl Regional Court in July 2006, charged with culpable homicide for having run through three signals at Red. The court was told that

---

[223] *Sify News*, Chennai, India, 4th January 2003

[224] *Outlookindia.com*, 4th January 2003

[225] https://eparlib.nic.in/bitstream/123456789/759683/1/lsd_13_12_20-02-2003.pdf

[226] Ibid, 8th January 2003

Matthee had had 27 years' service at the time of the accident and on that day had been on duty for 10 hours and 50 minutes when the collision occurred. [227]

In further evidence, the court was told that nine of the ten passengers who had died had fallen through the floor of the coach when its undercarriage collapsed.

Driver Ferdinand Jooste of the freight train testified that immediately after the collision he had sent his assistant ahead around a curve to protect the opposite running line, while he had gone back to find what had happened. Continuing, he explained that he had been taking a cargo of mealies (milled white maize – a South African breakfast dish) from Table Bay Harbour to Boland. His train had been put onto a side line for another passenger at Brackenfall, then proceeded almost to Muldersvlei when it had been run into. He had informed Muldersvlei of the collision over the radio and then waited for assistance. [228]

The defence alleged that the deaths of the passengers had not been the fault of Driver Matthee but of Metrorail for permitting rolling stock in poor mechanical condition to remain in service. There was also medical evidence that the driver had suffered two so-called fainting fits in the previous 18 months, linked, it was said, to the death of his wife from cancer, and the defence alleged that he had fainted a third time immediately before the collision. The magistrate asked if, in view of these previous episodes of fainting, the driver had not been negligent in continuing to operate a train. Defence counsel John Van der Berg replied that in view of medical assurances that the driver was in good health, he had not been culpable in continuing to operate a train.

The trial ended in early December 2006 when magistrate Anna-Marie Immelman found that negligence had not been proved and Driver Mathee was acquitted. [229]

A least seven people reportedly died when a crowded passenger train, the *Fajar Utama Ekspres*, from Palembang to Bandar Lampung, collided with a stationary coal train just short of Bandar Lampung, on the island of Sumatra, Indonesia, on Thursday 19th May 2005. Police Chief Col. Abdul Muis told a local radio station that, "At least seven people have been found dead," though another source stated that there were only four deaths. The police chief added that there were hundreds of injured, many of whom had been taken to local hospitals and clinics, while the radio station reported that at the time of their report, rescuers were still working to free survivors from the wreckage. [230] [231]

Five passengers died and 30 were injured when passenger train K 127 from Xi'an to Changchun collided with the rear of a freight train at Shenyang, China, in Liaoning province, derailing five coaches. The collision, on the Changchun – Dalian line, occurred at 19:52 on Sunday 31st July 2005. A passenger related, "We were riding in an air-conditioned carriage. When the accident happened it shook violently and some of the passengers in the sleeping car fell from their upper berths. Then the lights in the coach went out."

Railway Minister Liu Zhijun instructed Vice Minister Hu Yadong to proceed to the scene to organise the rescue operation and the reopening of the line. The minister also requested the railway department to open an investigation into the accident. Governor Zhang Wenyue of

---

[227] *Mail and Guardian*, Johannesburg, South Africa, 26th July 2006
[228] *News 24*, Durban, South Africa, 25th July 2006
[229] Ibid, 6th December 2006
[230] *BBC News*, London, England, 19th May 2005
[231] Wikipedia – Rail transport in Indonesia

Liaoning province likewise went to the scene. [232]

As in the Charlottedale collision in South Africa three and a half years earlier (see above), a signal failure, due to a theft of signalling cables, presumably for the copper they contained, was a factor in the accident. [233]

---

[232] Wikipedia – List of rail accidents (2000-2009)

[233] https://web.archive.org/web/20050804001109/http://www.trains.com/Content/Dynamic/Articles/000/000/006/074powwx.asp

CHAPTER 5

SIDELONG, HEAD-ON, DOUBLE AND FOULING COLLISIONS

Trains can come into collision with one another in a variety of ways. Rear collisions, that is, ones involving one train running into the back of another, trains meeting one another head-on on a single track and trains colliding on rail/rail level crossings, are detailed in other chapters. This chapter will deal with other types of collision such as those involving trains running into one another at junctions, meeting head-on on multiple tracks and colliding with vehicles left in sidings but which end up foul of running lines.

Five people died and more than 50 were injured in the second train accident to occur in Bangladesh in three days, when an express train bound for Dhaka was rammed by another train coming from the north-eastern city of Sylhet outside of Chittagong station on Tuesday 16th April 1991. [234]

At about 07:58 on Sunday 4th August 1991, Northern Railway train No. 1144, the Up Chhapra – Gwalior Mail, failed to stop at the Up Outer signal at Sonik station, Uttar Pradesh, India, on the Lucknow – Kanpur section of the railway station, and collided head-on with a Down Military Special, which had just arrived on the adjacent Down Loop line. Eleven people died in the accident and 39 were injured, 21 of them seriously.

The Commission of Railway Safety report on the collision made three important recommendations:

(a) Drivers should pass psychological tests prescribed by the Research Design and Standards Organisation before being put on Mail/Express trains.

(b) Action was necessary to provide adequate cadres of duly selected, trained and tested drivers for passenger-carrying trains.

(c) Seating-cum-luggage rakes (SLRs) with luggage compartments at either end should be employed where reversal of SLRs was difficult. [235]

The *Super Express* of Pakistan Railways from Sargodha to Karachi collided with a goods

---

[234] *The Guardian*, London, England, 17th April 1991

[235] Government of India Ministry of Civil Aviation and Tourism Commission of Railway Safety Annual Report for 1991-92

train that was being shunted across its path in the station at Piran Ghaib, Pakistan, on Thursday 29th August 1991. Three of the carriages of the express were derailed and five or six people died and at least 83 were injured. [236]

An overcrowded passenger train from Pointe-Noire to Brazzaville was in head-on collision with an Ogooué Mining Company freight train loaded with timber at about 02:00 on Friday 6th September 1991. The accident occurred on the Chemin de fer Congo – Océan at Mvoungouti station, Republic of Congo, 30 kilometres from Dolisie. The engine and two coaches of the passenger train fell down a 50-metre-deep ravine following the collision. Both train drivers died[237] and some reports stated that 150 passengers had lost their lives, while a probably more reliable report stated that 104 bodies had been recovered and another 29 people were missing, and 300 more were injured. It was, in any case, the most serious accident to occur in the country.

The passenger train appears to have been stationary when the freight train ran into it at the entrance to the station. The line was steeply graded and the accident was reportedly due to the excessive speed of the freight train, a problem with its brakes due to the dilapidated state of the locomotive and a braking error on the part of its driver.[238] The cause of the accident remains unclear, however. A conductor, presumably of the freight train, who had jumped from its engine, spoke at any rate, on condition of anonymity, of a brake failure, but Jean Goma, the director of the railway company, denied that there had been any problems with the brakes, stating that full checks had been made on both trains before they started their journeys. Transport Minister Jacques Okoko suggested instead sabotage, stating, "I'm not accusing anyone, but I think this act could have been committed by enemies of the Congolese people who don't want to see the transition {to multi-party democracy} succeed," adding that he was "perplexed" at the assertion of a brake failure. [239]

President Denis Sassou Nguesso visited the scene of the disaster and declared a week of mourning. Soldiers and Red Cross workers were still busy at the scene on Saturday. [240]

The victims were taken to hospitals in Pointe-Noire and Dolisie, which were overwhelmed by the magnitude of the disaster. Surviving victims and their beneficiaries finally received their first compensation for the accident 32 years later, in 2023. [241]

Melun, France, was the scene of a major disaster at 06:29 on Thursday 17th October 1991, when Société nationale des chemins de fer français (SNCF) train No. 9602, a Nice – Paris Lyon sleeper service, consisting of four coaches and several flat waggons for passenger cars, was involved in a near head-on collision with a freight train from Corbeil. As a result of the impact the leading sleeping car of train No. 9602 was telescoped into its engine. The passenger train was reportedly only travelling at 40 km/h at the time, the freight train was travelling at about 65 km/h. It was in the leading sleeper car that all the 16 initially reported fatalities occurred, apart from the driver of the freight train. Almost 60 others were injured, 11 seriously, among them the driver of the passenger train, who was only cut free in mid-morning and had to have a leg amputated, though he was stated to be out of danger. The passengers in the second, third and fourth coaches of the sleeper train were said to have alighted after the accident and stared in stunned silence at the sight in front of them. There had been about 120

---

[236] *Express and Echo*, Exeter, Devon, England, 29th August 1991

[237] *Winnipeg Free Press*, Winnipeg, Manitoba, Canada, 7th September 1991

[238] Wikipedia – Mvoungouti train accident (1991)

[239] *The Lethbridge Herald*, Lethbridge, Alberta, Canada, 9th September 1991

[240] *The Odessa American*, Odessa, Texas, United States, 8th September 1991

[241] Author's personal notes, source mislaid

passengers on train No. 9602 in all, many of them returning from a late holiday on the Côte D'Azur. Had the accident happened at the height of the holiday season, the train would have been carrying many more passengers.

Prefect Gérard Deplace activated the emergency plan at 07:25 as soon as the magnitude of the disaster became evident. A command post was set up, a firefighters' tent served as a chapelle ardente (chapel of rest or mortuary chapel) for the bodies of the victims, who had initially been lined up along the track, while 150 firefighters from Seine-et-Marne, Essonne and Paris worked alongside 140 police officers, 25 gendarmes, and 25 doctors for 36 hours at the scene, together with fire engines, rescue vehicles and other heavy vehicles, 30 in all, an army of ambulances and three helicopters.

The death toll apparently rose to 24 on Friday, with six of the injured succumbing in hospital and the recovery of two further bodies from the wreckage. [242]

President Mitterand ordered an immediate investigation into the accident and Transport Minister Paul Quilès stated that he expected speedy results from the inquiry. [243]

At the beginning of March 1992 the transport minister announced that the inquiry had determined that the accident had been due to human error; Driver Christian Roye of the freight train, who was killed instantly, had been overrunning signals. It was theorised that he might have been taken ill.[244] [245] The tachograph recovered from the freight locomotive indicated that the train had been travelling at 85 km/h when its driver had encountered a Yellow signal 1,500 metres from the junction. He had then reduced speed to 65 km/h but afterwards had failed to further slow down until he applied the brakes in emergency at the Red signal, only 150 metres short of the fouling point. [246]

Two passenger trains collided at Zaoqiao station, Taiwan, on the Zhunan – Changhua line in Miaoli County, at 16:00 on Friday 15th November 1991. Zaoqiao was a crossing place on the single line. Train No. 1, a Chu-Kuang Express, southbound and heading for Kaosiung, was just entering the station on the main line and in the opposite direction northbound train No. 1006, a Tze-Chiang Express, en route to Taipei, was approaching and had entered the auxiliary line (crossing loop). The northbound train failed to stop, ran through the loop and re-entered the single line before train No. 1 had cleared it, striking the antepenultimate coach of the southbound train. The collision reportedly occurred near a tunnel.[247] The last three coaches of the southbound train and the first three coaches of the northbound train (both were 100 series electric multiple units) were wrecked and overturned. According to some accounts the trains had both been travelling at 100 km/h when they collided.[248] Thirty people died and 112 were injured. The casualties included a number of Japanese tourists, among them Congressman Hideaki Oshiro.

The post-accident rescue operation continued through the night and into Saturday, illuminated by floodlights, while a cold winter wind blew. The accident and emergency department at the local hospital was overwhelmed by the number of victims and some had to be sent elsewhere.

---

[242] *Medicine Hat News*, Medicine Hat, Alberta, Canada, 18th October 1991
[243] *The Times*, London, England, 18th October 1991
[244] *The Daily Telegraph*, London, England, 18th October 1991
[245] https://www.sudouest.fr/faits-divers/il-y-a-30-ans-la-catastrophe-ferroviaire-de-melun-fait-16-morts-6572015.php
[246] *European Stars and Stripes*, Darmstadt, Germany, 19th October 1991
[247] Ibid, 17th November 1991
[248] *The Capital*, Annapolis, Maryland, United States, 16th November 1991

Deputy Director Yeh Hong-shui of the Taiwan Railway stated after the accident that, "The northbound train may have entered the single line before the southbound train had completely left it." [249]

The Railway Safety Committee (RSC) of the Taiwan Railway Administration immediately opened an inquiry into the accident, which was completed in only a fortnight. This was basically an internal inquiry by the railway itself, as there was no independent investigating body until 2018. The committee checked the tachograph of the engine and the records of the signalling centre, checked the maintenance of the signal and conducted tests at the site. At the close of the investigation the committee concluded that the accident had indeed been due to the northbound train overrunning signals, and that in some manner Driver Su Jinkun of train No. 1006 had imagined that he had seen a Green signal. Because of the curved approach to the station, the view of the signal was not as good as it might have been. When the train ran past the signal it should in any case have been stopped by an automatic train stop (ATS) system, but the ATS had been inoperative and hence of no use.

The report of the RSC only extended to three pages and it made no recommendations. [250]

Driver Su and Taiwan Railway Director Chen Shifang (who had held his position for only four months at the time of the accident) were subsequently both prosecuted for negligence. Chen was acquitted but Driver Su was sentenced to 4 years' imprisonment. [251] [252]

Eastern Railway train No. 335, the UP Vishwabharti Fast Passenger, collided head-on with No. 235 Down Diesel goods train at Pichkurir Dhal station, West Bengal, India, at 20:10 on Tuesday 25th February 1992, on the broad-gauge section of the Sahibganj Loop of the Howrah Division of the railway. Five people died in the accident and nine were hurt, four seriously.

Some minutes prior to the collision the signals had been cleared for the goods train to proceed but its crew had refused to do so because they had already exceeded their working hours. The goods train was then cancelled but the block instrument could not be restored to normal and the facing points leading from the single line to the loop line, on which the goods train was stood, were not reversed. The passenger train was then accepted in the opposite direction on a paper line clear. The intention was for that train to be piloted beyond the Up Outer and Home signals into the station but it failed to stop at these, was diverted into the Loop line and the collision ensued.

The action was determined to have been caused by the incorrect setting of the facing points and the failure of the passenger train to stop at the signals for the pilotman.

The Commission of Railway Safety report on the accident contained three important recommendations:

(a) The utilisation of staff engaged in train operation beyond scheduled working hours should only be resorted to in exceptional circumstances.

(b) Drivers, switchmen, stationmasters, assistant stationmasters and gatemen should be adequately counselled on train passing duties and should be frequently booked for obligatory transportation training. The Railway should suitably formulate the syllabus and periodicity of such training.

(c) Frequent ambush checks should be conducted in cases of signal failures where OP/T-27

---

[249] *Ironwood Daily Globe*, Ironwood, Michigan, United States, 16th November 1991

[250] https://www.kansai-u.ac.jp/Fc_ss/center/study/pdf/bulletin013/bulletin013_07.pdf

[251] Wikipedia – 1991 Miaoli train collision

[252] Wikipedia – Eisenbahnunfall von Zaoqiao

were issued from the last stopping station in the rear to check:

(1) Whether or not the driver followed the required speed restriction at the defective signal.

(2) Whether or not the driver was hand signalled at the foot of the defective signal. [253]

At 05:13 on Tuesday 3rd March 1992 *Latvia* express passenger train No. 4 from Riga to Moscow collided with freight train No. 3455 coming in the opposite direction at Podsosenka, Russia, in the Tver oblast. Podosenka was a crossing place on the single-track Rzhev branch of the Oktyabrskaya Railway and the trains collided head-on at No. 1 points at the end of the crossing loop. In charge of the engine of the passenger train were Driver G.T. Morozova with Assistant Driver O.V. Efimov. Train No. 4 was running three minutes early. It received a Yellow signal at the entrance to the crossing loop, while signal No. Ch1 at the far end of the loop was, of course, at Red, pending the arrival of the freight train, but the passenger train failed to stop, ran past the signal at Danger and a collision ensued. Fire broke out following the accident in the locomotives of the two trains and spread to at least two of the carriages of the passenger train. Thirty-five or 36 passengers and five or eight railwaymen died, including the two engine crews, while 108 people were injured. Four locomotives, four passenger coaches, eight freight waggons, a set of points and 23 metres of track were destroyed and the line was blocked for 15½ hours.

Two hundred soldiers, six armoured personnel carriers, four all-terrain vehicles and four military helicopters took part in the rescue operation, which was made more difficult by the location of the accident in a wooded and swampy area.

The only eyewitness to the disaster was Signalman Tsvetkov, who was clearly not responsible for it. He reported that after making four unsuccessful attempts to contact the train crew of the passenger train by radio he had left his box and attempted to stop the train by waving a red hand lamp at it. When this still had no effect and the express had been only 500 metres from the end of the loop and was still not slowing down, he had returned to his box, presumably to replace his signals against the freight train. As he had done so he heard the brakes finally being applied on the passenger train, then the noise of the collision.

Two investigations were opened into the accident, one by an independent examiner and the other by a ministerial commission. It was noted that the train crew had only been on duty for four hours, after what should have been adequate rest, and there was no obvious reason why they should have failed to obey the signals.

The independent inquiry theorised that there had been a failure of the brakes of the express due to an accumulation of ice in the brake pipe.

The ministerial commission, on the other hand, concluded that there had been no problem with the brakes themselves but that the train crew had simply failed to apply them until too late, perhaps (the author's supposition) because they had dozed off in the early hours of the morning, which would explain why they had also failed to respond to the signalman's radio calls or Red lamp. [254]

Southbound express passenger train No. 211 from Nanjing to Canton, with 1,300 passengers, collided head-on with freight train No. 1310 at Wulidun station, Hunan province,

---

[253] Government of India Ministry of Civil Aviation and Tourism Commission of Railway Safety Annual Report for 1991-92

[254] https://www.parovoz.com/phpBB2/viewtopic.php?f=24&t=5438&start=0&st=0&sk=t&sd=a

China, before dawn on Saturday 21st March 1992. Both locomotives, five passenger coaches and eight freight waggons were derailed, overturned and struck a lineside railwayman's residence. A photograph showed a derailed carriage lying partly on the roof of a single-storey brick apartment building, with the other end of the carriage on top of a second carriage. Only parts of the wall of the apartment building remained standing.

Of the twelve known fatalities three were train crew, one was a peasant struck by the wreckage and others were railwaymen's relatives who had been in the apartment building. Another five people were missing, believed dead. It was not immediately known if any train passengers were among those who had lost their lives. Another 34 people were hurt, eight of them seriously.

Railway Minister Li Senmao went to the scene to organise the rescue operation.

The cause of the accident was under investigation. [255]

At 22:30 on Monday 17th August 1992 a passenger train and a freight train were in collision at Kazichene, Bulgaria, about 15 kilometres east of the capital, Sofia. The passenger train was carrying 520 holidaymakers home from the Black Sea resort of Burgas, many of them children. Eight carriages of the passenger train were derailed. Eight people died and about 60 others were injured.

The rescue operation following the collision faced considerable difficulties, as a newspaper report related. 'Rescue workers struggled for hours in the dark and with no crane to free passengers trapped in the mangled wreckage, BTA {Bulgarian Telegraph Agency} said. One of the derailed cars was tipped on its end and half-buried in the ground near the track, AP {Associated Press} photographer Ivo Hadzhimishev said from the accident site early Tuesday. Officials said 57 of the injured were hospitalised and at least seven were in critical condition. Doctors appealed for blood donors.'

The accident arose, according to the Transport Ministry, because the freight train had run through a Red signal and entered the main line, where the passenger train had slammed into it at a speed of about 100 km/h. State television indicated that the drivers had been dismissed from their jobs and arrested and that 21 senior railway officials had also been dismissed. Why so many management officials were deemed responsible for the accident was not stated. Railway Minister Alexander Andropov said that he might also tender his resignation, though he must first discuss the matter with Cabinet colleagues. [256]

The investigation into the accident found that Driver Krasimir Todorov of the freight train, having finished a cup of coffee, had alighted from his engine, leaving it in charge of Assistant Driver Valentin Krastev. The latter had been drunk and in his intoxicated state ran past a signal and brought the engine foul of the running line on which the passenger train was approaching. Both engines, five passenger coaches and two freight waggons were derailed in the resulting collision.

After a prolonged trial at Sofia in 1998 Assistant Driver Krastev was sentenced to 12 years' imprisonment, Driver Todorov received eight years and Conductor Angel Botev five years, the latter because he had failed to exercise proper supervision over the train crew. [257]

An express train ploughed into a locomotive that was on a test drive near Semnan, Iran, 160

---

[255] *European Stars and Stripes*, Darmstadt, Germany, 23rd March 1992

[256] *Daily Herald*, Arlington Heights, Illinois, United States, 19th August 1992

[257] https://slivenmedia.com/news/nay-tezhkite-zhelezopatni-katastrofi-v-bulgaria-ot-1969-g-do

kilometres east of Tehran, on Monday 18th January 1993. At least eight people died and 16 were injured according to a first report.[258] A later report spoke of 24 fatalities and 45 injuries.[259]

The main line of the Northern Indiana Commuter Transportation District (NICTD) was basically double track but at Gary, Indiana, United States, the two lines were gauntleted (interlaced) so as to cross a narrow box girder bridge over two other railroad lines. The transition from double to what was effectively a single line was protected by colourlight signals at each end of the bridge. In mid-January 1993 the bridge was the scene of a sidelong collision that killed seven passengers and injured 95.

On the morning of Monday 18th January 1993 two-car eastbound NICTD train No. 7 left Chicago, Illinois, for South Bend, Indiana, at 8:45 a.m., operated by Motorman David Riordan. At Hammond, Indiana, the conductor notified the motorman that there were no passengers on board who wished to alight at Clark Road station, a little-used flag stop where passengers wishing to alight informed the conductor and those wished to get on flagged down the train. This station was about half a mile short of the gauntleted bridge.

Motorman Riordan told the National Transportation Safety Board inquiry into the accident that when he had left Hammond, travelling at 70 mph, signal No. 621, a quarter mile before Clark Road station, had changed from Yellow (Approach) to Green over Red (Clear). Approaching Clark Road level crossing, adjoining the station, he had reduced speed to 40 mph, confirmed that there was no one there waiting to join the train, and had seen that signal No. 601, protecting the bridge, was also showing Green over Red. Accelerating for the climb to the bridge he had glanced at his watch, then, "When I looked back up, the signal appeared dark." Standing up to get a better view, he had seen that in fact the signal was Red, had applied the brakes in emergency and come to a stand with the train about a foot foul of the westbound track. A westbound train had then been coming off the bridge and they had collided.

In the opposite direction, two-car westbound train No.12, operated by Motorman Willard Blewett, had left South Bend at 8:05 a.m. for Chicago. A third car was added to the train at Michigan City, a brake test was carried out, and the train left there at 8:45 a.m., four or five minutes late. A deadheading (off-duty) collector/brakeman boarded the train at Ogden Dunes and entered the driving compartment. The two railwaymen had called out the signals to one another; Nos. 592 and 602, the Approach and Home signals for the bridge, equivalent to signals Nos. 621 and 601 for eastbound trains, were clear. As the train moved onto the bridge they saw train No. 7 approach and then pass signal No. 601. The collector/brakeman immediately vacated the cab, the motorman applied the brakes in emergency and the trains collided.

An unidentified person dialled 911 to report the accident at 9:37 a.m.

Of the 57 people on train No. 7 four passengers were killed, five were seriously hurt and 32 were slightly hurt. Fourteen passengers and two crew were unhurt. Of the 140 people on train No. 12 three passengers were killed, one passenger was seriously hurt, 54 passengers and three crew were slightly hurt and 79 passengers escaped injury.[260]

Passenger Steven Lewin, a jewellery maker from Valparaiso, told a reporter, "All of a sudden I heard this big, loud explosion and I was thrown from side to side." He had suffered a minor back injury. He explained that he had been sat in the second car of his train and that when he had looked into the first car he saw a body in the aisle, adding, "It looked to me like

[258] *Western Morning News*, Plymouth, Devon, England, 20th January 1993
[259] *International Herald Tribune*, Paris, France, 20th January 1993
[260] National Transportation Safety Board report

they had been decapitated, and I got ill and had to turn around." [261]

Passenger Margaret McNeill related, "The glass was flying. It was a tremendous impact." She was unhurt, as was passenger Jane Mitchell, who said, "We hung on, and there were people flying around."

The varying reactions of the passengers after the collision were described by passenger Abir Khater. "There was panic and people were saying, 'get out, get out,' while other people were saying, 'sit down, just sit down.'"

Fire Chief Ben Perry declared, "The walls were ripped open, creating an effect like a big razor blade." [262]

The NTSB noted that trains on the line were governed by timetable, track warrants, train orders, and the signal indications of an automatic block system.

The NTSB ruled out any problems with the track, the rolling stock of the trains or the signalling.

Motorman Riordan, the NTSB noted, stated that he had been wearing sunglasses at the time of the accident but tests showed that these should not have affected his ability to distinguish signal aspects. Nor should the position of the sun have interfered with signal observation. The motorman must therefore have been inattentive for some reason. On previous trips his train had been stopped at signal No. 601 on almost half of the journeys for train No. 12, so it could not have been the case that the signal was always clear for his train. Whatever the reason, his failure to stop at signal No. 601 clearly indicated that he had been inattentive.

Both Motorman Blewett and the collector/brakeman on train No. 12 had stated that they had seen train No. 7 as their train entered the bridge and the collector/brakeman had exclaimed, "He's still coming." As train No. 12 had exited the bridge the collector/brakeman had cried, "They're still moving," and, "We're going to hit," and had run back into the train. The NTSB concluded that had Motorman Blewett reacted more quickly and braked hard at once, he could have reduced the impact of the collision even if he could not have entirely prevented it.

The fitness of both motormen therefore came into question, as did the adequacy of testing of the fitness of employees for duty.

The accident report went on to examine in considerable detail the crashworthiness of the vehicles of the two trains.

The topic of Positive Train Separation, which the NTSB stated had been on its 'Most Wanted List' since its inception in September 1990, was also mentioned.

The report reached eight conclusions:

1. No anomalies or deficiencies had been evident in the train operation (except train handling), track, signal system, or mechanical equipment. Neither engineer had been impaired by drugs or alcohol or lack of rest. Both engineers had been experienced in and qualified to perform their duties.
2. The engineer of train No. 7 had been inattentive to his duties when he passed the Approach indication displayed by signal No. 621 and the Stop indication displayed at signal No. 601. Because of his inattentiveness, he had failed to stop at signal No. 601, which caused his train to foul the westbound track.
3. Had the engineer of train No. 7 acted immediately when he had perceived a dark signal and applied the emergency brakes, as he should have, train No. 7 would have proceeded past

[261] *Greensburg Daily News*, Greensburg, United States, 19th January 1993
[262] *Herald Bulletin*, Anderson, Indiana, United States, 19th January 1993

signal No. 601 but would have stopped short of where it fouled the westbound track.

4. At the time the engineer of train No. 12 had realised that train No. 7 had passed the Home signal, the two trains might have been too close for the engineer of train No. 12 to stop his train and avert the collision; however, if he had applied his brakes at this time, the severity of the accident would have been mitigated.

5. Although the Safety Board had been unable to conclusively determine whether the engineer of train No. 12 could have taken action to prevent the accident, the evidence showed that the actions he had taken were neither timely nor appropriate.

6. The use of collision energy absorption structures in the corner post assemblies of these railcars would have reduced the collision impact intrusion and might have prevented or substantially reduced the number of fatalities and serious injuries.

7. The collision between the two trains could have been prevented had a positive train separation system been in place and operational.

8. The performance of both engineers in this accident raised questions about the adequacy of procedures used by the railroad industry in determining fitness for duty.

The NTSB determined that the probable cause of the accident had been the inattentiveness of the engineer of train No. 7, resulting in his train passing a Stop signal and partially blocking the westbound track. Contributing to the severity of the accident had been the failure of the engineer on train No. 12 to take timely action to slow or stop his train before the collision. Contributing to the severity of the injuries had been the breach of the passenger compartment in the lead cars of both trains.

The NTSB made several recommendations.

To the Federal Railroad Administration:

In cooperation with the Federal Transit Administration and the American Public Transit Administration, it should study the feasibility of providing car body corner post structures on all self-propelled passenger cars and control cab locomotives to afford occupant protection during corner collisions. If feasible, it should amend the locomotive safety standards accordingly.

To the Federal Transit Administration:

It should cooperate with the Federal Railroad Administration to study the feasibility of providing car body corner post structures on all self-propelled passenger cars and control cab locomotives to afford occupant protection during corner collisions.

To the American Public Transit Association:

(a) It should cooperate with the Federal Railroad Administration to study the feasibility of providing car body corner post structures on all self-propelled passenger cars and control cab locomotives to afford occupant protection during corner collisions.

(b) It should develop improved procedures for determining fitness for railroad personnel on safety-sensitive positions.

To American Short Line Railroad Association:

It should develop improved procedures for determining fitness for railroad personnel on safety-sensitive positions. [263]

The gauntleted bridge was replaced by an ordinary double-track structure in 1997. [264]

[263] National Transportation Safety Board report

[264] https://railroad.net/what-is-the-status-of-the-gary-gantlet-bridge-t56772.html

A passenger train on its way from Nairobi to Malaba, near the border with Uganda, was waiting to enter the station at Visoi, Kenya, not far from Nakuru, on Saturday 31st July 1993, when it was in collision, either head-on or sidelong, with a goods train. Five passengers died in the accident. [265] [266]

A mistake by a signalman, who routed a passenger train and a freight train onto the same line, resulted in a collision at Zifta, Egypt, on Sunday 12th December 1993. Ten people lost their lives and 76 were injured, 25 of them seriously. [267]

Two passenger trains collided head-on in Arjawinangun station, West Java, Indonesia, at 02:30 on Friday 10th May 1994. The *Senja Ekonomi* from Surakarta to Jakarta was standing in the station when the *Tegal Arum* train from Jakarta to Tegal entered the station on the wrong line and travelling at full speed. Eight people were killed outright and two were fatally injured; all had been in the leading car of the *Senja Ekonomi* train. At least 50 others from both trains were injured, many seriously. Those who lost their lives included married couple Darsan and Tanya, and their 14-year-old daughter Tanti. On the following Monday their son Timbul Jaya, 6, was still recovering from his injuries and as yet unaware that he was now an orphan.

While promising a full investigation into the accident, Transportation Minister Haryanto Dhanutirto pronounced that it had been due to the driver of the *Tegal Arum* train running past a signal, adding that, "He might have fallen asleep at that time." He stated that the signalling and the brakes of the train had been working correctly.

A passenger on the train stated that it had been running faster than usual since it started from Jakarta and that, "It only stopped when it crashed."

Rony Wahyudi, head of the Cirebon office of Perusahaan Umum Kereta Api (abbreviated to Perumka, the state railway company), asserted, "There is no excuse. The driver had already had a day's rest. Apparently he did not use his time properly."

Both trains were badly damaged but as the accident occurred on a minor alternate route there was little disruption to traffic. [268]

On the afternoon of Saturday 20th August 1994 a northbound passenger train from Sfax collided with a southbound passenger train from Tunis at or near Kalla-Seghira, Tunisia. Fire broke out following the collision and a number of the victims of the accident were burnt to death in the wreckage before they could be released. Eighteen people reportedly died and 74 were injured. The rescue operation was still in progress hours after the accident. The injured were taken to hospitals in Sousse and Monastir.[269] A government delegation, which included the transport and health ministers and was led by Interior Minister Abdallah Kallel, went to the scene of the accident. [270]

On the night of Friday 13th January 1995 the Simanta Express, bound with Khulna,

---

[265] Author's own notes, source mislaid
[266] http://www.nationaudio.com/News/Daily/Nation/250399/News/News17.html
[267] *Courier Express*, DuBois, Pennsylvania, United States, 13th December 1993
[268] *The Jakarta Post*, Jakarta, Indonesia, 23rd May 1994
[269] *The News Herald*, Panama City, Florida, United States, 21st August 1994
[270] The *Daily Press*, Victorville, California, United States, 21st August 1994

collided, probably head-on, with a stationary local train in the station at Hili, Bangladesh. Both trains were crowded with passengers. State-run television broadcast that there were many casualties, though they could not give a figure. Reporters for the mass-circulation *Janakantha* and *Ittefaq* newspapers spoke of 100 or even 150+ fatalities, though rescuers stated at the same time that they had only recovered 30 bodies, with more believed to be trapped in the wreckage. About 300 were injured. It was not known how many passengers had been on the trains and no official death toll was announced. Bangladesh Railway spokesman Mostafa Jamil stated that at the time of his report road ambulances were already at the scene and a relief train with medical supplies was on its way. He added that two engines and four coaches were badly damaged. [271] [272]

A major disaster occurred in the station at Kalubathan, Bihar, India, on the Asansol – Dhanbad electrified broad-gauge section of the Asansol Division of the Eastern Railway, at about 16:22 on Thursday 1st June 1995. An Up EC BKSC goods train had been received on the Up Loop line at the station and was followed by train No. 3151, an Up Sealdah – Jammu Tawi express, which should have been received on the Up Main line, but the facing points were not reset and instead it was diverted into the rear of the goods train. Forty-five passengers were killed and 175 were injured, 56 seriously.

The Commission of Railway Safety made a good number of recommendations in relation to the disaster:

(a) A specific competency certificate should be issued to the Chief Signal Inspector and Signal Inspectors (SIs) for working in a section with panel interlocking and route relay interlocking (RRI). This certificate should be given after a proper oral examination as well as practical tests.

(b) Whenever an electric signal maintainer (ESM) was permitted to work in a panel-interlocked RRI station he should first work with his predecessor for a minimum period (to be specified by the Railways) so that he became fully familiar with the circuitry in the relay room, the panel room, and in the field. Before he was given independent charge he should be tested by at least an assistant signal and telecommunications engineer (ASTE). An adequate number of electric signal engineers should be kept ready after such training for required relief arrangements during the leave of regular incumbents and in emergency.

(c) Nobody other than the authorised ESM and SI should be permitted to work in the relay room, panel room and field and their work should have the prior approval of the concerned ASTE, who would give details of the works to be done by them. In addition, these works should be done in the presence of the maintenance SI and/or ESMs, who should ensure that the circuitry that governed the movement of the train was not disturbed and that disconnection memos were issued when needed.

(d) Short-cut methods for repairing signals and so on should not be permitted. For this, a special drive and regular surprise checks, including night checks, should be carried out by the inspectors and officers of the signal and telecommunications and operating departments. The inspecting officers and inspectors should also correlate the inspection notes, notes of footplate inspections (by officers and inspectors of all departments), signal failure registers, disconnection memo register, OP/T-27 books, joint points and crossing inspections registers, individual inspection records of signalling installations being made by inspectors and officers (with actual time spent for each item). Such analysis would be helpful in pinpointing the discrepancies in various inspections and detecting cases of notes

[271] *The Register Herald*, Beckley, West Virginia, United States, 15th January 1995

[272] *The Orange County Register*, Santa Ana, California, United States, 14th January 1995

and entries made that were not in conformity with actual site conditions.

(e) It is recommended that a proforma be evolved for recording the measurements and joint observations after an accident, in the relay room (for all the relays, cable racks, cables, junction arrangements, power supply installations, battery chargers, generators, batteries, with connections and leads going to and coming from in these installations), panel board, track circuits, point machines (with points), junction boxes, location boxes, axle counters and signals including earthing arrangements and available voltage and current at end points.

(f) A detailed proforma of the overhead equipment-cantilever (OHE) equipment including earthing arrangements at the site of accidents should be prescribed.

(g) After an accident, photographs for the signal and telecommunications (S&T) installations such as panel, relay room, and so on and other items relating to the OHE, track and rolling stock should be taken under the guidance of the officers of each of these departments to enable joint observations and recordings to be adequately supplemented.

(h) Detailed instructions for the sealing of the relay room and panel room after an accident should be issued.

(i) After an accident the traffic should not be resumed (except to save life) until a safety certificate jointly signed by the representatives of the traffic, S&T, engineering, mechanical and electrical departments had been given, clearly stating that all the prescribed joint observations had been recorded, required photographs had been taken and the necessary tests had been done to ensure safety and normal working.

(j) The through running speed should be restricted to 15 km/h when works not requiring a disconnection memo were being done in the relay room in panel-interlocked or RRI stations.

(k) A proper procedure order for the exchange of advice regarding defective signals and panel defects between the cabin stationmaster, stationmaster (platform) and section controller should be issued so that there was no confusion at any stage and the responsibility of each member of staff was clearly defined.

(l) At a particular station, the stationmaster (deputy station superintendent) should sit in the cabin only so that action for the complete arrival of trains, signal failure registers, disconnection memos, the issue of OR/T-27 forms, exchange of signals, and so on, were better coordinated. If required, additional staff for commercial duties should be posted for the platform office.

(m) In the Eastern Railway Supplementary Rules it stated that points set for a Loop line, after the arrival of a train on that line, should be reset for the Main line (if that line is clear). It was suggested that the Supplementary Rules should be modified to state that the change of points should be done under the exchange of Private Numbers between the cabin stationmaster, the stationmaster (platform) and Control. A maximum time limit for such resetting of the points might also be specified, keeping complete arrival checks and train parting in view.

(n) It should be ensured that the officers and staff not only carried out their scheduled inspections but also that the prescribed details were covered in these inspections and that there was prompt compliance with them.

(o) The plans for electrification works as well as signalling works should show both signalling and OHE layouts detailing cable layouts, earthing, bonding, and so on. These plans should be signed by officers of the S&T as well as the electrical departments of Open line and Construction Electrification (where applicable).

(p) The inspection of the Chief Signal and Telecommunications Engineer, Chief Signal Engineer and other officers of the S&T department from headquarters should also include checks on relays, cables and panels during some of these inspections, so that the message went down the line that headquarters officers were also monitoring these items.

(q) It should be ensured that Siemens relay front covers, including the group relays, were sealed and the aspects were specifically checked by inspectors and officers.

(r) S&T staff taking the relay room key should record the details of all the works (with a time schedule), which they planned to do in the relay room before the key was handed over to them by the stationmaster.

(s) The Railway should issue the detailed schedule of maintenance works for the relay room, giving periodicity for each schedule. [273]

At 5:39 p.m. on Friday 16th February 1996 a collision and fire occurred on the main line of CSX Transportation (CSX) between Washington, D.C. and Brunswick, Maryland, at Georgetown Junction, Maryland, United States, a mile west of the station at Silver Spring. It was a head-on collision involving National Railroad Passenger Corporation (Amtrak) passenger train No. 29 (the *Capitol Limited*), from Washington to Chicago, Illinois, and Maryland Rail Commuter (MARC) passenger train No. 286 from Brunswick to Washington. As a result of the collision eight passengers and three crew on the MARC train were killed and 11 of its remaining 12 passengers were injured. Of the 182 passengers and crew on the Amtrak train, 29 were injured. The passengers who died had all been participants in the federal Youth Corps job training scheme for disadvantaged young people, who had been on their way home to Washington for the weekend from the training centre in Harpers Ferry, West Virginia. The railwaymen who died, all CSX employees, were Engineer Richard Orr, 43; Conductor James Majors, 48; and Assistant Conductor James Quillen, 53. [274]

The accident was investigated by the National Transportation Safety Board (NTSB).

The main line in the area was double track, signalled for operation in both directions on each track and controlled by a dispatcher in Jacksonville, Florida, designated the AU dispatcher. Operation was by timetable, the direct traffic control (DTC) block system and the signal indications of a traffic control system (TCS). Colour position light signals were provided, capable of showing four different aspects:

Proceed (two vertical Green lights below a White marker light).

Medium Clear – speed limited to 30 mph through points connections, then proceed (two vertical Green lights above a White marker light).

Approach – prepare to stop at next signal (two diagonal Yellow lights below a White marker light).

Stop (two horizontal Red lights, no marker light).

MARC train No. 286 was a "push-pull" service, consisting at the time of the accident of a passenger coach control cab car leading, two coaches and an engine at the rear, thus it was in push mode. It had left Brunswick eastbound at 4:30 p.m., running on Track No. 2. Two passengers joined it at Kensington station, a flag stop a mile short of Georgetown Junction. As the train approached Kensington station the engineer of westbound MARC train No. 279 on Track No. 1 overheard Engineer Orr on train No. 286 acknowledge signal No. 1124.2, about 1,000 feet short of Kensington station, which must have been displaying Approach.

Westbound Amtrak train No. 29 consisted of two diesel units and 15 coaches and had left

[273] Government of India Ministry of Civil Aviation and Tourism Commission of Railway Safety Annual Report for 1995-96

[274] *The News*, Frederick, Maryland, United States, 19th February 1996

Washington at about 5:25 p.m. It had also been running on Track No. 2 from Washington Union so as to overtake CSX freight train No. K591, which was stopped short of Georgetown Junction at 5:28 p.m. The dispatcher had set the route for the Amtrak train to cross from Track No. 2 to Track No. 1 at Georgetown Junction, in front of both the freight train and the MARC train approaching in the opposite direction on Track No. 2.

As an aside, in the opinion of the author this typifies American railway working, sacrificing safety for convenience. On British railways, at any rate in former times, on double-track lines each track was reserved for trains travelling in one direction, and an Up train, say, would not be crossed over to run on the Down line even if there were four slower trains in front of it on the Up line and none timetabled on the Down line for 10 hours. The signalling in any case would not be set up for such a move. Less convenient but safer.

The approaching Amtrak train received a Medium Clear indication at signal WAS-2 and was just negotiating the crossover to Track No. 1 when it was struck by the MARC train.

The event recorder on the MARC train indicated that it had accelerated after leaving Kensington station, slowed for a speed restriction, then accelerated again before Engineer Orr had applied the brakes in emergency about 18 seconds before the collision. This, together with placing the locomotive in forward mode (it was operating in reverse while propelling), had reduced its speed from 66 to about 38 mph on impact. The whole of the MARC train and both diesel units and the leading eight cars of the Amtrak train (only one of which had been carrying passengers) were derailed. Fuel from a ruptured tank on the leading engine of the Amtrak train was sprayed onto the control cab car of the MARC train and ignited at once.

The crew of freight train No. K591 witnessed the accident and notified the dispatcher, while about a dozen members of the public called the fire service at about the same time. The first emergency services reached the scene at 5:46 p.m.

Ten passengers in the leading car of the MARC train, nine of them Youth Corps participants, were able to escape from it and gave evidence to the NTSB inquiry. The report noted, 'they stated they had known nothing about the operation of the emergency doors or windows. Some stated that they did not see any emergency placards explaining the use of emergency exits and that no crewmember had given any emergency announcements before the accident. One student remarked that he had seen the emergency placards but had paid no attention to them. The survivors noted that about half of the students were asleep before the collision. One stated that he had been seated in the rear section of cab control car 7752 when the conductor had run from the front of the car shouting, "Brace yourself! Get on the floor!" The student had dropped to the floor and held onto the seat on which he had been sitting. He said that windows had shattered during the collision, that the car was on fire afterwards, and that he could feel the heat from the fire on his back. He reported that he crawled to the rear of the car and tried unsuccessfully to open both side doors. He saw light through an opening between the cars and climbed through the space.

'One student, who was sitting next to an emergency window on the last seat in the right rear section of control cab car 7752, described the smoke as extending from about 2 feet above the floor to the car ceiling. This student said that he did not have time to open the window and that he believed he could escape faster by using the door through which he had entered the car.

'Another student stated that the conductor with another person came from the front of the car shouting, "everybody run to the back" and the conductor had reached the midpoint of the car when the collision occurred. The student reported that after the collision, he was thrown between the seats, the lights went out, smoke came into the car, and that other students were screaming and running to the rear of the car. He saw students unsuccessfully attempting to open the rear side doors, and he explained that because of the fire, no one attempted to open the emergency windows. He slid to the floor because he could not see or breathe with the smoke. He observed that while he was on the floor, one person exited "off the train through a hole or crack in the train." He followed another person through that opening and ran down a hill to a large residential building. He estimated that he escaped within 5 minutes.

'One student, who had been standing and talking to another student, stated that he had seen two crewmembers running rearward through the car from the control compartment and had heard them shouting, what he understood to be, "aboard, aboard." He noted that the two crewmembers appeared to be expressing shock and disbelief that another train was approaching on the same track. He said that he saw the conductor jump behind one of the seats and that he himself crouched behind and held onto the last seat on the left side of the car in anticipation of the collision. He recounted that he proceeded immediately after the crash to the rear door exits, that he felt the heat from the fire on his back, and that smoke quickly filled the car. He reported that he looked "high and low" for "handles or gadgets or something to open the doors" but smoke obscured his vision and that he could find nothing to open the doors. He added that he pounded with this hands on the door windows but the glass would not break. Other students were running toward him, coughing from the smoke, and yelling for someone to open the doors. He also heard banging and kicking noises that suggested other passengers were attempting to open or break windows.'

All the passengers who had escaped from the leading car had done so via an opening formed in the collision between the first and second cars. By the time firemen arrived, it had been clear that no one who was still in the leading car could have survived.

The signalling was of course checked after the collision, especially eastbound signals Nos. 1142.2, which should have been displaying Approach for the MARC train, and signal No. EAS-2, at Georgetown Junction, which should have been displaying Stop. The signalling was found to be in order and the visibility of the signals had been good.

Analysing the accident, the NTSB stated that there was no evidence of equipment failure, problems with the track, employee fatigue, nor of any of the train crews having used alcohol or drugs. It had been snowing at the time of the accident but no other train crews had reported problems seeing signals.

The NTSB theorised the most likely scenario was that Engineer Orr had observed signal No. 1142.2 at Approach but subsequently had been distracted by the Kensington station stop, radio conversations with the engineer of train No. 279, defect detector broadcasts and perhaps discussions with the conductor and assistant conductor over loss of job seniority due to a CSX reorganisation. With these distractions he had probably forgotten about the Approach aspect that he had received before stopping at Kensington and had assumed that he had a clear road through Georgetown Junction, where it was unusual to be stopped. The conductor and assistant conductor had been in the control cab but had taken no action to correct the engineer's mistake.

The NTSB again raised the subject of Positive Train Separation.

The safety standards of passenger coaches relating to rapid exit from them in an emergency such as a fire were examined in detail, also the crashworthiness of the cars of the MARC train and of the fuel tanks of locomotives.

The NTSB reached 28 conclusions regarding the disaster:

(1) The train equipment and the track had functioned as designed and the AU dispatcher had conducted the dispatching activities properly. Neither the three MARC train 286 crewmembers nor the two Amtrak train 29 locomotive crewmembers had been impaired by alcohol or drugs. All the train crewmembers had been in good health, had no evidence of fatigue, and had been experienced in and qualified for their duties.

(2) The weather conditions had not impaired the ability of the MARC train 286 crewmembers to distinguish the indication of Kensington signal 1124-2.

(3) The signal system had functioned as designed.

(4) The MARC train 286 engineer had apparently forgotten the signal aspect that required him to be prepared to stop at Georgetown Junction, due to interference caused by various events, including performing an unscheduled station stop, which had occurred between the presentation of the Approach aspect at signal 1124-2 and the Stop signal at Georgetown Junction.

(5) Neither the conductor nor the assistant conductor while in the cab control compartment appeared to have effectively monitored the engineer's operation of MARC train 286 and taken action to ensure the safety of the train.

(6) Had the Federal Railroad Administration required the recording of the train crewmembers' voice communications, the essential details about the circumstances of this accident could have been provided.

(7) Had the Federal Railroad Administration and Federal Transit Administration required the CSX Transportation Inc. to perform a total signal system review of the proposed signal changes that included a human factors analysis within comprehensive failure modes and effects analyses, this accident might have been prevented.

(8) Federal funds granted for the signal modifications on the CSXT Brunswick Line to accommodate an increase in the number of Maryland Rail Commuter trains had not ensured that the safety of the public was adequately addressed.

(9) Without a separate collection database specific to commuter rail inspections and accidents/incidents, it had been difficult for the Federal Railroad Administration to evaluate its own effectiveness of inspections and to identify problematic trends.

(10) The Federal Railroad Administration reliance on the need for increased vigilance of wayside signals and special actions in operating rules, such as the crew communication rule of emergency order 20, had not adequately safeguarded the public.

(11) Had a train control system that could utilise the cab signal equipment on the Maryland Rail Commuter cab control car been a part of the signal system on the Brunswick Line, this accident might not have occurred.

(12) A fully implemented positive train separation control system would have prevented this accident by recognising that MARC train 286 was not being operated within allowable parameters, based on other authorised train operations, and would have stopped the train before it could enter into the unauthorised track area.

(13) The emergency egress of passengers had been impeded because the passenger cars had lacked readily accessible and identifiable quick-release mechanisms for the exterior doors, removable windows or kick panels in the side doors, and adequate emergency instruction signage.

(14) The absence of comprehensive Federal passenger car safety standards had resulted in the inadequate emergency egress conditions.

(15) A need existed for Federal standards requiring passenger cars be equipped with reliable emergency lighting fixtures with a self-contained, independent power source when the main power supply had been disrupted, to ensure passengers could safely egress.

(16) Prescribed inspection and maintenance test cycles were needed to ensure reliable operation of emergency windows in all long-distance and commuter rail passenger cars.

(17) The exterior emergency door release T handles for the MARC cars had neither been in place nor accessible to firefighters because no requirements for their maintenance or accessibility had existed.

(18) The catastrophic rupture of the Amtrak unit 255 fuel tank in the collision with the MARC cab control car 7752 had released fuel, which sprayed into the interior of the cab control car and resulted in the fire and at least 8 of the 11 fatalities.

(19) Because other commuter passenger cars might also have interior materials that might not have met specified performance criteria for flammability and smoke emission characteristics, the safety of passengers in those cars could be at risk.

(20) The Federal guidelines on the flammability and smoke emissions characteristics and the testing of interior materials did not provide for the integrated use of passenger car interior materials and, as a result, were not useful in predicting the safety of the interior environment of a passenger car in a fire.

(21) Even though the Montgomery County Fire and Rescue Services personnel had responded promptly to the emergency, they could do nothing to save any of the accident victims because passenger coach cab control car 7752 had already been completely engulfed in flames when the first firefighter had arrived on scene.

(22) The Montgomery County Emergency Management Agency disaster plan had lacked procedures for responding to railroad passenger train accidents, such as simulating the accident response with coordinated management, which could have emphasised the importance of being familiar with passenger cars and of coordinating activities between the Montgomery County Fire and Rescue Services, the Maryland Rail Commuter, and the CSX Transportation Inc.

(23) The confusion during the initial emergency response had resulted because the CSX Transportation Inc. and Maryland Rail Commuter had lacked a formal emergency management plan to follow.

(24) The CSX Transportation Inc. personnel operating Maryland Rail Commuter passenger trains had not been adequately trained to understand and, therefore, execute their responsibilities for passengers in emergencies.

(25) The lack of appropriate training for emergency responders in the areas of emergency planning, coordination and communications, rescue methods, inaccessible terrain along railroad property, familiarity with railroad equipment, and disaster drills might become a recurrent problem for other emergency response organisations unless a national effort was made to address emergency response training for railroad accidents.

(26) The MARC train 286 engineer's use of the reverser during the emergency brake application had resulted in a marginally increased stopping distance for MARC train 286.

(27) The Federal Railroad Administration had not addressed the use of computer-aided dispatching system records to provide information for the identification and evaluation of potential safety-related trends for corrective action.

(28) The CSX Transportation Inc./Maryland Rail Commuter system had lacked a comprehensive safety oversight to ensure the safety of the commuting public.

The NTSB stated that the probable cause of the disaster was the apparent failure of the engineer and the train crew, because of multiple distractions, to operate MARC train 286 according to signal indications and the failure of the Federal Railroad Administration, the Federal Transit Administration, the Maryland Mass Transit Administration, and the CSX Transportation Inc. to ensure that a comprehensive human factors analysis for the Brunswick Line signal modifications was conducted, to identify potential sources of human error and to provide a redundant safety system that could compensate. Contributing to the accident was the lack of comprehensive safety oversight on the CSX Transportation Inc./Maryland Rail Commuter system to ensure the safety of the commuting public. Contributing to the severity of the accident and the loss of life was the lack of appropriate regulations to ensure adequate emergency egress features on the railroad passenger cars.

The NTSB made a good number of recommendations.

To the Federal Railroad Administration:

(a) It should amend 49 Code of Federal Regulations Part 229 to require the recording of train crewmembers' voice communications for exclusive use in accident investigations and with appropriate limitations on the public release of such recordings.

(b) It should require comprehensive failure modes and effects analyses, including a human factors analysis, for all signal system modifications.

(c) It should develop and maintain separate identifiable data records for commuter and intercity rail passenger operations.

(d) It should require, in the interim of a positive train separation control system being available, the installation of cab signals, automatic train stop, automatic train control, or other similar redundant systems for all trains where commuter and intercity passenger railroads operated.

(e) It should require the implementation of positive train separation control systems for all trains where commuter and intercity passenger railroads operated.

(f) It should require all passenger cars to have easily accessible interior emergency quick-release mechanisms adjacent to exterior passageway doors and take appropriate emergency measures to ensure corrective action until these measures were incorporated into minimum passenger car safety standards.

(g) It should require all passenger cars to have either removable windows, kick panels, or other suitable means for emergency exiting through the interior and exterior passageway doors where the door could impede passengers exiting in an emergency, and take appropriate emergency measures to ensure corrective action until these measures were incorporated into minimum passenger car safety standards.

(h) It should issue interim standards for the use of luminescent or retroreflective material or both to mark all interior and exterior emergency exits in all passenger cars as soon as possible and incorporate the interim standards into minimum passenger car safety standards.

(i) It should require all passenger cars to contain reliable emergency lighting fixtures that were each fitted with a self-contained independent power source and incorporate the requirements into minimum passenger car safety standards.

(j) It should provide promptly a prescribed inspection and maintenance test cycle to ensure the proper operation of all emergency exit windows as well as provide that the 180-day inspection and maintenance test cycle was prescribed in the final rule.

(k) It should require that all exterior emergency door release mechanisms on passenger cars be functional before a passenger car was placed in revenue service, that the emergency door release mechanism be placed in a readily accessible position and marked for easy identification in emergencies and derailments, and that these requirements be incorporated into minimum passenger car safety standards.

(l) It should require that a comprehensive inspection of all commuter passenger cars be performed to independently verify that the interior materials in these cars met the expected performance requirements for flammability and smoke emissions characteristics.

(m) It should update 49 Code of Federal Regulations Part 228.17, Train Dispatcher's Record of Train Movements, to include the same parameters for electronic recordkeeping of the dispatcher's record of train movements.

To the Federal Transit Administration:

(a) It should revise the grant application process to require a comprehensive failure modes and effects analyses, including a human factors analysis, be provided for all federally funded transit projects that were directly related to the transport of passengers.

(b) It should cooperate with the Federal Railroad Administration for requiring, in the interim of a positive train separation control system being available, the installation of cab signals, automatic train stop, automatic train control, or other similar redundant systems for all trains where commuter and intercity passenger railroads operated.

(c) It should cooperate with the Federal Railroad Administration for requiring the implementation of positive train separation control systems for all trains where commuter and intercity passenger railroads operated.

(d) It should cooperate with CSX Transportation Inc. in the development and installation of a positive train separation control system where Maryland Rail Commuter equipment operated on CSX Transportation Inc. tracks.

To the CSX Transportation Inc.:

(a) It should develop and install a positive train separation control system on track segments that had commuter and intercity passenger trains.

(b) It should develop and implement a formal emergency management plan that contained procedures specific to employee responsibilities and interaction with emergency response agencies and other transportation entities.

(c) It should develop and implement, in cooperation with Maryland Mass Transit Administration/Maryland Rail Commuter, a complete training agenda for all CSX Transportation Inc. passenger traincrews that provided experience in the correct use of emergency equipment, in emergency communications procedures, and in passenger evacuation and assistance in an emergency and also included the distribution of a comprehensive employee guidance manual.

(d) It should conduct, in cooperation with Maryland Mass Transit Administration/Maryland Rail Commuter, the Baltimore County Emergency Management Agency, the City of Baltimore Emergency Management Agency, the Metropolitan Washington Council of Governments, the Jefferson County Commissioners, and the Berkeley County Commissioners, periodic disaster drills to assess their emergency management plans, to reinforce and evaluate their emergency training, and to test the communication with the organizations.

(e) It should inform all operating train crewmembers of the circumstances of this accident and emphasise the crew responsibility while in the operating compartment for the safe operation of the train.

(f) It should inform its engineers of the circumstances of this accident and caution them not to use the reverser during emergency brake applications for those trains on which the use of the reverser would eliminate the dynamic braking, thus increasing the stopping distance.

To the Maryland Mass Transit Administration:

(a) It should cooperate with CSX Transportation Inc. in the development and installation of a positive train separation control system where Maryland Rail Commuter equipment operated on CSX Transportation Inc. tracks.

(b) It should develop an emergency plan that would provide a detailed description of emergency response procedures as well as a protocol to coordinate activities with the emergency response organisations and other transportation entities when an accident occurred.

(c) It should develop and implement, in cooperation with CSX Transportation Inc., a complete training agenda for all CSX Transportation Inc. passenger traincrews that provided experience in the correct use of emergency equipment, in emergency communications procedures, and in passenger evacuation and assistance in an emergency and also included the distribution of a comprehensive employee guidance manual.

(d) It should conduct, in cooperation with the CSX Transportation Inc., the Baltimore County Emergency Management Agency, the City of Baltimore Emergency Management Agency, the Metropolitan Washington Council of Governments, the Jefferson County Commissioners, and the Berkeley County Commissioners, periodic disaster drills to assess their emergency management plans, to reinforce and evaluate their emergency training, and to test the communication with the organisations.

To the US Department of Transportation:

(a) It should review the testing protocols within the various modal administrations regarding the flammability and the smoke emissions characteristics of interior materials and coordinate the development and implementation of standards for material performance and testing with the Federal Railroad Administration and the Federal Transit Administration.

To the Federal Emergency Management Agency:

(a) It should include in its training at the U.S. Fire Administration National Fire Academy a curriculum that addressed the needs of State and local emergency management agencies to respond to a major railroad accident and that familiarised emergency response organisations with railroad equipment and appropriate rescue methods for railroad accidents.

To the Governor and the General Assembly of the State of Maryland:

(a) It should instruct and empower an appropriate State agency to provide continual, effective, and independent safety oversight of all aspects of the Maryland Rail Commuter operations.

To the Association of American Railroads:

(a) It should assist the railroad industry with the development of positive train separation control systems through a continuing review of non-railroad technology and assess its adaptability to railroad communication-based control systems.

(b) It should assist the railroad industry with the development of positive train separation control systems by acting as a clearing house for information on the status and results of pilot projects and by disseminating that information to the railroad industry and the Federal and participating State transportation organisations.

(c) It should assist the railroad industry with the installation and operation of positive train separation control systems by maintaining industry standards to ensure open architecture and an interoperability of equipment for train control systems.

(d) It should inform its membership of the circumstances of this accident and caution them not to use the reverser during emergency brake applications for those trains on which the use of the reverser will eliminate the dynamic braking, thus increasing stopping distance.

To the Montgomery County Emergency Management Agency:

(a) It should develop comprehensive procedures for responding to railroad passenger train accidents and include these procedures in its disaster plan.

To the Baltimore County Emergency Management Agency, the Baltimore City Emergency Management Agency, the Metropolitan Washington Council of Governments, the Jefferson County Commissioners, and the Berkeley County Commissioners:

(a) They should conduct, in cooperation with the CSX Transportation Inc. and the Maryland

Mass Transit Administration/Maryland Rail Commuter, periodic disaster drills to assess their emergency management plans, to reinforce and evaluate their emergency training, and to test the communication with the organisations.

To the American Short Line Railroad Association, the Brotherhood of Locomotive Engineers, the United Transportation Union, the International Brotherhood of Teamsters, and the American Public Transit Association:

(a) They should inform their membership of the circumstances of this accident and caution them not to use the reverser during emergency brake applications for those trains on which the use of the reverser will eliminate the dynamic braking, thus increasing stopping distance.

Also, as a result of its investigation, the National Transportation Safety Board reiterated the following recommendations.

To the Federal Railroad Administration:

(a) It should promulgate Federal standards to require the installation and operation of a train control system on mainline tracks that would provide for positive separation of all trains.

(b) It should conduct, in conjunction with the Association of American Railroads, the General Electric Company, and the Electro-Motive Division of General Motors, research to determine if the locomotive fuel tank could be improved to withstand forces encountered in the more severe locomotive derailment accidents or if fuel containment could be improved to reduce the rate of fuel leakage and fuel ignition. Consideration should be given to crash or simulated testing and evaluation of recent and proposed design modifications to the locomotive fuel tank, including increasing the structural strength of end and side wall plates, raising the tank higher above the rail, and using internal tank bladders and foam inserts.

(c) In conjunction with the Association of American Railroads and the Railway Progress Institute, it should establish a firm timetable that included, at a minimum, dates for final development of required advanced train control system hardware, dates for an implementation of a fully developed advanced train control system, and a commitment to a date for having the advanced train control system ready for installation on the general railroad system.

To the General Electric Company:

(a) It should conduct, in conjunction with the Federal Railroad Administration, the Association of American Railroads, and the Electro-Motive Division of General Motors, research to determine if the locomotive fuel tank could be improved to withstand forces encountered in the more severe locomotive derailment accidents or if fuel containment could be improved to reduce the rate of fuel leakage and fuel ignition. Consideration should be given to crash or simulated testing and evaluation of recent and proposed design modifications to the locomotive fuel tank, including increasing the structural strength of end and side wall plates, raising the tank higher above the rail, and using internal tank bladders and foam inserts.

To the Electro-Motive Division of General Motors:

(a) It should conduct, in conjunction with the Federal Railroad Administration, the Association of American Railroads, and the General Electric Company, the research to determine if the locomotive fuel tank could be improved to withstand forces encountered in the more severe locomotive derailment accidents or if fuel containment could be improved to reduce the rate of fuel leakage and fuel ignition. Consideration should be given to crash or simulated testing and evaluation of recent and proposed design modifications to the locomotive fuel tank, including increasing the structural strength of end and side wall plates, raising the tank higher above the rail, and using internal tank bladders and foam inserts. [275]

Northeastern Railway Up passenger train No. 583 collided head-on with a stationary Gorakhpur Cantonment Special goods train at Domingarh station, Uttar Pradesh, India, on the Gorakhpur Cantonment – Gonda broad-gauge, single-line section of the Lucknow Division of the railway, at 17:24 on Thursday 18th April 1996. The accident was a very serious one, resulting in 54 deaths and 84 cases of injury, 47 of them serious.

The passenger train had overrun the Up Home signal at Danger and entered the Loop line to collide with the goods train.

The Commission of Railway Safety made several recommendations in the aftermath of the disaster:

(a) A GS {general sitting – non-air-conditioned second-class unreserved} coach manufactured in 1966, which had been involved in the collision, had corridors at each end where considerable passenger casualties had occurred. All such coaches should be removed from passenger service and replaced with the latest design of coaches, which had two toilets at each end.

(b) The standard train composition of ten coaches on train No. 583 Up should be restored, especially considering the severe overcrowding of this short-distance train. This could be done even by utilising surplus second-class sleeper coaches, which were available on the railway. In addition, the manufacture of new second-class GS coaches should be increased.

(c) A passenger had been riding in the locomotive of train No. 583 Up. The Railway should take preventive measures to ensure that passengers were not allowed to travel on the engines of passenger trains.

(d) The availability of speed-recorder charts and mechanical speedometers should be ensured. The engines of all goods engines should also be fitted with speed-recorder charts and mechanical speedometers. [276]

On the night of Sunday 14th July 1996 a passenger train and a freight train collided near Thénia, Algeria, about 40 kilometres from Algiers. Seven people died and 107 were injured, 15 of them seriously. [277]

About 50 people died on about Friday 8th August 1997 in a collision between a passenger train and a goods train at a location 300 kilometres from Baghdad, Iraq, according to a single brief internet report. [278]

On Friday 19th September 1997 an express passenger train came into sidelong collision with a freight train at Southall, England, on a line owned since privatisation of the railway system by Railtrack Great Western. The accident was a serious one, resulting in the death of seven passengers and the injury of 139. It was the subject of an inquiry by the Health and Safety Executive, conducted by Professor John Uff QC, FREng, whose report ran to 312 pages.

---

[275] National Transportation Safety Board report

[276] Government of India Ministry of Civil Aviation and Tourism Commission of Railway Safety Annual Report for 1996-97

[277] *The Telegraph*, Edwardsville/Glen Carbon, Illinois, United States, 15th July 1996

[278] ArabicNews.com internet report August 1997

The passenger train, 1A47, was a high-speed train (HST) consisting of front and power cars with seven coaches in between. On the morning of the day of the accident it had been worked from the depot at Old Oak Common to London Paddington by Driver James Tunnock, who discovered that it had two faults. One was that the buzzer providing communication between the guard and the driver was not working, so that the guard was not able to indicate to his driver that the train was ready to start from a station and this had to be done instead by the station staff. The second, more serious fault, was that the driver found, when he arrived at Paddington, that he could not cancel the automatic warning system (AWS) in power car No. 43173. He thus had to isolate it in order to release the brakes. At Paddington he reported the problems to Operations Supervisor George Barnfield and to Control at Swindon but not, as he should have done according to the Rules, also to the signalman.

Driver Tunnock then worked the HST as train 1B08, the 07:00 Paddington-to-Swansea express. For this journey the leading power car was No. 43163, in which the AWS was working. The same HST set then formed train 1A47, the 10:32 return working from Swansea to Paddington, now, with the reversal of direction of the train, with power car No. 43173, with its isolated AWS, at the head of it. Driver Tunnock also drove train 1A47 as far as Cardiff Central, where he was relieved by Driver Larry Harrison. Driver Tunnock impressed upon his colleague the fact that the AWS had been isolated and also affixed a note to the dashboard of the power car, 'AWS ISOLATED REPAIR BOOK FULL.'

Driver Harrison had not previously driven a train with the AWS inactivated. There were differing views on whether the AWS was an essential safety device or merely an aid to drivers. The train was also equipped with a Driver's Safety Device (DSD), which required the continual depressing of a foot pedal and the DSD also incorporated a Driver's Vigilance Device, whereby a warbling sound was emitted every minute, requiring the release and re-depressing of the pedal within three seconds to forestall a brake application. Finally, the train was fitted with Automatic Train Protection (ATP), which would have applied the brakes should a train pass an adverse signal without any action on the part of the driver, but as neither driver was yet qualified to use it, the ATP was not switched on.

The journey to London went normally to Reading, with Driver Harrison being accompanied between Swindon and Reading by Railtrack employee Tim Mayo, engaged in a signal-sighting exercise. The train left Reading on the Up Main line and speed was correctly reduced to 100 mph for several temporary speed restrictions, imposed due to the condition of the track. After the last restriction, west of Slough, speed was increased to 125 mph. Approaching the site of the accident at Southall East Junction, 8¾ miles from Paddington, the last signals on the Up Main were Nos. SN 298, rather more than 11¼ miles from Paddington, SN 280, 1,113 yards further on (the accident report alternates between imperial and metric units; this account will give imperial units throughout); SN 270, 1,155 yards further on; and SN 254, 1,673 yards beyond that. This last signal was 448 yards short of the point of collision.

When train 1A47 approached, signal No. 298 was Green, No. 280 was Double Yellow, No. 270 was Single Yellow and No. 254 was Red. Driver Harrison failed to react to either the Double Yellow or Single Yellow warnings, though he did finally apply the brakes after that, probably when signal No. SN 254 had come into view, showing Red. This signal was at Danger because English Welsh and Scottish Railway freight train 6V17 was crossing from the Down Relief line into Southall Yard, which was on the Down side of the Down Relief line (the lines here were paired by type rather than by direction so crossing from the Down Relief line to the yard on the Down side involved crossing both the Up, on which the express was approaching, and the Down Main lines). The freight train consisted of a locomotive and 28 empty waggons and was being operated by Driver Alan Bricker. Fortunately for the latter his engine was already across the Up Main but the waggons were not, so a collision was unavoidable. The passenger train was still travelling at between 60 and 80 mph and the freight train at 21 to 25 mph in the opposite direction when they met. The power car of train 1A47 struck the seventh waggon of the freight train a glancing blow, then collided with the eighth and following waggons. The collision occurred at 13:15.

The accident was immediately noticed in the new Integrated Electronic Control Centre (IECC) at Slough, which went by the name of Slough New signal box, though it bore little resemblance to a proper, traditional box. Signalman Forde there had manually set up the route for the freight train to pass from the Down Relief line into the yard at Southall, whereby the Up Main signals had been necessarily placed at Danger and the Automatic Route Setting (ARS) equipment had been prevented from setting up any conflicting route. Signalman Forde saw from his visual display unit (VDU) that train 1A47 was overrunning signals, then that multiple track circuits were being shown as occupied, indicating the collision. He had then manually placed all the signals in the area at Red.

Driver Bricker told the inquiry that he had become alarmed at the speed of the approach passenger train, had been able to see that it was braking hard and that, nevertheless, there would be a collision. He had tried to accelerate out of the way, though with no hope of success. At the moment of collision the brake pipe between the waggons had been severed and their brakes came on automatically. The leading coach of the HST was derailed to the left, the second coach struck the freight waggons and was itself struck by the third coach, being bent into a U shape, and the third coach also struck a waggon. The next two coaches were derailed; the rear two and the rear power car remained on the line. As might be expected the shock of the collision was felt throughout the train, so that even Senior Conductor Abdul Khanghauri, in the rear coach, was thrown to the floor and briefly stunned.

Driver Harrison told the inquiry that he recalled passing signal No. SN 298 at Green and "whizzing through Hayes." He remembered putting some paperwork in his bag and the next thing he could recall was seeing signal No. SN 254 at Red with, about a mile ahead, the engine of the freight train "at a funny angle" and realising to his horror that there must be a collision. His first thought had been that the signal had been put back 'in his face' and he had applied the brakes, possibly first a full service application and then an emergency application. Realising that he could do no more, he had then retreated into the interior of the power car and escaped with cuts and bruises.

How the driver came to miss the signals was unclear. It was possible that he had dozed off momentarily (a microsleep) after last cancelling the DVD. Another possibility suggested was that he had weighed down the DSD to enable him to briefly leave his seat, though his driver's bag as recovered after the accident was not heavy enough to do so. The possibility of him having experienced a microsleep seemed more likely.

Many local people called the emergency services, which had already been alerted from Slough New signal box. Two police in a patrol car were quickly on the scene and the first ambulance arrived at 13:24. Fifteen road ambulances eventually attended, together with an air ambulance and the fire service, who were able to quickly extinguish a small fire in the front power car. [279]

Newspaper accounts described the accident in detail. 'The 10:32 Great Western express eased out of Swansea precisely on time for the three hour journey to London. At the start, only about 30 passengers had joined the train. When it reached the first stop, at Cardiff, 20 minutes later, there was a crush of passengers waiting on the platform. Many were journalists who had worked late into Thursday night covering the devolution vote {a referendum to establish a National Assembly for Wales}. The train became so crowded that not everyone could get a seat. Mark Cole, a BBC reporter, said: "Some were standing in the aisles, Others were perched

---

[279] The Southall Accident Inquiry Report

on the edge of seats, or sitting at the ends of the carriages." Some moved forward to empty seats in the first-class carriages. Stewards pushing refreshment trolleys found their paths blocked by the throng of passengers. As the express accelerated to about 125 mph many of those on board, Mr Cole included, were dozing.

'The train reached Reading on tine and had pulled out for the final stretch towards Paddington when it braked violently. It was 1.20 p.m. Celia Ellacott, a Channel 4 journalist, said: "When the train slowed down and started juddering, the thought flashed through my mind that we might crash, then suddenly it just happened. It was terrifying."

'One businesswoman said: "It jumped the tracks and cantilevered on to its side. The windows seemed to shatter. Debris from the other train started to hit us. Dark, black smoke filled the carriage. People were very scared."

'Mr Cole woke up as the train "seemed to pick up speed and within seconds had left the track – glass was flying through the windows. The carriage I was in veered off. It caught fire. There was smoke everywhere. People started to panic."

'Nick Sutton, a BBC researcher, was in a lavatory cubicle and was thrown back against the wall. An acrid smoke engulfed the carriages. A guard's voice came over the intercom with a warning. Outside, power cables were crackling as they fell beside the windows, flashing above the carriages. The guard {Senior Conductor Khanghauri} said, "Don't run. Don't jump out." He asked passengers to move towards the back of the train. In the darkness, people were crying. Luggage had been thrown across the compartments, hitting travellers and making escape more difficult. Some had cuts from flying glass. Others helped them to staunch the flow of blood. Passengers stood in the dark, queueing to be released. The doors were still locked. Mr Cole said that eventually the only way out was by pushing the remaining glass from a window, climbing up and jumping the 10ft to the ground as the carriage rested on its side: "People were in a bad state, and the doors would not open. We had to climb out of the windows and jump down. A lot of people had been cut."

'Chris Dixey, 50, an engineer from near Canterbury, in Kent, was in the third crashed carriage, which was on its side. About 30 people were with him, many of them elderly: "There were a lot of people shouting, then one man said to people to stay in their seats, because it was on its side." Mr Dixey tried to break a window, but it would not smash. "There was a flash of fire, then it went out, but I was worried about the fire," he said. "Eventually we got a door open. There were some plain-clothes policemen at the side of the track, They helped the old people out. They were very good."

'Ms Ellacott said: When we got out we were told to mind the overhead cables. We climbed out and looked forward and saw that the engine was in flames."

'As the first survivors hauled themselves through the broken windows and slithered down the outside of the carriages, they were confronted by the sight of the bodies that had been flung from the ruptured carriage. Mr Sutton said: "As I walked off the train, I saw a body lying by the side of the tracks. No one was touching it." His shirt was ripped and there was blood all over him. Mr Cole said bodies were lying beside the track: "I saw two or three fatalities. They were on the track by the first-class carriage. The engine was on its side and on fire. There was great concern that the train was not safe at that time."

'Local residents were running to help. Manjit Singh, who lives in Park Avenue, backing on to the railway line, reached the trackside with a friend, Tony Malr. They saw the driver of the express, sitting stunned beside the shattered diesel cab. "I grabbed him and asked if he was all right. He said, 'What has happened?' He was all black from the smoke. From the second carriage were cries of 'Help, help, help.' There were a lot of injured people, everyone was just saying, 'Help us, help us.' We were trying to get people out from the rear carriages and they had to go past these bodies. People got hysterical. We got curtains and things that were lying around and covered up the heads and faces."

'Rajinder Vagarwall, a coach company owner, described how he and four neighbours tore down a fence to get to the injured. "We pushed down the wire-mesh fence to get to the track. We went to the first carriage and people were shouting and screaming, but they were too badly stuck to get out so we had to go to the second. There we helped out three Chinese people – two men and a woman. They were extremely upset and shaken. There was also an old lady of 80 and her sister, who were too shaken to come out at first. We managed to guide them out and took them to our house."

'As fire crews and paramedics descended on the crash site, the local major disaster plan went into action. At Ealing Hospital, a mile from Southall station, the casualty department was cleared, wards were emptied, operating theatres were put on standby and all operations were cancelled as Dr Michael Rudolf, the medical director, and his staff waited for the injured. "We had been warned by police that it was carnage out there, similar to that of an air crash," he said. At the Middlesex Hospital, a spokesman said they had been told to expect large numbers of the less seriously injured passengers: "It's something we practise for."

'The walking wounded were led through a park, many carrying luggage they could salvage. Some were wrapped in orange blankets and taken to a school. Harjeet Singh Jaglev, 37, a housewife, said: "People were letting the injured into their houses and getting out the first-aid kits. Some ladies were screaming and crying." Her son Rajdeep, 12, a pupil at Villiers High School, where many of the injured were taken, said: "I had just got back to school after lunch when there was an announcement from our teachers telling us there had been an accident. Some of the pupils stayed to help the injured people, but the younger ones were told to go home." A group of 30 16-year-olds gave tea and biscuits to the walking wounded, acting as volunteer assistants to staff and emergency workers. The rest of the school's 1,200 pupils went home. The school, which is just 500 yards from the scene of the crash, is a designated emergency centre. Dai Jones, a senior member of staff, said he was enormously pleased with the way his pupils coped sensitively with the injured: "They were marvellous. I am very proud." [280]

The accident report noted that the track at Southall was near the end of its life but was in good condition. The signalling was examined very carefully after the accident and signal No. SN 270 was found to be somewhat misaligned, but not to such a degree as would affect its visibility.

The question as to whether the freight train should have been crossed in front of the express had been raised in the press. Signalman Forde had had to make a rapid decision regarding train regulation, with the movement of ten other trains to consider, and decided by instinct and experience to cross the freight train over as he had. An analysis showed that out of 30 possible routing options only one caused less delay than the one he had adopted and he could not have been expected to choose that option in the time available to him. Since privatisation there had also been the question of penalty payments having to be made should the freight train have been unnecessarily delayed. The report discussed the train regulation policy at length.

The accident report went on to consider driver competence and training, then discussed the isolation of the AWS equipment. It emerged that the AWS had first failed at Oxford the evening before, when the HST had been about to work train 1A85, the 19:47 Oxford to Paddington. Driver Allan Taylor had isolated it, like Driver Tunnock, and reported the matter to Signalman Richard Parker in Oxford panel box, though the latter had not understood that the AWS was out of action and so had not reported this to Control. The train had been worked to Paddington without incident and thence back to Old Oak Common depot, where Driver Taylor had failed to note the fault on either a Defect Report or an Incident Report form. The problem with the AWS had not been detected during the overnight maintenance at Old Oak

---

[280] *The Times*, London, England, 20th September 1997

Common. There was no record of the telephone call that Driver Tunnock had made at Paddington to Swindon Control on arrival in London to report the driver/guard communication and AWS problems, which had happened during the change of shifts at Swindon Control and must have been mislaid, so when his train arrived at Swansea he had been surprised to find that there was no fitter there to attend to the problems. Driver Tunnock had then rung Swindon Control again from Swansea. The controllers not being at their positions at that moment, the call had been taken by Information Controller Sandra Hallett, who had only been working there three weeks. She had not understood the significance or meaning of the message and it had not been passed on. Driver Tunnock had then telephoned Landore depot, only a short distance away, and said that he had told the fitters there of both problems. They had duly arrived and fixed the driver/guard communication but that was all. In any case the train would have had to be taken out of service to fix the AWS problem. An alternative would have been to turn the HST on the triangle of lines outside of Swansea, so that the power car with the operative AWS was leading, but there was no procedure for this.

Contamination of the AWS switch by an unknown chemical was the cause of the failure.

The inquiry was delayed because Driver Harrison was subsequently charged with seven counts of manslaughter. Seven months later, Great Western Trains (GWT), the company operating the HST, were charged with corporate manslaughter, no individual being named, while both GWT and Driver Harrison were charged with offences under the Health and Safety at Work Act. Finally, at the beginning of July 1999, the corporate manslaughter charge against GWT was dropped, as were all charges against Driver Harrison. GWT pleaded guilty to the Health and Safety at Work charges and was fined £1,500,000. The inquiry was then resumed.

The report went on to discuss the evolution of opinion regarding AWS, which for a long time had been regarded merely as an aid to drivers. With the reduction of collisions due to signalling errors with modern signalling and with faster trains, the proportion of accidents due to driver error had increased. With these developments there had been a change of opinion from AWS being merely an aid to drivers to it being an essential safeguard.

AWS had its limitations, because a driver could acknowledge a warning of a caution signal and then continue at unreduced speed, while on suburban lines where trains often encountered a series of Yellow signals and resulting AWS warnings, so that drivers could become conditioned to automatically cancelling the warning signals. The Rules as to whether a train could continue to run with its AWS isolated were also ambiguous and susceptible to different interpretations.

The report discussed the situation regarding the introduction of Automatic Train Protection (ATP), which had it been in use would have prevented train 1A47 overrunning signals and hence the accident. The installation of ATP on the main line concerned was begun by British Railways in 1991 but at the time of privatisation in 1994, there had been concerns at its cost. Problems caused by the fragmentation due to privatisation, with the ATP equipment itself, with driver training on ATP and the coming introduction of the cheaper Train Protection and Warning System (TPWS), meant that at the time of the accident there was only a 50% chance of an ATP-designated train having the ATP in use, and no chance at all of it being operative if the driver was not trained in its use (as neither Drivers Tunnock nor Harrison were). By the end of the inquiry, ATP was in use on all trains.

The inquiry resulted in 93 recommendations:

(1) All parties in the rail industry should co-operate in the collection of evidence to support reliable research into human behaviour studies relating to driver performance. Railtrack should co-ordinate this work and TOCs {Train Operating Companies} incorporate the results into training programmes.

(2) Evidence should include that to be provided by CIRAS {Confidential Incident Report and Analysis System} and from On-Train Data Recorders used to monitor driver

behaviour. ASLEF {Association of Locomotive Engineers and Fireman – the train driver's union} in particular should give their full support to such an initiative.

(3) Simulators should be introduced for driver training and for the observance of driver behaviour.

(4) Driver training should include driving in abnormal situations permitted by the Rules and specifically driving with AWS isolated to the extent so permitted, including the use of simulators.

(5) Testing of driver competence and knowledge of Rules should be extended to cover application of the Rules to practical situations, including all abnormal driving situations permitted by the Rules.

(6) Drivers should be encouraged to report all actual or suspected faults, whether through formal fault reporting Ongoing procedures or through CIRAS.

(7) Railtrack together with ATOC {Association of Train Operating Companies} should establish a national qualification and accreditation system for drivers including centrally held records to be available to the current employer.

(8) Railtrack and ATOC should monitor the transfer of drivers between operators and the numbers of drivers trained by each TOC and consider whether there were any safety implications involved.

(9) Current Rules governing drivers' permitted daily and weekly working hours should be reviewed in light of current research into human behaviour.

(10) Railtrack must ensure that Rules and Group Standards applicable to operators, including drivers, were clear and unambiguous. In particular, Railtrack should urgently complete the review of operating Rules to ensure they were workable in the privatised, fragmented industry.

(11) The use of more than one document (whether Rules, Group Standards or otherwise) to cover a single operational issue should be avoided, save where proper reasons existed for use of multiple sources.

(12) All train-borne safety equipment should be clearly designated as to whether or not it was vital to the continued running of the train.

(13) AWS was to be regarded as vital to the continued running of the train.

(14) Clear procedures for steps to be taken on failure of any train-borne safety equipment should apply nationally, subject only to such company variation as was fully justified.

(15) All parties must emphasise the need to comply with the Rule Book and must not condone departures from the Rules.

(16) Railtrack and TOCs must impress on drivers and other staff the need to use formal fault-reporting procedures where the Rules so provided, and that the duty to report a fault must be performed personally and the report delivered to the person or body identified in the Rules, and not to any other person or body.

(17) Fault-reporting procedures should be reviewed and made as simple and convenient to use as practically possible. They should include provision for an acknowledgement and an explanation if relevant.

(18) Failure to provide forms, defect-repair books or other means of reporting faults should be regarded as a disciplinary offence.

(19) Appropriate procedures for receiving and making an automatic record of verbal reports should exist in all control centres, similar to the facilities installed by GWT in 1998.

(20) Level and quality of training for information controllers should be reviewed.

(21) Controllers' posts in Railtrack and TOCs should be designated as "safety-critical" as defined in the Railways (Safety Critical Works) Regulations 1994.

(22) Fault databases (including RAVERS {Rail and Vehicle Records System}) should ensure that repeat faults were logged and that a 28-day history of defects was available to managers and maintenance staff.

(23) Databases should be programmed to determine and highlight statistically significant trends in faults reported and to display such information to managers and maintenance staff.

(24) GWT should maintain full records of competencies for all maintenance staff as required by job descriptions and safety responsibility statements.

(25) GWT should regularly monitor the workload of all maintenance staff.

(26) A current and detailed list of items required to be inspected for each examination should be prepared for and used by maintenance teams.

(27) Documentation for the A-Exam should require ATP reset and self-test.

(28) Maintenance staff should be provided with a flow-chart to show the derivation of all sources of repair work, to include RAVERS (with check on repeat items) and appropriate structure for Request, Repair Book and Maintenance Control items.

(29) Maintenance procedures should require checking of the history of reported defects, including repeat faults, and the taking of appropriate action.

(30) An improved AWS test box, capable of detecting faults not revealed by the magnet test, should be provided as standard issue at all maintenance depots.

(31) Efforts should be concentrated on ensuring that AWS and other train-borne safety equipment did not fail in service through preventable causes. This should include regular replacements of equipment, maintenance of full service records and provision for full traceability of repairable parts and components.

(32) Contractual ownership and other rights in AWS equipment must be clarified and defined.

(33) ATOC and Railtrack should monitor the supply of new parts and components to ensure continued availability on an indefinite basis, including the introduction of improved components.

(34) Railtrack should ensure that the alignment and sighting of signals was confirmed at the time of commissioning, both from the signal and from the track, and appropriate records made, including photographs.

(35) Railtrack should ensure that checks on alignment and sighting of signals were made at least annually, and at a greater frequency to be determined on the basis of errors found.

(36) Railtrack should review all maintenance contracts to ensure that all parties were aware of what checks were included and which were excluded.

(37) Railtrack should ensure that any further proposed change of regulation policy was preceded by a risk assessment.

(38) There should be a review of Condition H11 of the Track Access Conditions, which should make clear that no regulating decision was to be made on the basis of protecting commercial interests. Safety and security must be paramount considerations.

(39) Railtrack should review their Level 1 Policy Statement in the light of any amendment to Condition HI l.

(40) More Level 2 and 3 Policy Statements should be introduced, having due regard to any relevant risk analysis.

(41) Railtrack should review the operation of ARS to consider whether more Green signals should be booked ahead of higher speed trains, and generally whether the speed and length of trains was adequately taken into account.

(42) Any further review of LRM {Layout Risk Method} should take into account Dr Murphy's risk analysis.

(43) HMRI {Her Majesty's Royal Inspectorate) should keep under long-term review the effect of speed on numbers of casualties in rail accidents.

(44) A review should be carried out by ATOC, with input from all interested bodies, on the ways in which internal safety features might be modified and standardised to provide the best practicable means of emergency exit under accident conditions, including vehicles lying on their side, to include the provision of emergency lighting and standardised public announcements.

(45) The review should consider dates and means for the introduction of identified improvements to existing stock.

(46) A single body should be empowered to specify common standards for safety features in the interior of passenger vehicles and to identify and approve types of vehicles and/or operators to which particular standards are to apply.

(47) The design of coaches should be such that internal doors could be easily opened in a crash situation, in darkness and irrespective of the attitude of the vehicle; and that hammers intended for breaking windows could be easily located in the same conditions.

(48) Safety briefings or other appropriate means of communicating safety information to passengers should be adopted, including pointing out safety notices to passengers. ATOC should monitor the methods adopted by TOCs and issue guidance documents after a suitable trial period, including recommendations for different types of journey.

(49) A design study and risk assessment should be carried out to determine whether freight wagons could be designed with less aggressive features without detriment to their primary function.

(50) Consideration should be given to the most appropriate form of coupling for freight trains, to minimise damage in the event of collision, including a risk assessment.

(51) No recommendation was made concerning crumple zones in passenger carriages, save that the matter should be given attention by the Ladbroke Grove Inquiry.

(52) Train crews should be given improved training and briefing on emergency actions, including a practical evacuation.

(53) Standards for evacuation of passengers should be proved by practical exercises using typical groups of passengers and train crew and repeated on a regular basis to be approved by HMRI.

(54) Consideration should be given to modification of the design of OHL {Overhead Line Equipment} structures to improve their response to accidents, if achievable without detriment to their primary role.

(55) All trains should be fitted with data recorders. All data recorders should record speed, time, location, brake application and AWS cancellation, and should be simple and speedy to download.

(56) Consideration should be given to developing a cheaper form of data recorder for

retrospective fitment where this would allow earlier fitting.

(57) Steps should be taken to put in place means to resolve inter-company issues relating to research and development at all levels. Specifically, the following issues must be addressed.

(58) Rights (including ownership) and obligations in all equipment added to vehicles, together with lineside equipment upon which its operation depended, must be defined in legally enforceable terms.

(59) The above recommendation to include review of Safety Cases and franchising arrangements and consideration of action by the Rail Regulator.

(60) Consideration should also be given, for the purpose of the above recommendations, to enlarging or re-organising existing inter-company groups, including considering whether the Railway Group should include ROSCOs {Rolling Stock Companies} and component suppliers.

(61) S&SD {Railtrack Safety and Standards Directorate} together with HMRI and other bodies having responsibility for accepting or approving new equipment or stock should review their procedures with a view to reducing delay and introducing fast-track procedures where possible.

(62) There should be a review of progress on implementing the Recommendations of the DTp {Department of Transport} Review Committee Report (chaired by Sir David Davies) published in September 1996

(63) One or more System Authorities should be created to oversee the specific development of any new project on the railways and to oversee continuation of work on existing projects, including AWS and ATP.

(64) Future R&D {Research and Development} must be the subject of rigorous programming, cost-projections and funding arrangements, including reliable contingencies. R&D funding must be on a cross-industry basis, irrespective of whether individual TOCs decided to fit new technology.

(65) Development of ATP should be managed and funded in future through a System Authority having broad industry representation and support.

(66) ATP should be maintained in a fully operational state on Great Western lines currently fitted, until replaced by an equally effective train protection system.

(67) GWT and Railtrack should consider extensions to the present coverage of ATP.

(68) GWT, Railtrack and HMRI should consider whether trains with AWS isolated could run normal services where ATP is fitted and operational.

(69) All parties in the industry must ensure that paper-based procedures did not become divorced from reality. This should include senior managers maintaining a direct knowledge of the situation in railway workplaces.

(70) Paper-based audits should be backed up by unplanned inspections and other direct observation of the work under review.

(71) Steps should be taken to ensure that all risk assessments were rigorous and that those initiating risk assessments were appropriately qualified and informed. Steps include giving attention to HSE {Health and Safety Executive} Guidance Notes covering risk assessment, which should accordingly be kept under review.

(72) Consultation procedures and times involved in revision of Group Standards or the introduction of new Group Standards, should be reviewed by the Inquiry to be held into

safety procedures.

(73) The technical investigation of serious rail accidents should be controlled by HMRI save in exceptional cases of suspected crime which was unconnected with the running of the railway.

(74) HMRI should ensure that a single, thorough and definitive technical investigation was carried out, to include the recording of all appropriate factual data, the collection of physical evidence from the scene of the accident and decisions as to handing the site back to the rail companies.

(75) Standing contracts for the provision of consulting services by recognised railway experts should be amended to make provision for HMRI to require any appropriate individual to provide expert services for the immediate accident investigation, including the services of any appropriate laboratory or testing house.

(76) HMRI should decide, irrespective of any ongoing criminal investigation, what data or information was to be passed on to rail companies for rapid action.

(77) The primary forum for deciding upon appropriate recommendations following an accident should be the Rail Industry Inquiry (RII). Procedures for holding such an inquiry at the earliest possible date should be strengthened and should include the presentation by HMRI of their investigation.

(78) Procedure for conducting an RII should be reviewed. This should include ensuring that the RII panel was independent of all parties having an interest in the accident.

(79) Consideration should be given to whether procedures could be adapted to make any RI1 accessible to the public, save where the needs of confidentiality otherwise require.

(80) Nothing should be permitted to delay the opening of an RII nor the completion of their Report and Recommendations.

(81) Consideration should be given to whether an additional HSC {Health and Safety Commission} independent accident investigation body should be created to take over the accident investigation functions of HMRI under Recommendation 73

(82) Existing protocols between Railtrack, BTP {British Transport Police} and HMRI should be reviewed in the light of the above Recommendations.

(83) Any subsequent Inquiry directed under the Health and HSC Safety at Work Act. 1974 should involve all parties in the rail industry who might have an interest in Recommendations to be made, through the involvement of representative groups including ATOC.

(84) Passenger representation at such inquiries should not be limited to those involved in the immediate accident. Consideration should be given to enlarging the role of CRUCC {Central Rail User's Consultative Committee} and the provision of appropriate funding for their full participation in Inquiries.

(85) Responses to the S&SD paper "The Future of Accident Investigation in the Railway Industry" should be taken into consideration in applying the foregoing Recommendations.

(86) Steps should be taken to upgrade the role of Rail Incident Officer and to ensure that the person so designated had sufficient authority and standing for the task in hand, bearing in mind the tensions that could develop in the early stages of an accident response.

(87) Consideration should be given to means of speeding up the process of earthing and isolation of traction current following an accident on an electrified section of line.

(88) Routes for evacuation away from an accident should take into account the need to avoid distressing scenes.

(89) Further consideration should be given to the sensitive handling of persons rescued from accidents, including whether they should be sent onward by train.

(90) More effective means of liaising with hospitals and casualty gathering areas should be considered.

(91) Identification of victims should be speeded up and information released to relatives at the earliest possible time.

(92) Casualty bureau procedures should be reviewed in order to ensure that they remained open for as long as required and that adequate telephone facilities were available.

(93) Post-accident de-briefing procedures should be reviewed to ensure that combined de-briefings were held between all involved Railway Industry and Emergency Services groups. [281]

On Friday 10th October 1997 an express passenger train collided with a freight train stopped in the station at Shingia, Bangladesh, near Jessore. Twelve people died and 30 were injured. [282]

Just before midnight on Christmas Eve, Wednesday 24th December 1997, the *Rohi Express*, bound for Rawalpindi, collided with another passenger train in the station at Rustam Sargana, Pakistan. There was a thick fog at the time of the accident. Both trains were fairly well-filled with holiday travellers and 32 people reportedly died and about 40 were injured. Emergency services worked through the night under the glare of floodlights to extricate the injured and the services of the Pakistani Air Force were called in so that their helicopters could take casualties to hospital. Medical Supervisor Javed Omar at Jhang Hospital said that most of the injured had been discharged after receiving treatment but four were more seriously hurt. [283]

Other, possibly later reports stated that the number of dead had increased to 35 and of injured to 100. The other train was a southbound local service from Sargodha to Multan and the trains had collided head-on, because the express had not been diverted to the other running line through the station. Witnesses stated that villagers had been the first outsiders on the scene and had torn at the wreckage with their bare hands to reach screaming passengers. [284]

Passenger Rehmet Bibi, 40, related, "Suddenly there was this big bang and people started crying and screaming." She was among the injured.

Prime Minister Nawaz Sharif ordered an immediate inquiry to be opened and on Thursday afternoon, investigators were taking statements from witnesses at the scene, while rescuers were still working to recover several bodies with the help of breakdown cranes, which had by then arrived. [285]

Early on Monday 5th January 1998 a passenger train collided with a shunting freight train, which was performing shunting operations near Karachkum in northern Tajikistan. The

---

[281] The Southall Accident Inquiry Report

[282] https://www.ejrcf.or.jp/jrtr/jrtr15/pdf/topics.pdf

[283] *BBC News*, London, England, 25th December 1997

[284] https://www.deseret.com/1997/12/25/19353320/train-crash-in-pakistan-takes-35-lives/

[285] *Syracuse Herald-Journal*, Syracuse, New York, United States, 26th December 1997

passenger train had been on its way from Urgench in neighbouring Uzbekistan to Andizhan in Tajikistan. Fire broke out after the collision and four carriages were destroyed. Different reports stated that at least six people had died and 20 were injured, and that 15 had died and 26 had been injured. [286] [287]

A passenger train and a freight train collided near Tiefa, Liaoning province, China, on Sunday 3rd January 1999. Twenty-two people were killed outright and two more died later. A railway source stated that another 23 people had been hurt, while an independent source stated that 100 people had been hurt and that ten of them were in a critical condition. Both trains had belonged to the state-owned Tiefa Mining Bureau; the passenger train was a service used to convey miners to and from their work. [288]

A major catastrophe occurred in India at 01:55 on Monday 2nd August 1999 on the Northeast Frontier Railway when train No. 4056, the Down *Brahmaputra Mail* from Dibrugarh to New Delhi, came into head-on collision with train No. 5610, the Up *Avadh Assam Express* from New Delhi to Guwahati, in the station at Gaisal, West Bengal. The official report stated that the accident had resulted in fatalities to 286 people, while another 359 were injured, 156 of them seriously.

The mistake that led to the catastrophe was made at the adjoining Kishanganj station. Three of the four lines in the station were temporarily out of use in connection with the enlargement of the layout, with all trains using the remaining track. Because of the engineering work the points had to be cranked over manually instead of being operated from the station panel. Upon leaving the station for Gaisal, train No. 5610 should have been switched back to the Up line at Kishanganj East cabin but the wrong set of points was manually reset, with the result that it left there instead on the Down line. Driver B.N. Roy failed to notice that his signals were out of position and the station staff failed to observe that the Up signals had not reverted to Danger after the train had passed them.

In the collision the engine of the *Avadh Assam Express* reared several metres into the air and the engine of the *Brahmaputra Mail* rammed into several coaches of the Assam train. Fire then broke out in the wreckage and killed many of the injured trapped in the wreckage, the blaze also spreading to the station buildings. The magnitude of the disaster overwhelmed the emergency services. Many of the rescue vehicles had a 14-hour road journey from Calcutta to reach the scene, so that some victims who might have been otherwise saved succumbed, while the hospitals at Kishanganj and Islampur were likewise unable to cope with the number of victims. Heavy rain on the day after the disaster helped to extinguish the fires and rescuers then worked to retrieve the bodies from 12 mangled coaches. Most of the corpses were unrecognisable.

Unofficially it was claimed that the true number who had died in the accident was over 1,000, including 90 soldiers. Both trains had been grossly overcrowded and there had been many passengers travelling without tickets, who were not included in the official figures.

Railway Minister Nitish Kumar tendered his resignation after the accident. [289]

The accident was the subject of an inquiry held by Justice G.N. Roy, a retired high court judge, who in the course of his 16-month investigation cross-examined 106 witnesses. He

[286] *The Medicine Hat News*, Medicine Hat, Alberta, Canada, 6th January 1998
[287] *The Orange County Register*, Santa Ana, California, United States, 6th January 1998
[288] https://www.emerald.com/insight/content/doi/10.1108/dpm.2000.07309bac.003/full/html
[289] Wikipedia – Gaisal train collision

found much to criticise in his report, in which he placed the responsibility for the accident on a number of different railwaymen. The person with the greatest responsibility, he concluded, was Assistant Stationmaster Ram Naryan Singh at Panjipara, the intermediate station between Kishanganj and Gaisal, where the *Avadh Assam Express* had apparently halted for 11 minutes on the wrong line, before continuing towards Gaisal. The other railwaymen to blame, Justice Roy considered, were Driver B.N. Roy and the assistant driver of the *Avadh Assam Express*, Driver B.C. Bardhan and the assistant driver of the *Brahmaputra Mail*, Assistant Stationmaster (Cabin) S.P. Chandra, Cabinman Allauddin and Pointsman Jagdish Ram in Kishanganj East cabin, and Cabinman Gulabchand Gupta in Panjipara cabin.

In addition to the errors that had led up to the disaster, the station staff at Gaisal had failed to at once inform the Division Headquarters of the Railway at Katihar that there had been a major disaster. The judge remarked, "In fact, the primary report that they sent said that the accident was due to a powerful bomb blast inside the Avadh Assam Express, which was carrying a number of defence personnel from Assam to the war-front at Kargil."

The judge observed, "It was a clear case of human failure. The cabinmen, pointsmen, assistant station masters and station masters failed in their duties to either detect or correct the wrong allotment of tracks to the two trains, resulting in the collision." Reiterating that this type of failure could not be called sabotage, he said, that the Commission had been aimed at finding whether the human failure itself was an act of 'omission or commission'. "And I found that it was gross omission." He noted that though Gaisal itself was at the 'receiving end' of errors committed by operating staff at Kishanganj and Panjipara, Assistant Stationmaster Chandra at Gaisal had also shirked responsibility by fleeing from the scene after the accident.

Judge Roy recommended the overhauling of the Northeast Frontier Railway administration, which he said was suffering from 'low morale' due to heavy casualties of railway personnel, as a result of insurgency activities in the North East. He added that the administration also lacked finances to run the system smoothly. He also suggested that radio wireless devices should be installed to connect the station staff with train drivers, that there should be colour signalling to inform the station staff whenever a train was on a wrong line and optical fibre connections to spruce up communication links. In addition, the remote station in North Dinajpur district had not had the resources to mobilise immediate rescue operations after a disaster of the magnitude it had witnessed." The Kishanganj SDO {Sub-Divisional Officer} admitted during the inquiry that timely arrival of more gas cutters could have lessened the casualty figure considerably," the judge observed. [290]

During its long existence, from 1835 to 1948, and despite its extensive network, the Great Western Railway of England had very few really serious disasters, but its successors, first British Railways, Western Region, and after 1994 Railtrack Great Western, were not so fortunate, with bad accidents in the next 50 years at Milton in November 1955, West Ealing in December 1973, Taunton in July 1978, and Southall in September 1997. Then on 5th October 1999 at Ladbroke Grove, just outside London Paddington, one of the most serious accidents of the 20th century occurred.

Between the terminal platforms at London Paddington and Ladbroke Grove Junction, about two miles to the west, there were six running lines, numbered 1 to 6, all signalled by this date for bi-directional working, while beyond Ladbroke Grove there were just four principal running lines, the Up and Down Main and the Up and Down Relief, paired by type. The signalling was controlled from an Integrated Electronic Control Centre (IECC) at Slough, with Solid State Interlocking (SSI) and a computerised Automatic Routing System (ARS). The

---

[290] https://m.Rediff.com/news/2001/apr/02train.htm

running signals were four-aspect colourlights.

On the morning of the accident train 1K20, a three-car Class 165 Turbo diesel multiple unit (DMU) forming the 08:06 departure to Bedwyn, operated by Driver Michael Hodder, 31, departed from Platform No. 9 at Paddington and was routed by the ARS onto Line 4 as far as Ladbroke Grove. From there its intended route was from Line 4 via a crossover to Line 3, thence onto the Down Main line, via two crossovers, the first to take it onto the Up Main line and the second to take it from the Up Main onto the Down Main. Meanwhile in the opposite direction, travelling on the Up Main line, was the 06:03 Inter City high-speed train (HST) train from Cheltenham to London, 1A09, consisting of eight coaches with a power car at each end, operated by Driver Brian Cooper, 52. The Up HST had priority at Ladbroke Grove Junction, so having received Green indications at signals SN 17 at the end of Platform No. 9 and at signal SN43, the Bedwyn train would have received a Double Yellow indication at signal SN 63, a Single Yellow indication with a junction feather at Signal SN 87, for it to cross from Line 4 to Line 3, and then a Red indication at signal SN 109, where it should have halted until the Cheltenham train had passed through on the Up Main line. The Bedwyn train was provided with the Automatic Warning System (AWS), the warning horn of which had sounded at signals SN 63, SN 87 and SN 109 and each time had been cancelled by Driver Hodder and a brake application thereby forestalled. Having passed signal SN 109 at Danger the Bedwyn train was diverted onto the Up Main and at 08:06 came into an almost head-on collision with the Cheltenham train at a combined speed of about 130 mph (210 km/h). The leading car of the Bedwyn train was almost destroyed, while diesel fuel from its ruptured fuel tank started a fire that particularly affected coach H of the HST, immediately behind the leading power car. Driver Hodder, Driver Cooper, 23 passengers on the Bedwyn train and six passengers on the Cheltenham train were killed or fatally injured, while 417 people were injured, some very seriously. [291]

'Passengers had been folding away their newspapers, prompted by the familiar landmarks, as the 6.03 Great Western express sped towards journey's end. Mark Dwyer stood to retrieve his briefcase from the overhead rack and noticed the giant clock on the Pall Mall Deposit building was nudging towards 8.10. A moment later he was pitched headlong along the aisle as a colossal explosion reverberated through the carriage which, he remembers, seemed to rear up in the air. "The compartment thumped back on to the ground and toppled on its side and everyone was tumbling over one another," he said. "A seat came flying at me. There was this deafening sound of metal tearing. People were screaming and the train seemed to roll on for ever."

'The 24-year-old accountant, his face pitted with glass and blood, struggled to his feet as a tongue of flame flicked past the window of the carriage, which lay on its side against the grass verge of the track. Trails of acrid smoke began filtering through the carriage as he joined others, lying on his back trying to kick-out side windows to escape. "I couldn't even tell which way up we were. You couldn't see anything through the smoke, but it was getting hotter inside and people began to panic." A hundred yards away, supermarket staff, who had just opened the West London Sainsbury's store that sits above this busy stretch of track at Ladbroke Grove, shuddered with the force of the collision, fearing their petrol station had exploded. Those who rushed outside described how the bright autumn sunshine was eclipsed by rolling clouds of black smoke from the railway line.

'Dazed and bloodied, the first survivors to clamber through a gash ripped in the roof of one compartment stumbled alongside twisted wreckage trying to make sense of what had happened. There were shouts for them to avoid touching power lines draped across the carriages, some of which were now engulfed in orange flame. Small bonfires erupted around their feet as pools of diesel oil ignited.

[291] The Ladbroke Grove Rail Inquiry

'It took several minutes for Fiona Whiton to realise that another train had buried itself into the side of the express. The secretary, 23, begins to cry as she recalls seeing a woman's face pressed against a window of the Thames Turbo that had just set off for Great Bedwyn. "Her hair was on fire. I couldn't make out what she was saying. Her face was contorted in pain and she beat on the glass with her fists. I couldn't do anything to help. Someone shoved a coat over her head to douse the flames. The smoke was getting thicker and I could hardly breathe and didn't know which way to turn. I was afraid of being run down by another train." Only then did she notice that her fawn jacket was dripping with blood from an 8 in. cut along her right arm.

'Walking wounded, faces blackened with soot, were led up a steep incline to the supermarket forecourt where they sat at picnic benches and on blue metal chairs. Some were led into the cafeteria as boxes of tissues, kitchen rolls and first-aid kits were grabbed off shelves. Many were shaking so badly that they could barely grip offered plastic cups of sweet tea.

'Some were too shocked to speak. Others swopped stories of their escape and those with mobile telephones handed them around so that survivors could contact loved ones or explain to office managers why they wouldn't be in. There were so many that police had to call for double-decker buses to ferry them to hospital. Those on the upper deck could look down on to the carnage they had escaped. Some sobbed silently as they gazed at the mangled remains.' [292]

The accident was the subject of a public inquiry by Lord Cullen, the report of which stretched to 298 pages.

With the death of Driver Hodder in the collision, the key witness who could have perhaps explained how the accident had happened was of course unavailable, though a number of important points were evident. Firstly, he had had very little experience as a driver, having only qualified as such a fortnight previously. An assessment of his situation-handling skills had not been made during his training, nor had he been warned about previous incidents when signal SN 109 had been passed at Danger (so-called SPADs in modern terminology) – eight times in six years.

On the morning of the disaster it had been a fine day and the low sun had reflected off the signals. The driver of a previous train that morning stated that "all the signals right the way across were lit up like a Christmas tree" on gantry 8, that carrying signal SN 109. It seemed that Driver Hodder had been deceived by a reflection of the sunlight on the Yellow aspect of signal SN 109 into believing that it was showing Yellow rather than Red and that he could therefore proceed.

The line had been re-signalled in the early 1990s for bi-directional working, with the signals placed on gantries over the lines because of the limited space between the tracks. So that drivers could see the top, proceed, aspects sooner, the signals, rather than having their four aspects displayed vertically in the normal fashion (Yellow over Green over Yellow over Red) had been modified into a 'reverse L' formation, with the Red aspect to the left of the other three aspects. In addition, the overhead electrification of the line in 1994 for the new Heathrow Express service had left further obscuring of the signals and the Red aspect of signal SN 109 would have been the last signal aspect on gantry 8 to have become visible to Driver Hodder. The inquiry noted that new or altered signals and those with multiple SPADs should have been reviewed by a signal-sighting committee but this had not been done. Four separate groups had been set up to deal with the problem of SPADs, with overlapping functions, but they had not achieved very much.

The inquiry was also critical of the training Driver Hodder had received, the fact that he had received no warning regarding SPADs and the shortness of his training period, plus the fact that only 13 days after qualifying he had been permitted to operate a train on the

[292] *The Times*, London, England, 6th October 1999

notoriously difficult approach to Paddington, whereas under the British Railways training regime, after qualifying he would have been only permitted to drive trains on less complex routes for the first two years. Consideration had also not been given to the fact that Driver Hodder had had no previous railway experience.

The inquiry noted that although the Bedwyn train had been equipped with AWS, the alerts it had given could be cancelled by the driver. Had an Automatic Train Protection (ATP) system been fitted and working, it would have applied the brakes automatically and stopped the train. National introduction of ATP had been recommended after the 1988 Clapham disaster but later had been abandoned because the safety benefits had been considered insufficient to justify the cost. The Cullen inquiry agreed that a cost-benefit analysis could not support the introduction of ATP.

There had been no flank protection provided at Ladbroke Grove Junction. Had there been, the Bedwyn train would have been routed onto the Down Relief line rather than the Up Main line. Flank protection had probably not been provided because at the time that the layout had been re-modelled, the installation of ATP was thought to be imminent.

The inquiry noted that as soon as Signalmen Dave Allen, James Hillman and Tom Siddell, on duty in Slough IECC, had realised that the Bedwyn train had overrun signal SN 109, they should have sent an "emergency all stop" signal to the drivers of all trains via cab secure radio (CSR). Such a signal had been sent, but only after the Bedwyn train had been about 200 yards past the signal and possibly not before the actual collision. The signalmen, supported by Signalling Manager Heinz Winters, stated that their understanding was that they should first wait to see if the driver was stopping of his own accord before resorting to the "emergency all stop signal." The signalmen had not been trained in the use of CSR nor had they ever used it in response to a SPAD. Lord Cullen was not impressed, stating that: "The evidence strongly indicated that, despite all the written instructions, SPADs were regarded by signallers as a matter of driver error. That showed not only a dangerous complacency but a lack of collaboration in the management of safety. The evidence of Mr Winters and Mr Siddell in particular suggested that signallers at the Slough IECC did not fully regard themselves as a line of defence against the consequences of a SPAD, which could be fatal."

He added: "While I have expressed criticisms of the reaction to the SPAD on 5 October 1999 the root causes of the approach which Mr Allen took, and of his understanding of what he was expected to do, lay in the way in which the operation of the control centre was managed, including the instruction and training of signallers. It is surprising that the deficiencies to which I have referred were not picked up by senior management."

The Cullen report made no less than 89 recommendations:

(1) The system for the reception of information about missing persons, casualties and survivors should be computerised. It should be possible for information received to be entered directly into the computer and for information from it to be provided, to the extent appropriate, to callers. There should be a set procedure for the returning of a call.

(2) Computerisation should be extended to all police forces, so that the information collated by each was readily available to all others.

(3) The police service, in co-operation with the emergency services, should use their best endeavours to ensure that common telephone numbers were issued for the use of members of the public who were seeking to give or obtain information about persons who had, or might have, been involved in a major incident.

(4) The Railway Group should review emergency planning, including liaison with the emergency services, arrangements for the aftercare of survivors and the provision of support and facilities for the bereaved and injured.

(5) Where a material change to track or signalling or both was proposed, there should be an express consideration of all relevant safety issues by an analysis of the material factors, if necessary by means of a risk assessment. This should be done on a holistic basis at the design concept stage and repeated at defined stages up to and including full implementation.

(6) Railtrack procedures, and the actions of management to enforce them, should be directed to ensuring that:

(a) A recommendation that was accepted was implemented according to a defined timescale.

(b) The person to whom a recommendation was allocated for implementation was required to report periodically the action which had been taken, the state of progress and the reasons for any delay.

(c) The monitoring of the implementation of a recommendation was assigned to an identified individual whose duties were clearly defined, whether by job description, formal instruction, or training or a combination of these methods.

(d) The person to whom monitoring was assigned was required to ensure that the recommendation was implemented according to a defined timescale.

(e) A recommendation should not be abandoned unless, exceptionally, this was shown to be fully justified to the person to whom monitoring was assigned.

(f) Any management system to which the recommendation related was altered to align it with the recommendation.

(g) The effectiveness of a recommendation was audited after its implementation.

(h) Full records were kept of all recommendations and their state of progress.

(i) There was a system for the central tracking of recommendations which were directed to Railtrack Line and those which, either immediately or thereafter, were directed to one or more of the Zones.

(7) Consideration should be given to extending sub-para (i) of Recommendation 6 to recommendations which were directed to one or more of the Train Operating Companies (TOCs) and others.

(8) Railtrack should ensure that the risk assessments and any consequent actions required under Group Standard GK/RT 0078 in respect of the signals in the Paddington area were carried out as soon as possible.

(9) Railtrack should conduct a safety examination of the layout over 0-2 miles from Paddington Station so as to satisfy Her Majesty's Railway Inspectorate (HMRI), if necessary by a risk assessment and additional measures, that it was safe for operation at current speeds and to current traffic arrangements. Such a safety examination should be repeated before the implementation of any change which was or might constitute, in the opinion of the HMRI, a material change of circumstances.

(10) No change should be made in the direction of running on line 3 or in the current speed limits on any of the lines out to two miles six chains from Paddington Station unless and until the following had been done to the satisfaction of the HMRI, namely:

(a) A risk assessment had demonstrated that the change could be implemented in safety, and, if this could be achieved only if certain measures were taken, what these measures were.

(b) Such measures had been implemented and shown to be effective.

The risk assessment should take account of the following possible measures, inter alia:

(a) The conversion of four-aspect to three-aspect signals.

(b) The addition of flank protection at SN109 and elsewhere if appropriate.

(c) The installation of standard, simple, non-distracting and consistent means of line identification.

(d) The alteration of the height, configuration and mounting of signals.

(e) The installation of an additional gantry to the east of Portobello Bridge for carrying Down signals previously carried on gantry 8. The risk assessment should be carried out by persons independent of Railtrack and in accordance with usual standards and the best available methods. It should take account of human factors that might affect the actions of drivers and signallers, and any risks which the carrying out of any of these measures might create.

(11) Signallers and drivers should jointly attend away days and other training processes to develop their mutual understanding.

(12) Thames Trains should increase the frequency of the briefing of drivers with a view to ensuring that each driver had a face-to-face meeting with his or her driver standards manager at least monthly, if not more often, and safety should be the first item on the agenda of these meetings.

(13) The adoption by TOCs of the teaching and practice of defensive driving was endorsed.

(14) The TOCs should review the effectiveness of the systems in place to deliver the required level of driver competence at least once every three years and should retest the driver against the revised systems at the same frequency.

(15) The Association of Train Operating Companies (ATOC) study on the central licensing of drivers should be progressed expeditiously.

(16) ATOC should consider the application of National Vocational Qualifications (NVQs) to the driver licensing scheme presently under their consideration.

(17) The development of a culture within the industry in which information was communicated without fear of recrimination, and blame was attached only where this was justified, was commended.

(18) Thames Trains and other TOCs should ensure that their driver training and testing programmes adequately reflected the need for specific, relevant and validated criteria. Drivers should be tested against these criteria, and a definite pass standard should be established. Consideration should be given as to how often drivers should repeat key steps in their training before submitting themselves for testing.

(19) Further research should be carried out to develop the understanding of human factors as they related to train driving.

(20) The safety audit process should be strengthened, and the quality of communication during the process should be improved.

(21) An organisation whose activities were being audited should disclose all material and relevant information to the auditor in regard to the area of the activity which was being audited.

(22) The standard on signal sighting should require that explicit consideration was to be given to the readability of a signal. It should be made clear that the fact that a signal complied with a minimum requirement was not of itself to be taken as meaning that it was adequate.

(23) The standard on signal sighting should deal explicitly with the additional time required

for the reading of certain signals, including (but not necessarily limited to) those mounted on gantries.

(24) Human factors experts should be involved in the revision of the standard on signal sighting.

(25) The reference to "very short duration" in the standard on signal sighting should be clarified.

(26) Areas where ambiguity in the meaning of "very short duration" might have caused, or might still cause, problems should be identified. There should be a retrospective review of all locations where this might be the case, so that appropriate action might be taken.

(27) The expression "overhead line equipment" in the Group Standard on signal sighting should be clarified by the statement that it referred only to wires and droppers.

(28) The standard on signal sighting should define acceptable limits to the temporary obscuration of a signal, subject to the overriding right of a signal sighting committee to determine whether the nature and extent of the interruption in the individual case was such that the sighting was unacceptable.

(29) The standard on signal sighting should explicitly define the cab sight lines within which signals must be positioned by reference to the envelope governing the position of the driver's eye, which was specified for each particular rolling stock.

(30) The report by W.S. Atkins, "Initial Study of Signal Sighting Practice on Railtrack Infrastructure", Issue 1, 6 March 2000, was commended.

(31) Railtrack, in consultation with the TOCs, should examine the availability of signal sighters to meet the expected workload and take all necessary steps to ensure an adequate supply of trained signal sighters and an adequate range of skills.

(32) It should form part of Railtrack's safety management system that it was the responsibility of senior Zone operating and signal engineering management to decide whether the recommendations of a signal sighting committee under the Group Standard on SPADs were to be implemented and, if not, what alternative measures were to be taken, and, in either event, that the relevant measures were implemented.

(33) The Group Standard on SPADs and its associated documentation should be reviewed to ensure that there was no presumption that driver error was the sole or principal cause, or that any part played by the infrastructure was only a contributory factor.

(34) The use of the word "disregard" in the Group Standard on SPADs and its associated documentation should be reconsidered.

(35) Persons who investigated, and made recommendations as a consequence of, SPADs should be trained in the identification of human factors and in root cause analysis. Their competence in these areas should be formally recorded and renewed by refresher courses. The analysis of SPAD data should be specifically directed to eliciting the part played by human factors and assessing the significance of the hazards against which the signals which had been passed at Danger were intended to afford protection.

(36) The instructions for signallers as to their response to a SPAD should be:

(a) Clarified.

(b) Set out in a single set of instructions, while if there were matters specific to a particular area, they should be covered by separate local instructions.

(37) The instructions for signallers should state explicitly that the signaller was expected, in the event of a SPAD, to make an assessment and to take action immediately.

(38) The instructions for signallers should provide a set of options, including the use of the CSR (where it was available) either to send an emergency stop message to a particular

train or a general stop message. This range of options should be supported by full and regularly repeated briefing as to the type of circumstances in which each option was or might be appropriate.

(39) Railtrack should institute a system whereby all signallers in the signal box (or centre) were briefed by their line manager following a SPAD in their area, and there was appropriate dissemination of information which might be of assistance to signallers elsewhere.

(40) Railtrack should ensure that the reports which were made to the Zone about a SPAD should include a report by the signaller as to the actions taken by him or her and the reasons for such actions.

(41) The use of simulators in providing fully effective training of signallers in dealing with emergencies was endorsed.

(42) Railtrack and the TOCs should take steps to ensure that signallers and drivers obtained a full appreciation of the nature and demands of each other's work.

(43) Railtrack should review the work done by signallers to identify all non-essential tasks and eliminate them from the work performed by them while they were in charge of a workstation.

(44) A supervisor should be employed on a continual basis to ensure that the workstations were operated in the most effective way.

(45) Signallers should take the opportunity from time to time to practise the controlling of train movements.

(46) Railtrack management should set out the criteria for allowing signallers, in exceptional circumstances, to exceed the maximum of 72 hours of work per week, and ensure that these criteria were, and continued to be, correctly applied.

(47) There should be a unique alarm for SPADs, which should sound until it was turned off.

(48) The speed with which signallers could take action to move points in an emergency should be improved.

(49) There should be a study of the possibility of the automatic replacement of a signal to Danger where a SPAD has occurred and the layout was such that there was a significant danger of collision.

(50) Subject to satisfactory risk assessment, an arrangement should be made whereby, when a train which was fitted with the CSR passed a signal at Danger, an audible warning automatically sounded in the cab.

(51) There should be a national system of direct radio communication between trains and signallers.

(52) Signallers, managers and maintenance staff working at IECCs should be instructed as to the need to preserve CSR data disks in the event of a SPAD taking place.

(53) The enhancement of the cabs on HSTs to improve driver protection along with energy absorption and compatibility with other vehicles, and the enhancement of measures for the retention of bogies on the coaches of HSTs, should be considered, subject to an assessment of feasibility, costs and benefits, with a view to possible retrofitting.

(54) The current standard for crashworthiness in respect of new vehicles should be reviewed in the light of the crash at Ladbroke Grove with respect to the objectives referred to in Recommendation 53.

(55) In the case of Turbos, the enhancement of end pillar weld connections, the possible

enhancement of crashworthiness by weakening the ends and strengthening the saloon of the cars, and the fitting of shear-out couplers and anti-overriding devices should be considered, subject to an assessment of feasibility, costs and benefits, with a view to possible retrofitting.

(56) The current standard for crashworthiness should be reviewed, in light of the crash at Ladbroke Grove, in order to ensure that there are adequate measures for safeguarding survival space.

(57) In the case of new vehicles constructed of aluminium, consideration should be given to:

(a) The use of alternatives to fusion welding.

(b) The use of improved grades of aluminium, which were less susceptible to fusion weld weakening.

(c) The further development of analytical techniques.

(58) The revision of the Group Standard for crashworthiness should be pursued with particular reference to:

(a) The design requirements for more realistic scenarios.

(b) High-speed accidents.

(c) Dynamic verification testing.

(59) The enhancement of the security of seating in Turbos and of tables in HSTs should be considered, subject to an assessment of feasibility, costs and benefits, with a view to possible retrofitting.

(60) Comprehensive market research in regard to safety-related measures should be carried out in order to take account of the views of informed passengers.

(61) The following measures should be considered with a view to enhancing protection against fire:

(a) A review of Group Standards in respect of improved crash resistance of fuel tanks.

(b) Consideration of the feasibility of reducing fuel inventories and of utilising smaller fuel tanks.

(c) In respect of frontal impacts, consideration of the repositioning of fuel tanks away from the leading ends of trains from behind bogies wherever this is practicable.

(d) Avoidance of placing fuel tanks in exposed and vulnerable locations.

(e) Examination of the use of additives to reduce the propensity of a fuel to atomise.

(f) The employment within fuel tanks of internal flexible linings or a honeycomb construction.

(g) Consideration of the most appropriate material for fuel tanks.

(h) Recognition of the need for supporting theoretical and experimental work in respect of the foregoing.

(62) The scope of Schedule 1 to the Railway (Safety Case) Regulations 2000 should be extended so as to include explicitly the arrangements which the duty holder had established in regard to facilities, instructions and signs for the escape of persons in an emergency.

(63) The provisions in the schedule as to evacuation and escape should be supported by adequate guidance from the Health and Safety Executive (HSE).

(64) The code of practice on public information on train safety and emergencies should be

kept up to date.

(65) So far as was feasible, the safety information issued to passengers and the means by which they could be evacuated or escape from a train should be standardised.

(66) A system should be established for the collection of human factors information pertinent to issues of passenger safety following rail accidents.

(67) Passengers should be given general safety advice both before and after they had boarded their train.

(68) Expert assistance should be obtained on the advice that should be given to passengers as to what to do in the event of there being a known threat of serious danger to them in remaining on board.

(69) The provision on board of explanatory information about the emergency facilities of individual trains was endorsed.

(70) The use of on-board announcements to draw attention to safety information was endorsed.

(71) The requirement for emergency signs to be luminous should be made retrospective.

(72) So far as is feasible, emergency signs on all trains should be capable of being understood by passengers without the necessity to read text.

(73) There should be research with the aim of arriving at a system of signage which is common to all trains in Great Britain.

(74) Research should be carried out into the means of safeguarding emergency lighting systems from disablement by the forces involved in sudden deceleration.

(75) The provision of "snap wands" should be considered as a supplementary means of providing lighting in an emergency.

(76) In the case of every coach (on any train) with internal doors that slid in the same direction, one of the following should be carried out by 31 December 2003:

(a) The coach should be fitted instead with opposite-handed internal doors.

(b) The coach should be fitted instead with double-leaf internal doors.

(c) A panel in the door should be rendered removable so as to enable passengers to pass through. The above was subject to the proviso that if the HMRI were satisfied, on application by the TOC concerned, that it was not practicable for that change to be achieved within this period, they might grant a deferment for an appropriate period in which the work was to be done.

(77) The staff-only doors on all trains should have an override device to enable them to be used by passengers in an emergency.

(78) Signage primarily in the form of pictograms similar to those used on aircraft, and depicting the correct operation of emergency door mechanisms, should be developed. The signage should conform to current human factors standards on signage and be displayed prominently adjacent to each door and beside the door release mechanisms, as well as within the carriage. The mechanisms should be provided with artificial illumination to highlight their location at all times, with a backup power supply in case of an emergency.

(79) The daily routine check of every train should include confirming that all ladders could readily be used. A mechanism to enable ladders to be released quickly should be devised and fitted.

(80) There should be a thorough review of the adequacy of the number of, and signage relating to, emergency hammers. This should include the provision of means of

illuminating the location of hammers in an emergency, with a backup power supply in case of emergency.

(81) There should be research into the feasibility of, and risks associated with, removable windows, the adequacy of windows as a means of emergency egress, the number of dedicated windows which were necessary and the provision as to the maximum distance between each passenger and a bodyside door or emergency exit.

(82) Tests should be carried out into the practicability of building emergency hammers into the passenger alarm system so that they could be released only after an alarm had been activated.

(83) The incorporation of escape hatches in existing carriages should be the subject of feasibility and risk assessment and the provision of escape hatches in new carriages should likewise be considered.

(84) All members of the on-board train staff (including persons working under contract) should be persons who had been trained in train evacuation and protection.

(85) The possibility of installing on driver-only trains a telephone by which passengers could communicate with the signaller in the event of the driver being killed or incapacitated should be studied.

(86) The feasibility of a "roaming" communication system for train staff should be examined.

(87) The possibility of remote broadcasting from outside the train, where it was not already available, should be investigated.

(88) The availability on trains carrying passengers of the items of emergency equipment mentioned in the standard on emergency and safety equipment should be unrestricted.

(89) A review of compliance with the above recommendations should be conducted on behalf of the Health and Safety Commission (HSC) within six months of publication of this report, and further reviews should be put in hand as necessary thereafter. The HSC should publish the outcome of such reviews. [293]

A Rangoon to Mandalay express travelling at 80 km/h collided head-on with a Pegu local train near Ywathagyi, Myanmar, at about 22:00 on Friday 17th December 1999. Between 70 and 100 people were believed to have died in the accident.[294] The location of the accident was also given as having been near the station at Toegyaung Galay, 11 kilometres from Rangoon. The overnight express was reportedly stopped waiting for a signal to change. Several coaches were derailed as a result of the collision. News of the accident was suppressed by the State Peace and Development Council, the military government at the time, and it was not reported by state media. [295]

Eleven people died and 32 were seriously injured when passenger and freight trains collided at Hinda, Republic of Congo, 10 kilometres north of Pointe Noire, on the Chemin de fer Congo – Océan, on Wednesday 12th April 2000. An incorrectly placed set of points had diverted one of the trains onto the wrong line. Most of the victims were traders travelling from Pointe Noire to Dolisie.

Train services over the line had only been partly resumed in January after having been

[293] The Ladbroke Grove Rail Inquiry

[294] *BBC World News*, London, England, 22nd December 1999

[295] https://www2.irrawaddy.com/article.php?art_id=1730&page=17

suspended for most of the previous year due to an ongoing civil war. They were due to be fully resumed later in April and it was feared that the resumption might be delayed due to the accident. [296] [297]

Two trains collided head-on at Mekar Sari, Indonesia, about 35 kilometres south-west of Jakarta, on Monday 1st May 2000. Hundreds of locals rushed to the scene to assist in the rescue operation, while less usefully, a nearby toll road was blocked when numerous car drivers stopped and got out of their vehicles to gaze at the crumpled engines and carriages. Four passengers were killed more or less instantly and a fifth died soon afterwards. Another 21 passengers were injured.

Railway spokesman Zainal Abidin stated that, "We are still investigating the cause of the accident but there is a possibility that a malfunctioning signal light caused it."

Three people had died in an accident a fortnight earlier on 18th April. [298]

At least eight people died and 22 were injured in a collision early on Friday 2nd March 2001 in the station at Nacavala, Mozambique. Two goods trains were involved, one of which was stationary when the other crashed into it head-on after being diverted onto the wrong line. The victims were apparently unauthorised passengers who were struck or crushed by falling containers as a result of the collision. As well as the dead and injured there was heavy damage to equipment, especially to the stationary locomotive, which was derailed.[299] Most of the unauthorised passengers, who would have been riding with the complicity of the train crew, would have been street sellers who rode the freight trains with their goods so as to keep prices low.

An inquiry into the accident found the crews of both trains responsible for it, that of the Nampula to Nacala for approaching the crossing place at excessive speed and that of the stationary train for failing to reverse the points so as to direct the approaching train onto the other track. [300]

A serious collision occurred at Pécrot, Belgium, on the morning of Tuesday 27th March 2001. The accident occurred on line No. 139, a double-track electrified route between Leuven and Ottignies, used by local and long-distance passenger trains and international freight trains. Single-line working, with trains running in what was normally the wrong direction on the other running line, was quite common in this section, especially at weekends.

The accident occurred between the stations of Wavre, in the Walloon part of Belgium, where the population speak either French or Walloon (a language similar to French) and Leuven, where the population speak Flemish, a language very close to Dutch. As will be seen below, this was to be a factor in the accident. One of the trains involved in the accident was a two-car electric multiple unit (EMU), which had formed train No. 2358, the 08:07 passenger working from Leuven to Wavre and was then booked to return as an empty coaching stock (ECS) to Leuven. The other train was No. 9458, the 08:32 passenger train from Leuven to Louvain-la-Neuve, formed of two two-car EMUs coupled together.

About 08:35, a freight train left Wavre for Leuven on the correct running line. Then at

---

[296] *BBC World News*, London, England, 13th April 2000

[297] https://allafrica.com/stories/200004130041.html

[298] *The Janesville Gazette*, Janesville, Wisconsin, United States, 1st May 2000

[299] *Panafrican News Agency*, 2nd March 2001

[300] AIM News headlines, 6th March 2001

08:41, the Wavre signalman noticed that the ECS had departed northwards from track No. 4 towards Leuven on the wrong line and against the Starting signal at Red. The Wavre signalman immediately attempted to contact his counterpart in Leuven to warn him and hopefully have the passenger train stopped, but here the difficulty arose that the former signalman only spoke French and the latter only spoke Flemish, so there was difficulty in getting the message understood. At this time, at any rate, there was no requirement for Belgian railwaymen to be bilingual. The situation was thus different from that pertaining in Switzerland, where four languages are spoken (French, German, Italian and Romansh) and where railwaymen were required to be able to converse in at least two of them. How much this difficulty in communication played a part in the accident, it is hard to be sure. It was highlighted in Press reports on the accident but if the passenger train had left Leuven on time it would have done so nine minutes before the Leuven signalman received the warning anyway; too late, it would seem, for him to have done anything.

At 08:43 the Wavre signalman attempted to have the current switched off, which, both trains being electrically powered, would have saved the situation. Here, another complication arose, for the Wavre signalman was unable to contact the electrical control centre at Namur directly but had to do so via Brussels, causing a further fatal delay.

According to some reports Brussels control also attempted, unsuccessfully, to warn the drivers of the two trains by radio but was unable to do so because of poor reception, due to the hilly terrain and lineside vegetation. Another report states, however, that track-to-train radio had not been installed at the time of the accident. Also lacking at this date was an automatic train stop at the Starting signal; this system would not begin to be installed until 2006.

The young driver of the ECS had only been operating trains since January and had already featured in an incident in which he had overrun a Stop signal at Danger, for which offence he had been suspended for two weeks. There was also evidence that he had stopped his train alongside or even slightly in advance of the Starter when his train had arrived at Wavre and that he might have then inadvertently set off upon receiving the tip from his conductor that the station work was complete. The conductors of the ECS and another train had also, without giving notice, swopped duties, which might have been another factor in the accident.

Having wrongly left Wavre, the ECS had reached a speed of 85 km/h at the moment of the collision, while the passenger train was moving at 83 km/h. The ECS had run 8 kilometres to the point where the trains met, passing over several level crossings, the barriers of which were not lowered. The current was reportedly finally cut off at 08:47 and the trains collided at 08:49, the front car of the ECS mounting and destroying the leading car of the passenger train. Both drivers, one conductor and five passengers lost their lives and 12 people were injured. [301]

Anne-Marie Choumuerthe witnessed the collision from her home in Pécrot. She related, "I heard a train braking and then saw them one on top of the other. There was a noise like two tin cans crashing together. I went out to see what had happened. I heard and saw people at the carriage windows screaming."

Another villager, Bernadette Kounen, had also heard a loud bang and seen clouds of black smoke. She said on Belgian radio, "I went out to see where this could be coming from because I thought that it was a gas explosion, but when I saw the trains, which were tangled up, it was awful. They {the passengers} were shrieking, that was terrible. Some people managed to get out but they were in very bad shape."

Ameile Jassont lived in a house close to the track. She said, "I saw a black cloud of smoke rise up. The train was wobbling back and forth. It all happened so fast."

---

[301] *Today's Railways*, No. 66 (June 2001), pg. 7

A teenage passenger on the train had not at first realised there had been a collision. "I thought that it was just a derailment," she explained. "It was only when I got out that I saw that the train had been torn apart. I could hear a woman screaming. People were trying to open the windows with hammers, which did not work. It was only when I got out that I became aware of what had really happened. I had a very narrow escape." [302]

A court case subsequently followed, in September 2004, in which it was accepted by the national railway company, the Société nationale des chemins de fer belges, that the accident had resulted from human error on the part of the driver of the ECS. Illogically. It would seem Judge Phillipe Ridelle cleared the deceased driver of criminal responsibility, even though it was his mistake that had been the principal cause of the accident, while fining the railway company 100,000 euros, which sum it was stated would be invested in increased safety measures. [303] [304]

At 03:45 on Sunday 2nd September 2001 the engine of the *Cirebon Ekspres* was being attached to its train in Cirebon Kejaksan station, West Java, Indonesia, and was foul of the incoming line from Jakarta when it was struck by the incoming *Empu Jaya* train from Jakarta to Yogyakarta, carrying 400 passengers. The force of the impact pushed the engine of the *Cirebon Ekpres* back some metres and it then overturned, blocking the running lines.

At least 41 people died in the collision and another 62 were hurt. Driver Suwanto and Assistant Driver Sudarto of the *Empu Jaya* train were amongst those who died, while the driver of the *Cirebon Ekpres* was extricated from his crushed locomotive and rushed to Gunung Jati hospital in Cirebon. The injured were distributed between Gunung Jati, Pelabuhan and Ciremai hospitals, all in Cirebon.

One of those who lost their lives was a 29-year-old man named Nuryanto, who was very shortly to be married. His friend, named Sito, explained that Nuryanto had been on his way from Jakarta to Cirebon to complete some paperwork in connection with his forthcoming wedding in Jakarta. His fiancée, named Supuroh, aged 25, who came from Cirebon but lived in Jakarta, was informed of the accident by a phone call from Gunung Jati hospital and made her way there, where she cried, "Where is Mas Nur...where is Mas Nur?" Then she fainted. On recovering she exclaimed repeatedly, "I can't believe it... I can't believe it."

"I can't believe it myself," said Sito.

Hundreds of rescuers worked through the night using hacksaws and ropes to release the injured. Heavy-lifting equipment to shift the two locomotives and two wrecked coaches had still not arrived, reportedly, by 15:00.

Transportation Minister Agum Gumelar made his way to Cirebon upon learning of the accident. He promised that a full investigation would be made into it and estimated that it would cost the railway company 30,000,000,000 rupiah, including the cost of delays to other trains.

Sabotage was ruled out as a cause of the accident. Stationmaster Bijak Filsadjati told a reporter, "The accident happened because the train entered the station without waiting for the signal."

PT Kereta Api Indonesia spokesman Gatot Wibowo likewise stated, "It looks as if it was human error... the *Empu Jaya* ignored signals and came into the station without reducing speed."

---

302 *BBC News* Europe, London, England, 27th March 2001

303 Wikipedia – Pécrot rail crash

304 https://mx-schroeder.medium.com/lacking-linguality-the-2001-p%C3%A9crot-belgium-train-collision-76542028e81d

Bijak added that, "All the trains which pass through the station {about 80 daily} have been diverted via the southern route. The *Cirebon Ekpres* passengers, who could not leave because of the accident, have had their money refunded, including the reservation fees." He said that all the medical expenses of the injured would also be covered.

A post-mortem conducted on the body of Driver Suwanto showed no trace of alcohol or drugs, nor any evidence of a heart attack. That his train had been running late and he had fallen asleep was one possibility initially being considered, given the hour at which the collision occurred. [305]

The true cause of the accident, revealed by the Komite Nasional Keselamatan Transportati accident investigation board, was different. It had not been the result of driver error, as had been thought. Instead, an escape of carbon monoxide (the newspaper report says dioxide but monoxide is more likely) from a faulty exhaust pipe of the diesel locomotive, which was situated directly in front of the driving cab, had caused Driver Suwanto to lose consciousness and very likely had already killed him before the collision. Assistant Driver Sudarto had presumably likewise been poisoned. [306]

Less than four months after the Cirebon accident, at about 05:00 on Christmas Day, Tuesday 25th December 2001, the *Empu Jaya* train was involved in a second major disaster, again a collision, at Ketanggungan Barat station, West Java, Indonesia, not very far away from Cirebon. The *Empu Jaya* train, on its way from Jakarta to Yogyakarta, was in collision with the *Gaya Baru Malam* train, bound for Jakarta. Both trains were packed with passengers. The *Empu Jaya* train, in which most of the casualties occurred, consisted of an engine, 11 coaches and a restaurant car. The *Gaya Baru Malam* train, which must have been a multiple unit, consisted of a power unit that included a passenger portion, nine coaches and a restaurant car.

Again, the number of casualties was heavy, with 45 people losing their lives and dozens injured. Officials stated that a number of passengers had been trapped in the wreckage and smashed windows in order to escape.

Setiyo Raharjo, director of the national search and rescue service, told reporters, "We are bringing steel cutting equipment as many people are trapped inside."

A nurse at the general hospital in the town of Brebes, 20 kilometres away, confessed that, "We are having trouble handling this accident," while Sergeant Sugeng of the Brebes district police said that his men had been helping in the rescue work, as well as investigating the accident. [307]

An investigation into the collision determined that it had been due to the *Empu Jaya* train overrunning signals and thus colliding head-on with the stationary *Gaya Baru Malam* train 147 metres beyond the stop signal that should have protected it. Driver Mohammad Toat, who survived the accident, though he was seriously injured, told the police that the brakes of his train had failed, preventing him from stopping. An examination of the brakes after the accident, however, showed that they were working normally.[308] As the accident had again occurred in the early morning, perhaps this time there was a possibility that Driver Toat had dozed off. A further case of carbon monoxide poisoning was presumably ruled out.

On the evening of Saturday 16th March 2002 train No. 80 Down, running non-stop

[305] *The Jakarta Post*, Jakarta, Indonesia, 3rd September 2001
[306] Ibid, 10th December 2001
[307] *BBC News* Asia Pacific, London, England, 25th December 2001
[308] https://www.liputan6.com/news/read/26384/tabrakan-ka-di-brebes-murni-kesalahan-manusia

between Lahore and Faisalabad, was diverted into a loop line at Qila Sattar Shah station, Pakistan, in order to cross train No. 15 Up, the *Jarnawala Express*, bound for Shorkot. The driver of the Lahore train apparently entered the loop at too high a speed and/or braked too late so that he was unable to pull up before the train rammed the earth barrier at the end trap siding. The engine and leading bogie (carriage) in which there were 44 passengers overturned and another carriage was derailed. [309]

The above initial report on the accident spoke of seven dead and 15 injured, who were taken to hospital in Sheikhupura.

A presumably somewhat later report stated that five coaches had been derailed, two of which were smashed, while the other three were badly damaged, and that the number of dead had increased to ten. [310]

A passenger train from Sylhet to Chittagong collided head-on with a stationary passenger train in Azampur station, Bangladesh, at about 04:30 on Sunday 27th October 2002. Three coaches of the colliding train were derailed and at least eight passengers were killed and 30 were injured. The injured were treated at Akhaura hospital, where doctors said that at least five of the injured were in a critical condition. One doctor told a reporter, "We will try and send some patients to Dhaka as they need specialised treatment and surgery." [311]

Railway spokesman Mostafa-e-Jamil stated, "Primary investigations suggest that the accident occurred as the train carrying about 600 passengers approached the station, ignoring a signal. Four railwaymen have been suspended, including the driver and his two assistants. The accident could have been more serious had the train been moving more quickly. The derailed and damaged coaches have been removed and the tracks reopened to restore communication between Sylhet and Chittagong." [312]

At about 23:00 on Saturday 21st December 2002 two freight (engineering?) trains collided in what was possibly a rear collision at or near Sharatnagar, Bangladesh. One of the trains overturned as a result of the collision and seven railwaymen died and 25 were injured, while the line between Ishwardi and Sirajganj was blocked for seven hours. Both trains were stated to have belonged to Ircon International, an Indian railway construction firm. [313]

At about 03:10 on Thursday 21st April 2005 the Western Railway train No. 9168 Up *Sabarmati Express* from Varanasi to Ahmedabad was diverted onto a loop line and collided with the stationary Up JNPT Conraj goods train on a loop line at Samlaya Junction station, Gujarat, India. Fourteen passengers, Driver Rai Singh and Assistant Driver Kishan Lal were killed[314] and 40 passengers were seriously injured and 110 were slightly hurt. It was quickly apparent that the accident had been due to human error rather than sabotage, General Manager Z.A. Ansari stating, "The accident was avoidable. The signal locking system was disturbed, which led to the tragedy."

Passenger Meenu Singh told a reporter: "We were asleep when we felt heavy jerks two or three times. Soon the coach overturned and people inside started shouting for help and crying.

---

[309] *Dawn*, Karachi, Pakistan, 17th March 2002
[310] *Kuwait News Agency*, Kuwait City, Kuwait, 17th March 2002
[311] *BBC News*, South Asia, London, England, 27th October 2002
[312] *Dawn*, Karachi, Pakistan, 28th October 2002
[313] https://www.voabangla.com/a/a-16-a-2002-12-22-3-at-94338649/1383944.html
[314] www.hinduonnet.com/thehindu/thscrip/print.pl?file=2005042211000100.htm&...

There were some army personnel in the compartment who soon found the gate, which was jammed. The gate was opened and they helped us come out."

Guard Pyarelal Meena of the goods train, a Container Corporation of India (CONCOR) container train bound for the Jawaharlal Nehru port terminal,[315] had a lucky escape. He related, "It was 3.10 a.m. when I saw the Sabarmati Express coming on Track 3 loop line. My train was on Track 4. However when the train came closer, it changed tracks and came on to the one my train was on. I showed the red light, but in vain. I lost all hope that I would survive, I was in a shock." The guard, standing at the door of the van, was thrown 15 or 29 metres by the force of the collision. He continued: "Didn't know where I was. I just realised I was alive when I opened my eyes. My hands and legs were injured and my body was aching." He had then walked to the station to report the accident and was taken to a hospital in Vadodara. "Now, I want to go home and meet my two children and wife," he added. [316]

Gujarat Chief Minister Narendra Modi went to the scene, followed by Railway Minister Lalu Prasad, who later announced compensation of 100,000 rupees for the next of kin of the dead and 15,000 rupees for the injured. The railway minister also visited the injured in hospital at Vadodara, where protesters stoned his car. [317]

A newspaper account written several days after the accident explained how it had occurred. The events leading up to the collision had begun with a problem with the interlocking in the early hours, which the linesman had been called out to attend to, but the proper procedures had not been followed, leading to a situation where the signalman was able to clear his signals for one route while the points were out of correspondence and diverted the train to another. The newspaper wrote: 'Only the signal maintainer is authorised to do any work at the points and that too with the permission of the stationmaster. As per the protocol, in case of a fault in the interlocking, the signal maintainer should approach the stationmaster for permission to work on the points. The stationmaster issues a memo to the signal maintainer and gives him the key to the relay room. Only then can he work on the points or authorise the helper to work at the points. "The stationmaster gave the keys of the relay room without following the procedure. There is no entry to certify that the keys were taken," said {General Manager) Ansari. As per the norms, the stationmaster has to personally inspect the work and can signal for movement of the train only after he is fully satisfied with the job. However, in the case of the 9168 Up local, no such precaution was taken. According to sources, the keys to the relay room were with the helper and he had access to it whenever he wanted. That is why the helper, who is not qualified to work at the points, was at the job.' [318]

Assistant Stationmaster Ramesh K. Vankar, Switchman Jaisingh Lala, in the B cabin Helper Budha Punak and Linesman Balwantsinh Jensinh Baria were held to be directly responsible for the accident and arrested on murder charges according to one report, while six other railwaymen were found to be less directly responsible. The murder charges were later dropped.[319] Another report stated that six railwaymen had been arrested (it states five but lists six names) and while agreeing on the names of the linesman and helper, named the others as Stationmaster Y.S. Wanker, Senior Section Engineers K.S. Gohil and B.R. Meena, and Signal and Telecommunications Engineer A.K. Luna.[320] A third report stated that Station Superintendent I.P. Sharma had been 'awarded' compulsory retirement, as would the other

[315] www.hinduonnet.com/thehindu/holnus/0022005042211757.htm
[316] www.indianexpress.com/print.php?content_id=68925
[317] http://news.webindia123.com/news/printer.asp?id=78717&cat=India
[318] http://cities.expressindia.com/fullstory.php?newsid=126207
[319] http://news.webindia123.com/news/printer.asp?id=183359&cat=India
[320] http://cities.expressindia.com/fullstory.php?newsid=163120

nine railwaymen at the close of departmental proceedings. [321]

The Commission of Railway Safety found that the disaster had resulted from the clearing of the Reception and Dispatch signals without the correct setting of points B9(N) having been ensured and with there having been unauthorised interference with the interlocking circuit. It made 17 recommendations:

(a) The Western Railway should ensure the implementation of the instructions contained in their letter No. SG.2A/21 dated 27.4.2005.

(b) Cards with check list/troubleshooting charts should be prepared for attending to signal failures. Separate cards should be prepared for points, signals, track circuits and so on. Staff should take the relevant cards while proceeding to attend to a failure and should proceed with items sequentially up to their authorisation and then consult the signalling inspector.

(c) It must be ensured that problems due to illiteracy did not hamper the work by keeping maintenance staff in such combinations that at least one person in a group could maintain records and prepare messages.

(d) The railway should issue instructions to signalling staff that the testing of signals should be done by getting physical conditions created and not by interference with relays, circuits and terminals.

(e) The use of wires for testing purposes should be stopped and probes fixed with adequate lengths of wires and jacks should be provided in relay rooms.

(f) It should be ensured that data logger signals were thoroughly checked at the time of installation.

(g) The checking of the springing back of fixed heel switches should be included in the items for checking in the joint inspection of points and crossings by signal and permanent way officials.

(h) Supervisors should be counselled to take note of the use of vague or wrong terms in the working and repeatedly counsel staff to use standard and clearly defined terms.

(i) The railway should ensure that works requiring disconnections were undertaken only after a disconnection message had been issued and the disconnection had been allowed.

(j) The section controller should be debarred from interacting directly with the switchmen in end cabins provided with telephone connections to Control, except during an emergency. During normal work, instructions from Control should only be via the stationmaster.

(k) The Railway should ensure compliance with Subsidiary Rule 5.12(5)(b)(v) by the random checking of records with data loggers and should take disciplinary action against erring staff.

(l) The Railway should ensure compliance with Subsidiary Rule 3.68(1)(c) by the random checking of records with data loggers and should take disciplinary action against erring staff.

(m) The Railway should implement the procedure order contained in their letter No. SG.215/46 Vol. III dated 20.5.2005.

(n) The Western Railway Administration should issue orders for designating authorised officers for issuing and amending Station Working Rules.

(o) The Railway should arrange the inspection of all works, including the change from

---

[321] www.newkerala.com/news.php?action=fullnews&id=54407

mechanical to electrical working of points, commissioned in the last three years, and take corrective action for points where the rod of the switch was projecting more than 50 mm from the switch extension bracket. The Railway should ensure that in future, such works conformed to the Research Design and Standards Organisation's (RDSO's) drawing and specifications.

(p) The Railway should seek the RDSO's decision on the extent of deviation which could be allowed to remain in position of the switch extension bracket in already completed works, from the RDSO's drawing No. SA 9101 and take further action as per RDSO's decision.

(q) The Railway Board might consider directing the RDSO to take up for study the provision of electropneumatic brakes on all types of coaches. [322]

Three trains were involved in a major catastrophe at Sarhad station, Pakistan, seven kilometres from the city of Ghotki, in Sindh province, at about 03:40 on Wednesday 13th July 2005. The events that led to the accident began, according to Sukkur railway Deputy Superintendent Chaudhry Nazir Ahmed, when train No. 24 Down, the *Quetta Express*, from Lahore to Quetta, developed a problem and stopped in the station for attention. While it was stationary there it was run into from behind by train No. 16 Down, the *Karachi Express* from Lahore to Karachi. The signal had been cleared, the deputy superintendent explained, for the *Quetta Express* to depart but had been misread by the driver of the *Karachi Express* as applying to his train. As a result of this first collision a number of carriages were derailed and thrown onto the opposite running line, where they were struck by the *Tezgam Express*, from Karachi to Rawalpindi. It was on the *Tezgam Express* that most of the casualties had occurred. The deputy superintendent stated that Driver Ghulam Mohammad and Fireman Noor Hussein of the *Karachi Express*, and Driver Din Mohammad, Fireman Mohammad Yousuf and Guard Sabir Hussain of the *Quetta Express* had all been killed. The driver, fireman and guard of the *Tezgam Express* had all been injured and the guard of the *Karachi Express* was missing. Another railway official stated that Pointsman Muneer Ahmed should have diverted the *Karachi Express* onto a loop line around the *Quetta Express* but this had not been done. The pointsman also lost his life in the accident, while a gateman was missing.

Dr Ali Nawaz and Dr Sattar Leghari informed a newspaper by telephone that the number of dead in the accident was 164. They stated that the bodies of 126 victims had been brought to Ghotki hospital, four to the Combined Military Hospital (CMH) at Pano Aquil, five to Mirpur Mathelo, five to Sukkur, and 24 to other places. Railway authorities, however, would admit only to 128 fatalities. There had been about 2,000 passengers in all aboard the three trains. The bodies brought to Ghotki hospital were placed in a newly constructed hall that lacked air-conditioning and some also had to be put in tents, while the hospital was running short of medicines to treat the injured.

Railwaymen, police, Army jawans (soldiers) and local residents worked for many hours amongst the wreckage. Villagers had rushed to the scene upon hearing the noise of the double collision, while Major-General Ahmed Nawaz Saleem Mela stated that troops had gone to the scene as soon as news of the accident had been received. They had sealed off the area and Army helicopters had flown the most badly hurt to nearby hospitals and to Multan, while over 100 road ambulances, many of them hospital vehicles, had also been employed. A newspaper reporter who went to the scene wrote, 'The crash site was strewn with decomposed bodies. Rescue workers used electric cutters to pull out bodies and wounded passengers from the smashed carriages. Decomposed parts of the bodies were collected in bags for proper

[322] Commission of Railway Safety Annual Report 2005-06

identification. Thousands of people had gathered at the site, which posed difficulties in the rescue work. Though the army, rangers and police personnel had been deployed, people were still pouring in.'

Passengers Riaz Ahmed and Ramzan, from Bahawalpur; Zahoor Ahmed, from Sahiwal, who was injured; and Farzana Awan described how they had been asleep when he was woken by the sound of a loud explosion, followed by a severe jerk. Luggage had fallen from overhead racks, causing injuries. Then there had been another massive explosion and total chaos reigned. "We heard people screaming for help and saw blood everywhere, then we lost consciousness."

A passenger named Khurram, aged 13, from Wazirabad, who was injured, related that he and his sister had been awake at the time. A younger sister, aged 2, had been sleeping with their mother, Naseem Begum. Khurram said that he and his older sister both fainted when they saw that their mother had been killed. Their father lost both of his legs in the accident.

Another passenger, Mohammad Amin, said, "I woke up at the noise of a huge bang and then there was a big jerk and smoke all over the place. There was total darkness...I hit the floor and fainted."

Uninjured passengers were forwarded by road in air-conditioned buses and coaches while relief trains were sent from Rohri and Khanewal. Twelve coaches of the *Tezgam Express*, four economy-class coaches of the *Quetta Express* and three coaches (two economy and one lower/AC) of the *Karachi Express* were damaged in the accident. [323]

Worried Karachi relatives and friends of people thought to have been on the trains hurried to the station there, to be told that officials had no lists of passengers. Many families hired private cars to go to the scene and look for loved ones, many of whom had been holidaymakers from Karachi. Police official Aga Mohammed Tahir commented, "It's a very gruesome situation," while a witness told a news agency, "It's a painful scene. There are bodies scattered all over. People are crying. Fathers are looking for children, husbands for their wives and brothers for their sisters."

Some were lucky. Waseem Riaz, from Chakwal, wrote, 'Today early in the morning the first news I heard was this railroad accident. My uncle was in the *Karachi Express* with his family. My father called me up and was crying and not even able to talk {about} what happened. The news took the soul out of me, as last Sunday he was here visiting us after several years. Luckily he and his family survived and we were relieved after three hours of the worst ever suffering.'

Others said that they were deterred from travelling by rail in the future. Imran Ashan Mirzal from Lahore wrote, 'I have travelled on this route several times. I am really shocked and horrified. I will certainly not travel on trains any more. I can well imagine that the government will conduct superficial inquiries and after four weeks this accident will be treated as a bygone matter. Pakistan has to learn a lesson with this tragic accident. Well organized and business oriented railways can generate enough money to invest in technological upgrading.' [324]

An inquiry into the catastrophe placed the principal responsibility for it on the deceased Driver Mohammad and Fireman Hussein of the *Karachi Express*, who had overrun the Sarhad Down Outer Home and Home signals at Danger. A contributory factor was that the brakes of two of the coaches of the *Karachi Express* had been isolated, so that it could not pull up as quickly as it should have been able to when its crew had seen the tail lamp of the *Quetta Express* in front of them.

[323] *Dawn*, Karachi, Pakistan, 14th July 2005
[324] *BBC News*, London, England, 19th July 2005

No less than eight other railway officials were also found to have been at fault and in May 2006 it was announced that Sarhad Relieving Stationmaster Iftikhar Hussain, Lahore Train Examiner Zulfikar Ali, Ghotki Signal Maintainer Mavo and Sukkur Section Controller Niaz Ali had been dismissed; Ghotki Signal Sub-engineer Mohammad Sharif had had his contractual appointment terminated; Sukkur Assistant Signal Engineer S. Haider Hussain was still having his case considered; and Rawalpindi Senior Train Examiner Tasleem Khan and Ghotki Signal Inspector Abdul Rashid had had their increments withheld for two years. [325]

[325] *Dawn*, Karachi, Pakistan, 19th May 2006

CHAPTER 6

SINGLE LINE COLLISIONS

The operation of trains in opposite directions over a single line of rails carries the obvious danger of head-on collisions between them. By 1991 there were many hundreds of miles of single-track railways throughout the world, especially in places where the density of traffic did not justify the provision of a second line. In Great Britain and some former colonies such as India, the working of some sections of single line sometimes still required drivers to be in possession of a physical token or equivalent, the tokens being contained in instruments at each end of the section, which were electrically locked with one another so that only one token could be removed at a time. In North America, until recently some lines had still been worked on the train order system, whereby a central dispatcher issued orders to train crews via local operators. More recently in North America and elsewhere a system of Centralised Traffic Control was used on busier lines, again with a central controller but with train movements being controlled by lineside signals, eliminating the need for operators. On quieter lines systems known as Direct Train Control (DTC) and Track Warrant Control (TWC) had been introduced from 1986, where instructions were issued directly either by radio from dispatchers to train crews (DTC) or using pre-printed forms (TWC), again eliminating the need for operators. The track warrant system is described in more detail below in connection with the 1991 Fowler accident and the 1997 Devine accident. On many lines in Europe and elsewhere the crossings of trains were arranged locally by telephone between adjoining stations, with the aid of fixed signals, while countries such as New Zealand had switched, like the United States, to using track warrants. Various other methods of single-line control have also been used, such as tokenless block and on short, dead-end branches 'one engine in steam' working with a train staff.

Collisions during temporary single-line working, where one of a pair of double tracks was out of use, are also included in this chapter.

In what was probably a single-line collision, what were described as two passenger trains, though one at least was a mixed train, collided west of Ciudad Serdán, Mexico, on the Mexico City – Veracruz line of the Ferrocarriles Nacionales de México, before dawn on Tuesday 26th February 1991. At least five people died and over 50 were injured. More serious consequences were avoided because two freight waggons loaded with toilet paper at the head of one of the trains had helped to absorb much of the impact. [326]

In tune with a worldwide trend, Japanese National Railways had been privatised in 1987, after which some unprofitable branch lines had been closed, while others continued to operate

---

[326] *Playground Daily News*, Fort Walton Beach, Florida, United States, 27th February 1991

as so-called third-sector railways, subsidised by local governments. Among the latter was the 14.7-kilometre-long, single-track Shigaraki line from Kibugawa station on the West Japan Railway (JR West) Kusatsu line. The Shigaraki Kogen Railway (SKR), as the branch became known, had a number of intermediate stations that were not crossing places, limiting its capacity to one train hourly in each direction. Normally only SKR passenger trains used the line, shuttling between Kibugawa and Shigaraki, but in May 1991 the World Ceramic Festival was being held at Shigaraki, meaning much-increased traffic over the line and its use by through JR West trains. In order to handle this increased traffic load, an intermediate crossing place was established at Onodani Signal Station, about 9.1 kilometres from Kibugawa.

On Tuesday 14th May 1991 JR West special eastbound express passenger train No. 501D left Kyoto at 09:30 for Shigaraki, five minutes late, and carrying 716 people, though its seating capacity was only 252. At 09:44 the signalling staff at the Kameyama Centralised Traffic Control (CTC) Centre noticed that the Kyoto train was late and activated the Direction Priority Lever (DPL) to place the westbound Starting signal at Shigaraki at Red. The DPL had been installed by JR West in 1990 so that priority could be given to Shigaraki-bound trains, apparently without informing the SKR. Train No. 501D regained some time with short station stops and left Kibugawa only two minutes late at 10:18.

In the opposite direction, train No. 534D was due to depart from Shigaraki at 10:14 but the Starting signal there remained Red because of the operation of the DPL. At 10:24, without communicating with Onodani, the SKR business manager (another source states it was the operations manager) at Shigaraki instructed the motorman of train No. 534D to proceed, despite the Starter remaining at Red. In such a situation a device known as a Faulty Departure Detector (FDD) should have placed the eastbound Starter at Onodani at Red. The same situation had occurred on 8th and 12th April and again on 12th May, when trains had similarly been allowed to leave Shigaraki against the Starter and the FDD had functioned properly and held opposing trains at Onodani. On 14th May, though, the FDD failed to work, and the eastbound Starter at Onodani remained Green. The motorman of train No. 501D was rather surprised that there was no opposing train for him to cross at Onodani, but the Starter was Clear for him, so he set off. The trains collided between Onodani Signal Station and Shigarakikyu-ato (an intermediate station on the branch, not the same place as Shigaraki) at about 10:35. Forty-two people died and 614 were injured. There had only been 15 people on train No. 534D, of whom 12 died, including the business/operations manager who had ordered it to depart. The SKR lost five of its 20 staff in total and two of its four cars; the motorman of the JR West train survived, though he was seriously injured. The ceramic festival was cancelled after the disaster and the line remained closed until 8th December. Onodani Signal Station was abolished and JR West stopped working through trains over the branch.

An investigation determined that two factors had led to the disaster. Firstly, the staff at Shigaraki had switched from the normal Regular Blocking System (relying on the fixed signals) to the Alternative Blocking System (resorting to hand signals) without bothering to inform the staff at Onodani Signal Station.

Secondly, the FDD system had failed to work. The Ministry of Transport Railroad Bureau stated merely that, "It is assumed that the signal circuit was not connected properly at the time, but the cause is undetermined."[327] Another source states that as train No. 534D had left Shigaraki a signal engineer had been inspecting relays, under the supervision of the track and signalling manager, and his interference had caused the failure of the FDD.

After the accident the SKR apologised to each of the families of the deceased, in contrast to JR West, who claimed that they had done nothing wrong.

---

327 https://www.shippai.org/fkd/en/cfen/CA1000607.html

In 1999 Osaka District Court fixed compensation for the victims at 5,500,000,000 yen, originally to be shared equally by the SKR and JR West. JR West appealed the court's decision and eventually the SKR was ordered to pay 70% and JR West 30%.

In March 2000 Otsu District Court found the Shigaraki duty stationmaster, the signal engineer and the track and signalling manager guilty of professional negligence causing death and injury. The sentence of this court is not recorded. [328]

In 2002 JR West was found guilty of negligence in Osaka High Court because they had been aware that SKR staff had previously permitted trains to proceed against Red signals but had taken no measures to stop the practice. [329]

On Friday 30th August 1991 eastbound Burlington Northern Railroad (BN) freight train No. 602 collided head-on with westbound BN freight train No. 603 on the single line at milepost 85.55, between Fowler and Naismith, Montana, United States.

Eastbound train No. 602 (No. 2275 East) was en route from Whitefish, Montana, to Laurel, Montana, via Shelby, Montana, and consisted of three diesel units, 20 loaded and 14 empty cars, in charge of an unnamed conductor with Engineer James Henderson and two unnamed brakemen. At 5:08 p.m. the second-shift branch line dispatcher in Seattle, Washington, who had taken over from the first-shift dispatcher at 3:45 p.m., issued track warrant (TW) No. 8860 by telephone to the conductor of this train, authorising it to proceed from east yard limits at Shelby to Ledger, about 20 miles. Fowler and Naismith are places between Shelby and Ledger.

Westbound train No. 603 (No. 6905 West) had left Great Falls, Montana, at 1:50 p.m. From Conrad, Montana, it consisted of six diesel units, 57 loaded and 28 empty cars, in charge of Conductor Meyers with an unnamed engineer and Brakemen Michael Holst, 39, and Richard Strissel, 23. [330]

In the resulting collision Engineer Henderson of train No. 602 and Brakemen Holst and Strissel of train No. 603 were killed and the other five railwaymen were all more or less injured. The accident was, perhaps, an especially hard blow to farmer Allan Strissel, father of Richard, who had himself lost both his arms trying to unclog a hay baler only two months previously. [331]

The conductor of train No. 602 testified that upon seeing the other train he had shouted a warning and jumped. His shout was credited with saving the lives of the two brakemen of his train, who had been riding in the trailing units. The brakemen described the conductor of the other train, No. 603, as having appeared to be in shock after the collision, while the conductor said the same of the brakemen. Doubtless they all were. The engineer of train No. 603 spoke of applying the brakes, shouting, "Jump," to his conductor and doing so himself. As he crawled away from the train he had seen the "diesel tanks exploding," a smoke cloud and fire.

A rancher herding cattle with his family heard the impact of the collision and going up a hill saw the accident and a man waving his arms for help, whereupon he had returned to his home two miles away and dialled 911 for the emergency services. Other farmers, seeing the smoke and suspecting a farm fire, had also dialled 911. Fire, police and ambulances arrived within 15 minutes.

The National Transportation Safety Board (NTSB), who investigated the accident, noted

[328] https://www.ejrcf.or.jp/jrtr/jrtr33/pdf/f26_edi.pdf

[329] Wikipedia – Shigaraki train disaster

[330] National Transportation Safety Board report

[331] *The Daily Interlake*, Kalispell, Montana, United States, 3rd September 1991

that the track warrant (TW) system for working single lines had been introduced in 1986, replacing, on many lines, the old train order system, whereby a dispatcher issued orders to a series of operators, who in turn gave them to train crews. To establish a meeting point between opposing trains under the TW system, the train dispatcher issued a TW directly to one train usually by radio, which included instructions for the conductors to enter onto a preprinted form, with a box that, at a minimum, stated the "proceed from" and "to" limits being granted and a box that instructed it to "clear main track at last named point". The train travelling in the opposite direction was issued a TW that included, at a minimum, the "proceed from" and "to" limits being granted and an instruction to "hold main track at last named point" (the "to" limit granted). The train instructed to clear the main track at the station where the meet was to take place, was not authorised to proceed on the main track beyond the initial switch (entrance points) of the siding without additional TW authority. The train instructed to hold the main track was not authorised to proceed beyond the switch (exit points) at the end of the "to" siding until additional TW authority was issued. This system protected the two opposing trains, and the physical passing of the trains was accomplished. A computer was used by the dispatcher to generate orders and had an inbuilt logic to prevent the issuing of conflicting or lap orders (an order authorising one train to proceed from A through B and C to D, at the same time as another order authorised an opposing train to proceed from C to B, thus giving the two trains the right to the line between B and C at the same time). What the system did not ensure was that the TWs it produced were identical to those transmitted by the dispatcher to the train crews.

The NTSB pointed out that the transmission of orders under the old train order system had followed a strict ritual, whereby amongst other things station or siding names, train numbers and times were pronounced and spelt out letter by letter; the orders were always issued first to the train whose movement was restricted and were repeated back to the dispatcher by the different operators. No order was "complete" and in effect until each operator had repeated the order back correctly to the dispatcher.

The branch line dispatcher at Seattle, Washington, sat alone in a small, air-conditioned room. Thirty radio channels were available for use, there were six video screens, a printer and three keyboards, one each for computerised track warrant control (CTWC), centralised traffic control (CTC) and COMPASS (a computer inquiry system for train movement information, consist lists and so on). Clocks on the wall showed Pacific and Mountain times. On the day of the accident, first shift branch line Dispatcher Nancy K Jaeger had controlled 33 of the 55 daily trains, issuing three train destination lineups, 15 CTC track and time limits and 28 TWs (seven of which the computer had, however, rejected because of spellings, conflicts or unrecognised directions).

After working as a clerk and train order operator, Dispatcher Jaeger had begun work as a dispatcher at Glendive, Montana, in January 1980, but after only a few days had been overheard by the chief dispatcher to be issuing conflicting orders on an unsignalled single line and had been taken off dispatching, returned to her previous work. She had begun work at Seattle on 1st January 1991 but mostly had been employed as assistant chief dispatcher (an administrative post) and only six days as branch line dispatcher. She was therefore not very familiar with CTWC and said that the job needed a lot of listening and concentration.

At 2:55 p.m. first-shift (early turn) Dispatcher Jaeger issued TW No. 8851 by voice radio to Conductor Meyers of train No. 603 at Dutton, Montana. The destination of the train as set out on the computer was Ledger.

The conductor was not sure what the dispatcher had said and asked, "West yard limits, Shelby?"

"Yes," the dispatcher replied.

Conductor Meyers then repeated, "West yard limits, Shelby," to Dispatcher Jaeger and she "okayed" it, although it read Ledger on her computer. Both trains thus had track warrants for the stretch between Shelby and Ledger.

The radio conversation between Dispatcher Jaeger and Conductor Meyers went as follows:

Dispatcher: *Seattle Branch Dispatcher answering.*

Train No. 603: *Dispatcher. This is 603. We'd like to get a warrant to get out of Dutton.*

Dispatcher: *OK. Let's make it TW 8851, eight-eight-five-one, to 6905, six-nine-naught-five, West, w-e-s-t. You're at Dutton now?*

Train No. 603: *Roger. We're stopped at the east end of Dutton. Or the west end, I mean.*

Dispatcher: *Dutton, Dutton, D-u-t-t-o-n,*

Item number 2, t-w-o. Proceed from Dutton, D-u-t-t-o-n, D-u-t-t-o-n to (nine-second pause) Ledger, L-e-d-g-e-r, on main, m-a-i-n, track.

Item number 7, s-e-v-e-n. Not in effect until arrival of 7825, seven-eight-two-five East, e-a-s-t, at Dutton, D-u-t-t-o-n.

Item number 8, e-i-g-h-t. Hold main track at last named point.

Item number 15, one-five. Protection as prescribed by Rule 99 not required for following trains on main track, and okay to repeat.

Train No. 603: *TW number 8851, dated August 30th 1991, to 6905, six-nine-oh-five, West, w-e-s-t, at Dutton, D-u-t-t-o-n.*

Item number 2, t-w-o. Proceed from Dutton, D-u-t-t-o-n, to west, w-e-s-t, yard limits Shelby, S-h-e-l-b-y, on main track.

Item number 7, s-e-v-e-n. Not in effect until arrival of 7825, seven-eight-two-five East, e-a-s-t, at Dutton, D-u-t-t-o-n.

Item number 8, e-i-g-h-t. Hold main track at last named point.

Item number 15, one-five. Protection as prescribed by Rule 99 not required for following trains on main track.

Dispatcher: *8851 is okay 13, BK, 1455, one-four-five-five, Dispatcher N-K-J.*

Train No. 603: *TW 8851, eight-eight-five-one, is okay at 1455, one-four-five-five, Dispatcher NKJ.*

Dispatcher: *All right, thank you.*

Train No. 7825 East (several diesel units coupled without cars) was already at Dutton when train No. 603 arrived and its engineer heard part of the above radio conversation but not all.

Interviewed in hospital after the accident, Conductor Wesley stated: "So we stopped there and got a track warrant to get out of Dutton and go to Shelby. And we had a little bit of trouble getting the warrant. I could hear him, and I said to the engineer, 'What is it he is copying?' He (the engineer) said, 'I thought he said Shelby.' And I said, 'Well, that's what I thought he said.' I says, 'Oh, it don't matter, because I'm going to have to repeat it back and spell it to him.'"

The conductor contended that ultimately it was the responsibility of the dispatcher that the track warrant was accurately transmitted: "My understanding of it was, though, that the dispatcher was in control of where you had authority to move. They controlled your movement of where the train went. It was my understanding that the dispatcher was in charge of that because they had to okay it. So they had to give you permission to do what you were going to do. They had to okay it and then after they had okayed it, gave you a time. And then after that you had a track warrant."

The train No. 603 engineer agreed, saying, when asked whose responsibility it was that the TW had been transmitted correctly: "I assume the dispatchers since they're giving us the information and then checking to see if we received it properly."

Asked if he relied on the dispatcher to verify the accuracy of the TW he replied: "Yes. She's—or the train dispatcher is putting up the information, has the computer in front of their face. And we're, you know, repeating back what we heard. This happens all the time. If there's a mistake, then the dispatcher tells you what you have incorrect and you correct it. Then you're trying to get an okay after that."

The dispatcher training manager noted 15 errors in the above transmission between the dispatcher and the conductor, most of which concerned the non-use of the words "over" and "out."

The conductor testified that he had expected train No. 602 to be at or near Shelby when he had copied TW 8851 because of an earlier derailment near there. He could not explain why he had associated TW 8851 with Shelby other than he thought he had heard the dispatcher say "west" as in "west yard limits," a description that would apply to Shelby but not to Ledger, which was merely a siding with no yard.

After the accident, an audit of TWs issued on 27th and 28th August disclosed 18 errors in 135 TWs (incomplete line feedback, incomplete dates or times, wrong locomotive unit numbers, wrong or misspelt locations and so on). Eleven of these errors had been corrected by the dispatchers themselves. Of the remaining seven Dispatcher Jaeger, who had been on duty for 9½ of the 48 hours, had made five.

The NTSB analysed the accident, stating that the weather, the equipment, the track, alcohol and drug use had not been factors. It described the circumstances of the accident, the fact that the conductor had been unsure as to what the dispatcher had said and that he had relied on the dispatcher to correct his feedback, if he read it back incorrectly.

The first shift dispatcher had been working her third straight day on the branch line position. Her skill using the CTWC had been improving but she had still been struggling with it. Even for an experienced dispatcher, the branch line job was busy and in addition to having to redo TWs rejected by the computer she had been trying to keep the trains moving. Thus, she had failed to detect the TW error by the conductor of train No. 603.

Given her previous very brief 1980 employment as a dispatcher at Glendive, which had been terminated, and her more recent errors, it seemed likely that she lacked the aptitude to be a dispatcher.

The NTSB identified five errors that might have led to the fatal misunderstanding:

(a) The voice intonation used by the first shift dispatcher.

(b) The accepted procedure of simultaneously formulating and transmitting a TW from the computer screen piecemeal.

(c) Failure to use proper radio procedure.

(d) Failure of the conductor and engineer to request repeat or clarification of the destination, as outlined in BN operating rules, when they had not clearly heard it.

(e) Failure of the first shift dispatcher to recognise and correct the wrong destination during crew repeat back.

The NTSB reached 14 conclusions regarding the accident:

(a) The weather, the train equipment, the track, and alcohol or illicit drug use had neither caused nor contributed to the collision. The operations personnel had been qualified to perform their duties in accordance with BN procedures.

(b) The train No. 603 conductor had relied on the read back rather than asking for a repeat of the track warrant 8851 authority limit.

(c) The first shift dispatcher had failed to detect and correct the train No. 603 conductor repeat of the track warrant 8851 authority limit.

(d) Burlington Northern Railroad had yet to implement an employee selection programme as stated in Safety Recommendation R-83-43, which would ensure that individuals considered for safety-critical positions, such as dispatchers, were capable of fulfilling the position requirements.

(e) The first shift dispatcher's piecemeal method of computerised track warrant formation and transmission, as well as the dispatcher's and the train No. 603 crew's poor radio procedure practices, had not been isolated events. Such practices had been known to exist by Burlington Northern Railroad management and fostered an environment in which the misunderstanding of the track warrant 8851 authority limit occurred.

(f) Had either the Federal Railroad Administration or Burlington Northern Railroad followed up the concerns and recommendations about radio procedures from the National Train Dispatcher Safety Assessment 1987-88, this accident might not have occurred.

(g) In the absence of a clearly understood track warrant 8851 authority limit, the train No. 603 conductor had "heard" what he had expected to hear.

(h) The first shift dispatcher's error in not correcting the train No. 603 conductor's read back had resulted from several problems that had combined to reduce her attention when track warrant 8851 was issued.

(i) The first shift dispatcher had recognised her need for refresher training on the computerised track warrant control system before commencing the branch line position on 27 August; however, Burlington Northern supervisors discouraged such requests. Burlington Northern Railroad had no recurrent computerised track warrant control operating training and radio skills programme for shift dispatchers.

(j) The first shift dispatcher's one comfort break during her shift might not have been sufficient to maintain the required mental ability for a complete shift, and Burlington Northern Railroad had made no effort to periodically schedule relief or extra dispatchers on positions.

(k) Burlington Northern Railroad management had neither provided personnel specifically to relieve, help or monitor dispatcher, nor had it adequately trained supervisors in the use of computer equipment.

(l) Burlington Northern Railroad had an inadequate quality control programme and an inadequate dispatcher training programme.

(m) Had an advanced train control system been involved and working in the accident area, the accident would probably have been prevented.

(n) The development of a practical positive train separation system had not progressed as quickly as it should have.

The NTSB determined that the probable cause of the accident had been the poor communication practices of the first shift dispatcher and the train No. 603 conductor and the Burlington Northern Railroad's failure to establish an adequate management oversight and quality control programme in its train operation. Contributing to the accident had been the combined failure of Burlington Northern Railroad and the Federal Railroad Administration to follow up on the dispatching concerns and recommendations of the Federal Railroad Administration *National Train Dispatcher Safety Assessment 1987-88.*

The NTSB made a series of recommendations:

To the Burlington Northern Railroad:

(a) It should accelerate and implement an objective identification, selection, and screening programme for dispatcher candidates to determine their suitability to successfully perform as dispatchers.

(b) It should require dispatchers to complete formulation and entry of a computer-acceptable track warrant before transmitting it to the train crew.

(c) It should implement a programme to teach and enforce proper radio procedures for dispatchers and train crews so that compliance with applicable Federal and railroad rules would be met.

(d) It should establish within its qualification programme a formal requalification programme for dispatchers and supervisors who intermittently worked such positions.

(e) It should evaluate each dispatcher position to determine the adequate number of breaks necessary to maintain optimum mental ability and provide adequate personnel and equipment resources to adequately monitor, test, evaluate, help and relieve shift dispatchers.

(f) It should establish a dispatcher audit/quality control programme and implement an effective periodic dispatcher testing programme.

To the Federal Railroad Administration:

(a) It should follow up on the concerns and recommendations made to railroads in its safety assessments and request a response to provide closure on the safety problems uncovered.

(b) In conjunction with the Association of American Railroads and the Railway Progress Institute, it should establish a firm timetable including, at a minimum, dates for the final development of Advanced Train Control System hardware, dates for the implementation of a fully developed Advanced Train Control System, and a commitment to a date for having the Advanced Train Control System ready for installation on the general railroad system.

To the Association of American Railroads:

(a) In conjunction with the Federal Railroad Administration and the Railway Progress Institute, it should advise its membership of the facts and circumstances of this accident and encourage them to implement and install an Advanced Train Control System.

To the Railway Progress Institute:

(a) In conjunction with the Federal Railroad Administration and the Association of American Railroads it should establish a firm timetable including, at a minimum, dates for the final development of Advanced Train Control System hardware, dates for the implementation of a fully developed Advanced Train Control System, and a commitment to a date for having the Advanced Train Control System ready for installation on the general railroad system. [332]

At 8 a.m. on Tuesday 17th September 1991 two Norfolk Southern Railway (NS) freight trains collided head-on on the single line just west of Knox, Indiana, United States. One

[332] National Transportation Safety Board report

railwayman was killed in the accident; another was seriously hurt and four were slightly hurt. The accident formed the subject of a report by the National Transportation Safety Board (NTSB). The line was worked by Centralised Traffic Control, supervised by a dispatcher at East Wayne, Indiana.

One of the trains involved in the accident was eastbound No. 629, consisting of three diesel units and 95 cars loaded with molten sulphur, in charge of Conductor Mark McKee with Engineer Eric Goff and Brakeman Leval Barney Jr. This train had left Calumet Yard, Chicago, Illinois, at about 5:20 a.m. destined for Ft Wayne, India. Initially the dispatcher instructed its crew to meet two westbound freight trains at Thomaston Siding; later, on learning of the location and speed of train No. 629, he changed his mind and advanced it another 10 miles to meet the westbound ones at Knox Siding instead, where it was to take the siding. Train No. 629 duly passed Thomaston Siding and was approaching the signal designated in the NTSB report as 'C,' about two miles short of the beginning of Knox Siding, which was displaying Approach Diverging (reduce speed to 25 mph to enter the siding). Engineer Goff told the NTSB inquiry that his train had been travelling at about 46 mph at that point and he had applied the brakes to slow to 25 mph. Then, unexpectedly, the signal had changed to Restricting (reduce speed to 15 mph and be prepared to stop short of any obstruction) and he had increased the braking. Shortly after passing the signal, the headlight of a westbound train had appeared, whereupon all three crew had jumped, escaping with slight injuries. Train No. 629 was travelling at about 25 mph upon collision. [333]

The other train was No. 277. This had left East Wayne Yard, Ft Wayne at 5:22 a.m. for Chicago. It consisted of a locomotive and seven loaded container-on flat cars (COFC) double-stack cars, in charge of Conductor Stephen R. Sustek, 47, with Engineer Richard A. Marks, 40, and Student Engineer P.R. Rice, 30.[334] Part of the way through the journey, and well before Knox Siding, Engineer Marks had handed over operation of the train to Student Engineer Rice, the engineer standing behind the student most of the time thenceforth. Student Engineer Rice stated that he had overheard a conversation between the dispatcher and another train but had not known which train or what was said. Approaching Knox he had reduced speed to 21 mph to comply with the 30 mph restriction through the town, then begun to increase speed again. Signal 'A' (NTSB designation), at the approach to Knox Siding, he had read as clear and he had assumed that Engineer Marks, stood behind him, must also have seen it. He and the engineer had not called out the signal to one another as required by Rule No. 34, though he thought the conductor had, and that the conductor had also remarked, "I guess we'll meet them at Thomaston." Conductor Sustek, who was seriously injured, denied saying this. According to the student Signal 'B' (NTSB designation), at the far end of Knox Siding, had also been clear (displaying Green over Red), though the NTSB determined that it must have been displaying Stop (Red over Red). Again the student engineer and engineer had not called out the signal to one another, though the student engineer stated that the conductor had called it as clear. The student engineer said that the train had been travelling at 35 mph as it had re-entered the single line and he had not noticed that it had run through the points.

Conductor Sustek testified that all the signals before Knox had been called out by the crew. He had not called out either signal at Knox nor made any remark about Thomaston, as at that time he had been in the washroom. When he returned from there they had been back on the single line and shortly afterwards he and the student had seen a light, which at first they had thought was a reflection of the sun on a metal building, then realised was the headlight of an oncoming train. The student engineer had applied the brakes and jumped; he and the engineer had not had time to do likewise and they had sat on the floor and braced themselves (Engineer Marks was killed).

---

333 Ibid.

334 *Anderson Herald-Bulletin*, Anderson, Indiana, United States, 19th September 1991

A track foreman witnessed the crash and notified the dispatcher at 8:05 a.m., when the latter called the police, medical personnel and fire service. When the firemen arrived at 8:20 a.m. they found that the wreckage was on fire and that one train was carrying hazardous materials (burning sulphur generates irritant and corrosive sulphur dioxide gas). Thirty families within a two-mile radius were evacuated from 9:10 a.m., returning in the evening, by which time the fire had been extinguished. The four slightly injured crewmen were quickly taken to hospital but the badly injured conductor was not found until 11:10 a.m. The body of the engineer was recovered after the fire had been put out.

Three diesel units and five tank waggons were destroyed in the collision. Only one event recorder out of four, on the rear unit of train No. 629, survived the collision. The cost of the accident was estimated at $3,476,788.

The NTSB determined that there had been no problems with the track or the traffic control (TC) system, while the TC computer log showed that train No. 277 had passed signal 'B' at Red and had run through the switch (points) at the west end of Knox Siding. There had clearly been a lack of vigilance on the part of the crew of train No. 277 in running past signal 'B'. Also, had they been monitoring radio communications they should have heard the crew of train No. 629 announcing signal 'C' showing Approach Diverging. According to the rules the engineer and conductor were jointly responsible for the safety of the train.

The crew of train No. 629 should have reduced speed to 15 mph upon signal 'C' changing to Restricting.

Had the East Wayne dispatcher not arranged for the trains to meet at Knox Siding, involving a slight delay to train No. 277, the more important train, there would have been a much greater delay to train No. 629 at Thomaston Siding. The dispatcher's decision to have the trains meet at Knox instead of Thomaston had therefore been correct.

Tests showed that signal 'A' should have been visible to the student engineer or anyone stood behind him for about 22 seconds. Signal 'B' should have been visible from the student engineer's side for 90 seconds. The headlight of train No. 629 should have been visible for about 14,000 feet.

The testimony of the student engineer conflicted, inasmuch as he stated that he had not noticed the conductor leave his seat, while the conductor stated that he had been in the washroom when the train was at Knox.

After the crash the conductor reportedly asked the student engineer, "What really happened? ...Was you bullshit?"

The student engineer had replied, "We were bullshit."

The NTSB determined that train No. 629 had been operated correctly and even if it had been stopped dead after sighting train No. 277, the collision would still have occurred, because the latter train had been accelerating.

Knowing that the student engineer had had a college (university) education and that he had been a top pupil at the NS school, Engineer Marks might have been overconfident of the student's ability.

Student Engineer Rice had made 11 trips through Knox but said that he was unfamiliar with the territory. The NTSB commented that his unfamiliarity should have made him more vigilant, as should the engineer and conductor. Because of the long hood of the engine signal 'A' could only be seen from the (student) engineer's side.

The crew of train No. 277 might not have been expecting a meet at Knox. They had only overheard part of the conversation between the dispatcher and the crew of train No. 629, because No. 277 was a priority train, and because they had been routed via the main track at Knox.

Because of their positions, the conductor and engineer had probably not been able to see signal 'A' but they had not adequately supervised the student engineer and the signal aspects had not been called.

The NTSB reached six conclusions regarding the accident:

(a) The engineer and the conductor of train No. 277 had failed to properly supervise the student engineer while approaching the siding at Knox.

(b) Train No. 277's crew had demonstrated a lack of vigilance and crew coordination in failing to stop at signal 'B'.

(c) Train No. 277's crew could have seen signals 'A' and 'B', and the sight distance had been sufficient for proper train operations.

(d) The crew members of train No. 277 could have stopped in time to avoid the accident, using either normal or emergency braking, had they seen and responded appropriately to either signals 'A' or 'B'.

(e) Changing Rule No. 34 to require that crews call signals over the radio should help to prevent future accidents.

(f) Before the accident, Norfolk Southern and the Knox area communities had not implemented adequate measures to ensure on-scene coordination.

The NTSB determined that the probable cause of the accident had been the failure due to lack of vigilance, coordination, and discipline of the crew of train No. 277 to comply with the signals at Knox. Contributing to the accident had been the inadequate supervision of the student engineer by the engineer and conductor.

The NTSB only made one recommendation:

To the Norfolk Southern Railway company:

(a) It should review and revise its programmes for train crew supervision, locomotive cab discipline and training of student engineers in light of the circumstances of this accident and make necessary improvements. [335]

The single-line Ferrovie dello Stato branch from Ciampino to Velletri, only 15 kilometres south of Roma, suffered a single-line collision at 17:45 on Monday 27th January 1992. Train No. 7217 from Velletri to Roma, consisting of three cars, was running a few minutes late. It had just started from the small station at Casabianca, Italy, which was not a crossing place, when it was struck by fast train No. 7220 from Roma to Velletri, travelling probably at 60 km/h. Driver Gabriele Giammattei, 43, of train No. 7217 and passengers Constantino Rodu, a Romanian, 40; Claudio Milletti, 37; and Alberto Zaccagnini, 34, were killed outright, Driver Tommaso Cocuzzoli of train No. 7220 was trapped in the wreckage of his cab and died at 21:00 in the arms of his son Paolo before he could be released, while Driver Romano D'Antini of train No. 7217, who was likewise trapped in the wreckage, was found to be dead when he was released by firemen an hour later. Driver Marco Vittori of train No. 7220 was seriously hurt but managed to say, as he was being taken to the operating theatre at San Giovanni Hospital in Roma: "The Ciampino stationmaster signalled me away with his hand, a signal to proceed, gave me a Green light. When I saw the other train in front of me I couldn't do anything." The trains were crowded with commuters and students and about 140 passengers were injured and were taken to hospitals in Roma, Marino, Albano, Velletri, Frascati and elsewhere.

---

[335] National Transportation Safety Board report

Railway cleaner Alessandro Calvani, 50, was on the train bound for Roma and escaped with a sprained neck and shock. He told a reporter: "I think the fault lay with the Ciampino stationmaster. Or at any rate whoever gave the other train a Green light to proceed before we had arrived there. That is the only possible explanation. There is only one track on this damned line. I was on my way back to Roma and was sat just behind the driver's cab, opposite the second door of the coach. The crash was extremely violent; it was as if I was in the middle of an earthquake. The train was split in two. I heard people screaming and they were jumping into the air. Then I found myself on the ground and I couldn't understand what was happening. There were two women near me. Their legs were stuck under the seats, they weren't moving and they didn't even complain. The firefighters told me that they had died. It's a miracle that I'm here to say what happened."

Natalina Fazzolari, the wife of the former Casabianca stationmaster, was stood at the window of the station house and watched the train from Velletri begin to pull out of the station after picking up four or five passengers. "When it restarted," she said later, "it had not travelled more than thirty metres when there was a loud bang and it started going backwards. I saw it reappear as if its driver had put it in reverse. What happened then was terrible. The passengers were coming out of the carriages screaming like madmen. They were crying, 'it's a bomb. A bomb has exploded.' They were looking for shelter. Dozens of them came to our house. They all wanted to telephone home. So sad, they were as pale as death. Some of them were bleeding."

Signora Fazzolari's neighbour, Mario Siciliano, the wife of another railwayman, added, "The train coming from Roma was going fast. It is usually travelling at 60 km/h when it reaches our station. The other train was also moving. Something must have happened at Ciampino. Usually the train coming from Velletri, on the single track, calls at Casabianca and then continues to Ciampino before the other train comes from there."

At 17:30 on the evening of the accident Stationmaster Manlio Novelli at the junction station at Ciampino had gone off duty, leaving the station in charge of Porters Sossio Dolce and Alfredo Valente. When train No. 7220 from Roma had arrived shortly afterwards Porter Valente was sat drinking a cup of coffee in the station bar. All of a sudden he saw the train setting off again down the Velletri branch, although train No. 7217 had yet to appear, because Porter Dolce had cleared the Starting signal for it to proceed. Running from the bar, with the cup of coffee in his hand, Valente asked Dolce where the train from Velletri was, at which point the latter realised too late what he had done. Dolce then picked up the telephone in an attempt to get train No. 7217 held at Cecchina station, down the branch, but was told that it had left there two minutes earlier. The intermediate stations between Ciampino and Cecchina were unstaffed and there were no intermediate signals which could be replaced to Danger, nor was there any means of contacting either of the train drivers directly. There was a gateman at a level crossing, which Dolce then rang, but again he was told that the train had already passed there. The line was electrified but there was insufficient time to contact the electric control centre in Roma to have the current cut off. All that remained was to wait in despair for news of the now inevitable collision.

Upon news of the collision reaching Ciampino Porter Dolce left the station and wandered around the streets of Ciampino. At 23:00 he turned up at the home of a friend, Bernardo Dolfrancesco, the mayor of Colfelice, and confided to him what had happened, then at 03:00 the next morning he surrendered himself to the carabinieri at Castelgandolfo. He was taken to prison at Velletri and risked a sentence of ten years' imprisonment for manslaughter.

Meanwhile, the railway unions had organised strikes in support of Porter Dolce and in protest against the lack of modern safety features on secondary lines such as the Velletri branch, while large sums were being spent elsewhere on glamorous high-speed lines. [336] [337]

A passenger train and a freight train collided head-on near Dorud, Iran, on the Tehran – Ahvaz line, on Saturday 23rd May 1992, in what was probably a single-line collision. Twenty-six people lost their lives and 28 more were injured. The cause of the accident was not immediately known and services on the line were suspended pending an investigation. [338]

Less than five months after the Casabianca accident (see above) Italy suffered another serious single-line collision, this time near Caluso, on the line between Chivasso and Ivrea. This second accident happened at 15:02 on Wednesday 10th June 1992 and occurred on a line that at the time was not part of the Ferrovie dello Stato, but instead was operated by the Regimento Genio Ferrovieri, a branch of the Italian Army dealing with railway transport. The line was therefore equivalent to the Longmoor Military Railway in England, with the difference that it was available for use by ordinary civilian passengers. A mistake in the signalling made by two of the soldiers operating the line led to the accident.

The trains involved in the accident were diretto service No. 2449 from Torino to Aosta and accelerato (stopping) train No. 10370 from Aosta to Chivasso. The line was single-track with crossing places, and the day of the accident the trains collided about 150 metres on the Aosta side of Caluso station, at the entrance to Sant' Antonino tunnel.

By the late evening there were known to be six dead, plus 17 injured in hospital at Chivasso, 11 in hospital at Ivrea and another three in the Centro Traumatologico Ortopedico (CTO) Hospital in Torino. The passengers taken to the CTO Hospital were Mariella De Melas, 34, who had been trapped between two seats and whom firemen had only been able to extricate at 22:15, and Corporal Gerardo Cardillo, one of whose feet had to be amputated to free him. The third patient at the CTO Hospital was Driver Roberto Mortaretti of the diretto. Driver Giovanni Innocenti, 22, of the accelerato, was also amongst the injured, though he was not too badly hurt. The six who lost their lives were Driver Mauro Guscelli, 45, of the diretto; and passengers Angela Bertletti, 29, from Pavone; Anna De Lucchi, 20, and Elisabetta Borello Ferraris, 45, both from Ivrea; Mery Delfina Frati, 40, from Hone, in the Aosta Valley; and Dino Glarey, 50, from Cogne, Valdostano.

All the victims had been on the diretto, which had been filled with commuters, university students and gamblers bound for the casino at Saint Vincent. There were also two 17-year-old foreign students who were staying in Aosta: Esmana Villehnina from Ireland and Lisa Cruckshane from Australia; they were frightened but not badly hurt. Some of the surviving passengers told reporters of their experiences. Minnie Memi, 21, from Saint Vincent, was a student studying languages, on her way home from the Università di Torino and sitting in the middle carriage of the diretto. She said: "I felt a strong brake application, followed immediately by a terrible roar." Although she had suffered a 'whiplash' injury she said: "I was lucky because I was facing towards the rear of the train. The person facing me was thrown to the floor. The first moments were terrible. People were screaming and trying to get out of the coaches. There was confusion and above all the fear of fire. We looked out of the window and saw a locomotive and a carriage destroyed. Smoke was coming out of the tunnel and we were scared."

Alessandro Morandini, 21, from Verrès, who had been with Minnie, related: "They were terrible moments. After a few attempts we managed to force a door and get out."

---

[336] *L'Unità*, Roma, Italy, 28th and 29th January 1992

[337] *La Stampa*, Torino, Italy, 28th and 29th January 1992

[338] *The Orange County Register*, Santa Ana, California, United States, 24th May 1992

Some of the students made their way to the front of the train to try and release the injured, while other passengers made their way back to Caluso station to raise the alarm. Dozens of ambulances, police and firemen were on the scene in a few minutes and two helicopters from the air rescue service were also sent. The uninjured passengers waited in the waiting room of Caluso station until they were collected in relatives' cars or were taken to their destinations in replacement buses.

As usual, several separate investigations were begun into the accident, in this case by the Ministry of Transport, by, apparently, the Ferrovie dello Stato, although the line was not at the time part of the state system, and by the judiciary. The first interrogations were carried out in a room at Caluso station, with the questioning at length, by Ivrea public prosecutor Bruno Tinri, of Stationmaster Renato De Vellis and Station Inspector Luigi Mancini, both serving soldiers.

The line was worked by a form of the apparato centrale elettrico a itinerari (ACEI, central electrical apparatus with itineraries) system, which allowed the remote control of points, signals, level crossings and so on from a panel, together with a manual electric block system.

Following the investigations, blame for the accident was placed on Sergeant Pasquale Mancini, 24, in charge of the signalling at Caluso, and his colleague at Candia Canavese, the next station north of and on the Aosta side of Caluso, Corporal Major Roberto Giunta, 23. The two soldiers faced penalties of up to 12 years' imprisonment for causing a railway disaster or up to 10 years for multiple manslaughter.

Exactly how the accident had happened was determined by the investigation. Accelerato No. 10370 had left Aosta on time at 13:05 and was due at Chivasso at 15:08. In the opposite direction diretto No. 2449 had left Torino Porto Nuova at 14:20 for Aosta. The trains were timetabled to cross one another at Rodallo, the next station on the Torino side of Caluso, at 14:57. However, on the afternoon of the accident a violent thunderstorm had put the signalling system out of action for some time, while the accelerato had lost ten minutes on its journey. There was still telephone communication between the stations, and Giunta and Mancini decided to amend the crossing of the two trains to Caluso in view of the fact that the accelerato was running late. At 14:55 Giunta at Candida informed Mancini at Caluso that the accelerato had arrived at Candida but forgot to report that it had set off again from there for Caluso. Not having been informed by Giunta that the accelerato was on its way, Mancini did not clear his segnale di protezione (Home signal) at the exit from Sant' Antonino tunnel, so the accelerato came to a stand at the Red signal there at 14:59, 200 metres from Caluso station. Based on his telephone conversation with Giunta only a few minutes earlier, Mancini should in any case have kept his signals at Danger against the diretto, but instead he cleared them, and the latter train, which was not booked to stop at Caluso, passed through the station at reduced speed at 15:02 and entered the single line. Rounding a curve, its drivers saw the accelerato stopped in front of them and applied the brakes in emergency to somewhat reduce the effect of the collision, which occurred almost at once.

As after the January accident at Campino, the railway unions organised strikes in the aftermath of the collision to protest against the lack of modern safety technology on the line, as well as the lack of investment and cuts in the labour force. Meanwhile, Agostino Pertruzzelli, president of the transport users' committee for the Chivasso – Ivrea line, called for its doubling and electrification, plus its transfer to the Ferrovie dello Stato, while questions were asked in Parliament regarding the accident. [339] [340]

In February 1994 Giunta and Mancini were each sentenced to two years' imprisonment for causing the accident. The court also ordered an investigation into whether other parties were

---

[339] *L'Unità*, Roma, Italy, 11th and 12th June 1992

[340] *La Stampa*, Torino, Italy, 11th, 12th, 13th, 14th, 16th and 19th June 1992

to blame but nothing came of it.

The relinquishing of Army control of the line, ruled out soon after the accident on the grounds that the military needed somewhere to train soldiers in railway operation, occurred at the beginning of 2001, when it became part of the Ferrovie dello Stato. At about the same time the method of operation was altered to Centralised Traffic Control (CTC), while electrification between Chivasso and Ivrea was completed in 2006 and is due for completion to Aosta at the end of 2025. The line remains single track. [341]

A fairly common problem on modern railways is the theft of copper signalling cables, given the relatively high price obtainable for the metal. It was two such thefts in Czechoslovakia in the summer of 1992, which led indirectly to a single-line collision. On 6th June, 30 such cables were stolen by an unknown person, affecting the signalling of trains on the Budkovce – Drahňov – Vojany line. With the aid of temporary modifications, some of the signalling functions concerned could be maintained but after a further cable theft at Drahňov, just before midnight on 17th June, all the electronic protection normally provided was lost. In order to keep the trains running they had to be offered and accepted over the telephone, with no safety controls to prevent trains simultaneously entering a single-line section from each end at the same time.

Trains continued to be signalled safely via telephone messages until early on the morning of Monday 3rd August 1992. At about 04:00 on the day, Signalwoman Alžbeta at Budkovce had train No. EPn 61 159 bound for Maťovce at her station. According to her statement, as this train had approached her office she positioned herself outside and stopped it by exhibiting a Red hand signal, as she had to give its driver a written Op order in relation to the failure of some barriers, presumably automatic crossing barriers. The driver had then whistled and assuming that he was about to stop, she had returned to her office, prepared the Op order and also prematurely cleared her Starting signal for the train to proceed. Having written out the order she was leaving her office to hand it to the driver when she found that having seen the Starting signal go to Green, the train was accelerating away. Unable to stop the train she had then returned to her office to report its departure to Signalman Miloš at the next station at Drahňov, to be told by him that train No. Mn 81 400 was already on its way to Budkovce, so there would likely be a collision between them. As the line was electrified, her last hope was to switch off the current, but before this could be done the trains had collided at Kilometre 32.4, at 04:07. Six railwaymen died in the collision. Drivers Vojtech Tóth (43) and Ján Bak (28), Assistant Driver Pavol Kičinko (57), and Conductors Ladislav Brziak (49), Štefan Koneval (37) and Peter Smolák (33). It is not known who was on which train. The damage resulting was estimated at between 3,000,000 and 6,000,000 koruna).

Neither of the train crews had apparently realised the danger of a collision until the last moment, when the driver and conductor of train No. Mn 81 400 at least had attempted to escape into the engine room of their locomotive. This was doubtless because the single-track, standard-gauge line on which the trains were both running was paralleled by a broad-gauge track, which had been laid from Uzhhorod in Ukraine as far as Košice in Czechoslovakia to permit the transhipment of goods, especially iron ore, to and from Russia without a break of gauge.

At the inquiry into the accident Signalwoman Alžbeta insisted that train No. EPn 61 159 had been accepted by Drahňov, which Signalman Miloš denied, and he counterclaimed that train No. Mn 81 400 had been accepted by Budkovce. There were no outside witnesses to the telephone conversations between the two but an examination of the train register at Budkovce

---

[341] https://www.risvegliopopolare.it/il-10-giugno-1992-il-pauroso-incidente-ferroviario-tra-caluso-e-candia-canavese-di-doriano-felletti/

revealed certain discrepancies, while there was a marked difference in the handwriting in it before and after midnight. The investigation concluded that the principal responsibility for the accident lay with Signalwoman Alžbeta at Budkovce, who had failed to offer and have train No. EPn 61 159 accepted by Drahňov, while Signalman Miloš there had failed to notify Budkovce before the departure of train No. Mn 81 400 from his station. However, but for the theft of the signalling cables, the Starting signal at Budkovce could not have been cleared with the other train in the section. [342] [343]

The *El Purépecha* express passenger train of the Ferrocarriles Nacionales de México, which ran between Mexico City and the port of Lázaro Cárdenas, was involved in what was probably a head-on collision on a single line, at a remote location somewhere near Lázaro Cárdenas, in Michoácan state, Mexico. Ten or possibly more people died in the accident, which was blamed on 'poor co-ordination of timetables.' The accident happened on the afternoon of Friday 7th August 1992. There had been 300 people on the express, and apart from those who died, about 270 of them were more or less injured. The collision occurred near the La Villita reservoir and rescuers evacuated many of the casualties by boat, while helicopters were also used in the rescue work. [344] [345]

On Tuesday 18th August 1992 regional passenger train No. R 6411 collided with freight train No. 71154 at Km. 22.26, near Grünbach station, Austria, on the Schneebergbahn. Seven people died in the accident and 51 were injured. The same two trains had collided at Km. 20.77 on 3rd June 1992 and there had also been a previous collision on the line, involving different trains, in September 1989. The August accident was blamed on the motorman of the regional train failing to wait to cross the freight train at Grünbach, as set down in the working timetable. The line was not provided with Streckenblock but after this third accident, Grünbach and other stations were provided with Starting signals. [346] [347] [348]

Early on the morning of Thursday 1st October 1992 freight train No. 10507, consisting of an engine and 31 waggons destined for Turkey, arrived at the yard at Plovdiv, Bulgaria. The engines were changed and the train continued its journey with Driver Yordon Tomov and Assistant Driver Petar Stoimenov. At 05:26 the freight train arrived at Harmanli station. In the opposite direction train No. 292, an international express from Istanbul, Turkey, to München, Germany, was due to pass through Harmanli at 06:05. Stationmaster Kolyo Kolev at Harmanli testified that the freight train was too long to fit into the crossing loop (possibly more waggons had been added to it at Plovdiv) and so it had to be divided.

Meanwhile, the international express had arrived at the next station at Lyubimets. The Starting signal there was Red and Driver Nikolay Kanev of the express telephoned the signalling centre at Plovdiv, to be told that he could proceed. A minute later (?) Driver Tomov

---

[342] https://vkport.sk/2019/08/04/historia-zeleznicna-nehoda-budkovce-3-8-1992/
[343] https://dolnyzemplin.korzar.sme.sk/c/22976290/za-tragediou-v-budkovciach-boli-aj-ukradnute-prepojky-dodnes-ich-nevedia-ustrazit.html#:~:text=augusta%201992%20sa%20v%20kilometri,a%20Peter%20Smol%C3%A1k%20(33)
[344] *Port Arthur News*, Port Arthur, Texas, United States, 7th April 1991
[345] *Daily Herald*, Arlington Heights, Illinois, United States, 10th August 1992
[346] https://www.schienenweg.at/forum/index.php?thread/40822-zusammenst%C3%B6%C3%9Fe-2143-5047-in-den-1990ern/&pageNo=2
[347] Wikipedia – Liste von Eisenbahnunfällen in Österreich
[348] https://www.parlament.gv.at/dokument/XVIII/AB/3580/imfname_443756.pdf

set off from Harmanli towards Lyubimets without permission. His train had reached a speed of 70 km/h when it collided on the single line with the express, which was itself travelling at 62 km/h, near Biser village, about midway between the two stations. Both locomotives and seven vehicles of one or another train were destroyed in the violent collision. Driver Tomov, 35, of the freight, Driver Kanev, 36, and Assistant Driver Dimitar Demirev, 29, of the express; (spare?) Driver Ivan Malyov, 29; Head Conductor Nencho Penchev, 42, and Conductor Georgi Georgiev, 32, also of the express; and a Norwegian passenger named Ralf Liene were all killed. The fact that the leading vehicle of the express was a mail van probably helped prevent more passenger deaths.

The investigation into the accident revealed that Driver Tomov had been drunk, with a post-mortem blood alcohol level of 1.85%. It was not known when he had been drinking but it emerged that he had attended a pre-journey medical examination the previous evening three and a half hours early. Assistant Driver Stoimenov, who survived the collision (was he, perhaps, left behind when the train left Harmanli?), testified that the engine cab had smelt of alcohol and that, "Generally, Yordan doesn't like to be silent – he talks while on the road. However this morning he didn't speak, he was thoughtful. There was a bruise on his forehead above his left eyebrow, and he told me that he had had a fight with someone, but he didn't say who." [349]

A major disaster occurred in the North Eastern Railway zone of Indian Railways on Thursday 15th July 1993. Train No. 5553 Up, the *Mithilanchal Express*, and train No. 334 Down, a Nirmali – Samastipur passenger train, collided with one another between Haya Ghat and Thalwara stations, Bihar, India. Twenty-two passengers and the driver of train No. 334 died in the accident and 80 people were injured, 39 of them seriously.

The *Mithilanchal Express* consisted of YP Class 4-6-2 steam locomotive No. 2631 and eight all-steel coaches.

Train No. 334 consisted of YG Class 2-8-2 steam locomotive No. 3195 and seven coaches.

The single-track, metre-gauge line between Haya Ghat and Thalwara stations was worked on the Absolute Block System with Neal's Ball Token instruments. Two-aspect semaphore signals were provided. Haya Ghat station had Standard III interlocking with Warner (Distant), Outer, Home, Starter and Advanced Starter signals. Thalwara station had modified non-interlocked signalling with only Outer and Home signals.

The accident was investigated by the Commission of Railway Safety.

Guard A.K. Srivastawa of train No. 334 stated that his train had arrived at Thalwara at 09:29. After five minutes the starting bell had rung, he had seen his driver being handed a token in its holder, his driver had whistled and he had given a Green signal for the train to start. After a few minutes' journey his train had slowed and a sudden jerk had thrown down him and Permanent Way Inspector A.K. Sinha, who had been with him in the van. A second jerk had thrown him out of the van. He was hurt but had handed his detonators to Guard R.R. Berma to protect the train, then despite his injuries attended to the injured. His train had been moving at 10 km/h when the collision had occurred at 09:38. His driver had been killed and his fireman seriously injured.

Guard Ramendra Jha of the express testified that his train had arrived at 09:10. After getting a paper line clear his driver had attempted to start at 09:14 but due to a problem with the vacuum brakes it had not been able to get away until 09:32. At Km 24/1-2 the vacuum had

349 https://telegraph.bg/telegraphplus/novini/istoriiata-pomni-stotici-ubiti-i-raneni-v-rodnite-zheleznici-13-zloveshti-incidenta-vzeha-zhivota-na-116-dushi-369744

suddenly dropped to zero and as he had gone to the door of his van to look out, its door had opened automatically, throwing him down, there had been a loud noise and the train had stopped suddenly at 09:38. Getting out, he had seen that there had been a collision, protected his train and finding that the other guard had advised Control, he had attended to the injured. He put the speed of his train at the moment of collision at 40 km/h. At Kishanpur and Haya Ghat the passengers, mostly milkmen, used to open the clappet valve resulting in a fall in the vacuum, closing it before he reached the coach concerned. At Haya Ghat his driver had not whistled, even after the signal had been lowered, whereupon he (the guard) had gone to the assistant stationmaster's office and told the latter to send line clear, after which his driver had started. He concluded that train No. 334 had come on a wrong line clear.

Assistant Stationmaster (ASM) Jayant Kumar Sirivastawa at Haya Ghat stated that the block/token instrument to Thalwara had been out of order when he commenced duty at midnight, with its handle in the Train Coming From (TCF) position. Traffic was being worked with paper line clear tickets. He had given Line Clear for the express to Rambhadrapur, it had arrived at about 09:05 and he had requested Line Clear from Thalwara on the Control phone in the presence of Section Controller K.N. Thakur, because the phone connected to the block/token instrument had not been working properly. Stationmaster (SM) R.L. Das at Thalwara had duly given him Line Clear with Private Number 53. The express had then suffered three or four episodes of vacuum trouble before it had finally started at about 09:35. About 09:45 stationmaster Das had requested Line Clear for train No. 334, whereupon he had asked if the express had arrived there complete. A man had shortly afterwards arrived at the station with the news that there had been an accident. At first he had assumed that the express had been derailed, then the section controller had informed him that there had been a head-on collision. He had asked the controller how train No. 334 could have left Thalwara, the controller had asked stationmaster Das, and the latter had said that train No. 334 had left with a token, delivered to its driver by the electrical signal maintainer (ESM). He (Sirivastawa) had then had to flee the station to escape the attention of angry passengers.

Assistant Stationmaster Binay Kumar Thakur at Thalwara said that he had taken charge there at midnight when the block/token instrument to Haya Ghat had been out of order and paper line clear working had been in operation. This had still been the case at 08:00 when Stationmaster Das had taken over. The instruments had not failed previously in the three months he had been at the station. Beats were being received at Thalwara from Haya Ghat but not by Haya Ghat from Thalwara. The block phone had been in order.

Assistant Stationmaster L.P. Singh at Laheria Sarai, the next station north of Thalwara, stated that the working between his station and Thalwara had been normal. Electrical Signal Maintainer I P.K. Jha had ordered Electrical Signal Maintainer III (ESM) (Batteryman) Mohammad Ashiq to proceed from Laheria Sarai to Thalwara on train No. 334 and he presumed that the latter had done so.

Pointsman R.P. Singh at Thalwara stated that upon instruction from SM Das he had been set the facing points at the Haya Ghat end of the station for train No. 433 Up to enter the station on line No. 1. That train had arrived, he had reset the facing points and returned to the station, giving the key to the points back to the stationmaster. Rather later, the stationmaster had informed him that train No. 334 Down, which had crossed train No. 433 Up at Laheria Sarai, would be arriving and proceeding towards Haya Ghat, which it had duly done. Later, angry members of the public had appeared at the station and he fled. He had been aware of the failure of the block instruments and had not been able to see the ESM arrive at his station from where he had been at the points.

Pointsman R. Ray at Thalwara described dealing with train No. 433 Up and SM Das then advising him and Pointsman Singh that train No. 334 Down would be received on line No. 1 and then proceed to Haya Ghat, which it had done. Like his colleague he had had to flee the station after the accident to save his life. He had seen the ESM arrive on train No. 334 Down. When he had returned from setting the facing points at the Laheria Sarai end of the station for train No. 334 Down ESR, Ashiq had been opening the back cover of the token instrument to Haya Ghat and Casual Labour Safaiwala/Porter Lalan Kumar Jha had told him (Ray) that the ESM had given him (Jha) a token to give to the driver of train No. 334 Down, which the safaiwala had done. The pointsman said that he had known nothing of the Up express. (Safaiwala is an Indian term for a person employed in collecting or removing rubbish.)

Section Controller Thakur at Samastipur stated that he had learnt of the collision at 09:45. ASM Sirivastawa at Haya Ghat had reported to him the arrival and departure times of train No. 5553 Up and the departure time of train No. 433 Up had been reported by ASM Thakur at Thalwara. After that, no one had contacted him regarding any train movements and he had not spoken to SM Das at Thalwara the whole day. In the presence of the controller, ASM Sirivastawa at Haya Ghat repeated that he had obtained Line Clear for train No. 5553 Up from SM Das at Thalwara on the control phone and ASM Thakur denied that he had reported the departure time of train No. 433 Up to Control, saying that he had boarded that train at Thalwara after finishing his duty.

Chief Controllers K.K. Sharma and U.C. Tiwari described their organisation of the relief measures after the accident.

Traffic Inspector Sudhir Kumar Sinha at Darbhanga spoke of going to the accident site. He said he had been aware that ESM Ashiq, called to Thalwara to correct a fault in the block/token instrument, had extracted a token from it. He went on to describe the routine for station inspections in the area under his jurisdiction.

Station Superintendents R.K. Singh at Banmankhi and Kolahal Prasad at Darbhanga (who had been a passenger on train No. 334 Down) gave evidence.

Cabinman Ganga Prasad in Haya Ghat B (or North) cabin at the Thalwara end of the station related that he had received Private Number 40 for Line Clear for train No. 5553 Up on Line No. 1 from the station and had given Private Number 19 to the station. He had received Private Number 53 for the Advanced Starter but that signal had not been lowered as the ASM had not taken off the slot and the train departed on a paper line clear. He had understood from the ASM that train No. 5553 Up was to cross train No. 334 Down at Thalwara.

Cabinman Devendra Prasad Singh in Haya Ghat A (or South) cabin, at the Rambhadrapur end of the station, related that he had exchanged private numbers with the ASM regarding Line Clear for train No. 5553 Up and for its arriving complete. The ASM had advised him that train No. 5553 Up was to cross train No. 334 Down at Thalwara. Private numbers had been exchanged with the ASM but he could not recall their numbers.

Assistant Operating Managers K.K. Rana and S.S.P. Sinha at Samastipur gave evidence.

Stationmaster Das testified that he had come on duty at Thalwara at 08:16 on the day of the accident, taking over from ASM Thakur, who had informed him that trains Nos. 433 Up and 334 Down would cross at Laheria Sarai as per Control, that the ESM would be coming down on train No. 334 Down to rectify the fault in the block/token instrument and that train No. 334 Down would cross train No. 5553 Up at Haya Ghat. He had then been busy selling tickets and ASM Thakur, in his presence, had informed both the controllers and the safaiwala that trains Nos. 433 Up and 334 Down would cross at Laheria Sarai, while train No. 334 Down would cross train No. 5553 Up at Haya Ghat, the latter train having been delayed there by passengers interfering with the vacuum. Before the arrival of train No. 334 Down he had tried to talk to both the section controller and Haya Ghat but interference on the phone had prevented him from doing so. ESM Ashiq had arrived on train No. 334 Down, had hurriedly repaired the block/token instrument and had extracted token No. 19 and handed it to Safaiwala Kumar to give to the driver of train No. 334 Down. He had reported the failure of the block/token instrument 15 hours before the accident, which he blamed on the carelessness of the Samastipur section controller, ESM Ashiq, ASM Thakur and ASM Sirivastawa, pointing out that the accident would not have occurred had there been Starting and Advanced Starting signals at Thalwara station. He denied having given Private Number 53 for train No. 5553 Up. The Line Clear token had been given without his knowledge. He had not recorded the failure of the block/token instrument because it had got dark at 17:00 and he had been busy issuing paper line clears, which involved a lot of writing, and in issuing tickets.

Safaiwala Jha stated that upon the instruction of SM Das he had lowered the signal for train No. 334 Down and replaced it afterwards, had collected the token from that train and given it to SM Das and that SM Das, rather than ESM Ashiq, had given him the token for the forward section to Haya Ghat to give to the driver of train No. 334 Down. He had known nothing of train No. 5553 Up.

Electrical Signal Maintainer Jha at Laheria Sarai deposed that he had learnt of the failure of the Thalwara – Haya Ghat block/token instruments at 17:00 on the day before the accident but had been occupied with jobs at Laheria Sarai. ESM Ashiq had, he maintained, been with him at Laheria Sarai at 09:20 on the day of the accident and had not gone to Thalwara.

Electrical Signal Maintainer Ashiq agreed that he had never gone to Thalwara. Pointsmen Ray (here spelt Roy in the report) and Singh repeated that they had seen ESM Ashiq at Thalwara, the ESM again denied having gone there.

Chief Signal Inspector S.P. Srivastawa at Darbhanga stated that he had been informed of the failure of the Thalwara – Haya Ghat block/token instruments at 07:00 on the day of the accident, also of several problems at Laheria Sarai, and had instructed ESM Jha to deal with the Laheria Sarai matters first.

Several railwaymen gave evidence as to the rescue work.

The driver of train No. 5553 Up and the firemen of the two trains had all been seriously injured in the accident and were only able to give statements a fortnight later.

Driver Grade A Atma Prakash of the express related that his train had been delayed 17 or 18 minutes at Haya Ghat by a vacuum disturbance, then he had left there with a paper line clear ticket. Because of a curve he had only seen the other train when it was 200 to 250 metres away, when he had shut off steam and applied the brakes. His speed had then been about 50 km/h.

Fireman Lakshman Rai of the express corroborated the evidence of his driver.

Fireman M. Baitha of train No. 334 Down stated that they had left Thalwara after receiving token No. 19 there. When they had seen the other train about 400 metres away his driver had shut off steam and applied the brakes, reducing the speed of their train from 30-40 km/h to 10-15 km/h upon impact.

Commissioner of Railway Safety V.S. Dutta detailed tests done and observations made.

The commissioner fixed the time of the collision, as stated by both guards, at 09:38.

The speed of the trains, the commissioner fixed at 50 km/h for the express and 15 km/h for train No. 334 Down.

The commissioner detailed the events leading up to the collision. The Thalwara – Haya Ghat block/token instruments had failed. This failure had not been recorded at either station but was registered in the Control office. ASM Sirivastawa at Haya Ghat alleged that using the Control phone he had obtained Private Number 53 for train No. 5553 Up. SM Das at Thalwara and Section Controller Thakur denied this, though the section controller stated that Haya Ghat had given the arrival and departure times for train No. 5553 Up. From a perusal of the private number book it was clear that Private Number 15, rather than 53, had been given for train No.

5553 Up. The section controller had also stated that the arrival and departure times of train No. 433 Up had been given by ASM Thakur, who had left on this train for Dharbanga at 09:06.

What had probably happened, Commissioner Dutta theorised, was that ASM Sirivastawa or his staff at Haya Ghat had been listening on the Control phone to obtain Line Clear permission and the private number for train No. 5553 Up from Thalwara, at the time that Thalwara had been giving the timings for train No. 433 Up to Control. SM Das at Thalwara had been busy issuing tickets, so ASM Thakur, going off duty and waiting for train No. 334 Down, had lent a hand to his colleague by reporting train No. 433 Up to Control. During this time the ASM or another staff member at Haya Ghat had sought permission and the private number for train No. 5553 Up and ASM Thakur had repeated 53, which the ASM Sirivastawa at Haya Ghat had taken as the private number for train No. 5553 Up, dispatching the train with that number and giving the number to Control. The Control phone, the commissioner noted, should not be used for private numbers as a number of persons were often on the phone exchanging information and confusion could arise. ASM Sirivastawa should instead have used the block phone.

Both ESM I Jha and ESM 111 Ashiq alleged that the latter had remained at Laheria Sarai but a number of witnesses had seen ESM Ashiq at Thalwara, that he had attended to the block/token instrument, had extracted token No. 19 from it without any cooperation from Haya Ghat and had handed it to the porter to give to the driver of train No. 334 Down.

Examination of the block/token instrument at Thalwara showed that the ESM's lock for its back door had been unlocked, that the SM's lock had been defective and that a broken contact would have meant that the instrument was permanently out of operation. It was clear, therefore, that a token could only have been extracted by the technical staff bridging the contacts. It was likewise clear that ESM Ashiq had come to Thalwara. SM Das denied having been aware of the despatch of train No. 334 Down before it had left his station but had recorded its despatch in the station register and it seemed clear that he had come to an understanding with the ESM to make a temporary connection between the block/token instrument contacts, so as to allow a token to be extracted and to avoid the hassle of having to issue a paper line clear ticket. Train No. 334 Down had then been hurriedly despatched. Only after the train had left at 09:32 had the SM asked for the private number from Haya Ghat and then learnt that train No. 5553 Up had left there. The drivers of both trains had thus had authority to enter the single line, could only have unexpectedly seen one another at a short distance around a curve and had not been responsible for the collision.

In conclusion, Commissioner Dutta determined that the collision had been due to:

(a) The Thalwara SM, with the assistance of the ESM, obtaining a token from the instrument without obtaining permission supported by a private number from the Haya Ghat ASM and giving it to the driver of train No. 334 Down.

(b) The Haya Ghat ASM hearing the number 53 on the Control phone, wrongly assuming that it was a private number for the despatch of train No. 5553 Up and despatching the latter train with a paper line clear ticket.

Three officials, the commissioner stated, were primarily responsible for the accident:

(a) SM Das at Thalwara for allowing ESM Ashiq to open the block/token instrument in contravention of General Rule No. 14.12 (i) (e) and extracting a token and giving it to the driver of train No. 334 Down without having obtained permission and a private number from Haya Ghat, in contravention of paragraph 18 (i) of Appendix D and Southern Railway Rule No. 14.13 (i).

(b) ESM III Ashiq for opening the block/token instrument without written permission from the SM and extracting a token without the knowledge and cooperation of Haya Ghat.

(c) ASM Sirivastawa at Haya Ghat for not obtaining Line Clear on the block phone or following the correct procedure using the Control phone, leading to his wrongly accepting 53 as the private number for train No. 5553 Up. He had contravened paragraph 18 (iii) of Appendix D and Southern Railway Rule No. 14.13 (i).

Secondary responsibility, the commission stated, rested on ASM Thakur at Thalwara for answering the Control phone and answering an inquiry from Haya Ghat with, "53."

In addition, the commissioner listed a number of staff as being blameworthy:

(a) ESM I Jha at Laheria Sarai for wrongly alleging that ESM Ashiq had not gone to Thalwara and for deficiencies in the block/token instrument at Thalwara, such as defective locks.

(b) Chief Signal Inspector Srivastawa at Darbhanga for ineffective inspections and supervision.

(c) Traffic Inspector Sinha at Darbhanga for ineffective inspections and supervision.

(d) Section Controller Thakur at Samastipur for failing to ensure that the Thalwara and Haya Ghat ASMs had followed the correct procedures in exchanging Line Clear messages and Private Numbers.

Commissioner Dutta made several remarks and recommendations:

(a) The failure of the block/token instrument had not been recorded in the Failure Register or formally reported for over 16 hours. Similar non-recordings of failures had happened previously and were not safe. All failures should be recorded and the instruments not brought back into use until the failure had been rectified and the repair noted in writing.

(b) Signal failures should be recorded in the SMs Diary by outgoing and incoming SMs at shift changes.

(c) There had been a casual substitution of paper line clears for token working and the ESM had only learnt of the failure at Thalwara, 4 kilometres away, 14 hours later, by chance. Instructions for action during a failure of the block/token instruments should be as under:

(1) Concerned signal and telecommunications staff should attend to failures by the first available means by day or night.

(2) Entry should be made in red ink in the train register by every shift ASM.

(3) A board with the legend "Block Suspended" should be put on the instrument handle.

(d) The police had only given permission for the removal of the damaged coaches 27 hours after the accident, after the attentions of a police photographer. This was too long. The Railway should write to the state government to issue directives to the police in order that the wreckage could be removed with minimum loss of time.

(e) North Eastern Railway rake links should be reviewed to see that carriage and waggon examination and maintenance was done at shorter intervals and long lie-overs without examination were avoided. [350]

Western Railway train No. 1593 Down, a Kota – Bhopal passenger service, and a N/SRE Up goods collided head-on at Km. 132/0-2, between Chhabra Gugar and Bhulon stations, Rajasthan, India, on the Kota – Ruthiyai single-line, non-electrified, broad-gauge section of the Kota division of the railway, at 21:19 on Tuesday 21st September 1993. The goods train

[350] Government of India Ministry of Civil Aviation (Commission of Railway Safety) Railway Accident Investigation Repot on Head-on Collision of 5553 Up Express with 334 Down Passenger Between Haya Ghat and Thalwara Stations of North Eastern Railway on 15th July 1993.

had passed through Bhulon station without authority and collided with the passenger on the single line with catastrophic results, 78 people being killed and 88 injured, 26 seriously.

The listing of all the recommendations made by the Commission of Railway Safety for a number of accidents at one place towards the end of the 1993-4 annual report, with no clear indication of which recommendation(s) apply to which accidents, has made it difficult to be certain which specific recommendations were made with regard to this accident, but it is believed to have been the following:

(a) Brake power certificates for air-braked trains should only be issued in the prescribed format. All the particulars on the certificate should be fully and legibly filled in on all its copies.

(b) Fresh brake power certificates or revalidation of old certificates should be done, keeping in view all the norms laid down by the Railway Administration. Rigorous interpretation should be done on whether the integrity of a rake was affected or not.

(c) Facilities for the intensive examination of all air-braked trains at nodal points should be established as a priority.

(d) Deviations from extant instructions, such as not offering trains for examination when due, allowing trains to run with invalid brake power certificates and allowing excess load on trains, should only be permitted under exceptional circumstances and then only after the personal approval of an administrative-grade officer of the operating department.

(e) The running of 'trial loads' should be permitted only after the issue of written instructions covering all aspects of the trial.

(f) While checking drivers, the importance of the brake continuity rest must be driven home. The duties and procedures for monitoring the drivers of different modes of traction should be streamlined. A record of monitoring, based on an objective pattern of questions and observations, should be brought into vogue.

(g) The brake vans of air-braked trains should be provided with reliable pressure gauges.

(h) Intensive examinations should be made at the originating stations to ensure 100% brake power.

(i) The responsibility for getting freight trains examined should be passed on to a foreign Railway/Division if the trains passed through more than one Railway/Division.

(j) Undue pressure should not be exerted by Control on the crew and station staff to start a train, unless the crew was satisfied of the availability of adequate brake power in the train.

(k) The speedometers supplied by M/s Autometers India Ltd should be closely monitored for their performance and reliability. [351]

Two Kereta Api Indonesia passenger trains collided between Depok and Citayam stations, Indonesia, in a single-line collision, during the morning rush hour on Tuesday 2nd November 1993. The magnitude of the accident varied according to different reports, with one source stating there were at least 69 dead and 77 injured,[352] while another said that there were only 11 dead, with 110 injured.[353] The Department of Transportation announced that there were at least 18 dead, with 200 injured, adding that the earlier higher figure had probably been due to

[351] Government of India Ministry of Civil Aviation and Tourism Commission of Railway Safety Annual Report for 1993-94

[352] *The Chronicle Telegram*, Elyria, Ohio, United States, 2nd November 1993

[353] *The Gleaner*, Kingston Jamaica, 3rd November 1993

double-counting and that some people thought dead were found to be alive.[354] The collision occurred at 07:45 as people were travelling to work.

A housewife named Mulyana, who lived near the scene of the accident, told a reporter that, "It sounded like a bomb. I was very scared." One of the electric trains was reportedly travelling at 60 km/h; the speed of the other was not stated. The work of freeing the victims took a considerable time and at noon a five-year-old child and an elderly woman were still trapped in the wreckage. The two closest hospitals were overwhelmed by the number of injured, many of whom had to be laid in corridors or outside because of the lack of room.

A misunderstanding between the responsible officials at each end of the single-line section caused the accident. The pengatur perjalanan kereta api (PPKA) (signalman) at Depok allegedly despatched an eight-car electric train to Citayam without having had it accepted by his colleague at that station, and at the moment when another train from Bogor for Jakarta had already left Citayam for Depok. The trains met on a curve at Ratu Jayam so that neither driver had much warning of the other train. As the train from Depok was almost empty and thus lighter than the heavily loaded one from Citayam, the former was crushed and forced upwards by the latter in the collision. As well as being crushed some of the passengers in the Citayam train were thrown out of it by the impact. Those who died included both motormen and both conductors; the latter were probably accompanying the motormen in their cabs.

Both station officials subsequently stood trial, charged with causing the accident through negligence, and both were found guilty and sentenced to three years' imprisonment.

The Jakarta – Bogor main line on which the accident occurred was subsequently doubled throughout. [355]

Two Kansas City Southern Railway freight trains collided head-on on a curve at 13:40 on Thursday 10th February 1994 just north of Anderson, Missouri, United States. The trains involved were Extra No. 705 South, from Pittsburg, Kansas, to Siloam Springs, Arkansas, a local freight train consisting of a diesel engine and 13 cars, and Extra No. 741 North from Heavener, Oklahoma, to Pittsburg, Kansas, consisting of three diesel units and 81 cars. The southbound train was travelling at an estimated 26 mph and the northbound at an estimated 19 mph when they met. Five crew members were injured in the collision. Four were taken to the Freeman-Neosho Hospital and the fifth, who was more seriously hurt, was conveyed to St John's Regional Medical Center in Joplin, Missouri. Flammable naphtha being carried in three tank waggons on one of the trains was not spilt and did not ignite but about 100 gallons of diesel fuel were lost from a tank of one of the engines. [356] [357]

The National Transportation Safety Board (NTSB) issued a brief report on the accident. It noted that crews of both trains had been issued with joint time and track authority to operate in the area at the same time by the dispatcher at Shreveport, Louisiana, and that both crews had been aware that the other crew had the same orders. The dispatcher had not issued a meet order for the two trains, nor, said the NTSB, had he been required to do so. He had, however, told the crew of train No. 705 South that he expected they would arrive at Anderson before the northbound train. The crew of train No. 741 North had not heard this conversation and the dispatcher had not discussed the meet with them. The train crews had not contacted one another, nor had they been required to do so. In such a situation the regulations required the trains to move at restricted speed, not exceeding 20 mph, and prepared to stop within half of

[354] *Xenia Daily Gazette*, Xenia, Ohio, United States, 2nd November 1993
[355] Wikipedia – The 1993 Jabotabek Ratujaya Commuter Line Collision
[356] *Courier News*, Blytheville, Arkansas, United States, 11th February 1994
[357] *The Joplin Globe*, Joplin, Missouri, United States, 12th February 1994

the range of vision of the engineers.

The NTSB reached five conclusions:

(a) The entire train crews had failed to proceed at restricted speed.

(b) The dispatcher had given out insufficient information.

(c) The operating rules had been inadequate.

(d) The procedures for train operation had been inadequate.

(e) The rules had been inadequate (it is not quite clear how this differs from (c); the report merely states, 'Inadequate rule – company operator/management.').

The NTSB determined that the probable cause of the accident had been: 'The inadequate operating procedures that permitted an overlap of train movement authority on the same track without any provisions for train separation, and the failure of both train crews to comply with restricted speed rules and slow their trains to a speed that would allow them to stop within ½ the range of vision.' [358]

Just over a fortnight after the Anderson accident, two Illinois Central Railroad freight trains collided head-on on the single line at Flora, Mississippi, United States, at 13:45 on Saturday 26th February 1994. Involved in this accident were train Extra 6152 South from Memphis, Tennessee, to Jackson, Mississippi, consisting of two diesel units and 116 cars, and Extra No. 6131 North from Jackson, Mississippi, consisting of two diesel units and 116 cars. (The destination of this train is stated in the official report of the accident to have been Flora but it seems likely that this is a misprint, as Flora was merely an intermediate siding, i.e. crossing place,) The northbound train was travelling at an estimated 34 mph and the southbound one at an estimated 25 mph when they collided. Upon seeing that a collision was inevitable the crews of both trains had leapt from their respective locomotives and in doing so, Engineer Jerry Plunk of the northbound train was fatally injured. Conductor R.E. Crowley of the northbound train and Engineer Q.B. Gray and Conductor Bill R. Gandy of the southbound train were all injured. Flora Fire Chief Tommy Quattlebaum told a reporter that after the accident, "There was a lot of debris. Engines were stacked on top of one another." [359]

The National Transportation Safety Board (NTSB) investigation into the accident determined that the crew of Extra No. 6131 North had been issued with a permit to proceed to Cynthia, a siding between Jackson and Flora siding, and only to proceed north of Cynthia after they had met Extra 6152 South there. Somehow, the crew of the northbound train had the impression that the southbound train had already arrived in the yard at Jackson and without questioning the dispatcher in distant Chicago, Illinois, they had immediately proceeded beyond Cynthia, covering another nine miles before meeting the opposing train. A further finding of the investigation was that the conductor of train Extra 6152 South had tested positive for marijuana, with a urine level of 23 ng/ml.

The NTSB reached three conclusions regarding the accident:

(a) The crew of Extra No. 6131 North had failed to comply with the track warrant.

(b) The crew of Extra No. 6131 North had been inattentive.

(c) A member of the crew of Extra No. 6152 South had been under the influence of drugs.

The NTSB determined that the probable cause of the accident had been: 'The failure of the crew of Extra No. 6131 North to comply with the track permit which instructed them to stop

[358] National Transportation Safety Board report

[359] *Laurel Leader-Call*, Laurel, Mississippi, United States, 28th February and 1st March 1994

at Cynthia until Extra 6152 South had arrived.' [360]

The third single-line collision to occur in the United States in less than six months was also the most serious of the three. It happened at Marathon, Texas, at 09:05 on Monday 25th July 1994. One of the trains involved was Southern Pacific Transportation Extra No. 9719 East, from Long Beach, California, to Houston, Texas, consisting of five diesel units and 42 cars, in charge of Conductor Eddie J. Hansen with Engineer Milby Winn. The other train was Extra No. 9713 West, from Memphis, Tennessee, to Long Beach, California, consisting of seven diesel units and 25 cars, in charge of Conductor James M. Davis with Engineer F.H. "Rick" Bonnell. The eastbound train was travelling at an estimated 39 mph and the westbound at an estimated 25 mph (the maximum authorised speed) when they collided on a curve. Fire broke out following the collision and the four train crew, who all hailed from Sanderson, Texas, were killed, their bodies being trapped in the wreckage. Various types of hazardous materials being carried on both trains were not involved in the fire and there was no spillage of them. Most of the cars of the trains were not derailed or damaged but in the words of the local sheriff, "All of the engines were burnt plumb to a crisp. They were all jumbled together." [361] [362]

The National Transportation Safety Board (NTSB) conducted its usual investigation into the accident. It determined that Extra No. 9719 East had passed an Advance Approach signal at 64 mph with its dynamic brakes applied. Half a mile further on, two automatic brake applications had been made but the brakes had then been released a mile in the rear of the succeeding Stop signal, which the train had passed while still travelling at 43 mph. The brakes had then been applied again, reducing its speed to 39 mph upon impact with the westbound train.

Although all four of the train crew had died in the accident, its cause was not difficult to determine. It transpired that changes in the crew lineup had resulted in the crew of Extra No. 9719 East being called up for duty four turns early. As a result, Conductor Hansen had only had 4 ½ hours' sleep in the previous 51½ hours and Engineer Winn had only had 4½ hours' sleep in the previous 35 hours. It was obvious that as a result, the engineer and conductor had been very fatigued and they had probably either been asleep or close to it and hence had overrun the signals.

The NTSB concluded that:

(a) The engineer and conductor of Extra No. 9719 East had failed to comply with signal indications.

(b) The two railwaymen had been fatigued due to lack of sleep.

(c) The two railwaymen had been inattentive.

The NTSB determined that the probable cause of the accident had been: 'The failure of the crew of Extra No. 9719 East to control their train and observe signal indications due to reduced alertness consistent with falling asleep, and the failure of SP to provide accurate train lineup information for crews reporting for duty.' [363]

One would have thought that having been called to work the train with so little rest, the engineer and conductor could have refused to do so with good reason.

Northbound Queensland Rail freight train No. C531 from Moolabin to Townsville collided

[360] National Transportation Safety Board report

[361] *Odessa American*, Odessa, Texas, United States, 26th July 1994

[362] *Del Rio News-Herald*, Del Rio, Texas, United States, 26th July 1994

[363] National Transportation Safety Board report

head-on with southbound freight train No. C740 from Townsville to Acacia Ridge at about 20:06 on Thursday 28th July 1994. The accident occurred on the electrified single-track North Coast Line between Beerburram and Elimbah, Queensland, Australia, about a kilometre south of Beerburram. Both drivers, the only train crew, were badly enough hurt to be admitted to hospital. The engine of the southbound train, No. C740, was too badly damaged to be repaired and the line was blocked for 22 hours. An inquiry found that the Automatic Train Control system had operated as designed and that train No. C740 had run past signal BM 18 at Danger. [364]

Several narrow-gauge railways in the Harz region of Saxony-Anhalt state, Germany, were taken over from Deutsche Reichsbahn by the private Harzer Schmalspurbahnen (HSB) company at the beginning of February 1993. One of these lines, the Harquerbahn, suffered a single-line collision only a few months later, on the afternoon of Sunday 21st August 1994.

On the day in question, downhill passenger train No. 8934 should have waited at Drängetal for uphill passenger train No. 8937. In some manner the driver and conductor of train No. 8934 overlooked the crossing and proceeded, whereupon the trains collided near the Thumkuhlental bridge, between Drängetal and Steinerne Renne station, at kilometre 52.4. Because the collision occurred on an 80-metre-radius curve, with the forest further reducing visibility, the train crews only saw one another at the last moment and there was barely time for either driver to brake before they met. Both engines were derailed and several carriages partially telescoped one another or overturned. No one was killed but 31 people were injured, the two engine crews and some of the passengers seriously.

Following the accident the HSB equipped the line with simplified signalling and axle counters and introduced working by Signalisierter Zugleitbetrieber, with train movements being authorised by a central controller. [365] [366]

Just over a month after the Harquerbahn accident there was another single-line collision in Germany, this time in Schleswig-Holstein on the Altona-Kaltenkirchen-Neumünster Eisenbahn. This accident occurred near Bad Bramstedt station and was more serious than the previous one, with six fatalities and about 80 injuries, some serious.

Since 1957 the line had been worked by a form of direct traffic control, with train drivers being given permission by a central controller via train radio to proceed from one crossing place to another.

The trains involved in the accident were both LHB VT 2E diesel railcars. One was southbound from Neumünster, the other was travelling in the opposite direction and they were intended to cross one another at Bad Bramstedt. Between them they were carrying about 80 passengers and both were in charge of experienced drivers. The accident occurred because the northbound train, having stopped at Bad Bramstedt station, set off again from there without its driver having contacted the controller and before the southbound train had arrived. It had proceeded about 500 metres when the trains met on a curve at about 15:24, when each train was travelling at about 60 km/h. Both drivers died in the accident, while another fatality was the 14-year-old daughter of the driver of the southbound train, who had been riding with him in the cab.

---

364 https://www.chapelhill.homeip.net/FamilyHistory/Other/QueenslandHistory/QUEENSLANDRAILWAYDISASTERS.htm

365 https://www.selketalbahn.de/meldung2019/s1908231.php

366 https://chroniknet.de/chronik/zugunglueck-eisenbahnunfall-deutschland-1951-heute

A large number of rescuers took part in the relief operation, which was made more difficult by the location of the collision on a high embankment. The last survivor was removed at 19:00 and the last body at about 20:30.

The accident was the most serious in the 110-year history of the train company. Subsequently, the company equipped the line with a train protection system with fixed signals and train stops, which applied the brakes should a train pass a signal at Danger. The reason for the driver's error could not be determined. The public prosecutor concluded that the accident had been due to human error. [367]

For most of its existence the former London Brighton and South Coast Railway line from Hurst Green Junction, near Oxted, to Uckfield and Lewes, had been double track but in 1989/90 most of it had been singled, including the stretch between Hever and Ashurst. It was on this stretch, near the station at Cowden, England, on what had become by then with privatisation Railtrack South, that a single-line collision occurred at 08:27 on the foggy morning of Saturday 15th October 1994. Following the singling, the line was provided with colourlight signals, provided with the Automatic Warning System (AWS), and was controlled by the signal box at Oxted from an entrance-exit (NX) panel under Track Circuit Block regulations.

The trains involved in the accident were the 08:00 Up passenger service from Uckfield to Oxted (2E27), in charge of Guard Johnathan Brett-Andrews with Driver Brian Barton, and the 08:04 Down passenger working from Oxted to Uckfield (2E24), in charge of Guard Robert Boyd with Driver David Rees. Both were Class 205 diesel electric multiple units. The Down train had left Cowden station and travelled 300 metres when it collided with the Up train. Driver Rees in the Down train and Guard Brett-Andrews, Driver Barton and two passengers, Mr Raymond and Mrs Moira Pointer of Crowborough, in the Up train were killed, while 11 passengers and Guard Boyd were slightly hurt and/or suffered from shock. The leading cars of each train were wrecked, that of the Up train suffering particularly badly, and the second cars badly damaged.

The accident was the subject of a comprehensive inquiry by the Health and Safety Executive, conducted by Major C.B. Holden.

Relief Signalman S.W. Webb at Oxted stated that he had set the route for the Down train to proceed as far as signal No. OD59 and for the Up train as far as Signal No. OD58. The former signal was at the Uckfield end and latter signal was at the Oxted end of the long Ashurst loop, Signal No. OD58 protecting the entrance to the single line thence to Hurst Green Junction. When the Up train entered the Ashurst loop from the Uckfield end he had then cleared his signals for the Down train to continue to Uckfield. He then set about preparing his breakfast, when the sounding of an alarm had warned him that something was amiss and he saw from his panel that points No. 532 at the Oxted end of the Ashurst loop (now referred to as Blackham Junction) were flashing "out of correspondence" and that tracks circuits LF and LD between Ashurst and Hurst Green were both occupied, indicating that the Up train had entered the single line already occupied by the Down train. Realising that a collision was almost unavoidable, he had telephoned Control at Croydon at 08:28 and informed them to call the emergency services. There was nothing more he could do, as there were no fixed signals in the area that he could replace to stop one or both trains, nor was there any means of communication between his box and either of the trains.

Guard Boyd of the Down train testified that there had been a slight delay in starting from Oxted, after which the journey to Cowden had been normal. Shortly after leaving Cowden the brakes had come on in emergency, there had been a 'crunch' and the train had stopped.

[367] Wikipedia – Eisenbahnunfall von Bad Bremstedt

Looking out of the window, he saw that his train was derailed but had not realised that there had been a collision, thinking instead that his train must have struck something. Going forward and finding the driver's cab empty, he had assumed that his driver had gone forward to protect the opposite running line and after reassuring his passengers, had gone back to Cowden station, where he met Rail Operator V.D. Hodges.

Rail Operator Hodges said that he had alighted from the Down train at Cowden station to take up duty there. He had noticed the noise of the collision but finding nothing wrong at the station, dismissed it. At 08:35, the Up train being late, he had telephoned the Oxted supervisor to enquire where it was, to be told that it was at Ashurst. Guard Boyd had then appeared and informed him that the Down train had struck something. Neither of them knew the telephone number of Oxted box so they had to obtain it from the Oxted supervisor, when they had learnt of the collision.

Relief Rail Operator W. Burton stated that he had travelled on the 06:48 train from East Croydon to Edenbridge (between Oxted and Hever) to take up duty there. Guard Brett-Andrews and Driver Barton, both of whom he knew, had been on this train and the guard had been in the driver's cab both at East Croydon and at Edenbridge.

Passenger D.M. Stone related that he had boarded the Up train at Buxted (the first station out of Uckfield) and had seen a man in uniform alongside the driver as the train approached there.

Dr D.A. Lawton related that she had boarded the Up train at Crowborough. The fog had been patchy and dense in places. Obviously a regular traveller, she said that she had been a little surprised not to see the Down train at Ashurst. She had not noticed any bang when the train had run through the points after leaving Ashurst. There had been no fog at that point, and she thought that but for the curve and trees, she would have then been able to see Cowden station up ahead. She had not noticed any brake application before the collision, certainly not an emergency one.

Deputy Controller K. McManus at Croydon spoke of receiving a telephone call from Signalman Webb at Oxted, who had clearly been distraught. Having had the signalman repeat his message to check that he had it right, he had informed Control Manager R.P. Westwood. The two of them had been trying to work out the location of the collision using Ordnance Survey maps and a reference book, so they could inform the emergency services, when they had been advised of it by Kent Police, a passenger from one of the trains having called the emergency services from a nearby farm.

Control Manager Westwood stated that he had advised the police that Cowden station would be the most suitable place for a control room to be set up, while concerned for the welfare of Signalman Webb, had arranged for someone from Oxted station to be with him until he could be relieved.

Area Movements Inspector D.G. Morgan related that he had been 'on call' and on learning of the accident from Croydon Control, he had reached Oxted box about 09:00, where he checked how matters stood and assisted Signalman Webb until the latter could be relieved at about 10:30.

Signal Maintenance Assistant J. Collins stated that he had examined signal No. OD56 at 12:00. It had been showing Red. In the presence of three colleagues he had then opened the signal head and examined it inside, finding all in order apart from some contamination of the lenses.

Signal Maintenance Engineer V.G. McLellan testified that he had checked signal No. OD56 on the day after the accident and considered that it had been showing 'a reasonable red.' He considered the results of a test indicating that it had only 13.6% efficiency to be very questionable. He had examined the signal both externally and internally and checked its alignment. All signals were subject to quarterly and annual maintenance and drivers were able to report any problems with signal sighting.

Technician Officer A. Pearce stated that he had carried out quarterly services on the signal in February and July 1984, and the annual service in April. He had cleaned the signal during the quarterly services and on an extra visit on 12th October he had brushed away a few dead flies. On that day the signal had been visible from Ashurst station.

The signal head was subsequently removed and examined at Crewe. Mr D.A. Jack, head of the Optics Section of British Rail Scientifics there, said that he had found the bodies of a number of dead insects inside the signal lenses, which were contaminated with what proved to be fly excrement. This could only be removed by the equivalent of paint stripper, something not available to signal technicians in the field. There was also a deposit of tungsten from the lamp filaments. It was he who had judged the luminosity of signal No. OD56 to be only 13.6% of a reference lamp. He considered that in foggy conditions, the driver of an Up train would only have had one second to view the signal.

Detailed evidence was given of examinations of the SSI in Oxted box and of the AWS equipment, of examinations of the brakes of the Up train, and so on. All was found to be in order. Further evidence was given relating to the visibility of signal No. OD56 and of the somewhat undesirable interposition of Ashurst station between it and its Distant signal, No. OD56R.

As to the damage to the rolling stock in the collision, Mr J.H. Lewis of British Rail Research stated in relation to the leading vehicle of the Up train, 'In my experience I do not think that I have ever seen a vehicle quite so badly damaged in terms of the amount of bending and twisting of the metal that forms the underframe of the vehicle. The superstructure of the vehicle had been completely wiped off as it had been overridden by the Uckfield-bound train. There was virtually nothing left above floor level at all and the roof structure had been torn off completely and was straddling the two vehicles. The front bogie had come off and the rear bogie was still attached but derailed. Essentially all that was left was the underframe and the superstructure was just a pile of wreckage.' He considered the closing speed at collision to have been between 88 and 104 km/h.

Evidence was given in respect of Driver Barton and Guard Brett-Andrews. Supervisor A.J. Gavin stated that the driver and guard had appeared fit for duty when they signed on duty on the day of the accident.

Traction Inspector A.J. Bushell said that he had been satisfied with the general competence of the driver and guard and added that although Driver Barton was not that familiar with the line, he knew it well enough not to have become lost, even in foggy weather. He pointed out that fog was not a valid reason for the guard to be riding in the driver's cab; the only reasons were a failure of the driver's safety device and examination of the line. After passing over the AWS magnet in fog, the train should have been driven at no more than walking pace until the signal aspect could be seen.

Supervisor G.A. Tryon at Selhurst confirmed that there had been nothing unusual in the demeanour of the driver and guard on the day of the accident.

Trains Inspector L.J. Ward at London Victoria stated that Guard Brett-Andrews had been warned three times about riding in the driver's cab, including a final warning on 4th February 1994.

Subjects such as the signalling of the line, the provision of automatic train protection (ATP), the provision of cab-secure radio (CSR) and the strength of the rolling stock were discussed at some length in the report.

The inspecting officer concluded that it was clear the Up train had run past signal No. OD56 at Danger and that there had been no equipment malfunction. Signal No. OD56 had probably only been visible to the driver of an Up train for between 50 and 20 metres. Its intensity had, however, been no worse than that of a significant number of other signals on the system. The likely speed of the train at the signal was 60 km/h, giving the driver between one and three seconds to see the signal; insufficient time if there had been a distraction at the critical moment.

Major Holden considered that the Caution aspect given by signal No. OD56R and concomitant AWS warning had been forgotten during the subsequent stop at Ashurst station. The further AWS warning at signal No. OD56 had been cancelled in an automatic reflex action and the Red aspect of the signal, visible only very briefly, had then been missed. To have run past the signal, he stated, 'was blameworthy but understandable.' To have continued without having registered the signal was 'not only inexcusable and blameworthy but totally irresponsible.' Regardless of whether he or the guard had been driving the train, Driver Barton was in charge of the driving and so was totally responsible for the accident.

Had cab-secure radio been available, Signalman Webb would likely have been able to prevent the accident. The carrying of mobile telephones as a substitute for radio was unsatisfactory. Carrying one was not mandatory, they were unreliable, their battery endurance was low and there were black spots on the line where they did not work.

The presence of Guard Brett-Andrews in the driver's cab must be regarded as having been an unnecessary distraction, heightened by his acknowledged extroverted character.

Some form of ATS would have prevented the accident and it should eventually be installed throughout the national railway system, beginning with high-speed lines and major resignalling schemes.

The report considered the evidence of the inquest on the five who lost their lives in the accident, at the close of which the jury, at the direction of the coroner, returned verdicts of Accidental Death on Driver Barton and Guard Brett-Andrews and open verdicts on the other victim. It was suggested at the inquest that the AWS had been isolated and that Guard Brett-Andrews had been driving the train, but the inspecting officer concluded that there was no clear evidence that either was the case.

The report reached ten conclusions:

1. There had been no malfunction of the signalling equipment.
2. The Up train had passed Signal OD58 at Danger.
3. At the time of the accident the AWS of the Up train had been operational.
4. The driver of the Up train, Driver Barton, was wholly responsible for the accident.
5. Had cab-secure radio been available, the accident could have been prevented.
6. It was unlikely that the accident would have been prevented entirely by National Radio Network (NRN, a general-purpose radio telephone network) but it might have taken place elsewhere between Cowden and Ashurst and with less violence.
7. No perishable evidence (regarding signal No. OD56, the AWS equipment, the SSI, and so on) had been lost.
8. The Uckfield branch had been signalled in accordance with the standards in force at the time.
9. The line had been adequately and safely signalled.
10. The presence of Guard Brett-Andrews in the driving cab must be regarded as having

been an unnecessary distraction.

The report ended with 15 recommendations:

1. Railtrack should ensure that adequate facilities were in place so that the AWS was properly tested before a train entered service.
2. The instruction to report the isolation of the AWS should be stiffened to require that such reports should be made immediately, either by radio or from the first available telephone.
3. Railtrack should review what information was displayed at telephones.
4. Action should be taken to ensure that control room manuals contained up-to-date information.
5. A review should be made of laboratory procedures to ensure that all testing was subject to rigorous protocols and that proper standards were established against which such tests were undertaken.
6. All clocks or other devices in a signal box that showed or recorded the time should be properly synchronised.
7. In future, before any major change was made to a layout or signalling, a risk assessment should be undertaken.
8. The layout risk study should be completed urgently, say, within the next 6 months, and when it was to hand, Railtrack should propose for consideration by HM Railway Inspectorate a plan for retrospective assessments.
9. The emergency procedures manual that dealt with accident investigation should be revised to cover more than just the preservation of evidence and to emphasise the need for a comprehensive plan for all subsequent investigations to be made at the site.
10. The question of driver distraction in all its forms in the environment of the modern driving cab should be added to the Signals Passed at Danger Reduction and Mitigation (SPADRAM) research project).
11. The SPADRAM research programme should be completed with the utmost priority.
12. (a) As a minimum, all new high-speed railways should be fitted with a form of ATP.

(b) Every major resignalling scheme submitted for approval that did not have a specific proposal for the fitting of ATP should demonstrate that full consideration had been given to its fitting later, together with the risk assessment called for in paragraph 12(d).

(c) Every resignalling scheme should be so designed that the incremental cost of providing the selected form of ATP (which need not be universal throughout the railway network) was minimised.

(d) Every signalling or resignalling scheme that did not incorporate ATP should be accompanied by a risk assessment of the measures selected for the mitigation of SPADs until such time as ATP was fitted.

(e) Once a line was fitted with automatic train control, automatic train protection or some interim measure, no train in normal service that was unfitted to operate that system should be permitted, except under stringent conditions that would give the same level of security, access to that line.

13. (a) Railtrack should produce a timed and costed plan for the completion of CSR on the agreed routes and of NRN to its planned engineering target coverage.

(b) The railway should improve the capacity of NRN to handle emergency messages to and from trains.

14. In the absence of the complete replacement of Mark I rolling stock, which would be the preferred solution:

(a) An urgent programme of research into the practicability of improving the crashworthiness of older designs of rolling stock should again be undertaken.

(b) The results of this research should be discussed with HM Railway Inspectorate within one year, so that an agreed programme of implementation could be drawn up.

(c) The opportunity should be taken in the design of new rolling stock or major refurbishment of old rolling stock to provide accommodation for the installation at the time or later of ATC, ATP or other agreed SPADRAM measures.

15. A timed and costed programme, to be funded by the railway industry as a whole, for the fitting of all traction units having a planned life of more than 3 years with on-train monitoring recorders, should be prepared and submitted to HM Railway Inspectorate for consideration. [368]

Northern Railway train No. 391, a Bikaner – Jaipur passenger service, collided head-on with a tamping machine at Km. 66/9-10 between Borawar and Makrana stations, Rajasthan, India, on the Degana – Phalera non-electrified, single-line, broad-gauge section of the Jodhpur Division of the railway at 10:39 on Saturday 11th February 1995. Four people died in the accident and six were injured.

The passenger train consisted of a Class WDM-2 diesel locomotive and six coaches. It had left Borawar station with the correct Line Clear token for the section, while the tamping machine had entered the single-line section at the Makrana end without authority. As the author has not been able to access the accident report, exactly how this happened is not known.

The Commission of Railway Safety made a number of recommendations in the wake of the accident:

(a) The Down Warner and Outer signals at Makrana should be re-sited to provide the required visibility distance of 1,200 metres from the Outer signal.

(b) The instructions that during periodical signal inspections from the locomotive footplate and thence by line checks, lasting solutions to improve the visibility of signals, wherever restricted, were reiterated. The alternative of making routine remarks regarding matters such as the chopping off of branches was not sufficient.

(c) The instructions that the rules issued vide General Manager (Engineering) Northern Railway's letter No. 219-W/61/(UT)/Pt. V, dated 10.9.84, should be strictly followed in the working of all types of on-track machines were reiterated.

Paragraph 1227 of the Indian Railways Permanent Way manual should be supplemented to specify the distribution of duties and responsibilities between permanent way officials and the track machine operator, as per the rules framed by the Northern Railway.

(d) Suitable supplementary rules should be incorporated in the General and Supplementary Rules, particularly for the guidance of station staff, clarifying that:

(1) During the running of track machines light, the track machine official in charge will perform the duties of driver to whom Line Clear and signals have to be given.

[368] Health and Safety Executive report on railway accident near Cowden on 15th October 1994

(2) In case the machine is to work on a running line at a station or in a block section, the traffic block shall be requisitioned by the permanent way official. The sanction memo and token (if applicable) shall also be handed over to the permanent way official who will guide the movement of the M/C on the track. [369]

What was effectively a single-line collision occurred between Lokur and Danishpet stations, Tamil Nadu, India, on the Jolarpettai – Erode electrified section of the Palghat Division of the Southern Railway, at about 22:15 on Sunday 14th May 1995. Train No. 6019, the Down Madras – Kanyakumari express, collided head-on with an empty Up Super Jumbo goods train. A major catastrophe was the result, 49 passengers and five railwaymen losing their lives and 50 people being injured, 30 of them seriously.

The accident was a peculiar one, as there were, in fact, two lines between the stations involved, but rather than being worked as a conventional double line. they were worked, 'for administrative purposes,' whatever that might mean, as two single lines. Such a method of working is not exactly unique but is certainly most unusual. To the author, it is reminiscent of the working of the Brandling Junction Railway in England in the early days, when after a similar and fatal head-on collision in 1844, the working was altered to that of a normal double-line railway.

In the present case, the express had already entered what was termed the 'A' line at the Lokur end of the section when the goods train entered the same line at the Danishpet end. Both locomotives, eight coaches of the express and some of the waggons of the goods train were derailed and overturned in the resulting collision. Relief trains were sent to the scene from both Jolarpettai and Erode and the injured were taken to the Government hospital in Salem. The accident was apparently due to the driver of the goods train starting from Danishpet against the signals; he, together with the other members of the engine crews of both trains, was killed in the collision. The Danishpet signalman was suspended after the accident, possibly merely as a formality until the investigation had been completed. Not surprisingly, the disaster formed the subject of a prolonged discussion in Parliament. [370]

Only two brief recommendations were made by the Commission of Railway Safety regarding the accident:

(a) The twin-single-line operating system should be examined with a view to converting it into conventional double-line working.

(b) A system should be introduced whereby the Advanced Starting signal released the Starting signal. [371]

Two Danske Statsbaner regional passenger trains collided head-on at the south end of Jelling station, Denmark, where they should have crossed one another, on Tuesday 1st August 1995. The train from Herning arrived at the station first and having discharged its passengers, it then set off again, for unclear reasons, without having awaited the arrival of the opposing train from Frederica (Velje). Fortunately it had not gone far when the two trains met and both drivers applied their brakes before the trains collided at a closing speed of 50 km/h. There were no fatalities but 69 people were injured, of whom 54 were taken to hospital and 19 were

---

[369] Government of India Ministry of Civil Aviation and Tourism Commission of Railway Safety Annual Report for 1994-95

[370] https://rsdebate.nic.in/bitstream/123456789/161914/1/PD_173_16051995_28_p445_p488_11.pdf

[371] Government of India Ministry of Civil Aviation and Tourism Commission of Railway Safety Annual Report for 1995-96

admitted. The trains were badly enough damaged for one of them to have to be cut up at the scene. [372] [373]

Two people died and 23 were injured in a head-on collision between a passenger train and a freight train on the single line near Eisenstadt, Austria, on Friday 15th September 1995. The driver of the freight train, seeing the danger, was able to bring his train to a stand before it was struck by the passenger train, thus reducing the effect of the collision. [374]

At least 16 people lost their lives in a head-on collision between two passenger trains, presumably on a single line, at Kongarish, Bangladesh,[375] near Chandpur and 110 east of Dhaka, on the night of Monday 8th January 1996. A local train heading from the port of Chittagong to Chandpur collided with the *Sagarika Express* bound for Chittagong, on an open stretch of line with visibility extending for several miles. Both engines and four carriages were derailed in the collision and in addition to those killed over 100 people were said to have been injured. It was feared that more bodies might be found as civilian rescuers, aided by troops, scoured the wreckage.

A three-person committee was set up to investigate the accident. At least one railway official, who wished to remain unidentified, told a press reporter that communications along the line had been made difficult since Sunday by the theft of signalling cables. [376]

What was apparently a single-line collision occurred near Mwekera, Zambia, early on Wednesday 17th January 1996. Two passenger trains, crowded mostly with people returning home from visits to markets, were involved. By the following day the bodies of 31 victims had been recovered and it was feared that more dead remained in the wreckage. The accident was blamed on a signal failure. [377]

On the morning of Monday 9th December 1996 a passenger train, probably a diesel railcar, was in collision with a freight train consisting of a locomotive and 21 waggons on the single line between Hrubá Skála and Karlovice-Sedmihorky, in the Czech Republic. [378]

The accident occurred because Driver Petr Mitlehner had started away from the station at Hrubá Skála station with the passenger train, in which there were about a hundred passengers, without waiting for the signalman's instructions. He defended himself by saying that he believed there had been no one on duty at that hour. After a few minutes his train collided with the freight train, which he should have crossed at Hrubá Skála. An elderly woman was killed on the spot and 50 people were injured, about 12 people seriously.

In February 1999 Driver Mitlehner was given a two-year suspended sentence by Semila District Court. In September of the same year, the regional court in Hradec Králové overturned the original sentence and sentenced him to three years' imprisonment as well as banning him from operating the train for six years. In justifying the increased sentence, the judge stated, "The defendant had a huge responsibility to people and property and he failed in

[372] Wikipedia – Togulykken I Jelling
[373] https://www.toglyde.dk/ulykke.htm
[374] *Western Morning News*, Plymouth, England, 16th September 1995
[375] *The Times*, London, England, 10th January 1996
[376] https://www.upi.com/Archives/1996/01/09/Bangladesh-train-crash/5290821163600/
[377] *Cedar Rapids Gazette*, Cedar Rapids, Iowa, United States, 19th January 1996
[378] https://cesky.radio.cz/zpravy-utery-10-prosince-1996-8055305

this capacity. The basic and principal cause of most railway accidents is gross violation of rules and regulations, as was the case in this instance." In justifying the severity of the sentence, the judge stated that he was surprised the defendant did not even know of the so-called notebook timetable he had at his disposal. "It gives me goosebumps to think that passenger trains are driven by people who don't know the basic regulations," the judge said.

Driver Mitlehner, 51, admitted his responsibility for the accident and sincerely apologised to the relatives of the deceased and the survivors of the accident, while adding, after the trial, that he was certain that others involved in running the train should have been in the dock with him. [379]

At 16:00 on Monday 10th February 1997 a passenger train from Bayamo to Habana, carrying many military conscripts from the eastern provinces, collided with a light engine belonging to the Sugar Ministry at Caguasal, Cuba, 10 kilometres west of the city of Ciego de Ávila. Fourteen people were initially reported killed and 173 injured, 15 of them extremely seriously. Radio Rebelde stated that 40 of the injured had been admitted to the Provincial Hospital in Ciego de Ávila.[380] The number of dead appears to have later climbed to 18.[381] An initial inquiry attributed the accident to the passenger train having 'violated the right of way' of the engine. [382]

Two Union Pacific Railroad freight trains collided head-on at 10:52 p.m. on Sunday 22nd June 1997 near Devine, Texas, United States. With only freight trains involved, the accident was not a major disaster but the engineer of one train, the conductor of the other and two transients riding on one of the trains lost their lives; the remaining conductor was seriously hurt and the other engineer was slightly hurt. The accident occurred in un-signalled so-called 'dark territory,' with train movements and meeting places on the single line being controlled by a dispatcher, who issued track warrants to the crews of trains by radio. About a dozen trains in each direction passed over the line daily. The National Transportation Safety Board (NTSB) investigated the accident.

The NTSB again described the procedure for issuing track warrants, given in its report on the 1991 Fowler collision. To establish a meeting point between opposing trains, it stated, the train dispatcher issued a track warrant to one train that included, at a minimum, the "proceed from" and "to" limits being granted and a box that instructed it to "clear main track at last named point". The train travelling in the opposite direction was issued a track warrant which included, at a minimum, the "proceed from" and "to" limits being granted and an instruction to "hold main track at last named point" (the "to" limit granted). The train was instructed to clear the main track at the station where the meet was to take place and was not authorised to proceed on the main track beyond the initial switch (entry points) of the siding without additional track warrant authority. The train instructed to hold the main track was not authorised to proceed beyond the switch (exit points) at the end of the "to" station until additional track warrant authority was issued. This system protected the two opposing trains, and the physical passing of the trains was accomplished. When one train arrived at a crossing place before the other, the dispatcher could either wait until the other train had arrived before issuing a further track warrant, or alternatively he could issue a track warrant to the first train to arrive at the crossing place that it could proceed further upon the arrival of the opposing

379 https://www.idnes.cz/zpravy/cerna-kronika/soud-poslal-strojvedouciho-na-tri-roky-do-vezeni.A_990927_172327_krimi_jpl

380 *El Nuevo Herald*, Miami, Florida, United States, 12th February 1997

381 *The Medicine Hat News*, Medicine Hat, Alberta, Canada, 7th November 1997

382 *Sedalia Democrat*, Sedalia, Missouri, United States, 10th February 1997

train (conditional track warrant control authority). A computer was used by the dispatcher to generate orders and had an inbuilt logic to prevent the issuing of lap orders (an order authorising one train to proceed from A through B and C to D, at the same time as another order authorised an opposing train to proceed from C to B, thus giving two trains the right to the line between B and C at the same time – in earlier times the dispatcher's nightmare).

The trains involved in the accident were No. 5981 North from Laredo, Texas, to Baton Rouge, Louisiana, consisting of three diesel units, 83 loaded and 11 empty cars, and No. 9186 South, from Chicago, Illinois, to Mexico, consisting of two diesel units and 62 loaded cars. At 2:30 p.m. on the day of the accident train No. 5981 North was given a track warrant by the Austin subdivision second-shift (late turn) dispatcher at Omaha, Nebraska, to proceed from Laredo, Texas (all the following places are also in Texas), to Callaghan, which it did. A southbound train was crossed there, then the dispatcher authorised the train to continue to Gardendale. At Gardendale, a second southbound train was crossed and here the U.S. Border Control removed two suspected illegal immigrants from the train. Gardendale was left at 6:45 p.m. The train proceeded to Derby, where eight ballast cars were added to it, thence to Melon, where the engineer left the train to purchase food at a store. At 9:18 p.m., at Melon, the second-shift dispatcher instructed the train to proceed to the siding at Gessner. At 10:30 p.m. the third-shift (night turn) dispatcher then issued a track warrant for the train to proceed from Gessner to San Antonio, after the arrival of train No. 9186 South at Gessner.

As train No. 5981 North reached Devine, 27.6 miles north of Melon but still 11.9 miles short of Gessner, Engineer Brian W. Bolner and Conductor Terry Yarbrough saw a bright glow on the horizon and wondered what it could be. A little later it became obvious that it was the headlight of an approaching engine. Engineer Bolner then placed his brakes in emergency and jumped, while Conductor Yarbrough, who was killed, either also jumped or was thrown from the engine.[383] Engineer Neil A. Wilhelm of the southbound train was also killed, while Conductor Randy Dennis on that train was critically injured. [384]

At 9:56 p.m. the second-shift dispatcher had issued track warrant No. 8261 to train No. 9186 South to proceed to Gessner, and the train left San Antonio at 10:10 p.m. At 10:28 p.m. the third-shift dispatcher issued track warrant No. 8289 to the train. This second track warrant contained four instructions (called 'elements' in the NTSB report):

(a) Proceed from Gessner to Melon on main track.

(b) Not in effect until arrival of 5981 North at Gessner.

(c) Hold the main track at the last named point {Melon}.

(d) Flag protection not required against following trains on the same track.

The second and third-shift dispatchers changed shifts between 10:15 and 10:20 p.m. and discussed the situation regarding what trains were where at that time. It was reasonably busy. At 10:20 p.m. the third-shift dispatcher spoke via the radio to the crew of an Amtrak passenger train and at 10:24 p.m. to the crew of another freight train. This call ended at 10:26 p.m. when the dispatcher called Conductor Dennis of train No. 9186 South and read out the instructions given in track warrant No. 8289, but crucially omitted the item: 'Not in effect until arrival of 5981 North at Gessner.' Conductor Dennis repeated back the instructions, when the dispatcher failed to notice the omission and okayed the order at 10:29 p.m. The dispatcher then radioed the track warrant to train No. 5981 North, this time correctly (the radio calls were all recorded).

Acting on the incorrect information received from the dispatcher, train No. 9186 South thus proceeded beyond Gessner without waiting for train No. 5981 North and the trains

[383] National Transportation Safety Board report

[384] *Texas City Sun*, Galveston, Texas, United States, 25th June 1997

collided on a road overbridge at Devine. The two units of the southbound train fell from the bridge to the road below following the collision, while the three units of the northbound train, although they did not fall from the bridge, were badly damaged. Freight cars from both trains piled up behind the engines and a fierce fire broke out in the wreckage from spilt diesel fuel.

The NTSB noted that several dispatchers reported that the most difficult time for a dispatcher was during the first 30 minutes of the shift, when a dispatcher was "trying to assimilate everything" and mentally planning the operation of the territory during that shift. At this time, the relieving dispatcher had to process the information just received from the briefing of the departing dispatcher, prioritise the tasks that were to be performed during the current shift, and respond to radio transmissions received from train crews that could not be responded to during the shift changeover. In addition, supervisors often called the dispatchers during the first 10 to 15 minutes to be apprised of operations on the territory. The dispatchers at Omaha operated inside a station with waist-high walls partitioning one dispatcher from another. Several dispatchers, including the third-shift dispatcher, stated that radio conversations between other dispatchers and operating crews could be overheard, that adjacent dispatchers also conversed with each other about train movements and other operational matters, and that conversations and other noise were greatest during shift changeover. Intercoms were available for communications but were generally not used. Many dispatchers reported that they found the noise distracting and believed that glass barriers to muffle sounds would decrease the noise and provide a better working environment.

The third-shift dispatcher reported that he usually directed between 15 and 20 trains on the three subdivisions in his territory. The UP indicated that, on average, dispatchers dealt with 21 telephone and radio calls each hour, talking to train crews, track maintenance workers, and dispatch centre supervisors. It added that when discussing track warrant information, dispatcher and train crew conversations typically took about 2 minutes. According to the Federal Railroad Administration (FRA), one dispatcher stated that he spent about 6 hours of his shift on the telephone and radio. According to dispatchers interviewed, because of the high workload on several territories, some dispatchers issuing track warrants might be mentally or physically attending to their next task and not concentrating on the read-back communication from the train crew. The dispatchers also stated that newly qualified dispatchers were involved in territories of high operating demands. Several dispatchers reported experiencing increased stress while on duty and specifically cited the following as contributing factors to the elevated stress levels: increased workload due to increased train volume and congestion, the lack of support from management in recognising the dispatching challenges, the management requirement that they work on scheduled off days, and the frustrated attitude of the traincrews. The UP management reported that some of the recent concerns expressed by the dispatchers were volume-related issues. Dispatchers stated that the workload continued to increase after the Devine accident due to increased train volume. Their duties focused on moving trains in and out of sidings, dispatching vans to pick up crewmembers who had reached the maximum hours-of-service time, dispatching relief crewmembers, and responding to an increased number of phone calls from the field personnel.

The NTSB report went on to review in detail the training of dispatchers, the injuries sustained by the various crew members, the rescue operation and so on.

Analysing the accident, the NTSB determined that neither the weather, the track, nor the train equipment had contributed to the accident. None of those involved had been overtired, nor had any of them consumed alcohol or drugs. The computer-generated track warrant No. 8289 had been correct. However, the third-shift dispatcher had failed to convey accurately the information on the track warrant to the conductor of the southbound train or to detect his error when the information was read back to him.

The dispatcher had been working as such since August of the previous year and had made no previous mistakes. The NTSB remarked that the track warrant verification procedure had no redundancy procedure to confirm that the read-back of the transmission had been accurate. The dispatcher's error had been made during the first 10 minutes of his shift, as were approximately 30 per cent of all dispatcher violations.

The NTSB commented, 'Many of the territories to which less experienced dispatchers are initially assigned, such as the Austin subdivision, have nearly doubled in train volume since the early 1990s, when they might have been more easily dispatched because of the fewer trains operating. Such territories often pose operational challenges to even the most experienced dispatchers. Veteran dispatchers reported that under conditions of high-operating demands, less experienced dispatchers might issue track warrants while mentally or physically attending to their next task and not concentrating on the read-back communication from the train crewmembers. The FRA noted during its safety audit of the {dispatching centre} that dispatchers working under high-workload conditions were not consistently monitoring the computer screens during read-backs of track warrants because of other task demands, which included answering the telephone, communicating with other dispatchers, and reading line-ups and performing transfers with their relief shift dispatchers. Some dispatchers, as a result, might forgo safe dispatching practices in an attempt to manage the high-operating demands.'

The report went on to consider matters such as management oversight of the dispatchers, the distractions suffered by the dispatchers due to the noise generated by the conversations of other dispatchers, the additional risk posed by the use of conditional track warrants (after-arrival orders), the emergency response to the accident, the crashworthiness of the locomotives and the crashworthiness of the event recorders (three of the five of these on the diesel units had been destroyed in the collision and/or fire).

The subject of Positive Train Separation (PTS), which automatically intervened should an engineer disregard a signal, was then raised and the NTSB noted it had been advocating it since 1970.

The NTSB listed 17 conclusions regarding the accident:

1. Neither the weather, the train equipment, nor the track had caused or contributed to the collision. The computer-generated track warrant 8289 information had been correct. Both operating traincrews had been in compliance with the requirements specified in the Hours-of-Service Act and qualified to perform their duties; no crewmember or dispatcher fatigue was indicated. Neither alcohol nor drug use appeared to have been a factor in the accident.
2. The third-shift dispatcher's failure to accurately issue track warrant 8289 to train 9186 South and his failure to detect and correct the 9186 South conductor's repeat of the track warrant authority limit had resulted in the crew receiving an incorrect track warrant that allowed the opposing trains 5981 North and 9186 South to operate on the same track in opposite directions through Devine on June 22, 1997.
3. The third-shift dispatcher had not communicated the accurate information in track warrant 8289 to the crew of train 9186 South.
4. The Union Pacific Railroad dispatchers' elevated workload at the beginning of shifts might have contributed to the disproportionately greater number of dispatching violations occurring during this time.
5. Some Union Pacific Railroad apprentice dispatchers might not have been adequately prepared to be placed and operate safely in territories of high-operating demands, immediately after completing the training program.
6. The Union Pacific Railroad might have jeopardised safe dispatching operations by qualifying unprepared apprentice dispatchers and assigning less experienced dispatchers to territories of high-operating demands.

7. Because the Union Pacific Railroad had not met its 5-year experience standard for on-the-job (OJT) dispatcher trainers, complying with the higher standard of a minimum 10-year experience level for OJT dispatcher trainers might not have been achieved.

8. The third-shift dispatcher's failure to communicate the information in track warrant 8289 accurately to the 9186 South traincrew and to verify the accuracy of the read-back information had resulted from operational shortcomings at the Harriman Dispatch Center {in Omaha, Nebraska}.

9. Some Union Pacific Railroad corridor managers had not consistently provided appropriate technical support to the train dispatchers.

10. Although no evidence was found that adjacent noises in the dispatching area had contributed to the third-shift dispatcher's inattention to the track warrant 8289 information in the Devine accident, a dispatcher's performance might be affected by unnecessary, avoidable sound distractions.

11. Had the Union Pacific Railroad after-arrival system in dark (non-signalised) territory operations not been used in the Devine accident area, the opposing trains 5981 North and 9186 South would not have been occupying the same block of track.

12. The Federal Railroad Administration had failed to develop dispatcher standards and needed to accelerate the establishment of regulatory standards for train dispatchers.

13. The Federal Railroad Administration surveillance and enforcement of compliance with Federal regulations at the Harriman Dispatch Center before the Devine accident had been inadequate and ineffective.

14. Although the local emergency response had been timely and adequate, the lack of readily available fire suppression foam equipment showed a need for additional firefighting equipment to mitigate such significant fire events.

15. Based upon the estimated speeds of the locomotives at impact, the loss of survival space in the locomotives, and the severity of the massive fire, the collision in Devine had not been survivable for crewmembers or anyone occupying the locomotive equipment at the time of the impact.

16. Had the event recorders been designed to withstand crash forces and fire exposure, the three destroyed would have survived and could have provided data for the investigation.

17. Had a positive train separation control system been installed and working in the Devine accident area, the two trains would not have been allowed to enter the same block of track travelling in opposite directions and, as a result, the head-on collision on June 22nd, 1997, would not have occurred.

The NTSB determined that the probable cause of this accident had been the failure of the third-shift dispatcher to communicate the correct track warrant information to the train crew and to verify the accuracy of the read-back information, because the Union Pacific Railroad management had not established and implemented workload policies and operational procedures to ensure a safe dispatching system and the Federal Railroad Administration had failed to provide standards and oversight in all aspects of train dispatching operations. Contributing to the accident had been the lack of an installed positive train separation control system that would have prevented the trains from colliding by automatically intervening in their operation because of inappropriate actions being taken.

The NTSB made a series of recommendations:

To the Union Pacific Railroad:

(a) It should evaluate its dispatcher training program and make necessary revisions to place greater emphasis on all safety-critical activities, including procedures used to issue and confirm track warrants.

(b) It should conduct an audit of its train dispatching operations to identify specific factors that could lead to dispatching errors and include in the audit an assessment of dispatching errors that occurred during or shortly after shift changes or because of improper radio procedures.

(c) It should conduct an audit of its train dispatchers' activities to evaluate the current workload and make necessary changes to dispatcher operations to distribute workload based on the individual dispatcher's qualifications, ability, and experience.

(d) It should examine the circumstances in which its policy to require a minimum 5 years of experience to qualify as an OJT dispatcher trainer was not followed and take action to ensure that its qualification policies were followed.

(e) It should develop and implement a comprehensive program to select and train experienced dispatchers to serve as dispatcher trainers.

(f) It should evaluate and determine the technical expertise required of corridor managers and make the necessary changes to ensure that corridor managers were qualified to provide proper dispatching assistance to the train dispatchers.

(g) It should identify all distractions, evaluate their effects on dispatchers, and take action to establish a working environment conducive to safe dispatching operations.

(h) It should discontinue permanently the use of after-arrival orders in dark (non-signalised) territory.

To the Federal Railroad Administration:

(a) It should revise 49 Code of Federal Regulations 220 to address track warrants and other current railroad operating practices.

(b) It should require railroads to discontinue permanently the use of after-arrival orders in dark (non-signalised) territory.

(c) It should develop and establish dispatcher selection and training standards, dispatcher trainer standards, and workload limits for dispatchers by January 1st, 2000.

(d) It should evaluate its surveillance and enforcement activities at dispatching centers and take appropriate corrective actions to ensure that Federal oversight was adequate and effective.

(e) Working with the railroad industry, it should develop and implement event recorder crashworthiness standards for all new or rebuilt locomotives by January 1st, 2000.

To the Texas Railroad Commission:

(a) It should develop a system that would make fire suppression foam equipment readily available to emergency management agencies and local rural fire departments for the fighting of hazardous materials fires.

Also, as a result of its investigation, the National Transportation Safety Board reiterated the following recommendation:

To the Federal Railroad Administration:

It should promulgate Federal standards to require the installation and operation of a train control system on main-line tracks that would provide for positive separation of all trains. [385]

---

[385] National Transportation Safety Board report

Two trains collided head-on, presumably on the single track, at an unstated location on the Kinshasa – Matadi Railway in the Democratic Republic of Congo on Monday 20th October 1997, due, it was stated, to a 'faulty signal.' Twelve people died and at least 10 were injured and admitted to several hospitals in Kinshasa. The report on the accident stated that it was the third of its type to occur in the country that year, the second being a collision involving a goods train and a line-inspection car near Tadidi Mosi a fortnight earlier, and the first at an unstated date having occurred at Lufuntoto, about 100 kilometres from Kinshasa, in which at least 100 people had died. It has not been possible to trace any information about this apparently earlier major catastrophe and it seems likely that it refers to a tragedy that occurred on Sunday 4th May 1997, when the bodies of about 100 Rwandan Hutu refugees were found aboard a train after it had arrived at Kisangani, Zaire, having either been trampled to death or else suffocated in grossly overcrowded coaches. Awful though this was, it was not, of course, a train accident as such. [386] [387]

Two passenger trains collided head-on in a ravine at Parchal, Portugal, between Estômbar and Portimão, Portugal, at 18:50 on Saturday 8th November 1997. The force of the collision caused the coaches of one train to climb on top of the other and engine No. 1215 was badly damaged. Three of the approximately 80 passengers and one of the drivers were killed and 14 people were hurt.

Regional train No. 5806, the 18:05 service from Tunes to Lagos, should have waited at Estômbar station to cross train No. IR 981, the 18:20 service from Lagos to Real de Santo António, but it did not. It was believed that the driver of the regional train had forgotten that he had to cross the other train there, as this was only the case on Saturdays. [388]

Less than a month after the Portugal accident there was a single-line collision in Luxembourg, on Monday 1st December 1997. The guard of train No. 2012, an electric multiple unit, gave his driver the "right away" signal to indicate that the station work was complete, whereupon the driver started away from Hollerich station against the Starting signal at Red and entered the single line. Before the train had progressed very far, it was in collision with an empty coaching stock consisting of engine No. 3616 and seven carriages. Fortunately the driver of the latter had seen what was about to occur and was able to brake almost to a standstill before the impact. No one was killed but 80 passengers were injured, 20 seriously. This was the second single-line collision suffered by the Société Nationale des Chemins de Fer Luxembourgeois in 1997; in the previous one on 6th April, a driver had died. The earlier accident had also been caused by a driver running past a Red signal. [389]

Three people died and 49 were injured, of whom three were said to be in a critical condition, when two trains collided on the single line between Shirbin and Bilqas, Egypt, on the line between Cairo and Damietta in Mansura province, on the evening of Saturday 10th April 1999. Both trains were crowded with workmen returning home.[390] One of the trains was thrown into a field by the force of the collision and caught fire, while the other was badly damaged. A woman passenger who did not wish to give her name related that the crash had

[386] *allAfrica.com*, 23rd October 1997
[387] *Daily News-Record*, Harrisonburg, Virginia, United States, 5th May 1997
[388] *Today's Railways*, Sheffield, England, March 1998
[389] Ibid
[390] *BBC News*, London, England, 11th April 1999

been "like a bomb," and said that after the fire had started, "I heard people screaming as if they were being fried like lambs."

From his hospital bed, badly burned and swathed in bandages, Driver Mohammed Amin Sayed Ahmen stated, "I was given permission from the signalman to proceed along the line. About 50 metres away I saw another train coming from the opposite direction. I attempted to pull up but there was no time. My engine and the leading coach were pushed off the line. My engine and the leading coach caught fire. I was able to break a window and jump out." [391]

Initially it was thought that Driver Ahmed Saad Abdel-Meguid, in charge of the other train, had lost his life in the accident but it transpired that he had been unhurt and attempted to leave the scene. He was later arrested and faced charges of negligence and with, as a news agency report mysteriously put it, 'travelling without a signalman in a rural area.' In all, 17 railwaymen, including superior officers, were suspended for negligence. The exact mechanism of the accident is unclear from the news reports, however, it appears that telephone communications between the stations had failed and it may be supposed that the rural area signalman referred to in the news report was in fact a pilotman, who in such circumstances was required to accompany each train through the section, or at any rate to be at the end from which a train was authorised to depart, if a second train was to follow in the same direction. Due to a mix-up, a train must have been allowed to depart from one end without the pilotman while he was accompanying or had authorised a train to depart from the other end. [392]

On Tuesday 4th January 2000 southbound passenger train No. 2302 left Trondheim, Norway, on time with 75 passengers at 07:45 for Hamar. It consisted of a Class Di 3 diesel locomotive and three coaches, in charge of Conductor Jan Anders Arneberg with Driver Stig Juliussen. Due to waiting for connections, it was 21 minutes late leaving Røros but time was then regained and on departure from Rena station the delay had been reduced to seven minutes. In the opposite direction northbound passenger train No. 2369, a two-car Class 92 diesel multiple unit, had left Hamar station for Rena at 12:30 with just 10 passengers. It was in charge of Conductor Erik Stårvik, 31, with Motorman Egil Lodgaard, 58. At Rustad station one passenger boarded the northbound train and it left there at 13:07, now with nine passengers as two had previously alighted, plus its motorman and conductor. At 13:12½ the trains collided head-on on the single line at kilometre 182.75, at Åsta, Norway.

The trains met on a gentle curve. It was cold (-5 degrees Celsius) and there was fresh snow on the ground. With the curve and trees bordering the line on either side it was estimated that the drivers could only have seen one another for four seconds, when their trains were 200 metres apart. The tachograph from the northbound train indicated that it had been travelling at between 80 and 90 km/h at the moment of the collision with no indication of a brake application. The tachograph of the southbound train was destroyed in the collision and subsequent fire, but it had covered 7.1 kilometres from Rena in five minutes and 20 seconds, giving it an average speed of about 80 km/h. The maximum permitted line speed was 90 km/h.

In the collision the engine of the southbound train climbed over the front car of the northbound one before overturning down the embankment, while its leading coach was half-demolished. Coaches 2 and 3 of the southbound train were little damaged. The front car of the northbound train was totally destroyed and about 1,500 litres of diesel fuel from the ruptured tanks of both trains ignited after the collision.

Motorman Lodgaard, Conductor Stårvik and six passengers in the northbound train and Driver Juliussen and 10 passengers in the southbound train lost their lives, while 67 people

---

[391] *Janesville Gazette*, Janesville, Wisconsin, United States, 11th April 1999

[392] *Daily News*, Estherville, Iowa, United States, 12th April 1999

were injured. At least two passengers in the leading coach of the southbound train were still alive and conscious when rescuers arrived 40 minutes after the collision but were trapped and died in the fire before they could be released. All the passengers in the two rearmost coaches of the southbound train survived, as did 22 of those in the leading coach, whereas only three passengers from the northbound train survived.

The first report of the accident came from local resident Torbjørn Tollersrud, who rang the police at Elverum just half a minute after the collision, to say that he had heard a tremendous bang and that a plume of smoke was visible. He added that he was sure something had happened on the railway line. The police in turn alerted the Elverum fire brigade at 13:17 and the Elverum AMK (Akuttmedisinsk kommunikasjonssentral, Emergency Medical Communications centre) at 13:18. By 13:27 hospitals in Gjøvik, Hamar, Kongsvinger, Tynset and Oslo had also been alerted to the accident.

The first rescuers to arrive at the scene of the accident, at 13:25, were lensmann (police sergeant) Per Ravnkleven and lensmannsbetjenter (police constables) Henning Farstad and Bjørn Fredrik Monné. They found that two trains had collided and that they were on fire. The passengers in the two rearmost carriages of the southbound train had already alighted from it and some of them were active in rescuing passengers from the leading coach and in ineffectual attempts to fight the fire with extinguishers. The first ambulance arrived from Åmot at 13:30 and the first fire engine at about 13:32.

The firemen concentrated their efforts on saving the lives of the passengers who were trapped in the wreckage, but the intensity of the diesel-fed blaze frustrated their task. For example, the use of foam had little effect, for with the very high temperatures it simply evaporated. Eventually, attempts to free several of the victims had to be abandoned and all their efforts were concentrated on saving one woman. With the use of hydraulic equipment, she was eventually freed at 17:00, apparently not too badly hurt, and was flown by air ambulance to hospital at Ullevål. In all, about 600 people were involved in the rescue operation. [393]

A television station in England reported, 'A British skier among survivors of a Norwegian train crash has told how he and seven friends dragged out passengers as flames engulfed their carriage. Ben Stephenson, 30, was among a party of British skiers in their 20s and 30s who were inside the train travelling towards the city of Hamar when it collided with an oncoming locomotive. At least seven people died in the collision with more still unaccounted for.

'The Foreign Office confirmed on Tuesday night that eight British nationals were on board when the two passenger trains collided near the town of Lillehammer. They had been on a New Year skiing trip and were on their way to Oslo for the return flight home on Thursday. Among the British party were Fiona Donaldson and Richard Stephenson of Oxford, although the other six also have connections with Oxford, either through work or through studying at the university. Of the eight, three were taken to hospital in Elverum suffering minor injuries; two were discharged on Tuesday night after treatment.

'Mr. Stephenson, from Chester, said he was faced with a race against time as fuel sprayed into the wreckage from a crumpled locomotive. He said a number of passengers were obviously dead inside his group's carriage after the collision, but others were trapped behind seats and tables unable to escape. "We were just sitting on the train when without warning there was an almighty bang and the train just rocked on to its side and burst into flames," he said. Inside the carriage there were those who were obviously not alive, but others were trapped by the wreckage and diesel was spraying all over the place. The fire was coming down the train very quickly and we needed to get people out before they were burned alive. We had to scramble out of the

[393] *Åsta-ulykka* by Jan Erik Kregnes

broken window on to the track, where we were helped by local villagers."

'Mr. Stephenson said he and his friends, who all escaped serious injury, were approaching the end of a seven-day skiing trip over the New Year. He and fellow survivors, most of whom lost their luggage in the fire, were taken to nearby homes where they were wrapped in blankets and given hot drinks. "We have nothing at all left, everything was burned including our passports," he said. The group has since been moved to a hotel near the crash scene. The Foreign Office has sent a consular official to the scene to help with arrangements to bring the party home. A spokesman said he hoped they would all be able to fly home later on Wednesday.' [394]

The Rørosbanen, the line on which the accident occurred, was worked by Centralised Traffic Control (CTC) with a control centre at Hamar and colourlight signals. According to the timetable, train No. 2369 should have waited at Rustad until 13:10, while from the signal indications logged onto the panel at Hamar the starting signals had been displaying Green for southbound train No. 2302 at Rena and Red for northbound train No. 2369 at Rustad. It was also clear afterwards that train No. 2369 had run through the points at the end of the crossing loop at Rustad, which had been set for the arriving southbound train. The line was not at the time of the accident equipped with Automatic Train Control (ATC) to prevent a train starting against a Danger signal, nor was there an audible alarm in Hamar panel to alert the signalman there that the northbound train was overrunning signals. There was a visual alarm but at 13:07 Signalman Nybakken at Hamar had been preoccupied with traffic on the busier main line between Hamar and Eidsvoll, so that he did not observe the alarm until 13:11½. There was also no train radio provided on the Rørosbanen. It would have been possible to get in touch with the drivers or conductors of the trains using mobile telephones, but the relevant telephone numbers were either recorded on the wrong list or were not recorded at all. After some delay the mobile number for the engine of the southbound train was obtained from DROPS Nord (driftsoperative senteret – the traffic operation centre in Trondheim) and an attempt was made to call its driver by Signalman Nybakken at 13:18, but by then, of course, the trains had already collided. At almost the same moment a telephone call was received by Hamar from a railwayman, Torbjørn Digre, who had been a passenger on the southbound train, reporting what had happened.

After the Tretten disaster of February 1975, described in a previous volume of this series, which had likewise been due to a driver entering a single line against a Starting signal at danger, Norges Statsbaner (NSB), the state railway company, had begun installing ATC on the electrified main lines of the country, beginning in 1979. The installation of CTC and ATC on all NSB lines was recommended but as a secondary line, which might anyway shortly face closure, the Rørosbanen remained without it at the beginning of 2000, mainly because of the cost. A train radio system known as Scanet was also restricted to lines deemed to be profitable. Finally, in September 1997 the NSB had changed the regulations so that only the driver and not both the driver and conductor, as before, was responsible for ensuring that the Starting signal was clear before the train set off. The railway inspectorate was not happy about this change, especially as regards the special conditions on the Rørosbanen, and protested to the Ministry of Transport, without the old procedure being reintroduced.

The commission set up to investigate the disaster reached 14 conclusions:

(1) Neither driver error nor a wrong-side signal failure could be excluded as a cause of the accident.

(2) The alteration in the operating procedures by which the driver alone was entrusted with ensuring that the Starting signal was Clear before departure had never been

---

[394] http://news.bbc..../591103.st

approved by Tilsynet (the railway authority) but had continued, nevertheless. Thus, just the driver of train No. 2369 had been responsible for checking the indication of the Starting signal at Rustad.

(3) When the CTC system had come into use, with control of train movements from Hamar, the togekspeditørene (local train dispatchers) had been withdrawn. Had they been retained, the risk of an undetected wrong-side signal failure would have been reduced.

(4) If the scheduled train crossing points had still been denoted in the working timetable and the regulations had required the driver to contact the signalman if there was any alteration in the crossing place, even if the Starting signal was Green, there would have been a barrier against either a wrong-side signal failure or a misreading of the signal by a driver.

(5) Had ATC been installed on the line, it would have prevented the driver of the northbound train from passing the Starting signal at Danger at Rustad.

(6) Because the Rørosbanen was not electrified there had been no possibility of the signalman stopping the trains by turning off the current in an emergency.

(7) The lack of audible alarms in the Hamar control centre had prevented the signalman from being immediately alerted that two trains were on a collision course.

(8) In the absence of audible alarms there should have been clear rules stipulating how frequently the Hamar signalman should monitor each portion of the line.

(9) Had train radios been provided, the Hamar signalman could have at once contacted the trains when he had observed what was happening on his panel and so prevented the collision.

(10) There were no regulations relating to the reporting, reception or storage of mobile telephone numbers at Hamar control centre.

(11) There were insufficient regulations and routines regarding shift changes.

(12) There was a lack of rules and regulations regarding the exchange of information between DROPS Nord in Trondheim and Hamar control centre.

(13) Neither the NSB nor Jernbaneverket (the government agency responsible for the administration of the railway) had carried out a risk analysis in relation to safety concerns related to the alterations and delays of plans for the Rørosbanen.

(14) There had been no systematic review of safety on the Rørosbanen during the 1990s, with the result that serious safety deficiencies had been neither analysed nor followed up.

The commission made eight recommendations:

(1) Measures should be implemented to ensure that a proactive safety management was applied to the general railway operation.

(2) There should be a safety manager for railway operations with a direct line to senior managers.

(3) Responsibility for the follow-up and implementation of safety measures must be in the line.

(4) Jernbaneverket should intensify its efforts to develop an adequate and efficient internal control system in all parts of its operation.

(5) Competence, requirements and training plans should be devised for all staff with

safety responsibilities.

(6) A risk analysis should be conducted to assess the risks connected with railway operations.

(7) Measures should be implemented to increase staff motivation to report and give feedback on undesirable incidents in all parts of the organisation.

(8) Reports of unwanted incidents should be compared and systemised to determine if failures were being repeated and if they were safety-critical. [395]

Automatic train control was subsequently introduced on the line, together with train radios.

The annual report of the Commission of Railway Safety for 1999-2000 details a collision that occurred between Saharsa Junction and Sombarsa Kachery stations, Bihar, India, on the Mansi – Saharsa metre-gauge section of the Samastipur Division of the Northeastern Railway on Monday 6th March 2000. The circumstances of the accident are not entirely clear from the details given in the report but it apparently involved a head-on collision between Down Class YDM-4 diesel locomotive No. 6487 and an Up Patrol Special. The report found that the accident had resulted from 'overspeeding in slightly foggy weather, not continuously whistling from whistle board to unmanned level crossing No. 28, not lighting the head light, not using the VHF set and non-vigilance by the driver of the light engine, in condition of total failure of communication in the above block section and outside interference by persons on the light engine.' Five people died in the accident and 22 were injured, five of them seriously.

The commission made six recommendations in relation to the accident.

(a) The Railway should provide speed recorders on metre-gauge diesel locomotives and speedometer maintenance should be improved so that they remained in working order during a run.

(b) Rules for the use of VHF sets by drivers in emergencies such as a total communications failure in a block section should be framed and noticed.

(c) During a total communications failure, the engine used to open communications should use its flasher light, irrespective of whether it was day or night time, due to possible foggy weather on the run and as a warning to any engine approaching from the opposite direction.

(d) An assistant stationmaster (ASM) should be posted as controller only after working for at least three years as an ASM and having been found suitable to act as a controller.

(e) Drivers and diesel assistants must be psychologically tested at the time of their promotion and those not screened so far should be tested at once.

(f) The No. T-12 form giving Authority to Proceed without Line Clear should have counterfoils that could be kept as a record by the ASM. [396]

The *Tawangjawa II*, an economy-class passenger train from Jakarta, came into head-on collision on a single line with a freight train at about 06:30 on Sunday 10th September 2000, between Pelabuhan and Krengseng stations in Batang regency, Indonesia, about 100 kilometres west of Semarang. The passenger train, consisting of five cars, was reportedly

[395] https://ntnuopen.ntnu.no/ntnu-xmlui/bitstream/handle/11250/270820/528469_FULLTEXT01.pdf?sequence=1&isAllowed=y

[396] Government of India Ministry of Civil Aviation and Tourism Commission of Railway Safety Annual Report for 1999-2000

packed; the freight train consisted of an engine and eight waggons.

Twelve hours after the accident rescuers, police and local people were still working to free victims from the wreckage, while a breakdown train had arrived at the scene. Sergeant Major Haryo Diko Dewo, deputy chief of the Batang police traffic accident unit, told a reporter, "We're still searching for more victims among the train accident debris. Delays have been unavoidable." The injured were taken to Montongsari Hospital in Waleri, Kendal General Hospital, Kendal Islamic Hospital, Pekalongan General Hospital and Semarang Karladi General Hospital. At the time of the report there were 16 known dead and 'dozens of injured.' Only five of those who had died could be immediately identified.

One report stated that the PT Kereta Api Indonesia company shuttled executive passengers for the Jakarta-bound *Kamandanu* and *Senja Utama Semarang* trains by bus to waiting trains at Pekalongan. According to another account the *Senja Utama Semarang*, executive-class *Argo Bromo* and super executive-class *Argo Bromo Anggrek* trains were delayed until midnight and other trains by five to eight hours.

A preliminary investigation suggested that the *Tawangjawa II* had been permitted to leave Pelabuhan station before the freight train had arrived there. "It was this decision that most probably resulted in the head-on crash, But investigations are still under way," Sgt. Maj. Haryo said. [397]

A northbound National Railways of Zimbabwe passenger train from Bulawayo to Victoria Falls collided head-on with a southbound goods train from Hwange to Bulawayo on the single line between Intundla Siding and Ingwe Siding, Zimbabwe, about 60 kilometres from the town of Dete, at about 02:50 on Wednesday 11th October 2000. Twelve hours after the collision rescue workers were still struggling to release passengers from the wreckage. Firemen and police went to the site from Hwange, Bulawayo, Dete and Victoria Falls while a large number of people turned up at the two hospitals in Hwange in search of relatives and friends. The accident occurred in a remote area on the border between Hwange National Park and the Ngamo Forest land, accessible only by the use of four-wheel-drive vehicles, which took the injured and uninjured passengers to Dete, from where ambulances forwarded the injured to Hwange. The uninjured passengers were forwarded on buses from Dete to Bulawayo.

Mr Gideon Siyandalala of Binga, admitted to St Patrick's Hospital in Hwange with leg injuries and concussion, told a reporter that he had been asleep when a loud bang had awoken him. "I was confused and realised that I was trapped in my seat. I heard many people moaning in pain and the next thing I passed out. When I came to, some people were trying to extricate me from the train wreckage but I passed out again and the next thing I was in hospital."

Sixteen people died in the accident and 20 were injured. [398]

An inquiry was opened into the collision. The line was worked by Centralised Traffic Control (CTC) and a previous relatively minor accident in December 1999 had been blamed on the CTC operators and the train crews failing to abide by the regulations when trains were authorised to pass signals at Danger. [399]

Two trains collided head-on on the Coimbra – Serpins branch in Portugal at about 14:00 on Thursday 4th April 2002. One was a regular passenger service bound for Serpins and the other was a special carrying trainee drivers and Rede Ferroviária Nacional (REFER, the

---

[397] *The Jakarta Post*, Jakarta, Indonesia, 11th September 2000
[398] http://qzwhre.africao.../full.asp?articleid=4359&issue=35
[399] http://qzwhre.africao.../full.asp?articleid=4597&issue=36

Portuguese rail infrastructure company) staff. The accident happened between Lousã and Serpins and was blamed on the failure of the passenger to wait at Lousã for the special. Five people died, all REFER and Combois de Portugal (since privatisation the train operating company) employees. Several others were injured, including the passenger train driver, who was badly hurt. It was the first serious accident on the branch for over a century. The failure of REFER to install Centralised Traffic Control on the branch and the recent destaffing of Lousã station were blamed for the accident. [400]

Fuel shortages in Zimbabwe had forced many buses, taxis and private cars off the road in early 2003, so about 1,100 passengers crowded 13 economy carriages of a National Railways of Zimbabwe overnight train bound for Victoria Falls, early on Saturday 1st February 2003.[401] The shortage of fuel had led to black marketeering and some of the passengers were reportedly carrying petrol in jerry cans. Coming in the opposite direction was a freight train, whose cargo included tankers of an unspecified inflammable liquid.[402] At about 03:00 the two trains collided on the single line between Dete and Mambanje Siding in the Hwange mining district. The freight train was travelling at full speed when they collided on a curve, while the passenger train was picking up speed after a stop. Following the collision the old wooden-panelled coaches of the passenger train caught fire and most were gutted. Sleeping in the rear two coaches of the passenger train were members of the Lions Club charitable organisation, on their way to a conference in Victoria Falls. Unhurt in the accident and seeing that the front of the train was on fire, they were able to uncouple the rear two coaches and using knotted sheets made into a rope were able to pull them clear of the blaze. Later, an engine pulled them back to Bulawayo with the survivors.[403] About 50 people lost their lives in the collision, including the crews of both engines, while another 64 were injured. The bodies of some of the victims were burnt beyond recognition and early on, only 11 had been identified, amongst them presumably those of the engine crews. Other bodies were later identified using DNA testing.

President Robert Mugabe sent his condolences upon learning of the accident, Transport Minister Witness Mangwende blamed the disaster on human error, and Provincial Governor Obert Mpofu stated, "We must find out what is wrong in our transport network." [404]

Preliminary reports blamed the collision on a mistake made by the signalman at Thompson Junction, who was responsible for the 14-kilometre stretch between there and Dete, who had supposedly authorised the two trains to enter the single-line section at the same time.[405] The exact details of what happened are not available but since the previous November, power failures and inadequate maintenance had led to problems with the colourlight signalling, train crews being issued with written orders to proceed. The signalman was arrested after the accident and tested for alcohol consumption, presumably with negative results. [406]

Spain suffered a major disaster at 21:40 on the evening of Tuesday 3rd June 2003, involving Talgo express passenger train No. 226 from Madrid to Cartagena and freight train No. 83407 from Murcia to Albacete. The two trains collided on the single line near Chinchilla, Spain, with disastrous results.

---

[400] *Today's Railways* Europe 78

[401] *The Indiana Gazette*, Indiana, Pennsylvania, United States, 1st February 2003

[402] *Sun Journal*, New Bern, North Carolina, United States, 2nd February 2003

[403] Wikipedia – Dete train crash

[404] *European Stars and Stripes*, Darmstadt, Germany, 2nd February 2003

[405] www.herald.co.zw/index.php?id=30543&pubdate=2004-04-01

[406] *Times-News*, Burlington, North Carolina, United States, 3rd February 2003

Like 44% of Spanish railways at the time, the line on which the accident occurred was worked by telephone block, with trains being dispatched and crossing movements being arranged over the telephone by the stationmasters at the different stations. On the day of the accident the express train had arrived at Chinchilla on the main line from Madrid, at about 21:30. Just previously, the Chinchilla stationmaster had accepted the freight train from Navajuelos station on the Cartagena branch. At 21:36, for some unexplained reason, presumably due to a mental block or a moment of forgetfulness, the stationmaster had then cleared the Starting signal for the Madrid train to pass onto the single-track branch, although the freight train had yet to arrive. A minute later, the Talgo express started off towards Pozo Cañada, the first station on the branch. The stationmaster later insisted that he had not displayed his dispatch baton, without which the train should not have started, even with the signal cleared. Seeing what was happening, the stationmaster had run along the platform shouting and gesticulating in vain at the departing train.

The Talgo soon reached a speed of 180 km/h, while the freight train travelling towards it was moving at a speed of 146 km/h. Whether one or both engine crews applied the brakes in emergency at the last moment was not known as all were killed in the collision, which occurred at 21:40 three kilometres from Chinchilla. In the collision the heavy freight locomotive crushed the front of the Talgo train and in the process the fuel tank of the freight engine was ruptured, the spilt fuel ignited and the leading three coaches of the ten in the Talgo train were turned into an inferno. [407]

Nineteen people probably died in the accident: 14 passengers and five railwaymen. Three other passengers were still unaccounted for several days after the accident and might have died in the collision and their bodies been consumed by the fire. Another 46 passengers were injured, eight seriously (there had been between 79 and 82 on the express). Initially local residents were told to stay indoors and keep their windows closed as there were fears of a leakage of sulphuric acid fumes from the freight train. Later, Red Nacional de los Ferrocarriles Españoles stated that the tank waggon that had previously carried the acid had been empty. Most of the freight train was made up of TECO-type container waggons from Cartagena, destined for Vicálvaro. [408]

A passenger named Mercedes told Spanish radio: "I thought we were all going to die. We saw that the first three coaches were on fire. It was horrible."

A BBC News report wrote: 'Francisca Naranjo escaped from the buffet car where at least half a dozen charred bodies were later found. "I felt a terrible shock," she said. "We were all thrown to the floor, one on top of the other. I saw flames in the coach from where I'd come. I'm very lucky."

'Another traveller, Nieto Pinto, said also she had an incredible escape. "I saved myself because I was in the toilet at the moment of the collision," she said. "I reeled from side to side. When I tried to get out to find my husband there was a tremendous fire, everybody was screaming."

'One resident who rushed to the scene from his nearby home said he had been unable to help trapped passengers. "I couldn't get close to them…I can't forget that image," said Juan Carlos Garcia Morate. "I saw horrible scenes."

'By the time firefighters reached the scene, the intensity of the inferno meant there was no hope of survivors in the front part of the train. "From when we arrived we knew no one was alive, that was absolutely certain," said Antonio Peinado, head of the provincial fire service.

[407] https://mx-schroeder.medium.com/staffing-issue-the-2003-chinchilla-spain-train-collision-37b447da475b
[408] Federación Castellano Manchega de Amigos del Ferrocarril

"If there had been the slightest chance that someone was alive we would never have suspended the rescue operations." The heat from the blaze was so intense that rescue efforts had to be called off overnight on Tuesday.

'Some of the survivors who had already got out of the train remained seriously ill in hospital on Wednesday.' [409]

In the aftermath of the disaster a union official complained that: "We use nineteenth century technology for twenty-first century speeds. All that responsibility for just 1,000 Euros a month. They will probably send him {the Chinchilla stationmaster} to prison now." The *El Mundo* newspaper, which interviewed the official, agreed, writing: 'it was a human error which technology could have avoided. It is tempting fate to continue to have as a system of safety and traffic control on such a busy and profitable line the same manual system that was introduced in 1942 and which depends solely on the stationmasters who, if they make a mistake, have no way of correcting it.' [410]

The stationmaster, José Luis D.C., stood trial in 2006 charged with 19 counts of manslaughter and 46 counts of causing injury through negligence. His defence lawyer argued that he was a stable family man with two children, one of whom was handicapped, and that he had not been solely at fault, whilst he no longer worked on the railway. The court was lenient and the stationmaster escaped with a two-year suspended prison sentence and a ban on controlling trains for four years (which as he no longer worked on the railway anyway, was a little academic). [411]

Two passenger trains collided head-on on the Deutsche Bahn Crailsheim – Aschaffenburg line between Schrozberg and Niederstetten stations, Germany, at 12:03 on Wednesday 11th June 2003. Regional Express No. 19533, en route from Aschaffenburg to Crailsheim, was a diesel multiple unit (DMU), while Regional Express No. 19534, travelling in the opposite direction to Wertheim, consisted of a diesel engine and four coaches.[412] The trains met on a blind curve at kilometre 30.360 and as a result of the collision the diesel engine fell eight metres down an embankment, while the lighter DMU was crushed and overturned. Both drivers and four passengers were killed and at least 15 passengers were injured.[413] The passengers who died were all from the same family: a 39-year-old woman, her 10 and 12 year-old sons and her 5-year-old daughter. They had been on their way for a summer bike ride in the countryside. The husband and father of the family was seriously injured. [414]

The single-track line was controlled by Streckenblock with fixed signals, which had it been working normally, should have prevented the accident. The events leading up to the collision began with the passage through Niederstetten towards Schrozberg of freight train No. IRC 52245 from Aschaffenburg, bound for Crailsheim and onwards to Nuremberg. As it did so, a Vorsignalwiederholer (Distant signal repeater) malfunctioned, due, the Niederstetten fahrdienstleiter (signalman) wrongly assumed, to a problem with a level crossing. The level crossing was linked with the Streckenblock and a problem with the latter would mean that main running signals were locked. The Niederstetten signalman therefore agreed with his colleague at Schrozberg to suspend working of the Streckenblock until the problem was fixed, though had he realised it, he could have cleared the fault himself.

---

409 http://news.bbc.co.uk/1/hi/world/europe/2963572.stm

410 *The Guardian*, London, England, 6th June 2003

411 https://mx-schroeder.medium.com/staffing-issue-the-2003-chinchilla-spain-train-collision-37b447da475b

412 www.eisenbahn-webkatalog.de/news/print.php?newsid=483

413 www.Ananova.com/news/story/sm_789666.html

414 http://edition.cnn.com/2003/WORLD/europe/06/12/germany.traincrash.ap/

The two trains that collided were due to cross at Schrozberg at 12:00. The Niederstetten signalman allowed train No. 19533, which was running six minutes late, to leave his station for Schrozberg at about 11:55, apparently before train No. IRC 52245 had arrived at the latter place. The freight train then shortly did arrive at Schrozberg, and the signalman there, confusing it with train No. 19533, cleared it back to Niederstetten. The two signalmen then assumed that the section between them was clear and train No. 19534 was allowed to leave Schrozberg for Niederstetten. (The full details of how the accident happened are rather more complicated than this and also apparently involved a design fault in the signalling equipment at Niederstetten, but this is the best account that the author can manage, given his limited understanding of both German and the workings of Streckenblock.)

A large number of rescuers went to the accident scene, including firemen, the Red Cross, police, the Bundesgrenzschutz (Federal Border Guard) and an emergency chaplain. The injured were taken to hospitals in Bad Mergentheim, Crailsheim, Künzelsau and Schwäbisch Hall.

The 26- or 27-year-old Schrozberg signalman, whose error directly led to the accident, had served as signalman for about 15 months, either there or at neighbouring Blaufelden. The Eilwangen public prosecutor subsequently charged both signalmen with six counts of negligent homicide, causing bodily harm through negligence and endangering the safety of people travelling on the railway. In June 2005 Eilwangen Regional Court awarded the Schrozberg signalman a suspended sentence of 18 months imprisonment, while his colleague at Niederstetten was fined 4,800 euros. [415]

At 16:30 on Wednesday 11th August 2004 the *Bashkent Express*, with 153 or 171 passengers and nine crew, en route from Ankara to Istanbul, was in head-on collision with the *Adapazarı Express* from Istanbul to Adapazarı near Tavşancil, Turkey, in northern Kocaeli province. It was the third serious railway accident in the country in as many weeks.

'Rescue workers in blue and orange jumpsuits climbed through the wreckage near the village of Tavşancil in Kocaeli province, some 50 miles east of Istanbul, searching for more victims. They threw mangled pieces of metal out of the shattered windows and used axes and torches to cut through the wreckage. Armed troops were guarding the site. By nightfall, rescue operations had largely ended and emergency crews were using flashlights to search inside the wreckage for any bodies.'

A statement from the prime minister's office announced that six people had died in the accident and 85 had been injured. Another statement, however, indicated that there were only four deaths and 67 injured. Most of the injured had been treated and released from hospital by nightfall.

Local railwayman's union official Cemal Yaman explained that the *Bashkent Express* had failed to wait at a crossing point and had come back out onto the single line past a Stop signal at Danger. Deputy Prime Minister Abdullah Gul agreed, stating, "One of the trains passed a Red light. When the driver noticed he tried to reduce speed but unfortunately the accident occurred."

However, Conductor Hasan Yucedag, who suffered a broken arm and was speaking from his hospital bed, maintained that, "The signal was Green for us. It suddenly changed to Red as we were about to pass it." [416]

Passenger Mahmut Yanmis suffered cuts to his head requiring stitches and minor injuries to his arms. He told CNN Turk television, "I made my way through a broken window. As I

---

[415] Wikipedia – Eisenbahnunfall von Schrozberg

[416] *The Delphos Herald*, Delphos, Ohio, United States, 12th August 2004

was exiting the coach I saw the decapitated body of a conductor {driver?} trapped at the door of the locomotive. I saw the body of the passenger sitting next to me dead. Body pieces were scattered everywhere."

A middle-aged man at the crash site was screaming, "My God! My God! Ismail! Ismail!" presumably referring to a victim. [417] [418]

The government launched three separate inquiries into the accident. [419]

A brief and somewhat unclear report spoke of 'some' passengers dying and 'some' being injured in a head-on collision of trains between Kanbalu and Chatthin stations, Myanmar, on Tuesday 24th August 2004, the accident involving the San Thawda Down train from Myitkyina to the northern part of the country and the Mandalay No. 2 Up express. Possibly this was a single-line collision but the report also speaks of a train jumping the tracks, due to the negligence of the railway employees, without further explanation. The exact casualties in the accident were not known and the ruling military junta had not acknowledged that the accident had occurred. [420]

At about 11:45 on Tuesday 14th December 2004 Northern Railway train No. 9112 Down, the *Jammu Tawi Express*, collided head-on with train No. 1 JMP, a Jalandhar – Pathankot diesel multiple unit local train, on the single line between Bhangala and Mirthal stations, in the Punjab, India. Thirty-eight people died in the accident, including Driver Avtar Singh of the local train and Assistant Driver Jaspal Singh of the express. Another 21 people were seriously injured, including Driver Harjinder Singh of the express, and 19 were slightly hurt.

Speaking from his hospital bed, Driver Singh, 57, of the express told a reporter: "It was an awful sight. It was as if death was running towards me. I could not believe that the incoming train was on the same track. It was like death was approaching fast. I pulled the emergency brakes and the train slowed down, but I could not avert a collision as the distance was less. I regained consciousness and found I was trapped in the wreckage. I was rescued later. My entire body is hurting. Just a few minutes before we {he and Assistant Driver Singh} were talking to each other and the next moment he was dead. I could not even touch him or go near him." Driver Singh, who had 37 years' service on the railway, added that he had been running at top speed after receiving a Green signal at Mirthal. [421]

Passenger Neeta Mohindroo, who had been riding in coach S-6 of the express, and was visibly shaken, said: "I felt a violent jerk and the next moment I realised that everybody was jumping out to save their lives. As we came out we saw the passenger train engine on fire and twisted pieces of iron all around."

Farmer Tarsem Singh was working as usual in the paddy fields adjoining the railway. 'As usual he was waiting for the distant ripple in the wind that would signal the approaching throb of the Jammu Tawi Express. Today he saw the train approach from the direction it always came from and smash into the local train, "that should not have been on the track at that time. It was a deafening sound and the ground shook under me. Then came the clanging sound of metal, some sparks and a lot of dust. I could hear people howling and groaning and crying for help," Tarsem said.'

---

[417] *Sandusky Register*, Sandusky, Ohio, United States, 12th August 2004

[418] http://www.china.org.cn/archive/2004-08/12/content_1103692.htm

[419] https://eu.goupstate.com/story/news/2004/08/12/train-collision-leaves-six-dead-85-injured/29733043007/

[420] http://english.dvb.no/news.php?id=2961

[421] http://www.indianexpress.com/print.php?content_id=60857

Local resident Shyam Babbar related: "I saw smashed bodies in the damaged carriages. There was a woman and a small child next to her. Both were dead."

'Local people were the first to reach the accident spot and started rescue operations. People were seen wailing as the bodies, some of which were mutilated, were taken out of damaged coaches. The scene was even pathetic when bodies of women and children were extricated. Rescue operation was being conducted on a war footing with the help of police, army, railway personnel and locals. The operation was still continuing.

'The injured were taken to hospitals at Mukerian, Jalandahar and Pathankot.

'Rescue trains were arriving at the site, railway sources said. Two relief trains had been despatched from Amritsar.

'The Northern Railway authorities have issued helpline numbers for inquiries related to the accident. The numbers are 0181-2456072 and 0181-2456240 at Jalandahar, 011-23341074 and 011-23344128 at Delhi and 078-2213344 at Ahmedabad.

'Several trains plying on the Jalandahar – Pathankot section, including the 9112 Up Ahmedabad – Jammu Express, 2471 Swaraj Express, 6317 Himsagar Express and 9367 Malwa Express, were diverted via Amritsar – Pathankot route following the accident.' [422]

There had been about 700 passengers on the express, some of them Hindu pilgrims returning from the Vaishno Devi shrine. The number of passengers on the local train was not stated but was doubtless considerably less.

Railway Minister Laloo Prasad Yadav reached the accident scene in the evening and promised strong action against those responsible for what he rather melodramatically called "brutal murder." He stated: "Both trains were allowed to move on the same track by two stations despite the station heads having communication links. I don't consider this an accident. I consider this brutal murder. It is an irresponsible human error." [423]

Some articles published on the internet detailed individual stories resulting from the accident: 'The train tragedy near Mukhsar in Punjab left ten-year-old Mintoo orphaned with both his parents and two sisters killed in the accident. Mintoo, however, had a miraculous escape and suffered no injuries in yesterday's accident which killed 37 people. The boy was returning from Jalandahar with his family after attending a cousin's marriage in Pathankot in the DMU train which collided head-on with the Jammu Ahmedabad Express. His father Dalbir Singh, mother Neelamjit Kaur, and sisters Bajinder and Harpreet, were killed in the accident. The four bodies were brought to the Mukerian Civil Hospital last night where Mintoo was seen crying bitterly.

'Tragedy struck another four-year-old girl, Komal, who lost her mother and brother in the accident. Komal, along with her family members, had boarded the DMU train from Bhangla village, just two km from the railway station {actually only a level crossing} near Mansar where the incident occurred. Her mother, Bholi, and brother Pawan Kumar lost their lives in the accident. Komal, who was admitted to the Mukerian Hospital with injuries in her left eye and shoulder, was unaware of the death of her family members. She was repeatedly asking her relatives the whereabouts of her mother and brother.

'Railway Minister Lalu Prasad has announced a compensation of Rs. 5 lakhs each to the kin of the deceased out of which Rs. 1 lakh would be released immediately.' [424]

Crossing Keeper Gopal Das, 45, at Crossing No. 125 at Mansar village was credited with

---

[422] http://www.deepikaglobal.com/search_newsdetail.asp?newscode=85341
[423] http://www.telegraphindia.com/1041215/asp/others/print.html
[424] http://www.hinduonnet.com/thehindu/holnus/002200412151220.htm

having reduced the severity of the collision. He had been advised by the Bhangala stationmaster of the approach of the local train but had known nothing of the express until he had realised that it was also approaching his crossing. "The moment I realised that two trains were heading towards each other, I put red flags on both sides of the track to warn the train drivers. While the driver of the Jammu – Ahmedabad Express applied the brake, the DMU driver did not probably see the flag due to a sharp bend," he related.[425] The crossing keeper at No. 27 crossing had likewise been advised of the local train by Bhangala and when he had seen the express approaching from the opposite direction instead, he had telephoned Bhangala to ask what was happening but had not attempted to stop the express. [426]

Two days after the accident Minister Prasad told Parliament that one of the two stationmasters involved had been arrested and police were searching for the other, while the Section Engineer (Construction) and the Section Engineer (Telecom) had been suspended. [427]

It emerged that at 13:10 on the day before the accident the Neale's block token instruments between Bhangala and Mirthal stations had failed. Ironically this was due to damage to a cable in the course of engineering work in connection with the doubling of the line.[428] Between then and the time of the accident, recourse had been taken to the use of paper Line Clear tickets, which were issued to the trains as authority for them to proceed, with messages being exchanged by wireless between the stations. An official stated that at any one time, seven or eight messages could have been exchanged simultaneously, and this was where the fatal confusion, which had led to the accident, had occurred.

Assistant Stationmaster Varinder Kumar Malviya (who had been standing in at Bhangala for Assistant Stationmaster Amarjit Singh, on leave) and Stationmaster Ram Lal at Mirthal both disappeared after the accident but were apprehended within a day or two.[429] Both were charged with causing death through negligence under Section 304 A of the Indian Penal Code.

An investigation by the Commission of Railway Safety blamed the accident on gross negligence on the part of the construction, signalling and operating departments of the Punjab division of the railway. The construction department was found to have been at fault for having damaged the signalling cables and then failing to report the damage to the officials concerned. The signalling department was blamed for not having rectified the failure of the block instruments for 22 hours and the operating department for improper working of the paper Line Clear system. The heads of all three departments had been transferred and also faced disciplinary proceedings. [430]

The provisional report of the commission made three recommendations:

(1) The use of the Very High Frequency wireless sets presently installed at the stations, which were accessible to many railway officials, for obtaining Line Clear for trains, should be prohibited with immediate effect.

(2) The provision of a voice recorder in the Firozpur control office should be expedited.

(3) A route plan for underground cables laid along the track should be prepared and cable markers should be provided at specified intervals. [431]

---

[425] http://in.*Rediff.com*/cms/print.jsp?docpath=/news/2004/dec/15train1.htm
[426] http://timesofindia.indiatimes.com/articlesshow/960214.cms
[427] http://www.teamindia.net/news/index.php?action=fullnews&id=44690
[428] http://timesofindia.indiatimes.com/articleshow/958982.cms
[429] http://www.indianexpress.com/print.php?content_id=60876
[430] http://www.zeenews.com/links/articles.asp?aid==221017&sid=NAT
[431] http://pib.nic.in/release/release.asp?relid=6217

Ferrovie dello Stato Intercity passenger train No. 2225, which had left Verona at 11:39 on Friday 7th January 2005 for Bologna, collided head-on with freight train No. 59308 near the closed station of Bolognina di Crevalcore, Italy, at 12:53.

The engine of the passenger train was reportedly derailed as a result of the collision, the leading coach, carrying about 100 passengers, was penetrated by heavy steel girders being carried on the freight train, and the second coach was thrust up into the air.

"I can hardly speak. I am still in shock from the terror, the fear," a surviving passenger told a television reporter. "At first I thought that it was a tree trunk or perhaps a house, then I saw through the window on my right a carriage up this high, vertical," he related.

Another passenger, Steve Djaumen from Cameroon, said that the train had appeared to be travelling at normal speed and he had not noticed any brake application before the crash. He and others had been able to push open a door and climb out after the accident. "From the coach in front of ours, which was completely destroyed, we heard people screaming for help," he stated.

'Rescue workers and emergency services battled time and fog to get to the wreckage, which was strewn across a field between the small towns of Crevalcore and Camposanto, north of Bologna. Medics treated the injured on the roofs of the tangled carriages. Mud in the fields alongside the track made it difficult for ambulances to reach the disaster scene and the injured had to be taken on stretchers to roads. Fire brigade officials said thick fog in the flat countryside made it difficult to bring cranes close to the track. Italian railways and magistrates opened an investigation into the cause of the collision.'

'Six hours after the crash, rescue workers using blow torches and metal cutters were still trying to reach part of the wreckage where they feared more passengers might have been trapped. "We expect to work through the night," said national fire brigades chief Mario Morcone. "We have still to lift a part of the carriage that was sent flying 10 metres up before it fell and we hope we don't find other victims in there," he said.' [432]

It was Saturday night before the body of the last victim was recovered from the wreckage, bringing the number of dead up to 17, including Driver Vincenzo De Biase and Conductor Paolo Cinti of the passenger train and Driver Equizio Abate and Conductor Ciro Cuccinello of the freight trains. About 60 passengers were injured, five critically. The uninjured passengers were taken to Crevalcore by bus and later forwarded on another train.

Some of both the survivors of the crash and the rescuers suffered from shock and post-traumatic stress disorder, some unable to believe what had happened and others shocked into silence. Some of the passengers who escaped suffered survivor's guilt, believing that they could have done more to help those who had died.

The widow of Driver De Biase could not believe he could have been responsible for the disaster, saying, "He was an experienced railwayman who knew the tracks by heart."[433] The verdict of the investigation, however, was that her husband and Conductor Cinto had been to blame and that they had run past two Red signals in the dense fog and through a set of points back onto the single line, before the freight train had arrived at the crossing place. It was common practice, though not obligatory, for Control to notify train drivers of unusual meets with other trains or extra workings but because of a change of shifts, Driver De Biase had not been advised of the freight train. The inquiry believed that this had probably been the main factor in his failing to wait for the opposing train. Initial theories that he had been under the influence of alcohol or drugs or conversing on his mobile phone were dismissed. Unable to take any action against Driver De Biase, Bologna public prosecutor Andrea Scarpa

[432] www.reuters.co.uk/printerFriendlyPopup.jhtml?type=topNews&tsoryID=650482

[433] http://news.bbc.co.uk/2/hi/europe/4155073.stm

subsequently charged 10 members of the railway management with causing a railway disaster, involuntary homicide and involuntary wounding. The cases against them were all later dropped.

The railway unions and opposition politicians complained that secondary lines had been neglected to save money and a token 10-minute stoppage was followed a few days later by a 24-hour strike by railwaymen demanding safety improvements.

Following the disaster, parts of the track were doubled, while Sistema di Controllo della Marcia del Treno (SCMT, a cab signalling system) and Sistema Supporto Condotta (SCC, a driver information system) were both introduced. [434]

[434] Wikipedia – Crevalcore train crash

CHAPTER 7

RUNAWAY TRAINS AND VEHICLES

A variety of situations can lead to a train or part of a train running away out of control. If there are railway staff in attendance the train or engine can only run away if there is a failure of the brakes or if the brakes are incapable of holding it on a steep gradient. In the case of unattended vehicles, the action of gravity alone is sufficient to precipitate a runaway. In contrast to the situation on a road, a train or vehicles can sometimes run for long distances without mishap, an accident occurring only if and when the runaway meets with another train, comes to the end of the line, or reaches a curve too sharp for the runaway to negotiate at the speed at which it is travelling.

On Sunday 3rd March 1991 a heavily loaded mixed train, consisting of an engine, at least four passenger carriages and 11 goods waggons, set off from the station at Nacala, Mozambique. It shortly got into difficulty ascending the climb out of the port and after covering about eight kilometres, it stalled, its brakes were unable to hold it, and it began to run back down the hill. Continuing at increasing speed, it ran all the way back to its starting point, where it crashed with disastrous results. At least 96 people lost their lives and another 113 were injured as the coaches and waggons piled up on one another. Authorities feared that the number of dead could rise with the discovery of more victims in the wreckage. [435]

At about 13:30 on Monday 30th September 1991 a bogie flat rail (BFR) waggon escaped from Thiparthy station, Andhra Pradesh, India, on the South Central Railway, and began to run down the incline at its Miryalaguda end, until between the Up Home and Up Distant signals it came into collision with oncoming train No. 532, an Up Repalle – Secunderabad passenger service. As a result, four people lost their lives and 50 were injured, five seriously. Damage was estimated at 3,631,000 rupees.

The Commission of Railway Safety attributed the runaway to a failure of the railway staff, who had hand-shunted a BFR loaded with rails in the station yard, which had falling gradients, in contravention of the Station Working Rules.

The report of the Commission on the accident made 10 important recommendations:

(1) Train examinations should be undertaken by properly qualified train examiners and not by semi-literate fitters.

(2) The conditions of working and the lighting arrangements in the train examiner's yard at Moula Ali required improvement.

---

[435] *Desert Dispatch*, Barstow, California, United States, 5th March 1991

(3) The staff strength of train examiners should be augmented and vacancies filled up.

(4) The combined check by drivers and guards before the clearing of stabled loads was not being done effectively.

(5) The literacy of semi-literate drivers should be improved. Those not coming up to the required standard of knowledge should be taken off running duties. The performance of Senior Locomotive Inspectors should be monitored.

(6) The quality of the periodic overhaul at Lallaguda Workshops was very unsatisfactory and required improvement.

(7) The Operating and Mechanical Department officers were not doing their scheduled inspections. Several officers were not conversant with Policy instructions. They should be sent for frequent refresher courses.

(8) The Railway should publish an "Operating Manual" incorporating the various circulars issued from time to time.

(9) The General and Supplementary Rules should be suitably modified with regard to the duties of staff affecting the safety of train movement.

(10) The contract for the loading of rails into BFRs was not properly framed. [436]

A Belgian narrow-gauge tourist train known as *Li Trimbleu* was involved in an accident in late 1991. At about 14:35 on Saturday 5th October the train, which consisted of a diesel locomotive and three waggons, set off with a complement of 117 passengers, all adults and most of them pensioners, for the 100-minute journey from Blegny-Mine, near Liege, Belgium, through the countryside to Montroux. About 10 minutes into the journey, near Dalhem, the driver realised that there was a problem with the train. Bringing it to a halt, he left it to head for a nearby farm to telephone for assistance. Unfortunately he omitted to properly secure the train and it shortly began to run away, travelling 500 metres before it left the rails on a curve. Two of its carriages overturned and seven passengers died and 35 were injured, four seriously. An investigation found that the driver had been unfamiliar with the proper operation of the train brakes. The line did not reopen after the accident. [437] [438] [439] [440]

On Monday 18th November 1991 Ferrocarriles Nacionales de México locomotive No. 1133 was found by its crew to be defective. It was duly sent to Puebla for repairs and returned to service the next day, though a number of faults, including with its brakes, had not been corrected. Driver Antonio Sergio Castellanos Conde was that day, Tuesday 19th, ordered to work freight train No. 9130, consisting of 20 waggons loaded with over 2,000 tons of cement, sorghum and soya beans, from Esperanza to Tehuacán. At 11:00, after inspecting the engine and finding everything apparently in order, Driver Conde reluctantly agreed to work the train.

When the train reached kilometre 23, near Santiago Miahuatlán, its air brakes failed. This happened at about the worst possible point, as from there the line fell very steeply, so that the train then quickly began to gain speed and six of its waggons apparently became detached. Its

[436] Government of India Ministry of Civil Aviation and Tourism Commission of Railway Safety Annual Report for 1991-92

[437] https://www.arvia.be/archive/blegny-1991-10-05-article-de-presse-accident-du-trinbleu/

[438] https://www.lavenir.net/regions/liege/dalhem/2016/10/05/il-y-a-25-ans-la-catastrophe-du-trimbleu-faisait-7-morts-S5PPRSJYQNGX7ILIDJOYWA7E44/

[439] https://www.loquis.com/en/loquis/5420063/Li+Trimbleu+Train+Accident

[440] Wikipedia – Trimbleu train accident

crew attempted to uncouple the engine from the remainder of the waggons but failing to do so, Driver Conde and the garroterros (brakesmen), realising that they could do no more, jumped from the train, now travelling at 80 km/h, to save their lives.

At 12:43 the runaway train, reportedly travelling at 180 km/h, reached the station at Peñafiel, a suburb of Tehuacán, Mexico. The staff there watched as the driverless train continued its mad run for a few hundred more metres before leaving the rails at the junction of Avenida José Gard-Crespo and Avenida de Las Américas, opposite Benavente School. The 100-ton engine 'rose into the air and rolled, dragging the waggons with it, which crushed one against the other, forming a cloud of cement dust, earth and black smoke,' accompanied by the screams of people in the vicinity. The accident happened at the worst time of day, for the streets were crowded with parents going to pick up their children from Benavente School, shoppers, buses and cars. Two full-size buses, a crowded microbus and about a dozen cars were flattened by the engine and waggons of the train. A Chinese restaurant and a car repair shop were also struck and badly damaged, the owner of the car repair shop being killed, along with many others. Bruno Cortez Ramirez, an aide to the mayor of Tehuacán, told Reuters news agency that 50 bodies had been recovered and that there could be many more, while other officials said that more than 100 injured had been taken to Hospital 15 of the Instituto Mexicano del Segurto Social, the nearest to the scene. Cortez added that he was very angry that the crew of the runaway train had failed to advise the town of the danger,[441] though whether they would have been in a position to do so seems debatable. Driver Conde and the four brakesmen felt it wise to make themselves scarce after the disaster and two days later, Tehuacán justice official Herve Ponciano Mukul stated, "So far, it has been established that the drivers are on the run and all their likely hiding places have been searched. When they are found they will be charged with manslaughter, aggression and damage to the national communications."[442] According to one report, the five railwaymen were never heard of again. [443]

Unofficially the death toll in the disaster was said to have been over 100; the official figures were that 32 people had lost their lives at the scene and another 9 had later succumbed to their injuries.

There were widespread protests against the railway after the disaster, exacerbated when photographs of the mutilated bodies of some of those who had died were published, with threats to tear up the tracks when they had been repaired. To this, the government responded that anyone who sabotaged the railway would be arrested and punished. The railway was eventually rerouted, while the Avenida de Las Américas was renamed Boulevard Héroe de Nacozari. [444] [445]

Few details are available relating to an accident that occurred between Gymuri (previously called Leninakan) and Nalband, Armenia, on the evening of Thursday 27th May 1993, but it apparently involved a collision between a passenger train and seven runaway freight waggons. Gymuri Mayor Levon Adamian stated that 30 people had died and 48 had been injured. Interior Minister Tatul Safrian agreed with the mayor on the number of fatalities but stated that there were 60 injured, 12 seriously. The cause of the accident was under investigation. [446] [447]

---

[441] *Los Angeles Times*, Los Angeles, California, United States, 20th November 1991
[442] Ibid, 22nd November 1991
[443] https://primeralinea.com.mx/web/noticia/15382
[444] https://www.icadtotal.com/blog_de_tehuacan/historicas/1991-el-trenazo-de-tehuacan/
[445] https://elmanana.com.mx/nacional/2024/6/6/el-trenazo-de-tehuacan-la-tragedia-ferroviaria-mas-sangrienta-de-mexico-125423.html
[446] *The Odessa American*, Odessa, Texas, United States, 29th May 1993
[447] *The Lethbridge Herald*, Lethbridge, Alberta, Canada, 28th May 1993

A shortage of locomotives in Norway was a significant factor in a runaway accident there, which occurred at Nordstrand, on Sunday 3rd October 1993. In order to alleviate the lack of engine power, Norges Statsbaner (NSB) hired nine Class NS 6400 locomotives from the Netherlands.

On the day of the accident an Oslo Sentrum to Ski passenger train, formed of a type 69 electric multiple unit set, was standing in Nordstrand holdeplass when it was struck by NS 6400 engine No. 6454, travelling at 75 km/h. As a result, five passengers in the local train lost their lives and 16 people were injured, four of them seriously. Those injured included both drivers.

The investigation into the accident determined that when the Dutch engine had left the yard at Lodalen, the controls for the brake cylinders had been left in the wrong position, with the result that its compressed air brakes had been inoperative, leaving only the electrodynamic brake and two parking brakes. The shunting engine had proceeded from Lodalen to Rosenholm holdeplass, where the parking brakes had been released by its driver, at which point it had begun to roll backwards down the gradient. On this type of locomotive the electrodynamic brakes would not work when it was running in reverse and the driver did not know how to stop his engine. His efforts to telephone Control at Oslo S to warn them of what was happening also failed.

The driver was subsequently charged with negligence but at his first trial, he was acquitted on the grounds that he had been working under stress, that his training on the Dutch locomotives had been inadequate, that the instruction manual for the engine had been translated from Dutch to Norwegian by an unauthorised person, that the text on the brake controls was in Dutch and that the settings of the brake valves were the opposite of those found on other NSB shunting engines.

The prosecution then appealed to the Supreme Court against the acquittal and the driver had to endure a further trial in 1996. This time, he was found to have been negligent but did not suffer any punishment or have to pay Court costs. The NSB, on the other hand, was fined 150,000 kroner for a violation of the Arbeidsmiljøloven (Working Environment Law) for the inadequate markings on the brake controls. [448] [449] [450]

The commission set up to investigate the accident made 10 recommendations:

(1) The stop cocks for the air and electrodynamic brakes should be marked in a responsible manner with Norwegian text.

(2) The stop cocks for the air brakes should be sealed in the open position.

(3) The working instruction should be upgraded and all drivers who operated freight trains should receive a better introduction to the air brake system's principle of operation and the significance of the stop cocks.

(4) The regulations and procedures for material hire should be determined.

(5) The rules for operator control in deliverance of the air brakes should be prescribed. It should be evaluated if there were special requirements for handling these brakes.

(6) The follow-up and control of set routines in relation to the checking out and insertion of traction engine should be prescribed.

(7) Emergency numbers should be established directly to Control via the mobile phone

448 https://web.archive.org/web/20080227220651/http://www.sikkerhetsdagene.no/kapitler/7Sklet.pdf

449 https://www.sintef.no/globalassets/upload/teknologi_og_samfunn/sikkerhet-og-palitelighet/rapporter/sintef-a3144.pdf#28

450 Wikipedia – Nordstrand-ulykken

network.

(8) The technical possibility of sending an acoustic warning system to Control if unintended movements occurred at a distant location should be evaluated.

(9) The training of personnel working with new material should be quality guaranteed.

(10) The possibility of expediting the expansion of the train radio system should be evaluated. [451]

A mixed train from Mandalay to Myitkyina was descending an incline as it approached Wuntho, Myanmar, when it suffered a brake failure, ran away and was derailed, on Friday 30th December 1994. One of its carriages fell into a ravine, several were overturned and one was left hanging from a bridge, according to an announcement from Burmese radio, monitored by the British Broadcasting Corporation. The statement added that 102 people had been killed and 53 injured in what was the most serious railway accident in the history of the country. Railways Minister U Win Sein and Deputy Health Minister Colonel Than Zin visited the accident scene. [452] [453]

On Saturday 24th June 1995 shunting operations were in progress at Čachnov station, on line No. 261 between Žďárec u Skutče and Svitavy in the Czech Republic. During these manoeuvres a brake van and three waggons loaded with scrap iron and timber (processed tree trunks) were left standing on the single-track main line, which fell steeply towards the next station at Krouna at gradients as steep as 20 per mille (1 in 50). The brakes were applied in the light van but no hand brakes were applied on the three waggons, which proved insufficient to hold the 123-ton weight on the gradient, so they began to run away.

Approaching from the opposite direction was a diesel railcar carrying 23 people, including its driver and presumably a conductor, on its way from Ždárec to Polička. There was no way to warn the railcar of the runaways and having passed through Krouna station and a request stop of the same name, the runaways, which had reportedly attained a speed of 100 and possibly 130 km/h, collided with the railcar between Krouna and Předhradi stations. The impact threw the railcar back 60 metres and most of it was demolished. Its driver and 17 passengers (or possibly 16 passengers and the conductor) were killed outright and another passenger died of her injuries a week later. Of the four survivors only one, a girl who chanced to be in the toilet at the rear of the train at the time, escaped unhurt.

One of the survivors, Mrs. Košnarová-Pražanová, lost her husband and was herself so badly hurt that it was a month before it was certain that she would recover. "It went through our train. I was crushed between the floors. I was lying on the floor of the passenger train and the floor of a waggon was above me. They only gave me a faint hope of pulling through, but I was young and healthy," she said. She had a fractured spine, burns, abrasions, bruises, hematomas and cerebral oedema. "My lungs were bruised, so I was on artificial ventilation, my femur was broken, and my nerves were damaged, so I can't control my left leg completely," she related. She receives a disability pension. Her one very slight consolation was that she was able to save a piece of film from their damaged camera, with the last pictures of her husband on their honeymoon in Spain.

The driver and conductor of the freight train were sentenced to 6½ and 5 years'

---

[451] https://ntnuopen.ntnu.no/ntnu-xmlui/bitstream/handle/11250/270820/528469_FULLTEXT01.pdf?sequence=1&isAllowed=y

[452] *The Indiana Gazette*, Indiana, Pennsylvania, United States, 1st January 1995

[453] *The Daily Gleaner*, Kingston, Jamaica, 1st January 1995

imprisonment for negligence; the sentence on the conductor was later increased to 8 years. Both were also banned from operating trains for several years. [454] [455] [456]

On the night of Wednesday 25th October 1995 (another source states Tuesday 24th) the *Galuh* passenger train, consisting of an engine and six coaches, from Jakarta Pasar Senen to Banjar, Indonesia, arrived at track 2 at Cibatu station. Then at 23:05 the *Kahuripan* train from Bandung to Kadiri, consisting of an engine and seven coaches, arrived there. The *Galuh* train was already running late when it arrived and it then developed a problem with its engine. In order to avoid further delay to the *Galuh* train and also to the *Kahuripan* train, the decision was made to combine the two for the onward journey from Cibatu as far as Banjar, and this combined train, now consisting therefore of two locomotives and 13 coaches, departed at 23:38. The line onwards was steeply graded, both uphill and downhill.

The long train, carrying 728 passengers, climbed from Cibatu to Cipeundeuy station, where it was not booked to stop. Beyond Cipeundeuy the line fell again at 25 pro mille (1 in 40) and it was here that the brakes of the train failed. Its speed increased uncontrollably until it reached the 100-metre-long Trowek Bridge near Tasikmalaya, at kilometre 241, where it left the track. Four carriages were thrown off to the right and one fell into a 10-metre ravine. Some of the passengers in other carriages panicked and in the darkness opened doors or jumped from windows and fell from the bridge into the ravine.

Villagers from Sarapat were the first on the scene and rendered what assistance they could. Fourteen passengers died in the accident and 71 more were injured.

Following the blockage of the line, the *Cisadane* train between Bandung and Madiun was cancelled and the *Badrasurya*, *Mutiara Selatan* and *Turangga* trains between Bandung and Surabaya were diverted via Cikampek and Kroya.

In the longer term, a compulsory stop for all trains at Cipeundeuy station, at the start of the descent, which had been in force for many years but had then been abolished, was reinstated, even for the *Argo Wilis* executive train, while Trowek Bridge and the station nearby of the same name were renamed Cirahayu Bridge and Cirahayu. [457] [458]

An initial report on an accident near Litvinovo, Russia, in the Kemerovo region of Siberia, 1,930 kilometres east of Moscow, stated that 50 people had died when four runaway cement waggons collided with a passenger train late on the night of Friday 31st May 1996 in a remote forest area. Three coaches of the passenger train were overturned and a rescue helicopter was sent to the scene, though efforts to provide assistance were hampered by the darkness. The brakes of the waggons had not been properly secured, resulting in them running away and completely smashing the leading vehicle of the passenger train in the ensuing collision. A second report stated that 17 people had been killed and 45 admitted to hospital, a number of whom were in a critical condition. [459] [460]

---

[454] https://www.idnes.cz/pardubice/zpravy/krouna-zeleznicni-nehoda-nestesti-vlaky-odskodneni.A200623_114851_pardubice-zpravy_mvo

[455] https://ct24.ceskatelevize.cz/clanek/regiony/pardubicky-kraj/dvacet-let-od-vlakove-tragedie-u-krouny-dnes-uz-by-k-ni-dojit-nemelo-130770

[456] https://chrudimsky.denik.cz/nehody/nehoda_vlaky_krouna20100623.html

[457] https://mugen.id/tragedi-trowek-1995/

[458] https://kumparan.com/wiji-nurhayat-1584020433391182608/mengenang-tragedi-cirahayu-24-oktober-1995-1uS9Sce0MAO/full

[459] *Port Arthur News*, Port Arthur, Texas, United States, 1st June 1996

[460] https://www.upi.com/Archives/1996/05/31/Russian-train-crash/5401833515200/

Half of a train, which had been divided, reportedly rolled back and was derailed in Kiazi, southern Zambia, on or about Saturday 9th April 1996. The death toll was stated to have reached 100. [461]

A disaster occurred in Brazil on Wednesday 18th September 1996, involving a heavy freight train and the *Expresso Barrinha* passenger train. The latter, train SUJ 0135, consisting of an engine and just two carriages, was stopped at a level crossing 700 metres short of the station at Japeri when it was run into from behind, at 08:16, by freight train FOR 0134. The latter train consisted of an engine and 18 waggons carrying coils of copper wire from the Belgo-Mineira company, weighing 1,784 tons, on its way from Dias Taváres, near Juiz de Fora, in Minas Gerais province, to Rio de Janeiro.

The freight train had left Dias Taváres at 21:48 on Tuesday for the Caju freight terminal in Rio. Its normal speed was 25 km/h but it got out of control descending the Serra Mario Belo, 16 kilometres from Japeri. It had arrived at Barra do Pirai, the last station before the Serra Mario Belo, at 05:50, and left there at 06:07. It is not quite clear from the newspaper reports how the accident occurred but it appears that the crew of the train had been instructed by Engineer Christiano Marcos at Barra do Pirai that they were not to bother with performing a brake test there, which would have taken 30 or 40 minutes, so as to avoid a late arrival at Rio. Chief Engineer Mario Antonio of the Rede Ferroviária Federal denied, however, that a brake test had not been performed. At any rate, descending the incline the train began to get out of control and it had reached a speed of 100 km/h by the time it reached Japeri. At 08:12 Drivers Cláudio Guimarães and Daniel Rodriguez Barreira radioed Control in Barra do Pirai to advise them that they were having a problem controlling the speed of their train but, as Controller Francisco José Ferreira told a reporter, "We didn't have time to do anything."

There had been about 100 passengers on the Expresso Barrinha. Some of them had seen the oncoming freight train in time to jump out. Both trains were derailed by the violence of the collision. Eleven people died at the scene, five passed away in hospital and another 39 were injured. The fire service were criticised by a newspaper for taking 20 minutes to arrive, which does not seem an exceptionally long time, while in the meantime several victims had been attended to by local residents. Nurse Elio Rodriguez Fortin, one of the first on the scene, said, "The noise was terrible. It sounded like a loud thunderclap. Many of the survivors were running about aimlessly, covered in blood and fainting as they tried to get away." Fifty firemen from Nova Iguaçu, Paracambi, Duque de Caixas and Rio de Janeiro took part in the rescue operation, as well as medical personnel and military police of the Queimados battalion.

One of the patients taken to the Hospital do Posse in Nova Iguaçu was a railwayman named Luís Fernando Guedes Conceiçáo, 47, who jokingly described himself as the "picotador" (puncher), presumably a ticket inspector or perhaps the conductor. Under observation he did not seem very badly hurt and described how he had never been in an accident before and had been scared. He was then taken into surgery with suspected internal bleeding at 17:00 and a few minutes later, he died. Another railwayman who lost his life in the accident was electrician António Jardim. Normally he caught the bus into work because it was quicker but this morning he had overslept and missed it, so had to resort to the train instead.

Passenger Icléia Matilde de Souza was discharged from the Hospital do Posse at 13:00, having been treated for a broken collarbone. She told a reporter, "I thought: 'My God, another train is going to run into ours.'" She had thought of trying to jump out of a window. Travelling with her had been her friend Marli, whom she had not seen after the accident. Icléia

[461] www.jrtr.net/jrtr08/topics.html

added, "It was God that saved my life."

Estevão Pedro da Costa, 34, from Paracambi, was a daily passenger on the *Expresso Barrinha* with his wife Maria de Lurdes Souza Pas. They did not see the oncoming freight train but remembered the terrible noise of the impact. Both briefly lost consciousness and were taken to a clinic in Japeri. Maria de Lurdes was quickly discharged but then went to the casualty department at the Hospital do Posse, because she was suffering from severe neck pain. With priority being given to more serious patients she had to wait for over three hours for an X-ray examination. Estevão was more badly hurt, suffering reportedly a fractured spine, collarbone and right foot. He related, "I'm in a lot of pain and I still can't remember exactly what happened." Estevão's brother-in-law, Jair do Amaral Lúcio, complained of the poor service on the line, with trains often being cancelled without warning, meaning a walk of four or five kilometres to a bus stop.

Isaac Popoutchi, president of the Rede Ferroviária Federal (RFF), promised that a commission of inquiry would be set up to investigate the accident. He noted that an accident of such a magnitude had not occurred on the RFF for 20 years, stated that it was hoped to reopen the line within a couple of days, and that he did not believe the accident would affect the imminent listing of part of the system on the Rio stock exchange. (As in other parts of the world, the state-owned railways of Brazil were in the process of privatisation at the time.)

Fernando Paulino, a railway union spokesman, accused the RFF of abandoning the Barra do Pirai – Japeri branch, the last in the state to still have a passenger service, stating that the RFF intended to end the service as soon as privatisation occurred. He complained, "They are only bothering about freight transport and not taking into account the people who need to use trains. The RFF have already closed the branch between Barra Mansa and Ribeirão Vermelho and want to close this one."

Union director Tânia Regina Ferreira added, "Everything is deteriorating. The government has decided to just get rid of the problem and await privatisation. Meanwhile, accidents keep happening. We continue to think that this damage is deliberate, to speed up the sale of the company."

The unions also denounced the job cutting, which had reduced the number of railway employees in the region from 10,800 in 1988 to only 4,900 in 1996, while the possibility that the freight train had been overloaded was also raised and several drivers had complained of brake problems. [462]

Pakistan Railways suffered a major catastrophe at 03:40 on Monday 3rd March 1997, involving a runaway train. It happened at Mehr Shah, near Khanewal, Pakistan. The 17-coach *Zulfikar Express* from Peshawar to Karachi got out of control approaching the station, where it was to cross a Karachi – Lahore express, and reaching the end of the loop without being able to pull up, it was diverted into the trap siding, at the end of which it became derailed in the sand drag.

A press report described the accident. 'Rescue workers carved through the twisted wreckage of a train with blow torches Monday trying to reach scores of injured screaming for help. At least 125 people – including 24 children – died in the accident, hospital workers said. Another 450 passengers were injured, 50 of them seriously, when the brakes failed on the express train and it derailed early Monday morning, Pakistan's state-run news agency said. The accident occurred in eastern Pakistan just outside of Khanewal, 240 miles south of Islamabad.

[462] *Jornal do Brasil*, Rio de Janeiro, Brazil, 19th September 1996

'Residents of Khanewal – the first to arrive on the scene – pulled bodies from the overturned cars and covered the dead in white shrouds. Shops closed and hundreds of people went to local hospitals to donate blood.

'The government called in the army to help with the rescue. For more than 12 hours, dozens of injured passengers were trapped inside five overturned railway cars as rescue workers painstakingly sliced through the wreckage. "You can hear the people screaming for help," said rescue worker Shamshed Iqbal. "We are having trouble getting to them."

'Some railway cars rolled on top of others, making it difficult to reach the wounded, but by evening the last of them were believed to be free. Rescue workers, however, said more bodies could remain trapped inside.

'The 17-car train, carrying 1,500 people, was en route from Peshawar in Pakistan's Northwest Frontier province to Karachi. Omar Hyatt from Khanewal Hospital said 125 bodies had come in, many of them labourers who were heading to Pakistan's largest city in search of work. "The bodies are still arriving," Hyatt said. "Many of them are barely recognisable."

'Rail officials said the train's brakes apparently failed when it tried to stop to make way for a second passenger train that was leaving Khanewal. When the driver realised the brakes had failed he pulled onto a track for runaway trains. The train slammed into a pile of sand at the end of the track and the first five cars overturned.

'Pakistan's train system is antiquated and dozens of accidents have killed nearly 2,000 people in the last decade.' [463]

Prime Minister Nawaz Sharif flew by helicopter to the scene of the accident and also visited the injured in hospital at Khanewal.

By the following day, with the deaths of some of the injured, the number of dead had risen to 136. [464]

Divisional Superintendent Sardar Saifullah Khan Qaiserani at Multan told reporters that had the express not been so diverted by Cabinman Allah Ditta, a worse accident would have resulted from it colliding with the opposing *Allama Iqbal Express*, which another railwayman stated had been due to cross the *Zulfiqar Express* at Jangal Mariala station, several kilometres further on.

Driver Khalid Islam of the runaway train, who survived the accident, admitted at the inquiry into it that on 25th February the divisional medical officer at Sukkur had told him that he was suffering from high blood pressure and had some symptoms of cancer, and that he should not undertake night duties. However, upon presenting the medical report to the divisional mechanical engineer at Sukkur the latter had requested that he perform his duties as before. There seems a possibility, therefore, that the driver had lost control of his train due to some sort of medical episode. Meanwhile, the examiners who had checked the train earlier denied that there had been any problems with the brake system. The inquiry was reportedly unable to reach any definite conclusion as to the cause of the accident. [465]

The failure of a coupling led to three carriages of a passenger train to run back down an incline, colliding first with a Jeep and a van, on one or two level crossings, injuring their occupants, then after travelling almost two kilometres, with another passenger train between Sucat and Alabang stations, Muntinlupa, the Philippines, on Monday 22nd September 1997.

[463] *Cedar Rapids Gazette*, Cedar Rapids, Iowa, United States, 4th March 1997
[464] *The Lethbridge Herald*, Lethbridge, Alberta, Canada, 5th March 1997
[465] *Dawn Wire Service*, Karachi, Pakistan, 8th March 1997

The second train of five coaches was considerably overcrowded, with some passengers riding on its roof, some of whom were injured when they jumped off, moments before the collision. One coach of one or the other train was derailed.

Passenger Maria Villamin, a 44-year-old garment worker who had been on the front train, said later, "Everybody was shouting and yelling in panic from the time that our train got detached until the train crashed." She related that a bump had preceded the coaches running away and that after the collision a number of passengers had been injured in their rush to escape.

Driver Jesus de Jesus of the second train told a reporter that when he had seen the runaway coaches he had tried to stop his train, but there had been no time.

Ambulances rushing to the accident scene were held up by rush-hour traffic, though on the other hand many car drivers were able to help the victims. Mayor Ignachio Bunye of Muntinlupa declared that the local hospital had been overwhelmed by the number of injured.

Seven people died in the accident and 220 were injured.

Police opened an investigation and indicated that Philippine National Railways officials might be charged with negligence. [466] [467] [468] [469]

Given that the passenger train was presumably fitted with some sort of automatic brakes, it is unclear why some of its coaches began to run away simply because the train became divided.

A major runaway disaster occurred in North Korea at an unstated date in the summer of 1997, on the line between Hyesan and Gilju (also spelt Kilju), which was steeply graded, especially between Baekam and Woonhung stations. Two trains were involved, the second of which was an overnight passenger service from Hyesan to Gilju. The first train stalled on the steep Baekam incline, whereupon the following passenger train came to a stand in its rear and its engine was detached from the coaches to assist the stalled front train up the bank. After initially remaining at a stand, the eight coaches then began to run back down the incline. As it was about 03:00, most of the passengers were asleep, while those who were awake at first did not realise that the train was not just restarting. When they did realise, a panic arose and some passengers jumped from the train.

The runaway coaches reportedly covered over 100 kilometres, which might be something of an exaggeration, and passed through 10 stations, with coaches becoming derailed on various curves, so that only two coaches remained by the time that the terminus at Hyesan was reached, where they crashed explosively. Warning of the runaway having been given from up the line, Hyesan station itself had already been evacuated. A good number of the younger and more agile passengers had saved their lives by throwing themselves from the train earlier on, so it was mostly elderly people, some of the younger women and children who suffered. No figures were given for the number of passengers killed or injured, though the casualties allegedly ran to several hundred.

North Korea was to suffer a probably even more disastrous runaway accident in November of the following year (see below). [470]

---

[466] *Sedalia Democrat*, Sedalia, Missouri, United States, 22nd September 1997
[467] *Daily News Miner*, Fairbanks, Alaska, United States, 22nd September 1997
[468] *The Medicine Hat News*, Medicine Hat, Alberta, Canada, 23rd September 1997
[469] *Pacific Stars and Stripes*, Tokyo, Japan, 24th September 1997
[470] http://nk.chosun.com/e.../news.html?ACT=detail&res_id=625

Early on the morning of Thursday 20th November 1997 Deutsche Bahn freight train KC 71153, consisting of a locomotive and 22 tank waggons of petrol and fuel oil, weighing 1,851 tonnes, was en route from Stendell to Nossen, Germany. Its route involved haulage by a diesel engine as far as Berlin-Grünau, where it would reverse direction and from where it would be hauled by an electric engine as far as Riesa. There, a diesel engine would be substituted for the final part of the journey over the non-electrified line to Nossen.

The train arrived at Berlin-Grünau late, the diesel engine was detached and electric engine No. 155 103 was attached to it. Perhaps because he was in a hurry, due to the train being late, its driver connected the screw coupling but omitted to connect the brake pipes and open the brake valves. He then charged the brake system until the pressure reached 5 bars, but failed to notice that he had achieved the required brake pressure very quickly, because only the air on the locomotive was being replenished.

In addition, a simplified brake test needed to be carried out in co-operation with a Zugforbereiter (train preparer) because the brake pipe connections had been broken and remade. The train preparer should have checked that the brakes on the rear waggon applied and released correctly, that the brake pipes were coupled throughout and that all the brake valves were open. Instead he found that the brakes on the rear waggon were not released, which he wrongly attributed to an overloaded brake line with a pressure greater than 5 bars. At this point, according to the regulations, the simplified brake test should have been aborted and a full brake test carried out instead, whereby the brakes of each waggon were released manually by operating the individual release cables, all the hose connections and brake valve settings being checked at the same time, and a check made that the brakes applied and released on each waggon. The problem with the brakes of the rear waggon should also have been notified to the driver. However, the train preparer had two other trains to deal with at the same time, so to save himself trouble he simply manually released the brakes of all the waggons. With the brakes not being recharged from the engine they were now inoperative throughout the train.

The driver alleged that the train preparer had then indicated to him that the brake test had been completed, which the train preparer denied. The Fahrdienstleiter (signalman) was advised by someone, it was never established who, that the train was ready to depart, the signals were cleared and the train set off, with the train preparer, if he had not notified the driver that the brake test was complete, having raised no objection to the train starting. It was possible that the driver had taken the "brakes OK" signal intended for another train as applying to his own.

With no other traffic on the line, the train was not checked on its journey as far as Elsterwerda, and so there was no need for its driver to use his brakes. The line also fell steeply between Hohenleipisch and Elsterwerda, so the train gathered speed on this section. Although the change of engines had originally been planned for Riesa, because the train was running late Control decided to make the engine change at Elsterwerda instead. This involved the train being diverted from the main line through a low-speed crossover onto track 5, and the Einfahrsignal (Home signal) was cleared to indicate "Reduce speed to 40 km/h, prepared to stop." Approaching Elsterwerda and finding adverse signals for the first time since leaving Berlin-Grünau, the driver applied his brakes, but obtained very little effect, because they were only working on his engine. Contacting the Elsterwerda signalman, the driver advised the latter that his brakes were not working but the signalman could do nothing to help him.

At about 06:35 the train attempted to negotiate the facing crossover, through which speed was limited to 40 km/h, at a speed of just under 90 km/h. The engine stayed on the track and came to a stand 177 metres further on. Behind it, the coupling snapped, 15 of the tank waggons left the rails and 13 of them caught fire. Petrol was sprayed several hundred metres, an engine shed and an engine stabled there caught fire and motor cars travelling along a road outside the station were pushed to the other side of the carriageway. During the efforts to fight the fire the Elsterwerda fire chief was crushed by a collapsing building and killed, a second fireman died of his injuries five days later and eight firemen and police officers were injured.

Efforts to extinguish the blaze were delayed because the firemen were initially unsure what chemicals were involved and whether water could be used, while there was also a shortage of water and a continued risk of further explosions. Quantities of petrol also leaked into the sewage system.

At noon several intact, un-derailed waggons were uncoupled and pulled out of harm's way and in the evening, after the fire had finally been extinguished, petrol was pumped out of four other waggons. The firefighting operation lasted for 34 hours.

The train driver and train preparer later received suspended prison sentences for negligence, while Deutsche Bahn paid out 1,200,000 euros to 122 victims of the accident. [471]

After leaving Rawalpindi, Pakistan, at 15:15 on Saturday 31st January 1998, train No. 6, the Down *Tezrao Express* from Peshawar to Karachi progressed 25 kilometres to Sihala, where it was signalled into a loop line to cross an opposing train. (The train is also referred to in some accounts as the *Zulfikar Express*, the name it bore during the government of President Zulfikar Ali Bhutto prior to his overthrow in 1977, but by this date it had apparently reverted officially to its former name.) According to surviving passengers the express had been driven at excessive speed and on reaching Sihala its driver apparently lost control of it, the train failed to stop at the end of the loop and its engine and four leading coaches were derailed in the sand drag at the end of the trap siding. A signal box and crossing gates were demolished before the train struck the Sihala Jamia Masjid (mosque), which stood by the lineside. Several railwaymen in the signal box and at the crossing gates saw what was about to happen in time to get clear. Driver Mohammad Hanif also jumped and saved his life but 12 people were killed outright or fatally hurt and about 20 others were injured. Fireman Mohammad Ramzan was amongst the dead. A worse accident was probably averted because having become stuck in the sand, the derailed coaches did not overturn.

A shopkeeper named Zaheer witnessed the crash. He told a reporter: "I saw the train moving and then I ran away. I did not know what had happened but clouds of dust covered the sky, bricks and pieces of iron were scattered over a wide area and the screams of men and women could be heard."

The level crossing with the Rawalpindi – Khutta road was blocked for some hours before breakdown cranes could clear it of wreckage. When the line had finally been cleared the express continued its journey with a new engine and rolling stock.

Prime Minister Nawaz Sharif expressed his sympathy with the victims, directed the Railway Minister Yagoob Nasir to provide assistance for the families of the deceased and medical facilities for the injured and ordered the opening of an inquiry into the accident. There were criticisms in the media, though, that even 28 hours after the disaster neither the railway minister nor the chairman of Pakistan Railways had gone to the scene of the accident.

Driver Hanif was later arrested, charged with negligence and overspeeding, though the railway authorities were reluctant to assign blame for the accident, pending the outcome of the inquiry, and it was stated that although Driver Hanif had complained at Rawalpindi of problems with the train brakes, his concerns had been dismissed. What was beyond dispute, was that in addition to the driver and fireman, because of the crowded state of the train, four passengers had been riding on the locomotive. General Manager Mohammad Afzal stated: "It

---

[471] Wikipedia – Eisenbahnunfall von Elsterwerda

is illegal because passengers are not allowed to travel with the driver and fireman." Only one of these four passengers, Sajid Ali, who had been riding on the right-hand side of the engine alongside its door and had probably jumped together with the driver, survived the accident. He related later: "Just before the crash, the driver told us to recite Kalma-a-Tayyaba {Islamic prayers, the first of which is: 'There is no deity but Allah, Mohammad is the messenger of Allah.') because he could not control the train. We had paid the fare to Lahore to the fireman because the train was overcrowded." [472] [473]

It will be noted that an almost identical accident had happened to the same train at Mehr Shah, less than two years previously (see above).

South Central Railway train No. 7336, the Manmad – Kachiguda Express, bound for Hyderabad, was stationary in Parli Vaijnath station, Maharashtra, India, at 01:10 on Friday 24th April 1998, when its rear coach was struck by some ballast waggons that had become detached from an engineering train and run back down an incline. Twenty-four passengers died and 36 others were injured. The rear coach of the express was badly damaged. The injured were rushed to Ambaijogai hospital. Twenty of them were discharged after treatment but 10 others were said to be in a serious condition.

Maharashtra Deputy Chief Minister Gopinath Munde announced an *ex gratia* payment of 50,000 rupees to the next of kin of the dead, and 5,000 rupees to the injured.

South-Central Railway general manager N.C. Sinha, along with the chief public relations officer, rushed to the spot to begin an investigation, while Union Railway Minister Nitish Kumar flew to the spot by a special aircraft to oversee the rescue and relief operations. [474] [475]

Eighteen people died and over 100 were injured, some seriously, when a passenger train from Blantyre to Makhanga carrying 250 people ran away and was derailed at Malabvi, Malawi, on Friday 22nd May 1998. [476]

The following year, a commission of inquiry into the country's worst railway accident, presided over by Judge Duncan Tambala, published its report, which in part found that, "Ultimately we hold the entire organisation of Malawi Railways (1994) Limited responsible and guilty of negligence for permitting a train with insufficient brake to depart Limbe on a downhill journey to Makhanga. We take the view that Malawi Railways must pay in full for loss of life, personal injuries and for loss of damage to properties."

According to the report, the locomotive, the brake failure of which was blamed for the crash, had only 30 percent brake power instead of the required 100 percent. The coaching stock, meanwhile, had only 16 percent braking power. Rail experts stated that coaches needed to have at least 90 percent braking power.

The commission attributed the accident to the falling gradients after Limbe township and the wet conditions of the rail track, which, it said, had aggravated matters, rendering the application of all the braking systems on the locomotive and the rest of the train useless.

The report added that, "Laxity by Malawi Railways workshop staff resulted in failure to meet recommended standards of both the locomotive and rolling stock. We hold the

---

472 http://www.best.com/~karachi/news/feb98/020298.txt

473 https://asianstudies.github.io/area-studies/SouthAsia/SAserials/Dawn/1998/07Feb98.html

474 https://m.*Rediff.com*/news/1998/apr/24mishap.htm

475 Government of India Ministry of Civil Aviation and Tourism Commission of Railway Safety Annual Report for 1998-99

476 *The Odessa American*, Odessa, Texas, United States, 24th May 1998

engineering services department of Malawi Railways responsible."

The report was welcomed by survivors of the accident who had been fighting for months for compensation.

"It is good that, finally, justice will be done. We lost a lot of property in that accident on top of footing hospital bills for a cousin of mine who sustained serious injuries," Moses Kamwendo, a resident of Blantyre, said. With even more reason to be grateful was Anne Yasini, who had been paralysed and confined to a wheelchair following the accident, and who was almost thrown out of a South African hospital for failure to meet her medical bill. [477]

Unauthorised passengers tampering with the brakes were blamed for the derailment of a passenger train No. 894 from Alexandria to Kafr el-Sheikh at Kafr el-Dawar, Egypt, about 30 kilometres from Alexandria, on Sunday 18th October 1998. The train was apparently diverted through a crossover while travelling at high speed, with the result that the engine and at least two of its coaches were derailed and overturned. The derailed vehicles demolished a wall and struck a war memorial and several shops, causing a heavy death toll of mostly townspeople, while passengers on the crowded train were mostly injured. A number of people at the station were also among the casualties.

Teenage Hany Roushdy, who witnessed the accident, related that the train had been running very fast. He said, "It did not seem to be able to stop. It went off the tracks and towards the town square," and that the noise of the crash had sounded like thunder.

Businessman Mustafa Attiya spoke of hearing girls screaming. "I turned around to look and saw the train shoot across and hit them," he declared.

Shortly after the crash there were chaotic scenes as wailing townspeople flocked to the scene, while riot police attempted to control them.

A television appeal was made for blood donors and Ismail Sallam, himself a practising surgeon, went to the scene. Many people flocked to the hospitals in response to the appeal while earlier, those at the scene had used their clothes to stem the bleeding wounds of the injured.

Railway official Mahmoud Marei suggested that roof-riding passengers had interfered with the brake pipe, perhaps accidentally obstructing it. [478] [479]

Health official Nabih Yousef Shaltout announced on Sunday that there were 47 dead and 101 cases of injury. On Monday the lifting of the last of the derailed coaches by a large crane revealed the bodies of another five victims. [480]

In 1998 just two passenger trains daily ran between the port city of Dancheon and the town of Geumgol, North Korea, departing from each end at 07:00 and 20:00. If things went well the journey took about three hours but there were often holdups, when the train could take two or three hours longer. On Thursday 19th November 1998, the 20:00 train from Dancheon ran normally until it had passed through Epa station, then the power failed (the line was electrified) and the train ground to a halt. Accustomed to this sort of thing happening, the passengers were not unduly concerned. After a halt of about an hour the train began to move

---

[477] https://www.ipsnews.net/1999/09/rights-malawi-victims-of-train-crash-to-be-compensated/#:~:text=The%20accident%2C%20which%20happened%20on,crash%20in%20Malawi's%20modern%20history.

[478] *BBC News*, London, England, 19th October 1998

[479] *The Orange County Register*, Santa Ana, California, United States, 19th October 1998

[480] https://www.deseret.com/1998/10/19/19407561/toll-rises-to-52-in-egypt-train-accident/

again, except that instead of going forward it began to go backwards. At first, the passengers assumed that the train was returning to Epa station for attention but when it ran through that station without stopping and continued its backward progress at increasing speed, its passengers became aware that something was seriously wrong.

In the panic that followed, some of the passengers jumped from the runaway train, while others held on to anything that they could for dear life. The mad rush of the train reportedly continued for half an hour, during the course of which, four of its eight carriages left the track on curves, one of them falling from a bridge into a river. Presumably due to a transition to rising gradients, the remainder of the train finally came to a halt only about two kilometres short of Dancheon.

Since 1994 North Korea had experienced an economic crisis due to a combination of financial mismanagement, natural disasters, international sanctions and the collapse of the Soviet Union. As a result, it was a time of mass starvation and many of the passengers on the ill-fated train were housewives from Dancheon travelling out into the country to exchange goods on the black market for potatoes.

The accident happened at about 01:30 on Friday 20th and it was about 08:00 before rescuers finally arrived at the scene. Nighttime temperatures were very low, down to about -10° Celsius, and it was thought that many of those from the derailed coaches who had not been killed outright had died from exposure before help arrived and would have survived had they been reached earlier. It was also alleged that there had been insufficient rescuers and that they had been poorly equipped. The bodies of the dead were lined up along the side of the track, arranged by the authorities by gender and age. The bodies of many other victims from the coach that had fallen into the river were thought to have been carried down the river to the sea. Railway officials put the number of dead at 500; unofficial estimates were that there had been 1,500 dead, which, if true, would make it one of the worst catastrophes in railway history, and about equal in terms of casualties to the Boxing Day 2004 catastrophe in Sri Lanka, when a train was struck by a tsunami (see later).

A survivor of the accident, who had been a passenger in one of the un-derailed coaches, wrote of subsequently resuming his journey to Geumgol on a freight train, accompanied by passengers going to search for relatives who had been on the accident train. Later he defected from North Korea and describing the catastrophe, wrote scathingly of Kim Jong Il spending millions of dollars in decorating his father's mausoleum and boasting of North Korea being a 'strong nation,' while yet not having sufficient rescuers or equipment to save his own people.[481] Why the train had run away was not stated; presumably it had suffered a brake failure for some reason, perhaps poor maintenance.

On the evening of Saturday 18th September 1999 an engine was engaged in shunting operations at Wah, Pakistan, when it developed a problem. Its crew alighted to try and fix the fault and while they were on the ground the engine began to run away down an incline towards the next station at Taxila. The driver then made a futile attempt to arrest the progress of the runaway engine by placing stones on the rails, which the engine simply crushed or pushed to one side.

After running some three kilometres and attaining an estimated speed of 80 or more kilometres an hour the engine came into collision with the six-coach *Chenab Express* from Peshawar to Karachi at 22:42. Published news reports state that the express was travelling in the opposite direction but as Wah station is on the Peshawar side of Taxila this cannot be the case and it would appear that either the express was proceeding from Karachi to Peshawar or

[481] https://www.dailynk.com/english/1500-dead-in-a-train-accident-in-d/

else that the runaway engine overtook and collided with the rear of the it. At any rate, two carriages of the passenger train were badly damaged and others were derailed.

Awakened by the sound of the crash, nearby residents hurried to the scene. The dead and injured were taken from the train and cars, lorries and horse-drawn carts were flagged down to take the injured to various hospitals, mostly to the Wah Military Hospital. Two of those taken to the latter place were Mohammed Aurengzheb and his wife, who was unconscious. Cradling her in his arms, Mohammed told a reporter, "I woke up with the crash and I began to look for my wife. I didn't know what had happened. I yelled her name." [482]

About 20 people died in the accident, among them the conductor of the express, and about 80 were injured. Most of the victims were from a party of 83 Afghan Ismaili refugees on their way to Karachi for resettlement by an aid organisation. [483]

A goods train from Senegal collided with a rake of eight carriages that had run away down a steep gradient on Monday 11th October 1999. The accident occurred at Kati, Mali, 15 kilometres from Bamako. Driver Minkoro Coulibaly of the goods train, who escaped injury, stated that he had seen the runaways when they had been only about 10 metres away. Three railwaymen travelling in a coach immediately behind the engine of the goods train were killed and others were injured. Fire apparently broke out after the collision, as a report stated, 'Rescue teams, including fire fighters, Red Cross staff and security forces, used burners, power saws and sledgehammers to retrieve bodies from the charred carriages.' An inquiry was opened to determine how the accident had happened. [484] [485]

Because the road to their homes was closed for repairs, about 80 children and their parents, who had just enrolled for a new school term at Padang, the capital of West Sumatra, Indonesia, chose to hitch a ride back on an eight-waggon coal train on Wednesday 28th June 2000. For many of them it proved to be a fatal decision because the brakes of the train failed as it was descending a gradient, it ran away and then derailed in the Batang Anai sub-district of Padang Panjang, about 50 kilometres north of Padang. At least 32 people, mostly children, were crushed beneath the derailed waggons and killed and another 21 were injured. [486]

A power failure was the indirect cause of a serious runaway accident in the metropolitan region of São Paulo, Brazil, on a line of the Companhia Paulista de Trens Metropolitanos (CPTM) on Friday 28th July 2000. When the power failed at 21:15, passenger train No. 145, an electric multiple unit of the 1100 series, en route from São Paulo to Francisco Morato, was stopped in the station at Perus. Behind it, in the preceding station at Jaraguá, was passenger train No. 217 of the 1700 series. The line fell 66 metres in the five kilometres between the two stations, thus with an average falling gradient of about 1 in 76.

When train No. 217 came to a stand in Jaraguá station its passengers were asked to disembark and Motorman Osvaldo Pierucci also alighted from his cab, according to his account in order to scotch the wheels of his train with wooden blocks, the power failure having rendered the other brake systems of the train inoperative. Student Motorman Selmo Quintal, 24, remained in the cab. The train then began to run away.

---

[482] *Sunday Herald*, Lethbridge, Alberta, Canada, 19th September 1999

[483] http://danger-ahead.railfan.net/reports/rep99/taxila_pkstan990918.html

[484] *allAfrica.com*, 12th October 1999

[485] *Danger Ahead! Historic Railway Disasters*, 13th October 1999

[486] *The Irish Times*, Dublin, Eire, 28th June 2000

Control was alerted of the runaway and reportedly attempted to get train No. 145 to leave Perus station. Three times the conductor signalled to his motorman to start but the train did not do so, as there was apparently a problem in closing the train doors. The station loudspeaker then instructed the passengers to alight. Before they had time to do so, the eight minutes it had taken for the runaway to cover the five kilometres between the stations had elapsed and it collided violently with the rear of the other train at a speed of reportedly 70 km/h. Nine passengers died and 115 were injured. Student Motorman Quintal, who had jumped from the runaway at some point, was also killed, while the station was demolished.

Replacement buses were provided for the passengers who normally used Linha 7 Rubi while it was closed.

The CPTM blamed Motorman Pierucci for the accident, stating that he had failed to secure his train. A committee was formed to investigate the accident but its findings were not made public.

Twelve years after the accident the CPTM was ordered by a court to pay compensation to the relatives of the deceased and to the injured. [487] [488]

Freight trains were approaching the station at Nvoungouti, Republic of Congo, from opposite directions on the morning of Wednesday 10th January 2001, when the brakes of one of them failed and they collided. Some of the waggons of one or both of them were then derailed and fell into a ravine. Although neither was a passenger train, many people were riding on them, and by Thursday afternoon there were known to be 30 dead and about 100 injured, of whom 53 were taken to hospital in nearby Dolisie and another 45 to the main hospital in Pointe Noire. Cans of fuel being carried on one of the trains caught fire in the accident and many of the injured suffered burns.

Transport ministry spokeswoman Therese Issabou stated that there were believed to be more bodies in the wreckage and by Friday, 38 had been recovered and another seven remained to be recovered from the waggons in the ravine. [489] [490]

At least 12 people possibly died and more than 25 were hurt in an accident that occurred at about 20:30 on Thursday 16th August 2001, when a Muzaffarpur – Sealdah passenger train collided with six waggons of a goods train, which had become detached due to a defective coupling and had run back down an incline. The accident happened near the station at Girdhaur, Bihar, India. Another source stated that there were less than 10 dead. [491]

Kenya suffered a major disaster in the early hours of Sunday 20th August 2000 when nine waggons of a freight train carrying petrol (five) and liquefied petroleum gas (four) from an oil refinery at Mombasa to the capital, Nairobi, ran away. After running some distance they collided with a stationary waggon, which had reportedly been deliberately placed in their path to stop them, at Athi River. What was described as a tremendous explosion followed the collision at 00:10 and a massive fireball then erupted and spread over a 100-metre radius. Killed outright, mostly while they were asleep, were 16 people, 13 of them Kenya Railways employees and their families. Two other victims were residents of a shanty town that adjoined

[487] https://viatrolebus.com.br/2020/07/ha-20-anos-dois-trens-se-chocavam-na-estacao-perus-da-cptm/

[488] Wikipedia – Acidente na estação Perus

[489] *The Lethbridge Herald*, Lethbridge, Alberta, Canada, 12th January 2001

[490] *The Daily Review*, Towanda, Pennsylvania, 13th January 2001

[491] *Rediff.com* 17th August 2001

the railway. Another 36 people were taken to local hospitals, many with severe burns. A relatively minor side effect of the disaster was that a Mombasa – Nairobi passenger train had to be terminated at Sultan Hamid, from where its travellers had to complete their journeys by road in hired matatus (private minibuses). [492]

Anthony Chira, whose own house suffered damage in the accident, said, "There was liquid fire in the air, which was dropping like it was raining fire. I thank God for being alive because some people were burnt beyond recognition. People were running and shouting with fire on their heads." Chira, who lives in a railway-owned house within the station's grounds, 50 yards from where the explosion took place, told The Associated Press that he was awoken around midnight by the "feel of heat." He then saw fire everywhere and heard explosions "again and again." Two of the three rooms in Chira's home, which he shares with his wife, three children and a railway employee, were gutted by the fire.

'More than 16 hours after the crash, houses and carriages were still smouldering as firefighters and railway engineers cleared up the site. The mangled carriages were spread across six lanes of track. The accident was caused by train cars reversing down the track out of control after being detached from a freight train.' [493]

'Philip Kioko, the father-in-law of a man who perished in the fire with his family, said the explosion was an instantaneous moment of horror. His son-in-law, Gideon Soi, died with his wife, Regina, and children. Kioko said the accident occurred while he was on duty manning the workshop shed. "I heard a loud explosion then saw a ball of fire leap into the dark sky. Momentary silence then engulfed the neighbourhood before the whole place turned into fire," he recalled.' [494]

The runaway waggons had been detached at a point some kilometres from Athi River because the load of the train had been too great for its engine to manage. A railwayman who did not wish to be named explained, "It was not expected that the waggons would roll back." [495]

A couple of days after the accident the number of dead had risen to 25 and Joseph Mbwiria, a spokesman at the Kenyatta Hospital in Nairobi, stated that most of the other injured patients there had at least 60 per cent burns, adding, "Many of them have internal and external injuries and have breathed a lot of hot air which has damaged their lungs." [496]

Stationmaster Joseph Muriuki Kathenwa at Athi River and Guard Hezekial Mukhwana of the freight train were subsequently charged with negligence in connection with the disaster, which finally claimed 35 lives.[497] It is not known whether the two railwaymen were convicted or, if so, what penalties were imposed.

Life for the survivors of the disaster was hard. 'One of them, 25-year-old Veronica Kabenge, was asked to leave the hospital last December 24 in spite of her poor health. "I was asked to take her home. The doctors told me that Kenya Railways, which was supposed to clear the bills, is yet to do so," her mother, Ms Agnes Kaswii, said. Ms Kaswii said she now uses a charcoal burner to keep her daughter warm in a single-room tin shack in Athi River. Her daughter is too ill to speak.

'In the neighbourhood, 25-year-old Dorris Atemo struggled to sit up and take a break from the lying position she had been in for several hours. She has had an adequate share of the tragedy, having lost her husband and two children in the inferno. Ms Atemo returned home on

[492] *allAfrica.com*, 20th August 2000

[493] www.*Ananova*.net/news/story/world_kenya-rail-accidents_940752.html

[494] *allAfrica.com*, 21st August 2000

[495] www.ble.org/pr/news/headline082100g.html

[496] www.ble.org/pr/news/headlines082200f.html

[497] *allAfrica.com*, 30th December 2000

October 29, a year after being admitted to hospital, but has fallen ill again. She now hopes that her father, Mr Jairus Zebedi Abweje, will eventually get a cheque from the corporation and take her back to hospital. Mr Abweje said, "I have been to KR headquarters many times in the past three weeks after my daughter fell ill again. At the Kenyatta National Hospital, they said they could not admit her before KR issues them with a cheque and that is what I have been trying to chase. I am worried because my daughter has eaten very little in the recent past." He hopes that his efforts will be rewarded one day soon.

'Three other victims – Ms Elizabeth Nduku, 39, whose hands were disfigured and who lost four children, Mr Moses Limera, 20, and Matthew Limera, 16 – said they were awaiting further help and more medical treatment. Moses and Matthew lost their elder brother in the inferno.' [498]

InterCity train No. 30 left Kandy for Colombo, Sri Lanka, at 06:30 on Sunday 13th January 2002. It consisted of a M6 type diesel electric locomotive and six coaches and was booked to make the 121-kilometre journey with only two intermediate stops, at Perandeniya Junction and Veyangoda.

Much of the Kandy – Colombo line was sharply curved and had steep gradients, so that although a speed of 90 km/h was permitted on the flatter straight sections, it was limited to 20 km/h over portions of the line, for example between Kadugannawa, 13 kilometres from Kandy, and Rambukkana, 26 kilometres from Kandy. Between Kadigamuwa, about four kilometres nearer Kandy, and Rambukkana, the line also fell very steeply at 1 in 44.

On the day of the accident the train left Perandeniya Junction after a minute's stop. It passed through Kadugannawa station at 06:57 and through Balana station at 07:08. So far it was under control but after passing the small station at Makehewala, in the section between Balana and Ihala Kotte, the driver noticed that the train speed appeared to be too high, while the vacuum had dropped to zero. As there was no speedometer on the engine he could not determine the exact speed but he instructed the assistant driver to apply the hand brake to steady the train. This was done, while the guard and the underguard, riding in two brake vans, also applied their hand brakes, all with no appreciable effect on the speed. Now beginning to panic, the driver applied the hydrodynamic brake on the engine. Despite this the speed still did not decrease and the train was now clearly out of control. It was now approaching Kadigamuwa, another tablet station on the single line, and the driver feared that there might be another train coming towards them, in the forward section, so it was a relief for him to see a member of the station staff standing on the platform, with the tablet holder for the next section. The tablet pouch for the previous section was thrown onto the station platform and the new pouch was seized with difficulty but the troubles of the crew were far from over, for now they were entering the steepest part of the line, which was also sharply curved. A potential head-on collision with another train had been avoided but near the village of Kirivallapitiya, not far short of Rambukkana station, the combination of the high speed of the train and the sharp curvature proved too much and it left the rails. The first news of the accident came from a report from a local person using a public telephone outside of Rambukkana station.

Fifteen people lost their lives in the derailment and over 200 were injured, many of them very badly.

The guard and the driver of the train gave evidence at the inquest on the victims of the disaster. Guard Sumathiratna Piyasekara testified, "I took over responsibility for the Intercity Express train at Kandy at about 5.30 a.m. on 13th January. I inspected the train well and it's not my responsibility to inspect the engine. The driver should take charge of the train at least

[498] *allAfrica.com*, 2nd January 2002

an hour before I get there. He has to check the oil, water and brakes of the engine before taking it from the yard. He should have done this in the yard. Therefore I don't know if he did so. The train was at the yard at about 6 a.m. There are three types of brake on the train. These are the vacuum, dynamic and air. Only the vacuum is indicated on the metre in my van. It was up to 18 inches. That is the normal level. We left Kandy at 6.30 a.m. The train stopped at Peradeniya and left there after tablets had been exchanged."

The guard stated that between Balana and Ihalakotte the train had begun to accelerate and exceed the normal speed. The vacuum had fallen to below ten inches. At this point he had realised that there must be a problem with the brakes and had applied the hand brakes, without effect. He had had no means of communicating with his driver.

Continuing, the guard said, "At Kiriwallapitiya, a mile short of Rambukkana, the train left the rails and toppled over into a paddy field. There is a curve here and a lineside sign indicates a speed limit of 25 km/h. All seven compartments of the train left the track, together with the engine. I was admitted to Kegalle Hospital and then transferred to Colombo General Hospital. I believe that if my driver had applied the brakes while changing 'notches' the tragedy could have been avoided."

Driver Katthigoda Kamkamamlage Lesli Ranjith related, "I was working on 12th January and finished at 12.40 a.m. on 13th January. I had to come on duty again that day at 5 a.m. so I told the officer in charge of drivers to bring the train from the yard to the platform as I had to get ready and arrange other equipment. Sometimes we do it. I reached the station at 6 a.m. I didn't check the engine. I believe the officer in charge checked the engine in the yard. There was no process for this to be recorded. The dynamic brakes were not working on the train. They are not normally working on any of the trains but we are not instructed to fail a train because of this. At Makehelwala the vacuum fell to zero. From there to Rambukkana there is a falling gradient and the train began to accelerate. My assistant shouted that there was a problem with the brakes as he was exchanging the tablets. The train derailed at the 53 mile post. The train should have stopped when the vacuum dropped to zero but it did not stop." [499]

A union representative pointed out that the driver might well have been suffering from fatigue after working continuously for three days. It appears, from his evidence, that the driver had only had a few hours' rest before taking out the accident train.[500] Vigilance devices had reportedly been removed from most locomotives because drivers found having to reset them every 30 seconds to forestall a brake application too arduous a task when working 12- or 15-hour shifts. [501]

The investigation into the disaster found that the brake power of both the locomotive and the coaches of the train had been deficient, as was indicated by the condition of some of the brake blocks of the carriages after the accident, which showed no indication of heavy braking. Four of the six coaches of the train had been overdue for their major service, one by more than three years. A vacuum brake test had not been carried out before the train had left Kandy, while the lack of a speedometer on the locomotive had also made it difficult for the driver to accurately judge his speed. [502]

On Tuesday 27th February 2002 a freight train consisting of an engine and 28 waggons was on its way from Sopron in Hungary via Eberfurth, Austria, where a reversal was necessary, to Wien Central. Meanwhile, coming in the opposite direction was a rolling road (RoLa) freight

---

[499] http://origin.island.lk/2002/01/28/news08.html
[500] www.dailymirror.lk/archives/dmr220102/News/defective.html
[501] www.dailynews.lk/20.../new19htm
[502] https://railpagesrilanka.blogspot.com/2012/11/

train consisting of an engine, a couchette coach for the lorry drivers, and 21 waggons loaded with their vehicles, en route from Wels in Austria to Sopron.

Unknown to the crew of the train from Sopron, though they realised that its brake performance was not as good as it might have been, the tap of the brake pipe was closed between the third and the second waggons from its rear. From Eberfurth, with the change of direction of the train, the closed brake pipe was now located between the second and third waggons from the engine, so that the brakes of most of the train were inoperative. A brake test should have been performed at Eberfurth, which should have detected the problem, but it was either not carried out at all or was not done correctly. The result was that when the train was approaching the station at Wampersdorf, Austria, on falling gradients, its driver lost control of it and it ended up colliding head-on with the RoLa train at about 15:00. [503]

The lorry drivers on the RoLa train were the main victims of the accident. By the evening the bodies of six of them had been found and rescuers stated that they could see the body of a seventh still in the wreckage. The 14 or 15 survivors were all injured; four of them were reportedly unconscious. Helicopters and road ambulances took the injured to local and Wien hospitals as 150 or more rescuers worked at the scene. Most of the lorry drivers were Hungarians; of the six recovered dead three were identified as Hungarians, two as Yugoslavs and one as a Turk. [504]

Driver Casaba Lanyi, presumably of the RoLa train, speaking from his hospital bed with his head bandaged, told state television, "I just had time to take a deep breath and then it was too late to do anything else before the collision."

Somewhat hampering the rescue work was the fact that the 'jaws of life' and similar equipment, which were suitable for use in road accidents, proved unable to cut through the thicker metal of the railway vehicles. One of the surviving lorry drivers was pinned in the wreckage for three hours before he could be freed. [505]

A major catastrophe occurred in Mozambique at about 05:00 on Saturday 25th May 2002. A mixed train from South Africa was descending a falling gradient between Moamba and Tenga and fairly close to the latter station, when it developed an unspecified mechanical fault. At this point the driver decided to divide it, leaving the passenger carriages, in which there were 600 people, behind on the bank while he proceeded with the goods waggons, loaded with cement, to Tenga station. To prevent the passenger carriages from running away down the incline after the front of the train had been detached he adopted the same ineffectual measure as had led to the Armagh catastrophe in Ireland, 113 years earlier – four large stones were wedged beneath the wheels of the passenger coaches. The front portion of the train was then detached and went forward; the makeshift scotches came loose or were dislodged in some manner and the carriages began to run away.[506] The gradient must have been pretty steep and the passenger train must have achieved a considerable speed, for the collision was a disastrous one. Initial reports stated that there had been 196 deaths and health officials were said to be pleading for blood donations for 400 injured.

A newspaper wrote, '"Dead bodies were lying all over," Antonio Cevere, who was travelling on the passenger train and was injured, told Radio Mozambique.

'Rescue workers with tractors and construction equipment were trying to free people trapped in the train's twisted wreckage, and officials said the death toll could rise further in

---

[503] Wikipedia – ÕBB Class 1044

[504] *BBC News*, London, England, 26th and 27th February 2002

[505] *Daily News-Record*, Harrisonburg, Virginia, United States, 27th February 2002

[506] *The Times Daily*, Florence, Alabama, United States, 27th May 2002

what was already the worst train disaster in the history of this impoverished southern African nation of 18 million people.

'President Joachim Chissano declared the accident a national emergency and called on people to donate blood to the injured and support those who had lost their families. Most of the dead and injured were thought to be Mozambiquan.' [507]

Another report stated, 'Mozambican television showed footage of a pile of bodies lying next to the train tracks and soldiers and police searching for more victims. "There are still people under there. When we got here, passengers were dying in our arms because we could not get them out quickly enough," a police officer told the television station. "Everything is destroyed. There are bodies that have been cut in half and we don't know how to match them up, there are severed limbs," said the officer, who was not identified.

'The majority of the injured were taken to hospitals in Maputo in private cars and many people died along the way.

'Transport Minister Tomas Salomao blamed the crash on the national railway company and said it would have to pay compensation. "Now we have to deal with funerals and the people who are being treated in the hospitals," he told state television. "Then we will establish the type of assistance people need."

'Antonio Libombo, an official of the national railway company, said he did not know what had caused the crash and refused to comment on Salomao's statement.' [508]

The above and some other reports indicated that the train was bound for Ressano Garcia with Mozambicans bound for South Africa on a weekend shopping trip but this does not agree with the report that the train was travelling from Moamba towards Tenga, which would imply that it had been travelling from Ressano Garcia towards Maputo.

Some of those who were involved in the disaster were, perhaps, surprisingly understanding. Severiano Matavele, 46, who had lost a brother-in-law, said, "I am not angry. This is something that can happen at any time." President Chissano, meanwhile, pledged that, "Justice will be done." [509]

The number of dead apparently subsequently rose to at least 205. [510]

A day short of a month after the Tenga accident, at 08:30 on Monday 24th June 2002, an even worse runaway catastrophe occurred between Igandu and Msagali, Tanzania. A passenger train carrying more than 1,200 people was on its way from Dar es Salaam to Dodoma, the capital, and on to Kigoma. It had passed Msagali and about 35 kilometres further on was climbing a bank when there was a problem with its brakes and it stopped just short of the summit of the climb. The driver got down, examined and adjusted the brakes, reboarded his engine and restarted. At this point the brakes apparently totally failed and the train began to roll backwards, at ever-increasing speed.

Surviving passenger John Maganga, 32, related that when the train had begun to roll back, "The driver left the engine and ran into the cabins telling people to close the windows and shouting that the train was out of control." It is not clear what the driver thought closing the train windows would achieve. He jumped off the train at some point and escaped with slight injuries.

The runaway train apparently attained a very high speed and after passing through two

---

507 *The Daily Times*, Farmington, New Mexico, United States, 27th May 2002

508 *The Intelligencer*, Doylestown, Pennsylvania, United States, 26th May 2002

509 *The Frederick News-Post*, Frederick, Maryland, United States, 28th May 2002

510 *Los Angeles Times*, Los Angeles, California, United States, 26th May 2002

stations it collided with a freight train proceeding at low speed towards Dar es Salaam. All but one of the 22 vehicles of the passenger train were derailed as a result of the collision.

Soon after the accident Dodoma Regional Medical Officer John Kitimba, who was in charge of the rescue operation, said, "At least 200 people are dead but we fear there could be more." In addition there was a large number of injured, journalist Daniel Musangya estimating that there were about 800 people who were badly hurt. Doctor Jacon Chembela at Dodoma Hospital described the situation there as desperate, with more than 200 victims with broken limbs, head and chest injuries. "We are receiving more people...we are trying, we are doing our best but we are stretched, with little equipment," he said. In a bid to help the situation Health Minister Anna Abdallah arranged for an aircraft carrying more doctors to be flown from Dar es Salaam to Dodoma.

Television reporter Betty Mkwasa told a news agency that more than 10 hours after the accident, as darkness fell, the sounds of people trapped in the wreckage, pleading for help, could still be heard, adding that the problem was that the rescuers lacked the heavy equipment required to cut through the twisted steel to free passengers.

By chance Parliament was in session at Dodoma at the time of the collision and Prime Minister Frederick Sumaye made his way from there to the accident scene, accompanied by Mark Mwandosya, the Transport Minister. Some of the injured were taken to hospital in ministerial cars. Back in Parliament later, the Prime Minister recounted, "The way the coaches are piled on top of each other; from far away you would think it was a multi-storey building." He declared two days of national mourning. [511]

A correspondent wrote a detailed account of the disaster for the British Broadcasting Corporation: 'A giant yellow crane hovered over the wreckage of the Tanzanian train - tugging at one of the crumpled carriages with a length of steel rope. Rescue workers crowded round watching the carriage jerk and lift away from the remains of another carriage to the sound of screeching metal and breaking glass. "Stop it there," a man shouted up to the crane driver. He gestured towards a dark recess under the mangled train. A grey-haired man was sitting upright, squashed awkwardly between the floor of one carriage and the side of another. He was dead and surrounded by personal belongings - a suitcase, broken bottles, envelopes, a plastic bag, squashed oranges and a child's handwritten exercise book.

'Nearby, a young man who said his name was Felix was hunting through another carriage. "I'm looking for my friend," he explained quietly. The two had been travelling on the busy passenger train to the nearby town of Dodoma - in the arid heart of Tanzania - from the coastal city of Dar-es-Salaam. Felix was limping from a wound to his shin. Like the others on board the train's 22 carriages, he had waited for 30 terrifying minutes as the runaway train gathered speed, racing down the steady incline until finally it crashed into a freight train.

'All around us crowds of green-uniformed soldiers and rescue workers - some wearing rubber gloves or even plastic bags for protection - sifted through the debris, a grim job. Body parts still littered the crash site. A hand protruded from beneath a brown carriage. Men walked past with stretchers, some covered, others showing charred flesh.

'"We drove through the night to get here," said Dr Angela Limu, at the closest hospital in the small town of Mpwapwa. She and a group of colleagues had come from Dar-es-Salaam to help out. "We brought all the equipment we could," she said. "But we are running out here. We need bandages, antibiotics, pins for setting bones." Behind her, in one of the hospital's five crowded wards, men lay stretched out shoulder to shoulder on the floor.

'"We have at least 400 patients here," Dr Limu said. "Children, women. Broken bones,

---

[511] *BBC News*, London, England, 24th June 2002

serious head injuries. Some of them need evacuating quickly to a better hospital."

'Norbert Shee was sitting upright on one of the ward's few beds. A big, stained bandage was fixed across the top of his head. Norbert is a civil servant working for the National Social Security Fund. He had taken the train to start a new posting in Tabora, north-western Tanzania. "I cut my head on something," he said, sounding slightly dazed. "At first we didn't realise there was anything wrong when the train started going backwards. Then it raced through a station," he went on, explaining that they had all begun to panic. "None of us tried to jump off," he said. "It was going too fast - when it stopped I saw my colleague. He was dying."' [512]

There were allegations of sabotage as a cause of the accident, carried out by trade union members who were angry at the proposed privatisation of the railway, but these were denied by the prime minister and others and were ruled out in the final report on the disaster, though this was not made public on account of "legal technicalities."

The final death toll was 288. [513]

Early on Wednesday 18th February 2004 a total of 51 waggons escaped from a siding at Abu Muslim station, Iran, at kilometrage 736 of the Garmsar – Mashhad line and began to roll down an incline. They were loaded with a variety of flammable and potentially explosive goods, including 17 with sulphur, 10 with cotton wool, seven with fertiliser and six with petrol. The runaways travelled some 40 kilometres without being detected before they were derailed on an embankment near Khayyam station (kilometrage 696), 22 kilometres east of the town of Neyshabur.[514] A fire then broke out in the wreckage, at about 04:00. Firemen arrived to extinguish the blaze while other rescuers searched the wreckage for possible victims, though none were found.

So far, it was an expensive accident in terms of property loss, which had also blocked the railway, but that was all. However, as usually happens in such situations, the derailment and resulting large fire were a spectacle, which attracted a crowd of local residents – ordinary people and some local politicians, mainly from the village of Deh-Yukeh – as well as some senior railway officials. By 09:37 firemen had almost extinguished the blaze, when without warning there was a tremendous explosion, of a magnitude estimated as having been equivalent to the setting off of 180 tons of trinitrotoluene. Locals thought that it was another earthquake (a major quake elsewhere in the country had killed 40,000 people only two months earlier) and seismologists recorded a tremor of 3.6 on the Richter scale at the time of the blast.

Homes in five nearby villages were flattened by the force of the explosion, Deh-Yukeh and Hashem-Abad being the worst affected. Ali Husseini, head of the local Red Crescent, stated that accommodation was being set up for those villagers who had survived the explosion.

A British Broadcasting Corporation account of the aftermath of the disaster wrote, '"The scale of the accident is very extensive and the damage seems to be more than initial estimates suggested," said local official Varik Barakche.

'"The whole town {of Neyshabur} is shocked by this accident," the editor of a local newspaper, Saeed Kaviani, told AP news agency. "Official vehicles mounted with loudspeakers are roaming the city calling for volunteers to donate blood." Mr Kaviani said dozens of people remain buried under the rubble of their homes in nearby villages.

The blast shattered windows more than 10 kilometres (6 miles) away and could be heard in

---

512 Ibid, 25th June 2002

513 https://allafrica.com/stories/200209160555.html

514 Wikipedia – Garmsar – Mashad railway line

the regional centre of Mashad, 75 km from the scene. TV pictures showed massive destruction, with derailed waggons and choking black smoke.

'Top city officials, including the local governor, mayor and fire chief, are believed to be among the dead. Irna news agency also said the head of the city's energy department had been killed and the director-general of the provincial railways was missing. An Irna reporter was also killed in the blast.

'Several hundred injured people have been taken to hospitals in Neyshabur and Mashhad. An official from the Natural Disasters Headquarters in Neyshabur said the hospitals had been overwhelmed with casualties.

'UN Secretary General Kofi Annan has offered United Nations assistance to those affected by the explosion. "The secretary general was distressed to learn of the casualties and damage," Mr Annan's spokesman said in a statement.

'The disaster comes at a time of political uncertainty in Iran, with controversial parliamentary elections to be held on Friday.' [515]

The day after the disaster, Ali Husseini announced that 309 bodies had been recovered, though there could still be more in the debris.[516] Another report put the number of dead at 315 but several days after the accident, Mohammad Maqduri, an official of Khorasan province, in which the accident had occurred, revised the figure down to 289. [517]

President Mohammad Khatami ordered an investigation into how the waggons had come to run away. An initial theory that an earthquake had started the waggons moving was quickly rejected and on 23rd February Khorasan Governor-General Hassan Rasuli also ruled out the possibility of sabotage by a disgruntled railwayman as a cause. Failure to properly secure the waggons seems likely to have been the cause.

Questions also arose regarding the transport together of the goods involved, none of which had been classified as 'dangerous' before the accident, and the failure to recognise the danger after the initial derailment and perhaps arrange an evacuation of nearby villages.

Train No. 1108 Down of the North Central Railway, the Down *Bundelkhand Express* from Varanasi to Gwalior met with an accident at Datia station, Madhya Pradesh, India, at 08:51 on Monday 3rd October 2005. The train apparently suffered a brake failure on its approach to the station, was diverted onto a loop line, ran into the trap siding at the end of the loop and demolished a signal cabin beyond the end of it. The engine and six coaches of the train were badly damaged. Thirteen died, including Driver Ramzan Khan, and 36 were injured. The train was travelling at an estimated 90 km/h when it reached the end of the loop line, through which speed was limited to 15 km/h. [518]

''Why is the train passing so fast,' was the question uppermost in the mind of Anil Trivedi, who runs a snacks shop near the crossing, the very instant before he heard a deafening sound and dust ballooned into the air. The dust took about five minutes to settle but by then heart-rending cries were renting the air. "Casting out business to the winds, we shopkeepers engaged in extricating passengers. But there was simply no means to access coaches piled one atop the other. The police and administrative official arrived soon, but railway authorities and the rescue team reached only about one-and-a-half hours later," he narrated.

[515] *BBC News*, London, England, 18th February 2004

[516] https://reliefweb.int/report/iran-islamic-republic/death-toll-iranian-train-derailment-rises-over-300

[517] https://reliefweb.int/report/iran-islamic-republic/iran-provincial-governor-rejects-possibility-sabotage-train-disaster

[518] http://us.Rediff.com/cms/print.jsp?dotpath=/news/2005/oct/03train.htm

'Another shopkeeper, Navneet Trivedi, said that vehicle drivers and pedestrians waiting at the crossing simply had no idea of impending doom. Several vehicles came under the train, he claimed.

'Instead of halting at Datia, the express actually picked up speed. The train suddenly changed track; the coaches began vibrating. I fell off my seat and passed out. After regaining consciousness I found myself lying near the tracks," said Kamlesh Soni, who hails from Bharuna Sumer in Uttar Pradesh's Hamipur district and is recovering in the local hospital.' [519]

The injured were taken to hospitals in Datia and Jhansi. Railway Minister Lulu Prasad stated that 5 lakh (500,000) rupees would be paid to the next of kin of the deceased and that jobs would be offered to a family member of the deceased. [520]

Madhya Pradesh Chief Minister Babulal Gaue alleged that the driver of the express had fallen asleep.[521] There was also a report that the driver had stated there was a problem with the engine and requested that a replacement be provided at Jhansi, but this had not been done. Assistant Driver Ram Singh had not been on the engine when it reached Datia, having moved to a sleeper coach at Jhansi, supposedly to answer a call of nature.[522] He and Guard Sundar Singh were suspended after the accident. [523]

Unfortunately it has not proved possible to access the full report of the investigation into the accident. It appears possible that a second engine was attached to the train at some point. An angle cock of the air brake system may then have been accidentally closed and the error not detected by performing a brake continuity test, but this is only speculation. [524]

In its annual report, the Commission of Railway Safety listed seven recommendations relating to the accident:

(a) Video cassette recordings of serious accidents should be prepared not by taking clippings here and there but by recording in a continuous manner, specifically showing the positions of coaches, engines, coach and engine undergear and clues that might in turn provide some information for the inquiring authority.

(b) The Railway Administration should constitute only one or two teams of senior supervisors from different departments to take the joint observations of tracks, coaches, engines, signals and so on and any other relevant information.

(c) It was desirable to have separate sets of drivers and assistant drivers to operate diesel and electric locomotives running passenger trains.

(d) Digital voice recorders with a memory of at least one week should be installed in the driving cabs of all locomotives to analyse the conversations between drivers, assistant drivers and station staff in case of need.

(e) Breathalyser and blood tests of the train crews and all train passing staff on duty at the stations at either end of the block section should be undertaken immediately after train accidents.

(f) Instructions to ensure that the type of locomotive attached/booked for a train was authorised to run at the maximum permissible speed of the train for the entire run to the next loco changing point should be reiterated. Whenever in an emergency it was necessary to attach/book a locomotive with a lower sanctioned speed, a caution order

---

[519] http://news.webindia123.com/news/printer.asp?id=128219&cat=India
[520] www.hindu.com/thehindu/holnus/001200510031350.htm
[521] http://ww1.mid-day.com/news/nation/2005/october/120149.htm
[522] www.indianexpress.com/print.php?content_id=79510
[523] https://www.irfca.org/articles/accidents-2004-05.html
[524] www.newindpress.com/NewsItems.asp?ID=IEH20051005130511&Page=H&Title=To...

should be issued to the drivers to observe the sanctioned speed.

(g) Instructions to ensure that a dead locomotive was escorted by a competent person not lower than an assistant driver before it was attached to any train/light engine should be reiterated. [525]

East Central Railway Up Gomoh – Chowpan passenger train No. 619, bound for Shaktinagar, was pulling out onto the main line from a loop line at Barwadih station, Jharkhand, India, in the Dhanabad Division of the railway, at 14:35 on Wednesday 9th November 2005 when it was struck by a runaway coal train (LMH Special Up goods) bound for Bhatinda in the Punjab. Four bogies (coaches) of the passenger train were derailed, together with the engine and 31 waggons of the coal train, while two passenger coaches were completely buried in coal. The coal train had overrun signals because its driver had lost control of it. The driver and assistant driver fled the scene after the accident. Nine passengers were killed outright or were fatally hurt and 26 were injured, eight seriously, and were taken to different hospitals. [526] [527]

The Commission of Railway Safety determined that the accident had been due to closure of the leading end angle cock of the third waggon behind the locomotive by 'unidentified miscreants.'

The commission made four recommendations:

(a) The instructions as contained in Slip No. 1 of General and Subsidiary Rules East Central Railway, 2005 Edition, in Subsidiary Rule 44.19 (i)(1)(o) regarding the provision of walkie-talkie sets in working order to the drivers and guards of all trains, passenger and freight, should be reiterated and implemented in the field. The responsibility of providing walkie-talkie sets to the guards of the train should be changed to the Operating Department instead of the Signal and Telecommunications Department, the Operating Department being the user department.

(b) Post loading Guard Driver Report checks based on Railway Board instructions No. 98/M(N)/951/12 dated 25.5.2005 by the drivers and guards of all trains should be ensured with proper recording at the station/siding point.

(c) The instructions contained in Chapter III of the Eastern Railway Accident Manual, Calcutta, 1982, relating to the preservation of clues and the recording of joint measurements in time without delay should be reiterated.

(d) The Railway Administration should consider the provision of a locking arrangement in miscreant-prone areas to make it difficult/inaccessible for miscreants to operate the handle of the angle cock. [528]

Twelve of the 57 waggons of a South Eastern Railway goods train carrying iron ore from mines at Roxy to a steel plant at Rourkela left the rails near the station at Bimlagarh, Orissa (Odisha from 2011), India, on the Chakradharpur Division of the railway, at 10:30 on Monday 5th December 2005, after the brakes of the train had failed. Twenty other waggons were described as having piled up upon one another. Although it was a freight train, it was used by what were described as 'tribals' to participate in the weekly haat (rural market) at Bimlagarh

[525] Commission of Railway Safety Annual Report 2005-06
[526] https://www.Rediff.com/news/2005/nov/09jhar.htm
[527] https://timesofindia.indiatimes.com/city/patna/7-dead-50-hurt-in-train-collision/articleshow/1290373.cms
[528] Commission of Railway Safety Annual Report 2005-06

in the absence of any other form of transport. About 70 tribals were riding in the brake van behind the two engines of the train. Twelve of them were killed in the accident, as were Guard N.L. Ho and three other railwaymen (gangmen). Driver L.D. Ram and Assistant Driver Xavier Soren were injured and were admitted to Bondamunda Railway Hospital and later transferred to Ispat General Hospital. Four other injured were taken to a hospital at Lahunipada. [529] [530] [531] [532]

The Commission of Railway Safety blamed the accident on the failure of the driver and guard to carry out a Guard Driver Report with due care in terms of Subsidiary Rule 4.31.93, with the result that the train had insufficient brake power. [533]

[529] www.outlookindia.com/pti_print.asp?id=339796
[530] www.telegraphindia.com/1051206/others/asp/print.html
[531] www.outlookindia.com/pti_print.asp?id=340415
[532] www.newindpress.com/NewsItems.asp?ID=IEQ20051210225059&Page=Q&Title=OR...
[533] Commission of Railway Safety Annual Report 2005-06

CHAPTER 8

BRIDGE FAILURES

Some of the most serious railway accidents over the years have resulted from the failure of bridges. Bridges may fail for a variety of reasons. The inherent design of the structure may be faulty, the materials used in its construction may be poor and inferior, the materials may deteriorate over time, or its foundations may be washed away by a stream or river that has become exceptionally swollen following abnormally heavy rainfall. With improvements in bridge engineering, failures due to faulty design had greatly diminished by the end of the 20th century, but accidents, including major disasters, continued to occur from causes such as bridge foundations being washed away. A relatively new danger, mainly though not entirely affecting road underbridges, would be so-called 'bridge bashing' by lorries or buses too high to pass beneath the structure, or occasionally by ships.

A possible major bridge accident that occurred near Lutterbach, Elsaß Lothringen, Germany, on Monday 13th March 1876 has only just come to light. A slow train, which had left Mulhouse at 5 a.m. for Strasbourg, was approaching Lutterbach and crossing a bridge over what was variously described as the River Ill and the River Doller, which had overflowed its banks, due to floodwater arising from heavy rain in the Alps and the Vosges Mountains. The swollen river undermined the bridge foundations and an arch of it collapsed, followed by a large portion of the remainder of the structure. According to the contemporary *Progres de L'Est* newspaper, the engine had reached the far side of the bridge but the remainder of the train fell into the river. The carriages of the train were, according to this account, thrown one upon the other, smashed to pieces and disappeared into the water. Passengers who had not been killed by the initial shock and fall of the bridge were trapped in the coaches and drowned or smashed against the rocks. Upwards of 30 bodies had soon been recovered but a number of other passengers remained unaccounted for.[534] [535] However, another source states that although the engine, tender, luggage van and two passenger coaches fell into the river, there had only been 13 passengers in all on the early-morning train and that only one had died, while several passengers and the conductor were badly hurt.[536] It is, of course, impossible after this length of time to know which account is closer to the truth.

Northern Railway train No. 4007, an Up Mughalsarai – Allahabad – Lucknow – Bareilly

534
https://paperspast.natlib.govt.nz/newspapers/CL18760616.2.6?items_per_page=100&query=bettini&snippet=false&sort_by=bySH
535 *The Cincinnati Daily Star*, Cincinnati, Ohio, United States, 18th March 1876
536 Eisenbahnunfall von Lutterbach

express, was derailed at Km. 152/11-12 between Allahabad and Prayag stations, Uttar Pradesh, India, on the Allahabad – Prayag – Pratapgarh – Lucknow section of the Lucknow Division of the railway at about 00:55 on Thursday 3rd January 1991. Fortunately the accident was not as serious as it might have been, but 12 people were hurt, one badly.

The derailment occurred at Bridge No. 174, a masonry and girder structure that had been struck by a lorry carrying an out-of-dimension load, shifting the bridge masonry and bending the rails. [537]

A major catastrophe occurred in Kenya on Saturday 30th January 1993 at a bridge over the Ndethia Geithia River near Darajani village and not too far from Mtito Andei, on the main line between Mombasa and Nairobi. The bridge collapsed beneath the train and the engine and its five leading carriages fell into the river at about 03:00, though the application of the brakes by the conductor reportedly saved the rear third-class coaches from following. The bridge dated from 1898 and was stated to have required few repairs; its foundations had been undermined by flash floods following heavy rain and a pier washed away. The accident occurred in uninhabited scrubland five kilometres from the Mombasa – Nairobi road. The name of the river allegedly translated as 'God Save Us.'

The flooding had damaged the communication system, delaying the reporting of the disaster, a fact that perhaps goes some way towards explaining complaints from survivors that it had taken 10 hours for rescuers to reach the scene. When they did arrive, 'military, police and railway security teams used power saws and axles to cut through coaches to retrieve trapped bodies.' After a while, 'The river's bank was littered with bodies retrieved by armed forces divers and police rescue units. Witnesses said some of the railway cars had been washed at least 500 yards from the bridge.' [538]

Early reports were that 140 people had died in the accident. By the following day it was known that there were 117 fatalities and Police Commissioner Philip Kilonzo stated that divers had worked through the submerged coaches and removed all the bodies and survivors. The search continued for the bodies of other victims, which had been washed downriver. Sixty bodies had been taken to the mortuary in Nairobi and the relatives of any of the 600 passengers on the train who were missing had been asked to go there in the hope of identifying them. At this point more than 180 people were still missing but it was thought that many might have wandered off into the bush after the accident. It was likely to be several days before the final number of dead could be known. A Canadian and at least one European were known to be among the dead. Some of the survivors reportedly saved themselves by squatting on the roofs of the submerged coaches. [539] [540]

The most serious accident in the history of the American National Railroad Passenger Corporation (Amtrak) up until at least July 2025, occurred at about 2:53 a.m. on Wednesday 22nd September 1993, at the Big Bayou Canot Bridge, near Mobile, Alabama, United States, on a line owned by CSX Transportation. It resulted in the death of 42 passengers and five railwaymen, serious injuries to four people and slight injuries to 107.

The events leading up to the disaster began at about 12:55 a.m. when the towboat *Mauvilla*, pushing six barges, left the National Marine Fleet on the Mobile River. It was

---

[537] Government of India Ministry of Civil Aviation and Tourism Commission of Railway Safety Annual Report for 1990-91

[538] *Orange County Register*, Santa Ana, California, United States, 31st January 1993

[539] *The Lethbridge Herald*, Lethbridge, Alberta, Canada, 31st January 1993

[540] *The Medicine Hat News*, Medicine Hat, Alberta, Canada, 1st February 1993

manned by a captain, Pilot Willie Odom, and two deckhands, the pilot having taken over the watch at 11:30 p.m. the previous evening. The *Mauvilla* was following a faster towboat, the *Thomas B. McCabe*. At the start of its journey the weather was clear but it then became increasingly foggy, to the point that it was impossible to see the head of the tow. At this point, the pilot reduced speed and decided to tie up to the riverbank until the fog cleared. The *Mauvilla* towboat continued its journey, the pilot looking for a tree he knew he would find at the Mobile-Tensas cut-off of the river, and which he could tie up to. However, unknown to the *Mauvilla* pilot, his towboat and barges had actually left the Mobile River and were proceeding up the Big Bayou Canot, a subsidiary waterway not normally used by commercial traffic, which was bridged by the CSXT railroad, this bridge having a vertical clearance of about seven feet. At some point, the pilot saw the bridge on his radar but he wrongly assumed that it must be another towboat that had been moored and had swung out from the riverbank. He was still trying to work out what was in front of him when he felt a bump and thought he had run aground. In fact, the leading port and starboard barges being pushed by the towboat had struck the south and centre piers of the through girder span of the bridge at about 2:45 a.m. The forward centre barge struck the girder span between the two piers, displacing it 38 inches. Awakened by the noise of the impact, the captain went to the wheelhouse and the crew were then occupied in attempting to retrieve two barges that had come loose in the collision, during the process of which the *Mauvilla* backed into the bank of the Big Bayou Canot and was stuck there for several minutes. At this point, none of the crew was aware that they were not still in the Mobile River.

Amtrak train No. 2, the *Sunset Limited*, had left New Orleans, Louisiana, at 11:34 p.m., 34 minutes late, for Miami, Florida. It left Mobile, Alabama, at 2:33 a.m., 33 minutes late, with 202 passengers and 17 crew. The journey from Mobile had been normal, with the engine crew calling out the signal indications over the radio. The assistant conductor remembered that the last signal called had been at Bayou Sara, at milepost 658.4. At about 2:53 a.m., the train, travelling at about 72 mph, struck the displaced girder on the Big Bayou Canot Bridge, at milepost 656.7, and was derailed. As a result of the derailment part of the bridge collapsed. The leading unit of the three pulling the train embedded itself in the mud of the bank of the bayou and the other two units, the baggage car, a dormitory car and two of the remaining passenger cars of the train, fell onto the bank or into the water. The four rear cars remained on the bridge. Fire from the damaged fuel tanks of the diesel units then broke out, gutting them and the baggage and dormitory cars behind them.

Engineer Billy Rex Hall, Assistant Engineer Ernest Lamar Russ and Assistant Engineer Michael Vinet, in the leading diesel unit, were all killed; Conductor Dwight Thompson and Assistant Conductor Gary Lee Farmer, who chanced to be in the penultimate car of the train at the time, survived. They said that there had been no warning of the accident and no brake application. The conductor attempted to radio his engineer but received no reply. At 2:56 a.m. the assistant conductor made a mayday call over the radio. This was picked up by the crew of CSXT train No. 579 at milepost 660.4, who forwarded it to the yardmaster at Sibert Yard, Mobile, from whence it was transmitted to the dispatcher at Jacksonville, Florida, and to the Mobile police. There was confusion for some 18 minutes as to the location of the accident, which the assistant conductor had thought was the Mobile River Bridge, rather more than three miles north of the actual accident site. The accident had occurred in a remote area of wooded swampland with no road access.

The captain of the *Mauvilla* had also made a mayday call at 3:05 a.m. that his towboat and barges had struck something but he thought they were still in the main Mobile River. At 3:07 a.m. he was told that a train was derailed and on fire but still was not sure that his boat had been responsible for the train accident.

Going forward, the conductor and assistant conductor realised the extent of the accident and began evacuating passengers, setting up a relay system, using passengers who could swim, to evacuate people from the submerged and partially submerged coaches. All the coaches of the train were of the double-deck type and the lower decks of the submerged coaches filled with water in seconds, giving passengers little time to escape. Six members of the onboard staff evacuated the burning dormitory car (the other two had been killed in the initial accident), were rescued by the crew of the *Mauvilla* and then assisted in rescuing the passengers. The crew of the *Mauvilla* pulled 17 passengers from the water and the crew of another towboat, the *Scott Pride*, pulled out 20. Later, Coast Guard helicopters went to the scene. The last survivors reached hospitals or hotels at 8:30 a.m. [541]

An account of the disaster, written a couple days after it had occurred, stated: 'Although the exact details of the crash are still under intensive investigation, survivors described the devastating event as a combination of falling, explosion, fire, and being thrown under water. According to witnesses, the train was proceeding normally across the eighty-year-old train bridge when a loud noise was heard. They then described a "falling feeling" and heard a large explosion. Within seconds, a large orange fireball was seen to engulf the area, presumably as one of the locomotives propelling the train blew up. Next the lucky survivors found themselves in the water or in a car that was rapidly filling with water. The bayou into which the train plunged was said to be about 12 feet deep.

'Shortly thereafter, heroes like Michael Dopheide, a 26-year-old law graduate, helped dozens of people from the submerged cars. Chaos reigned as numerous retired people attempted to escape both the flames and the water. Passengers helped each other by forming lines and passing people, hand to hand, to the shore. Others clung to large pieces of debris in the alligator and snake infested bayou that flows from the Mobile river. Aside from penlights and one person's small pocket flashlight, the only light that was available was the burning glow of the Amtrak engine that had exploded.

'As the first calls for help went out, it quickly became evident that the Saraland bridge wasn't easily accessible by roads; there were none nearby. Rescuers had to access the site by helicopter and boat. The U.S. Coast Guard was alerted, and H-60 helicopters that are normally used for rescue on the Gulf of Mexico were diverted to the rescue effort. Pilots and flight medics said that they were hampered in their initial efforts due to the fact that it was extremely dark, the sky surrounding the area was filled with dense smoke created by burning diesel oil used by the locomotives, and there was no place to land the helicopters.

'According to survivors, it may have been as much as 45 to 50 minutes before anyone arrived at the scene to begin the rescue efforts. Reportedly, the first units at the incident were the crew of a Coast Guard helicopter. Coast Guard Lt. Robin Stearett was quoted by the *Washington Post* as saying, "There was smoke and fire everywhere...one train car was totally underwater...it was obvious that there were going to be injuries...it was devastating...it was the most catastrophic wreck I have ever seen." It remains unclear whether the train was on fire prior to entering the water, largely because no members of the lead engine train crew have been found alive.

'Rescuers were dispatched from several nearby jurisdictions, many bringing boats of various descriptions to the scene. A command and control unit was reportedly established at the Scott Paper Co. parking lot, which was located three miles downstream from the actual crash site. The paper company also dispatched two work barges to the site, which were used to act as dive platforms and to store the recovered bodies of the deceased. Pictures taken by local television stations showed the bodies of several people, lined up on the barges, waiting to be taken to a temporary morgue that was established at a nearby local lumber company. A Coast Guard spokesperson said that within 30 minutes of activation that the Coast Guard had eight

[541] National Transportation Safety Board report

helicopters and six boats enroute to the scene.

'Early reports said that there were many survivors but didn't ascertain the number of fatalities. Helicopter crew members described the perilous process of extracting people from the water by winch, in a dangerous, dark, and smoky environment. As boats began to arrive on the scene, they said that many people were clinging to various objects in the water. Several passengers reportedly assisted others to swim to the shore. One survivor described several passengers forming a "human chain" that was passing children and elderly to the safety of the shore.

'Shortly, it became evident that rescue divers would be needed to enter the cars that were partially or totally submerged. Divers were requested from several emergency agencies, and more than twenty soon arrived at the scene. Divers that were already working described the work as "tedious and demanding," as they moved foot-by-foot, searching through the murky water that filled the coaches. Everything had to be done by touch, due to the fact that the water of the bayou was filled with mud and silt. The divers reported finding victims still strapped into their seats, and another trapped in a locked bathroom. They also lamented the deaths of children, found frozen in the horrible positions of their deaths.

'Soon the Mobile County Sheriff's Flotilla was activated and dozens of volunteer boaters and rescue divers began to arrive on the scene. Volunteers began to appear on the banks bordering the bayou. Seemingly, everyone that heard about the tragedy had come to help. Emergency medical service personnel and ambulances were dispatched from every nearby community. Reportedly, some emergency agencies were delayed in the response due to the remoteness of the accident site and confusion in communications between those on-site and those still responding.

'EMTs and Paramedics from the Mobile Fire Department and other agencies described the victims as "unbelievably calm," considering the magnitude of the incident. Numerous reports of passengers helping passengers continued to be received. Medics described the passengers as suffering from a variety of injuries, including joint, head, neck and back pains. Others were described as suffering from exposure, having spent an extended period in the water. Still others were reportedly dazed and in shock. Some patients were also described as having burns and smoke inhalation. Helicopters began ferrying the injured to local hospitals.

'By 8.00 a.m. and full sunlight, it was obvious that several people were still missing and that there were numerous fatalities. The grim task of body recovery began, as most of the living were already transported to a number of hospitals. 37 people were found essentially unhurt and loaded on another train that Amtrak brought to a site adjacent to the crash site. 10 people, including the crew of the lead locomotive, were still thought to be missing. The search went on all day, with hundreds of rescuers and support personnel participating. As the end of the rescue occurred, the investigation of the cause began in earnest.' [542]

The accident was the subject of a comprehensive investigation by the National Transportation Safety Board (NTSB). Its report reached 24 conclusions:

(a) The train crew's qualifications and the condition of the track, signals, and train equipment had not contributed to the accident.

(b) The Sunset Limited, travelling about 72 mph, had derailed after striking a girder that had been displaced about 38 inches when the towboat Mauvilla, pushing six barges, had struck the Big Bayou Canot railroad bridge.

(c) Had the pilot used the radar as visibility deteriorated, he could have observed the intersection formed by the Mobile River and the Big Bayou Canot and could have

[542] www.emergency.com/amtrkcrs.htm

avoided turning his tow and barges into the wrong waterway.

(d) The pilot had exercised poor decision-making under the prevailing conditions by continuing to navigate when he could not find a tree to which to tie up, by continuing his approach to an unidentified object, and by not summoning help from other crewmembers.

(e) The Warrior & Gulf Navigation Company had not provided adequate radar navigation training to the pilot, nor had it provided him with a compass.

(f) The Warrior & Gulf Navigation Company's written evaluation form had not fully identified and assessed those skills critical to vessel operation, limiting its value as a management tool for ensuring safe vessel operations.

(g) The American Waterway Operators, Inc. should encourage member companies to incorporate into towboat operator evaluations a practical method of assessing proficiency in navigation, including the use of radar.

(h) The Coast Guard standards for inland towing vessel operator licensing were inadequate because they did not address the radar skills necessary for the safe operation of vessels in restricted visibility.

(i) All uninspected towing vessels should be required to carry appropriate navigational devices, including charts, in the wheelhouse, and all such vessels, except those operating in very limited areas, should be required to have a radar installed.

(j) Delays in obtaining samples from vessel crewmembers, which had prevented conclusive determination of whether alcohol was a factor in this accident, could continue to be a factor in marine accidents because Coast Guard regulations applicable to post-accident alcohol testing did not specify the time limits within which such testing must be conducted.

(k) As evidenced by information provided by his employer and a statement from his doctor, the pilot had been able to control his diabetic condition satisfactorily since 1982 and the disease most likely had not precluded him from operating the Mauvilla safely.

(l) The Warrior & Gulf Navigation Company should establish procedures that encouraged towboat operators to inform management when they were taking medication, determine whether such medication might affect their performance of duties, and arrange for a qualified relief, if necessary.

(m) The transportation industry needed to disseminate information and educational materials that stressed the relationship between the use of medications (over-the-counter and prescription) and an employee's fitness for duty.

(n) The damage to the bridge discussed in repair plans issued on November 1, 1979, was probably due to a barge strike, which might have occurred as a result of high water and winds generated during Hurricane Frederic.

(o) Currently available data pertaining to bridge-vessel collisions was insufficient to allow the US Department of Transportation to develop a bridge risk assessment methodology or to allow the railroads to systematically identify vulnerable bridges and take action to increase protection for them.

(p) The development of a national risk assessment program for railroad bridges required the full cooperation and support of the railroad industry.

(q) The US Department of Transportation needed to develop a national risk assessment program for determining bridge vulnerability to vessel collisions.

(r) In order to carry out a national risk assessment program for highway and railroad bridges, the transportation regulatory agencies needed a standard methodology so that they could determine the vulnerability of the nation's highway and railroad bridges to collisions from marine vessels, formulate a ranking system for identifying those bridges at greatest risk, and provide guidance on the effectiveness and appropriateness of protective measures.

(s) Bridges vulnerable to impact from commercial marine traffic should bear unique, readily visible markings so that waterway and bridge users were better able to identify bridges involved in accidents when reporting such mishaps to emergency responders, and a list of bridge identification markings should be published periodically in a national register of bridges.

(t) Considering the remoteness of the accident site, the weather conditions, and the limited modes of transportation available, the emergency response had been well-coordinated and effective.

(u) Amtrak had not had an effective system of apprising passengers of safety features, and some passengers had been unaware of safety information.

(v) As Amtrak had not equipped its passenger cars with portable lighting, passengers had been at a disadvantage in evacuating the train.

(w) Emergency responders had been at a disadvantage because they had been unable to obtain an adequate passenger and crew list from Amtrak until the next day.

(x) If the Mobile County Emergency Management Agency had held drills to simulate a train accident, the incident commander might have known about Amtrak's procedures for accounting for passengers, and CSXT might have had the correct telephone number for the Coast Guard.

The NTSB determined that the probable cause of the derailment was the displacement of the Big Bayou Canot railroad bridge when it was struck by the *Mauvilla* towboat and tow as a result of its pilot becoming lost and disoriented in the dense fog because of:

(1) The pilot's lack of radar navigation competency.

(2) The Warrior & Gulf Navigation Company's failure to ensure that its pilot was competent to use radar to navigate his tow during periods of reduced visibility.

(3) The US Coast Guard's failure to establish higher standards for inland towing vessel operator licensing.

Contributing to the accident was the lack of a national risk assessment program to determine bridge vulnerability to marine vessel collision.

The NTSB made a series of recommendations to various bodies.

To the US Department of Transportation:

(a) It should convene an intermodal task force that included the Coast Guard, the Federal Railroad Administration, the Federal Highway Administration, and the US Army Corps of Engineers to develop a standard methodology for determining the vulnerability of the nation's highway and railroad bridges to collisions from marine vessels, to formulate a ranking system for identifying bridges at greatest risk, and to provide guidance on the effectiveness and appropriateness of protective measures.

(b) It should require that the Federal Railroad Administration and the Federal Highway Administration, for their respective modes, used the methodology developed by the intermodal task force to carry out a national risk assessment program for the nation's railroad and highway bridges.

(c) It should require the modal operating administrations to develop and disseminate bulletins, notices, circulars, and other documents that called attention to the need for an employee reporting procedure concerning use of medication (over-the-counter and prescription) while on duty and urged the transportation industry to develop and implement informational and educational programs related to this subject.

(d) It should consider the use of RACONS, radar reflectors, and other devices to make bridges more identifiable on radar.

To the US Army Corps of Engineers:

(a) It should cooperate with the US Department of Transportation in developing a standard methodology for determining the vulnerability of the nation's highway and railroad bridges to collisions from marine vessels, formulating a ranking system to identify bridges at greatest risk, and providing guidance on the effectiveness and appropriateness of protective measures.

(b) It should promote, in cooperation with the US Coast Guard, the development and application of low-cost electronic charting navigation devices for inland rivers.

To the US Coast Guard:

(a) It should amend 46 CFR 4 and 16 to specify the time limits, not to exceed 8 hours, within which employers must conduct post-accident alcohol testing.

(b) In consultation with the inland towing industry, it should develop radar training course curricula standards for river towboat operations that emphasised navigational use of radar on rivers and inland waters.

(c) It should upgrade licensing standards to require that persons licensed as Operators of Uninspected Towing Vessels held valid river-inland waters radar observer certification if they stood navigation watch on radar-equipped towing vessels and require that employers provided more specific evidence of training.

(d) It should require that all uninspected towing vessels carried appropriate navigational devices, including charts, in the wheelhouse.

(e) It should promote, in cooperation with the US Army Corps of Engineers, the development and application of low-cost electronic charting navigation devices for inland rivers.

(f) It should require that radar be installed on board all uninspected towing vessels except those operating within very limited areas.

(g) It should require that all bridges vulnerable to impact by commercial marine traffic bore unique, readily visible markings so that waterway and bridge users were better able to identify bridges involved in an accident when they reported such accidents to emergency responders.

(h) It should periodically publish a list of bridge identification markings in a national register of bridges.

To the National Railroad Passenger Corporation (Amtrak);

(a) It should develop and implement a uniform system to effectively apprise passengers of information pertaining to safety features.

(b) It should develop and implement procedures to provide adequate passenger crew lists to local authorities with minimum delay emergencies.

(c) It should equip cars with portable lighting for use by passengers in an emergency.

To the Federal Emergency Management Agency:

(a) It should encourage local authorities to conduct emergency drills that simulated transportation accidents involving railroad operations.

To The American Waterways Operators, Inc.:

(a) It should recommend that member companies equipped their tugs and towboats with suitable navigation devices, including charts.

(b) It should assist the Coast Guard in developing a curriculum for a training course on river radar navigation.

(c) It should recommend that member companies incorporated into towboat operator evaluations a practical method of assessing proficiency in navigation, including the use of radar.

To the Warrior & Gulf Navigation Company:

(a) It should require that company towboat operators completed a recognised training course on river radar navigation after the curriculum for such a course had been developed.

(b) It should establish a training protocol that required company towboat operators to demonstrate proficiency in use of radar, compasses, and charts and incorporate into towboat operator evaluations a practical method of assessing proficiency in river navigation techniques, including use of radar.

(c) It should equip all company towboats with a suitable compass, a complete, up-to-date set of navigation charts for the waters over which the vessel is intended to operate, and other appropriate navigational devices.

(d) It should establish procedures that encouraged towboat operators to inform management when they were taking medication, to determine whether such medication might affect the performance of their duties, and to arrange for a qualified relief, if necessary.

To the Association of American Railroads:

(a) It should immediately begin to collect data on vessel collisions with railroad bridges from its members and, if appropriate, take steps to increase protection for bridges identified as vulnerable.

(b) It should cooperate with the US Department of Transportation in developing a national risk assessment program for railroad bridges.

To the American Short Line Railroad Association.

(a) It should immediately begin to collect data on vessel collisions with railroad bridges from its members and, if appropriate, take steps to increase protection for bridges identified as vulnerable.

(b) It should cooperate with the US Department of Transportation in developing a national risk assessment program for railroad bridges. [543]

At about 18:45 on Friday 11th August 1995 Österreichische Bundesbahnen international passenger train IC 566 *Niederösterreichische Tonkünstler* from Wien Westbahnhof, Austria, to Lindau, Germany, was heading towards Bludenz on the electrified Arlbergbahn and had reached a bridge over the Masonbach stream near Innerbraz, Austria. Immediately before the arrival of the train there had been a violent thunderstorm accompanied by heavy rain. This had converted the stream into a mudslide, which swept away the bridge directly in front of the train and also brought down the overhead line equipment. Power to the train was thereby cut off but the train was too close to stop and the front Class 1044 locomotive and leading three

[543] National Transportation Safety Board report

coaches fell from the bridge. The fourth coach overturned on the line and the rear three coaches and another locomotive at the end of the train remained on the line. The train driver and three passengers lost their lives. According to the tachograph from the rear engine the brakes had been applied about 50 metres from the bridge.

The alarm was raised at about 19:00 and in the difficult conditions it took rescuers 20 minutes to get to the scene, with a local mountain rescue team being the first to arrive. The search for possible victims continued into the darkness with fears that some might have wandered off or been swept away by the flood. Search dogs were used in the hunt for victims. The injured had to be carried down the track to waiting helicopters and ambulances. In addition to the four people who died, over 100 of the approximately 220 passengers on the train were badly hurt.

An investigation indicated that no one had been at fault for the accident, which had not been predictable. A number of bridges on the line were later fitted with pressure sensors, which would trigger an alarm if a landslide or avalanche struck them. [544]

An express train bound for Seoul and diverted due to a landslide was derailed and overturned on a bridge at Goisan, South Korea, on about Sunday 24th September 1995. One passenger died and 120 others were injured. Three days of heavy rain had weakened the foundations of the bridge. [545]

A record 6.14 inches of rain fell in Charlotte, North Carolina, United States, between midnight and 2 p.m. on Wednesday 23rd July 1997 and the resulting floodwater caused the failure of the Sugar Creek Trestle. As a result, the engine and five of the 90 cars of a CSX Transportation coal train fell into the creek. The train crew were able to jump clear in time but 2,500 gallons of fuel were lost from the diesel engine. [546] [547]

South Central Railway train No. 7234, a Sirpur Kaghaznagar – Secunderabad – Bhagyanagar passenger train, was derailed at Bridge No. 194, between Asifabad Road and Repalliwada stations, Andhra Pradesh, India, on Monday 28th August 2000. Four passengers were injured, two seriously and two slightly, while damage caused was estimated at 11,124,000 rupees. The Commission of Railway Safety investigated the accident and found that a flash flood following heavy rain, combined with the breach of a poorly maintained railway-affecting irrigation tank, had resulted in the undermining of the foundations, tilting and damage to two bridge piers on the Down line. The Commission noted that: 'The unsafe conditions were created due to inaccurate calculations of design discharge scour depth and design depth of foundation as well as disregard of instructions for monsoon patrolling and posting of stationary watchmen.'

The Commission made seven recommendations:

(1) Regular inspection and maintenance of railway affecting tanks {RATs} should be ensured by the State government.

(2) The compliance with instructions on monsoon patrolling and stationary watchmen should be ensured. These should be test-checked by higher officers during their inspections.

---

[544] https://mx-schroeder.medium.com/bridgeless-the-1995-braz-austria-derailment-6be6f15c5d60

[545] www.jrtr.net/jrtr06/topics.html

[546] *The Intelligencer*, Doylestown, Pennsylvania, United States, 24th July 1997

[547] *The Telegraph*, Alton, Illinois, United States, 24th July 1997

(3) Man-days lost due to patrolling should be recouped and the posts sanctioned filled up so that these were available at the beginning of the monsoon.

(4) Pre-monsoon instructions issued by the headquarters should give a timetable for compliance, which should be ensured by the heads of departments.

(5) The adequacy of the waterways of the bridges in the South Central Railway, particularly on the ex-Nizam's Railway, should be investigated with revised 'C' values of the Dickens formula {an empirical formula used to estimate the peak run-off of water from a catchment area}.

(6) The Railways should ensure that the guard's equipment as per the standards specified was always available in all passenger trains.

(7) The RAT signs should be prominently displayed in the case of railway affecting tanks upstream as per extant instructions. [548]

A possibly major catastrophe involving a bridge occurred in Myanmar at about 04:10 on Friday 1st June 2001. Passenger train No. 41 Up from Mandalay to Myitkyina was crossing the Sindaw (or Hsintaw Creek) Bridge (No. 159) between Nyanug Khin and Htantabin villages,[549] when, as an eye-witness reported, "While the train was crossing over the bridge, the bridge just collapsed and the train engine and four coaches went directly into the river."

Local sources stated that the structure had been in very poor condition and further weakened by recent flooding. The rainfall in the country had been the heaviest for almost a decade. The same source added that although the poor condition of the structure had recently been reported to the authorities by an inspector, nothing had been done regarding the report.

The following day, state-run television confirmed that an accident had occurred not far from Mandalay; no casualties were mentioned. Transport Minister Pan Aung and local authorities went to the accident site and told rescue workers not to disclose the number of victims. According to the passenger list there had been 280 travellers on the train but survivors said that there had also been many other passengers riding without tickets. There were unconfirmed rumours that over 300 lives had been lost. [550]

On Friday 22nd June 2001 Southern Railway train No. 6602, the Mangalore – Chennai Mail, was crossing Bridge No. 924 over the Kadalundi River, between Parappanangadi and Kozhikode stations, Kerala, India, when the structure collapsed and three coaches towards its rear, a Mysore reservation coach, a ladies' coach and a general compartment, fell into the swollen river, while three other bogies (coaches) were derailed. The train had left Mangalore at 11:15 and the accident happened at about 17:00.

A newspaper report highlighted the story of one unfortunate family caught up in the accident. 'Middle-aged Hilaya pulled her husband and son from the jaws of death under such conditions when very few would have shown the courage and presence of mind. But unfortunately her courage and warm hug could keep her son 12-year-old Ashid alive only for six more days after that nightmarish Friday night. And her husband Ashraf is still battling for life in the ICU. With Ashid also succumbing to his injuries at the Medical College Hospital in Kozhihode on Wednesday, the toll in the rail mishap involving the Mangalore – Chennai Mail

[548] Government of India Ministry of Civil Aviation and Tourism Commission of Railway Safety Annual Report for 2000-2001
[549] http://rebound88.tripod.com/01/jun/03.html
[550] https://www2.irrawaddy.com/article.php?art_id=3349

at Kadalundi touched 50.'

The final number of dead was probably 52, with about 240 injured.

'Ashraf, hailing from Kannookkara near Badagara of this district was going to a relative in connection with a death, along with his wife and four sons when the mishap occurred. The S-6 coach. in which they were travelling, was one of the ill-fated bogies which submerged into the Kadalundi river in the incident. Seeing her husband and son Ashid trapped in the mud, she recovered the sense of mind to pull them out. The other sons, Harshad, Ramshid and Shahid; and Hilaya suffered only minor injuries. But the condition of Ashid and his father was serious since they were brought to the hospital. The condition of Ashid worsened further Wednesday morning and he succumbed to his injuries. Doctors said he died due to cerebral haemorrhage. The condition of Ashraf is still critical, hospital sources said. Ashraf was running a small tea shop at Paris street here' [551]

Some controversy surrounded the cause of the accident. A team of Indian Navy divers acting on behalf of the Commission of Railway Safety advised that one of the bridge piers had failed beneath the weight of the train. This had resulted in a girder slanting to the left, causing the coaches to fall to the left in the direction of travel. The bridge inspection also revealed a second breakage and it appeared that the Railway did not have a proper system of underwater bridge inspection.[552] Engineers from the Regional Engineering College in Kozhikode disputed the bridge pier failure theory, stating at a press conference that, "There is a suspicion in some quarters that the broken pier is the villain of the tragedy, This argument is untenable. All the piers are equally old and presumably in the same condition. If the broken pier can be repaired and replaced and trains run (which the Railway is planning to do) are not the other piers equally likely to collapse?" It was suggested that a derailment or at any rate a coach wobbling had preceded the bridge collapse. [553] [554]

On Friday 14th September 2001 a freight train on the Ferrocarriles Chiapas-Mayab, carrying 180 tons of cement, fell from a low bridge into the Chalaca River on the southern Pacific coast of Mexico, close to the border with Guatemala. Floods due to heavy rain had undermined the foundations of the bridge. The results were not serious, though two train crew needed treatment in a local hospital. There had been a number of previous derailments on the line, implying that it was not in good condition. [555]

Tropical depression Delfina was associated with torrential rain, which led to severe flooding in northern Mozambique in early January 2003. The scouring of the foundations of a bridge over the Meluli River at Mutivaze, 50 kilometres east of Nampula, led to its collapsing beneath the weight of a freight train on 5th January. The train driver was killed and two other people were injured. [556] [557]

Canadian National Railway (CN) eastbound freight train No 356-51-14 left Prince George, British Columbia, for Edmonton, Alberta, at 06:35 on Wednesday 14th May 2003 in charge of

---

551 *Indiane Express*, Noida, Uttar Pradesh, India, 27th June 2001
552 *The Hindu*, Chennai, Tamil Nadu, India, 8th July 2001
553 Ibid, 24th July 2001
554 www.flonnet.com/fl1815/18151250.htm
555 *The New Mexican*, Santa Fe, New Mexico, United States, 15th September 2001
556 https://reliefweb.int/report/lesotho/southern-africa-complex-food-security-crisis-situation-report-8-fy-2003
557 http://www.zana.gov.zm/news/viewnews.cgi?category=1&id=1042304173

Conductor Ken LeQuesne with Engineer Art McKay. At about 12:30 it was due to cross a westbound train at McBride, British Columbia, Canada, but due to engine trouble it had been reported as running an hour late. Just after 13:00, the conductor of the westbound train tried to contact the eastbound one. Getting no response, he spoke to the controller, who had last heard from the train at 11:40 when it had been near Loos, British Columbia. The controller then made 12 unsuccessful attempts to contact the train between 13:23 and 13:35. At 13:39 and 13:44 the controller requested the assistance of the Tete and Fraser subdivision track supervisors respectively, to try and locate the train, while at 13:54, the conductor of the westbound train at McBride and a track maintenance employee set off westwards in a taxi for the same purpose. A column of black smoke at Mile 7.9 (McBride was Mile 0.0) indicated to the latter two employees where the train was. There was no road access to closer than about half a mile to the location of what was obviously an accident, so the conductor instructed the track maintenance employee to return to McBride in the taxi to report it while he made his way to the site on foot.

Arriving at the accident site, the conductor found that a bridge had collapsed, the locomotives and a number of cars of the eastbound train had fallen into a gully and they were on fire. There was no sign of the train crew and the intense blaze made it impossible to approach the locomotives. Control were updated on the situation. The Royal Canadian Mounted Police (RCMP) were advised of the accident at 14:08 and the ambulance service at 14:15. The Provincial Forest Fire Service had already been advised of the fire by local residents at 13:35, a ground crew of firefighters was despatched at 14:44 and a firefighting helicopter at 14:55.

Conductor LeQuesne and Engineer McKay of the eastbound train were killed. Two locomotives, five cars loaded with lumber and the entire bridge were destroyed by fire.

The Transportation Safety Board of Canada (TSB) investigated the accident.

The TSB report noted that train No. 356-51-14 consisted of two diesel locomotives, 68 loaded cars, four empty cars and 13 residue tank cars, thus 85 cars in all. Event recorders had been provided on the locomotives but they had been too badly damaged in the accident to provide any information.

The track at the site of the accident consisted of 36-pound continuous welded rail, with wooden ties and crushed tock ballast and was in good condition. The bridge that failed had been an open-deck timber pile trestle across a dry gully. It had 26 spans, was 293 feet long and had a maximum height of 26 feet.

The locomotives had fallen 25 feet from the bridge into the gully, followed by the leading four cars of the train. The fifth car was derailed and the remainder of the train had remained on the track. Spans 7 to 26 of the bridge had been destroyed in the derailment and fire and spans 1 to 6 had been damaged by the fire. There was no indication of an emergency brake application on the approach to the bridge. There was likewise no indication of dragging equipment, which could have derailed the train, while a broken rail found at the west end of the bridge appeared to have fractured because of the stresses caused by the derailment.

The inspections and maintenance of the bridge were detailed. The bridge had been rebuilt in 1969 and repaired between 2001 and 2002. It had had annual inspections in 1999, 2000, 2001 and 2002, the last named having been on 25th September of that year.

Subjects such as the capacity (load-bearing ability) of the bridge, expressed as a Cooper E loading (after Thomas E. Cooper, who introduced it in 1894) and communication with the train (only possible over a range of 2 to 3 miles with the portable radios which the train crew had) were discussed in the report.

Analysing the accident, the TSB noted that train operation did not appear to have been a factor in the accident. The absence of any indication of braking demonstrated that the train crew had had no warning of any problem with the bridge.

Two possible scenarios for the accident were explored. The first was that the bridge collapse had been precipitated by the derailment of the train due to a broken rail at the approach to it. Factors such as the finding that the rail had broken due to rapid overstress, the fact that the rail had been in good condition, the lack of a large amount of cumulative traffic or of particularly cold weather made this scenario, the TSB concluded, highly unlikely.

The second scenario was a failure of a bridge component. It was theorised that some bridge members had failed in the vicinity of bent 15 (the bent of a trestle bridge was defined in the report as the braced framework of timber piles – the vertical supports of the structure. The horizontal structures that hold the members of a bent together are called caps). The bent had failed either beneath the weight of the locomotives of the accident train or under a preceding train. Internal rot had been present in the cap of span 15 and also in the stringers, these being the longitudinal timbers that rested in turn on the caps and spanned between the bents. The stringers transfer the axle load of the train from the bridge deck, composed of the rails and ties (sleepers), to the bents.

The TSB report listed a number of conclusions:

(a) The two crew members on train 356-51-14 had been fatally injured when the lead locomotive had toppled off the bridge and been destroyed by fire.

(b) It was most likely that cap 15 had been crushed under the weight of the lead locomotive when its bearing capacity had been exceeded, leading to the bridge caving in and the subsequent derailment.

(c) As the load shared by stringers 5 and 6 of span 15 had increased due to the internal rot of the adjacent stringers, the reaction load transferred to cap 15 had been concentrated over a smaller area, thus exerting increased stresses over the existing void in cap 15.

(d) The condition of cap 15, identified as reject in the 1999 detailed report, had not been reflected in subsequent inspection reports. Therefore, its continuing deterioration had not been reassessed.

(e) The failure to identify the urgency and the severity of the condition of the bridge had not been recognised, despite subsequent inspections, because of shortcomings in the inspection, assessment, planning, and maintenance process.

(f) As a result of heavy workload, and overlapping duties during job transitions, the Planning and Inspection (P & I) Engineer had relied on the inspectors' overall assessments and most recent visual inspection reports, which had not indicated any deficiencies on the bridge. Therefore, the severity and urgency of the condition that had been identified in 1999 had not been recognised.

(g) Even though Canadian National (CN) Standard Practice Circulars (SPCs) acknowledged the use of occasional audits to ensure the quality of work of inspectors and their immediate supervisors, the audits that had been carried out had not been effective. Because there had been no Transport Canada (TC) audit of work procedures, there had been no opportunity to identify the deficiencies associated with such bridge inspection and maintenance procedures.

(h) Reliance on an overall assessment of a system might be appropriate when the system had built-in redundancy, as in the case of the stringers or piles. However, when the system's safety relied upon single components, such as a cap, an overall assessment would not identify an unsafe condition and would give a false indication that the system was safer than it actually was.

(i) When the condition of defective components was not updated on each inspection report, the report no longer accurately reflected the actual condition of the structure, which could lead to miscalculating the load-carrying capacity of a bridge.

(j) The practice of assigning equal loads to all stringers, including helper stringers, could lead to overestimating the carrying capacity of a bridge, thus increasing the risk of overloading and failure of the internal stringers.

(k) Unless careful analysis of detailed inspection reports was carried out as a matter of course, and capacity ratings were performed using the remaining sound timber sections, the severity of bridge conditions might not be fully appreciated.

(l) In an environment where workload was high and reliance was placed on the capabilities of inspectors and the quality of their reports, the effectiveness of audits was essential to assess the adequacy of practices and their variance from company procedures.

(m) Had the territory been equipped with a system that communicated the location of the train at frequent intervals, such as the Global Positioning System, or frequent calls had been made between the train and the rail traffic controller, the train's location and status could have been recognised earlier, prompting a quicker response.

(n) Based on the available evidence and subsequent analysis, neither train operations nor a mechanical failure had contributed to this occurrence, nor was it likely that track failure had been a factor.

(o) The complete destruction of the bridge and all its related components, the loss of the locomotive event recorder data, and the lack of complete and comprehensive bridge maintenance and inspection records had impeded the investigation efforts. Even though the calculated capacity rating of the structure had been lower than the equivalent rating of some of the equipment handled over the bridge, no failure had occurred between 1999 and the time of the accident because there had been an adequate safety margin associated with the timber unit shear stress.

(p) The observed variance between company procedures (SPCs) and actual field practices in conducting inspections, using the condition rating, and the underestimating of the severity of the condition of components with internal rot, suggested that the training on bridge inspection (both on-the-job and workshops) provided to employees had not met its aim of enhancing the knowledge of all inspectors across the CN system and ensuring cross-system uniformity of the inspection and reporting process.

The TSB report made two recommendations:

(1) Canadian National should verify the condition of its timber bridges and ensure their continued safety with effective inspection and maintenance programs.

In response, CN indicated that, as of April 2004, all timber bridges had been inspected in full compliance with CN Standard Practice Circular 4000, Inspection of Steel, Timber and Concrete Bridges, and that Bridge Summary Reports (five-year plans), identifying required repair and maintenance activities, were current. Also, as of June 2004, it had put into place a comprehensive, computerised Bridge and Culvert Condition System (BCS) to capture bridge and culvert condition information electronically from the field and to store the data in a centralised database. Bridge components were ranked from 1 (poor) to 9 (new), which was intended to facilitate improved identification and prioritisation of required repairs. The inspection process was automated, from inspection planning and scheduling, to inspector and Planning and Inspection (P & I) Engineer sign-off. Both inspector and P & I Engineer sign-off activities were stored in the database. With the new BCS system, all rankings and comments noted from previous inspections would be brought forward when the next inspections were generated. As of July 2004, all inspectors had been trained in how to use the BCS and all inspectors had undergone a one-week bridge and culvert inspection refresher training in January 2005, while CN intended to provide a four-day comprehensive refresher training program every four years. BCS was part of CN's overall Bridge Management System (BMS), which was still under development. The compilation of the bridge and culvert inventory had been completed in 2002, and the BCS was implemented in 2004. CN expected to complete the planning module in 2005. It would assist in automating the prioritising and scheduling of bridge maintenance and repair, and the tracking of completed work activities. Until the BMS was completed, the regions would continue to use their Bridge Summary Reports. As of December 2003, the completion of all timber bridge maintenance and repair work had been reported by e-mail by the Regional Bridge Supervisors to Bridges and Structures (B & S) Engineers. In July 2003, Transport Canada's (TC) Pacific Region had met with CN officials to discuss the bridge program. Following the meeting, TC had conducted a review of CN's inspection and maintenance records to ensure that CN had verified the condition of its bridges.

> (2) Transport Canada should incorporate in its compliance reviews a comparison of railway working procedures and practices with railway inspection and maintenance records.

In response, TC indicated that it was revising its existing inspection and monitoring programs to integrate monitoring and inspection activities of railway infrastructure, equipment, and operations into the Safety Management System (SMS). The SMS approach would require that TC examine each railway company's standards and practices with respect to the safety management of its infrastructure, as well as audit company inspection and maintenance records. By incorporating compliance reviews with the comparison of working procedures and practices and by examining railway records, TC would have a better opportunity to discover gaps in railway management of bridge condition assessment and repair. It was aggressively developing an audit program that focused on all railway structures. [558]

CN was subsequently charged under the Railway Safety Act with failing to ensure that maintenance and inspection records had been kept on the bridge. It pleaded guilty and was fined $75,000 by a court in Prince George, British Columbia, in December 2005. [559]

As many as 13 waggons of a 59-waggon South East Central Railway goods train fell into the Hasdeo River from a bridge between Korba and Kusmunda stations, Chhattisgarh, India, at about 01:45 on Wednesday 16th July 2003. The waggons were loaded with coal. No one was hurt. [560]

At about 07:20 on Friday 6th May 2005 pier No. 4 of Bridge No. 256 between Nukaka and Opoutama, New Zealand, suffered a catastrophic failure and collapsed beneath work train No. 60, which was being piloted across it at the beginning of a two-day work programme to underpin said pier. As a result, a 60-tonne crane and spans Nos. 3 and 4 of the bridge fell into the river. The ten members of the work gang had fortunately moved off the bridge before the train started over it and no one was hurt. The line was closed for several months while the crane was recovered and four new bridge piers and five new bridge spans were constructed.

The bridge on the Palmerston North – Gisborne line had been built in 1922 and consisted of six 60 ft steel gate girder spans resting on hardwood timber piers.

The accident was the subject of a thorough investigation by the Transport Accident

---

[558] Transportation Safety Board of Canada report
[559] *The Lethbridge Herald*, Lethbridge, Alberta, Canada, 8th December 2005
[560] http://sify.com/printer_friendly.php?id=13201231&ctid=2&lid=1

Investigation Commission (TAIC), whose findings are summarised below:

(a) Pier 4 of Bridge 256 had catastrophically failed under the weight of work Train 60 because most of the 10 piles making up pier 4 had been severely weakened by an infestation of marine borers.

(b) The usual signs of marine borer infestation in the piles around the low tide mark had been masked by the water salinity layering effect caused by the slow-running nature of the Nuhaka River, but the relatively warm estuarine environment was typical of a marine borer habitat, and Bridge 256 should have been regularly inspected for an infestation.

(c) The level of engineering experience in the rail system had not been sufficient to support the inspection and maintenance of the rail infrastructure in the years leading up to and at the time of the bridge collapse.

(d) A lack of engineering resource had contributed to:

(1) The misdiagnosis of the cause of the piles sinking on pier 4.

(2) The incorrect assessment of remedial work to correct the sinking pier.

(3) Poor appreciation of the effect of using a heavy crane in working mode to effect repairs to a weakened bridge.

(4) The assumption that the cause of the pile settlement was the same as the longstanding scour that had led to a bearing problem at a nearby bridge.

(e) The structures inspector had not had sufficient engineering expertise to oversee the underpinning project safely and make a decision on whether to close the bridge at any stage.

(f) The weight and axle loadings of work Train 60 had been within the original design limits of Bridge 256 but had exceeded the failure point of the aged and weakened structure.

(g) The reduction in the replacement of timber components in rail bridges during successive periods of ownership and structure of the rail system, together with a reduction in the frequency and quality of bridge inspections, had increased the risk of a catastrophic failure beyond what would normally be considered safe.

(h) An appropriate level of regulatory intervention in the rail industry should have picked up and acted on the warnings in two separate independent reports that the rail infrastructure would be at increased risk unless material-replacement programmes were improved and engineering expertise was at least maintained.

Because Ontrack, which owned and managed the rail network, had already taken suitable safety actions, the TAIC did not issue any recommendations. [561]

A cyclonic storm affected Andhra Pradesh state in southern India at the end of October 2005. House collapses and drownings resulted in the deaths of at least a dozen people and flash floods caused the partial failure of a railway bridge near Bitragunta station on Friday 28th of that month and the capsizing of a South Central Railway goods train. This minor accident, in which no one was hurt, was the prelude to a major catastrophe the following day, from the same cause. [562]

At about 04:20 on Saturday 29th October 2005 the South Central Railway *Delta Fast*

---

[561] Transport Accident Investigation Committee report

[562] www.hinduonnet.com/thehindu/thscrip/print.pl?file-2005102916420100.htm&date=200...

Passenger from Rapalle to Secunderabad was derailed between Valigonda and Ramannapet stations, Andhra Pradesh, India (the location is now in Telangana state). The accident happened at Bridge No. 61, over the Valugodu culvert near Gollapalli village. The incessant rain had caused at least three irrigation reservoirs to fail; the resultant floodwater had washed away the approach embankment to the bridge and the bridge had collapsed. The engine and seven of the 18 coaches of the express fell from the bridge into the water. According to one account the *Narasapur Express* had passed the site without incident just 20 minutes earlier.

Almost 250 Army and Railway Police Force personnel took part in the rescue operation, assisted by local people and the services of two Indian Air Force helicopters and six boats. The injured were transported to Nalgonda and Secunderabad hospitals. [563]

There had been about 1,000 passengers on the train, many of them going home for the annual Diwali Hindu festival of lights. The toll of the accident was therefore high and at least 116 people died, some reports putting the figure at 123 dead. Driver Samuel of the express was amongst the fatalities. Assistant Driver Sreenivasa Rao, on the side of the engine that remained above the water, survived. Twenty-six passengers were stated to have been seriously hurt, and 77 passengers and one railwayman (probably Assistant Driver Rao) slightly hurt.

A newspaper wrote: 'A Koteswara Rao of Guntur lost his 10-year-old son Sekhar. They were coming to a relative's house here {Hyderabad} to celebrate Diwali. He ran in circles looking for his son but in vain.

'"It is the saddest day of my life. I was helpless. The scene of my wife being washed away will continue to haunt me forever," said P. Anandam, a resident of Balanagar near here (Hyderabad). He was returning home with his wife Krupa Rani after visiting some relatives in Gunhur.

'Rafiq of Gunhur, his pregnant wife Haseen and his sister-in-law Sultana came here for Eid shopping. "We had just finished Sahar (pre-dawn meals for Ramzan fast) when we heard a loud noise," said 28-year-old Rafiq. Some army personnel saved the couple. However, 12-year-old Sultana was not so lucky.

'All 18 compartments of the train were packed with passengers. Introduced recently, Delta Fast Passenger is popular among passengers from many parts of the coastal district of Guntur for its relatively low fares. The fare for reserved classes on this train, which covers 400 km distance, is just Rs. 75, about half of the fare on express trains.' [564]

Another newspaper had the stories of other passengers: '"I don't know how it all happened. I was in the fourth coach from the engine. Suddenly there was water in the compartment and it kept rising. I just caught hold of the fan on the roof and kept my head high," said Ventakeshwara Rao, a resident of Guntur. He received injuries on his left arm and chest. His elder brother, Veeranarayana, who was also in the same compartment was injured in the head and leg. They were among the 15 injured passengers who were shifted to the Railway Hospital at Lalaguda here. Gopi Natsiah (36), a pastor of Khanpur in Adilabad district, believes he was saved because of his prayers. He got up at 4 a.m. and was reading the Bible before the accident occurred. "The entire coach was flooded. It is a miracle that I escaped with minor injuries," he said from the hospital bed.' [565]

'South Central Railway General Manager Thomas Verghese subsequently stated that the bridge had been intact at 01:00 when Patrolman K. Sankaraiah had been at it. The patrolman had then set off towards Ramannapet station and on his return towards the bridge later had

---

[563] www.Rediff.com/cms/print.jsp?docpath=/news/2005/oct/29train.htm

[564] www.newkerala.com/news.php?action=fullnews&id=44611

[565] www.hinduonnet.com/thehindu/thscrip/print.pl?file=2005103015721200.htm&date=200...

himself been washed away by the floodwaters and injured. [566]

The inquiry into the disaster determined that 3.2 centimetres of rain had fallen on the day before the accident. The discharge capacity of the culvert was about 2,000 cusecs of water. On the day of the accident the quantity had been 51,000 cusecs. [567]

On the last day of the year the Secunderabad police arrested Senior Section Engineer K. Venkateswarlu, who had been in charge of the Chityala subdivision on which the accident had occurred. The arrests of other railwaymen were expected to follow; it being alleged that negligence on the part of railwaymen had led to the disaster.[568] There were allegations that the bridge had collapsed at about 01:00, so its failure should have been discovered in time to prevent the disaster, and that another railwayman, Gangman Bondaiah, had not been on duty, while the railway complained that they had not been immediately notified by the local authorities of the failure of the irrigation tanks. The bridge had been a modern one as the line had only opened about 1990. [569]

The Commission of Railway Safety issued six recommendations relating to the disaster:

(a) The linear waterway including the opening for the existing railway bridges and the design for the overflow surplus for railway affecting tanks should be per laid down Codal provisions/bridge rules by Railway Administration and Minor Irrigation Department of the Andhra Pradesh Government as jointly signed by the Chief Bridge Engineer, South Central Railway, Minor Irrigation and Chairman, State Committee of Engineers, Andhra Pradesh Railway affecting and Bridges, vide paragraph 4.01 of the Draft Manual of Instructions for Railway Affecting Tanks and works. Andhra Pradesh government should approve the return period for calculating maximum design discharge and the degree of outflow surplus weir for Railway Affecting Tanks as 50 years. The Railway Administration and the Minor Irrigation Department of the Andhra Pradesh government should modify the existing railway bridges and the overflow surplus weir on a time-bound programme basis.

(b) The instructions contained in the Andhra Pradesh state government order No. G.O. Ms No. 903 Revenue (V.O.) department, dated 7.11.1990, should be reiterated to all concerned revenue/irrigation officials up to the level of Village Administration for informing the Divisional Railway Engineers or the nearest railway stationmaster, gangmen/gatemen at the nearest track location of any mishap of unusual occurrence of floods, cyclones or other natural calamities including the likelihood of a breach of any railway affecting tank in time. For this purpose the High Flood Level/Maximum Water Level for the surplus weir shall be considered as a danger level and prominently displayed for the nominated person to take immediate action.

(c) For Railway Affecting Tank bridges, the pucca flooring including curtain and drop walls along with the pitching of the formation up to 0.3 m above the danger level for a length on either side of the bridge should be decided by the Railway Administration based on the height of the bank, the distance of the Railway Affecting Tank from the railway bridge, etc., should be provided.

(d) C & T boards should be provided on the approach to Railway Affecting Tank bridges as provided in the Blue Book on Monsoon instruction, 2003, issued by the South Central Railway.

(e) The Railway Board might be impressed upon the necessity for providing some

---

[566] www.newkerala.com/news.php?action=fullnews&id=45303
[567] www.ndtv.com/addons/printpage.asp?id=82112&callid=1&caption=undeined
[568] http://news.webindia123.com/news/printer.asp?id=207137&cat=India
[569] https://www.irfca.org/articles/accidents-2004-05.html

electronic/electrical gadgets for assessing the flood levels of vulnerable bridges and connecting them with the signals of approaching stations or with the stationmasters so that in case of the water rising beyond a certain level the signal was actuated or the alarm raised in the stationmaster's room.

(f) Walkie-talkie sets should form a part of the monsoon patrol equipment for both stationary watchmen and mobile patrolmen. The patrol charts for the entire section should be prepared and on receipt of a weather warning report the patrolling of the section by the mobile patrolman should be similar to the system as laid down in the blue book on monsoon instructions with exchange of the patrol book by them/signing of the patrol book by the assistant stationmaster/deputy station superintendent of the concerned stations. [570]

[570] Commission of Railway Safety Annual Report 2005-06

CHAPTER 9

OPENING BRIDGE ACCIDENTS

Accidents involving opening bridges appear to have occurred in only a handful of countries, with the majority occurring in the United States. It seems that most countries managed to a large degree to avoid building such bridges, which from a railway point of view are a constant source of expense and worry. Robert Shaw states that the early railroads on the eastern seaboard of the United States were forced by opponents of the new form of transport to build such bridges over even insignificant waterways where they were never really needed.[571] Only a very small number of accidents involving opening bridges occurred between 1991 and 2005.

On probably Wednesday 2nd March 1848 a 'luggage' train arrived at an open drawbridge at Weedon, England, on the London and North Western Railway. Although the proper signals were exhibited to stop the train, its driver failed to notice them in time to pull up and at least part of the train fell from the bridge onto a boat loaded with bricks, which was passing beneath it at the time, sinking the boat. No one was killed but eight passenger trains were delayed and their travellers had to walk on planks from one side of the bridge to the other, which was probably a rather unnerving experience, at least for the more timid amongst them. [572]

At about 6:28 a.m. on Saturday 23rd November 1996 eastbound National Railroad Passenger Corporation (Amtrak) train No. 12, the *Fast Mail*, from Washington, D.C. to Boston, Massachusetts, was derailed as it was crossing the Portal Bridge, a swing bridge across the Hackensack River in Secaucus, New Jersey, United States. Both engines and all 12 cars of the train were derailed and the engines and four leading vehicles went down an embankment into some marshy ground. The derailed train also fouled the opposite running line and was struck by Amtrak train No. 79, *The Carolinian*, from New York to Charlotte, North Carolina. No one was killed but two deadheading railwaymen were seriously hurt and 42 passengers and railwaymen were slightly hurt. Apart from one passenger who was slightly hurt all the injured had been on train No. 12. [573]

The accident was investigated by the National Transportation Safety Board (NTSB).

The NTSB determined that at 4:00 a.m. on the day of the accident the bridge operator on duty at Portal Bridge had received a call from a tugboat requesting that the bridge centre be

---

[571] *A History of Railroad Accidents, Safety Precautions and Operating Practices* by Robert B. Shaw, pg. 187

[572] *The Welshman*, Carmarthen, Wales, 10th March 1848 (Reproduced in *L & N W R Society Journal* Vol. 10, No. 1)

[573] *The Intelligencer Record*, Doylestown, Pennsylvania, United States, 25th November 1996

swung open to allow the vessel to proceed north up the Hackensack River. The bridge operator had duly notified the Amtrak train dispatcher, who had unlocked the bridge at 4:03 a.m. in preparation for the opening. The bridge operator testified that the first steps in the bridge-opening sequence had been completed normally. After he had activated the switch to lift the mitre rails (the transition rails between the running rails on the bridge approaches and the movable portions of the bridge), the operator had not received a control panel indication that it was safe to proceed, because Amtrak had removed the rail position indication circuitry for the mitre rails. In accordance with standard operating practice, the operator had then used a bypass switch to continue the bridge-opening sequence. When he had attempted to swing the bridge open, he found that the swing span would not rotate. Feeling the bridge shake and vibrate, the operator had then aborted the opening. The accident investigation determined that either prior to or during the attempt to lift the mitre rail for the north rail on the west end of track 1, both side bars joining the mitre rail to a longer running rail had broken. As the lift rod had pushed up on the mitre rail, the mitre rail and the running rail had separated at the joint. The lift rod had continued to lift the heel of the mitre rail, but the toe of the rail had remained seated in the bed plate for the stationary rail on the bridge approach span. When the bridge operator had attempted to swing the movable span, the mitre rail had hung in the bed plate on the stationary span, preventing the bridge from opening. When the bridge operator had then aborted the opening, the lift rods had lowered the mitre rails, but the heel of the hung mitre rail had not seated properly. Instead of falling back into the bed plate, it had come to rest on top of the broken side bar sections attached to the running rail. The 5-inch difference in elevation between the two tracks had created a ramp, which would derail the next train to cross the bridge on track 1. The rail position detection circuitry, which would have indicated on the bridge operator's control panel that the rail was not seated properly, had been removed in 1987. Because electrical continuity was maintained across the rails despite the misalignment, the signals governing the Portal Interlocking had displayed a clear indication. Amtrak procedures in effect at the time did not require the bridge to be physically inspected after a failed opening and before clearing the bridge for train traffic. About 5:40 a.m., an Amtrak electrician had arrived to troubleshoot the bridge and to determine why it had failed to open. Although he had walked onto the bridge from the well-lit west end, he had not noticed the misaligned rail. The electrician had still been attempting to determine why the bridge had not opened properly when train No. 12 derailed on the bridge and sideswiped passing train No. 79.

Regarding the removal in 1987 of the circuitry, which would have informed the bridge operator that there was a problem, the NTSB noted that problems with excessive vibration on the Portal Bridge had begun almost immediately after the installation of Conley mitre rails on the Portal Bridge in 1985. One result of the vibration had been that the limit switches installed to detect the position of the mitre rails and relay this information to the bridge operator's control panel had failed repeatedly and could not be economically maintained. To correct this problem, Amtrak had removed the limit switches and associated mitre rail indication circuitry in 1987. Even though Amtrak had purchased a new type of heavy-duty circuit controller to replace the limit switches, the new devices had never been installed on the Conley mitre rails. When the Conley rails had been replaced by mitre rails of Promex design in 1992, no detection circuitry had been installed, even though the mechanism had provisions for such devices. The rail position detection circuitry had served the important safety function of reporting to the bridge operator that the rails were properly lifted before an attempt was made to open the bridge and also were firmly seated after the bridge was closed. In fact, the bridge interlocking mechanisms would not allow a bridge opening or closing operation to proceed unless either the system received a safe indication from the detection circuitry, or the bridge operator used a bypass switch to allow the operation to continue without a safe indication. With no detection circuitry in place, bridge operators had always used the bypass switch, even though, without a visual inspection of the rails before and after each opening (which Amtrak had not required), the bridge operator had no way of knowing whether the attempt to raise or lower the mitre rails had been successful. This explained why, on the morning of the accident, the bridge operator had attempted to swing the bridge centre span even while the toe of the mitre rail on the west end of the north rail of track 1 had still been seated in the bed plate on the approach span. The torque applied to the mitre rail as the bridge span attempted to rotate might, in fact, have bent the rail slightly and contributed to the failure of the heel of the rail to align properly when the lift rods were retracted. After he had aborted the bridge opening, the bridge operator had assumed that the rails were safe for train traffic because the signal system had shown a clear indication. But the signal system had indicated only that the rails had made a complete electrical circuit and that the toe of the mitre rail was positioned properly; the system could not and had not indicated that the heel of the mitre rail was properly aligned.

The NTSB commented that if Amtrak management had understood the important safety function served by the rail position detection circuitry, it had done nothing to compensate for the removal of the system. An appropriate course might have been to require a visual inspection of the mitre rails to confirm that they were completely lifted before opening and were completely seated afterwards. This practice could have been continued until the detection circuitry could be replaced. In a post-accident memorandum to Portal Bridge operators and Spuyten Duyvil Bridge signal maintainers on 30th January 1997, Amtrak had instituted such visual inspections. The NTSB was concerned, however, that for almost 10 years, Portal Bridge operators had been allowed to assume that a critical procedure in the opening or closing of the bridge would always be successful and that no confirmation, electronic or visual, was required before using a bypass switch. The NTSB concluded that if Amtrak management had had in place on the Portal Bridge a functioning rail position detection system or procedures that required visual confirmation of the proper positioning of all mitre rails, the accident probably would not have occurred. The NTSB believed that Amtrak should ensure that current or future mitre rail installations on Portal Bridge were equipped with a position detection/indication system which provided the maximum protection possible and which was interlocked with other bridge systems to prevent the bridge from being opened or cleared for train traffic until the position of the mitre rails could be confirmed to be safe.

The NTSB listed its conclusions regarding the accident:

(a) Fatigue, drugs, weather, and the signal system had not been causal or contributory factors.

(b) The design, the materials, and the operation of the mitre rail system in place on the Portal Bridge at the time of this accident had made the side bars susceptible to fatigue cracking and had thus led to the side bar failure, which had precipitated the accident.

(c) The welding performed on the side bars had been inadequate and inappropriate as a permanent repair and had served to concentrate stress on the already fractured areas of the side bars.

(d) The weld repairs could have been adequate as a temporary fix had a detailed and repetitive inspection program been established to ensure continued safe operation until permanent repairs or replacements could be made.

(e) Amtrak management had been aware of failures in mitre rail side bars at least 10 months prior to the derailment, but because the company had erroneously considered cracked or broken side bars to be a maintenance issue rather than a safety issue, it had not made replacements or permanent repairs that could have prevented the accident.

(f) Amtrak management had not developed and implemented adequate mitre rail inspection procedures to identify defects in all components of the mitre rail assemblies on the Portal Bridge.

(g) Had Amtrak, when it had first learned about the cracked side bars on the mitre rails, revised its mitre rail inspection procedures to include raising the mitre rails for inspection, the accident might have been prevented.

(h) If Amtrak management had had in place on the Portal Bridge a functioning rail position detection system or procedures requiring visual confirmation of the proper positioning of all mitre rails, the accident probably would not have occurred.

(i) Had the accident resulted in more serious injuries, the confusing communication of the accident location by the Amtrak police dispatcher and the resulting delay in emergency response could have resulted in additional risks to train occupants.

(j) The traction motor current data had not accurately indicated the throttle position and, therefore, use of the data for this purpose by Amtrak had not met Federal Railroad Administration requirements for monitoring and recording train throttle positions.

(k) Amtrak's use of a multiplexer to monitor and record both traction motor current and cab signal on a single channel of the event recorder had been inappropriate and ineffective and, as a result, the NTSB investigators had found it impossible to determine the cab signal indications in this accident.

(l) If the entire event recorder systems, including sensors, wiring, etc., in Amtrak locomotives 910, 901, and 930 had been thoroughly tested during their most recent 60-day inspections, the incorrect module configuration would likely have been found and corrected, and the traction motor current data retrieved after this accident would have been useful in determining the pre-accident cab signals received by the train crews.

(m) Amtrak management had failed to foster an environment that promoted adequate inspection, maintenance, and repair of the mitre rail assemblies on the Portal Bridge and to permanently correct defects in the mitre rail side bars, which had been discovered 10 months before the accident.

(n) If Amtrak had been required to meet Federal standards for inspection and maintenance of the special trackwork on Portal Bridge, the defects in the mitre rail side bars might have been detected and repaired before they could cause a derailment.

(o) The results of the Federal Railroad Administration movable bridge survey would be beneficial to the railroad and rail rapid transit industry in preventing accidents similar to the derailment on Portal Bridge.

The NTSB determined that the probable cause of the accident was the failure of Amtrak management to foster an environment that promoted adequate inspection, maintenance, and repair of the mitre rail assemblies on the Portal Bridge and to permanently correct defects in the mitre rail side bars that had been discovered 10 months before the accident. Contributing to the accident had been:

(a) The failure of the Federal Railroad Administration to develop track inspection standards for special trackwork and to periodically inspect such track as part of its oversight responsibilities.

(b) Amtrak's removal of the mitre rail position detection circuitry without installing replacement circuitry or implementing procedures to compensate for the loss of this safety-critical system.

The NTSB ended its report as usual with its recommendations.

To the National Railroad Passenger Corporation (Amtrak):

(a) It should perform a comprehensive stress analysis of the design of any mitre rail assembly currently in use or intended for use on the Portal Bridge to identify critical areas of high cyclic stress and ensure that the mitre rail design adequately accommodated these cyclic loads.

(b) It should continue to monitor the safety of special trackwork on its movable bridges and ensure that its special inspections were adequate and of sufficient frequency to detect failures or potential failures involving all components of all its special trackwork.

(c) It should develop and put procedures in place to ensure that any failures or potential failures noted during these inspections were corrected before they developed into safety hazards.

(d) It should ensure that current or future mitre rail installations on the Portal Bridge were equipped with a mitre rail position detection/indication system that provided the maximum protection possible and was interlocked with other bridge systems to prevent the bridge from being opened or cleared for train traffic until the position of the mitre rails could be confirmed to be safe.

(e) It should review the training of its police dispatchers and ensure that dispatchers were trained to correctly identify all Amtrak locations to emergency response agencies.

(f) It should perform a thorough test of the entire recording system on every locomotive equipped with an event recorder to ensure that cab signal data records could be easily and positively identified and evaluated.

(g) It should conduct a comprehensive internal management review of the circumstances of the accident to determine why several layers of Amtrak management had failed to act in a timely fashion to correct a known hazardous condition on the Portal Bridge and make the management or procedural changes necessary to ensure that conditions affecting the safety of rail operations were given the highest priority.

To the Federal Railroad Administration:

(a) It should inform the railroad industry that traction motor current was not a valid indicator of throttle position, and the requirement to record throttle position contained in 49 Code of Federal Regulations 229.5(g) could not be met by recording traction motor current.

(b) It should ensure that all operators currently using traction motor current as a substitute for throttle position modified their event recording systems to monitor and record throttle position directly.

(c) Pending the results of its Railroad Safety Advisory Committee Event Recorder Working Group and its implementation of suitable requirements concerning event recorder system maintenance, it should require that microprocessor-based event recorders equipped to perform self-tests be subject to the testing and inspection procedures currently applicable to all other types of event recorders.

(d) It should expand the scope of its track safety standards to include special trackwork such as movable mitre rails and ensure that the condition and operation of special trackwork were included, when appropriate, in all Federal Railroad Administration track inspections.

(e) It should provide, in full or summary form, the results of the Federal Railroad Administration movable bridges survey to all railroads and rail rapid transit agencies.

To the Association of American Railroads:

(a) Pending the results of the Federal Railroad Administration (FRA) Railroad Safety Advisory Committee Event Recorder Working Group and the FRA's implementation of suitable requirements concerning event recorder system maintenance, it should advise its member railroads of the need to test and inspect all microprocessor-based event recorders equipped to perform self-tests in accordance with those procedures outlined in 49 Code of Federal Regulations 229.25(e)(2), which currently applied to all other types of recorders, to confirm proper event recorder function.

To the American Short Line Railroad Association:

(a) Pending the results of the Federal Railroad Administration (FRA) Railroad Safety Advisory Committee Event Recorder Working Group and the FRA's implementation of suitable requirements concerning event recorder system maintenance, it should advise its member railroads of the need to test and inspect all microprocessor-based event recorders equipped to perform self-tests in accordance with those procedures outlined in 49 Code of Federal Regulations 229.25(e)(2), which currently applied to all other types of recorders, to confirm proper event recorder function. [574]

A minor accident involving a swing bridge occurred at Clatskanie, Oregon, United States, at about 4:30 p.m. on Thursday 2nd September 2004. A Portland and Western Railroad train was performing a switching manoeuvre prior to departing from the sidings of the Stimson Lumber Company when an engineer inadvertently pushed a flat car loaded with lumber onto the adjoining draw, which was open at the time. Work was in progress on the Friday to retrieve the car and its load from the Clatskanie River, after which damage to the track and bridge would be assessed. [575]

Part of a northbound Canadian National Railway freight train carrying new automobiles became derailed as it was crossing a trestle swing bridge over the Fraser River at Burnaby, British Columbia, Canada, on the evening of Monday 5th December 2005. Four of the 39 cars of the train left the rails. Three of them remained upright on the bridge but the fourth fell into the water. No one was hurt. The trestle was subject to a weekly inspection and the cause of the derailment was not immediately known. [576]

---

[574] National Transportation Safety Board report

[575] http://news.statesmanjournal.com/article_print.crf?i=86174

[576] *The Lethbridge Herald*, Lethbridge, Alberta, Canada, 7th December 2005

CHAPTER 10

TRAIN ACCIDENTS INVOLVING BRIDGES

Bridges could feature in train accidents in one of two ways. Either the bridge itself could collapse for a number of reasons such as faulty design, faulty construction or because its foundations had been damaged; or a train could fall from a bridge, the structure itself remaining intact. Occasionally an overbridge could be struck by a derailed vehicle and fall onto a train. Bridge failures have been dealt with in the penultimate chapter; this chapter will concern itself solely with accidents where the bridge itself was not at fault. It may seem perverse to list such accidents separately, instead of in the chapters dealing with primary causes of the accidents, the procedure adopted elsewhere in this book, but the fortuitous presence of a bridge has such a potential for transforming what might have been only a minor accident into a major disaster that it has been decided to treat such accidents separately, in the same way as accidents occurring in tunnels have been given a chapter of their own.

A major environmental spill resulted from the derailment of the engine and seven tanker waggons of a Southern Pacific Transportation (SP) freight train from Los Angeles, California, to Pasco, Washington, at about 9:50 p.m. on Sunday 14th July 1991. The accident occurred on a bridge at Cantara Loops, California, United States, about six miles north of Dunsmuir. One of the tank cars contained metam sodium, a liquid weed- and tree-stump-killing chemical, some 19,500 gallons of which were lost from it into the Sacramento River. The chemical flowed down the river into the Shasta Lake, killing thousands of fish. About 360 people had to be treated for skin and eye irritation or respiratory problems from methyl isothiocyanate and other breakdown products of the chemical and some 200 residents living alongside the river were evacuated. A number of people suffered prolonged effects from exposure to the chemical and by the beginning of October the SP had paid out over $1,000,000 in compensation.

Referring to the effects of the pollution, Banky Curtis of the Department of Fish and Game stated: "all aquatic life in the river between the spill area and Shasta Lake has been destroyed." An unknown amount of plant life along the river banks was also killed. Curtis added: "This is one of the largest inland spills in California history. It appears that every living thing in the water was killed. It will probably take quite a bit of time, at least several years, maybe more, to re-establish the food chain. From a fishery standpoint the river has been eliminated." [577]

The derailment was blamed by SP Superintendent Bob Melbo on "stringlining," an effect occurring when the train tried to straighten as it rounded the 14-degree curve at the Cantara

---

[577] *The Orange County Register*, Santa Ana, California, United States, 16th, 17th, 18th, 20th, 21st, and 25th July and 7th October 1991

trestle.[578] Stringlining is defined as: 'Excessive lateral forces caused by heavy draft loading, resulting in wheels lifting over the low rail or the rail rolling over.' [579]

At about 12:15 p.m. on Tuesday 19th November 1991 part of an 89-car CSX Transportation freight train became derailed and fell from a bridge over the Salt River near Shepherdsville, Kentucky, United States. Twenty-eight cars left the rails, 14 of which either fell 40 feet from the bridge into the water, onto the riverbank or piled up on top of one another on the bridge. Cars carrying corn syrup and building materials caught fire; close to them and derailed was a car carrying potentially explosive methylene diphenyl diisocyanate. Another car in the train was carrying about 90 cluster bombs but it was not close to the fire and was not derailed. A tanker car carrying propylene oxide was also not directly involved. About two-thirds of the 80-year-old girder bridge collapsed.

Because of the danger of an explosion, Governor Wallace Wilkinson declared a disaster area around the scene of the accident, authorising the use of the National Guard, and 1,000 people were evacuated at least overnight as a precaution. No one was hurt. State Fire Marshal Rodney Raby decided that the safest option was to let the fire burn itself out, stating, "If we put personnel down there we have no assurances we'd get them out."

The accident was blamed on a dustcart that had collided with the bridge on the approach to the riverbank, knocking the track out of alignment. [580] [581]

Falling debris and a railroad car damaged a Greyhound bus and an automobile travelling along Interstate 90 at 4:15 a.m. on Christmas Eve, Tuesday 24th December 1991, after a Burlington Northern Railroad freight train was derailed at the west end of the Latah Creek Bridge, near Spokane, Washington, United States. Two cars fell from the bridge; a refrigerated car was left hanging from it and about a dozen cars were derailed but remained upright on the track. A tyre of the bus was punctured and the automobile windscreen was smashed but no was hurt.[582] [583] The derailment was subsequently blamed on a broken track joint. [584]

The derailment of 14 cars of a 57-car Burlington Northern Railroad freight train at about 2:30 a.m. on Tuesday 20th June 1992 on a bridge over the Nemadji River near Superior, Wisconsin, United States, led to one car falling into the river and a leakage of benzene from it. Witnesses spoke of a huge white cloud of chemical that hung over the harbour separating the cities of Superior and Duluth, Minnesota.[585] The spillage led to the evacuation of 30,000 people in Duluth and 20,000 people in Superior. Wisconsin Governor Tommy Thompson declared a state of emergency in Douglas County, which included Superior. At least 25 people were taken to hospital suffering from the effects of exposure to the toxic fumes and emergency crews worked to contain the spillage. Fish in the river were killed but the number

---

[578] https://eu.mtshastanews.com/story/news/2020/07/15/from-our-archives-cantara-loop-spill-was-29-years-ago/112681966/

[579] https://www.csx.com/index.cfm/about-us/company-overview/railroad-dictionary/?i=S#:~:text=Stringlining,or%20the%20rail%20rolling%20over

[580] *Bluefield Daily Telegraph*, Bluefield, West Virginia, United States, 20th November 1991

[581] *Times-News*, Burlington, North Carolina, United States, 20th November 1991

[582] *Fairbanks Daily News Miner*, Fairbanks, Alaska, United States, 24th December 1991

[583] *The Daily Inter Lake*, Kalispell, Montana, United States, 24th December 1991

[584] *Walla Walla Union-Bulletin*, Walla Walla, Washington, United States, 29th December 1991

[585] *Leader-Telegram*, Eau Claire, Wisconsin, United States, 30th June 1992

was not known, nor was the cause of the original derailment apparent. [586]

A spectacular and tragic accident occurred on Christmas Day, Monday 25th December 1995, when the *Estrella de Gibralfaro* express from Barcelona to Malaga and Seville ran into a landslide and was derailed near Las Correderas, Spain, in the Despeñaperros national park. The engine and leading coach of the train reared into the air and fell from a bridge into the gorge below, killing the two drivers. Eight other people were injured, one seriously. [587]

A major catastrophe occurred in India on the South Eastern Railway at 17:42 on Sunday 14th September 1997. Train No. 8033, a Down Ahmedabad – Howrah express, was derailed on the approach to Bridge No. 46 over the Hansdev River, between Naila and Champa stations, Madhya Pradesh, on the Howrah – Nagpur broad-gauge, main-line, electrified section of the Bilaspur Division of the railway. Five of the coaches of the train were derailed and one of them, in which most of the casualties occurred, fell into a dry ditch adjoining the river bed. The ditch was reportedly 45 metres deep and the coach was crushed in its fall. Eighty-eight people lost their lives in the disaster and 400 more were injured, 150 of them seriously. [588]

A press report on the accident stated, 'Villagers from surrounding areas helped rescue workers carry out the injured on stretchers, and several private aid groups contributed ambulances and supplies, railway official said. Several factory workers from a nearby industrial area rushed in hydraulic cranes to remove from a 150 foot ditch into which the train had fallen.

'Bajrang Kedia, a journalist travelling on the train, said a railway officer on the scene told him that the train was travelling at 18 mph {30 km/h} when the crash occurred, although the recommended speed was 6 mph {10 km/h} because of repair work being done on the tracks. The official said the train's engineer applied the brakes when he saw a caution flag, but it was too late, according to Kedia, an editor at Navbharat, one of India's largest Hindi newspapers. The train stopped with a jerk and the five carriages derailed, Kadia said. Three of the carriages were crammed with passengers, police and witnesses said.' [589]

A day or so after the disaster, Railway Minster Ram Vilas Paswan stated that it had been due to a mistake by a permanent way gang. A minor track problem could have been dealt with by making a temporary repair but instead the foreman in charge had chosen to completely replace a 10-metre length of track, without informing his superiors, and the work had continued after it should have finished at 17:30. The protection of the track work had also presumably been inadequate. After the accident the whole crew of 13 workers had fled and on Tuesday, 11 of them were still at large, including the foreman. Police Inspector General Surendra Singh stated that the two gangmen who had been apprehended would be charged with acts of negligence resulting in death, with other charges likely to follow.[590] The accident is thus reminiscent of the Staplehurst disaster in England, 132 years earlier.

Two hundred passengers on a mixed train in Bangladesh had a lucky escape on Thursday 4th November 1999 when part of the train was derailed and fell from a bridge into the Buri Teesta River. Five waggons loaded with rice fell from the bridge and another waggon and a

---

[586] *The Hutchinson News*, Hutchinson, Kansas, United States, 1st July 1992

[587] https://efs.efeservicios.com/en/foto/accidente-ferroviario-despenaperros/8000091499

[588] Government of India Ministry of Civil Aviation and Tourism Commission of Railway Safety Annual Report for 1997-98

[589] *European Stars and Stripes*, Darmstadt, Germany, 16th September 1997

[590] *The Orange Leader*, Orange, Texas, United States, 16th September 1997

passenger coach were derailed but stayed on the bridge. No one was hurt. [591]

Eastbound Canadian National Railway (CN) freight train No. M-365-21-20 had left Parent, Quebec, Canada, at 21:27 on Sunday 21st May 2000, with three locomotives and no less than 159 cars, 128 of which were loaded. While it was travelling at 29 mph it experienced an undesired brake application at about Mile 9.5, near Cressman, Quebec. After the train had stopped, its conductor went back and found that 13 gondola cars loaded with wood chips had derailed. Subsequently it was discovered that nine more cars further back in the train were also derailed and that three of them had fallen from a bridge over the Saint-Maurice River. Two of these were tank cars, which were punctured but they were fortunately empty apart from residue. Four hundred feet of track were damaged and two spans of the bridge were badly damaged. No one was hurt.

The Transportation Safety Board of Canada (TSB) investigated the accident. Its report contained seven findings:

(1) The train had derailed when a low rail in a curve had failed as a result of a vertical split head (VSH) in it.

(2) The VSH had not been a recent development; however, it was not detected because it had probably not been present long enough. The last rail test had been done about seven months previously.

(3) The special measures taken by CN to control rail breaks had been insufficient because, once VSHs were initiated, they propagated rapidly.

(4) Accumulated tonnage data could provide an additional indicator of rail condition and permit a more effective risk assessment and rail testing planning more appropriate to the track characteristics.

(5) Since inspectors had to perform many visual inspections concurrently, visual signs such as those indicating a VSH could escape their attention and were not always detected in bi-weekly inspections.

(6) The derailment of the 72nd to 80th cars had been a result of the first derailment at the head of the train.

(7) A train configuration that would take into account the dynamics of the train would allow more effective control of the forces generated and would reduce the risk of derailment in the event of undesired braking.

The TSB noted that after the accident, Transport Canada (TC) had increased its monitoring on the Saint-Maurice Subdivision and had sent a notice to CN regarding the track condition. Furthermore, the CN had reduced the speed of trains on certain sections of track and was regularly inspecting the track between Fitzpatrick and Parent. It intended to replace 21 miles of track on that subdivision in 2002.

TC and the railway industry had agreed to form a working committee, including union representatives. This would determine if changes to the Railway Track Safety Rules and to the methods used by the industry were required and would examine issues such as rail testing frequency and parameters to evaluate track conditions.

As for train configuration, TC recognised that by taking into account the train dynamics, the forces generated would be more effectively controlled in the event of undesired braking, thereby reducing the risk of derailment. Several stakeholders, including TC, had started

---

591 *Danger Ahead! Historic Railway Disasters*, 26th April 2002

discussions to initiate a research project on the subject. The goal of this project, entitled "Train Dynamics Simulator," was to create a simulator to assess the dynamics parameters on train movements to determine the impact on trains, cars and track. As a result of the project, new regulations could be established and action could be taken to reduce the risks related to train dynamics. [592]

The Bagouko Bridge, in Mali, which had collapsed following the derailment on it of a goods train on 21st July 2001, was brought back into use on 31st August of the same year. While it had been out of use, the price of commodities in Kayes, Mali, had soared, because they could not be transported by rail from Dakar or Bamako. [593] At least 40 staff had worked on the restoration of the structure, most of them being from the Société nationale des chemins de fer du Mali but including a dozen from the Société nationale des chemins de fer du Sénégal in the neighbouring country. (The former Chemin de fer Dakar – Niger had been split into the two companies upon the independence from France of Mali and Senegal.) [594]

A TransNamib goods train of eight waggons carrying cement bound for Tsumeb was derailed on a bridge 16 kilometres north of Omaruru, Namibia, at 06:10 on Tuesday 18th September 2001 and one waggon fell from the bridge into the river below. No one was hurt but the cost of the derailment was expected to be considerable. The cause of the accident was not immediately clear. [595]

At about 04:50 on Friday 9th August 2002 the driver of a lorry loaded with a 4.2-metre-high container attempted to pass beneath a railway underbridge with a maximum clearance of only 3.75 metres near Taichung station, Taiwan. The bridge was struck with considerable force and the impact was sufficient to distort the track. At about 05:20, train No. 2051, the empty stock of the *Tze-chiang Express* from Taichung to Changhwa arrived at the bridge, travelling at 90 km/h. Due to the distorted track it was totally derailed, its nine cars demolishing several posts carrying the overhead line equipment and damaging the station platform and a building.

"To my shock I discovered that the rail track had deformed into an "S" shape about 100 metres before we reached the bent rail. I pulled the emergency break and the train lost control and derailed. We didn't receive any warning," Driver Su Shih-shui said, after he had been sent to hospital. He needed three stitches. Fortunately no one was killed in the accident but the train conductor and the lorry driver were also injured. [596] [597]

The *Argo Dwipangga* express, filled with travellers returning to Jakarta from Solo, after having celebrated the Muslim Eid al-Fitr holiday in their home towns, was derailed in the Prembun district, Java, Indonesia, late on Monday 9th December 2002. The engine and eight of its 11 coaches fell down an embankment into rice fields in Tersobo village. Six people died and 29 were injured. The dead and injured were taken to hospitals in Kebumen and Purwarejo.

---

592 Transportation Safety Board of Canada report

593 www.subsaharaninformer.com/news/news.asp?cat=6&subcat=61&subregion=3&country=316&ne...

594 http://www.subsaharaninformer.com/news/news.asp?cat=6&subcat=61&subregion=3&country=316&ne...

595 http://www.namibian.com.na/2001/September/news/0114C8D90B.html

596 *TaiwanNews.com*, 10th August 2002

597 https://www.taipeitimes.com/News/front/archives/2002/08/10/0000159681

In addition, at least 10 trains between Jakarta and Surabaya had to be cancelled after the accident. Other trains for the south coast, including the *Gajayana*, *Mutiara Selatan* and *Turangga* were diverted via Malang and were many hours late arriving at their destinations.

As in the Taiwan accident exactly four months earlier, the derailment had been caused by a circumstance outside of the direct control of the railway. At the point of the derailment the line crossed a road by an overbridge with a maximum safe negotiable height of 2.4 metres. Immediately before the arrival of the train, a lorry that was 2.6 metres high attempted to pass under the bridge, struck and damaged it and distorted the track going over it. Police Chief Adj. Sr. Com. Prasta Wadyu Hidayat announced that the lorry driver, named as Ahmad Saefuddin, had been taken into custody for questioning.

Spokesman Zainal Abidin for PT Kereta Api Indonesia, the state railway company, said that it was hoped to reopen the line by Wednesday. Meanwhile, the company would refund the fares of passengers whose trains had been cancelled. [598]

Eleven people died when a freight train ran into a landslide and was derailed shortly after leaving Matadi, Bas Congo province, Democratic Republic of Congo, on Wednesday 26th November 2003. Ferdinand Derouaux of the transport ministry stated that two vehicles of the train had left the track and fallen from a bridge over the Mpozo River.[599] The previous day, about 200 people had died in a ferry disaster on Lake Mai-Ndombe, while there had been a serious air crash and road accident only the previous Saturday. [600]

At about 14:00 on Saturday 1st March 2003 a narrow-gauge (762 mm) tourist train consisting of a diesel locomotive and four coaches got out of control descending a long falling gradient and became derailed on a small concrete bridge on the 72-kilometre-long line Alishan Forest Railway in Taiwan, the final derailment occurring near Alishan station. One of the coaches fell from the bridge into a ravine and two others were left hanging perpendicular to the track. Seventeen passengers died and 173 were injured, 102 seriously. There had been about 150 people on the train.

Surviving passengers related that they had feared that their train would crash because of the speed at which it had been travelling. An unidentified woman whose face and hands were streaked with blood told a television reporter, "The cars suddenly left the rails and fell over. The car was filled with people. Some were standing. I looked at my daughter and saw her internal organs. I don't know where my daughter is now." [601]

A minor additional accident happened when a helicopter evacuating the injured crashed after its tail caught in some trees. No one was killed in this second mishap. Other rescuers went to the scene on foot and their efforts were hampered by the steep and thickly wooded mountain slopes through which they had to carry stretchers.

Wu Li-mei, 32, a woman passenger on the train who was a hospital administrator, was subsequently given a commendation for performing cardiopulmonary resuscitation on five injured passengers, thereby saving the lives of two children. [602]

The Ali Mountain, which the line served, was a popular tourist destination, especially at weekends.

---

[598] *The Jakarta Post*, Jakarta, Indonesia, 11th December 2002

[599] https://www.thenewhumanitarian.org/news/2003/11/27/185-die-ferry-collision

[600] https://www.voanews.com/a/a-13-a-2003-12-01-33-drc-s-67297682/380864.html

[601] http://edition.cnn.com/2003/WORLD/asiapcf/east/03/01/taiwan.crash.ap/

[602] www.taipeitimes.com/News/taiwan/archives/2003/03/08/197201/print

Conductor Su In-fu and Driver Tsai Cheng-sen were briefly taken into custody after the accident. It appeared that they had failed to conduct a brake test, or at any rate had not done it properly, before beginning the descent of the mountain, with the result that the brakes of the train had failed to operate properly, it had run away and had left the rails on a curve.[603] Subsequently, the two railwaymen, Co-driver Liu Po-yueh and Train Inspector Wen Fu-ming were charged with manslaughter.[604] It emerged that the accident had resulted from the brake valve between the engine and the leading coach having inadvertently been left closed, a fact not discovered by any of the railwaymen, and with the brakes only working on the engine the train had predictably not been controllable on the long descent. The four men subsequently stood trial for involuntary manslaughter and causing injury through negligence. They were all found guilty and the train inspector was sentenced to three years' imprisonment and the other three to two-and-a-half-year terms. [605]

South Central Railway train No. 7201, the *Golconda Express* from Guntur to Secunderabad, was involved in a major disaster at Warangal station, Andhra Pradesh, India (since 2014 in Telangana state), at 10:25 on Wednesday 2nd July 2003. The express was booked to stop at the station, instead of which it ran through it without halting, was derailed in a sand drag at the end of a loop line and its engine and two leading carriages fell from an underbridge onto the road below, crushing two autorickshaw taxis, two scooters and several vegetable sellers. Twenty-one people lost their lives in the accident, including five occupants of the autorickshaw, 16 others were seriously hurt and eight were slightly hurt. (Another source states that there were 22 dead and 100 injured.) Thirteen or 14 of the dead were passengers, the others being people on the road below. A banana vendor named Suresh had a lucky escape, due to a delay from a customer. "But for that lady, I would have died," he said. [606]

Assistant Driver Sammaiah stated after the disaster that he and his driver had realised that the brakes of the train were not working properly about two kilometres before the station. Speaking from his hospital bed, he said, "I tried to contact the stationmaster over the walkie talkie but there was no response. By that time all the controls had gone down to zero. It was a mechanical failure in the engine." [607]

Driver Mohammed Rashid was also injured and both he and his assistant fled the scene after the accident, the two railwaymen being admitted to the Mahatma Gandhi Memorial Hospital four hours later.[608] The driver related that he had first become aware of the brake failure at Nekkonda, 30 kilometres short of Warangal, when he had repeatedly attempted to warn the Warangal stationmaster of the fact but with no response. He added that he had managed to reduce the speed of his train from 110 km/h to 75. In response to the driver's allegation, six officials were suspended: Divisional Mechanical Engineer Kishan Singh, Assistant Engineer R. Upendra Kumar, Assistant Electrical Engineer Manjit Singh, Senior Section Engineer Syed Asan Basha, Permanent Way Inspector C. Narasimha Reddy and Senior Section Engineer D. Chukka Lingam. These suspensions were related to the apparent failure of the officials at Warangal to react to the driver's warning and also to an alleged failure to check the train's brake system before it had left Guntur. [609]

The provisional report of the Commission of Railway Safety blamed the accident on 'the

---

603 http://edition.cnn.com/2003/WORLD/asiapcf/east/03/02/taiwan.crash.ap/

604 http://stratstimes.asia1.com.sg/storyprintfriendly/0,1887,174346,00.html

605 www.etaiwannews.com/Taiwan/2003/12/18/1071712477.htm

606 www.keralanext.com/news/index.asp?id=12678

607 www.Ananova.com/news/story/sm_796260.html?menu=

608 *Newindpress.com*, 3rd July 2003

609 www.deepikaglobal.com/latestnews.asp?ncode=3758

delayed application of the brakes.' [610]

At about 06:00 on Tuesday 7th October 2003 an Iarnród Éireann bulk cement train from Castlemungret to Waterford was derailed as it was crossing a viaduct over the River Suir at Cahir, Eire. Following the derailment, the rear twelve waggons fell through the deck of the viaduct to the river and its bank below. No one was hurt but the waggons were wrecked and the viaduct suffered substantial damage.

The accident was considered sufficiently serious to warrant a full formal investigation by the Department of Transport (DoT). The subsequent inquiry was not able to state definitively what had caused the accident but concluded that shortcomings in the timber deck structure of the viaduct and shortcomings in the associated inspection and maintenance regime were the most significant causal factors. Had these had been fit for purpose, the inquiry determined, it was unlikely that the accident would have occurred. The inquiry made a number of recommendations relating to viaduct and track maintenance, the performance of two-axled cement waggons and speed limits. [611]

Iarnród Éireann was not very happy with all the conclusions of the DoT report, considering that the speed of the train, which had been slightly high but which the DoT had not thought important, had been a principal factor in the accident. [612]

Westbound Canadian National Railway freight train A-403-21-07 was derailed at 16:17 on Saturday 7th February 2004 as it was crossing a bridge over the frozen De Sud River at Montmagny, Quebec, Canada. Twenty-seven cars of the 94 in the train left the track, of which at least 12 fell from the bridge into the river, while three spans of the bridge were destroyed and the other six spans were damaged. A man who had been on the walkway of the bridge was slightly injured. One of the derailed cars contained chlorine but there was no leakage. [613]

The accident was investigated by the Transportation Safety Board of Canada (TSB), which determined that the derailment had been due to the hunting (unwanted oscillation) of an empty wood chip gondola car. The TSB also noted that the train journal indicating the location of two cars containing chlorine in the train had not been accurate, so that firemen had had to check their location visually (only one of the two chlorine cars was derailed). In summary, the TSB report stated that:

(a) The A-end of car CN 879261 had derailed first, at Mile 76.02, and had been dragged from that point to the public crossing at Mile 77.62, where the truck had been dislodged. The derailed car had struck the girders of the bridge at Mile 77.8, resulting in the derailment of the following cars.

(b) A wheel lift initiated by high-speed truck hunting had occurred as the run-in of train slack had taken place following the initiation of locomotive dynamic brakes, while the train was travelling at a speed of 58 mph. (The speed limit for freight trains was 60 mph.)

(c) CN 878 and CN 879 series wood chip gondola cars were prone to truck hunting when operating empty in trains at speeds in excess of 50 mph, presenting a risk of derailment.

(d) The risks of derailment due to truck hunting presented by CN 878 and CN 879 series wood chip gondola cars had not been identified as being similar to other gondola cars of

---

610 www.hinduonnet.com/thehindu/holnus/02291812.htm

611 Department of Transport (Eire) inquiry

612 www.timesonline.co.uk/printFriendly/0,,1-1507-1817747-1507,00.html

613 www.canada.com/co.../printstory.asp?id=B6BFA8EC-CD45-49C6-829F-647F33B2550

similar dimensions that were known to be prone to truck hunting.

(e) Despite the concerns expressed by the Board after the Mont-Saint-Hilaire accident and the reinforced Transportation of Dangerous Goods Regulations, inaccurate train journal information had continued to place first responders and the public at risk.

(f) Given the freshness and pattern of the wear marks and the distortion of the centre pin holes, the absence of mishandling of the car by the customer, and the presence of curves and turnouts at the yard at Rivière-du-Loup, it was unlikely that the car had been off-centre before Canadian National had taken the car at the interchange point.

(g) Loss of wheel-to-rail contact occurred more frequently than existing research indicated, suggesting a need for further research into the impact of worn wheel/worn rail profiles on truck hunting onset speeds.

Regarding the danger of derailments due to hunting with wood chip gondola cars and similar, the RSB stated that, 'Transport Canada may wish to assess the performance of the CN wood chip gondolas, and other similar equipment on other railways, when operated empty at high speed.'

Regarding the inaccurate journal, the Railway Safety Inspector ordered that: CN Rail must ensure that all trains operating were accompanied by accurate consists or train journals before trains departed yards and/or terminals under the following terms and conditions:

(a) It must verify that the actual cars on the train were accurately reflected on the train journal as to the number of cars and their sequencing, including the location or presence of dangerous goods, before departing a yard or terminal.

(b) It must verify the makeup of the train as per item (a) whenever cars were added or set off en route.

(c) A written record of each verification including the name and title of the person verifying the same must be maintained for the duration of this order. [614]

On the evening of Sunday 28th March 2004 a Union Pacific Railroad freight train was derailed as it crossed a bridge over Soldier Creek, near Silver Creek, Kansas, United States, at about 6:30 p.m. Thirty-eight cars fell into the creek and two more were derailed but the three diesel units stayed on the line and the engineer and conductor were unhurt. The train had been on its way from Wyoming to St Louis, Missouri. Work began at once to remove about 5,500 tons of coal and metal from the creek and begin construction of a temporary bridge.[615] The latter was complete by early on Thursday 1st April, the first train using it at 3 a.m. that day. The cause of the derailment was not immediately clear. [616]

At about 4 a.m. on Monday 3rd May 2004 a southbound Union Pacific Railroad freight train from Houston, Texas, to Mexico was changing tracks on a bridge over the San Antonio River at San Antonio, Texas, United States, when it was struck by a northbound Union Pacific freight train from Mexico to Houston. As a result of the collision both locomotives and 12 cars of the southbound train were derailed, the engines and five cars ending up in the river. The engineer of the southbound train suffered a broken ankle; its conductor sustained a broken wrist and a homeless person who had been sleeping under the bridge was also hurt. The derailed cars had been carrying beer, furniture and detergent, so no hazardous chemicals were

---

[614] Transportation Safety Board of Canada report
[615] www.ble.org/pr/news/pf_headline.asp?id=9939
[616] www.cjonline.com/stories/040104/bre_derailment.shtml

involved but 5,000 gallons of diesel fuel escaped from the engines into the river. The cause of the accident was being investigated. [617] [618]

Konkan Railway train No. 2620, the Up *Matsyagandha Express* from Bangalore to Mumbai, was involved in a major disaster at 06:10 on Wednesday 16th June 2004 when it ran into a rockfall and was derailed between Veer and Karanjadi stations, Maharashtra, India. Following the derailment the locomotive and part of the train fell from a bridge. Fourteen or fifteen people lost their lives and about 80 were injured (an early report spoke of 20 and 25 dead). Local villagers, alerted by the noise of the derailment, were the first on the scene and initiated the rescue work. The injured were taken to hospitals in Mahad and Raha. [619]

A news report highlighted the stress and worry of relatives waiting for those who had been on the train. 'Varman Shetty, who uses the train often to reach Mangalore, was expecting his wife and daughter. "I do not know anything about their fate. They have a mobile phone with them but I am not getting through," he said.

'Vivek Karmath said: "I was expecting my aunt and uncle and cousins for my marriage. I hope they are OK. I pray to God to bring them safely to Mumbai."

'Sunita Shetty came to Kurla terminus after receiving no information at the Panvel or Thane stations in Mumbai's suburbs. "I am just hoping for the best," she said.' [620]

For some, there was a happy ending. 'A tearful Bharati Sawant was waiting to hear about her family friend P. Vasudevan. A ticket-checker, he was supposed to have boarded the train at Madgaon. He was due to be on duty in the *Shatabdi Express* on Wednesday. "He called me from his mobile phone at 10 p.m. on Tuesday, but the line got disconnected. It is 4 p.m. and he hasn't contacted us yet," she said. Vasudevan's wife was pregnant, and she had not been informed. The tension eased only late in the evening when Vasudevan called up from Chiplun. By sheer luck, he had not boarded the *Matsyagandha Express*. [621]

The train had been travelling at 90 km/h at the time of the derailment. Its driver and assistant driver survived the accident but were badly hurt. From his hospital bed Driver Manohar Damodar Kkarmagar, 56, struggling to speak because of his injuries, said, "I just could not prevent the accident. I only saw the obstruction when my train was 100 metres from it. Had I been able to apply the brakes earlier the accident would have been less serious. As soon as my engine hit the boulder there was a huge thud and the train was derailed. Everything collapsed…" [622]

Attention naturally focused after the disaster on its similarity to the even more serious accident on the Konkan Railway, which had occurred between Rajapur Road and Vaibhavwadi stations in June of the previous year (see Other Derailments chapter). In both cases, the trains had been derailed through running into landslides after days of monsoon rains. Konkan Railways publicity officer Vaishali Patange admitted that various safety measures proposed after the previous derailment had not yet been introduced, though he promised that they would be operational within a few days. These included:

(a) A real-time instrumentation-based monitor that could detect minute slippages within the soil/boulder.

---

[617] *Del Rio News Herald*, Del Rio, Texas, United States, 3rd May 2004

[618] *The Hays Daily News*, Hays, Kansas, United States, 3rd May 2004

[619] http://ushome.redcliff.com/cms/print.jsp?docpath=/news 2004/jun/16train.htm

[620] www.newindpress.com/Print.asp?ID=1EL20040616055607

[621] http://cities.expressindia.com/fullstory.php?newsid=87991

[622] http://web.mid-day.com/news/city/2004/June/85817.htm

(b) An anti-collision device.

(c) Indigenously developed inclinometers (sensors to detect soil variations).

(d) Regular two-hour patrols of the lines.

Meshes to catch rock falls were already in place in cuttings with a height of 12 metres or greater above the tracks, including at the accident spot, but the boulder had fallen from below this height. [623]

The Commission of Railway Safety preliminary report on the accident concluded that it had occurred due to the failure of the east side slope of the Ambivali Cutting resulting in a fall of boulders and soil, which had obstructed the line. It noted that to prevent a recurrence in 2003-4 the Konkan Railway had mandated speed restrictions and patrolling at vulnerable areas and had installed warning devices (Raksha Dhaga), electronic pendulum-inclinometers, electronic rail gauges and seismographic monitors, constructed reinforced cement concrete retaining walls and gabion walls (cages filled with rocks), and had undertaken micropiling, rockbolting and netting of cuttings and tunnels over the previous five years (at a cost of 236,800,000 rupees in 2003-4). Since the second accident further inclinometers were to be installed at a cost of 15,300,000 rupees, safety nets were now to be installed in cuttings 5 metres or higher, pilot engines were to be run in front of mail and express trains between 19:00 and 07:00, speeds would be restricted in vulnerable areas and these would be patrolled during the monsoon season. [624]

Seven of the 13 cars of a Finger Lakes Railroad freight train, travelling at only five to eight miles an hour, were derailed at about 1:45 p.m. on Thursday 8th July 2004 at Auburn, New York, United States, and two boxcars, containing paper rolls from the Solvay Paperboard works, fell from a bridge into the Owasco Outlet stream, large rolls of industrial paper ending up in the water. No one was hurt. [625]

---

[623] http://web.mid-day.com/news/city/2004/June/85812.htm

[624] http://pic.nib.in/release/release.asp?re1id=2543

[625] *The Post-Standard*, Syracuse, New York, United States, 9th July 2004

CHAPTER 11

TUNNEL ACCIDENTS

In his classic work on British railway accidents, *Red for Danger*, L.T.C. Rolt remarks on the acute trepidation felt by Victorian travellers when they were passing through a tunnel, and their oft-repeated pleas to the railway authorities that all tunnels should be artificially illuminated, pleas which, probably for practical reasons if nothing else, fell upon deaf ears. He goes on to point out that these fears were by no means unjustified, especially in the period before the introduction of the absolute block system, when trains were merely spaced by time intervals. If a train should chance to break down in the open there was a good chance that the crew of a following train might be able to see it in time to pull up, even if there was insufficient time for the guard of the first train to go back and protect its rear. In the smoke- and steam-filled darkness of a tunnel there was no such hope. Furthermore, should a serious collision occur in a tunnel the confined space would likely result in the wreckage being hideously compressed, and should a fire break out in a long tunnel provided with ventilating shafts, the latter could act as chimneys. In addition, the rescue work would be hampered by the darkness and the difficulty of working in a confined space where the use of breakdown cranes was impossible, and the only resort was to drag the wreckage bodily out of the tunnel. Operation of the railways a century and a half later was of course much different from that in Victorian times, but the dangers of tunnels remained, as shown by continuing bad accidents and major disasters within them.

A passenger train from Segovia to Toledo was stopped at a Red signal in a tunnel near Chamartin station, Madrid, Spain, at 09:10 on Monday 28th January 1991 when it was rammed from behind by another passenger train from El Escorial. Motorman Rogelio Almería Gil of the colliding train was found to be dead on admission to Ramón y Cajal Hospital and 17 people were injured to various degrees. An inquiry determined that Motorman Gil had exceeded the permitted speed at five points during the journey. [626] [627]

Fire broke out in a Zurich RER (Réseau Express Régional) passenger train at about 20:40 on the evening of Tuesday 16th April 1991 in Hirschengraben Tunnel, Switzerland, not far from Zurich Stadelhofen station. The fire had probably originated in the penultimate coach of the train. It was discovered after the train had left Zurich Museumstrasse station for Stadelhofen but the train radio had not been working so its conductor could not notify his driver. For reasons not immediately apparent, the train had then stopped in the tunnel, whereupon the crew were able to advise Zurich station of the emergency, using a tunnel

[626] https://elpais.com/diario/1991/08/09/madrid/681737056_850215.html
[627] *The Daily Inter Lake*, Kalispell, Montana, United States, 28th January 1991

telephone. The first firefighters were on the scene within six minutes and in total 130 firemen, 35 rescue workers, 13 doctors and Rega (Swiss air rescue) staff attended the accident. Fifty-eight passengers were taken to hospital and admitted and another 140 passengers were able to evacuate the tunnel by their own efforts. Four of the injured were still in the hospital intensive care unit the following day, suffering from smoke inhalation. [628]

Two days after the accident it emerged that the train had stopped in the tunnel because a woman passenger had operated the emergency alarm. Spokesman Urs Neuenschwander of the Chemins de fer fédéraux suisses (CFF) emphasised that she could not be found fault with for having done so, but that the CFF was considering advising passengers that the alarm should not be used in a tunnel if possible. Meanwhile, the police were investigating evidence that suggested that an act of arson had caused the fire. [629]

A piece of luggage caught fire aboard a southbound Wuchang – Canton passenger train as it entered the Dayaoshan Tunnel in Guangdong province, China, at about 00:15 on Sunday 18th August 1991. The conductor quickly extinguished the fire but not before smoke had filled the carriage, causing the passengers to panic. Someone pulled the emergency alarm, bringing the train to a halt before it had left the tunnel. Dozens of the passengers then jumped from the train and a number of them, 10, 12 or 15 according to different reports, were struck and killed by a freight train passing on the other running line, while others were injured. [630] [631]

The Lexington Avenue line of the New York City Transit Authority suffered a serious accident at 12:28 a.m. on Wednesday 28th August 1991. Because of engineering work, a 10-car express train from Woodlawn, in charge of Conductor David Beerram, which had left there at 11:38 p.m., six minutes late, was being diverted at 14th Street-Union Square station, New York, United States, from the Fast line to the Local line through a crossover with a 10 mph speed restriction. Instead of its speed being reduced to that figure at Union Square, the train was still moving at an estimated 50 mph when it reached the crossover and a major derailment occurred, the leading car of the train striking a steel column supporting the tunnel roof, having its roof sheared off and being cut in half. The second car was folded in two, the third car was split in half and folded around a support beam and the rear fourth and fifth cars of the train were badly damaged. Despite the late hour, there were about 215 passengers on the train, five of whom were killed and 161 were seriously hurt.

After the accident passenger Albert Webb, 30, related that the train, "was going so fast, the next thing you knew it was 'boom.'" He added that he had seen a baby thrown from its mother's arms.

Joseth Ricketts, another passenger, told a reporter, "It was like an explosion. Some people from another car got thrown into ours. There was a lot of smoke. The car was just shredded. There were wires hanging down. People were screaming." [632]

Passenger Jerry Yambl, 22, stated that at a previous stop the motorman had, "let two cars from the front pass the platform and opened the doors. But there was nothing, no platform," while passenger Olalekan Ogundele said simply, "This guy never slowed down." [633]

Motorman Robert E. Ray, 38, who had been operating the train, went missing for some

---

[628] *Le Journal du Jura*, Biel Switzerland, 18th April 1991

[629] *Le Nouvelliste*, Sion, Switzerland, 20th April 1991

[630] *European Stars and Stripes*, Darmstadt, Germany, 19th August 1991

[631] *Hutchinson News*, Hutchinson, Kansas, United States, 19th August 1991

[632] *Altoona Mirror*, Altoona, Pennsylvania, United States, 28th August 1991

[633] *The Intelligencer Record*, Doylestown, Pennsylvania, United States, 30th August 1991

hours after the accident but was apprehended by police heading for his home in the Bronx district at 6:30 a.m. At 1:30 p.m. that afternoon he was tested for alcohol and was found to have a level of .21. A level of .10 was considered legally drunk in New York at the time. He told police that he had drunk three beers after the accident but it was pointed out that this amount of alcohol would not have been sufficient to give such a result. An empty vial of crack (cocaine) was also found in the motorman's cab, though it was not clear if Ray had consumed the drug. The motorman was immediately arrested and charged with five counts of manslaughter. [634]

It subsequently emerged that Motorman Ray had arrived 15 minutes late to sign on at Woodlawn, looking dishevelled, with bloodshot eyes and wearing tennis shoes, despite which Dispatcher Percival Hossack had permitted him to commence his shift and he had taken out the accident train, the departure of which had been delayed because of the motorman's late signing on. Mosholu Parkway, the first stop, the motorman had overrun by five car lengths, and Bedford Park Boulevard-Lehman College, the next stop, he had overrun by one car length. At each of these stops Conductor Beerram had admonished the motorman but had done no more. As Ray had continued to drive erratically, he had been repeatedly warned about his operation of the train both by Conductor Beerram and by Conductor Steve Darden, who had been riding 'on the cushions' at the end of the shift. According to the regulations Conductor Beerram should have stopped the train and reported his motorman's condition to Control, but clearly he, and, come to that, Dispatcher Hossack, had been in a difficult position, for had they reported Motorman Ray the latter would have been in deep trouble and probably would have been dismissed, while one does not like to squeal on one's mates. Obviously, had the dispatcher or conductor realised that the journey would end in disaster, they would have acted differently.

Directly after the accident the transit authority announced that it would introduce random, surprise testing for drugs and alcohol of subway and bus drivers. Mayor David Dinkins felt bound to state, "Please let us not suggest that crackheads are driving the trains in New York."

Motorman Ray later admitted that he had drunk heavily on the day before the accident because he had been depressed that his ex-girlfriend would not allow him to see their two children and that he had fallen asleep just before the accident. When he had come to after the accident (in which he had escaped injury) he remembered that there had been smoke all over the place and had thought to himself, 'What the hell has happened?' and then, 'Oh God, I must have had an accident.'

In late 1992 Motorman Ray stood trial and was found guilty of five counts of second-degree manslaughter and 26 counts of assault. He was sentenced to five to 15 years for manslaughter and one to seven years for assault, the sentences to run concurrently. He was released in April 2002. [635]

On Tuesday 26th November 1991 at least one person died and 60 were injured when a train caught fire at an unstated location on the Metroul București underground system in Bucharest, Romania. [636]

At about 10:30 on Saturday 7th December 1991 British Railways, Western Region Sprinter passenger train 1F08, the 07:00 service from Portsmouth Harbour to Cardiff, ran into the rear of High Speed Train (HST) 1B10, the 08:30 Inter City from London Paddington to Cardiff.

---

[634] Ibid.

[635] Wikipedia – 1991 Union Square derailment

[636] *International Herald Tribune*, Paris, France, 27th November 1991

The collision occurred towards the Newport (Welsh) end of the 4-mile, 628-yard-long Severn Tunnel. No one was killed but the driver of the Sprinter and 185 of its passengers were injured, the driver and five passengers seriously.

This was probably the first collision in the history of the tunnel, which had opened in 1885. Up until 1969 the line through it had been signalled between East and West signal boxes with semaphore signals and special block instruments. In that year, the boxes were abolished and track circuit block working was introduced, with full track circuiting through the tunnel and colourlight signals. However, it was found that the wet conditions within the tunnel made the use of track circuits impracticable and there were numerous right-side failures when the circuits became wrongly energised, putting the protecting signals back to danger, though no train was in the section. Thus, in 1987 the track circuits were replaced by axle counters, which had electronic detectors that counted the number of axles of a train entering the tunnel and held the protecting signal at Danger behind it until the same number of axles had been registered by another detector at the far end of the tunnel. The last protecting signal for Down trains was N 164 at Ableton Lane, 1,584 yards east of the tunnel; the next signal was N 168R, about five miles in advance near the Newport end of the tunnel. The signalling was controlled from Newport panel, with the equipment for the axle counters being housed in a relay room at Severn Tunnel Junction, near the Newport end of the tunnel.

The accident was investigated by the Health and Safety Executive.

Operations Manager Andrew Hancock told the inquiry into the accident that the events leading up to it had begun with the failure of the electronic remote-control equipment (time division multiplex) on the afternoon of Thursday 5th December. Operation of the line through the tunnel had then been switched to 'through routes,' whereby the signals operated automatically, clearing after the previous train had cleared the signal ahead. Whilst 'through routes' operation was in force, the Newport signalman had no indication on his panel of the position of trains or of the signal aspects. This had continued until about 14:00 on Friday 6th December, when the driver of a Down train had telephoned the Newport signalman to state that signal N 164 was at Red, when the signalman knew it should not have been, as the previous train had already cleared the section. The signalman had then arranged for a handsignalman to be stationed at signal N 164 to 'flag' trains past that signal upon their being advised by the panel that the section was clear, together with a tail lamp man at signal N 168R to advise the panel signalman that trains had passed that point complete. This arrangement had continued until 09:30 on the morning of the accident and had prevented delay from the driver of each Down train stopped at signal N 164 having to wait two minutes and then leave his cab to telephone the signalman.

Driver David Robertson of the HST testified that signal N 164 had been at Danger. There had been no handsignalman there and after stopping at it and waiting one minute he had alighted and telephoned the panel signalman, who had told him that there was a problem and instructed him to pass the signal and obey all others. He had then sounded his horn and proceeded at caution, prepared to stop short of any obstruction. His train had thus continued through the tunnel, travelling at about 20 mph, until he had suddenly felt an impact and the brakes of his train had come on. Knowing that he had not run into an obstruction, he had assumed his train must have been derailed, until going back, he found that it had been run into from behind.

Driver Stephen Carpenter of the Sprinter declined to give evidence at the inquiry on the advice of his lawyer. At the internal railway inquiry he stated firmly that signal N 164 had gone to Green as his train had approached it, whereupon he had proceeded through the tunnel almost at line speed, until he had seen the tail lamps of the HST ahead.

Guard Brian Nicholl of the Sprinter stated that his train had slowed to a walking pace approaching signal N 164, then accelerated, from which he had assumed that the signal, which he could not see from his position in the train, had cleared. The impact of the collision had thrown him down. On getting up and receiving a response from his driver that he was all right, he had alighted from his train, broken the tell-tale emergency wire in the tunnel and subsequently advised the signalman from a tunnel telephone that there had been a serious crash and the emergency services were required. He said that his train had slowed but not stopped at signal N 164, then had travelled through the tunnel at about 50 mph.

Signalman Kyril Morgan in Newport panel related that when he had begun duty at 06:00, his mate had told him that there was 'through routes' working through the tunnel and that the axle counters were disconnected. There had been a change of handsignalmen and tail lamp men at 06:00 and he had arranged for Signalman Terry Wood from Newport panel to take over as tail lamp man and Trainee Signalman Mark Ellott from Bristol to take over as handsignalman, then had changed the two men's positions to save Signalman Wood from having to travel over the Severn Bridge to the Bristol end of the tunnel. He had assumed that Signalman Ellott would have the flags and detonators used by the man he had relieved, but at about 09:07, Driver Forder of train IF06 had telephoned him to say that the handsignalman was not properly equipped. Signalman Morgan said that he had advised Bristol panel of this, assuming that they would send some flags down, but instead they had instructed him to withdraw the handsignalman at 09:30. After this had been done, trains 1B08 from Paddington to Swansea and 1B10 had been cautioned in turn at signal N 164 and had passed into the tunnel, then the Sprinter, IF08, had been described by Bristol panel. At about 10:35, he had received visible and audible indications that the tell-tale emergency wire in the tunnel had been broken but had taken no action until he received a telephone call from the driver of train 1B10, closely followed by one from the guard of train IF08, informing him that there had been an accident. He did not recall any telephone messages from the Signal and Telecommunications (S&T) staff at the Severn Tunnel Junction relay room regarding the resetting of the axle counters.

Signal and Telecommunications Engineer Michael Schofield stated that the remote-control system had failed at 09:15 on Thursday 5th December. This had been corrected at 10:15 but had failed again at 13:42. It could not be corrected that day and due to a lack of staff, the job had been postponed until Sunday 8th December. Then a driver had reported signal N 164 to be wrongly at Red at 14:02 on Friday 6th December. This had been assumed a failure of the Down line axle counter. It had not been possible to reset the counter and it had been assumed that the fault must lie with the trackside count-out heads. It had then been decided to tackle this job on the Saturday. This second failure had been unrelated to the first failure of the time division multiplex, which had not necessitated the resetting of the axle counter.

Engineering Technician Andrew Sperring, Engineering Assistant Jonathan Napper, Signal Section Manager/Supervisor Trevor Cantle, Signal and Maintenance Supervisor Vivien Daniel, Principal Technician Officer David Richard Croke, Principal Technician Officer (PTO) Stephen Andrews, Engineering Technician Raymond James Waters, Maintenance Support Engineer Peter Day and others gave evidence as to what was done in the Severn Tunnel Junction relay room on the morning of the accident. Their evidence is far too voluminous to be reproduced here, in addition to which much of it is highly technical in nature.

Before stating his conclusions regarding the accident the inspecting officer, Mr. R.J. Seymour, listed a number of errors he considered related to the collision and the emergency response to it:

(a) The failure of the Down line axle counter had apparently been initiated by Technician Sperring mistakenly withdrawing and replacing disconnection links in the Severn Tunnel Junction relay room. The latter believed that he had only touched a telephone circuit but the links were not clearly identified and he was not a telephone engineer.

(b) The S&T technicians, lacking a properly laid-down procedure for identifying faults, had failed to locate the problem at an early stage. Even the two PTOs, who were senior technicians regarded as experts in axle counters, had not discovered the fault until their unproductive journey to check the detector heads.

(c) There had been a casual attitude towards the resetting of the axle counters as evidenced by the fact that neither the signalman nor the technicians had taken the trouble to keep a record – as they should have done – of the resetting of the Up axle counter on the morning of the accident.

(d) The signalman had made a well-intentioned but improper decision to change the initial postings of the handsignalman and tail lamp man; although the use of a handsignalman had been a convenience rather than a necessity, the decision once taken should have been observed to the letter of the Rule.

(e) One of the fire extinguishers in the cab of the Sprinter, which had been unsecured, had almost certainly been responsible for Driver Carpenter suffering a fractured skull.

(f) The activation of the tunnel tell-tale alarm had been treated as an indication of a routine failure rather than as a serious emergency.

(g) The failure to see that the pagers of the rescue team had been in working order had led to an unnecessary delay in assembling a team, while a manager and a telephone were engaged in calling the team out individually.

(h) There had been a failure to ensure that the tunnel emergency phones were switched to Sudbrook as soon as the rescue operation began.

(i) There had been a delay in moving the trains in and out of the tunnel during the rescue operation.

(j) The start of Mr Day's site investigation had been unnecessarily delayed by the unsatisfactory state of the wiring diagram kept in the Severn Tunnel Junction relay room.

The inquiry reached three conclusions:

(a) The accident had resulted from an unaccountable error on the part of either the Sprinter driver or of the S&T technicians in the Severn Tunnel Junction relay room or possibly, though it seemed most unlikely, by an intermittent wrong-side signal failure, which had left no trace and had not recurred.

(b) The observance of the rules laid down for the operation and maintenance of the tunnel signalling, and in certain respects the rules themselves, had fallen short of an acceptable safety standard.

(c) Although the plan for dealing with major emergencies in the tunnel had been drawn up with care and was broadly satisfactory, the state of preparedness of the emergency provisions made by the operator had failed to meet the expectations of the professional emergency services and had caused serious delay to the rescue operation.

The inspecting officer made 11 recommendations:

(a) The operator should review the training of all concerned with maintenance and faulting work on the Severn Tunnel axle counters; there should be clear, comprehensive instructions and effective monitoring to ensure that the instructions were at all times observed.

(b) Provision should be made, whenever axle counters were used, for their resetting to be placed under the overriding control of the signalman, using a reset button in the signal box.

(c) Improved means should be provided for positively isolating axle counters from the signalling system during maintenance and fault-finding work.

(d) The British Railways Track Circuit Block Regulations should be revised and extended where appropriate to take account of the differences between track-circuited and axle-counter sections, especially under fault conditions.

(e) The operator should ensure that all disconnection links on vital signalling circuits were clearly labelled to identify their purpose.

(f) The operator should review the railway emergency procedures for the tunnel area, including the Newport panel and the shorter tunnels on the English approach, to ensure that both the directives and the means of achieving them were clearly laid down and that the state of preparedness was regularly monitored and recorded.

(g) The operator should proceed as quickly as possible to implement those improvements in the emergency equipment and procedures which had been agreed with the emergency services. Other procedural recommendations should also be implemented as soon as possible in consultation with the emergency services concerned. Recommendations for additional emergency equipment or facilities should be implemented to the extent that they were justified by the quantitative risk assessment, when finalised.

(h) The quantitative risk assessment should be an integral part of the safety case for the tunnel under the Railways Act 1983.

(i) The operator should review the arrangements for 'through routes' operation in the tunnel in order to find a means of providing the signalman with panel indications of the presence of trains in the tunnel during 'through routes' working, irrespective of the nature of the failure for which such working was initiated.

(j) The operator should ensure that all fire extinguishers carried on rolling stock were firmly secured in place and were located in such a position that their accidental dislodgement would not endanger the train crew or passengers.

(k) Although it was by no means certain that the accident would have been prevented by a system of automatic train protection (ATP) the operator should ensure that its plans for the installation of ATP, which automatically brought a train to rest in the event of a driver passing a signal at Danger, extended to the Severn Tunnel at the earliest possible date.

(The various references to 'the operator' stem from the fact that British Railways was in the process of being privatised at the time the report was made.)

Although the investigation into the accident was unable to determine its cause with certainty, to the author the likelihood of an experienced driver passing a colourlight signal at Danger in daylight, at the very moment that S&T technicians were interfering with the axle counters, is too much of a coincidence and makes it highly likely that an accidental resetting of the axle counters had resulted in a wrong-side signal failure, leading to the collision. [637]

On Saturday 12th June 1993 a 24-waggon freight train carrying petrol caught fire and exploded in a tunnel near Yanan, Shanxi province, China. Eight people died and the whole train was destroyed. [638] [639]

---

[637] Health and Safety Executive report

[638] *The Daily Telegraph*, London, England, 17th June 1993

[639] *New Castle News*, New Castle, Pennsylvania, United States, 16th June 1993

A serious collision occurred on the morning of Saturday 31st July 1993 on the international line linking Italy and Switzerland via the Simplon Tunnel. It happened not in the Simplon Tunnel itself but in the 2,266-metre-long Rio Rido-Preglia Tunnel, one of nine bores on the Italian side, between Domodossola station and the Simplon Tunnel proper. Two international passenger trains were involved. The *Italia Express* from Dortmund Hbf, Germany, to Roma, Italy, was stopped in the tunnel at a Red signal when it was run into from behind by the following *Galilei* express from Paris Lyon to Firenze (Florence), Italy, at 05:20. A third train from France was stopped in time to avoid a second collision and was subsequently shunted into a siding at Varzo station.

One passenger, Nicole Mule, 48, an Italian from Sicily who had been returning from working in Germany, lost his life in the accident and 60 others, mostly from the front train, were injured. The majority of the victims were Italian migrant workers returning home for the holidays but a number of Germans and some French, Dutch and American nationals were also hurt. The engine crew of the colliding train were unhurt.

The Chemins de fer fédéraux suisses stated that the *Galilei* express had stopped briefly due to a power outage caused by a thunderstorm, then it had restarted to collide at relatively low speed with the train in front. It was theorised that the storm might have caused the signals to malfunction but the two Swiss train crew were being interrogated by an investigating judge in Verbania. [640]

A major disaster occurred in Spain at 18:40 on Monday 2nd August 1993 in a tunnel on a single-track, narrow-gauge line of the Ferrocarriles Españoles de Via Estrecha (FEVE). Two trains were involved, a freight train from Oviedo to San Esteban de Pravia, carrying 540 tonnes of sodium carbonate and other goods, and a passenger train from Soto de Luiña to Oviedo. The freight train should have waited at Santa Maria de Grado station for the passenger train, instead of which it left there before the passenger train had arrived. The latter train left the station at Vega de Anzo on time and collided with the freight train in a tunnel after travelling 440 metres. A fire, resulting from the ignition of diesel fuel from the engines of one or both of the trains, broke out following the collision, leading to what was described as a 'living hell' in the tunnel. All those in the leading coach of the passenger train perished, as presumably did both drivers. In all, 12 people died and seven more were injured. At least one passenger, María Luisa Larraldi, had a lucky escape, as shortly before the collision she had felt dizzy and moved from the front to the rear of the train.

An account of the accident written 20 years later told the stories of some of the rescuers. Former police officer Paco González lived only 80 metres from Vega de Anzo station and was at home at the time. "We heard a loud bang," he recalled, and looking towards the tunnel from the veranda of his house he could see flames coming from its portal. With other residents he had hurried to the scene. "You don't see the danger, you just want to help," he said. "We got a woman and her daughter out, but when we tried to go back, it was impossible, everything was burning. It was so bad."

Lorry driver Ángel Fernandez, aged 33 at the time, had been on his way back to Léon after unloading coal at the Soto de la Barca power station. "I heard a huge explosion and the first thing I thought was that a tyre had burst," he said. He parked his lorry and met some locals who were sure that a train had crashed. With a young car driver he hurried down the line towards the tunnel. "We immediately met a driver {the conductor of the passenger train?} who told us that the train was derailed, that there were people in the train and that we must help them. I didn't think of the danger. We pulled out three or four people. A young woman

---

640 *L'Impartial*, La Chaux-de-Fonds, Switzerland, 2nd August 1993

aged 16 or 18 had part of her hair burned and had lost the flesh from one of her shoulders, so that the bones were visible. Her eyes pierced me. I've spent many years thinking about that girl and another person I saw in the burning car. They've been on my mind for the past 20 years," he related. The silence of the victims was also fixed in his memory. "No one spoke. The people we were pulling out didn't complain, they said nothing. They were in shock." After the fire brigade and other rescuers had arrived, an ambulance driver had accompanied him to the station platform, where he had sat down. "And when I went to get up again I couldn't walk. I had hurt my leg." He had refused to go to hospital as he had to continue to Léon with his lorry. Twenty years on he was anxious to meet the young car driver and the girl with the piercing eyes. "I think she survived," he added.

It was thought that the driver of the freight train had misunderstood a signal from the Santa Maria de Grado stationmaster for him to draw his train forward a little, perhaps to clear a set of points, as authority for him to proceed to Vega de Anzo. Once the freight train had started there were no train radios to communicate with its driver, nor was there any other means of stopping either train. [641] [642]

On Monday 28th November 1994 the brake van of a Central Railway Baroda – Kolhapur open-path goods train (one running in an unoccupied slot in the timetable to reduce congestion) became derailed at Km 110/0 of the Ghat section between Karjat and Khandala, Maharashtra, India. It was re-railed, only to derail again at Km 120/8-9, within Tunnel No. 25-C (otherwise Monkey Hill Tunnel, 2 kilometres, 304 metres long and at the time of the accident, the longest bore in India. With the construction of many new tunnels in recent years, by 2025 it had been relegated to the position of only being the 30th longest). The brake van was again re-railed and the train moved on but was stopped at the Khandala Home signal close to the far portal of the tunnel, which was Red. At this point the leading engine of the train was outside of the tunnel but the train was still within it. The signal was then cleared and the driver attempted to move off, only for the brake van to become derailed for a third time. The driver of the Kalyan breakdown train, which was following the goods, mistook the cleared Home signal as applying to his train, increased speed and collided with the rearmost of three banking engines at the back of the goods train. Two railwaymen lost their lives in the accident, 34 were injured, five of them seriously, and damage was estimated at 4,717,000 rupees.

The Commission of Railway Safety noted that the goods train had not been properly protected in the rear when its van had become derailed, while the breakdown train had not observed the rules of operation on a blocked section of line with restricted visibility.

The Commission made several recommendations:

(1) Breakdown trains in the Bombay Division called to work in ghat sections should be equipped with better cold cutting equipment, smoke/fume extractors, breathing apparatus, fire-retardant dresses for staff, water tankers (with pumps), trollies for mobility and hydraulic platforms (to work inside the tunnel). Breakdown staff should be given special training in firefighting techniques.

(2) Better ventilation arrangements in Tunnel No. 25 C such as jets/exhauster fans should be planned and executed by the Central Railway.

(3) The present lighting arrangements inside Tunnel No. 25 C needed reviewing, with the issue of suitable instructions for improved lighting.

(4) No brake vans should be permitted on goods trains in the Down direction over the ghat

---

[641] https://www.elcomercio.es/v/20130801/asturias/peligro-solo-quieres-ayudar-20130801.html

[642] Federación Castellano Manchega de Amigos del Ferrocarril

section between Karjat and Lonavla. However, arrangements should be made at Lonavla for the proper attachment of brake vans on Down goods train at the end of the ghat journey.

(5) The present composition of 40 Bogie Covered waggons (BCNs) on the North East and South East ghats in the Bombay Division needed review. This should be restricted to 30 BCNs (in normal cases) with exceptional loads up to 35 BCNs.

(6) Reliable inter-communication apparatus between the leading and banking engines of goods trains on the ghat sections should be provided for better co-ordination. Competent Diesel Loco Inspectors should be inducted into the Bombay Division to improve the calibre and working knowledge of diesel crews. [643]

An accident occurred between St Clair West and Dupont stations, Toronto, Ontario, Canada, at 18:02 on Friday 11th August 1995. A Toronto Transport Commission (TTC) subway (underground) train known as Run 34 was stopped at a Red signal, which had failed, in the tunnel short of Dupont station. While the train was still stopped at the signal it was run into from behind by a second train, Run 35. Three passengers, all women, died in the accident, 30 people, including Motorman Robert Jeffrey of the colliding train, were taken to hospital and about 100 were slightly hurt. [644] [645]

The line was controlled by colourlight signals. Some were automatic and showed the usual Red, Yellow and Green aspects, while there were also two-aspect Home or interlocking signals, the upper aspects of which displayed the same aspects as the automatic signals, the lower aspect displaying Red if the upper aspect was Red, otherwise it displayed Yellow if the points were set for a diverging route and Green if the road was set straight ahead. To complicate matters, the signals were also used to enforce train speeds, using a so-called grade timing system. There was an additional Lunar White aspect below the main aspects, which when illuminated, indicated that the next signal was only Red because of a speed restriction – the grade timing – and would clear as the train approached it. For the St Clair West – Dupont section, with a series of automatic signals and grade timing, a driver might expect to receive, when the section ahead was clear, a Yellow over White aspect, followed by a series of Red over White aspects, each clearing to Yellow over White upon the train's approach, with the final Red signal clearing to Green.

When Run 34 left St Clair West, Home signal No. SP77/X38 at the Dupont end of the station had been displaying a Yellow over Green aspect without a Lunar White because the next signal, No. SP71, was held at Red behind Run 34. After the collision Motorman Jeffrey claimed that signal No. SP77/X38 had been showing Yellow over White (an impossible aspect for that signal) and that he could not recall any signal aspects beyond that. He had then run past signals Nos. SP71 and SP65 without reducing speed and into the rear of Run 34, stopped at signal No. SP53 short of Dupont station, which was still unnecessarily at Red.

Train stops were also provided, which should have applied the brakes of the train if a signal was passed at Danger, these stops consisting of a T-shaped arm adjoining the right-hand rail designed to strike a trip cock on the leading car of a passing train to apply the brakes. The Ericsson train stop used was designed to have the trip arm placed in the centre of the track and as actually installed, failed to work properly due to wear and tear on the rails and train wheels, when the train was rounding a curve at speed. Hence, the train stops had failed in their

---

[643] Government of India Ministry of Civil Aviation and Tourism Commission of Railway Safety Annual Report for 1994-95

[644] *The Winnipeg Free Press*, Winnipeg, Manitoba, Canada, 13th August 1995

[645] *The Lethbridge Herald*, Lethbridge, Alberta, Canada, 14th and 17t August 1995

purpose on the day of the accident. [646]

Another factor in the accident was that Motorman Jeffrey, who had previously been a TTC bus driver, had only been operating the train for his second day, after 12 days of training. He was therefore very inexperienced. On his first day as a motorman he had also overrun a Red signal, this time triggering the train stop. [647]

Presumably because it occurred on a subway line, the accident does not appear to have been investigated by the Transportation Safety Board of Canada but it was the subject of a coroner's inquest, whose jury made 18 recommendations:

(a) The Railways Act of 1950 should be reformed to provide oversight over the TTC.

(b) The TTC should submit to an independent safety audit every two years, beyond that done by the American Public Transportation Association, the new agency to be set up under the revised Railway Act.

(c) A Due Diligence Checklist of deficiencies identified by the TTC should be completed.

(d) The providence and Metro should commit to a "State of Good Repair" funding policy with repairs taking priority over new works; future capital funding should be based on the State of Good Repair. (The jury considered that underfunding since the mid-1980s had contributed to the deterioration of the subway system and had jeopardised the safety of the TTC.)

(e) Improvements should be made to the Operations Training Centre, beginning with the hiring of an accredited adult training specialist. The updated training package should at least include realistic pass/fail grading, annual refresher courses for all operators (motormen), with emphasis on the meaning of the signalling and route supervisory accompaniment for at least one day of complete runs after completion of training.

(f) The updated Operations Training Centre should include a suitable subway simulator.

(g) There should be a comprehensive review of the signalling system with an emphasis on the following areas: Lunar White aspects, the removal of signal identification markers, the consistent placement of signal aspects, progressive speed controls, the expansion of Intermediate Point Headway Control to identify headway and train separation, the activation of all trip valves and the relocation of the reset.

(h) The auto-key facility should be eliminated and the raising of the trip arm immediately after the train had passed should be implemented.

(i) There should be advanced implementation of a new subway communications system.

(j) There should be improved communication within the TTC.

(k) A new Transit Control Centre should be brought into use, which included the updating of the skills of the TTC staff.

(l) The current Transit Control Centre should be responsible only for train operations with an adjoining facility dealing with intrusion alarms, facility maintenance, public information, media relations and related matters.

(m) A five-yearly emergency response exercise should include "everybody." There should be annual reviews by the Safety Department.

(n) Predictive and preventative maintenance should be improved with computer assistance where applicable.

---

[646] Wikipedia – 1995 Russell Hill subway accident

[647] *The Winnipeg Free Press*, Winnipeg, Manitoba, Canada, 4th January 1996

(o) There should be a review of equipment procurement with respect to quality control. The procurement of cheap equipment such as the Ericsson train stops was both a misuse of resources and a serious safety issue.

(p) There should be traceable design criteria and standards for track, signals and subway cars. There should be no modifications to anything without approval of the design review authority.

(q) Train motormen and the TTC must identify signal malfunctions by signal identification numbers. There should be a review of the discipline system to allow for signal malfunctions. Motormen reporting for duty should meet an inspector so that "State of the road" information could be passed on. Rookie motormen should not be scheduled together.

(r) The Chief Coroner's office should convene an annual press conference to update all parties on the implementation of the above recommendations.

Subsequent to the accident the TTC repaired the defective train stops and motormen were required to see signals change from Red to Yellow before passing them in grade timing sections. The signals were also modified so that those displaying Red aspects only because of grade timing now showed flashing Red aspects. The various recommendations of the coroner's jury were adopted and other improvements were made. [648]

A catastrophic fire occurred on a passenger train serving the underground (subway) system of Baku, the capital of Azerbaijan, at about 18:00 on Saturday 28th October 1995. The accident occurred between Ulduz and Nariman Narimanov stations, very shortly after the train had left the former. Passenger Tabil Guseinov, 45, said, "As soon as the train entered the tunnel I saw a flash. Then the flames enveloped the train car, there was a sound of breaking glass, and the lights went out. People started breaking windows to get out. We were starting to suffocate."

Passengers in the fifth (rear) car were apparently the first to smell smoke, then passengers in the fourth car observed white smoke, which shortly turned black and became irritating. After the train had only travelled 200 metres, it stopped in the tunnel. Accounts of the accident do not make it entirely clear why the train had stopped. Possibly it was because a passenger had operated the emergency alarm. After the disaster, a spokesman stated that it would have been better if the motorman had continued slowly to the next station. [649]

Passenger Gennady Nikkiferev related, "The train went on for a bit and then stopped. The driver came back but couldn't open the doors. Then the lights went out. We started to suffocate from the smoke and we realised we had to open the doors. The car was packed and when we opened the doors, half the people just fell out on top of one another."

Manish Gurbanov, 53, a passenger in the second car, explained that he had managed to escape by climbing through a ventilation duct. "I got through the tunnel by grabbing a cable on top of the tunnel but they say a lot of other people were electrocuted. People were dying all over the rails." [650]

When the train stopped the motorman had alighted from it and used a telephone in the tunnel to advise Control of the problem, amongst other things asking them to cut off the electricity. This would have had the effect of preventing the emergency doors of the train from opening but it did save more passengers, for example, Mr Gurbanov above, from being

---

[648] Wikipedia – 1995 Russell Hill subway accident

[649] *The Intelligencer Record*, Doylestown, Pennsylvania, United States, 30th October 1995

[650] *The Winnipeg Free Press*, Winnipeg, Manitoba, Canada, 30th October 1995

electrocuted. The driver had then assisted in the evacuation by opening (unlocking?) the internal doors towards the ends of the train. The external doors could be manually opened but the procedure to do so was difficult and instead passengers smashed windows, which had the unfortunate effect of letting more smoke into the cars. It was only 200 metres back to Ulduz station, as opposed to 2,000 metres forward to Nariman Narimanov station, but as the fire at the rear of the train grew in intensity, the fleeing passengers had no option but to go forward. Another factor making their escape rather more difficult was the restricted dimensions of the tunnel bore, which was only 5.6 metres high and 5 metres wide. Additionally, 15 minutes after the fire had begun, the ventilation system of the tunnel was switched to exhaust mode, drawing the smoke in the direction of the fleeing passengers. Further problems in the evacuation were the reduced visibility and a trench between the rails. The tunnel was illuminated, but the light bulbs in the tunnel roof became covered by the smoke until they gave out little light, while the central trench forced passengers to walk close to the tunnel walls, holding on to cables and each other's clothes. In addition, no medical treatment was available for the passengers who arrived at Nariman Narimanov as all the firefighters and doctors had gone to Ulduz. [651]

A newspaper account described the aftermath of the disaster: 'Survivors said Monday that it took rescuers and firefighters more than two hours to reach them after the blaze erupted. "There was a flash of light," 26-year-old Ilham Gamidov recalled from his hospital bed. People smashed the windows of the carriage, trying to escape." Gamidov said he helped a woman and one of her two children clamber out of the carriage. The woman's stockings had caught on fire, badly burning her legs. "Her other child had to remain behind," he said quietly.

'In another bed in the Fifth Clinical Hospital, a skinny 12-year-old boy, Aziz Maradov, leaned against his pillows, his chest wheezing from the effects of the toxic smoke. "I was so scared," he said. Maradov was reunited with his 17-year-old brother Mardan, in the hospital and had been told that his mother was in another hospital. But out of earshot a nurse confided the truth. "She is dead. He'll know soon enough," the nurse said.

'Sweating with fever and hoarse from the poisonous fumes, a 33-year-old teacher, Kirami Khankishiyeva, told her story staccato-fashion, then fell silent, her brown eyes wide and staring. "There were screams and everyone was trying to jump out of the train. I was alone," she said, recalling how she stumbled more than a mile down the dark tunnel, choking from the fumes, to the next station.

'As survivors battled their injuries, citizens of the Azerbaijani capital mourned the dead.' [652]

Officially the number of fatalities in the fire was 289, including 28 children. However, mortuary staff spoke of handling 303 bodies, while the Taran news agency quoted medical staff as saying that there were no less than 337 dead. On Monday 30th, 62 victims remained in hospital, many of them being in a serious condition. Another 207 were less badly hurt.[653] The bodies of the majority of the dead were found inside the train, including the children, some of them having been crushed or trampled to death in the panic to escape. Forty bodies were found outside the train. The majority of those who had died had succumbed to the lethal fumes generated by the fire, especially carbon monoxide and possibly hydrogen cyanide, from the burning synthetic material of the coaches.

The most likely cause of the fire was stated to have been an electrical fault in a traction motor of the fourth car, which had caused arcing and possibly itself had stopped the train. There was some suspicion, though, that the fire had resulted from an act of sabotage, and two

---

651 https://railsystem.net/baku-metro-fire-1995/

652 *European Stars and Stripes*, Darmstadt, West Germany, 1st November 1995

653 *The Winnipeg Free Press*, Winnipeg, Manitoba, Canada, 30th October 1995

large holes were reportedly found in the side of one of the cars. Nothing was substantiated regarding the sabotage claim, however.

One of the rescuers, a soldier named Çingiz Babayev, who also lost his life in the fire, was posthumously awarded the title of National Hero of Azerbaijan.

The government investigation into the catastrophe agreed that it had resulted from an accidental electrical fault, while commenting on the 'outdated Soviet equipment.'

The motorman of the train and the Ulduz station traffic controller were subsequently sentenced to 15 and 10 years' imprisonment respectively for criminal negligence by the Supreme Court of Azerbaijan. [654]

The first serious fire in the Channel Tunnel between England and France since its opening occurred on the evening of Monday 18th November 1996. HGV Shuttle train No. 7539, carrying lorries, was the service involved. It consisted of an engine, a coach for the passengers (lorry drivers) with open lattice-type waggons carrying the lorries themselves and was carrying 31 passengers and three crew. As it entered the tunnel from the French end at 21:48 it was seen that it was on fire, its driver was so advised and in accordance with the policy at the time he was told that his train would be run into a siding at the far end of the tunnel, where the fire could be dealt with. The speed of the train was then limited to 100 km/h and the doors of the crossover links leading to the tunnel running in the opposite direction were closed to prevent the spread of smoke. However, five minutes later the driver became aware of a fault that was likely to derail the train and he brought it to a stand at 21:58, now 19 kilometres inside the tunnel and opposite a door leading to a central service tunnel. At this point, smoke blocked the view of the service tunnel door and entered the coach, while for some minutes the supplementary ventilation system failed to work properly. The power supply to the electric train also failed, preventing it from moving. Finally, the ventilation system was reconfigured, whereupon it cleared the smoke and the passengers and crew were able to escape into the service tunnel. All were admitted to hospital in France for observation. [655]

'Sonia Matczak, 25, who is eight months pregnant, said: "The train stopped and I smelt burning plastic. Then I saw grey smoke curling up from under the doors and I knew there was a serious problem. I was not afraid at first, but when I realised we were trapped, I began to get very frightened. I really thought we were never going to get out of there. It was really strong, very acrid, and we knew something terrible had gone wrong."'

The passengers praised Conductor Emile Grard, 43, who had distributed dampened napkins for the coughing lorry drivers to clamp to their mouths. "He was on the intercom to the driver the whole time, so he never covered his mouth. That's why he inhaled so much smoke," said lorry driver Peter Hattis, 50, from Swansea.

Conductor Grard said: "Normal procedure would be to keep on going to reach the other side as fast as possible, but when that would risk human life the train must be stopped. I decided we had to evacuate the train. The idea was to find a way into the service tunnel by following the luminous arrows which are marked along the whole length of the tunnel. But at first the smoke was too dense and it was too dark to see anything. It was pitch black down there and we were about 19 kilometres from the French tunnel exit. The smoke was toxic and I told everyone to get paper serviettes and soak them in water and hold them over their faces. I could hardly swallow. It will always come back to me. It really only took about two minutes to get into the service tunnel. There are access points all along the way, and we found that we were very close to one. We have carried out exercises plenty of times for just this sort of thing

[654] Wikipedia – 1995 Baku Metro fire
[655] Wikipedia – 1996 Channel Tunnel fire

and it worked very well. In fact, when it all started some of the passengers thought it was just a drill. Unfortunately, I had to confirm for them that it wasn't."

Mme Matczak was initially admitted to intensive care because of her advanced state of pregnancy but was later given a clean bill of health. [656]

Firemen from both England and France fought the blaze, which was finally extinguished at 11:15 the next morning. Damage to the tunnel was extensive, with about 500 metres of it being affected. The HGV shuttle service was suspended and a restricted service of other trains used the adjoining tunnel until repairs were completed, the damaged tunnel being re-opened and the HGV shuttle service restored in mid-May 1997.

Three investigations into the accident were opened by Eurotunnel, the French judicial authorities and the joint British/French Channel Tunnel Safety Authority (CTSA). A number of problems were found. The usage of the HGV shuttles had been approved with the plan that, in the event of fire, the shuttle was to either continue out of the tunnel with the whole train or to detach the locomotive and passenger coach. Neither of these happened, as the fire had damaged the train such that it had been forced to stop, and the overhead line equipment had then failed four seconds after the train had come to a stand. The concept of an 'unconfirmed alarm' meant that the incident had not been treated seriously by Control until five minutes after the train had entered the tunnel on fire. The CTSA report recommended that all alarms should be treated seriously. In addition, the French crossover doors and one of the piston relief duct doors had failed to close properly, allowing smoke to enter the other running tunnel. Furthermore, the Control staff were overwhelmed, having not been sufficiently trained for an emergency, and they were using procedures and systems that were complex. This had led, for example, to the supplementary ventilation system being brought online late and running incorrectly for 15 minutes. Following the accident, additional staff were placed on duty in the control room. Eurotunnel's policy of attempting to drive trains through the tunnel in the event of an onboard fire, with a backup plan of uncoupling the locomotive and amenity coach and driving out of the tunnel, was abandoned and replaced by plans to bring trains to a controlled stop and evacuate the occupants into the service tunnel. Liaison between Eurotunnel and emergency services was improved with joint exercises and exchanges of personnel between the British and French fire brigades, so that each had experience with the other's operational procedures, while communications were also improved. [657]

At about 13:00 on Tuesday 1st July 1997 a freight train hauling 30 Peugeot car transporters stalled after entering the 2,009-metre-long Serre La Voute South Tunnel, between Exilles and Salbertrand stations, Italy, on the international main line between Torino in Italy and Modane in France. Drivers Antonio Bernardo di Meana and Renato Maria Tareo found that they were unable to restart their train or to extinguish the fire that had broken out in their engine. All they could do was apply all the train brakes and scotch its wheels to prevent it running backwards, before making a hurried exit on foot from the bore, as the tunnel was rapidly filling with smoke. The alarm having been raised, firemen from Torino and Susa hurried to the scene together with two ambulances, which took the drivers to hospital in Susa, both suffering from serious smoke intoxication.

Black smoke spouting from the tunnel prevented the firemen from entering it for some time and the fire destroyed eight car transporters and 60 cars, as well as causing considerable damage to the overhead line equipment. Eventually, in the evening, it was possible to push the train out of the tunnel, so that the firemen were able to deal with the blaze. Traffic was not

---

656 *The Times*, London, England, 20th November 1996

657 Wikipedia – 1996 Channel Tunnel fire

greatly affected as single-line working was opened up using the parallel North Tunnel. [658]

In Italy, Sunday 23rd May 1999 was the last day of the Serie A football season and Salernitana, the Salerno team, were playing away at Piacenza. The game between them ended in a draw, a result that meant the relegation of Salernitana to Serie B. Some 1,500 away fans left Piacenza at the end of the game, returning home on train No. 1681, consisting of an engine and 16 carriages. The return journey should have begun at 20:04 but disturbances from the fans meant that its departure was delayed until about 23:00. More disturbances occurred during the journey: fire extinguishers were emptied, seats were ripped up, windows smashed, stones collected during a stop at Bologna were thrown at buildings along the line and the communication cord was pulled frequently. Twelve police officers accompanying the train were insufficient to deal with the problem.

Early on Monday 24th May the violence culminated in the train being set on fire in the 10-kilometre, 265-metre-long Santa Lucia Tunnel, between Nocera Inferiore and Salerno, hence close to the train's destination, possibly by the use of a smoke bomb. According to the Salerno police commissioner this was done to distract the attention of the police, whom the fans knew would be waiting for them at the end of the journey. The rioting fans also pulled the communication cord again and according to one report, the train was forced to halt in the tunnel and firemen had to tackle the fire there. Another account states that the driver was able to pull the train slowly out of the bore.

The fifth carriage of the train was the worst affected; presumably this was where the fire had been started. A number of the fans jumped from the train to escape the blaze and four fans lost their lives. They were named as Giuseppe Diodato and Simone Vitale, both 23; Ciro Alfierti, 16; and Vincenzo Lioi, 15. Another 20 fans and two police officers were injured.

Fans Raggaele Grillo, 20, and Massimo Iannone, 22, were subsequently sentenced to eight years' imprisonment for manslaughter and arson, while Francesco Fiammenghi received a lesser sentence of eight months for causing damage to the train. [659] [660]

The exemplary safety record of the Shinkansen high-speed lines in Japan was slightly marred by an accident that occurred in the 8,488-metre-long Fukuoka Tunnel on Sunday 27th June 1999, on the Sanyo line of the West Japan Railway. As a train was passing through the tunnel a mass of concrete fell from its roof, damaging both the train coaches and its pantograph, though no one was hurt. Services over the section involved were stopped for about four hours. The accident was blamed on the formation of a crack within the concrete of the tunnel roof, at a so-called cold joint (a seam where an initial batch of concrete had hardened and set before a second batch was added). All 142 tunnels on the line were examined by a hammering test after the accident and any faulty areas found were removed.

Then on 9th October 1999 another mass of tunnel fell from the roof of Kitakyushu Tunnel on the same line, this time without involving a train. Here, the fault was found to be the non-removal of a 'pouring mouth' made during the construction process, resulting in a crack between it and the side wall, into which water had leaked, leading to the eventual collapse of a section of the tunnel lining as a result of temperature changes and vibration. Another, more extensive examination of all 142 tunnels was then carried out.

Improvements in the inspection and repair of concrete structures were introduced after the two

---

658 *La Stampa*, Torino, Italy, 2nd July 1997

659 *L'Unità*, Roma, 25th, 26th and 30th May 1999

660 Wikipedia – Santa Lucia Tunnel Massacre

accidents, which were blamed on defects in the original construction of the tunnels due to rushed work, plus a shortage of materials leading to the use of unsuitable sea sand. [661]

Another relatively minor tunnel accident occurred in Japan on Sunday 28th November 1999. A freight train from Nagoya to Sapporo was passing through the 1,232-metre-long Rebunhama Tunnel, between Rebun and Ogishi on the Muroran line, when its driver noticed an obstruction on the line and applied his brakes in emergency. The obstruction was five pieces of concrete, weighing in total about two tonnes, which had again fallen from the tunnel roof, and which were sufficient to derail the train. Like Fukuoka, Rebunhama Tunnel was quite a new one, dating from 1975. The accident was blamed on the spalling (cracking and delamination) of the concrete. [662] [663]

The most serious railway accident by far in the railway history of Austria in terms of casualties occurred not on the main-line system but on the Gletscherbahn Kaprun 2 funicular railway. This single-track railway had only opened in 1974 and ran from the town Kaprun to a location on the Kitzsteinhorn Mountain, from where skiers could descend the mountain slopes or proceed further via a cable car. The 946-mm-gauge railway climbed first on a long viaduct and then through the 3,300-metre Kitzsteinhorn Tunnel, ascending from the base station at a height of 911 metres to the Alpincenter station at a height of 2,450 metres. The line was only 3.9 kilometres long and had a maximum gradient of 500‰ (1 in 2); there was a crossing place in the middle of the tunnel. The railway was well-used by skiers.

At 09:02 on the morning of Saturday 11th November 2000 an ascending train (the *Kitzsteingrams* – Kitzstein goat) started from the Kaprun end of the line with 161 passengers and a conductor, while a descending train (the *Gletscherdrache* – Glacier dragon) started at the same time from the Alpincenter station with a conductor and a single passenger. Fire broke out in the rear control cab of the *Kitzsteingrams* shortly after it had begun its journey and when it was 532 metres within the tunnel the fire had developed sufficiently to damage the hydraulic brake lines, bringing the train to a halt in the tunnel. The *Gletscherdrache* was likewise halted in its descent. At 09:06 the conductor of the *Kitzsteingrams* radioed the control centre and reported the fire. Instructed to open the train doors and report back, he presumably attempted to do so but the hydraulically controlled doors would not open. The fire shortly burnt through a 16kV power cable running through the tunnel, cutting off communication with the control centre, plunging the train into darkness and also causing a total blackout at the Alpincenter.

At 09:10 the train doors were finally opened manually, after the conductor had unlocked them. He and the majority of the passengers then attempted to make their way up and out of the tunnel, stumbling over one another in the pitch darkness. Meanwhile, at the Alpincenter two employees had noticed the smoke early on and managed to persuade most of the people there to leave by various exits. Unfortunately, by leaving the doors at the centre open they worsened the fire by creating a chimney effect, sucking the poisonous smoke up through the tunnel. Most of those attempting to escape by climbing up the tunnel asphyxiated before they had even reached the intermediate crossing place; they all died, as did the conductor and single passenger from the *Gletscherdrache* and an employee and two skiers in the Alpincenter, a total of 155 people in all.

[661] https://www.shippai.org/fkd/en/cfen/CD1000130.html

[662] https://www.ejrcf.or.jp/jrtr/jrtr23/pdf/Topics.pdf

[663] https://www.researchgate.net/publication/323004457_Major_Functional_Risks_for_Operation_and_Maintenance_of_Tunnelling_Projects

The train windows were ironically of shatter-resistant acrylic for safety reasons and thus were difficult to break, but a handful of passengers managed to smash them with skis. Among them was an experienced volunteer fireman who was on a skiing holiday, and he, very aware that hot air rises, was able to convince 11 fellow passengers to follow him and go down the tunnel past the fire, rather than going up it towards the Alpincenter. This group of 12 eventually stumbled out of the south portal of the tunnel, coughing and exhausted but alive.

With the electric power having failed, having seen smoke and unable to contact either train, the staff at the Alpincenter had called the fire brigade at 09:12. The Red Cross arrived at the lower station before the firemen; when the latter arrived they evacuated the station for fear that the stalled train might run back down the incline, though it did not do so. At 11:34 an unconscious victim was retrieved by the firefighters from the Alpincenter; he was the only person whom the firemen could save. At this time and for some time afterwards the heat and smoke in the tunnel itself made it impossible for them to enter it from either end.

Those who lost their lives included 19-year-old German woman Sandra Schmitt, at the time the World Champion freestyle skier. Another person who died was Kevin Challis, a British ski instructor, who had been going up on the previous train with his eight-year-old daughter Siobhan but had given up his seat on that train to an elderly man. His daughter took the earlier train and survived. Those who died included 92 Austrians, 37 Germans, 10 Japanese, 8 Americans and 4 Slovenians.

One of the 12 passengers who had escaped from the burning train was cameraman Hans Fuchs. A newspaper wrote, 'Herr Fuchs was in the rear carriage when the lights went out. The passengers, cramped and uneasy, looked at each other nervously. "We assumed that the emergency electricity supply would spring into action, but then suddenly there was this vicious smoke and somebody shouted: 'The train's on fire, get out,'" he said. The burly man, in his late forties, smashed a window with his ski stick and propelled himself through the splintering glass. Then he turned left, staggering in confusion through the smoke to the valley end of the tunnel. The fire was being drawn upwards, sucked up as if in a chimney, eating everything in its path, melting metal. Almost gasping their last, Herr Fuchs and 11 others managed to cheat the fumes and the flames. Behind them were piles of bodies.'

A number of people were lucky enough to miss the train: 'Katja Liesmann, an advertising executive from Frankfurt, celebrated her 24th birthday on Friday night. It was a rowdy party with her father, a ski tour instructor, and a group of friends. Early on Saturday the alarm clock sounded, but she fell back into a doze, thereby missing the fatal funicular.

'It was a bright autumn morning on Saturday, and with the prospect of the first snowboard party of the season, there was little to hold people back from a trip up the mountain. Arnold Hochstasffl took his son Thomas, 15, and a 13-year-old friend to the valley base station and made his way down the mountain. On his way home, he noticed that the quiet village was being drowned by ambulance sirens and the overhead clatter of helicopters. "I was touched by fear, the deep fear that comes with knowing something bad has happened," he said yesterday. For hours he tried unsuccessfully to call his son by mobile telephone. Only in the late afternoon was he sure that the boys were alive: they had missed the train.

'So did a group of Slovenes. The tourists were waiting for one of the group who, to their collective irritation, was spending a long time in the station lavatory.' [664]

Various theories were offered to explain how the fire had started, such as a short-circuiting battery or even arson by a passenger. The investigation into the catastrophe determined that a faulty fan heater was to blame. The heater was not of a design intended to be used in a moving vehicle, notes on its packaging stating explicitly that it was intended for household use and to

---

[664] *The Times*, London, England, 13th and 14th November 2000

be free standing. An overheating protection device had also been removed. As the temperature of the heater had increased to too great a degree, a mounting of the heating element had broken off and the element had jammed against its wooden casing and caught fire. A small leak of hydraulic fluid had then also caught alight, the fuel lines had melted and the loss of hydraulic pressure had applied the brakes, stopping the train in the tunnel. Investigators found the same faulty heater and leaking hydraulics in one of the control cabs of the *Gletscherdrache* (the heaters in the control cabs at the other ends of the two vehicles were of a different, safer type).

Other deficiencies had contributed to the fire, for example there were no smoke detectors, and in the tunnel there was no smartphone reception, while although fire extinguishers were provided they were located in sealed units in the conductor's cabs, inaccessible to passengers. The skiers themselves possibly had also contributed to the severity of the fire as their outfits were reportedly highly inflammable, while aerosol sprays were often carried by them to wax their boards, in defiance of a rule prohibiting the transport of volatile and flammable materials.

Sixteen individuals, comprising company officials, technicians and government inspectors, subsequently appeared in court charged with negligence, in connection with the accident. The department responsible for ordering the heaters stated that they had specified that a different, safer type of heater should be purchased. Other defendants pleaded that the funicular, not possessing an engine, did not fall within the definition of a train and therefore the use of a room heater was not ruled out. The fact that the heaters had had to be partially disassembled and their safety seals had been removed was not bought out in court. The manufacturer's handbook clearly stated that the heaters should not be used in any location if the safety seals had been removed. The same excuse, that the funicular vehicles were not trains, was used to defend the lack of emergency exits, there having been no emergency brakes, the inaccessible fire extinguishers and so on.

After a two-year trial Judge Manfred Seiss ruled that there was insufficient evidence of criminal negligence and acquitted all the defendants. A separate prosecution of the manufacturer of the heaters was not then proceeded with.

The railway was not reopened after the accident but was replaced by a gondola lift. [665] [666]

Eastbound CSX Transportation (CSX) freight train No. L-412-16, consisting of two locomotives, 31 loaded and 29 empty cars, left West Baltimore Yard at 2:37 p.m. on Wednesday 18th July 2001, destined for Philadelphia, Pennsylvania, and then Oak Island, New Jersey. It was in charge of Conductor Chad Cadden, 27, with Engineer Edward Brown, 52.[667] Its journey shortly took it through the 8,700-foot-long Howard Street Tunnel, beneath the streets of Baltimore, Maryland, United States, and it entered the tunnel travelling at 23 mph. At 3:08 p.m. an uncommanded emergency brake application occurred while the train was still in the tunnel and it stopped about 1,850 feet short of the east portal. Unbeknown to its crew, the train had become derailed behind the engines. Engineer Brown was unable to contact the dispatcher using the train radio, which had been damaged in the derailment, but at 3:13 p.m. Conductor Cadden was able to contact the Baltimore trainmaster using his cell (mobile) phone. The engines were then driven out of the tunnel.

About the time that the engines were leaving the tunnel, the security director of a hotel above the bore dialled 911 to report a disturbance and a little later he also telephoned the CSX communications centre and told them that he had heard a "strong rumbling," which he

---

665 https://mx-schroeder.medium.com/caught-in-the-dark-the-2000-kaprun-glacier-railway-fire-b4ff46a7175

666 Wikipedia – Kaprun disaster

667 *The News*, Frederick, Maryland, United States, 19th July 2001

suspected was related to the tunnel. His message was forwarded to the CSX dispatcher, who found from the Baltimore trainmaster that the train was carrying hazardous materials and the emergency services were informed.

A tank car carrying inflammable tripropylene was punctured in the derailment, losing 28,600 gallons of the chemical. A fire ignited, and adjoining cars carrying paper and wood products provided ample fuel for the blaze. The resultant fire and smoke prevented the fire brigade from entering the tunnel for a considerable time. In addition, another derailed and damaged car lost 2,554 gallons of hydrochloric acid.

It was decided by the incident commander that there was no immediate threat of a catastrophic explosion or any release of toxic vapours and so only a limited evacuation of the city was ordered. Then, just after 6 p.m., it was discovered, due to falling levels at water treatment plants, that a large water main adjoining the tunnel had burst, it was believed as a result of the derailment and fire. About 14 million gallons of water were lost.

It was not until 7:42 a.m. on Monday 23rd that the incident commander was finally able to declare the site under control.

The National Transportation Safety Board (NTSB) carried out a detailed investigation of the accident but was unable to determine a cause for the derailment beyond stating that the most likely cause of it was an obstruction on the track, combined with changes in track geometry.

The NTSB report on the accident contained four recommendations.

(a) To CSX Transportation:

(1) It should maintain historical documentation of maintenance and inspection activities affecting the Howard Street Tunnel.

(2) It should take the necessary action to enhance the exchange of information with the city of Baltimore on maintenance and construction activities within and in the vicinity of the Howard Street Tunnel.

(b) To the city of Baltimore, Maryland:

(1) It should take necessary action to enhance the exchange of information with CSX Transportation on maintenance and construction activities within and in the vicinity of the Howard Street Tunnel.

(2) It should update and revise its emergency preparedness documents to include information on hazardous materials discharge response procedures specific to tunnel environments, as well as infrastructure information on the Howard Street Tunnel. [668]

Two passenger trains collided head-on in the 1,154-metre-long Biogna Tunnel, France, on the single line through the Alps between Nice, France, and Cuneo, Italy, at 11:15 on Monday 27th January 2003. There were only five passengers on the French train from Nice, which was close to Tende, its destination, but there were an estimated 200 on the Italian train. The passengers were thrown to the floor by the force of the impact and luggage fell from overhead racks. Italian passenger Paolo Branconi related: "Everywhere, people were screaming. Three people managed to get out of the car and call for help."

About 150 rescuers took part in the relief operation, with four helicopters flying overhead. The passengers from both trains were taken to the village of Saint-Dalmas de Tende where they were given food and drink before continuing their journeys by bus. The motorman and

[668] National Transportation Safety Board report

conductor of the train from Italy were killed and four passengers were seriously hurt. [669]

The exact cause of the accident was not immediately clear but it was attributed to human error. [670] Later, it transpired that it had resulted from an error by the stationmaster at Breil-sur-Roya, who had “forgotten” the Italian train. In early 2005 a Court sentenced him to a one-year suspended prison sentence, while two maintenance staff were sentenced to nine and six months respectively and the Société nationale des chemins de fer français was fined 40,000 euros, for manslaughter and negligence. [671]

The main headline of *The Korea Herald* on Wednesday 19th February 2003 was a shocking one. It read: ‘Subway arson attack kills 130.’ Even this figure proved to be well too low, for the final number of deaths was 192, with another 151 injured.

At about 09:30 on Tuesday 18th February 2003 train No. 1079 pulled into Songhyeon station on the Daegu underground system. Daegu was the third-largest city in South Korea, after Seoul and Busan. Among the passengers who joined the train at Songhyeon was Kim Dae-han, aged 56. Kim had suffered a stroke in 2001, which had left him partially paralysed and unemployed, while he had also been diagnosed as suffering from a manic depressive illness. Dissatisfied with the treatment he had been receiving, he had ten days earlier threatened to set fire to a psychiatric hospital. When he boarded the train, he had been in possession of a duffel bag in which were two milk cartons (another source states a shampoo bottle), filled with flammable liquid, probably petrol or possibly paint thinner.

After the train had left Banwoldang station at about 09:53, Kim had begun fumbling with the contents of his duffel bag and a cigarette lighter, to the alarm of the other passengers, who had attempted to disarm him. In the ensuing struggle the contents of one of the cartons or the shampoo bottle were spilt and they caught fire just as the train was pulling into Jungangno station. With his back and legs ablaze, Kim managed to escape from the train, as did many of the passengers around him, but in two minutes the fire had spread to all six cars of the train. Its rapid spread was helped by the presence of flammable insulation material between the aluminium shells of the cars, vinyl and plastic materials in the seat cushions and strap handles, and heavy plastic floor matting, which produced clouds of thick black smoke as it burnt. The motorman of the train, Choi Jeong-hwan, 31, failed, however, to immediately realise the gravity of the situation.

Meanwhile, approaching the station from the opposite direction on the other line was train No. 1080. Although the motorman of train No. 1079 had yet to realise that there was a problem, it was apparent that something was wrong to Control, who could see smoke accumulating in the station on their closed-circuit television monitors, so they radioed Motorman Choi Sang-yeol on train No. 1080 and advised him to enter the station cautiously, because there was a fire there. This train duly entered the station about four minutes after train No. 1079 and came to a stand alongside it. The doors of train No. 1080 were briefly opened, then were closed again, apparently in an effort to keep the smoke out. A little later, an automatic fire-detection apparatus shut down the power supply, preventing either train from leaving the station.

Three times while his train was stopped in the station, Motorman Choi Sang-yeol made announcements over the intercom, advising his passengers to sit tight while he attempted to contact Control to try and sort out the problem. Control then came back to him and told him, “Quickly, get away…shut down your engine and go!” At this point, the motorman opened the

[669] www.guardian.co.uk/worldlateststory/0,1280,-2354936,00.html
[670] www.guardian.co.uk/worldlateststory/0,1280,-2357204,00.html
[671] *Today's Railways* Europe 117

door of his cab and fled but tragically, in doing so, he had removed the master key, which was in accordance with the regulations but had the effect of shutting down the battery-operated power supply to the train doors, so that they could not be opened by the passengers. The 79 people still on the train were then trapped and all perished in the fire. Local media reported how some of the trapped passengers had made frantic calls to their loved ones on their mobile phones. A mother told her daughter, "Follow the other passengers and make you way off the train to safety," to which her daughter had replied, "Mum; the doors won't open."

Various factors made the disaster worse. For example, there were no fire extinguishers provided on the trains, no water sprinklers, nor was there emergency lighting at the stations. There was an emergency ventilation system but it proved inadequate to deal with the large amount of smoke generated. Many passengers became disorientated in the dark, smoke-filled surroundings and died of smoke inhalation while vainly searching for an exit. There was a massive influx of emergency personnel, reportedly over 1,300 in total, and the fire was extinguished at 13:38 but it was another three and a half hours before rescuers were able to enter the station, due to the toxic smoke.

Many of the victims were burned beyond recognition and DNA testing had to be used to identify them. Three remained unidentified even then. Those who had lost their lives had mainly been students or else young women working in department stores, which normally opened at 10:30.

Lee Dal-shik, 57, was missing his 19-year-old daughter, Hyun-jin. He said, "She disappeared yesterday when she went on the subway to meet a friend. I won't believe she's dead until I see her body or at least her remains," he said. He supplied authorities with her dental records and a hairbrush for DNA analysis. [672]

Various safety measures were introduced following the disaster. The plastic fabric of the trains was removed, they were sprayed with fire-resistant chemicals and manual door handles were fitted, while flashlights and gas masks were provided at stations or on trains.

Kim was of course arrested and faced multiple charges of murder and arson. The prosecutor and at least some of the relatives and friends of those who had died called for the death penalty for him. In the end, on account of his apparent remorse and his mental condition, the Court instead sentenced him to life imprisonment. A year later, he suffered a further stroke and died.

Nine employees of the Daegu Metropolitan Transport Corporation also faced criminal charges. Motorman Choi Sang-yeol was charged with involuntary manslaughter for having failed to evacuate his passengers in a timely manner instead of repeatedly telling them to remain where they were. To compound matters, as noted above, he had then abandoned his train and, in accordance with the regulations, taken with him the master key, which had had the effect of making it impossible to open the train doors and so trapped his passengers in the burning carriages. He received the stiffest sentence of the nine railwaymen – five years' imprisonment. Motorman Choi Jeong-hwan was presumably charged with the same offence; he received a slightly more lenient sentence of four years' imprisonment. The penalties imposed on the other seven railwaymen were not stated. [673] [674]

---

[672] *BBC News*, London, England, 19th February 2003

[673] https://www.koreaherald.com/view.php?ud=20231011000686

[674] Wikipedia – Daegu subway fire

CHAPTER 12

LEVEL CROSSING COLLISIONS

Wherever roads and railways intersect, they must do so either by means of a bridge or on the level. Bridges have the advantage of allowing road and rail traffic to operate entirely without interference from one another, but they are much more expensive to provide, especially in flat countryside, due to the need to provide extensive earthworks for approach embankments or cuttings as well as the cost of providing the span itself. Where underbridges are provided there is often also the need to build retaining walls and there can be problems with drainage.[675] The provision of level crossings, especially if, as in many parts of the world, they are unstaffed, is the much cheaper alternative, but where they exist, they introduce an additional source of danger to both road and rail users. Robert B. Shaw in his outstanding book on American railway accidents, *A History of Railroad Accidents, Safety Precautions and Operating Practices*, writes that he has excluded ordinary grade (i.e. level) crossing collisions from consideration on the grounds that they are not proper train accidents, inasmuch as trains usually suffer much less harm from such accidents than the road vehicle involved. This series of volumes, on the other hand, will attempt to include, amongst others, all such accidents that resulted in five or more fatalities, even if, as was often the case, none of those who died were train passengers or railwaymen. As in previous volumes, all accidents involving school buses have also been included.

The First World War had provided a great stimulus for the development of the internal combustion engine and the number of motor vehicles in use, and consequently of level crossing collisions involving them, greatly increased following the war. The surge in ownership of motor vehicles continued in many richer countries after the Second World War and the removal of the most dangerous level crossings, or at least the provision of more sophisticated crossing protection, did not keep pace, so there was a substantial increase in serious level crossing accidents in post-war years, which has continued up to the present. This chapter is therefore a long one.

Mrs Clara Muir Laing, 46; her children Irene Hilda Meale, 9, and Jean Muir Laing, 18 months; and her grandchildren Robert Murray, 4, and Philip Murray, 2, were all killed when the 1936 sedan in which they were riding was struck by a diesel-hauled passenger train at about 17:15 on the afternoon of Thursday 21st August 1952 on Cootamundra Road Crossing, Temora, New South Wales, Australia. Mrs Laing was returning home to Dimiseer, about 16 miles from Temora, from a visit to her daughter, Mrs Albert Murray, who had given birth to a daughter in Temora District Hospital on Wednesday night.

---

[675] *Backtrack*, Volume 24, pg. 88.

The train carried the car about 300 yards down the track. The four children were thrown from it and their bodies were so badly disfigured as to be barely recognisable. Mrs Laing was trapped beneath the front of the locomotive, which was lifted off the rails. The car was completely wrecked, parts of it being scattered along the line. A steam engine took the train on to Temora. No one on the train was hurt. [676]

Five teenagers returning from a surfing trip to the coast were killed when a diesel-hauled passenger train bound for Adelaide, carrying about 100 passengers, collided with their car on a level crossing at Middleton, South Australia, on the evening of Monday 10th October 1966. Most of the victims were believed to have lived in the Adelaide area. The car was carried about 70 feet and smashed surfboards from its roof were hurled about the track. It took over an hour to clear the wreckage and the train was delayed nearly four hours, possibly because it needed a replacement engine. [677]

Five people died and 31 were injured in a collision between a freight train and a bus on a level crossing in the suburbs of Taipei, Taiwan, on Monday 21st June 1971. [678]

A southbound National Railroad Passenger Corporation (Amtrak) passenger train running on Illinois Central Gulf Railroad tracks and travelling at 75 mph collided with a station wagon on a level crossing with Highway 67 near Godfrey, Illinois, United States, on the night of Friday 16th February 1973, carrying it 270 feet along the line. Godfrey teenagers Ronald Auer, 16; Rick Camp, 15; Randy Davis, 16; Mark Deeder, 16; and Tim Woods, 16, were all killed. Roald Auer owned the vehicle and was believed to have been driving it at the time. The crossing lacked barriers but was provided with flashing warning lights. A witness stated that the station wagon had initially stopped at the crossing but had then proceeded across the line, its driver clearly believing that he could get over in front of the train. No one on the train was hurt and it proceeded to St Louis after a short delay. [679] [680]

At 08:27 on the morning of Wednesday 20th July 1979, a 740-ton freight train approached Galabovo Crossing, Bulgaria, three kilometres from Simeonovgrad. At the same time, a "Chavdar" bus with 50 passengers was also approaching the crossing. The crossing keeper was not at his or her post and the barriers were not lowered. The bus driver did not notice the approaching train, although the train driver was sounding his horn, and a terrible collision followed. The steam locomotive crashed into the bus and dragged it 50 or 60 metres. In addition, the fuel tank of the bus was ruptured and exploded and flames engulfed the passengers. Thirteen people were burned to death at the scene and five died later in hospital. Another 30 were injured. [681]

A pickup truck carrying eight teenagers and driven by a 16-year-old boy failed to give way to a Union Pacific Railroad coal train travelling at about 50 mph at a level crossing, on the evening of Friday 11th January 1991. The train struck the truck, killing six of its occupants and

---

676 *Daily Advertiser*, Wagga Wagga, New South Wales, Australia, 22nd August 1952
677 *The Sydney Morning Herald*, Sydney, New South Wales, Australia, 11th October 1966
678 *New Nation*, Singapore, 21st June 1971
679 *Evening Democrat*, Fort Madison, Iowa, United States, 17th February 1973
680 *Wisconsin State Journal*, Madison, Wisconsin, United States, 18th February 1973
681 http://www.railwaypassion.com/forums/index.php?topic=1410.60

injuring the other two. The accident happened at Fort Gibson, Oklahoma, United States, and all the victims had been pupils of Fort Gibson High School. The 15-car train dragged the truck about 125 feet. Those who lost their lives were Amy Baird,15; Chad Daugherty, 17; Mandy Fowler, 15; Jeremy Michael Johnson, 16 (the driver); Pattie Perryman, 16; and Carrie Walker, 15. Of the other two teenagers, Linda Duvall was stated to be in a serious condition and Thomas Jackson was in a fair condition. Rescuers from Muskogee fire department had to wade through waist-deep water to get to the wrecked vehicle and the rescue took an hour and a half. At least seven road ambulances and an air ambulance went to the scene.

The girls, all sophomores (second-year students), had been returning from Linda Duvall's birthday party when their car had broken down and Johnson had given them a lift in his truck. The crossing was only marked by a sign. [682]

On about Sunday 17th February 1991 a Moscow-Ivano-Frankovsk express collided with a bus on an unprotected level crossing near Marshyntsi, in western Ukraine, USSR. Ten bus passengers, two of them children, were killed outright and the bus driver succumbed to his injuries in hospital. Three other bus passengers were admitted to hospital with serious injuries. [683]

A train collided with a car on an open level crossing at Columbus, Mississippi, United States, on Thursday 14th March 1991, killing the 32-year-old woman driving it, her two children and two related children, all four aged between 7 and 12. A fifth related child was taken to hospital critically injured. Chief Deputy Butch Howard of Lowndes County Sheriff's Department stated, "Apparently the vehicle approached the railroad crossing, saw the train, stopped, and then tried to go in front of the train." [684]

At about 09:05 on Tuesday 19th March 1991 South Eastern Railway train No. 407, an Up Shahdol – Katni passenger working, collided with a bus on the 'B' Class manned level crossing at Km 935/8-9 between Ghunghuti and Birsinghpur stations, Madhya Pradesh, India, on the Anuppurkatni broad-gauge, double-line section of the railway. The results were disastrous for the bus passengers, 27 of whom were killed, while another 52 were injured, 15 of them seriously. The barriers had been lowered for the passage of the train but the bus had smashed through them into the path of the train. No one on the train was hurt. [685]

Six people died in a level crossing accident at Jesenské, Slovakia, on Sunday 31st March 1991. [686]

A goods train collided with a matatu (a private minibus used as a share taxi) on a level crossing with Jogoo Road, Nairobi, Kenya, on Thursday 9th May 1991, killing seven people. [687]

---

[682] *The Joplin Globe*, Joplin, Missouri, United States, 13th January 1991
[683] *The Lethbridge Herald*, Lethbridge, Alberta, Canada, 18th February 1991
[684] *New Castle News*, New Castle, Pennsylvania, United States, 15th March 1991
[685] Government of India Ministry of Civil Aviation and Tourism Commission of Railway Safety Annual Report for 1990-91
[686] European Union Agency for Railways: List of fatal railway accidents 1990-2009 (more than five fatalities)
[687] *Daily Nation on the Web*, 'Ngai Ndethya' the worst yet, 25th March 1999

There was apparently a, presumably minor, level crossing collision involving a train and a school bus near Billimari, New South Wales, Australia, on Tuesday 4th June 1991. [688]

Eight people lost their lives in a level crossing collision at Corinth, Greece, on Saturday 29th June 1991. [689]

On Sunday 28th July 1991 a Caminhos de Ferro Portugueses (CP) express from Vilar Formoso to Lisboa collided with a car on an unguarded level crossing near the village of Lapa do Lobo, Portugal, or in railway terms at kilometre 106.66 of the Beira Alta line.[690] The car was wrecked and its five occupants, one of them a six-year-old child, were killed. The victims were locals from nearby Canas de Senhorim.

The accident reawakened discussion regarding the danger of level crossings. There were five unprotected crossings in the vicinity of the accident, 434 crossings on the 250-kilometre-long line and in the country as a whole, 2,390 crossings were unprotected, 689 had crossing keepers and 141 were automated. CP had commissioned a study to improve the safety of the crossings between Pampilhosa and Vilar Formoso but, as always, the problem was cost. Martins da Costa of the CP pointed out that it cost 20,000,000 escudos to automate each crossing and given the financial situation of the railway, they could not afford it. The provision of crossing keepers was similarly ruled out on the grounds of cost – who was going to pay their wages? [691]

At least nine people were killed and 20 were injured when a passenger train and an army lorry collided on a level crossing between Kaliningrad and Svetlogorsk, USSR, on Saturday 3rd August 1991. Five of those who died were passengers on the train and four were soldiers on the lorry. The electric train was derailed and two of carriages overturned. The lorry driver had attempted to cross in front of the train when the engine of his vehicle had unexpectedly died. [692]

A train collided with a minibus on Monday 2nd September 1991 in Lome, the capital of Togo. Seven people died and a dozen were injured. [693]

Six people died and 12 were injured when the driver of a Volán bus disregarded signals at a level crossing in fog and collided with a freight train between Vámosgyörk and Hort, Hungary, on Thursday 7th January 1992. The train was derailed and then struck by an express train. There was some indication that the crossing barrier had failed to descend and the road had been icy. The bus driver was amongst the fatalities. [694] [695] [696]

---

[688] Author's personal notes – source mislaid

[689] European Union Agency for Railways: List of fatal railway accidents 1990-2009 (more than five fatalities)

[690] Ibid

[691] *The Anglo-Portuguese News*, Lisbon, Portugal, 1st August 1991

[692] *Playground Daily News*, Fort Walton Beach, Florida, United States, 5th February 1991

[693] *The Chronicle Telegram*, Elyria, Ohio, 7th September 1991

[694] *The Daily Telegraph*, London, England, 8th January 1992

[695] European Union Agency for Railways: List of fatal railway accidents 1990-2009 (more than five fatalities)

On Wednesday 12th February 1992 a bus carrying 67 wedding guests got stuck on an unguarded level crossing at Chindwara, Madhya Pradesh, India. Neither the bus driver nor the passengers noticed an approaching passenger train, which struck the bus, killing 10 of its occupants and injuring 37. [697]

Juan Carlos Lemos was driving a Ford Taunus car containing 10 boys aged between 10 and 12 to a football match on Tuesday 12th May 1992 when his car was struck by a train on an open level crossing on the General Mitre line near Pueblo Andino, Argentina. Eight of the boys were killed more or less instantly in the accident and Lemos himself, who was also the father of one of the boys, succumbed to his injuries in hospital a fortnight later.

A police investigation subscribed the accident to the lack of barriers at the crossing and to the fact that it had been foggy at the time. Following the accident, locals erected new warning signs at the crossing and were active in keeping vegetation in check, to ensure that motorists had a clear view up and down the line but the railway, having carried out a traffic survey, continued to maintain that the crossing was not sufficiently busy to require barriers.

In 1998 a federal court in Rosario ordered Ferrocarriles Argentinos to pay compensation of about 1,500,000 pesos to the families of the victims. [698]

Perhaps more than usually tragic was a level crossing accident that occurred on Wednesday 20th May 1992 at Jarrenwisch, Germany. On that day, the son of a 44-year-old man was celebrating his sixth birthday with a party and his father had agreed to collect some of his son's friends from three neighbouring villages to attend the celebration. As the minibus in which the father and six children were travelling reached the crossing, situated on the Heide – Büsum line, the man for some reason disregarded the warning lights and bells, the vehicle was struck by express railcar No. 4769 and was wrecked. All seven of its occupants were killed. [699]

An intercity ANS bus from Padang, the capital of West Sumatra, to Jakarta, was in collision with a passenger train on a level crossing at Banjaratu village, Indonesia, on the evening of Tuesday 23rd June 1992. Of the 64 people on the bus, 32 were killed, including a six-month-old baby, and the other 32 were injured, 17 of them seriously. The bus was dragged about 60 metres by the train and then fell into a shallow ravine, about four metres deep.

The circumstances of the accident were unclear. Some witnesses stated that the gates of the crossing had not been closed despite the train being close at hand. Surviving bus passengers, on the other hand, maintained that the bus had broken down on the crossing due to a mechanical problem. [700]

A mistake by a bus driver was apparently the cause of a major disaster on Wednesday 16th September 1992 at an unstated location in the Kaliningrad oblast, Russia. The bus was broken to pieces by a train and parts of it were carried 600 metres. Seventeen people died and another

---

[696] https://www.origo.hu/itthon/1899/12/20030508alegsulyosabb

[697] *The Odessa American*, Odessa, Texas, United States, 14th February 1992

[698] https://www.pagina12.com.ar/1998/98-04/98-04-25/rota1.htm

[699] *Katastrophen der deutschen Bahnen Teil II Chronik 1845-1992* by H.J. Ritzau, pg. 274

[700] https://www.upi.com/Archives/1992/06/23/T-rain-crash-claims-many-lives-in-Indonesia/3406709272000/

two were badly hurt. All the victims were on the bus. [701]

Just before 17:00 on Saturday 26th September 1992 a light engine, travelling at about 100 km/h, collided with a minibus carrying a band of German tourists on a level crossing in the town of Agárd, Hungary. Fifteen of those on the bus were killed outright, one was fatally hurt and the other three, one of them its driver, were injured. The engine driver told Hungarian radio that he had braked when he saw the bus, but had been unable to avoid the collision. The crossing was protected only by red warning lights and possibly bells. After the accident the bus driver explained that he had been unable to distinguish the flashing warning lights due to the setting sun shining in his eyes. [702]

One person died and 14 were injured, two seriously, when a school bus and a coal train collided at a level crossing at Conara Junction, Tasmania, Australia, on Monday 7th December 1992. Nine teenagers and six adults were aboard the minibus at the time, of whom three adults – the male bus driver and another man and a woman – were flown by helicopter to Hobart for treatment. The woman died later in the Royal Hobart Hospital there. The other injured passengers suffered broken bones, cuts and bruises, and were treated at the accident scene by two doctors and paramedics. They were then taken by a fleet of ambulances to Launceston General Hospital for further treatment.

Police could not explain why the southbound minibus had smashed into the front of the north-bound train. Warning lights at the crossing were activated.

The chartered bus had been carrying George Town High School students and their parents on a four-day excursion to the state's south.

The accident forced the closure of the Midland Highway, the main link between Hobart and Launceston, for more than four hours, while wreckage was cleared.

The owner of the Conara store, John Brazier, who was one of the first on the scene, described the accident as a "holocaust". Mr Brazier, whose shop is 300 metres from the crossing, said that a customer had run in and told him of the crash. "It was like a holocaust when we got there. Straight out of a war movie," he related. Mr Brazier said the driver of the minibus appeared to have been flung through the window and catapulted back into the vehicle, where he had remained trapped for more than one hour. Another six people had been pinned in the bus until they were cut free. Mr Brazier said he rushed back to the store and brought blankets for other shocked passengers who sat or lay beside the road crying or dazed. "Many of them were just young kids and they were crying," he said. "They also had a lot of cuts and bruises and didn't have a clue what was going on." Mr Brazier added that the warning lights at the crossing, which lacked barriers, had still been flashing when he arrived at the scene.

George Town High School Principal Peter Cox said the group had intended staying at Oyster Cove near Hobart before spending several days at Bruny Island looking at Aboriginal sites. He said the bus had been chartered for the trip. [703]

Six people died in a level crossing accident at Chełmce, Poland, on Sunday 18th October 1992. [704]

---

[701] *The Lethbridge Herald*, Lethbridge, Alberta, Canada, 16th September 1992

[702] *The Southern Illinoisian*, Carbondale, Illinois, United States, 27th September 1992

[703] *The Canberra Times*, Canberra, Australian Capital Territory, Australia, 8th December 1992

[704] European Union Agency for Railways: List of fatal railway accidents 1990-2009 (more than five fatalities)

On Sunday 31st January 1993 a grossly overloaded bus was struck by a train on an unprotected level crossing between the small towns of Xinlitun and Gaotiashan, in Liaoning province, China. Although the bus had a capacity of 60 passengers, it was carrying 94 at the time of the accident. No less than 66 people were killed in the accident and 28 injured. All of those who lost their lives and all but one of those who were hurt had been on the bus; the exception was the train driver, who was slightly hurt. The train had been on its way from Chifeng to Dalian. According to the Xinhua news agency, investigators blamed the bus driver for the accident, stating that he, "was not attentive at the time and did not abide by regulations." [705]

A school bus was struck by train No. 7817 on a level crossing near Alsónyék, Hungary, on Friday 12th February 1993, killing 11 children and bus driver János Szellenheim, while the other 18 children were injured. The only slightly fortunate feature of the accident was that because of a flu epidemic, there had only been 29 children on the bus rather than the usual 68. According to witnesses the bus driver had stopped on the approach to the crossing but failed to notice the approaching train. He was aware of the passing times of the trains and might have thought it had already gone by; the morning was also foggy. More importantly, he had received no warning from the automatic crossing equipment that a train was approaching. The train driver and conductor both saw the bus, which like the train was heading towards Bátaszék, and the driver whistled for the crossing as the bus turned left off main road No. 55 to take the lane that passed over the railway, on its way to pick up a pupil from a farm. When the train driver had realised that the bus was not stopping, he had applied his brakes in emergency. The train struck the vehicle broadside at 07:16, cutting it in two and carrying the front of it 150 metres down the line, while the rear part fell into a ditch. The sound of the collision alerted local residents and the rescue operation began almost at once. The younger pupils, aged between 6 and 10, had in the main been sat at the front of the bus and it was they who were mostly killed, while their older fellow pupils sat towards the rear survived. No one on the train was hurt.

The crossing was protected by warning lights but possibly lacked barriers. However, the batteries for the crossing equipment had been stolen on 29th January and no spares were immediately available. An order advising train crews that the warning equipment at the crossing was inoperative had, however, been issued to trains at Kiskunhalas station, this order containing a provision that the speed of all trains must be reduced to 15 km/h over the crossing. There were some irregularities in the issuing of the order to train No. 7817 and despite its provisions, the train approached the crossing without slowing up at all, at a speed of 78 km/h.

Driver Gyula Zörényi and Assistant Driver Zsolt Appl subsequently found themselves in court. The driver admitted that on four occasions during the previous two days he had also passed over the crossing without reducing speed. He was found guilty, sentenced to two years' imprisonment and banned from driving trains for four years. The assistant driver was acquitted due to lack of evidence. On appeal to the Supreme Court, the driver's sentence was increased to three years, while the assistant driver's acquittal was made absolute. An internal inquiry also found a chief engineer to have been at fault for the delayed battery replacement, but he was not taken to court.

Improvements were made to the equipment of automatic level crossings after the accident to make the batteries thief-proof, while inventor Sándor Hédl designed a device that could be incorporated into road surfaces to warn motorists that they were approaching a level crossing.

A further measure taken after the accident was the reopening of a primary school at

[705] *The Daily Times*, Farmington, New Mexico, United States, 1st February 1993

Pörbölyi, closed in a school reorganisation the 1970s, so that younger pupils no longer had to commute by bus to neighbouring Bátaszék. Older pupils still made the journey, but their bus was no longer routed over the crossing. [706]

A railcar collided with a car on a level crossing near Irivo station on the Douro line in Portugal on Monday 1st March 1993. Six occupants of the car died and three more were injured. [707]

One child was killed and six others were seriously injured when a train collided with a school bus carrying 14 handicapped children on a level crossing on the outskirts of Athens, Greece, on Tuesday 16th March 1993. A police report indicated that the bus driver had ignored the red warning lights at the crossing. [708]

At about 3:15 p.m. on Wednesday 17th March 1993 National Railroad Passenger Corporation (Amtrak) passenger train No. 91 (the *Silver Star*), on its way from New York, New York, to Miami, Florida, was in collision with a loaded gasoline (petrol) road tanker on the level crossing with Cypress Creek Road (otherwise NW 62nd Street) in Fort Lauderdale, Florida, United States, on a CSX Transportation line. The train consisted of a locomotive and 11 coaches, while the articulated road tanker consisted of a 1991 three-axle Freightliner model tractor and a 34-ft-long aluminium semi-trailer, containing at the time of the accident 8,500 gallons of petrol. Cypress Creek Road was a six-lane divided highway. Due to roadworks, on the day of the accident the left-hand lane of the road approaching the crossing from the direction in which the tanker was coming was closed off, and closer to the crossing the centre lane was also closed off, so that the traffic on all three lanes was funnelled into just the right-hand lane.

Lorry driver Barson DeFrance, 51, had reported to work at the Port Everglades terminal of the Amerada Hess oil company at 2 p.m. on the day of the accident and had left their depot driving the tanker at 2:30 p.m., bound for Margate, Florida, only two miles from Fort Lauderdale. Because two lanes over the crossing were closed, traffic over the crossing with the single-track railway line was congested, with the result that the petrol tanker came to a stand with its tractor unit on the crossing, close to but not on the line, because of stationary vehicles ahead. The crossing bells then began to sound and the crossing barrier came down onto the roof of the tractor. As the train approached the whistle board 1,528 feet from the crossing, its engineer saw a small white car on the crossing itself with the road tanker behind it. He made a service brake application, blew his horn, and the car moved forward off the crossing. The road tanker, the front of which had initially only been just over two and a half feet over one rail, then began to slowly move forward as well, at which the engineer applied his brakes in emergency and sounded his horn again. The train had been travelling at 59 mph and was still travelling at 35 mph when it struck the rear of the tanker. Witnesses related that the collision had ruptured the tanker, spilling petrol, which had immediately ignited, and the fire had engulfed nine road vehicles. The lorry driver and five people, occupying three of the other road vehicles, were killed, while 15 motorists were injured. Nineteen people on the train (11 of its 118 passengers and 8 crew) were slightly hurt. The luggage and dining cars of the train were badly burnt, with serious consequences for some train passengers: 'Emma Ebby wiped away tears as she boarded a bus to Miami station after the crash. She was moving from Chicago to Miami and her possessions were in the luggage car that burned. "I lost everything I own," she said.'

---

[706] Wikipedia – Pörbölyi bus accident
[707] Wikipedia – Acidentes ferroviários en Portugal
[708] *The Guardian*, London, England, 17th March 1993

Witnesses to the accident spoke to reporters. '"He couldn't back up and he couldn't go forward," said Barbara Freeman, a witness to Wednesday's accident. "Then the crossing gate did come down, but it came down on the cab of his truck."

'"You couldn't do anything, it happened so fast," said Junior Ayala, who was stopped at a red light a block away. "You couldn't even get close; it was so hot. I could see people burning."' [709]

'A Broward County sheriff's deputy in his patrol car at the time pulled the driver from a burning vehicle behind him, said Sonya Friedman, a Fort Lauderdale police spokeswoman. "He tried to rescue another, but couldn't because of the fireball," she said.

"The explosion was unbelievable," said witness Tim Dahma.

'Cars and small trucks were scorched and scattered as if they had been in a tank battle.' [710]

Greta Machado, 19, of São Paulo, Brazil, said that from the next-to-last car of the train, "We couldn't feel the crash but we could see it. We could see the flames and the fire." [711]

The National Transportation Safety Board (NTSB) investigated the accident. It noted that had the engineer made an emergency brake application as soon as he had seen the tractor of the road tanker, he could have stopped the train at the crossing. However, at that point he could not have seen that it was a petrol tanker and the board observed that it would not be prudent for an engineer to place his train in emergency every time a road vehicle entered a crossing after the warning system had been activated. Placing his train in emergency could cause a derailment, or injure passengers, crew or passers-by.

The lorry driver had been experienced and had driven articulated lorries for 18 years. He had received a number of safety awards, though he had also committed several traffic offences, such as speeding and following another vehicle too closely. There was no evidence that he was fatigued and post-mortem tests for alcohol and drugs had been negative. When he had initially stopped under the gate (barrier), he had still been about five feet clear of the crossing but from his position in the cab, he might have thought that his vehicle was already foul of the line. Witnesses, including the train engineer, stated that he had then moved forward, then stopped, perhaps to change gear, before moving forward again. Tests showed that had he not hesitated in this manner, he would have just cleared the crossing in time. There was no evidence that he was a risk-taker.

There were 4,100 public road level crossings in Florida, experiencing about 100 accidents a year. They were numbered in order of safety, with 1 being the most dangerous. On this scale, Cypress Creek Road was number 15. A project to widen the road over the crossing from six to eight lanes had begun in 1992, expected to take 540 days, until January 1994. The accident had occurred on the 241st day of the work.

Witnesses described the congestion, with two of the three lanes over the crossing being closed. No one connected with the roadworks had consulted the railroad about traffic congestion over the crossing.

The NTSB reached six conclusions in relation to the accident:

(a) The emergency response had been timely and effective.

(b) Had the lorry driver remained in his position beneath the crossing gate, as described by witnesses, he would have avoided the accident.

(c) Had the lorry driver not hesitated and driven across the line when the crossing gate had

---

[709] *Alamogordo Daily News*, Alamogordo, New Mexico, United States, 18th March 1993
[710] *Roswell Daily Record*, Roswell, New Mexico, United States, 18th March 1993
[711] *The New Mexican*, Santa Fe, New Mexico, United States, 18th March 1993

first descended, he might have been able to clear the track in time.

(d) The taper and shift design of the work zone had caused congestion at the railroad/highway grade crossing.

(e) Neither the County project manager, the design engineer, nor the contractor had adequately considered the traffic congestion or the resulting obstruction of the railroad/highway grade crossing.

(f) Once congestion had developed, neither the County project manager, the design engineer nor the contractor had taken adequate measures to ensure safe traffic control in the work zone near the railroad/highway grade crossing.

The NTSB determined that the probable cause of the accident had been the inadequacy of the precautions taken by the Broward County project manager, the design engineer, and the contractor, which had resulted in traffic congestion at the railroad/highway grade crossing, and the lorry driver's decision to cross the railroad track though the warning system had been activated.

The NTSB made several recommendations.

To the Federal Highway Administration:

(a) It should include in Part VI of the *Manual on Uniform Traffic Control Devices* minimal standards on channelisation of work at track zones to minimise traffic congestion on railroad/highway grade crossings.

(b) It should incorporate guidance in its construction and maintenance training courses, which addressed work zones near railroad/highway grade crossings.

To the American Trucking Associations Inc.:

(a) As part of its safety programme, it should notify its members of the circumstances of the accident and alert them to the potential dangers when approaching work zone areas adjacent to railroad/highway grade crossings.

To Amerada Hess Corporation:

(a) It should develop and implement a grade crossing awareness programme, including participation in the Operation Lifesaver Program, to instruct its lorry drivers in the potential dangers at railroad/highway grade crossings. [712]

A goods train collided with a school bus on a level crossing in Bulawayo, Zimbabwe, on Saturday 20th March 1993. The rear of the bus was struck as it was carrying Northlea High School students from a sports event at the White City Stadium. One pupil was killed outright and three were fatally injured. Fifty-two others were less badly hurt. [713]

A school bus taking children, mostly aged 11 and 12, and their teachers on an outing to Yingkou, Liaoning province, China, was struck by an oil tanker train on an unprotected level crossing on the foggy morning of Friday 30th April 1993.[714] Of the 74 occupants of the bus 36 were killed, including at least 28 children and its driver, and the other 38 were injured. There were no reports of anyone on the train being hurt.[715] Many schools in China organised outings

[712] National Transportation Safety Board report

[713] https://www.chronicle.co.zw/never-argue-with-a-train-6-deaths-54-injured-in-66-accidents-from-january-to-august/

[714] *The Cedar Rapids Gazette*, Cedar Rapids, Iowa, United States, 1st May 1993

[715] *European Stars and Stripes*, Darmstadt, Germany, 2nd May 1990

and celebrations on May Day, one of the country's few public holidays. [716]

Twelve people died and the same number were injured when a train collided with a tractor in Urleasca, south-eastern Romania, on Sunday 1st August 1993. [717]

On Saturday 14th August 1993 a level crossing collision between Makó and Makó-Ujváros, Hungary, killed six people. [718]

A major level crossing accident occurred near Kosovo Polje, Kosovo, Serbia, on Monday 23rd August 1993,[719] when a bus carrying workers of the Kosovo Electrical Company was struck by a train. Twenty-one people on the bus were killed and over 30 were taken to the medical centre in nearby Pristina, four of whom were critically hurt.

After the accident, the bus was found to have been in working order but there was no indication that its driver had braked on the approach to the crossing, implying human error as the cause of the accident. The aftermath was complicated by inter-ethnic recriminations between Serbians and Albanians. Žarko Katić, the Serbian Minister of Traffic and Communication, stated that the crossing met all safety requirements, while the manager of the Kosovo Electrical Company claimed that it did not and that its lack of protection was illegal, as although it had signs it lacked gates (this word is given as 'ramp' in an internet account of the accident but gates or barriers is presumably the correct translation). Two buses should have been provided to take the early shift workers home from the factory, but for some undefined reason only one was provided and so it was overcrowded. It was also late and thus arrived at the crossing at the same time as the train. There were allegations that the Albanian operator of the bus had boasted of being a fast driver, so that no one would survive if he had an accident; also that some of the injured had been taken to an 'illegal hospital' in Glogovac, while the unions accused the electrical company of not caring about the safety of its workers. [720]

Michael Herron, 24, was one of six people killed when his girlfriend's 1963 Buick LeSabre car was struck by a passenger train on a rural level crossing 10 miles east of Stratford, Ontario, Canada, at 7 p.m. on Saturday 4th September 1993. The other victims were named as Gordon Algera, 15, of Chatham; David Clarke, 23, of Kitchener; Jason Hanson, 24, of Kitchener; William Landsborough, 18, of Monkton; and Bonnie Rudow, 23, of Kitchener. All these places are in Ontario. It was thought that the group had been on their way to a party in Stratford. None of the car passengers had been wearing seatbelts and five of them were thrown from the vehicle when it went down an embankment. The driver had apparently braked at the last moment, perhaps when he caught sight of the train. [721]

The crossing was protected only by signs. Michael Herron's father, Paul, stated, "The tracks are bad because you can't see. You down a bit of a hill and you're in a bit of a gully and you can't see the train if you're going any kind of speed – not that I'm saying they're speeding."

No one on the Chicago-to-Toronto train was hurt and its 139 passengers were forwarded to

---

716 *Medicine Hat News*, Medicine Hat, Alberta, Canada, 30th April 1993

717 *The Daily Telegraph*, London, England, 2nd August 1993

718 European Union Agency for Railways: List of fatal railway accidents 1990-2009 (more than five fatalities)

719 *The Guardian*, London, England, 24th August 1993

720 http://www.scc.rutgers.edu/serbiandigest/101/t101-9.htm

721 *The Medicine Hat News*, Medicine Hat, Alberta, Canada, 7th September 1993

their destination on buses. Passenger Tom Rajewski, of Flint, Michigan, United States, told a newspaper reporter, "We barely hit it. All I saw was a car headed for the tracks. He hit it, that was it." [722]

Initial reports stated that ten people had died and 22 had been injured when a bus attempted to cross the line in front of a train in Morelia, capital of Michoacán state, Mexico, on the evening of Friday 5th November 1993. By Sunday, the number of dead had risen to 16, while another 24 victims remained in hospital. The bus driver, Daniel Sánchez Hernandez, was amongst those who lost their lives. [723] [724]

South Central Railway train No. 7034, the Up *Sahayadri Express*, collided with a school bus carrying pupils to a picnic on unmanned Crossing No. 11, between Alandi and Phursungi stations, Maharashtra, India, on Saturday 11th December 1993, causing numerous deaths and injuries. The accident occurred on the Pune – Miraj broad-gauge section of the Hubli division of the railway. No less than 39 people, probably all but one or two of them children, lost their lives and 44 people on the bus were injured, 19 seriously.

A member of staff at the Sassoon Hospital in Pune recalled the accident some years later. 'All the doctors rushed to the Casualty of Sassoon Hospital. We saw children aged between 6 and 12 with all sorts of injuries. I still remember those children in their colourful dresses, those lunch boxes, those small girls with bangles in hands and flowers in their hair, those faces of anxious parents who wanted to know whether their child was alive or dead. Doctors were trying to help these children and save lives. Some children were brought dead and some were breathing their last breath.

'Even after so many years when I remember the scene it sends shivers through my spine. I kept on wondering if any better medical aid could have been given to these children at the site of the accident.' [725]

The accident was the subject of a comprehensive investigation by the Commission of Railway Safety, which heard evidence from a number of witnesses. The statements of outsiders who more or less saw the accident happen and assisted in the rescue work were followed in their published report by the account of Shri Chandusingh Bhimasingh Rajaput, the driver of the ill-fated bus, who stated that he had been taking 74 pupils from Sri Nath Hari Purandhare Primary School from Pune to Theur and Ramdhara. There were no gates at the crossing, nor any signals on the road and he had had no warning of an approaching train. He had followed another bus over the crossing, stopping short of it just for a second. He had neither seen nor heard a train and so had proceeded over the line. He was aware that according to the Motor Vehicle Act he was obliged to stop before negotiating an unmanned crossing. He did not know if the crossing in question was manned or not. There were no warning boards, no speed breakers on the road and the condition of the road had been bad. His bus had 49 seats but according to the rules 80 or 90 small children could be carried. There were loudspeakers on the bus but no music had been playing and most of the children had been asleep at the time of the accident.

School peon (assistant) Dattaram Baraku Shinde testified that the children had been in a happy mood and singing loudly when they had reached the crossing (so not all had been asleep, as related by the bus driver). He did not know if anyone had noticed the approaching train. He did not think that the cleaner (who was killed) had said anything to the bus driver

---

[722] *The Lethbridge Herald*, Lethbridge, Alberta, Canada, 7th September 1993

[723] *Fairbanks Daily News-Miner*, Fairbanks, Alaska, United States, 7th November 1993

[724] *Times-News*, Twin Falls, Idaho, United States, 8th November 1993

[725] *The Times of India online*, 5th January 2001

about the train and did not know whether or not the bus had stopped at the crossing. This was because his (the peon's) attention had been entirely focused on the children.

Teachers Bhagavat Gitanjali Gajanand and Makashir Arundhathi had also been riding on the bus but could not add any new information.

Driver 'A' G.R. Navagaonkar, in charge of the engine of the colliding train, testified that he had had no problems in the journey to Phursungi, where the station signals, which were not far from the crossing involved, had been clear. He had sounded the horn continuously from the whistle board for Crossing No. 11. Two telegraph poles from the crossing he had seen a bus enter it and applied his brakes in emergency. That bus had just got across in front of his train but then to his dismay, a second bus had followed the first onto the crossing and a collision with it could not be averted. His train had been travelling at 90 km/h at the time of the brake application and had stopped 483 metres beyond the crossing.

Diesel Assistant Shaik Mohammed Amein corroborated the evidence of his driver.

Guard K.O. Chacko confirmed that his driver had whistled for the crossing. After the accident he said that he had delegated Assistant Guard S.G. Mapari to protect the rear of the train, had sent information of the accident to Phursungi cabin via another stationmaster who had been travelling on the train and then had rendered first aid to the injured.

Various other witnesses gave evidence.

Discussing the accident, Commissioner Y. Ganesan put the time of the accident at 11:24 and the speed of the train at 90 km/h. He noted that whether or not the driver of the express had seen the bus was irrelevant, as the train had the right of way. That the train driver had whistled for the crossing was confirmed by independent witnesses, though the noise of the bus engine and the singing children might have masked the sound of the train horn from the bus driver. It was also clear that the train had been braked as its driver had stated and that its brakes had been in good order.

Suggestions that trains should reduce speed on approaching every unmanned crossing were totally impracticable. The statement of the train driver that the first bus had had a narrow escape was accepted. It was clear that the train driver was in no way to blame for the accident.

The crossing had been re-sited to improve the visibility of trains. In its new position, visibility of a train coming from Kolhapur (as the accident train had been) was 560 metres and from Pune 700 metres. The visibility requirements for the crossing were thereby fulfilled. In addition, the signals at Phursungi station in both directions were clearly visible from the crossing. Factors such as the rather rough road approaches to the crossing could not have contributed to the accident.

Neither of the bus drivers had stopped short of the crossing to check that no trains were approaching, if need be by sending a helper ahead. The statement of the driver of the leading bus that he had seen no train could not be accepted and it was clear that he had had a narrow escape.

The commissioner concluded that the accident had resulted from the failure of the drivers of buses Nos. MTQ 5221 and MTO 1040 to observe the safety precautions laid down in Section 131 of the Motor Vehicles Act, 1988, and Rule No. 20 (i) of the Maharashtra Motor Vehicle Rules, 1989. Drivers Chandusingh Bhimasingh Rajaput of bus MTO 1040 and Laxman Baburao Kili of bus MTQ 5221 were therefore responsible for the accident.

Commissioner Ganesan made three remarks and recommendations:

(1) State Governments should gear up their enforcement machinery to ensure the strict observance of the regulations regarding road vehicles stopping short of unmanned crossings. The Railway Administration should co-ordinate with the police authorities the conducting of ambush attacks at unmanned crossings and track down offenders but the

primary responsibility should rest with the State Government rather than with the Railway.

(2) Most of the drivers of road vehicles were semi-literate and were not very clear as to the precise legal provisions regarding the use of unmanned crossings. State Governments might arrange for active propaganda regarding the matter through regional newspapers in local languages and perhaps also make use of television for the same purpose.

(3) Further work in removing unmanned crossings where it could be done without detriment to local residents should be pursued. Underbridge No. 22, for example, was only 463 metres from Crossing No. 11 and could be fairly easily modified to take the Loni – Randhara road, which at present used Crossing No. 11. [726]

Northern Railway Down express No. 4588 collided with a Jeep on manned engineering Crossing No. 64-C, between Hanumangarh and Manaksar stations, Rajasthan, India, at about 17:45 on Saturday 15th January 1994. Seven people in the Jeep were killed. [727]

A very brief report in a British newspaper stated that 27 people died when a train collided with a bus on a level crossing near Baghdad, Iraq, on about Monday 7th February 1994. [728]

São Marcos da Serra, Portugal, was the location of a level crossing accident at kilometre 274.94 of the Sul line on Thursday 17th February 1994.[729] The driver of a school van taking 14 children aged between eight and 14 to school early in the morning failed to observe the warning light at the crossing and drove onto it in front of the train. The driver and six of the children were killed. Police were reportedly working to free two children trapped in the wreckage, while nine people, children and railwaymen, were less seriously hurt. [730]

The matter of frequent level crossing accidents having occurred in the São Marcos da Serra area was subsequently raised in Parliament, with 13 reported fatalities in the previous 12 years and demands that the Caminhos de Ferro Portugueses install barriers at the two busiest crossings. The railway company replied that the traffic intensity at the crossings was not such as to justify such an expensive measure, whereupon the Portuguese Communist Party accused the company of putting money before human lives. [731]

Western Railway No. 310 Down passenger train collided with a lorry on Crossing No. 56 between Liliya Mota and Savar Kundla stations, Gujarat, India, on the Dhasa – Mahuva metre-gauge section of the Bhavnagar division of the railway, on Monday 21st February 1994. The crossing was unmanned. Eight people on the lorry were killed and 27 were injured, 13

[726] Government of India Ministry of Civil Aviation and Tourism (Commission of Railway Safety) Railway Accident Investigation Report on Collision of No. 7304 Up Sahayadri Express with Bus No. MTO 1040 at unmanned level crossing No. 11 at Km 17/11-10 between Alandi and Phursungi stations on Pune – Niraj broad gauge section of Hubli division of South Central Railway on 11-12-1993

[727] Government of India Ministry of Civil Aviation and Tourism Commission of Railway Safety Annual Report for 1993-94

[728] *The Daily Telegraph*, London, England, 8th February 1994

[729] European Union Agency for Railways: List of fatal railway accidents 1990-2009 (more than five fatalities)

[730] *The Gleaner*, Kingston, Jamaica, 18th February 1994

[731] Wikipedia – Acidentes ferroviários en Portugal

seriously. The accident was blamed on the negligence of the lorry driver. [732]

Ignoring the warning signals at a level crossing, a lorry driver entered it and came into collision with a train at an unstated location in central Taiwan on about Sunday 20th March 1994. As a result of the collision three coaches of the train were derailed and the lorry driver, train driver and seven train passengers were killed and 24 were injured. Many of the passengers had to be cut free from the wreckage of the coaches, [733]

A train collided with a lorry on a level crossing at Gaerias, Portugal (kilometre 101/71 of the Oeste line), at 12:50 on Saturday 30th April 1994. The force of the impact was sufficient to cut the lorry in two and three building workers were killed outright, two succumbed to their injuries in hospital and a sixth was seriously hurt. [734] [735]

Thirty-five people died when the *Narayandri Express* collided with a tractor trolley in the Nalgonda district of Andhra Pradesh, India, on Tuesday 3rd May 1994, on the South Central Railway. [736]

Jogoo Road Crossing, Nairobi, Kenya, the scene of a bad accident in May 1991, suffered another one on Thursday 19th May 1994 when an engine collided with a Nyayo bus, killing five people. [737]

Northeastern Railway train No. 5063, the Up *Avadh Express* from Gorakhpur to Bandra, consisting of an engine and 17 coaches, collided with a tractor trolley on manned Class 'C' Crossing No. 269, between Gonda Kachehari and Maijapur stations, Uttar Pradesh, India, at 17:15 on Saturday 28th May 1994. Twenty-seven people on the tractor trolley were killed and six were injured, two seriously. After the accident the gates were found to be open and their chain and lock were hanging down in a broken condition. [738]

Five people died and 29 were injured when a train collided with a bus on an unguarded level crossing at Granja, Portugal (kilometre 39.91 of the Póvoa line), on Sunday morning, 29th May 1994. According to the train driver, the bus had approached the crossing without decreasing speed, "and I was too close already to stop in time." The bus driver, who escaped death, lived 500 metres from the level crossing and had used it as a professional motorist for years. These circumstances might partly explain the accident. The bus usually arrived at the level crossing several minutes before the passing time of the Póvoa de Varzim – Famalicão train, but on this Sunday it was about five minutes late. There were only three Póvoa – Famalicão trains on Sundays.. The train hit the bus, which was carrying passengers to the

---

[732] Government of India Ministry of Civil Aviation and Tourism Commission of Railway Safety Annual Report for 1993-94

[733] *Western Morning News*, Plymouth, England, 21st March 1994

[734] European Union Agency for Railways: List of fatal railway accidents 1990-2009 (more than five fatalities)

[735] *The Anglo-Portuguese News*, Lisbon, Portugal, 2nd June 1994

[736] https://www.Rediff.com/news/report/fire1/20030515.htm

[737] *Daily Nation on the Web*, 'Ngai Ndethya' the worst yet, 25th March 1999

[738] Government of India Ministry of Civil Aviation and Tourism Commission of Railway Safety Annual Report for 1994-95

Senhora da Saúde festivities in Laundos, in the middle, and completely destroyed it. [739] [740]

Mystery surrounded the death of a woman and four children in a level crossing accident at 7:52 a.m. on Thursday 14th July 1994. Penny Corona, 27, had been returning home to Lodi, California, United States, after dropping off her boyfriend, Brian Martindale, at work at a Stockton construction business, when the accident occurred at Eight-Mile Road Crossing, north of Stockton. Witnesses told how she had manoeuvred her 1971 Plymouth station waggon around the lowered barrier of the crossing, past flashing warning lights and ringing bells, into the path of a Union Pacific Railroad (UP) freight train, which was then only some 150 feet away with its horn blaring. The train was travelling at 57 mph, 3 mph below the maximum authorised speed. It consisted of three diesel units and 60 waggons loaded with new automobiles. Despite an emergency brake application it took it almost a mile to stop. The car was pushed some 360 feet before it went down an embankment. Ms Corona, her daughters Natosha Hope, 7, and Kayla, 6, and her boyfriend's daughters, Lyne, 7, and Altina Marie, 6, were all killed instantly.

UP Area Manager Steve Smith commented, "She would have had to see the train. Maybe she thought it was going slower than it was."

California Highway Patrol spokesman Robert Whitmire said, "A person using good judgement would have been able to...tell they weren't going to make it." [741]

A Burlington Northern Railroad freight train from Laurel, Montana, to Fort Worth, Texas, travelling at 45 mph, collided with a 1970 Ford Maverick car on an open level crossing at Douglas, Wyoming, United States, at 8:04 p.m. on Tuesday 2nd August 1994. Struck broadside, the car was carried 70 yards and its driver, Noah Stavnes, 18, and passengers Jennifer Coziahr, 14, Tiffany Rabun, 14, Jeremy Stavnes, 15, and Ryan Wilson, 15, were all killed. It was not known if the car had stopped at the stop sign protecting the crossing or if alcohol was involved. [742] [743]

Vicror Manurung, 40, the driver of a Bemo three-wheeler, apparently thought he had time to cross the line in front of the *Cirebon Ekspres* at 16:30 on Wednesday 24th August 1994, despite the warning alarm sounding at the Kebun Singkong Crossing in East Jakarta, Indonesia. Even the lowered bamboo barrier did not deter him as he smashed through it. Manurung and seven of his passengers were killed in the resulting collision and the other three passengers were seriously hurt. The bodies of the dead were Cipto Mangunkusumo General Hospital mortuary and the injured were taken to Persahabatan Hospital, nearby.

Although it seemed clear that the accident had resulted from the recklessness of the car driver, the police questioned Crossing Keeper Sukadi, 50, and Stationmaster Mustakim, 39, in charge of Klender station, a short distance away. [744]

A train collided with a minibus that had stalled on a level crossing in the town of

---

[739] European Union Agency for Railways: List of fatal railway accidents 1990-2009 (more than five fatalities)
[740] *The Anglo-Portuguese News*, Lisbon, Portugal, 5th May 1994
[741] *The Bakersfield Californian*, Bakersfield, California, United States, 15th July 1994
[742] *The Daily Inter Lake*, Kalispell, Montana, United States, 3rd August 1994
[743] *Daily Times*, Delaware County, Pennsylvania, United States, 4th August 1994
[744] https://jawawa.id/newsitem/inter-city-train-hits-bemo-killing-eight-people-1447893297

Kebumen, Java, Indonesia, on Thursday 1st September 1994. The bus was carried 500 yards and 11 people on it, five of them from the same family, were killed instantly.

A school bus that had picked up children from the hamlets of Vers-chez-Perrin, Vers-chez-Savary and Corges, to take them to school at Payerne, Switzerland, had 14 pupils on board aged between 10 and 15 when it was struck by a train at the Glatigny level crossing in Payerne just after 07:00 on Friday 23rd September 1994 and dragged about ten metres, before ending up in someone's garden. An 11-year-old boy was killed in the accident and eight other pupils were hurt, none of them that seriously, though one, who had suffered a leg fracture, was taken to Lucerne for surgery. The other seven children, who included two daughters of the woman bus driver, were treated in the hospital at Payerne.

A witness stated, "Most of the children got out of the bus after the accident without panicking. I don't think they had realised what had happened to them."

An investigation was opened into the accident, which was apparently due to human error on the part of the railway staff. Just before the collision a freight train had left Payerne for Moudon and the barriers at the crossing had been lowered for it to pass. As soon as the freight train had gone, a shunting movement, involving an engine and three waggons, had also begun and had to pass over the same crossing but in the meantime, the barriers had been raised, whereupon the bus had started over the line, as had a car coming from the opposite direction. The shunting train struck both, though the car driver was unhurt.

The removal of the crossing had been proposed in the 1970s. However, because of the cost of substituting a bridge, nothing had been done by Vaud canton. [745] [746]

On the evening of Friday 23rd September 1994 James M. Allen, 47, and his wife Susan, 45, were driving four friends of their daughter Brooke to a football game at Heath High School, where the school was to play a match against Liberty Union High School. Brooke herself, an only child and a cheerleader at the game, had gone on earlier. At 7:22 p.m. the car reached an open level crossing at Baltimore, Ohio, United States, at the same moment as a Conrail freight train, which struck and demolished it. Mr and Mrs Allen, twins Laura and Lisa Willey, 15, and Allison Good, 14, were all killed. The fourth girl, Holly Weekley, 14, was injured and was said to be in a fair condition the following day.

The train, from Dunkirk, New York, to Dickinson, West Virginia, consisted of four locomotives and 103 empty coal cars and was travelling at 40 mph. Its leading engine was equipped with a new safety feature named "ditch lights," flashing headlights designed to attract the attention of motorists.

A Conrail spokesman stated that he was not aware of any accidents at the crossing in the previous five years, though a woman who lived near the crossing told a reporter that she remembered at least three within the six years that she had lived there. She added, "I heard the train whistle and then I heard a bump, but it just sounded like train cars bumping together. The only time we've heard a train stop like that is when there's been an accident."

A wooden crossbuck was the only warning at the crossing. Lights and gates were scheduled to be provided in May 1995. [747] [748]

---

745 *FAN-L'express*, Neuchâtel, Switzerland, 24th September 1994

746 *L'Impartial*, La Chaux de Fonds, Switzerland, 24th September 1994

747 *The News Herald*, Panama City, Florida, United States, 25th September 1994

748 *The Times-Reporter*, New Philadelphia, Ohio, United States, 25th September 1994

On the morning of Thursday 20th October 1994 a freight train collided with a crowded bus carrying workmen on an unguarded crossing at Shibali, a village a few kilometres outside of Beijing, China. The force of the collision was sufficient to shear the vehicle completely in two and at least 17 bus passengers were killed and 30 were injured, according to the *Beijing Evening News*. Other sources stated that at least 20 had been killed and only a handful of the 80 or 100 people on the bus had escaped injury. The bus was owned by the Beijing Internal Combustion Engine Group Corporation. [749]

With the death of all seven of the occupants of the road vehicle involved, the cause of an accident at an automatic half barrier level at Santa Fe de Modújar, Spain, near the town of Gádor, at about 09:10 on Thursday 17th November 1994 could not be established with absolute certainty. The vehicle involved was an ambulance, which had picked up patients from Illar, Alhabia and Albodhuy to take them to the Torrecárdenas health centre in Almeira on a routine visit. Checks carried out by Red Nacional de los Ferrocarriles Españoles after the accident indicated that the crossing equipment was working normally, implying that the ambulance driver, Manuel Pelayo López, 42, had manoeuvred his vehicle around the lowered barriers. It was struck by a light engine proceeding towards Huéneja to pick up a mineral train, carried 200 metres and its occupants were all killed instantly. The train driver was unhurt. It was foggy at the time of the accident with reduced visibility.

The ambulance passengers who lost their lives were named as Gregoria Maillo Cabello, 31; Francisca Ayala Valverde, 37; her son Alejandro Maartinez Ayala, eight months; Ana Maria Castellón Orta, 30; her daughter Milagros Larios Castellón, five months; and Manuel López Rodriguez, 52. Francisca had been the wife of Francisco Martínez, the mayor of Albodhuy. A newspaper reported that Julián Larios Fernández, the husband of Ana Maria, had sobbed after hearing of the tragedy, "The girl had to go to the health centre twice a week. I always told Ana not to use the ambulance but to take the car."

Despite the assertions of the railway that the crossing had been working correctly, locals were not so sure. Antonio Pelayo López, 27, brother of the dead ambulance driver, told a reporter, "He had been making this journey six or seven times a day for four years and was very careful behind the wheel. He would never have crossed the track when the barriers were down. If he did so, it must have been because the barriers were up." Another resident commented that sometimes the barriers could be lowered for 15 minutes and no train would appear, while at other times one barrier would be lowered and the other remain raised.

Gádor City Council declared three days of mourning for the victims. There were ten level crossings in the Gádor area, four with barriers and six without. Gádor mayor Eugenio González called them "death traps" and Rosario Peral, Almeira civil governor, promised to intercede to have them removed. [750] [751]

On Wednesday 9th November 1994 what was described as a 'medium sized' bus smashed through the gates of a level crossing at an unstated location in Honan province, China. Sixteen people died and 35 were injured, some seriously. [752]

[749] *The Ottawa Herald*, Ottawa, Kansas, United States, 20th October 1994
[750] *El Paso*, Madrid, Spain, 18th November 1994
[751] https://www.diariodealmeria.es/almeria/Terrible-accidente-ferroviario-Gador_0_283472153.html
[752] www.jrtr.net/jrtr04/topics.html

A Chevrolet Corsica car carrying five teenagers was struck by a Norfolk Southern Railway freight train on an open level crossing near Crittenden, Kentucky, United States, at about 10:25 p.m. on Wednesday 23rd November 1994. Four of the teens: Steve Chandler, 16; Jesse Hagar, 14; Bertha Toll, 14; and her sister Liana, 16, who was the car driver, were all killed more or less outright. Larry Rippentrop, 14, was very badly hurt and succumbed to his injuries the following Sunday.

The crossing was one of two, 350 yards apart, which led into the Grant Mobile Home Park, where three of the teenagers lived. The crossing where the accident occurred was protected only by white and yellow signs, each with black lettering, while the other crossing was fully protected by gates, warning lights and bells. Residents of the mobile homes stated that they had heard the engineer of the train blowing his horn furiously for the crossing, while the engineer himself stated that the car had pulled onto the tracks and stopped, then had just been pulling away again when it was struck. It seemed likely, therefore, that the inexperienced driver had stalled her vehicle.

An investigation disclosed that Liana Toll had only had a learner driver's permit for six days and according to the law, she should not have been driving the car, except in the company of a licensed driver. [753]

A lorry carrying nine police officers and a police official was struck by a train, presumably on a level crossing, at Sohag, Egypt, on Sunday 4th November 1994. The vehicle was thrown into the River Nile and all were killed. [754]

Kindergarten children riding in a 16-seater minibus were the victims of a level crossing accident at an unstated location on the Ibusuki – Makurazaki line in Kagoshima Prefecture, Japan, on Thursday 26th January 1995. A four-car passenger train struck their bus, killing two of them and injuring 12 others, some seriously. [755]

The circumstances of a level crossing collision involving a Florida Central Railroad train and a school bus at Apoka, Florida, United States, on Tuesday 7th February 1995, were disputed. Driven by Nora Hare, 52, the bus was carrying 32 Piedmont Lakes Middle School students at the time of the accident. As a result of the impact the bus was carried about 70 feet down the line before it then overturned onto its side. The results were not as serious as they might have been but the driver and two students, Adrianne Gray and Lilquincy Grimmage, both aged 12, were badly enough hurt to be admitted to hospital, while the other 30 pupils were treated for minor injuries before being released. The train had been travelling at 20 to 25 mph at the time.

Lt. Floyd Baker of Florida Highway Patrol told a reporter that both the bus and the train had been southbound when the bus had made a hard left turn onto a dirt road across the track at an unprotected crossing. "The report from the engineer said that he sounded his whistle and the bus driver turned into his path," the trooper said. Some of the pupils, however, said that the bus had stopped before crossing the track. Lt. Baker said that failure to render right-of-way was a misdemeanour. "If we find that the bus driver failed to stay stopped and render right-of-way to the train we will be charging her," he added. Ms Hare had been driving a school bus for 21 years.

---

[753] *Times-Tribune*, Corbin, Kentucky, United States, 25th, 26th and 28th November 1994

[754] *The Herald Bulletin*, Anderson, Indiana, United States, 5th December 1994

[755] www.jrtr.net.jrtr05/topics.html

Florida Central Railroad stated that it would open its own investigation into the accident. [756]

The National Transportation Safety Board (NTSB) noted that the warning bell and whistle of the train had been operating. The view of the train from the bus and vice versa was limited to about 540 feet; the bus was overdue for its annual inspection. The NTSB concluded that the accident had been due to the failure of the bus driver to give way to the train. [757]

At 07:05 on Saturday 18th February 1995 the northbound *Rápido de la Frontera* express from Temuco to Santiago collided with a bus on the Juan Egenau level crossing just north of the station at Rengo, Chile, about 140 kilometres south of the capital, Santiago. The bus had been chartered to take people from the capital to a summer resort near Rengo. Fourteen bus passengers lost their lives in the collision and another 33 were injured, some of them seriously.[758] The bus was dragged several metres by the train, badly damaging two small lineside houses in its path. Some of those who lost their lives had survived the initial collision only to be fatally electrocuted by electric cables brought down by the impact. No one on the train was hurt. [759]

Late on the night of Sunday 2nd April or early on the morning of Monday 3rd April 1995 a passenger train collided with a heavy lorry on an unprotected level crossing, probably at a location south of Prachuap Khiri Khan station, Thailand, in the province of the same name. Ten carriages of the train were derailed and the engine and three carriages caught fire. Eleven or 12 passengers, the train driver, a second railwayman and the lorry driver were all killed and at least 77 passengers were injured, 20 of them seriously. The bodies of many of the dead were burnt beyond recognition.

Railwayman Likit Thanthankal said, “At first I thought it was a bomb.”

His colleague Somsak Deephan added, “I heard a bang and when I managed to get up, there was fire and dust everywhere.” [760] [761]

An attempt by the driver of a 1983 Volvo sedan containing six teenagers to cross the line ahead of a Burlington Northern Railroad train early on Saturday 8th April 1995 resulted in a collision with it and the death of all of them. The tragedy occurred at Titan Road Crossing, near Castle Rock, Colorado, United States, and west of Highway 86. The train consisted of an unspecified number of diesel units pulling 116 coal cars and was travelling at 43 miles an hour. It was not clear how fast the car was travelling. Although there were no barriers at the crossing it was protected by warning lights. The absence of skid marks indicated that the car driver had made no attempt to stop.

The engineer of the train had sounded his horn approaching the crossing. He told police he had seen “a brief flash of light as we hit something,” and the train brakes had then been applied in emergency. The car was dragged 167 feet, rolled over and burst into flame, burning the bodies of five of the victims so badly that at first it was thought that there was only one person in the vehicle. The sixth teen was thrown clear.

The victims were identified as Mike Budinger, age unknown; Mike De Boo, 17; Nathan

---

[756] *Daily News*, Fort Walton Beach, Florida, United States, 8th February 1995

[757] National Transportation Safety Board report

[758] *The Herald Bulletin*, Anderson, Indiana, United States, 19th February 1995

[759] http://members.fortubecity.es/treneschilefotos/e32.html

[760] *Medicine Hat News*, Medicine Hat, Alberta, Canada, 3rd April 1995

[761] *Rushville Republican*, Rushville, Indiana, United States, 3rd April 1995

Gladwin, 17 (the driver); Tom Maestas, 17; Beth Linton, 15, the victim who was thrown clear; and Tera Relfschneider, 16. All but Tom Maestas were students at Heritage High School; he had left Heritage earlier in the year for an alternative school.

Carol Dosmann, whose 17-year-old daughter had been dating one of the victims, said, "I don't know how the kids will cope with this. It's supposed to be a happy time with graduation."

Bob Dosmann said that his daughter should have been in the car but she had chosen not to accompany the others. "She wasn't there because she made the right choice," he commented.

His daughter reportedly remarked that the car driver was "doing stupid stuff like that all the time." [762]

Student Julie Vucich told a reporter, "I keep saying to myself. What if it was me? This should be a reality check for all of us. We can think we're indestructible but we're not."

The school planned to have grief counsellors on hand for when the pupils returned to school. Principal Jim Sekich said, "We're just devastated. We don't know what to expect."

Teacher Jill Ricard added, "It feels sick…It will hit hard because one student has friends, but six students have many friends."

It was not known if alcohol was involved in the accident but the teens had been returning from a party a few miles away.

A bus belonging to the Misr Menoufia Spinning and Weaving Company had just picked up textile workers from the towns of Shebeen El-Kom and Arab El-Raml to take them to the factory in Quweisna, Egypt, and was on the outskirts of the latter town in the Nile delta when it was involved in a disastrous level crossing collision at 06:50 on Saturday 15th April 1995. When the bus arrived at the crossing, two trains were also approaching it from opposite directions. As soon as the first train had passed the bus proceeded over the line, to be struck broadside by the second, an Alexandria-to-Cairo express. 'The wreckage of the bus, mixed up with shoes and clothing, was spread over 200 metres of track,' a newspaper reported. Forty-two people lost their lives in the accident according to first reports and another 45 were injured.[763] Six slightly injured victims were treated at the scene and the other 39 were taken to a nearby hospital. Although all of those killed had probably been riding on the bus, some of those injured were roof-riding train passengers who had been thrown off when the collision occurred. It took about two hours to clear the wreckage from the line.

There were conflicting reports as to how the accident had occurred. Witnesses stated that after the bus driver had stopped for the northbound train he had been flagged across by the crossing keeper. This scenario was contradicted, however, by a statement from the Interior Ministry, which indicated that although the warning bells at the ungated crossing had still been ringing, the bus driver had started across as soon as the first train had gone, ignoring the efforts of the crossing keeper to stop him. [764]

A minivan bringing a party of seven teenagers home from a religious retreat at a church at Cypress Hill Provincial Park, was in collision with a CP Rail freight train on a level crossing at Maple Creek, Saskatchewan, Canada, at about 4:30 p.m. om Sunday 28th May 1995. The eastbound train of rather more than 70 empty potash cars was travelling at 80 km/h at the time. The minivan was bound for Burstall, about 100 kilometres further north.

---

[762] *Gazette Telegraph*, Colorado Springs, Colorado, United States, 9th and 10th April 1995

[763] *Medicine Hat News*, Medicine Hat, Alberta, Canada, 15th April 1995

[764] *The Daily Inter Lake*, Kalispell, Montana, United States,16th April 1995

As a result of the accident Wally Baker Jn., 13; Blaise Bodnerchuck, 14; Brenda Lewis, 16; Mark Lewis, 14; and Trent Plato, 19, the driver, were all killed. The other two passengers, Kyle Bodnerchuck, 19, and Gerald Job, 15, were seriously injured and were taken to hospital, Kyle to the Royal University Hospital in Saskatoon and Gerald to the Regional Hospital in Medicine Hat. All were believed to be from Burstall, a village of about 500 people. Trent, who had been training to become a pastor, was one of the organisers of the retreat, which was sponsored by the Burstall Evangelical Mission.

Sergeant Bill Sparkes of Maple Creek Royal Canadian Mounted Police (RCMP) commented, "As unbelievable as it sounds, according to witnesses it's almost as if the driver never saw the train. Absolutely no alcohol was involved whatsoever. What we put it to is simply driver inattention. The crashing lights at the crossing were working. The train whistle did, in fact, go." He added that the accident was ironic, as, "They just spent the whole weekend as close to God as you can be." [765]

Continuing his statement, the RCMP officer provided the likely explanation for the accident, stating that Trent Plato had been following his father in another vehicle. The elder Mr Plato had heard the train whistling but decided that there was time for him to cross anyway and had done so. A few minutes later, looking back, he had found that his son was no longer following him, so he had gone back and found the accident. "He's actually blaming himself. It's pretty sad," Sgt Sparkes said. "It wasn't his fault. You hate to say it but it seems like it was inattentiveness that caused it." As to the train crew, who were in shock, he said, "It must be devastating to see the vehicle and not be able to do anything about it.

A witness to the accident was a man named Dave Ferguson, who had been pumping gas (petrol). He related that he had talked to the train engineer after the accident and the latter had appeared badly shaken. "He told me, 'I blew the whistle,' and, 'The lights were going. It's not my fault.' I told him, 'I know it's not your fault,' and then he started to cry," Ferguson said. [766]

CP Rail spokesman Don Bower stated that the train crew had been relieved and were being given counselling until they were judged fit to return to work. He remarked, "You're blowing your whistle and doing everything you can, hoping that someone… will realise there's a train coming." Cases of people trying to beat the train at crossings were disturbingly common. "For the two to three minutes it usually takes to wait for a train – it's just not worth it," Bower declared. Fifty-five people had died and 57 had been injured in Canadian level crossing accidents in 1994 and he noted that the numbers had remained static, despite the efforts of railway companies to increase safety awareness at crossings. He stressed that crossing signals should be obeyed as diligently as stop signs at road junctions and traffic lights, especially as the results of ignoring them could be so devastating. "The lights don't make you stop your vehicle. You stop it." [767]

A van drove onto Sg Tok Pawang Crossing, Malaysia, into the path of a train, on Tuesday 6th June 1995, resulting in a collision and the death of seven people in the van. A year later the van driver, S. Thachimana, aged 41 at the time of the accident, was sentenced to six months' imprisonment for negligence. He pleaded that the crossing gates had been open, that there had been someone standing near the guard hut, that neither he nor his passengers had heard the whistle of the approaching train and that bushes had obscured his view of it. His defence counsel submitted that negligence on the part of the crossing keeper in failing to close the

[765] *Medicine Hat News*, Medicine Hat, Alberta, Canada, 29th May 1995

[766] *Brandon Sun*, Brandon, Manitoba, Canada, 30th May 1995

[767] *Medicine Hat News*, Medicine Hat, Alberta, Canada, 30th May 1995

gates had caused the accident. [768]

A pickup truck taking a family of seven farm workers to their labours in an onion field on the morning of Thursday 8th June 1995 was struck by a train at about 5:30 a.m. The rural level crossing near Nyssa, Oregon, United States, was marked only by a stop sign. Sylvestre Perez, 47, the driver of the truck, apparently saw the approaching Union Pacific Railroad train and slowed up but then decided to beat the train over the crossing. Probably he was expecting a slow-moving freight train; instead it was an Amtrak passenger train travelling at about 70 mph. The impact tore the truck in two and wreckage and bodies were scattered for about a mile along the line. Sylvestre, his daughters Cecilia, 20; and Lidia, 23; his sister Sofia, 49; his nephew Victorino Trinidad, 28; Lidia's husband, Bernebe Ayala, age unknown; and Sofia's husband, José Luis Joaquin, 42, were all killed. The family were from Guerre state, Mexico, but had lived in Nyssa for seven years.

None of the 165 passengers or 13 crew on the train, the westbound *Pioneer*, en route from Chicago, Illinois, to Seattle, Washington, was hurt. The train was delayed about three hours by the accident.

The pickup truck had been in a convoy of vehicles taking about 40 workers to the field. Amador Lopez, the foreman of the workers, who had been in the leading vehicle, stated that he had crossed the railway before he noticed the train and had been about 1,200 yards further on at the time of the collision. He related, "It was daylight. I could see the train coming. I saw the train hit the truck." He considered that the crossing with a rural road, about two miles from the farming town, which was close to the Idaho state line, should have some form of protection. "It's only a lonely country road, but you have lots of migrant workers going over these crossings," he said. [769]

Six people lost their lives and seven more were injured in a level crossing accident between Győrszabadhegy and Nyúl stations, Hungary, on Tuesday 27th June 1995. [770]

Five people were killed outright and three were injured in a level crossing accident at Sorkifalud, Hungary, on Monday 11th September 1995.[771] The driver of a minibus carrying workmen went around the lowered crossing barriers and it was struck by a Zalaegerszeg – Szombathely passenger train. The bus driver was one of those killed. One of those injured succumbed eight days after the accident. [772]

At about 7:10 a.m. on Wednesday 25th October 1995 an eastbound Northeast Illinois Regional Commuter Railroad Corporation (Metropolitan Rail) (METRA) passenger train No. 624, running on a line of the Union Pacific Railroad, collided with a Transportation Joint Agreement (TJA) school bus on a level crossing with Algonquin Road in Fox River Grove, Illinois, United States. The bus was stopped at a T-junction with US Route 14, where the traffic lights were red, with its rear foul of the line. As a result of the accident seven of the 35

[768] *The Star online*, Petaling Jaya, Malaysia, 8th June 1996
[769] *Daily News-Miner*, Fairbanks, Alaska, United States, 9th June 1995
[770] European Union Agency for Railways: List of fatal railway accidents 1990-2009 (more than five fatalities)
[771] European Union Agency for Railways: List of fatal railway accidents 1990-2009 (more than five fatalities)
[772] https://www.origo.hu/itthon/1899/12/20030508alegsulyosabb

school bus passengers were killed and 24 were injured to various degrees. None of the three crew or estimated 120 passengers on the train was hurt. The accident was investigated by the National Transportation Safety Board (NTSB).

The inquiry into the accident noted that the regular school bus driver had been ill and the bus had been driven that day by Mrs Pat Catencamp, who had never been over the route before, either driving a bus or in a car. She had to enlist the help of one of the first pupils she had picked up to give her directions. By the time the bus reached the crossing, she had made all but one of her stops. She had halted the bus short of the crossing, switched on its hazard lights, opened the bus door and looked left and right, seeing no trains. Her passenger/guide had not seen any trains either. Closing the bus door, she had then pulled over the crossing, which was provided with full lifting barriers, warning bells and lights. [773]

Recounting the accident, Mrs Catencamp said, 'I proceeded across the tracks because the light {the traffic lights at the T-junction} was red and I knew that I had to go up there and trip the sensor in order for the light to go green to proceed through the intersection. It was like moments that it {the accident} happened.' She had not realised that the bus was stopped foul of the railway line, thinking that there was 'plenty of room,' and said later that, 'It never entered my mind that there wasn't enough room for that bus to fit.'

Most of the bus passengers said they had been talking to each other or sleeping during the ride. Some commented that the noise level on the bus "was not any louder than normal" and "was really quiet" and that "people were talking calmly, like in a restaurant, but not real busy." Eight passengers reported that the music broadcast from the bus AM/FM radio had not interfered with their conversations; nine passengers reported that they had heard the train horn. Four said that they had seen the crossing warning devices activate, and eight recalled seeing the train before impact. Another four passengers remembered standing up to move forward in the bus just before the impact. From the back of the bus, passengers were yelling, "a train is coming," "we're still on the tracks," "move the bus," and, "I think we're gonna get hit by a train." The passengers recalled that such comments had initially been made in a joking fashion, but as the train had come closer, the yelled remarks had become serious warnings. The bus driver reported that her passengers were making a noise; however, she had not understood that they were warning her that a train was coming and added that she had not heard the train horn, the crossing bell, or the sound of the gate striking the bus. She said that she had not been aware that a train was coming until it had struck the bus.

The passenger assisting the bus driver with the route recalled, although the bus driver could not, that both the two-way radio with the TJA dispatcher and the AM/FM radio had been transmitting. He said that he had heard the train horn just before the train struck the bus, and then he and the bus driver had looked up into the rearview mirror at the same time. They then had simultaneously exclaimed, "Oh!" He also reported that he had observed the crossing gate in the mirror after hearing the train's horn. He said that "… right before the train hit," he had become aware of the others on the bus yelling about the train coming.

Several motorists travelling on US 14 reported that the traffic lights at the junction with Algonquin Road had turned from green to yellow to red just before the collision. However, the bus driver and passengers said that they had never seen the traffic lights on Algonquin Road itself turn green. Others who witnessed the accident and who could observe the same traffic light as the bus driver could only state that it had been red before the accident and green afterwards.

Train No. 624 consisted at the time of the accident of a bilevel (double-decker) control car, followed by six bilevel passenger cars and a locomotive at its rear. Its engineer stated that it had left Crystal Lake on time at 7 a.m. The train had travelled eastbound at no more than 40

[773] *Daily Herald*, Arlington Heights, Illinois, United States, 27th and 28th October 1995

mph as it approached Cary, Illinois, the first station southeast of Crystal Lake. There he had slowed the speed of train 624 to between 15 and 18 mph, as required by a railroad timetable special instruction, to allow a westbound train from Chicago, which was in the station at the time, to discharge its passengers and to provide time for the departing passengers to cross the tracks. He then had a clear signal, which permitted him to operate the train at a maximum speed of 70 mph. The engineer said that train 624 had been travelling about 66 mph as it had crossed the Fox River Bridge (about 2,300 feet west of the Algonquin Road level crossing) when he had seen that the school bus had crossed the first track at "a very slow speed." He reported that the sun had been coming up over a hill on his left but it had not interfered with his view of the school bus. He considered that if the bus had continued moving at that low speed, it would have cleared his train. However, because it had been going so slowly, he had placed the train throttle in the idle position, made a full-service brake application, and sounded the horn. The engineer stated that when he had realised that the school bus had not cleared the second track, he had placed his train brakes in emergency application. He related that the bus had come to a stop with its rear on the railroad tracks in the path of the train.

Train 624 struck the school bus in its left rear side, and it rotated counterclockwise. The bus body and its chassis separated and came to rest with the chassis facing in a northwest direction and the body facing south (approximately 42 feet apart from each other). Gouge marks led up to the final rest position of the bus body, which had struck and knocked down the breakaway traffic light stanchion at the southeast corner of the US 14 intersection. Train 624 did not derail, and its cab control car stopped about 1,422 feet east of the point of collision.

By chance, the accident was witnessed by the Fox River Grove police chief, who had been sat in a police car nearby, and he immediately radioed for assistance via his portable radio. After train 624 had stopped, its engineer also radioed his train dispatcher and reported striking the bus. The Cary Police Department dispatcher received multiple 911 calls at 7:13 a.m., whereupon he had notified the Fox River Grove Fire Department and the Crystal Lake Regional Command Center. Located on Algonquin Road, only 350 feet southeast of the accident site, the Fox River Grove Fire Department had responded at 7:18 a.m. with an ambulance, a fire engine, four emergency medical technicians, and two paramedics. The Fox River Grove assistant fire chief had arrived on the scene almost immediately and acted as incident commander (IC), requesting the Mutual Aid Bo Aid Box Alarm System third alarm response, which he later upgraded to a fifth alarm response. The IC established a command post in the Algonquin Road and US 14 intersection and triage areas on both sides of the school bus. Shortly after the Fox River Grove Fire Department had arrived on the scene, 20 ambulances from 18 fire departments, as well as 2 helicopters, arrived and transported 32 of the injured passengers to medical facilities. The local hospital activated its disaster plan and dispatched a doctor to the scene at 7:27 a.m. An estimated 90 fire and emergency response personnel and the McHenry County Coroner's Office responded and were directed by a paramedic triage officer to assist in the treatment and transportation of victims. Additionally, officers from 12 police departments responded and assisted with the accident investigation and traffic control. In less than 90 minutes from the time of the collision all the seriously injured passengers had been transported to one of seven nearby hospitals.

The NTSB conducted its usual very thorough investigation into the accident. In its analysis it found that the weather at the time of the accident had been dry and clear and the rising sun had not interfered with visibility. There had been no problems with the maintenance of either the train or the bus. The engineer had been familiar with driving trains over the line and there was no evidence that he had been fatigued. The school bus driver had been operating buses for 8 years, had not been fatigued and had followed the correct procedure upon arriving at the crossing. Neither the engineer nor the bus driver had been under the influence of alcohol or illicit drugs.

Travelling at 66 mph, the train would have activated the crossing controls when it was 2,400 feet from the crossing, 24 seconds before the moment of the collision. The situation at the crossing was complicated by the existence of a controlled pedestrian crossing parallel to US 14 and between it and the railway crossing, which normally showed 'Walk' to pedestrians when the traffic lights were green for US 14. When the train activated the crossing controls there was a 12-second delay to allow pedestrians to clear their crossing, before the traffic lights on US 14 went to red and then those on Algonquin Road went to green. If the lights on US 14 had only just gone to green when a train activated the crossing controls, there was a further 3-second delay before the lights on US 14 went back to red, to prevent them flashing just momentarily to green in the face of a motorist. In this instance the lights on Algonquin Road would go to green only about three seconds before a train arrived at the crossing, little time enough for the driver of a vehicle stopped foul of the crossing to move clear of it.

Ten seconds before the collision, the NTSB noted, the engineer had sounded his horn, closed the throttle and made a brake application; the train was at that point travelling at 69 mph and 1,000 feet from the crossing. Six seconds before the collision, with the train moving at 67 mph and 600 feet from the crossing, the traffic lights on US 14 would have gone to red and those on Algonquin Road to green. A second later, 5 seconds before the collision, the engineer had made an emergency brake application.

The bus had stopped at the red traffic lights on the far side of the crossing and the crossing gate (barrier) had then descended upon its roof. At first the bus passengers had joked about this, then, as the train had come closer, they had become alarmed. The lights on Algonquin Road must have gone green for six seconds before the collision but the bus driver, looking back to the rear of the bus because of the noise made by the alarmed pupils, had not noticed them change. As stated earlier, the bus driver was unfamiliar with the crossing.

With a length of just over 38 feet, the bus was considerably longer than distance between the stop line at the end of Algonquin Road and the crossing, though the bus driver had not realised that her vehicle was overhanging the railway. In this respect the accident was similar to that at Fort Lauderdale, Florida (see above) where the lorry driver had failed to realise that he had initially stopped his vehicle clear of the line. No specific training was provided to school bus drivers regarding their vehicles having clearance at level crossings. Regular bus drivers, familiar with the limited waiting area beyond the crossing, normally stopped their vehicles short of it until the traffic lights at the junction with US 14 went to green.

The noise level in the bus from the alarmed passengers had distracted the driver and prevented her seeing the lights go to green. The radio speaker positioned next to the bus driver would have increased the noise level and prevented her hearing the warning horn of the train.

The routing of the school bus over the crossing had been unavoidable but bus drivers had not been warned of the hazards of the crossing, especially the short queueing area. Nor was the driving performance of substitute school bus drivers monitored.

Road widening at the intersection of US 14 and Algonquin Road in 1989 had reduced the distance between the stop line and the crossing from 28½ feet to 21 feet, insufficient for dump trucks, mobile homes, articulated lorries, buses and so on. No road signs warned motorists of the danger (after the accident, additional traffic lights were installed short of the crossing).

The time of between 2 and 6 seconds provided by the interconnected system of highway and railroad signals at the crossing was inadequate for safety, while there had been inadequate communication and misunderstandings regarding warning times between the railroad and the Illinois Department of Transportation (IDOT).

The NTSB reached 16 conclusions regarding the accident:

1. Neither the weather, the position of the sun, the track, nor the mechanical condition of

the train or of the school bus had either caused or contributed to the collision. The train engineer had been qualified to perform his duties and working in compliance with the hours-of-service requirements, and the bus driver had been trained and experienced to drive school buses. Neither alcohol nor drug use by the engineer or the school bus driver had been a factor in the accident.

2. Emergency response personnel had reacted promptly and acted effectively and efficiently at the collision site; the emergency response efforts had been well coordinated.
3. The guidance or training provided in the Illinois school bus driver training curriculum about vehicle positioning in relation to the roadway had been ineffective.
4. Had the school bus driver discerned the combined visual and audible warnings that a train was approaching, she might have had sufficient time to recognise the hazard and move the bus before impact.
5. The methods employed by the school district to identify and evaluate route hazards had been ineffective. Furthermore, had the school district ensured that all school bus drivers exchanged information about any identified route hazards, such as the short queuing area, the driver might have avoided the collision.
6. Had the regular and substitute school bus drivers been monitored during their morning routes, school officials might have been aware that the regular drivers had habitually stopped on the south side of the Algonquin Road grade crossing to wait for a green indication.
7. The Illinois Department of Transportation had not employed sufficient measures before the accident to prevent vehicles from encroaching on the railroad tracks while stopped at the north side of the grade crossing.
8. During the Safety Board tests, the railroad grade crossing signal system had provided 20 seconds or more warning time before the arrival of a train.
9. The installation and use of railroad and highway signal recording devices at interconnected/pre-emptive grade crossings could improve opportunities for highway and railroad personnel to determine if the signals were coordinated and operating properly.
10. The highway traffic signal hardware (heads, controllers, masts, posts, and loop detectors) had conformed to design standards and had operated as intended, but the highway signal system had not provided sufficient time for northbound traffic on Algonquin Road to clear the grade crossing.
11. The highway traffic signal system before the collision had provided a green indication for northbound Algonquin Road for 2 to 4 seconds based on the post-accident testing, or for 2 to 6 seconds based on the highway signal system programming.
12. Had a database containing grade crossing signal system information been available after this accident, the States could have more readily identified and then inspected specific crossings to ensure the signal systems posed no hazards.
13. The Illinois Department of Transportation had programmed its highway signal system without applying the minimum warning time information from the railroad.
14. Had an effective communication system existed between the Illinois Department of Transportation (IDOT) and the railroads about interconnected signal systems, IDOT, its contractors, and the Illinois Commerce Commission might have understood that the railroad had provided through the thumb wheel setting only the FRA minimum requirement of 20 seconds warning time before the arrival of a train at the grade crossing.

15. Had a coordinated programme to ensure effective communication between transportation modes about all aspects of grade crossing safety been in operation, the ineffective communication between IDOT and the railroad might never have occurred.

16. The passengers found outside the bus had sustained their fatal injuries during the initial impact sequence and not as a result of being ejected.

The NTSB determined that the probable cause of the collision was that the bus driver had positioned the school bus so that it had encroached upon the railroad tracks because of the failure of:

1) the Illinois Department of Transportation to recognise the short queuing area on northbound Algonquin Road and to take corrective action.

2) the Illinois Department of Transportation to recognise the insufficient time of the Green signal indication for vehicles on northbound Algonquin Road before the arrival of a train at the crossing; and

3) the Transportation Joint Agreement School District 47/155 to identify route hazards and to provide its drivers with alternative instructions for such situations.

Contributing to the accident had been the failure of the Illinois Department of Transportation and its contractors, the Illinois Commerce Commission, and the railroads to have a communication system that ensured understanding of the integration and working relationship of the railroad and highway signal systems.

The NTSB made a series of recommendations.

To the US Secretary of Transportation:

(a) It should develop a comprehensive and periodic railroad/highway grade crossing safety inspection programme to be conducted jointly by railroads and public entities and also require railroads and public entities to coordinate any changes to railroad/highway grade crossings before implementation.

(b) It should notify, in cooperation with the American Association of State Highway and Transportation Officials, the National Association of County Engineers, the American Public Works Association, the Institute of Transportation Engineers, the Association of American Railroads, the American Short Line Railroad Association, and the American Public Transit Association, railroads and public entities about the importance of exchanging information about railroad/ highway grade crossings.

(c) It should develop a common glossary of railroad/highway grade crossing terms and disseminate this glossary to railroads and public entities.

(d) It should develop a training program in the design and operation of railroad/highway grade crossings that included the interaction between rail and highway signal systems and require representatives of the railroads, public entities, and others who designed and maintained grade crossing signal systems to complete the training program.

(e) It should require the use and maintenance of railroad and highway traffic signal recording devices on all new and improved installations at railroad/ highway grade crossings that had active warning train detection systems and were interconnected/pre-empted to highway signal systems. These devices should record sufficient parameters to allow railroad and highway personnel to readily determine that the highway signals and railroad-activated warning devices were coordinated and operating properly. It should also require that the information from these devices was used during comprehensive and periodic joint inspections.

(f) It should require that existing recording devices for railroad and highway signals systems at interconnected/pre-empted grade crossings be retained or upgraded as necessary and require that these recording devices be maintained and that the information from these devices be used during the comprehensive and periodic joint inspections.

To the Federal Highway Administration:

(a) It should develop guidelines and amend the Manual on Uniform Traffic Control Devices for Streets and Highways to provide methods to delineate the area (zone) that a train, or its cargo, or both, might occupy on the track or tracks of a railroad grade crossing so motorists had visual reference points that enabled them to ascertain whether their vehicle was encroaching on the travel path of the train, or its cargo, or both.

(b) It should disseminate safety information, in cooperation with the National Highway Traffic Safety Administration and Operation Lifesaver, Inc., once guidelines were developed, to national, State, police, public service, and safety agencies for them to use in developing a training and education module which informed motorists how to recognise the area (zone) that a train and/or its cargo might occupy on the track or tracks of a railroad grade crossing.

(c) It should cooperate with the Federal Railroad Administration in the review and modification of the existing parameters of the National Highway-Rail Crossing Inventory to ensure that it met the needs of both railroad and highway users.

To the Federal Railroad Administration:

(a) In cooperation with the Federal Highway Administration, it should review and modify the existing parameters of the National Highway-Rail Crossing Inventory to ensure that it met the needs of both railroad and highway users and included, as a minimum, information on highway/railroad grade crossings having pre-emptive or interconnected signals. Once modified, it should review and update the information annually.

To the National Highway Traffic Safety Administration:

(a) It should determine what effect school bus sound attenuation materials had on the ability of a bus driver to discern both interior and exterior audible warnings.

(b) It should disseminate safety information, in cooperation with the Federal Highway Administration and the Operation Lifesaver, Inc., once guidelines were developed, to national, State, police, public service, and safety agencies for them to use in developing a training and education module that informed motorists how to recognise the area (zone) that a train and/or its cargo might occupy on the track or tracks of a railroad grade crossing.

To the State of Illinois:

(a) It should advise school bus drivers of the circumstances of this accident and provide the bus drivers with practical training about vehicle positioning on the road, especially at railroad/highway grade crossings.

To the Illinois Department of Transportation:

(a) It should review and modify the highway design for all railroad/highway grade crossings in Illinois to ensure that vehicles had adequate space and time to clear the crossing before the arrival of a train.

(b) It should train Illinois Department of Transportation personnel and contractors involved in the design, inspection, and maintenance of highway signals at railroad/highway crossings to ensure that they understood the integration and working relationship of the railroad and highway signal systems.

To the Transportation Joint Agreement School District 47/155:

(a) It should develop and implement a program for the identification of school bus route hazards and routinely monitor and evaluate all regular and substitute school bus drivers.

To the National Association of State Directors of Pupil Transportation Services:

(a) It should advise its members of the circumstances of this accident and provide guidance about vehicle positioning on the road, especially at railroad/ highway grade crossings.

(b) It should develop guidelines for the appropriate placement of radio speakers and use of radios on school buses and disseminate these guidelines to its members.

(c) It should advise its members to check their school buses and disable any radio speakers located immediately adjacent to school bus drivers' heads.

(d) It should encourage its members to develop and implement a program for the identification of school bus route hazards and to routinely monitor and evaluate all regular and substitute school bus drivers.

(e) It should advise its members to consider railroad/highway grade crossing accident histories or unusual roadway characteristics when establishing school bus routes.

To the American Association of State Highway and Transportation Officials:

(a) It should advise its members of the circumstances of this accident and, in cooperation with the US Department of Transportation, notify railroads and public entities about the importance of exchanging information regarding railroad/highway grade crossings.

To the National Association of County Engineers:

(a) It should advise its members of the circumstances of this accident and, in cooperation with the US Department of Transportation, notify railroads and public entities about the importance of exchanging information regarding rail road/highway grade crossings.

To the American Public Works Association:

(a) It should advise its members of the circumstances of this accident and, in cooperation with the US Department of Transportation, notify railroads and public entities about the importance of exchanging information regarding railroad/highway grade crossings.

To the Institute of Transportation Engineers:

(a) It should advise its members of the circumstances of this accident and, in cooperation with the U.S. Department of Transportation, notify railroads and public entities about the importance of exchanging information regarding railroad/highway grade crossings.

To the Association of American Railroads:

(a) It should advise its members of the circumstances of this accident and, in cooperation with the U.S. Department of Transportation, notify railroads and public entities about the importance of exchanging information regarding railroad/highway grade crossings.

To the American Short Line Railroad Association:

(a) It should advise its members of the circumstances of this accident and, in cooperation with the U.S. Department of Transportation, notify railroads and public entities about the importance of exchanging information regarding railroad/highway grade crossings.

To the American Public Transit Association:

(a) It should advise its members of the circumstances of this accident and, in cooperation with the U.S. Department of Transportation, notify railroads and public entities about the importance of exchanging information regarding railroad/highway grade crossings.

To Operation Lifesaver, Inc.:

(a) It should disseminate safety information, in cooperation with the Federal Highway Administration and the National Highway Traffic Safety Administration, once guidelines were developed, to national, State, police, public service, and safety agencies for them to use in developing a training and education module that informed motorists how to recognise the area (zone) that a train and/or its cargo might occupy on the track or tracks of a railroad grade crossing. [774]

Amtrak's *City of New Orleans* passenger train from New Orleans to Chicago collided with a pickup truck on a level crossing at Ponchatoula, Louisiana, United States, on the afternoon of Thursday 14th December 1995. The train pushed the truck about half a mile down the line before it stopped, with the truck still wrapped around its nose. Irma Brumfield, 54, the driver of the truck; Lawanda Patterson, 26; her daughter Alexius, 4; and her twin sons Michael and Mikel, 2, were all killed. It took emergency workers an hour to free the bodies from the wreckage.

The Illinois Central Gulf Railroad crossing was protected only by a stop sign. Police Chief Tim Gideon said that it was not known if the pickup driver had seen the train but probably, "she was just not paying attention."

The police chief and Mayor Julian Dufreche complained that although there was a theoretical speed limit of 35 mph through the town, trains passed through it at speeds up to 80 mph because the local ordinance was superseded by federal regulations. Amtrak spokesman Rob Borella said he did not know how fast the train had been travelling at the time of the accident, if its engineer had whistled for the crossing or if he had applied the brakes. None of the 126 passengers and 15 crew on the train was hurt. The train continued its journey after about three hours. [775]

Only four of the eight crossings in the town were equipped with warning lights and bells and the mayor said that the town had neither the money nor the rights of way to erect barriers at all of the crossings. [776]

Central Railway train No. 2925, the Down *Paschim Express* from Bombay Central to Amritsar, collided with a lorry on a manned engineering crossing between Asaoti and Ballabhgarh stations, Haryana, India, on the Mathura – New Delhi section of the Jhansi division at 10:02 on Sunday 31st December 1995. As a result of the collision two railwaymen, presumably the engine crew, and three labourers on the lorry were killed. The said labourers had reportedly forced open the crossing gates after temporarily immobilising the crossing keeper. [777]

A slippery, icy road was the principal factor in a level crossing accident near the village of Târguşor, Romania, about 30 kilometres north of the Black Sea port of Constanta, early on Friday 26th January 1996. Unable to stop on the treacherous road surface, the bus skidded, crashed through a crossing barrier and was hit by a passenger train, which dragged it some 600 metres. There had been 33 passengers on the bus, comprising 31 Romanians and two Turks. Ten of them and the Turkish bus driver were killed and 15 were seriously injured. The Romanian-owned bus had been on its way from Istanbul, Turkey, to the town of Brăila in

[774] National Transportation Safety Board report

[775] *The Orange County Register*, Santa Ana, California, United States, 15th December 1995

[776] *The Daily News*, Huntingdon, Pennsylvania, United States, 15th December 1995

[777] Government of India Ministry of Civil Aviation and Tourism Commission of Railway Safety Annual Report for 1995-96

eastern Romania. [778] [779]

Greatly reduced visibility, due to a blinding snowstorm that had affected much of the country, was a likely factor in two level crossing accidents that occurred in Egypt on Thursday 8th February 1996. In the earlier accident a train collided with a van near the town of El Rahmaniya, some 90 kilometres north-east of Cairo. Three women riding in the van were killed and seven were injured. No one on the train was hurt.

Later in the day a passenger train from Alexandria to Cairo collided with a lorry near the town of Kafr el-Zayyat, about 90 kilometres north-west of Cairo. This accident was more serious; many of the train's coaches were derailed and nine people died and 12 were hurt. Police blamed the lorry driver for failing to obey a signal at the crossing. [780] [781]

A serious level crossing accident occurred at Kutas, Hungary, at 12:04 on Monday 26th February 1996, on the electrified Dombóvár–Gyékényes main line (No. 41). Express freight train number 80202 from Kaposvár, bound for Gyékényes, had passed through Jákó-Nagybajom station a few minutes earlier. It consisted of engine No. V43 1152, operated by Driver János Szána, and 41 waggons and was travelling at 68 km/h at the crossing, within the permitted maximum speed of 70 km/h. Meanwhile, a bus driven by Béla Horváth had left Nagyatád at about 11:00. Train and bus reached the crossing at about the same time. Protection at the crossing was afforded by a pair of manually operated barriers but due to a mistake by Csabán Czillinger, the railwayman in charge of them, these were not lowered upon the approach of the train and so the bus was driven onto the line in front of it.

As a result of the violent collision that followed, the bus was cut in two. Its rear half was pushed off the track while the front half was carried several hundred metres, damaged beyond recognition and caught fire. The colliding engine did not escape unscathed either as it was damaged both by the impact and the fire and one of its bogies was derailed, while the pantograph was also damaged. Ten bus passengers and its driver were killed more or less outright, another passenger died soon after being admitted to hospital and a 12th person succumbed to their injuries a week later. Fourteen more bus passengers were seriously injured, as was Driver Szána. The passengers were largely a mixture of students and pensioners.

Dozens of police, ambulance, fire and local residents participated in the rescue operation. The first units were on the scene within 15 minutes of the alarm being raised. Traffic was suspended on the Jákó-Nagybajom–Beleg section while the accident was investigated and the line cleared, and international express trains were operated via the Gyékényes–Nagykanizsa–Székesfehérvár–Budapest route.

Crossing Keeper Czillinger was subsequently found to have been responsible for the tragedy and was sentenced to four years' imprisonment. The crossing was later provided with automatic half barriers. [782]

A train collided with a bus on a level crossing at Gavojdia, Romania, on Sunday 3rd March 1996. This accident was a little less serious than the January one at Târguşor but six lives were

[778] *The Daily Press*, Victorville, California, United States, 27th January 1996

[779] *Altoona Mirror*, Altoona, Pennsylvania, United States, 27th January 1996

[780] *The News Herald*, Panama City, Florida, Germany, 9th February 1996

[781] *The News*, Frederick, Maryland, Germany, 9th February 1996

[782] https://iho.hu/hirek/amikor-a-szili-kettetepte-a-kapos-volan-buszat-husz-eve-tortent-a-kutasi-tragedia-160313

lost and nine people were injured. The weather was clear at the time of the accident and the warning systems at the crossing were functioning normally (the report does not state what these were). The cause of the accident was being investigated. [783]

Seventeen or 18 people died and about 20 were injured when a train collided with a minibus carrying a wedding party on an unmanned level crossing about 5 kilometres from Jalpaiguri, West Bengal, India, on about Monday 12th March 1996.[784] There had been 44 passengers on the bus. The newly married bride and groom were both critically hurt and it was feared that some of the injured might succumb. [785]

A police spokesman said, "The train just banged into the bus. Those who escaped death were all very lucky."

At about 14:28 on Wednesday 3rd April 1996 Northern Railway train No. 433, which consisted of a three-car diesel multiple unit, had left Jullundhar City and after stopping at Jullundhar Cantonment was proceeding towards Chiheru, having attained a speed of 75 km/h, when it collided with a school minibus on Crossing No. C-72, Punjab, India. Nine people, presumably mostly children, were killed and two others were seriously hurt. The minibus had entered the crossing due to a failure of the crossing keeper, who despite having advised of the approach of the train, had not closed the gates. [786]

Thirty-three people died and another 24 were injured when a freight train collided with a bus on a level crossing near Kerogwe, Kenya, on Sunday 7th April 1996. [787]

At 6:55 a.m. on Wednesday 1st May 1996 a northbound Union Pacific Railroad train travelling at 48 mph collided with a Morehouse Parish Public Schools' bus on a level crossing in Morehouse Parish, Louisiana, United States. The engineer reported that he had seen the school bus approaching the crossing when his train had been half a mile away and had begun sounding his horn. In response the bus had slowed down but it had not stopped until its front wheels were between the rails. Just before the collision the bus driver had attempted to reverse his vehicle. Despite this, the train struck the bus, injuring 11 pupils and its driver.

The National Transportation Safety Board (NTSB) investigated the accident. It noted that there were obstructions that could temporarily block the view of the train engineer of the bus, and of the bus driver of the train. The view of the bus driver of a northbound train was also impeded by a fan in the centre of the bus windshield combined with the skewed angle of the crossing when the bus was 400 feet from it. The bus driver claimed that he had stopped at the crossing and opened the door to listen for a train, a statement disputed by the engineer.

The NTSB concluded that the performance of the bus driver had been "substandard." He had failed to stop his bus a safe distance from the crossing and had failed to hear the warning

---

[783] https://www.upi.com/Archives/1996/03/03/Six-Romanians-die-in-bus-train-accident/5521825829200/

[784] *The Guardian*, London, England, 13th March 1996

[785] https://www.upi.com/amp/Archives/1996/03/11/Indian-train-rams-into-bus-17-killed/9122826520400/

[786] Government of India Ministry of Civil Aviation and Tourism Commission of Railway Safety Annual Report for 1996-97

[787] Author's personal notes – source mislaid

whistle of the train. [788]

An Ernakulam – Kayamkulam train collided with a bus carrying a marriage party on an unmanned level crossing near Alapuzha, Kerala, India, on Tuesday 14th May 1996, killing 35 of them. [789]

Twenty-five people died when a passenger train bound for Allahabad collided with a tractor trolley on an unmanned level crossing near Varanasi, Uttar Pradesh, India, on Saturday 25th May 1996. [790]

A Ferrocarriles Nacionales de México passenger train bound for Nogales collided with a bus on a level crossing with the Avenida Santa Rosa in Mazatlán, Mexico, at about 22:00 on Friday 31st May 1996. Twenty people on the bus were killed outright, 14 died soon after in hospital and another 13 were injured. A four-year-old boy was the only person on the bus to escape injury. A preliminary investigation determined that because his radio had been on loud, the bus driver had failed to hear the whistling of the engine of the train, which he had also failed to see. Survivors stated that the bus had also been overcrowded and driven at excessive speed.

The train, consisting of an engine and three coaches, had been travelling at about 60 km/h when it hit the bus. Driver Elías Flores Peñas stated afterwards, "When I was about 100 metres from the crossing I saw a white and green Colosio bus coming from west to east, but assumed that it would stop, as the train has the right of way. Instead the bus began to climb the hill, it never stopped and when I was about 35 metres from the crossing I applied the brakes of the train in emergency, but it did not respond at once due to the weight of the engine and coaches. The train caught the centre of the bus and dragged it about 300 metres before it could be pulled up."

Five more of the injured died in the next day or so. [791]

Bus driver Juan Carlos Ramírez de los Ángeles survived the accident. He told the public prosecutor, "I remember making a last turn before the crossing that night and that I didn't have many passengers, about 15 seated. When I got to the railway I stopped as I always did, short of the tracks. I don't remember seeing anything and so I set off across the line. Suddenly I felt a movement to my side, that is a blow, then I don't remember any more."

The bus driver was subsequently sentenced to four and a half years' imprisonment, of which he served half. [792] [793]

On Friday 7th June 1996 a disastrous accident occurred between a Zambia Railways passenger train and an Ohan Transport Ltd petrol tanker on Lukanda Crossing, north of Kapiri Mposhi, Zambia. The lorry failed to give way to the train and there was a violent collision, as a result of which the petrol ignited, setting fire to the tanker, the locomotive and five coaches of the train. The lorry driver, the train crew and many train passengers lost their lives. Provincial Police Chief Nungu Sassasali said some hours after the accident that it was impossible to know

---

788 National Transportation Safety Board report

789 https://www.deccanherald.com/india/train-chronology-2501432

790 Ibid

791 *New Castle News*, New Castle, Pennsylvania, United States, 3rd June 1996

792 *Port Arthur News*, Port Arthur, Texas, United States, 2nd June 1996

793 https://www.noroeste.com.mx/mazatlan/a-20-anos-del-trenazo-en-el-crucero-de-la-santa-rosa-DCNO1014878

how many people had died as many of the bodies were badly burnt, while the carriages were still too hot to enter. There had been at least 40 passengers in each of the coaches. Twelve survivors were taken to the nearby town of Kabwe for treatment, most of them in a critical condition. The track and signalling at the crossing suffered significant damage.

The crossing was protected by warning lights and bells and with the lorry driver having been blamed for the crash, the railway subsequently claimed a substantial sum from the road transport company for damage to its locomotive, rolling stock, track and signalling, plus the loss of business while the line had been out of use following the accident. [794] [795]

A freight train collided with a bus on a level crossing near the mining city of La Oroya, Peru, early on Thursday 27th June 1996, killing 18 people and injuring 34. Surviving bus passengers said that their driver had tried to beat the train, which had consisted of two engines and 18 waggons. The accident happened in a flat, open area with good visibility. The force of the impact pushed the bus about 120 metres down the line and derailed both locomotives and four waggons. Mine workers assisted the police in rescuing the injured and recovering the bodies of the dead. [796]

A bus failed to give way to a heavy freight train at a level crossing near Teza, South Africa, a siding in the timber growing area of Kwa-Zulu-Natal province, on Tuesday 13th August 1996. At least ten people were killed and 34 were injured, seven of whom were flown to hospital by helicopter. [797]

An engine collided with a school bus on a level crossing near Mokry Batay, a few kilometres south of Rostov-on-Don in southern Russia on Thursday 26th September 1996. As the bus approached the crossing the warning lights were flashing, indicating the approach of a train, but it was only 15 minutes before school classes were due to begin, so the bus driver, Konstantin Sidelnikov, anxious to not get the children to school late, elected to accelerate over the crossing in front of the train. Instead, the diesel-hauled train caught the bus, dragging it 200 metres, crumpling it and cutting it almost in two.

By chance, driving behind the school bus and a witness to the accident, was the father of two of the children. He snatched his son and daughter from the wreckage and rushed them to hospital, along with the son of a neighbour, handed them over to the hospital staff and then collapsed from the shock. His daughter, Maria Mashukina, survived, but was left severely brain damaged. A report on the accident, written some years later, makes no mention of his son, implying that he was one of those who did not survive. Eighteen children died at the scene, three more passed away after reaching hospital and 16 children, three adults and the bus driver were hurt, nearly all of them seriously. Only one child, nine-year-old Lena Boldyreva, escaped unhurt. Running all the way home in a state of severe shock, in answer to her mother's questions as to what had happened she could only reply, "I've lost my schoolbag, I've lost my schoolbag."

The body of another child, Yulia Shevchenko, could not be found among the dead or the living, so her name was automatically added to the list of the dead and her family went into mourning. Hours after the disaster, it transpired out that the bloodied girl had been picked up

---

[794] *The Times*, London, England, 8th June 1996

[795] Supreme Court of Zambia Appeal No. 89/2013 Ohan Transport Ltd v Zambia Railways Ltd

[796] *Daily News-Miner*, Fairbanks, Alaska, United States, 28th June 1996

[797] *Pacific Stars and Stripes*, Tokyo, Japan, 15th August 1996

by a passing lorry driver and taken to a hospital in another district. Olga Shevchenko recalled years later, "The doctors there said that they had already covered my daughter with a sheet, having decided that she had choked to death on her own blood. But a nurse saw Yulia move her hand. My daughter was transferred to intensive care; she was in a coma. At the hospital they still believed that she had only a few days left to live. Then we brought our youngest daughter into the ward to say goodbye to her sister. She took Yulia's hand in her hands and held it. And in the evening Yulia regained consciousness." Over the next five years, the girl underwent 11 operations. Yulia had to learn to sit, walk, and talk all over again. But, having gone through all the pain, she still finished school, entered college, got married, and later became a mother herself. [798]

Five people lost their lives in a level crossing accident at Berlaar, Belgium, on Monday 7th October 1996. [799]

After a freight train had cleared a level crossing at Port Allen, Louisiana, United States, at 7 a.m. on Wednesday 4th December 1996 motorists began to cross the line, dodging around the crossing barriers, which remained lowered. The driver of an empty school bus followed suit and it was struck by a slowly moving Kansas City Southern Railway freight train, the horn of which was being sounded and its headlights were lit. The bus driver was injured.

The National Transportation Safety Board found that the substitute bus driver had been fully qualified, though at the time of the accident her commercial driving licence had been revoked because she had failed to respond to an administrative reclassification action. [800]

At about 07:26 on Boxing Day, Thursday 26th December 1996, Northern Railway train No. 2004, the Down *Shatabdi Express*, from New Delhi to Lucknow, collided with a minibus on manned engineering Crossing No. 148 'C' after passing through Maripat station, Uttar Pradesh, India, under clear signals. The train was travelling at 128 km/h when its driver saw the minibus crawling across the line. Five people on the minibus were killed. An unidentified minibus passenger had snatched the keys to the gates from the crossing keeper and opened them in the face of the train. [801]

Five people died and two were critically injured in a collision between a train and a van on a level crossing at Wellawatte, Sri Lanka, early in December 1996. [802]

At least five people died in a collision between a train and a tourist bus on a level crossing outside the town of Santa Cruz Tagolaba, Mexico, at about 05:30 on Saturday 18th January 1997. The bus was travelling on the Tehuantepec Transisthmic Highway, on its way from Veracruz on the Gulf Coast to Acapulco on the Pacific Coast. At least a dozen more were injured, many of them seriously, though the exact number was not immediately known. The

---

[798] https://rostov.aif.ru/incidents/details/124573

[799] European Union Agency for Railways: List of fatal railway accidents 1990-2009 (more than five fatalities)

[800] National Transportation Safety Board report

[801] Government of India Ministry of Civil Aviation and Tourism Commission of Railway Safety Annual Report for 1996-97

[802] *Daily News*, Colombo, Sri Lanka, 1st January 1997

bus had a capacity of 41. [803]

On Tuesday 21st January 1997 an express train collided with a school van on a level crossing at Gujar Khan, Pakistan, 60 kilometres south of Islamabad. Nine people lost their lives, six of them children. [804]

Eastern Railway train No. 3112 Down, the Lalquila – Sealdah express, collided with both a lorry and a bus on manned Crossing No. 19/A with the Grand Trunk Road between Bainchi and Simlagarh, West Bengal, India, at about 07:44 on Monday 24th February 1997. The accident, on the Howrah – Barddhaman electrified section of the Howrah division, resulted in ten deaths and 19 cases of injury, 15 of them serious. [805]

At about 12:00 on Sunday 9th March 1997 Central Railway Up passenger train No. 1556 from Khajuri to Guna had almost reached its destination when it collided with a bus at manned engineering Crossing No. 5 'C', between Tarwata and Guna stations, Madhya Pradesh, India, on the Bhopal division of the railway. Sixteen people on the bus were killed and 58 were injured, 45 of them seriously.

The Commissioner of Railway Safety made a series of recommendations in relation to the accident:

(a) Gate working instructions should be made as per guidelines given in S.R. 16.03.1 (b) and 16.03.2 (a) of the General and Subsidiary Rules, Central Railway, 1983.

(b) Proper fencing needed to be provided by replacing the existing wire fencing.

(c) The non-standard lifting barriers should be replaced by standard lifting (co-acting) barriers.

(d) Banners flags (instead of conventional hand flags) should be supplied to the gatemen who should be instructed to plant them ahead of the level crossing, in the direction of traffic, whenever the gate was open to road traffic.

(e) A speed restriction of 30 km/h for a distance of 100 metres on either side of the level crossing should be imposed for all trains during the daytime. During night working, all trains should stop short of the level crossing and be piloted by the pointsman, available on the train.

(f) A speed restriction of 20 km/h should be imposed for all road vehicles on both road approaches and monitored jointly by the rail and road authorities.

(g) Proper hoardings depicting a "rail-bus" collision and warning boards should be erected to indicate to the road users that they were approaching an "accident prone level crossing."

(h) The working instructions on all other manned crossings on the Guna – Khajuri section should be reframed as per the above recommendations. [806]

---

[803] *The Paris News*, Paris, Texas, United States, 19th January 1997

[804] *Western Morning News*, Plymouth, England, 4th March 1997

[805] Government of India Ministry of Civil Aviation and Tourism Commission of Railway Safety Annual Report for 1996-97

[806] Government of India Ministry of Civil Aviation and Tourism Commission of Railway Safety Annual Report for 1996-97

A southbound CSX Transportation freight train collided with a van on an open (so-called passive) level crossing near Kirkwood, Ohio, United States, just after 3 p.m. on Sunday 9th March 1997. Three of the occupants of the van were trapped in it and two were ejected; all were killed.

The victims were named as David Horner, 44; his wife Kathryn, 43, from Vanadlia, Ohio; and their children Jonathon, 18; Carmen, 17; and Preston, 13.

Harold Brandon, who lived only 100 feet from the crossing, spoke of hearing the horn of the train "screaming continuously" and then, "All of a sudden I heard a big thud." He said that he had been able to see the van upside down through the passing train cars and had called 911. He added that he considered the crossing especially dangerous because it sat higher than the rest of the road, while trees and other obstacles blocked the views of motorists. [807]

Women and elderly men riding on a bus and bound for a weekly farmers' market south of Seoul were the victims of a level crossing accident at about 09:00 on Monday 24th March 1997 in the town of Namwon, South Korea, about 200 kilometres from the capital. A passenger train struck the bus, killing at least 16 of them and injuring 20 more. An initial police investigation indicated that the bus driver had disregarded the warning signals at the crossing. [808]

A Burlington Northern Santa Fe Railway train collided with a possibly empty school bus on a crossing at Riverside, California, United States, on Friday 16th May 1997. The bus driver was killed. [809]

A nasty level crossing accident occurred at about 11:50 on the morning of Monday 8th September 1997 at Crossing No. 395 (also known as Noble Crossing) between the stations of Saint-Antoine-de-Breuilh and Saint-Foy-le-Grande, France. Involved in the accident were train No. 9796 from Bordeaux to Sarlat via Bergerac, consisting of a Class X 2100 diesel railcar No. X2149, and a road tanker loaded with 29,964 litres of petrol. The train was travelling at about 120 km/h when it reached the crossing, which was protected by lifting barriers, warning lights and bells.

Although the crossing with departmental road No. 936 was fully protected, it was situated on a double bend made by the road to cross the railway at right angles and had been the scene of 70 incidents and three collisions in the previous ten years. On the day of the accident, the lorry driver, M. Christian Le Breton, aged 47, for some unexplained reason failed to see the lowered barrier, as he told rescuers several times soon after the accident (his lawyer maintained later that he might have been taken ill). His tanker crashed through the barrier directly into the path of the train and was cut in two by the impact. Seconds later the contents of the lorry exploded and the ensuing fire engulfed both the train and a lineside house at the crossing that the lorry had been thrown against. Ten fire engines were sent to the scene, while air ambulances evacuated the most seriously hurt and 23 road ambulances attended the accident. It took 160 firemen two and a half hours to get the fire under control. [810] [811]

An unnamed passenger who had escaped unhurt told French television, "It was frightening. There was a crash and lots of smoke. I scrambled out of the train and then heard a

---

[807] *The Daily Herald*, Delphos, Ohio, United States, 10th March 1997

[808] *Pacific Stars and Stripes*, Tokyo, Japan, 26th March 1997

[809] http://www.stnonline.com/stn/sch.../rail_chart1.ht

[810] *Medicine Hat News*, Medicine Hat, Alberta, Canada, 9th September 1997

[811] *Sedalia Democrat*, Sedalia, Missouri, United States, 8th September 1997

sound like a bomb and saw flames."

Another passenger related, "The doors were jammed and we had to break the windows to escape." [812]

Twelve train passengers and its motorman were killed and 43 people were injured, ten of whom suffered serious burns. The lorry driver survived but was left partially paralysed.

Almost four years after the accident, in July 2001, the crossing was replaced by a bridge. The following year the lorry driver, who was also a former mayor of Allemans-du-Dropt, found himself in court in connection with the accident, was charged and found guilty of manslaughter and having involuntarily caused injury and was awarded a suspended sentence of three years' imprisonment. Raymond Guionie, his former employer, was ordered to pay 404,000 euros compensation, while the Société nationale des chemins de fer français (SNCF) was required to pay a token fine of one euro.[813] The same year a target of removing 15 level crossings annually was set up. [814]

An electric train collided with a bus carrying farm workers on a level crossing near Novomoskovsk, Ukraine, on Tuesday 7th October 1997, killing 18 of them. One bus passenger and its driver were seriously hurt; none of the passengers on the electric train was hurt. The crossing was equipped with warning lights and bells but lacked barriers. [815]

Shortly after leaving a station, a fast train bound for Beijing collided with a lorry loaded with 13 tons of ore from a mine on a level crossing north of Guangzhou, China, on Tuesday 28th October 1997. The collision broke the lorry in two and one part of it smashed into a small office. Six people (train passengers?) lost their lives and three more were seriously hurt. [816] [817]

A passenger train collided with a minibus in Kotabumi, Indonesia, on Tuesday 4th November 1997. At least 25 minibus passengers reportedly lost their lives and others were seriously hurt. [818]

Cuba suffered a disastrous level crossing collision in the town of Urbano Rey, at the Crucero del Rey crossing near the Urbano Noris sugar mill, on Thursday 6th November 1997. A 12-coach passenger train from Habana to Santiago de Cuba, travelling at 80 km/h, struck the centre of a bus, dragging it several metres. No less than 56 bus passengers and its driver were killed and another seven bus passengers were badly injured and admitted to a local hospital. The victims were all residents of Holguin province, in which the accident had occurred, apart from one person from Habana.

The accident was blamed on 34-year-old Juan Carlos Santos, the bus driver, who, it was stated, had driven past a stop signal and onto the ungated level crossing. The bus had been due

---

[812] *Western Morning News*, Plymouth, England, 9th September 1997

[813] https://www-leparisien-fr.translate.goog/faits-divers/peine-maximale-pour-le-routier-du-drame-de-port-sainte-foy-10-04-2002-2002973824.php?_x_tr_sl=fr&_x_tr_tl=en&_x_tr_hl=en&_x_tr_pto=sc

[814] Wikipedia – Accident de Port-Sainte-Foy

[815] *The Titusville Herald*, Titusville, Pennsylvania, United States, 8th October 1997

[816] *The Straits Times*, Singapore, 30th October 1997

[817] *Pacific Stars and Stripes*, Tokyo, Japan, 30th October 1997

[818] *Western Morning News*, Plymouth, England, 5th November 1997

shortly to stop in the town. [819] [820]

On Sunday 21st December 1997 a train collided with a bus on an unprotected level crossing some kilometres north of Bồng Sơn in Binh Dinh province, Vietnam. Two bus passengers were killed more or less outright and 13 more were hurt and taken to hospital, three of whom later succumbed to their injuries. Police indicated that the bus driver's view of the line had been obscured by a row of houses that had been erected too close to the tracks. [821]

At 12:10 on Tuesday 6th January 1998 a light engine ran into an autorickshaw on a manned level crossing between Yeswanthpur (now Yashavantapura) and Tumkur, Karnataka, India, in the Bangalore division of the Southern Railway, killing six people in the vehicle, which had entered the crossing due to the gate not being closed.

The Commissioner of Railway Safety made a number of recommendations in relation to the accident:

(1) (a) The engineering officers and staff had a feeling that they were not required to check equipment of the traffic manned level crossings or examine gatemen as to the rules. It should be clarified to all the railway administrations that examination of the gate equipment and gatemen in rules as provided in para. 914 of Indian Railways Permanent Way Manual (IRPWM) equally applied to the manned traffic level crossings.

(b) The IRPWM should be further enlarged to include a schedule of inspection of the traffic officers and the supervisors for the manned traffic IC gates. Presently this schedule was not provided in any rules or manuals. The station manager in whose jurisdiction a manned traffic level crossing was situated should be made responsible for the gate equipment, knowledge of the gateman in rules and for ensuring the proper maintenance of the gate with the help of the department responsible for the repair and maintenance of the level crossings.

(2) (a) At present there was no regular prescribed format for maintenance of Private Number registers at the manned level crossings not operated from the cabin/station and the same should be prescribed by the railway board for the uniform adoption on the Indian Railways. This register might have the following columns: Date, Train No., Likely departure, Time of exchange of Private Number by Station Manager (SM), SM's Private Number, Time of closing of the gate, Gateman's Private Number to SM, Time of opening of the gate, and Gateman's signature.

(b) The railway board had accepted the recommendation 9.3 of the Commissioner Railway Safety (CRS), Northern Circle, in his statutory inquiry on 'Dashing of 41J Down DMU with a minibus at level crossing No. C72 between Chiheru and Jullundhur stations of Northern Railway on April 3, 1996. It was, therefore, necessary that all the manned level crossings were provided with a wall clock or watch as the gateman's personal equipment so that he might record proper timings of Private Numbers of expected and actual times of passage of trains, times of opening and closing of the level crossings etc. The provision of these clocks/watches should be ensured within a positive time-frame, say by 31 3, 1999.

(c) In conformity with the provisions in 4.01 of the General Rules, it should be the responsibility of the Station Manager to communicate the standard time to the gatemen telephonically connected with his station every day after the same was transmitted to him at

---

819 *The Herald Journal*, Logan, Utah, United States, 7th November 1997
820 *El Pais*, Madrid, Spain, 7th November 1997
821 *Pacific Stars and Stripes*, Tokyo, Japan, 25th December 1997

16:00, and this communication should be under exchange of a Private Number. It would ensure that the same time was recorded for all the events from Control down to the field and level crossings and there would be no ambiguity about the different timings for the same event.

(3) It was also requested that all manned traffic level crossings should be interlocked and no non-interlocked level crossing should exist within the station limits and all these level crossings should invariably be equipped with lifting barriers along with the other provisions as in column (8) in item 5 in annexure 9/1 to para. 904 of IRPWM and also with the provision of approach warning hooters to alert road users of approaching trains. The railway administration should make a time-bound programme to convert existing non-interlocked manned traffic level crossings level crossings, to enhance the element of safety in train operation.

(4) (a) The recommendation of CRS, Central Circle, n 9.2 (vi) in his statutory inquiry report on ‘Collision of A38 Up EMU Local and a private truck in Mumbai division of Central Railway on Oct 11, 1997’ for evolving better boom locking arrangements on the lifting barriers to withstand tampering by outside elements is further commended for immediate acceptance and implementation on the Indian Railways.

(b) Even on the non-interlocked level crossings equipped with lifting barriers, the locking should be through a lever, as for the interlocked level crossings, which should be placed in the same platform as the winch. The key taken out from this lever after locking would considerably reduce the tampering, as presently the securing of the boom to the rest post was hardly followed by the gatemen.

(c) Two strong short chains with the rings at the end and sizeable locks should also be provided at all manned level crossings for use to secure and lock the level crossings for the safe passage of trains in case of the failure of the regular locking arrangements both at interlocked and non-interlocked level crossings. [822]

Sixteen pupils of the Beeville High School girls track team had a narrow escape when the rear of the school bus in which they were riding along with three games coaches was clipped by a Burlington Northern Santa Fe Railway freight train of 62 cars on a level crossing at Sinton, Texas, United States, on Saturday 28th February 1998. The team had been returning from the Tour of the Coast relays at Rockport when the accident happened at 9:45 p.m. The train was travelling at about 20 mph.

Following the accident, half of the occupants of the bus were taken to the Spohn Memorial Hospital in Corpus Christi and the other half to the Columbia North Bay Hospital in Aransas Pass. All were released after treatment, apart from one girl who was detained overnight with suspected concussion.

There were no barriers at the crossing but it was provided with warning lights and bells. The 76-year-old bus driver had stopped at the crossing but had then proceeded across the line, despite the approach of a train being indicated. It was believed that his view of the train had been obscured by a parked lorry. [823]

A southbound Burlington Northern Santa Fe Railway (BNSF) freight train consisting of six diesel units and 50 cars collided with a school bus on an open level crossing at Buffalo,

[822] Government of India Ministry of Civil Aviation and Tourism Commission of Railway Safety Annual Report for 1997-98
[823] *Corpus Christi Caller Times*, Corpus Christi, Texas, United States, 2nd March 1998

Montana, United States, at 7:34 a.m. on Tuesday 10th March 1998. The impact broke the bus apart and two brothers, Ben Petersen, 15, and Christopher Petersen, 13, were killed. As the full-sized bus had been at the start of its run there were only three other pupils on it at the time. All of them were injured, one of them sustaining a fractured skull, and they and the bus driver were taken to the Central Montana Medical Centre in Lewiston.

Upon news of the accident being received, Hobson high and junior high schools, the destination of the bus, were closed for the day. Speaking of the Petersen brothers, school superintendent John Baule said, "They were two young men who were well liked. They were good, decent kids, who got along with everybody." Speaking of the reactions at the school he added, "Our kids were very emotional, upset, distraught. Their reaction was immediate, one of grief and mourning. A lot of tears and hugs."

BNSF spokesman Gus Melonas stated that the engineer of the train had blown his horn for the crossing and applied his brakes when he had seen the bus on the crossing in his path. The train had been travelling below the 49 mph permitted speed for the line. [824]

Bus driver Granville Stuart Abbott, 36, was subsequently charged with failure to proceed safely over the crossing, failure to open the school bus door and listen for a train when stopped at the crossing and endangering the safety of the children. He stated that he could not remember looking left, the direction from which the train had come. [825]

A train collided with a farm truck filled with people on their way to a wedding on a level crossing in Chaiyaphum province, Thailand, on Friday 13th March 1998. Eleven people died and six were injured. The bodies of the victims were scattered along the line for 30 metres. The cause of the accident was not immediately clear. [826]

Southern Railway Up express No. 6323 collided with a Tata Sumo SUV on an unmanned level crossing between Kayankulam and Alleppey, Kerala, India, on Saturday 9th May 1998. Five people died. [827]

Five people died in a level crossing accident between Létavértes and Monostorpáyi stations, Hungary, on Wednesday 27th May 1998 when a Debrecen – Biharkeresztes passenger train collided with a minibus. [828] [829]

A minibus stalled on an unprotected level crossing in the town of Demak, Java, Indonesia, on Tuesday 2nd June 1998. Before it could get clear it was struck by a train, killing 17 of its occupants outright. [830]

On Sunday 7th June 1998 an empty Central Railway BCN goods train crashed into a

---

[824] *The Daily Inter Lake*, Kalispell, Montana, United States, 10th and 11th March 1998

[825] Ibid, 15th April 1998

[826] *The Joplin Globe*, Joplin, Missouri, United States, 14th March 1998

[827] Government of India Ministry of Civil Aviation and Tourism Commission of Railway Safety Annual Report for 1998-99

[828] European Union Agency for Railways: List of fatal railway accidents 1990-2009 (more than five fatalities)

[829] https://www.origo.hu/itthon/1899/12/20030508alegsulyosabb

[830] *The Daily Gleaner*, Kingston, Jamaica, 3rd June 1998

Tempo three-wheeler autorickshaw on manned 'C' Class engineering Crossing No. 241-A between Obaidullaganj and Itayakalan stations, Madhya Pradesh, India. Eight people lost their lives and one was hurt. [831]

The Commissioner of Railway Safety recommended that, 'Surprise checks should be done at level crossing gates to check if gatekeepers plant red banner flags (or in the alternative, use large size red flags, vertically hung) across the track before opening the level crossing for road traffic.' [832]

A northbound CSX Transportation freight train from Jacksonville, Florida, to Cincinnati, Ohio, collided with a van on an unprotected level crossing at Benton, Tennessee, United States, at about 4 p.m. on Monday 13th July 1998. The train pushed the van several hundred yards down the line before it could stop and rescuers had to cut the van open to free the victims. Pandora Garrett, who lived close by, told a reporter, "The top of the van was just folded in. They were putting pieces of it in a pickup truck."

Michael Varnell, 39; his wife Tammy, 32; and their children Davey, 8; Jessica, 8; and Kristin, 7; were all killed in the accident, which had occurred less than a quarter mile from their home. Roy Lowe, who saw the wreckage, commented, "I knew it didn't look like they were going to live."

CSX spokeswoman Jane Covington stated that the engineer had sounded his horn when he had seen the crossing. "By the time the engineer sees a car on the crossing it's too late." She said that the two train crew had not been hurt but, "As unfortunate as it is for the family, it's also unfortunate for the train crew. I'm sure they're experiencing some trauma as well." She added that the speed limit for the line was 60 mph and there was no indication that the train had been speeding. The crew had been interviewed and the "black box" on the locomotive would be examined. There had been no previous accidents at the crossing within recent memory. [833]

A Tamil Nadu State Transport bus travelling from Madurai to Salem and carrying 44 people was in collision with a Southern Railway Chennai – Madurai express No. 6711 on Class 'A' Crossing No. 36, near Murthipalayam, between Erode and Karur stations, Tamil Nadu, India,[834] at about 03:00 on Thursday 13th August 1998. Survivors stated that the driver of the bus, named Durai, had accelerated through the crossing as the gates were closing. The bus was carried 550 metres down the line before it fell into a field and disintegrated. By the following day 19 people had died, among them the bus driver and conductor, and the final numbers of casualties were 20 dead and 26 injured. [835] [836]

Crossing Keeper Kalandai told reporters that telephone communication between Karur Junction and his cabin had not been working for several days, so that in the daytime he had been obliged to close his gates upon catching sight of the train, while relying upon seeing the headlights of the engines at night.

The State Government announced compensation of 50,000 rupees for each of the families

---

[831] Government of India Ministry of Civil Aviation and Tourism Commission of Railway Safety Annual Report for 1998-99

[832] Letter from the Government of India Ministry of Railways (Railway Boad) to Chief Safety Officers of Indian Railways

[833] *The Daily Inter Lake*, Kalispell, Montana, United States, 14th July 1998

[834] www.tribuneindia.com/1998/98aug14/head2.htm

[835] *Danger Ahead! Historic Railway Disasters* report

[836] Government of India Ministry of Civil Aviation and Tourism Commission of Railway Safety Annual Report for 1998-99

bereaved in the accident and of 5,000 to 10,000 rupees for each of the injured. [837]

Recriminations between the Railways and Highways departments followed the accident. New barriers were provided at the crossing in place of chains but it still lacked an 'interlocking facility' (protecting signals locked with the gates) while there was a dispute as to whether 'speed breakers' (road humps) would have prevented the accident. The suggestion of replacing the busy crossing with a bridge had also been raised repeatedly. [838]

Another major level crossing accident occurred in India at about 12:30 on Monday 31st August 1998, this time on an unmanned crossing near Kheroda, in Rajasthan state. A minibus carrying members of the Raibari shepherd tribe from Mavli to Karavali was hit by a local train running on the Mavli to Bari Sadri metre-gauge line, with disastrous results. Thirty people on the bus were killed and the other 21 on it were injured.[839] Medical teams, police and railway officials from Udaipur, Ajmer and Bombay went to the scene of the accident. [840]

A train collided with a bus carrying about 70 people, who were members of the bride's family or friends, to a wedding in Shanxi province, China, on the morning of Sunday 13th September 1998. The force of the impact reportedly threw the bus over 200 metres from the road and 27 of those on it were killed and four more were injured. The accident happened on an unguarded crossing. [841] [842]

The new Asian republic of Turkmenistan suffered a major disaster on the night of Friday 18th September 1998 when a train collided with a bus near Keshi station, on the outskirts of the capital, Ashgabat (another source gives the station name as Vatutino and the name of the capital as Ashkhabad, this being a translation of its Russian name, used until 1991). The bus had been carrying a party of about 60 people returning from a wedding. About 40 bus passengers lost their lives in the accident, many of them women and children. It was not possible to immediately establish the exact number of fatalities because of the severe damage sustained by the bus. Authorities stated that a preliminary investigation indicated that the bus driver had been drunk. [843] [844]

Five young people aged between 18 and 20 lost their lives in a level crossing accident at Gol, Norway, on Saturday 19th September 1998. They had left a bar at 02:00 to visit the house of a friend, a route that took them over a level crossing protected by automatic half barriers. Their car apparently went around the half barriers and was struck by a freight train, which pushed it about 600 metres and killed its occupants. The westbound freight train had been stopped in Gol station nearby, awaiting the passage of a passenger train for Oslo, and it was thought that the driver of the car might have assumed the barriers had only lowered for the passenger train. It was not known for certain who had been driving but the owner of the car

---

[837] *The Hindustan Times*, Delhi, India, 14th August 1998
[838] *The Hindu online*, Chennai, India, 6th September 1998
[839] *Danger Ahead! Historic Railway Disasters* report
[840] www.tribuneindia.com/1998/98sep01/head5.htm
[841] *The Straits Times*, Singapore, 17th September 1998
[842] *European Stars and Stripes*, Darmstadt, Germany, 17th September 1998
[843] https://www.tehrantimes.com/news/16446/Forty-Feared-Killed-in-Turkmenistan-Rail-Bus-Crash
[844] *European Stars and Stripes*, Darmstadt, Germany, 20th September 1998

had reportedly only drunk water in the bar. [845]

The majority of the 65 passengers on an Andhra Pradesh State Road Transport Corporation (APSRTC) bus from Irikigudem to Miriyalguda on Friday 25th September 1998 were children on their way to take a supplementary examination in Miriyalguda. The rear of the bus was struck by a goods train on an unmanned level crossing at Bothalaparem, Andhra Pradesh, India, killing a number of passengers and injuring others. According to the police there were 25 fatalities, while a South Central Railway spokesman stated that there were 20 dead and the Road Transport Corporation put the figure at 19. About 30 others were injured. The Commission of Railway Safety report states that there were 18 dead and 20 injured. [846]

Dr Y. Satyanarayana of the APSRTC stated, "We are yet to ascertain the exact cause of the mishap. The steep terrain at the unmanned level crossing might have blocked the view of the approaching engine."

The injured were taken to hospitals in Nalgonda and Miriyalguda. [847]

Six young people on their way home from the Ladrillo Disco bowling alley in a Renault Clio car from Realicó, Argentina, had only travelled five kilometres when they reached an unprotected level crossing on National Route No. 188 at 05:50 on Sunday 11th October 1998. Javier Rodríguez, who was driving the vehicle, either failed to observe or attempted to beat a Ferroexpreso Pampeano freight train travelling from Intendente Alvear to Junin over the crossing. The train, which consisted of an engine and 29 waggons of sunflower oil, hit the car, dragged it a short distance and it was more or less destroyed. Rodríguez and passengers Ghimena Martín, Roxana Monge, Vanesa Peralta and Lorena Urtiaga, were all killed. They were all from Intendente Alvear and all aged between 18 and 25. One passenger, Paula Savaletta, survived the accident with very serious injuries and was hospitalised in a coma at General Pico. The train driver, Alejandro Ballester, 33, gave a statement to the police before resuming his journey. [848]

Four waggons loaded with gravel ran way down a gradient until they collided with a bus on its way from Khabarovsk to the village of Beryozovka, close to the former place, on an open level crossing at 19:00 on Monday 26th October 1998. The accident happened in the far east of Russia and close to the border with China. The impact tore the bus apart, it was carried about 200 metres and the waggons were derailed. Over 100 people were engaged in the rescue operation, with 28 ambulances and 50 vehicles from the Emergency Situations Ministry involved. Twenty-four people were killed and over 20 were injured, all but three of whom needed to go to hospital. It was clear that the bus driver, Anatoly Vaerennikov, who was one of those killed, could not have seen the danger in the darkness.

The waggons, belonging to the Dalspetsstroi company and stabled on a siding, had not, it was determined, been properly secured. They had run down a steep incline and reached an estimated speed of 80 km/h before striking the bus. Minister of Railways Nikolai Aksenenko accused the Dalspetsstroi staff of having committed an act of criminal negligence resulting in

---

[845] https://www.dagbladet.no/nyheter/de-mistet-sine-barn/65575621

[846] Government of India Ministry of Civil Aviation and Tourism Commission of Railway Safety Annual Report for 1998-99

[847] *The Indian Express*, Noida, India, 25th September 1998

[848] https://www.lanacion.com.ar/sociedad/cinco-muertos-por-un-accidente-en-la-pampa-nid113890/

loss of life in not properly securing the waggons. [849]

Multiple road vehicles were involved in a level crossing accident near Sango, Nigeria, on the evening of Saturday 31st October 1998. Four vehicles were reportedly hit by the train, including a bus, a pickup truck and car, most of the victims, according to eyewitnesses, being hurt when they were trapped in their rush to get clear. Eight people were believed to have been killed and others injured. The injured were brought to the Cornerstone Hospital in Ibadan and the bodies of the dead to the Adeyoyo State Hospital mortuary in the same place. [850]

A freight train collided with a bus on an unprotected level crossing in the Kalakla suburb of Khartoum, the capital of Sudan, late on Wednesday 11th November 1998. The bus was pushed more than 75 metres and it overturned several times. Four bus passengers were killed outright and 50 more were hurt, seven of them critically. [851]

Losing a wheel in some unexplained manner, a school bus became stranded on a level crossing near Wontergem, Belgium, on the line between Ghent and De Panne, on about Wednesday 9th December 1998. Fortunately a teacher had time to shepherd 50 pupils off the bus before it was struck by a fast passenger train travelling at 120 km/h. None of the children was hurt but six train passengers were slightly injured by flying glass and were taken to hospital. [852]

At about 04:45 on Friday 18th December 1998 the *Shimanta Express*, bound for Khulna, collided with a Chitra Paribahan overnight bus, which had left Dhaka at 22:30 the previous evening for Chuadanga. The accident happened on Kashipur Crossing, Bangladesh, not far from Mobarakganj station. Ten people were killed outright, three succumbed on the way to hospital and the condition of seven was said to be critical. Forty passengers in all were hurt, 15 of whom were taken to Jhenidah hospital, 14 to Kaliganj health complex and six to Jessore General Hospital. Those who died included Abdus Salam, 36, the bus driver; bus supervisor Raisuddin, 35; bus helper Altaf Hossain, 34; and Assistant Driver Mohammad Hossain, 45, on the train.

The line was blocked until 14:40 when a crane from Jessore removed the debris.

Survivors stated that the gates of the crossing had been open at the time of the accident and no crossing keeper had been on duty. [853]

Following an investigation by the railway company, Crossing Keeper Akbar Ali and Stationmaster Rustam Ali were suspended for alleged negligence. [854]

Crossing No. 3, between Bhuli and Tetulmart stations, Bihar, India, was the scene of an accident on Christmas Day, Friday 25th December 1998. Eastern Railway express No. 7054 and a bus carrying miners of the Bharat Coking Coal company from Bhooli colony to Bhansjoda colliery for their night shift were involved. The bus was dragged a kilometre, 11 bus passengers lost their lives and six were injured.

---

[849] http://vn.vladnews.ru/1998/iss179/text/news1.html
[850] *allAfrica.com*, 2nd November 1998
[851] *European Stars and Stripes*, Darmstadt, Germany, 13th November 1998
[852] *The Times*, London, England, 10th December 1998
[853] *The Daily Star*, Dhaka, Bangladesh, 19th December 1998
[854] Ibid, 20th December 1998

There were different accounts as to how the accident had happened. One stated that the vehicle had broken down on the crossing, another that the miners, fed up with the long wait at the closed gates for the train, had seized the keys to them from the crossing keeper and opened them themselves.[855] Ex-gratia payments of 4,500 and 5,100 rupees respectively were offered to the families of the dead and to the injured respectively. [856]

A mistake by a crossing keeper was the apparent cause of a level crossing collision that occurred about 15 kilometres from Sibi, Pakistan, in Baluchistan province, at about 10:30 on Friday 29th January 1999. A train collided with a road vehicle described as a "passenger van," killing ten people and injuring 12, two of whom later died. The crossing keeper was suspended after the accident, pending the results of an inquiry. [857]

The driver of a minibus carrying a party of student athletes, which was travelling along road No. 44 between Tiszakécske and Lakitelek, Hungary, disregarded the flashing warning lights and the bus was struck by a passenger train. Four young footballers and their coach, who was also the driver of the bus, were killed and a child was slightly hurt. The accident happened on Saturday 13th March 1999. [858]

At about 9:47 p.m. on Monday 15th March 1999 National Railroad Passenger Corporation (Amtrak) passenger train No. 59, carrying 207 passengers and 21 crew, operating over a line of the Illinois Central Railroad, collided with an articulated lorry on McKnight Road Crossing, Bourbonnais, Illinois, United States. Both locomotives and 11 of the 14 carriages of the train were derailed, 11 passengers died, 29 passengers and 5 railwaymen were seriously hurt, and 51 passengers and 4 railwaymen were slightly hurt. The National Transportation Safety Board (NTSB) inquired into the accident.

The NTSB report noted that lorry driver John Stokes, 58, had picked up his vehicle from Melco Transfer Inc. at Peotone, Illinois, at 8:30 p.m. He had been operating under a 60-day probationary licence, because his conditional driver's licence had been suspended for 90 days following his having committed three traffic offences within a year. He had then proceeded to the Birmingham Steel plant in Bourbonnais, where his trailer had been loaded with six bundles of 60-foot-long, ¾-inch-diameter steel reinforcing rods (rebar). The plant was only 650 feet from the crossing.

Train No. 59 was the *City of New Orleans*, en route from Chicago, Illinois, to New Orleans, Louisiana. After a brake test had been performed it had left Chicago at 8:03 p.m. and departed from Homewood, Illinois, at 9:27 p.m.

McKnight Road Crossing was protected by automatic half barriers, warning lights and bells.

Driver Stokes told the NTSB investigation into the disaster that when the crossing lights had begun flashing he had been "right on top of the track." He had not noticed the position of the gates (barriers). He said that he had been concerned at that point about applying his brakes too hard, for fear that any violent deceleration would cause the load on his trailer to shift forward and crush his cab, while a gentler brake application might result in his lorry coming to a stand on the crossing, in the path of the train. For a moment he had been uncertain whether

[855] *Danger Ahead! Historic Railway Disasters* report
[856] www.netguruindia.com/news/dec98%5Cn271298.html
[857] *Danger Ahead! Historic Railway Disasters* report
[858] https://www.origo.hu/itthon/1899/12/20030508alegsulyosabb

to stop or continue, then he had decided to "just floor it" and accelerate over the crossing in front of the train. He had seen that the light of the train "wasn't too far down the tracks then." At the moment of impact his lorry had been in sixth gear, moving at 20 mph.

A motorist who had witnessed the accident stated that when he had stopped at the warning lights of the crossing, the lorry had been approaching it. The gates had been descending as the lorry had entered the crossing and he saw the gate strike the right side of the trailer about a third of the way back and break off. Travelling at about 7 mph, the lorry had moved towards the middle of the two-lane road. Several other witnesses described what they had seen of the accident from different locations relative to the crossing.

The unnamed Amtrak engineer stated that he had sounded his horn when he had seen the lorry slowly trundling over the crossing. When he had seen that it would not clear the line in time, he applied his brakes in emergency. His train had been travelling at 79 mph and within a few seconds it had struck the rear of the trailer.

Kankakee County Police received the first 911 emergency call at 9:48 p.m. About three minutes later, a police officer arrived at the crossing and reported that a locomotive was on fire and carriages were derailed. The Kankakee County disaster plan was then put into effect. The fire brigade arrived at 9:52 p.m. and the first ambulance at 9:53 p.m.[859] The injured were taken to St Mary's and the Riverside hospitals in Kankakee, while one seriously injured patient was airlifted to the Loyola University Medical Centre in Maywood. There was confusion in establishing the number of passengers on the train and the firemen had considerable difficulty in dealing with the fire, which was not finally extinguished until dawn the following morning.

Newspapers carried the accounts of survivors of the accident. 'The train lights had been dimmed. There was nothing to see outside of the windows and 17-year-old Anna Fulson was preparing to sleep after a long ski trip. Then came a shuddering jolt. A screech of twisted metal. Finally, flames and cries of pain. "You could hear screaming," Miss Fulson said hours after clambering out of the Amtrak train that struck a semitrailer Monday night. "You could feel the car derailing...It was like in slow motion. Luggage was falling. And you were being slammed against the ceiling and walls." Miss Fulson was among 15 Mississippi teens and three chaperones returning from a 10-day Canadian ski trip when their train, Amtrak's City of New Orleans, derailed at a crossing 50 miles south of Chicago.

'Many in the ski group were riding in a double-decker car immediately behind the shattered sleeping car where all the dead were found. Leaking diesel fuel ignited and set that car on fire. After the lead locomotive hit the truck and derailed, the cars behind it also left the tracks and began to fold together like an accordion. One locomotive smashed through the sleeping car.

'"People were lying on the ground," said Melissa Watson, 14, one of the Mississippi group, who was playing cards in the club car when the train derailed. "You could hear children scream. 'Where is my mum?' You could hear parents scream. 'Where are my children?'"

'The crash occurred about 9.30 p.m. as many of the passengers heading from Chicago to New Orleans were preparing for sleep. Some were listening to Walkmans; others were gossiping. Many teens had put on T-shirts and sweatpants. Then it happened. Bump. Bump. Bang. Bang. A twisting motion. A hurtling crash that flung some of the passenger cars onto their sides. Then, total darkness. Someone flicked on a cigarette lighter every few seconds in the toppled teen's car. Boys in the group forced open an escape window above their heads. Boys began pulling out the girls, who used the train seats as makeshift ladders. One girl fumbled around to find her glasses among heaps of clothes and luggage.

---

[859] National Transportation Safety Board report

'"It was like your worst nightmare of what could happen on a train," said Christina Bomgaars, 15. "You can't imagine the terror of not knowing if you're above water, if you're on a cliff, or if you're going to roll over." The boys told them not to panic, but to move fast. And they did. "We had some idea they were scared our car was going to catch fire soon," said Miss Bomgaars, who recounted the night as she sat with her friends in the lounge of Lees Inn, a nearby hotel that had taken in crash victims.

'In another car toward the rear of the train, two women from Melbourne, Australia, on a month long journey across the United States, were watching a movie when the train suddenly screeched to a jarring halt. Donna Griffiths, a 28-year-old advertising executive, was tossed about in her seat but her passenger car remained upright. She had minor rib bruises but started handing out towels to people who were bleeding and blankets to children – items provided by rescuers. "There were horrible injuries," she said, recounting how she saw some people struggling to breathe, others bleeding profusely, and one little boy with part of his ear cut off. Her friend, Daniella Bauman, a 28-year-old graphics designer, feared another travelling companion might have been killed, but tracked down the woman at a hospital later. "It was better than winning the lottery," said a joyous Miss Bauman. "We just grabbed each other and hugged. It was such a beautiful moment. After seeing that horrible sight, I was just thinking, 'How could anyone survive.'"

Many survivors called family and friends to reassure them they were fine. Miss Bomgaars phoned her parents near Jackson, Miss., when she arrived at the hospital. She was not injured, though her experience shook her up. "It makes you so aware of how short life is and how quickly it can be gone," she said. "You don't have to do anything. You just have to be in the wrong place. God was watching over us."' [860]

The NTSB report went on to consider every aspect of the accident, including the rescue operation, the personal history of the lorry driver and the train engineer, the condition of the lorry and train, the track and signalling, the crossing installation and so on, also the results of various post-accident investigations.

Analysing the accident, the NTSB noted that there was no indication that the train engineer had been fatigued, had had any problems with his vision or hearing, or had been using drugs or alcohol. There had been no problems with the train or track, nor apparently with the lorry. Other than that it had been dark, visibility had been good. The lorry driver had tested positive only for prescription drugs prescribed for medical conditions such as hypertension and rheumatoid arthritis.

The statements of several witnesses and post-accident reconstructions did not support the lorry driver's statement as to the events leading up to the collision. Extensive tests showed no evidence of any malfunctioning of the crossing equipment and it appeared that the warning lights had begun flashing more than 20 seconds before the lorry had arrived at the crossing. Its driver had had little sleep in the days before the accident so it was possible that he had failed to notice the flashing lights, but from the evidence of witnesses it appeared that he had initially applied the brakes, indicating that he must have seen the lights but then had made an error of judgement in thinking that he could get across the line in front of the train.

The NTSB reached nine conclusions regarding the accident:

1. The following factors had not caused or contributed to the accident: the physical condition or actions of the engineer of Amtrak train 59; the mechanical condition of the train and the condition of the tracks; railroad operating procedures and policies; the mechanical condition of the truck-tractor semitrailer; the equipment making up Amtrak train 59; the weather; alcohol, drugs, or prescription medications; and the location of traffic control

[860] *Jacksonville Journal-Courier*, Jacksonville, Illinois, United States, 17th March 1999

signs on the approach to the grade crossing.

2. The accident-involved truck-tractor semitrailer had probably been travelling at between 7 and 14 mph at the time of the collision.
3. The grade crossing signal lights had begun flashing at least 26 seconds before the train's arrival at the McKnight Road grade crossing.
4. Based on the signal system tests and physical evidence, including evidence of the truck's position at the time of impact, both crossing gates had likely lowered as designed as the accident truck approached the crossing.
5. The truck driver had had ample time to safely stop his truck and avoid an accident, but likely as a result of fatigue, he had failed to respond appropriately to the signals and instead had decided to attempt to cross ahead of the train.
6. Melco Transfer, Inc., had failed to provide driver oversight sufficient to detect or prevent driver fatigue as a result of excessive driving or on-duty periods.
7. The flammability and smoke-generation properties of the train's interior materials had not contributed significantly to the spread of the fire.
8. Because of insufficient training in responding to railroad emergencies or inadequate/inappropriate resources, or both, the Bourbonnais Fire Protection District had not been prepared to respond effectively to a passenger train accident involving a significant diesel fuel fire.
9. The lack of railroad pavement markings had probably had no effect on the truck driver's driving behaviour.

The NTSB determined that the probable cause of the accident had been the lorry driver's inappropriate response to the grade crossing warning devices and his judgment, likely impaired by fatigue, that he could cross the tracks before the arrival of the train. Contributing to the accident had been Melco Transfer, Inc.'s failure to provide driver oversight sufficient to detect or prevent driver fatigue as a result of excessive driving or on-duty periods.

The NTSB made several recommendations.

To the US Department of Transportation:

It should provide Federal highway safety incentive grants to States to advance innovative pilot programs designed to increase enforcement of grade crossing traffic laws at both active and passive crossings.

To the Federal Railroad Administration:

For all railroads that installed new or upgraded grade crossing warning systems including crossing gates, and that were equipped with event recorders, it should require that the information captured by those event recorders included the position of the deployed gates.

To All Class I Railroads and all Regional Railroads:

For all their new and upgraded grade crossing warning systems that included crossing gates and were equipped with event recorders, it should ensure that the information captured by those event recorders included the position of the deployed gates.

To the National Railroad Passenger Corporation:

In fulfilling its Federal mandate to help prepare emergency responders to respond to an accident involving Amtrak equipment, it should emphasise to those responders the possibility that such an accident could result in large quantities of burning diesel fuel and urge them to be prepared to respond to this specific hazard.

To the International Association of Fire Fighters and the International Association of Fire Chiefs:

They should inform their memberships of the circumstances surrounding the emergency response to the March 15th 1999 grade crossing accident in Bourbonnais, Illinois, and of the need for responders to prepare for train accidents that might result in significant diesel fuel fires. [861]

In late September 2004 at Kankakee, Illinois, John Stokes was found guilty of the only two offences with which he could be charged: violating the law regarding how long lorry drivers could work without rest and failing to keep an accurate log book. He was sentenced to two years' imprisonment and in addition, his commercial driving licence was permanently revoked. [862]

The crossing was closed in January 2006. [863]

Train No. 5610, the Down Avadh – Assam express, had left Lucknow for Ghonda, en route from New Delhi to Gauhati, Assam, when it collided with a bus carrying a wedding party on Jhukia Crossing, Uttar Pradesh, India, at 20:21 on Tuesday 27th April 1999. Twenty-seven bus passengers were killed outright, 32 injured were admitted to Barabanki hospital, of whom five died, and 12 were admitted to Bahraich hospital, of whom three died.[864] Other reports stated that 37 people had died at the scene. The bus had a seating capacity of 52 but was carrying 74 people at the time of the accident. The accident occurred on an unprotected crossing at night. A police official stated, "Apart from the headlights of the bus and of a car that drove ahead of it seconds before the accident, there was no light there, no streetlight." As for the train, an official said, "The engine of the train is very badly damaged but we don't know yet if anyone on the train is injured." [865]

A train collided with a school bus on an unprotected level crossing at Podlog v Savinjski Dolini, Slovenia, on Thursday 13th May 1999. Four children aged 10 or 11 attending the Šmartne ob Paki or Bratov Latonja primary school (according to different accounts) were killed and 38 were injured, four of them seriously. A doctor in the hospital in nearby Celje stated, "The remaining 38 children were taken to hospital. Most of them have cuts and bruises and none of them is in a critical condition." [866]

The pupils had been returning from an educational trip to the Pekel Cave, a large karst (landscape formed from soluble rocks such as limestone) cave that was over 3,000,000 years old and included an underground waterfall. A former pupil, describing the accident 10 years later, wrote that it had been a very hot day, so that the children had all sat on the left-hand side of the bus with the curtains closed against the sun. Because of the closed curtains the children had not seen the approaching train, so there had been no panic. Possibly the bright sun and closed curtains had been factors in the failure of the bus driver to observe the train. The road over the crossing was too narrow for two vehicles to easily pass one another on it and as the bus had entered the crossing, a car had been about to enter it from the opposite side. The woman car driver had seen the train close at hand and had attempted to warn the bus driver of the danger but he had misunderstood her gestures as meaning that she wanted priority over the crossing and had proceeded instead.

---

[861] National Transportation Safety Board report

[862] *Daily Herald*, Arlington Heights, Illinois, United States, 29th September 2004

[863] *Chicago Sun-Times*, Chicago, Illinois, United States, 19th January 2006

[864] *Rediff online*, Mumbai, India, 27th April 1999

[865] *UTU News Digest* – Wednesday 28th April 1999

[866] *Medicine Hat News*, Medicine Hat, Alberta, Canada, 14th May 1999

The school bus driver was subsequently sentenced to five years' imprisonment for having caused the accident through negligence. [867]

The crossing was closed after the accident, with the road being diverted to another crossing that was equipped with automatic barriers. [868]

Five people were injured when a locomotive collided with a school bus on a level crossing at Beedenbostel, Lower Saxony, Germany, on Monday 7th June 1999. [869]

A level crossing south of San Miguel de Tucumán, Argentina, was the scene of a serious level crossing accident at 11:45 on Wednesday 16th June 1999, when a bus carrying about 45 passengers was struck by a freight train. The crossing was an unprotected one and it was foggy at the time. Dragged about 100 metres by the train, the bus was deposited in a sewage drain and badly damaged. Six of its passengers were killed outright, a pregnant woman passed away after admission to hospital and 26 passengers, some of them schoolchildren, and bus driver Ricardo Soria were injured. The train driver, Juan Carlos López, escaped unhurt. According to witnesses he had sounded his horn in an attempt to warn the bus driver, without success. He was given a routine test for alcohol after the accident.

Prosecutor Abraham Mussi went to the scene and collected information for his investigation. He provisionally blamed the lack of protection at the crossing for the accident. [870]

The so-called "Volvo train," a freight service carrying lorry cabs in containers from Göteborg to a factory in Olofström, was in collision with a car on an automatic half barrier level crossing at Veka, Sweden, east of Borås, on Tuesday 22nd June 1999. Weighing 310 tonnes and travelling at about 90 km/h, the train carried the car 500 metres and the woman driving the car and four children riding in it were all killed. Locals complained that at times the half barriers were down for long periods without any train appearing, tempting road users to zig-zag around them. [871]

Northeast Frontier Railway Down passenger train No. 30 from Kamakhya to Dhubri collided with two road vehicles on a manned level crossing between Moterhjar and Golakganj stations, Assam, India, on the Alipurduar division of the railway on Thursday 8th July 1999. Six people died and the same number were injured. The accident was blamed on the rash and negligent driving of a truck driver. [872]

Eastern Railway train No. SK 264, a Down Krishnanagar – Sealdah electric multiple unit, travelling at about 70 km/h, collided with a bus on a manned crossing, No. 66 B-E between Krishnanagar and Badkulla stations, West Bengal, India, on the broad-gauge Sealdah division

---

867 https://www.24ur.com/novice/slovenija/casovnica-tragedije-na-nasih-cestah-ki-jih-ne-bomo-pozabili.html?bl=0
868 https://www.vlaki.info/forum/viewtopic.php?t=3784
869 https://www.welt.de/print-welt/article585948/Bahnunglueck-Entgleister-Gueterzug-legt-Strecke-lahm.html
870 https://www.lanacion.com.ar/sociedad/siete-muertos-y-27-heridos-al-chocar-un-tren-con-un-omnibus-nid142440/
871 *Danger Ahead!*, Currents, 22nd April 2002
872 Government of India Ministry of Civil Aviation and Tourism Commission of Railway Safety Annual Report for 1999-2000

of the railway, at 04:54 on Tuesday 27th July 1999. Five people died and 22 were injured, 7 seriously.[873] The Commissioner of Railway Safety concluded that, 'The dashing of SK-264 DN EMU Local with the Bus took place due to the absence of interlocking between gate and gate signal and the lack of adequate safety precautions resulting {in the} gate being left in open position while the protecting gate signal was "Off."' [874]

A 37-year-old man, his wife and their four children, aged between four and 11, were all killed when a train struck the tractor on which they were riding on a level crossing at Owadów, Poland, on Tuesday 27th July 1999. [875]

Eight people, seven of them children, lost their lives when a train collided with a rural bus on a crossing in the town of Longaví, Chile, 300 kilometres south of Santiago, on Friday 13th August 1999. Police reports indicated that 18 other people on the bus were injured and were taken to the nearby regional hospital in Talca. A police investigation found that the bus driver had disregarded the warning signs at the crossing and had tried to cross the line in front of the train. Four adults and about 20 children had been travelling on the bus; the children were being taken to their school, though the report on the accident did not indicate if it was a school bus *per se* or if it was merely that the school children had formed the majority of passengers on the vehicle. [876]

Six members of a family were killed outright or fatally injured when the car in which they were riding was struck by a train on a level crossing at Honingnest Crossing, South Africa, north-west of Pretoria, late on Friday 27th August 1999. The victims were named as Rina Olivier, 52; her daughters Karien Swart, 22, and Santa, 20; her daughter-in-law Adrie, 25; and grandchildren Tiana, 6, and Clarise, 4. One of the women and one of the children survived long enough to be taken to Pretoria Academic Hospital but died on Saturday. Danio Swart, a two-year-old boy, was the only survivor of those in the car. The family had been coming from their home at Pienaarsrivier along the old Warmbaths road, on the way to a dance. Their husbands had been following in a separate vehicle and were the first on the scene. On Saturday afternoon the train driver was still in a state of shock and unable to describe the circumstances of the accident. [877]

On Tuesday 14th September 1999 a train collided with a school bus on an unprotected level crossing at Yibin, in Sichuan province, China. There had been 33 middle school children plus the driver on the bus, of whom six were killed, while the remaining 27 were all injured. [878]

Early on Thursday 16th September 1999 a passenger train from Surabaya to Jakarta collided with a bus on a level crossing at Pemalang, in central Java, Indonesia, 355 kilometres

---

[873] Ibid

[874] Abstract of report of Commissioner of Railway Safety on Dashing of SK-264 DN Krishna Nagar – Sealdah Local with a bus (No WB-63-0295) at manned level crossing No. 66/B-E at Km.97/3-4 between Krishna Nagar and Ranaghat single line electrified B.G. section of Sealdah division of Eastern Railway on 27.07.1999

[875] https://echodnia-eu.translate.goog/radomskie/najwieksze-tragedie-w-historii-na-drogach-regionu/ar/8762837?_x_tr_sl=pl&_x_tr_tl=en&_x_tr_hl=en&_x_tr_pto=sc

[876] *El Pais*, Madrid, Spain, 14th August 1999

[877] www.pta-online.co.za/pretnews/aug99/n280801.htm

[878] *The Salina Journal*, Salina, Kansas, United States, 17th September 1999

east of the Indonesian capital. The bus was dragged some 700 metres and 24 passengers, all on the bus, were killed and five more were badly hurt.[879] A local police officer suggested that the gates at the crossing had accidentally been left open. [880]

Southern Railway passenger train No. 251 from Chikjajur to Gumtakal collided with a private bus on a manned but non-interlocked level crossing between Chitradurg and Balenahalli, Karnataka, India, on the Chikjajur – Rayadurg broad-gauge single line on Monday 18th October 1999. Six bus passengers died, nine were badly hurt and two slightly hurt. The Commission of Railway Safety found the crossing keeper to have been at fault, as he had failed to close the gate on the north side of the crossing, allowing the bus to be driven onto the line. [881]

Rash driving on the part of a taxi driver was the cause of a bad level crossing accident at about 11 p.m. on Saturday 27th November 1999 at Compton, California, United States. The taxi was proceeding along Willowbrook Avenue, parallel to the Blue Line of the Metropolitan Transit Authority (MTA). Reaching the end of the avenue, the taxi had turned to the right into Greenleaf Boulevard and was driven past the lowered half barriers, flashing lights and ringing bells of a level crossing, directly in front of a southbound train travelling at 55 mph. The taxi was cut in two by the train, caught fire and four of its occupants were carried some distance down the track. All six of the people in the taxi, four men and two women, were killed; they were mostly apparently in their late 20s or early 30s but could not immediately be identified. The motorman of the train was the only person on it, so it was probably an empty coaching stock working. She escaped with slight injuries and was understandably badly shocked.

'"He was coming pretty fast. I think his intent was to beat the train," said Gabriel Castro, 29, of Compton, who was driving west on Greenleaf and had stopped to let the train pass. "From there on, you could hear all the rumble and see these sparks," Castro said. "I was just shocked. I saw the car blow up and I knew there was nothing I could do."

'Moises Rodriguez was inside his house watching television when he heard the collision. "Then we just heard like a loud explosion. You could feel the ground just shake and you could hear the car, pieces of it falling everywhere," said Rodriguez, 25, who lives near the intersection. The decapitated body of a woman was found in the front section of the car, Rodriguez said. Flames and sparks from downed wires that provide electricity to the commuter train prevented Rodriguez and other area residents from getting close to the wreckage.

'The accident derailed the train and closed the southbound tracks, which could reopen today. Metropolitan Transit Authority officials were trying to clear the northbound tracks so they could run trains in both directions. The MTA was sending passengers to commuter trains north and south of the accident scene by bus.

'The accident is the worst ever on the Blue Line, which started the agency's commuter rail service in 1990. It brings to 53 the number of people killed since the line opened. The MTA's other two commuter lines have seen just one fatality, a presumed suicide, MTA spokesman Ed Schannel said. "It is absolutely the most graphic illustration of the risk somebody takes in crossing railroad tracks when a train is oncoming," Schannel said. "It was an absolute

---

[879] *Daily News-Miner*, Fairbanks, Alaska, United States, 17th September 1999

[880] *International Herald Tribune*, Paris, France, 17th September 1999

[881] Government of India Ministry of Civil Aviation and Tourism Commission of Railway Safety Annual Report for 1999-2000

tragedy." [882]

It emerged that the taxi driver, Romaldo Gonzalez, 42, had two previous convictions for drunk driving, had been driving while his licence was suspended and had no permit to operate a taxi in Compton anyway. Tom Drischler, Los Angeles taxicab administrator, stated that there were dozens of taxi companies and thousands of vehicles operating illegally in southern California and that the matter was a serious problem. [883]

At 02:10 on Sunday 28th November 1999 train No. 7384, the *Maharashtra Express* of the Central Railway, en route from Nagpur to Kolhapur, collided with a Jeep on Crossing No. 29 at Nimblak, not far from Ahmednagar station, Maharashtra, India. The Jeep had been carrying sugarcane cutters from Junnar to Pathardi. The driver of the Jeep had reportedly attempted to drive through a small gap near the crossing, which had been permanently closed pending the construction of an overbridge, with traffic being diverted to a link road some distance away. Despite this, road users had continued to attempt to cross the line there rather than take the diversion and Gateman Vinayak Ramchandra Raut had been either posted at the crossing or else retained there after its closure to warn road users. As it turned out, he was unable to prevent the accident. The Jeep driver and two others were hurt and taken to hospital; 11 of his passengers were not so lucky and were killed.[884] [885] The workers were returning home after elections at the Vighnahar Cooperative sugar factory at Junnar. [886]

Early on the morning of Sunday 28th November 1999 the *Khyber Mail* from Peshawar to Karachi collided with a minibus owned by the New Khan Transport Company, which was on its way from Multan to Karachi, on a level crossing at Ranipur, Pakistan. As a result of the impact the bus was carried about a kilometre along the line. Nine bus passengers lost their lives and 18 were injured, six possibly critically.

Some witnesses stated that two buses had stopped on the crossing and their drivers had been talking to one another. It was also reported that the bus had been following a police van, that the police had forced the crossing keepers to open the gates, and that the train driver had run past a protecting signal at Red.

Departmental (railway) and judicial inquiries were opened into the accident and Driver Raja Dilnawaz and Crossing Keepers Darya Khan Jamro and Tagyal Khan were all taken into police custody for questioning. First information reports were registered by the police against the driver and Fireman Gulsher Mirbahar. [887] [888]

At about 21:30 on Saturday 18th December 1999 a lorry owned by cattle dealers, which had travelled from the northern state of Kano, in Nigeria, to the south-east of the country to purchase palm oil, was making its return journey. By the time it had reached the town of Otukpo in the centre of the country, it had picked up a large number of passengers. Here it came to a level crossing but despite being warned that a passenger train was approaching, the lorry driver elected to cross the railway anyway. The train duly demolished the lorry with

[882] *The Bakersfield Californian*, Bakersfield, California, United States, 29th November 1999
[883] *UTU News Digest*, 1st December 1999
[884] *Danger Ahead! Historic Railway Disasters* report
[885] Government of India Ministry of Civil Aviation and Tourism Commission of Railway Safety Annual Report for 1999-2000
[886] www.indian-express.com/ie/daily19991129/ige29136p.html
[887] www.paknews.org/cgi-paknews.cgi?news=99/November/29-11-1&key=accident
[888] www.dawn.com/1999/19991130/nat30.htm

disastrous results. According to Police Constable Sam Uduok at least 50 people who had been on the lorry lost their lives, though the exact number was not known, and only one survived. One account of the accident stated that the train was at least partially derailed but no one on it was apparently hurt. [889] [890]

At the end of December 1999, Hindu farm workers from the village of Sangaiahpatti, in the Karur district of Tamil Nadu, were on their way to Palani as part of their annual pilgrimage in the Tamil month of Margazhi. The men left first, travelling on foot, and their wives and children followed a fortnight later on Monday 27th, bringing food packets and pooja materials. (Pooja or puja is a Hindu devotional ritual.) It was a lorry carrying these women and children, plus a few elderly villagers, which was struck by a train on a level crossing at Oddanchatram, Tamil Nadu, India, on the evening of that day and dragged 200 metres down the line. The train was delayed nearly five hours by the accident.

Survivors of the accident related that the lorry had developed a problem with its gear box as it was negotiating the crossing, and stalled when frantic efforts by the lorry driver to restart it had failed. The pilgrims had not had much time to react and found it difficult to jump from the wooden plank at the rear of the lorry, so in desperation, a number of women had thrown their children from the lorry to save them. Thus, of the 14 dead, ten were women, all of them mothers. Three children aged two or three also lost their lives; the other fatality was presumably an elderly villager. Many of the victims were related to one another. About a dozen injured women and a good many children were admitted to Madurai and Dindigul government hospitals. There had been some 50 villagers on the bus. The accident happened at 19:45 and firemen and police had difficulty in locating the dead and injured in the darkness.

A senior official noted that visibility at the crossing had been good and blamed the lorry driver for attempting to use the crossing when he should have been aware that a train was due. Escaping injury, the driver made himself scarce after the collision and was being hunted by police. [891]

Southern Railway train No. 6707, a Coimbatore – Madurai Down express, struck a lorry on an unmanned level crossing between Oddanchatram and Akkaraippatti stations, Tamil Nadu, India, on the Dindigul Pallachi metre-gauge section of the Madurai division of the railway on Monday 27th December 1999. Sixteen people died and another 33 were injured, 13 of them seriously. The accident was blamed on the negligence of the lorry driver. [892]

A waggon (van?) bound for Sargodha was struck by a passenger train bound for Lala Musa on an unprotected level crossing at Chenab Nagar, Pakistan, on Sunday 2nd January 2000. Nine people died and 10 were injured. There was reduced visibility at the time of the accident due to fog. [893]

On Friday 28th January 2000 Mohinder Singh, driving a school van carrying pupils on their way to the Sacred Heart Roman Catholic convent school in Malout, Punjab, India, arrived at

---

[889] https://www.latimes.com/archives/la-xpm-1999-dec-20-mn-45841-story.html

[890] *The Orange County Register*, Santa Ana, California, United States, 19th and 20th December 1999

[891] *The Hindu online*, Chennai, India, 29th December 1999

[892] Government of India Ministry of Civil Aviation and Tourism Commission of Railway Safety Annual Report for 1999-2000

[893] www.dawn.com/2000/01/04/local27.htm

Crossing No. 27-C, which was unmanned, as a train from Bathinda to Sriganganagar was approaching. At first he reportedly stopped for the train, then he apparently changed his mind and thought that he could get across in time before the train arrived. Driver Heera Lal Verma reported that he had sounded his whistle when he had seen the van, then when the van had restarted after first stopping, he had applied his brakes but could only slow down the train somewhat. Five or six children died in the accident and four were hurt. The injured, who included the van driver, were taken to Dayanand Medical College and Ludhiana Hospital. [894] [895]

An *ex gratia* payment of 5,000 rupees was made to the next of kin of those killed and of 2,500 rupees to the seriously injured. [896]

Six children died and three were injured when a passenger train collided with a school van on unmanned Crossing No. C-27 at Theri in the Mukhtar district of Punjab, India, on Friday 28th January 2000.[897] According to the police, Mohinder Singh, the driver of the van, had initially stopped at the crossing when he had seen a Bathinda – Sriganganagar train approaching, had decided that he had time to cross in front of it, then realising that it was close at hand, panicked and applied the brakes rather than the accelerator so that the train had hit the rear of his bus. Driver Heera Lal Verma said that he had whistled and applied the brakes of his train when the bus had stopped on the crossing but had been unable to avoid the collision.[898]

Following the tragedy, it was announced that measures would be taken to convert half of the unmanned crossings in the Ambala Division of the Northern Railway to manned ones within a year, at a cost of 6 lakh (600,000) rupees each, including the salaries of crossing keepers and maintenance expenses. There were 300 unmanned crossings in total in the division, plus 345 manned ones. [899]

At least 13 people died and one person was hurt when a minibus carrying 19 people was struck by the *Argo Bromo* express, bound for Surabaya, on a level crossing in the town of Karawang, West Java, Indonesia, on Thursday 3rd February 2000. Ten people were killed outright, including the crossing keeper, and three were fatally injured and died on the way to hospital. The driver of the minibus was only slightly hurt and six minibus passengers escaped injury. According to witnesses the minibus had been driven past or through a crossing barrier. [900]

Driver Mohammad Khan and five pupils of Jinnah Cambridge Grammar Public School at Bhalwal lost their lives when the school bus (actually a Suzuki pick-up truck) in which they were riding was struck by train No. 12, the Down *Chenab Express* bound for Karachi, on an unprotected level crossing at Wilsonpura, Pakistan, at 07:50 on Saturday 12th February 2000. Another five pupils and a cleaner were badly hurt and were rushed to hospitals in Bhalwal and Sargodha.[901] The bus had been on its way from Phularwan town to Bhalwal at the time of the accident. Traffic on the Lala Musa – Sargodha line was blocked for three hours because the engine of the express was disabled by the collision and another had to be obtained before the

[894] *The Hindustan Times*, Delhi, India, 28th January 2000
[895] *The Tribune online*, Amritsar, India, 29th January 2000
[896] Ibid, 30th January 2000
[897] *Indiainfo.com*, Chronology of rail accidents in 2000
[898] www.tribuneindia.com/20000129/main2.htm
[899] www.tribuneindia.com/20000131/haryana.htm
[900] *UTU News Digest*, Monday 7th February 2000
[901] www.paknews.org/cgi-paknews/paknews.cgi?news=100/February13-2-5&key=accident

train could continue its journey. The driver of the express was blamed for the accident for having failed to whistle repeatedly at short intervals during thick fog, as per the regulations. [902]

A farm tractor pulling a trailer over an unprotected level crossing near Ksar el Kebir, Morocco, was struck by a train on the foggy morning of Wednesday 1st March 2000, killing at least 35 people and seriously injuring another nine. The train had been travelling between Casablanca, Rabat, and the northern city of Tangiers.

Sbai Hassan, secretary general of Larache province, in which the accident had occurred, stated, "All the casualties were among some 50 farmers on the trailer. None of the passengers on the train was hurt. We fear that the death doll will rise as most of the injured are in a very serious condition."

Moroccan television added that most of the victims had been women. [903]

Helicopters flew the dead and injured to a military hospital in Rabat, 180 kilometres to the south. The tractor driver and three young girls who had jumped from the trailer escaped injury.

Office national des chemins de fer (ONCF, the railway company) spokesman Mohamed Smouni announced that the accident was being investigated but it appeared to have been due to a combination of the fog and the imprudence of the tractor driver. [904]

A private bus carrying a wedding party, which was returning from the village of Muthi Bhambial to Naghled, was in collision with a train on an unmanned level crossing near Raja Ka Bagh, Himachal Pradesh, India, on Tuesday 7th March 2000. Despite being urged by his passengers to wait for the train, the bus driver had elected to beat it over the crossing. The bus was carried about ten metres. Ten people died and 29 were hurt. The injured were taken to Nurpur Civil Hospital, apart from three critically hurt who were brought to the Dr Rajendra Prasad Medical College in Dharamsala. Nearby villagers informed the police of the accident and began rescue work. [905]

At 6:18 a.m. on Tuesday 28th March 2000 CSX Transportation freight train No. S-213-25 from Cincinnati, Ohio, to Atlanta, Georgia, and Tampa, Florida, set off with a fresh crew from Ocoee Siding, Tennessee, where it had been stabled overnight. The brakes had been tested before the train departed. The train consisted of a locomotive and 33 loaded auto-rack waggons. It had travelled 9½ miles, its horn and bell had been sounded for several level crossings and it was moving at 51 mph when it approached Liberty Church Road Crossing, near Conasauga, Tennessee, United States, at about 6:40 a.m. (newspaper reports give the location of the accident as having been near Tennga, a location in the adjoining state of Georgia).

As the train neared the crossing the engineer saw a school bus, its yellow body illuminated in the sun, which had just risen, moving towards the line. He cried, "Hey! Hey!" and sounded the engine horn continuously, "to blow her down," despite which the bus did not stop but kept moving towards the crossing. The engineer then applied his brakes in emergency. The conductor, riding in the cab of the locomotive with the engineer, said that the bus driver had been looking straight ahead; the engineer remembered that she had looked towards the train at the last moment. They agreed that the door of the bus had been closed.

---

902 *Dawn*, Karachi, Pakistan, 13th February 2000

903 *BBC News*, London, England, 1st March 2000

904 *The Intelligencer*, Doylestown, Pennsylvania, United States, 2nd March 2000

905 www.tribuneindia.com/20000308/himachal.htm

The crossing was protected only by signs, with a standard crossbucks sign at the crossing itself, preceded by a diamond-shaped warning marked 'LIMITED SIGHT DISTANCE' accompanied by a sign reading '10 MPH' and further back, a circular railroad advance warning sign, indicating 'ONE LANE.' Both signs were painted yellow with black writing.

The Murray County, Georgia, school bus had picked up its first pupils at 6:35 a.m. It had then turned eastwards from Route 411 onto Liberty Church Road, Georgia, and normally picked up pupils along that road, which shortly crossed the state line into Tennessee, before proceeding along it over the crossing and rejoining Route 411.

Interviewed after the accident, bus driver Rhonda Cloer alleged that approaching the crossing she had been "coming down the hill, came to a complete stop," had looked both ways and opened and then closed the bus door. Checking in her rear-view mirror that all the children were seated, she had looked both ways again, seeing no train. She had then begun to cross the line and "it was like it fell from the sky."

A surviving pupil, who had been sat in the second row with her lap belt fastened, related, however, that, "The bus went down the hill and across the track without stopping, just the same way the driver did the day before, without stopping." It was only the pupil's second day riding the bus. She had not heard the train horn and could not recall if the bus windows had been open or closed.

In order to monitor pupil behaviour the bus had been provided with video recording equipment and a review of this after the accident indicated that, as the pupil had said, the bus had not stopped before crossing the line.

The train struck the bus between its axles, separating its body from its chassis, when the front of the body was in turn separated from the remainder and remained with the chassis, while the rest of the body was carried 190 feet from the crossing. The train was not derailed and stopped 1,900 feet from the crossing. Two of the pupils who were thrown from the bus and one who remained on it were killed, while one pupil who was thrown from the bus and two who had remained on it were seriously injured. The bus driver and one pupil (the one who gave evidence that the bus had not stopped at the crossing) were slightly hurt. Both had been wearing seatbelts, though the bus driver's belt had broken in the crash. Neither the engineer nor the conductor of the train was hurt.

A newspaper wrote, 'The train dragged the bus' battered yellow body about 100 yards. Some of the children were thrown to the gravel beside the tracks. Others were found bloodied inside.

'It "sounded like...thunder or a bomb blown up," said Joe Brown, who lives about 300 feet from the crossing. He said he found his best friend's daughter among the dead.

'John Watson, whose two sons had nissed the bus that morning, arrived with the first energency workers and pulled one badly injured girl from the wreckage. "I was shocked that she was still alive," he said.

'The crash is raising questions once again about the safety of rural grade level crossings. The one-lane crossing, just across the state line in Tennessee, had no warning lights or crossing arms, and the tracks curve through palm trees in hilly terrain, making it difficult to spot trains from the road.

'It was unclear if the driver stopped at the tracks – Georgia law requires school buses to stop at least 15 feet from the tracks before crossing – or took other precautions, such as opening the doors to listen for a train.' [906]

[906] *The Register-Herald*, Beckley, West Virginia, United States, 29th March 2000

The National Transportation Safety Board (NTSB) investigated the accident.

The NTSB report noted that the bus driver had tested negative for alcohol and illict drugs, as had the train crew.

A passing motorist had made a 911 call to report the accident at 6:41 a.m., initially stating that it had occurred at a crossing a half mile away because she was unfamiliar with the area, but giving the correct location in a further call at 6:45 a.m. Meanwhile, the conductor of the train had reported the collision to Control at 6:43 a.m. The first fire engine arrived at the scene at 6:50 a.m. Two helicopters, 17 firefighters and five road ambulances in total attended the accident.

All aspects of the accident are discussed in considerable detail in the NTSB report. Analysing the accident, the NTSB noted that school buses were amongst the safest forms of transportation in the country but the accident demonstrated that certain safety issues remained to be addressed.

It was noted that the weather had been clear and dry and although the sun had just risen, neither the bus driver nor the engineer had reported that it had interfered with visibility. There had been no reported problems with the train, the track, or the bus. The engineer and conductor had been well qualified and had acted correctly. Alcohol and drugs had not been factors in the accident. The visibility of trains at the crossing had been adequate.

Georgia and Tennessee law required school buses to stop at level crossings. Analysis of the videotape from the bus showed that although it had slowed to 15 mph approaching the crossing, it had not stopped. Had the bus stopped 15 feet from the crossing, its driver would have been able to see the train. Equally, had she turned off the radio and opened the bus door, she would have been able to hear the train horn. The bus driver's post-accident statement that she had in fact stopped at the crossing and had looked both ways was contradicted by the videotape, which also showed that she had previously driven over the same crossing without stopping eight times previously. It was likely that having negotiated the crossing previously without encountering a train, she had become complacent, and that the training she had received had not sufficed to alert her to the danger.

The inexpensive solution of placing Stop signs at all passive (unprotected) level crossings had been recommended in 1998 but so far only the states of Hawaii and Idaho had done this. Had there been a Stop sign at Liberty Church Road Crossing, the bus driver might have obeyed it.

With the bus door closed and the radio on, the bus driver had probably not been able to hear the train horn. Despite only being a year old, the bus had been fitted with a radio speaker right next to the driver's head.

An in-vehicle warning system might have alerted the driver to the presence of the train.

Some of the information on a grade (level) crossing inventory intended to identify the relative hazards of different crossings was found to be outdated and incorrect as regards matters such as train speeds and numbers of daily trains for 10 passive crossing locations checked by the Safety Board in the area.

Hazards such as passive crossings were not identified in the routing of school buses.

The passenger wearing the lap belt was the only one who had not been either killed or at least seriously hurt in the accident.

The accident had occurred in a rural area with little traffic. By chance, a passerby had witnessed the accident and called the emergency services and although she had not known its exact location, her description of the site had been sufficiently detailed for it to be identified, while in any case she had made a second call upon learning of the correct name of the accident location from a garage, and the train conductor had also reported the crash. The emergency response had thus not been delayed. Another time, the circumstances might have been less fortunate, a difficulty that would have been overcome had the bus been fitted with an automatic crash notification (ACN) system.

The NTSB reached 21 conclusions regarding the accident:

1. The weather, the mechanical condition of the school bus and train, and the condition of the track had not caused or contributed to the accident; the locomotive engineer and conductor had been qualified and had complied with Federal regulations; the sight distance along the tracks had been adequate for the school bus driver to see an oncoming train; and neither alcohol nor drugs had been a factor in the accident.
2. If the bus driver had followed the required procedures at the grade crossing, that is, if she had stopped at least 15 feet from the nearest rail, turned off her radio, and opened the door and window, she would have seen and heard the train and thus avoided the accident.
3. The driver had not stopped, had the radio on, and the door closed; thus she had difficulty in detecting the train horn and had probably been unaware of the presence of the train.
4. Although the driver had been educated on and knew the mandatory safety precautions at railroad/highway grade crossings, she had disregarded the required procedures and crossed the railroad tracks without stopping on the day of the accident and at least eight other times before the accident.
5. The Murray County School District had not monitored drivers or identified and corrected improper behaviour, thus missing the opportunity to observe this driver's behaviour at railroad/highway grade crossings.
6. Had a stop sign been present at the crossing, it would have reinforced to the driver the need to stop before crossing the tracks, likely prompting her to stop and not attempt to cross in front of the train.
7. If activated prior to a grade crossing, a switch that turned off all nonessential noisemaking components, including, but not limited to, the radio, could help drivers hear train horns and stop as necessary.
8. An in-vehicle warning system to alert drivers of the presence of an oncoming train would have provided the bus driver with additional information, likely prompting her to stop before crossing the tracks.
9. Though critical to assessing grade crossing hazards, the data in the Highway-Rail Grade Crossing Inventory had not always been accurate and complete.
10. Had Tennessee factored school bus use into its grade crossing hazard index, the accident crossing might have had a higher priority for receiving funds to install active warning devices.
11. The Federal Railroad Administration's Web-based Accident Prediction System or a State's grade crossing hazard index, as part of the school bus routing hazard identification program, could help school districts select the safest school bus routes.
12. The lap-belt-restrained passenger had sustained minor abdominal injuries as a result of the severity of the accident and the pivoting of her body about the belt; she had not been ejected due to the use and proper fitting of the available restraint.
13. Had the available restraints in the first row been appropriately worn in this accident and had they fitted properly, the two front-row passengers probably would not have been ejected during the accident sequence and the level of injury might have been reduced.
14. Components within the passenger compartment that were not fitted with energy-absorbing materials could contribute to serious injuries when passengers struck them

during an accident sequence.

15. The seriously injured passenger across from the impact area (seat 6A) had sustained his injuries due to incomplete lateral compartmentalisation, and all three passengers (seats 6A, 6F, and 8F) in the impact area had sustained injuries as a result of locomotive intrusion and exposure to high-impact forces.
16. The rearmost-seated passenger had sustained fatal injuries due to movement out of the seating compartment and subsequent impact with non-energy-absorbing interior components.
17. Even those passengers who were outside the area of impact and remained within their compartment could receive serious or fatal injuries due to impact with non-energy-absorbing components within a school bus.
18. Federal Motor Vehicle Safety Standard 222 exempted sidewall components and the sides of seat frames within the passenger compartment of school buses, thereby exposing passengers to these injury-producing components in lateral-impact collisions.
19. The driver's lap/shoulder belt webbing had failed due to the high forces applied to the webbing as the two parts of the school bus had separated and due to the large distance between the lap/shoulder belt anchor points.
20. In spite of a lack of street name signs or a street address, emergency responders to this accident had been able to efficiently coordinate their response operations and had provided timely, appropriate medical services and rescue operations, in part because of the 911 caller's prompt action and the emergency responders' familiarity with the accident area. Given the limited amount of traffic traversing Liberty Church Road, an emergency response might have been delayed if the passerby had not noticed the accident or if the train crew had been incapacitated.

The National Transportation Safety Board determined that the probable cause of the collision had been the school bus driver's failure to stop before traversing the railroad/highway grade crossing. Contributing to the accident had been the Murray County, Georgia, School District's failure to monitor bus driver performance and its lack of school bus route planning to identify hazards on routes and to eliminate the necessity of crossing railroad tracks. Contributing to the injuries of the school bus passengers outside of the area of intrusion had been incomplete compartmentalisation and a lack of energy-absorbing material on interior surfaces.

The NTSB made a number of recommendations.

To the States:

(a) In cooperation with the National Association of State Directors of Pupil Transportation Services, they should develop and implement a program of initiatives for passive grade crossings and school buses that included:

   (1) The installation of stop signs at passive crossings that were traversed by school buses, except where an engineering study showed that their installation would create a greater hazard.

   (2) The use of information about whether school buses routinely crossed passive grade crossings as a factor in selecting crossings to upgrade with active warning devices.

   (3) A requirement that all newly purchased and in-service school buses should be equipped with noise-reducing switches.

   (4) Enhanced school bus driver training and evaluation, including periodic reviews of on-board videotapes where available, especially with regard to driver performance at grade crossings.

(5) Incorporation of questions on passive grade crossings in the commercial driver's license manual and examination.

To the National Highway Traffic Safety Administration:

(a) It should implement rulemaking to prohibit radio speakers used for music or entertainment from being placed adjacent to drivers' heads in school buses.

(b) It should develop and incorporate into the Federal Motor Vehicle Safety Standards performance standards for school buses that addressed passenger protection for sidewalls, sidewall components, and seat frames.

(c) It should evaluate the feasibility of incorporating automatic crash notification systems on school buses and, if feasible, proceed with system development.

To the Federal Highway Administration:

(a) It should require States to update the Highway-Rail Crossing Inventory to accurately reflect current railroad operations.

To the Georgia Department of Education:

(a) It should require all school districts to disconnect radio speakers used for music or entertainment that were adjacent to school bus drivers' heads.

To the National Association of State Directors of Pupil Transportation Services:

(a) It should encourage its members to use the Federal Railroad Administration's Web-based accident prediction system or the States' hazard indexes for grade crossings when developing school bus routes.

(b) In cooperation with the States, it should develop and implement a program of initiatives for passive grade crossings and school buses that included:

(1) The installation of stop signs at passive crossings that were traversed by school buses, except where an engineering study showed that their installation would create a greater hazard.

(2) The use of information about whether school buses routinely crossed passive grade crossings as a factor in selecting crossings to upgrade with active warning devices.

(3) A requirement that all newly purchased school buses and in-service school buses be equipped with noise-reducing switches.

(4) Enhanced school bus driver training and evaluation, including periodic reviews of on-board videotapes where available, especially with regard to driver performance at grade crossings.

(5) Incorporation of questions on passive grade crossings in the commercial driver's license manual and examination.

(c) It should notify its members of how and why the driver's lap/shoulder belt had torn in this accident and of the potential consequences of large longitudinal distances between lap/shoulder belt anchor points.

To the School Bus Manufacturers:

(a) It should discontinue the installation in school buses of radio speakers used for music or entertainment that were adjacent to the driver's head.

The National Transportation Safety Board also reiterated a previous recommendation:

To the US Department of Transportation:

(a) It should develop and implement a field test program for in-vehicle safety and advisory warning systems, variable message signs, and other active devices; then ensure that the private entities who were developing advance technology applications modified those applications as appropriate for use at passive grade crossings. Following the modifications, it should take action to implement use of the advanced technology applications.

Gates were installed at the crossing in September in accordance with a Tennessee state law that required protection to be provided at a level crossing after a fatal accident. [907]

In January 2004 Rhona Cloer was sentenced to 90 days' imprisonment for causing the accident. [908]

Four people were killed outright, two died in hospital and three were hurt, when their van was hit by a train near Tuy Phước, Vietnam, in Bình Định province, on Sunday 7th May 2000. [909]

Northern Railway train No. 2, a mainline electric multiple unit passenger service, collided with a tractor trolley on manned Crossing No. 58-C between Block Hut 'N' and Makhanpur station, Uttar Pradesh, India, on the Tundla – Kanpur double-line electrified section of the Allahbad division of the railway on Tuesday 30th May 2000. Six people died in the accident and the same number were injured, three of them seriously. Some of the riders on the tractor trolley had intimidated and threatened the crossing keeper, whereupon he had yielded to them without even warning a switchman by telephone as to what he was going to do, or exhibiting a red light towards the approaching train.

The Commission of Railway Safety made nine recommendations in connection with the accident:

(1) At manned level crossings where the gates were normally closed to road traffic and there was telephonic communication, on lines with up to 80 trains daily, the assistant stationmaster (ASM) should obtain assurance that the crossing gates were closed before authorising a train to enter the section; over routes with more than 80 trains daily the gateman should obtain permission before opening the gates but the ASM should compulsorily give such permission after a maximum interval of 30 minutes, or 10 minutes after receiving a request from the gateman, for at least 2 minutes, if necessarily, by regulating the trains.

(2) Power should be delegated for the hiring of private vehicles for surprise checks by inspectors and officers of the engineering department, safety organisations as also traffic inspectors, to ensure that the level crossings were being operated as per the rules.

(3) The decision on the provision of a 70 cm opening instead of wicket gates on level crossings to permit a very large number of persons on cycles, scooters and motorcycles to pass through a level crossing after alighting should be reviewed.

(4) Instructions should be issued that *Line Clear* should be asked for and given only a few minutes, say up to 5 minutes, prior to the entry of a train into a section so that adequate time was available to permit level crossings being opened for the passing of road traffic.

(5) Train Actuated Warning Devices should be provided on priority instead of telephone communication at manned level crossings still not having telephone communication.

(6) The post of gateman should be upgraded to the grade of at least keyman so that the

[907] www.ble.org/pr/news/headline092600h.html
[908] *The Daily News*, Jacksonville, North Carolina, United States, 27th January 2004
[909] *Danger Ahead!*, Archive, May 2000

gatemen were picked by a process of selection and had to volunteer for this work for further promotions.

(7) Electric lighting should be provided at manned level crossings to improve working conditions at night time.

(8) The yardstick in terms of train vehicle units for the upgrading of crossings from Class C to B, B to A and A to Special should be stipulated. {Train vehicle units for a level crossing were the number of trains multiplied by the number of road vehicles over a crossing per day.}

(9) The instructions issued by the Railway Board regarding the granting of *ex gratia* payments in cases of accidents on manned level crossings to road users in cases of death or injury should be amended to permit such payments in every accident at a manned level crossing, prior to determining the cause. [910]

Eight people, seven of them Indian Congress workers, were killed on the spot when the Tempo Trax Jeep in which they were riding was in collision with train No. 12 from Miraz to Bellary on Crossing No. 59, between Banapura and Haligeri, Madhya Pradesh, India, at 11:35 on Saturday 3rd June 2000. The workers from Katharki village had been on their way to Haligeri 'to take part in the campaign for the zilla and taluk panchayat elections.' The Jeep had stalled on the unmanned crossing 'due to a mechanical snag.' Its driver, named Siddalingaiah, jumped clear at the last moment but was badly hurt and taken to hospital. The eighth person to die was described as a cleaner. Train services were delayed for over an hour. [911]

Asked why the crossing was not manned, officials stated that for a crossing to be manned it must be used by at least 6,000 Train Vehicle Units (number of trains x number of road vehicles) daily and at the most recent census in November 1997, the traffic intensity at the crossing had not reached this figure. There were over 24,000 unmanned crossings in the country as a whole. [912]

An express train from Nanning to Guilin collided with a tractor on an unmanned crossing in the Laibin district of Guangxi province, China, on Wednesday 7th June 2000. Ten people died and 13 were hurt. [913]

The flashing red lights protecting a level crossing at Klarenbeek, the Netherlands, failed to prevent a collision with a motor vehicle towing a caravan on Friday 16th June 2000. The driver of an Apeldoorn – Zutphen passenger train vainly tried to warn the road user by sounding his horn, then the train collided with the vehicle with sufficient force to cut it in two. The occupants of the vehicle – a man, a woman and three children – were not immediately identifiable but it was presumed that they belonged to one family. While the man and woman were thrown out of the vehicle, the children were carried 400 metres along with half of it by the train. The woman and children were killed outright and the man, who had been driving, died shortly afterwards. The front of the caravan was also destroyed and its contents were scattered over tens of metres of track. Due to the accident the line was blocked the whole

[910] Government of India Ministry of Civil Aviation and Tourism Commission of Railway Safety Annual Report for 2000-2001

[911] *The Hindu online*, Chennai, India, 4th June 2000

[912] www.timesofindia.com/060600/06mban21.htm

[913] http://dailynews.muzi.com/11/english/72105.shtml

afternoon; the passengers from the train were forwarded by bus. [914]

The super-fast *Vaishali Express* collided with a Jeep on an unmanned level crossing at Jasaoli village, between Siwan and Pachrukhi stations, Bihar, India, on Friday 30th June 2000, killing its five occupants. Four of the victims were members of the same family. [915]

A South Eastern Railway train collided with a tractor trolley on an unmanned level crossing between Muima road and Rajoli stations, West Bengal, India, early on Friday 28th July 2000. Six people were killed and 14 were hurt, all of whom had been riding on the tractor trolley, the driver of which, according to preliminary reports, had attempted to cross the line directly in front of the train. The injured were taken to Chandrapur Hospital. [916]

A freight train collided with a bus on a level crossing in Kisumu, Kenya, on Monday 17th August 2000, during the morning rush hour, killing six people. [917]

Only apparently very briefly reported on was a major disaster at Kabbah, Nigeria, on Friday 8th September 2000. A train travelling from the eastern to the northern part of the country collided on a level crossing with an oil tanker going along the Lokoja – Aboja road. No less than 40 persons were stated to have lost their lives, while another 15 were injured. The casualties, both dead and injured, were taken to the federal medical centre in Lokoja, the capital of Kogi province. [918]

A lorry belonging to the Jamaguri Tea Estate, carrying tea garden labourers, was struck by a Tinsukia to Lumding passenger train carrying about 500 passengers on an unmanned level crossing at Oting, Assam, India, at about 14:30 on Monday 2nd October 2000. Seven people, including the driver, was killed outright, four were fatally hurt and a dozen were injured and taken to the K.K. Civil Hospital in Golaghat. The train was slightly damaged but no one on it was hurt. [919]

Rabi Kota, the Golaghat district magistrate, stated, "An inter-state passenger train on the broad gauge was passing an unmanned level crossing when it hit a truck of the Jamuguri tea estate carrying about 30 workers. The truck was split into three parts, killing 11 persons and injuring 12 others." [920]

At 09:00 on Monday 23rd October 2000 the *Parahyangan Express* from Bandung to Jakarta collided with a lime truck on Perjuangan Street Crossing, Bekasi, Indonesia. The lorry was thrown against and badly damaged two motorcycles, then fell into a 50-metre-deep river. A police source stated that six people had been killed, while local residents claimed that there were 10 dead, though a PT Kereta Api Indonesia (railway company) spokesman would only

---

[914] *Nieuwsblad van het Noorden*, Groningen, the Netherlands, 17th June 2000
[915] *Danger Ahead!*, Archive, July 2000
[916] *Sawaal News*, India, 29th July 2000
[917] https://en.wikipedia.org/wiki/List_of_level_crossing_crashes
[918] *allAfrica.com*, 9th September 2000
[919] *The Times of India online*, 3rd October 2000
[920] http://news.indiainfo.com/2000/10/03/03train.html

confirm two deaths, two cases of serious injury and one of moderate injury. [921]

A crossing accident occurred between Kalanur and Lahli stations, Haryana, India, on the Rohtak – Bhiwani section of the Northern Railway, on a link road near the village of Basana, on Sunday 12th November 2000. A Bathinda-bound passenger train collided with a van on an unmanned level crossing at about 17:00. All five occupants of the van reportedly died on the spot and the van was badly damaged. Rail traffic was disrupted for nearly two hours. [922]

A freight train from Wedmadani to Khartoum collided with a minibus on a level crossing in the Mayo district of Sudan on Wednesday 15th November 2000. Nine people lost their lives and 15 were injured, five seriously.[923] The bus driver, who was amongst the injured, had attempted to cross the line despite the warning whistles of the train. Both the bus and train drivers were being questioned to determine the circumstances of the accident, the report added, after having indicated where the responsibility apparently lay. [924]

The driver of an overcrowded bus reportedly attempted to force it through a crossing as it was being closed to road traffic in the city of Surabaya, East Java, Indonesia, on Saturday 9th December 2000. As a result the bus was struck by a freight train, 11 people died and over a dozen were injured. The bus driver apparently escaped more or less unhurt, fled the scene and was still at large at the time of the report. [925]

On Monday 18th December 2000 a level crossing accident occurred in Iran involving a train, a lorry towing a trailer (or possibly an articulated lorry) and four cars. The lorry driver, travelling along the Tehran – Qom road, attempted to cross the line at Shahr-e Rey, south of Tehran, when the train was too close at hand and his vehicle was struck by it. Whether the cars were also struck by the train directly or whether the lorry was thrown into them was not stated. Whilst the lorry driver was able to jump clear at the last moment and thereby escaped, six occupants of the cars were killed outright and ten were injured, two of whom died in hospital. [926]

At about 06:00 on the morning of Monday 18th December 2000 a Transportación Ferroviaria Mexicana freight train collided with a bus on a level crossing in Santa Catarina, Mexico, killing 16 of its passengers and its driver, while a good number of other passengers were injured. At least a dozen passengers were ejected from the bus, which was carried 500 metres. Rescuers worked for two hours to free the victims from the wreckage.

Most of the passengers on the bus were on their way to factories in Monterrey. As in other countries, they lived outside of the city where the housing was cheaper and commuted to and from their jobs.

Several years later, the accident was the subject of a lawsuit involving the relatives of those killed and the injured on the one side and Kansas City Southern Industries (KCSI) on the other, lawyers arguing that the accident had resulted from the railway cutting corners by

---

921 www.tempo.co.id/news/2000/10/23/1,1,20,uk.html
922 *The Hindustan Times*, New Delhi, India, 13th November 2000
923 www.arabicnews.com/ansub/Daily/Day/001117/2000111704.html
924 *At Sudan Net*, 17th November 2000
925 *The Mercury News*, San Jose, California, 9th December 2000
926 Islamic Republic News Agency report, 19th December 2000

increasing train speeds without regard to safety, especially in Mexico where safety standards were in any case not as high as they were in the United States. It was argued that KCSI had refused to meet the cost of installing barriers or lights at the Avenida Cuauhtémoc crossing, or to place speed bumps on the road approaches, although the crossing was known to be dangerous and despite numerous official requests for them to do so. Nor had the speed of trains through Santa Catarina been limited.

One of the surviving passengers was Rubén Blancas Mata, aged 33 at the time of the accident, who several years later described his experiences of it. He related that he had made the 40-minute journey to his job in a fruit market in Monterrey every day. He had been dozing when the bus reached Santa Catarina, when screams had suddenly awoken him. "It happened very quickly. The train was coming fast," he said. He suffered a head injury in the accident and was knocked unconscious, as well as sustaining multiple fractures. He was several weeks in hospital and four years later was still suffering from the after-effects of the accident and unable to return to his previous job.

Housewife San Juanita Mirna saw a report of the accident on the television and immediately thought of her husband, who caught the Route 120 bus to his work on an assembly line in a Monterrey air-conditioning factory. She telephoned the factory to learn that he had not arrived there. "I knew right then, because he never misses work. It was horrible, so horrible," she said. She was faced with telling her children, aged 10 and 12 and at school at the time, and her adult daughter, that their father had been in an accident. "I didn't have the words to tell them what had happened," she said. Her husband was one of those who had died. She said that it was hard for her to live near the railway because it reminded her of her late husband. "I miss him so much," she added. Since the accident she and her children had to survive on a small pension. Like most of the other victims, they had been poor before the accident and now faced increased financial hardship. [927]

There had been another accident involving a bus on the crossing in November 2002, in which 33 people were slightly hurt. [928]

Twelve Ecuadorian nationals, employed illegally as migrant agricultural workers in Spain, lost their lives at 07:30 on Wednesday 3rd January 2001 when the nine-seater Citroen van in which they were riding to work on farms near Lorca, Spain, was struck by a passenger train travelling at 120 km/h. The other two people in the van, its driver and a 13-year-old girl, were injured, the driver critically. Wreckage of the van and the bodies of its occupants were strewn along 200 metres of the line. The three-car train, presumably a multiple-unit set, was slightly derailed and an elderly woman amongst its 25 passengers suffered from shock. The workers mostly belonged to two families who had moved to the area in the last two years. [929]

A Red Nacional de los Ferrocarriles Españoles (RENFE) spokesman pointed out that although there were no barriers at the crossing, it was properly marked with warning signs and the weather had been clear, so the approaching train should have been clearly visible. Dawn had, however, just been breaking, so visibility was probably not as good as it might have been. [930] [931]

The accident led to what was described as an 'unprecedented protest movement' from the

---

927 www.pitch.com/issues/2004-11-04/news/stline_print.html

928 www.sfgate.com/cgi-bin/article.cgi?f.../international1245EST6782.dtl&type=printable

929 https://www.lavozdegalicia.es/noticia/espana/2001/01/03/doce-ecuatorianos-mueren-arrollados-tren-paso-nivel-lorca/0003_358727.htm

930 *Ananova*, 3rd January 2001

931 www.cnn.com/2001/WORLD/europe/01/03/spain.accident.03/

immigrant community. None of the workers involved in the accident other than the van driver had had work permits. They were expected to work 10 hours a day for a daily wage of 4,000 to 5,000 pesetas (ESP) and their wages were often withheld from them. Their wages, when paid, were equivalent to £15 to £19 Sterling at January 2001 exchange rates or a maximum of £95 for a five day 10-hour week. The average pay for a male UK worker in April 2001 was £445, thus they were very poorly paid.

On about Monday 15th January 2001 five occupants of a car died when a train collided with it on a level crossing in Donetskoye, Russia, in the southern Urals region. [932]

An XPT passenger train from Sydney, New South Wales, to Melbourne, Victoria, collided with a white 1997 Holden Statesman car on Bells Line Road level crossing, near Gerogery, New South Wales, at 16:07 on Saturday 27th January 2001, killing five young men aged between 18 and 20 riding in the vehicle. Two wheels of the front power car of the train were derailed and it took 1,500 metres to stop, while wreckage of the car was scattered over a distance of 200 metres and caught fire. None of the 151 train passengers or six crew was injured but an ambulance took a distressed pregnant woman and another passenger suffering from shock to Albury Base Hospital, together with an eyewitness to the crash who also suffered shock. Buses took the remaining train passengers to the next station at Albury.

Trains normally travelled at 130 km/h over the crossing, which was provided with warning lights and bell. These were thoroughly checked after the accident and found to be working correctly. [933] [934]

The five men had been travelling from Wagga Wagga to Albury to watch their friend Nick Henderson play in his first rugby league match for the Melbourne Storm National Rugby League club.

It appeared that the driver of the car had attempted to beat the train over the crossing. A passenger on the train, Sam Durland from Melbourne, who was familiar with the crossing, told police that he had seen the car travelling south along the Olympic Highway towards the crossing. He recounted, "I thought, 'That car's not going to make it; it's not going to fit in front of the train.' They looked like they were going for it." Another passenger said the same thing. [935]

The crossing with the busy Olympic Highway had been the scene of previous accidents. There were sharp bends in the road on either side of the crossing so that the warning lights were not as visible to motorists as they might have been, while at sunrise and sunset, the sun might be in their eyes. No barriers were provided at the crossing on the grounds that because there was only a single track, there was no chance of a "second train coming" situation. Also, the teenage driver of the car was inexperienced.

Driver Keith Headland of the XPT suffered post-traumatic stress disorder as a result of the crash and stated at the inquest on the victims in July of the following year that he would never drive a train again. [936]

That the driver of the car had been trying to beat the train was hotly disputed at the inquest by the families of the victims and Coroner Carl Milovanovich stated, "In my view there is certainly one hypothesis that can be discounted immediately, and that is a suggestion that the

---

[932] *Ananova*, 16th January 2001
[933] *The Sun-Herald online*, Sydney Australia, 28th January 2001
[934] *News.cm.au*, 31st January 2001
[935] *The Sun-Herald online*, Sydney, New South Wales, Australia, 29th January 2001
[936] *The Age*, Melbourne, Victoria, Australia, 10th July 2002

boys were attempting to race the train. The fact that this version of events received so much publicity, and by innuendo given unwarranted credibility, was most unfortunate and most distressing for the families of the deceased. I wish to make it very clear to the family and to the media who will no doubt report this matter, that I, as coroner, have categorically rejected any suggestion that the boys were attempting to race the train."

The coroner continued, "The acute angle of the curve from a 100 km/h area leaves little room for braking if the bells or lights are not seen or heard or for a person who may not be familiar with that section of the road. The real tragedy in this matter is not whether the driver made an error of judgement, but that in this day and age when we all strive to reap the benefits of new technology, such as computers, advances in medicine, trains that travel at 160 km/h and even faster, we still have a 19th-century approach to level crossings on the basis that they are traversed by horses and carts."

The crossing was replaced by an overbridge, known as the "Five Mates Crossing" bridge after the five young men, in mid-December 2005 at a cost of $24,500,000. [937]

According to the evidence of the assistant driver of a Diva – Sawantwadi passenger train, rash driving by a motorist resulted in a level crossing collision on the Konkan Railway between Mangaon and Veer stations, Maharashtra state, India, on Friday 9th February 2001. He related that a Tata Sumo sports utility vehicle and a Premier Padmini car were negotiating an unmanned crossing, when one tried to overtake the other. The driver of the train had sounded his horn and then made an emergency brake application but could not avoid striking both vehicles. Ten people were killed, including a woman and two children, and four were injured and were rushed to Goregaon cottage hospital. The victims were all members of the same family. [938]

An Osthannoversche Eisenbahnen freight train collided with a school bus on a level crossing at Lübberstedt, Saxony, Germany, at 08:00 on the morning of Monday 26th February 2001. The were only two people on the bus, who were badly hurt. The crossing was protected only by St Andrew's crosses. [939]

A location given as Impala Mines, North West province, South Africa, not far from Rustenburg, was the scene of a serious accident at about 17:00 on Saturday 12th May 2001, when a train from Thabazimbi to Pretoria collided with a van carrying 11 people on a level crossing. By the following day six of the victims had died and the other five were reported to be in a critical but stable condition. All were mine employees. [940]

A train collided with a bus on a level crossing near the village of Ovrazhnoye, in the Kaliningrad oblast of Russia, on Wednesday 23rd May 2001. Fourteen people died and another 16 were injured in the accident, in which the bus was crushed beyond repair. The train escaped relatively unscathed and was not derailed. The Railways Ministry blamed the accident on the bus driver, who it stated had overtaken several vehicles stopped for the train and driven past a red light. It added that the driver had previously had his licence suspended for driving while intoxicated and at the time of the accident had been driving on a temporary permit. He

---

[937] Wikipedia – Gerogery level crossing accident
[938] www.centralchronicle.com/20010210/1002123.htm
[939] http://home.t-online.de/home/05820678-0001/ohe5.htm
[940] *SABCNews.com*, 13th May 2001

survived the collision and was taken to a local police station for questioning. [941] The bus passengers had mostly been elderly people going to their summer dachas. [942]

According to witnesses, a serious accident at Enugu, Nigeria, early on Sunday 27th May 2001 was caused by the absence of a crossing keeper at the point where Zik Avenue intersected the railway. A bus from New Market to Garki, carrying 42 passengers, mostly bound for services at various churches, was in collision with a passenger train consisting of an engine and five carriages. The bus was dragged some distance; four passengers and its conductor were killed and over 20 passengers were injured and taken to the University of Nigeria Teaching and National Orthopaedic Hospitals in Enugu. The crossing was equipped with barriers but there was no crossing keeper on duty at the time. After the accident, the train driver reversed his train to the station to report it before continuing to Port Harcourt. There was reportedly nearly a second accident involving another bus and a motorcycle when his train reached the crossing again, but a 14-year-old boy lowered the crossing barrier. [943]

Interviewed by a reporter, Nigerian Railways Corporation Eastern District Manager explained the apparent absence of a keeper at the crossing, adding that other vehicles had been stopped waiting for the train on either side of the crossing, when the bus driver had jumped the queue and attempted to beat the train. Furthermore, the train had not hit the bus, rather the bus had hit the train. He stated that there had been a crossing keeper on duty and when, for some unstated reason, he had been unable to lower the crossing barriers, he had stopped the road traffic with a red flag. After the accident the crossing keeper had run away, fearing that he would be lynched. The bus driver had also run away. [944]

As was often the case in India, it was a bus carrying a marriage party that was involved in a disastrous level crossing collision at Bahelia Buzurg Halt, between Gola Gokarannath and Rajagan stations, Uttar Pradesh, India, on Wednesday 30th May 2001. The bus had a seating capacity of 60 but was carrying 80 wedding guests when it was struck by Northeastern Railway Up train No. 175 from Mailani to Lucknow at 13:54. The reports do not indicate how many, if any, of the bus passengers were small children who would take up less room. Initial reports spoke of 16 dead, later accounts stated that there were at least 31 dead and another 49 injured, who were taken to hospital in Kheri. The victims, according to a police officer, were largely poor Muslims who worked on farms and did small jobs.

The crossing was unmanned. According to one report the bus had developed a technical snag and its rear wheel had stuck on the crossing, though another report stated that the bus driver had simply failed to see the inter-city express passenger train. Before it could be freed, if indeed it had been stuck, the bus was hit by the train and badly crushed. The bus driver survived the accident and reportedly fled. The train was able to continue its journey when the line was reopened at 16:00. [945] [946] [947] [948] [949]

---

[941] *The Garden City Telegram*, Garden City, Kansas, United States, 23rd May 2001

[942] *Ananova*, 23rd May 2001

[943] *AllAfrica.com News*, 11th June 2001

[944] Ibid, 25th June 2001

[945] *The Indian Express*, Noida, India, 30th May 2001

[946] *The Times of India online*, 31st May 2001

[947] *Indiainfo.com*, 31st May 2001

[948] *The Times of India*, Mumbai, India, 31st May 2001

[949] www.now-india.com/ge.../nownewsdetails.asp?newsid=7389&categoryid=6

Just after midnight on Monday 4th June 2001 an express train from Isparta to the Ægean port city of Izmir had not covered more than a few kilometres when it collided with a minibus carrying 15 people on a level crossing near Gonen, Turkey. Seven bus passengers were killed and the others were all injured, three seriously, and were taken to hospital in Isparta. No one on the train was hurt. The train driver was being questioned by police, though the accident was probably caused by the bus driver disregarding warning signals at the crossing. [950] [951]

A train collided with a fully loaded school bus on an open level crossing at Stromberg, Germany, near Mainz, on or about Wednesday 6th June 2001, injuring 17 children and the driver. Police were trying to determine how the accident had happened. [952]

About 4:39 p.m. on Sunday 1st July 2001 a woman and four children were killed when their car was struck by a Norfolk Southern Railway freight train on an unprotected crossing at Montpelier, Ohio, United States. The victims were named as Wanda Petre, 37; her children Amber, 14, and Chelsea, 11; her nephew Bradley Crontz, 12; and Chelsea Green, 10, a family friend. All were from Angola, 20 miles from Montpelier. No one on the train was hurt, while there were no survivors from the car. The automobile was carried 77 feet and the train stopped a quarter mile down the track.

The train was travelling from Fort Wayne, Indiana, to Detroit, Michigan, and consisted of an engine, 24 loaded and one empty articulated lorry trailers. It hit the 1993 Buick car broadside; the vehicle ended up on its roof and caught fire.

Lt. Mike Sharp of the State Highway Patrol stated that the engineer of the train had reportedly whistled repeatedly for the crossing, which was protected only by signs. He said it was not clear why Ms Petre had failed to see the train. "The investigation is pretty straightforward," the lieutenant added. "The train can only be in one direction at one point in time. We know where it was. We know where the car was." [953]

Township leaders had requested warning lights at the crossing two years before the accident but the crossing had not been deemed dangerous enough to warrant being given priority for receiving such protection. [954]

In an all too familiar scenario, at about 14:00 on the afternoon of Saturday 7th July 2001 the driver of a Toyota Hiace minibus thought that he could beat a train over a level crossing, despite the warnings of fellow drivers who had already stopped for the train, near the town of Mukono, Uganda. He failed, the freight train struck the minibus and eight people died and five were injured. No one on the train was hurt but a motorcyclist stopped at the crossing was struck by the wreckage of the minibus and injured. [955]

At 07:45 on Sunday 8th July 2001 a passenger train collided with a mini truck on an unmanned level crossing between Hathras City and Hathras Road stations, Uttar Pradesh, India, in the North Eastern Railway zone of Indian Railways. As a result of the accident six train

---

[950] *Ananova*, 4th June 2001

[951] http://w.../01-06-05.anadolu.htm

[952] www.germnews.de/archive/dn/2001/06/06.html

[953] *Pharos-Tribune*, Logansport, Indiana, United States, 2nd July 2001

[954] *Sandusky Register*, Sandusky, Ohio, United States, 3rd July 2001

[955] www.africast.com/article.php?newsID=18879

passengers and two people in the mini truck lost their lives and 11 people were injured. [956]

A Zhiguli car was struck by a Moscow-to-Lena train on a level crossing near the station at Chuna, Russia, on the evening of Tuesday 17th July 2001, killing its five occupants. The crossing lacked barriers but was equipped with warning lights, which the car driver had disregarded. [957] [958]

Travelling at about 130 km/h, an express train collided with a car on an automatic half barrier level crossing at Jutila, Finland, between Riihimaki and Lahti, on the evening of Monday 23rd July 2001. A freight train had already passed over the crossing in one direction and when the barrier had failed to rise afterwards, it was theorised that the car driver had assumed it had failed. It had apparently not occurred to him or her that, in the words exhibited on the notices at such crossings in Great Britain, it was probably a case of, 'Another train is coming if lights continue to show.' There was, indeed, another train coming from the opposite direction, No. 79, the 18:02 Helsinki – Iisalmi express. The car zig-zagged around the lowered barrier, was demolished by the express and three men and two women lost their lives. It was the second fatal accident at the crossing within a year and plans were already being worked on for its elimination. [959]

A party of policemen were on their way as reinforcements to Ghagrapar, where there had been an encounter with United Liberation Front of Asom militants, when the Maruti Gypsy four-wheel-drive vehicle in which they were riding was struck by a pilot engine preceding the *Kanchanjungha Express* of the Northeast Frontier Railway on a level crossing between Nalbari and Ghagrapar stations, Assam, India. Six policemen and a civilian were killed and two policemen were seriously injured in the accident, which occurred early on Tuesday 24th July 2001. The injured victims were taken to hospital in Guwahati. The crossing where the accident happened was provided with gates but there was no gateman on duty at the time of the collision and the gates were open for road traffic. [960]

A passenger train collided with a crowded bus in the town of Tegal, Java, Indonesia, on Wednesday 1st August 2001, killing at least 13 people and seriously injuring another five. One report stated that 19 had died. The bus had failed to give way to the train. [961]

Early in 2001 a lorry collided with a gate at Kotuwagoda Crossing, Sri Lanka, not far from Seeduwa station. Several months later, the damaged gate had still not been repaired and the failure of the railway to do so was a factor in a collision between a train and a bus at 11:45 on Thursday 16th August, though the warning bells and lights at the crossing were working normally, while there was also a crossing keeper. The latter described the events leading up to the collision. "When the warning bells rang, signalling the arrival of the train, I went to the Raddoluwa side, where the gate is damaged, with a red flag to signal the arrival of the train. Then a bus from Raddoluwa came in without slowing down and its driver tried to cross the

---

[956] https://archive.pib.gov.in/archive/releases98/lyr2001/raug2001/21082001/r2108200118.html
[957] *People's Daily*, Beijing, China, 18th July 2001
[958] http://www.rusnet.nl/info/cis-today/archive/01-07/18rn6.shtm
[959] www.saunala.../v2001.ht
[960] http://in.news.yahoo.com/010724/58/11w5z.html
[961] *Ananova*, 12th August 2001

line when the train was just 10 metres away."

Ten of the 60 passengers on the Central Transport Board bus from Raddoluwa to Colombo were killed, while 23 were injured and taken to Ragama hospital by passers-by and residents. The bus driver survived. He was given a breathalyser test by the police after the accident and this indicated that he had been under the influence of alcohol. [962]

After being struck by the train, the bus collided with a tree before it stopped. The passengers who died had all been seated towards its rear; they were six men, three women and a 12-year-old child. The bus driver and conductor both fled and faced arrest. The engine of the train was damaged in the collision and traffic over the line was temporarily suspended.

A letter to a Sri Lanka newspaper identified further dangerous features of the crossing as the presence of bus stops close to and on either side of it, which caused tailbacks of traffic over it when buses stopped to pick up passengers, and the very brief opening of the crossing gates between shunting movements over the crossing, which confused motorists. [963]

A party going from Beriar village to nearby Jhaur to mourn the death of a relative were involved in an accident when the three-wheeler in which they were riding was demolished by an Amritsar – Pathankot passenger train on an unmanned level crossing at 11:05 on Thursday 23rd August 2001. Nine persons were killed outright in the accident and at least one more succumbed in hospital. Most of the victims were women. The accident happened near Gurdaspur, Punjab, India. Guard Yog Raj Bhagat told a reporter that the remains of the three-wheeler had been carried almost a kilometre before the train pulled up. [964]

On Sunday 4th November 2001 a coal train was in collision with a chartered bus on a level crossing at Lopang, near Serang, Banten province, Indonesia. At least eight people died and a more than a dozen were injured, while a nurse in the emergency department at Serang hospital stated they had been told to expect more casualties. [965]

On the evening of Sunday 25th November 2001 train No. 4555, a Bareilly – Delhi express, collided with a vehicle variously described as a minibus and a van, on an unmanned level crossing between Kundarki and Machharyi stations, Uttar Pradesh, India, on the Northern Railway. At least nine people died and 10 were injured. [966]

A National Railroad Passenger Corporation (Amtrak) passenger train, operating over a line of the Burlington Northern Santa Fe Railway, collided with a blue 1989 Ford Aerostar minivan at about 4:15 p.m. on Wednesday 12th December 2001, on a level crossing with the little-used Poplar Avenue, on the outskirts of Shafter, California, United States. Travelling at about 79 mph the train, which was bound for Bakersfield and carried about 95 passengers, bore the van half a mile down the line and its driver and six passengers were all killed. The victims were all Mexican men who were coming home from work in the nearby walnut orchards. They were named as Mario Andrés Aguilar, 32, (the driver); Benigno Andrés Aguilar (his brother); Juan Jiménez, José Mario Muñoz, 25; Jesús Romero, 36; Héctor Guillermo Sarabia, 36; and Elroy Suárez, 41.

---

[962] *Daily News*, Colombo, Sri Lanka, 17th August 2001
[963] *The Sunday Times*, Colombo, Sri Lanka, 16th September 2001
[964] *Rediff online*, Mumbai, India, 23rd August 2001
[965] https://jawawa.id/newsitem/eight-killed-as-train-hits-bus-1447893297
[966] *Rediff online*, Mumbai, India, 25th November 2001

There were no barriers at the crossing but it was equipped with warning lights and bells, which were found to be working correctly after the accident. There was initially talk that Señor Aguilar had been drunk. His common law wife, Maria de Jesús Velez, pointed out, however, that the men had been returning from their work, where they were not allowed to consume alcohol, at the time of the accident, and this was confirmed by a post-mortem blood-alcohol measurement, which was found to be only 0.02%, the equivalent of one beer and well below the legal limit of 0.08%. Suspicions of drunk driving had been raised by the supposed smell of alcohol at the accident scene and by the fact that Señor Aguilar had been driving while disqualified because he had been convicted in both May 1999 and again in February 2000 for driving under the influence of alcohol. His driving record was therefore not good. Commenting on the fact that he had been driving while suspended, his wife said, "He knew he wasn't permitted to drive, but we had to. We have to work to pay bills."

It would still have been daylight at the time of the accident, though only half an hour before sunset. Possibly the visibility was not as good as it might have been at the time of the accident, or else Señor Aguilar had expected a slower-moving freight train and had thought he would have time to cross in front of the train.

Shafter resident Amanda Burton was driving along Highway 43 nearby and witnessed the accident. She related, "That train, it came out of nowhere. All you could see was smoke and glass and parts of van flying everywhere. The poor driver. All I could see was the driver waving his hands around for help. I've never seen a car wreck before. It was horrible."

A local newspaper highlighted the personal aspects of the accident. 'Velez said Andrés and her brother Guillermo, and her brothers in law, Jiménez and Benigno Andrés, were living in her house. She said the four men and their neighbours, Romero and Suárez, rode together to work in the walnut groves in Shafter. They were all working for labor contractor Rudolph Palormo. "I told Andrés that morning. 'Please come straight home, because I want us to go to Mass, because today is the day of the Virgin of Guadalupe,'" Velez said. But when Velez came home that afternoon after work, Andrés was not back yet. She said she called him on his cell phone but got no response. Later that afternoon, Palomo called her to tell her that the van might have been hit by a train.

'"When I went to the accident I saw it…it was my van," she said. Velez said her husband and her three relatives killed in the accident were from El Paraíso, in the Mexican state of Guerrero.

'Andrés had been in the United States for 13 years. The couple didn't have children together but Velez has two sons. Velez said Guillermo had been in the country for about four months and Jiménez for about a year. Guillermo didn't have children, but Jiménez had a wife and five children in Guerrero, she said. Velez said she is planning to send the bodies of Guillermo and Jiménez to their relatives in Mexico.

'Benigno Andrés Aguilar and his wife, Rocio, and two children came to the United States about a year ago, trying to find a cure for their sick son, Velez said. His son died less than a year ago, Velez said. Benigno Andrés Aguilar's wife was not available for comment. The family is planning to bury the Andrés brothers in Bakersfield.

'Despite negative reports about Andrés' driving record, friends and relatives defended him. "Mario was a very careful man. He was a very nice neighbor," said Romero's mother, Rosa Romero, who was at Velez's home, where the two comforted each other. Mario and my children loved each other, and I don't have a better neighbor than (Velez)," Rosa Romero said, hugging Velez.' [967]

---

[967] *Bakersfield Californian*, Bakersfield, California, United States, 13th, 14th and 20th December 2001

An attempt by the driver Abdurehim Mernet of a minibus to beat a train over an unmanned level crossing was the cause of a major level crossing accident near Artux, in the Xinjiang Uygur Autonomous Region of China, on Christmas Eve, Monday 24th December 2001. At about 11:15 train No. K-887 from Urumqi to Kashgar collided with the privately owned vehicle, which had reportedly stalled on the crossing. Twenty-eight people lost their lives in the accident near the Afghan border and 11 were seriously injured. Whether the bus driver survived the crash was not reported. The bus fell down a 15-metre-high embankment and its passengers were ejected from it, while the train sustained obvious damage, though it was able to resume its journey after about two hours. No one on the train was hurt. [968] [969] [970]

Three women and two men, the latter brothers, lost their lives when the Jeep in which they were riding was struck by a train on an unmanned level crossing near Kasu Begu village, Punjab, India, on the morning of Boxing Day, Wednesday 26th December 2001. The other five people riding in the Jeep were seriously injured and were admitted to hospital in Firozepur. The victims all came from nearby Ugoo village. [971]

On the night of Saturday 29th December 2001 six people died and 18 were injured when a Philippine National Railways (PNR) train collided with a jeepney packed with passengers on a level crossing in at Paloyon, Nabua, Camarines Sur province, the Philippines. The driver of the jeepney had entered the crossing apparently unaware of the approaching train, having overtaken several vehicles stopped waiting for it. The jeepney was crushed and dragged several metres. [972]

PNR District Manager Constancio Toledano stated that according to a report he had received, the jeepney had been overloaded, with passengers riding on its roof, and that most of the injured had been roof-riding passengers. He admitted that the crossing, which was only provided with luminous warning signs, was dangerous but explained that the dire financial situation of PNR ruled out the provision of full-time crossing keepers, though the addition of warning lights might be feasible. He called on the Nabua municipal government and the Departments of Transport and Public Works to fund the provision of crossing keepers. [973]

Fast train No. 1562 from Iasi to Brasov collided with a local bus on a level crossing at Frunzeasca, Romania, just north of the town of Tecuci, at 14:45 on Wednesday 20th February 2002. The bus was dragged several hundred metres and demolished. Eight people died in the accident and about a dozen were injured. Six badly injured patients were taken to the Anton Cincu Hospital in Tecuci and one was transferred to the Galati County Hospital. According to witnesses, bus driver Adam Valentin had ignored the warning lights and bells at the crossing. Unhurt in the accident, he was interviewed by the police and alleged that the engine of his vehicle had cut out as he was negotiating the crossing. [974]

---

968 *Ananova* report, 25th December 2001
969 *The Straits Times*, Singapore, 25th December 2001
970 *People's Daily*, Beijing, China, 28th December 2001
971 www.tribuneindia.com/2001/20011227/punjab1.htm
972 http://sg.news.yahoo.com/011230/1/288fa.html
973 *Inq7.net*, 31st December 2001
974 *Nine O'clock, The Web Edition*, Bucharest, Romania, 30th April 2002

Brazilian police stated that bus driver Lucivaldo Remigio Silva had probably failed to see a train when he arrived at a crossing near Espinosa in Minas Gerais state at 04:15 on Saturday 16th March 2002. As a result, the Araújo Tur coach from Jacobina to Montes Claros was struck by a freight train carrying sand and limestone(?), travelling along the Montes Claros – Bahia line. The train was travelling at about 50 km/h and it carried the coach at least 100 metres down the line. To make matters worse the fuel tank of the coach exploded upon impact. Five bus passengers were killed more or less instantly, one of whom was thrown through a window, a sixth succumbed soon afterwards and 25 passengers and the bus and train drivers were taken to hospital with trauma and burns. The bus driver was badly hurt, the train driver only slightly so, while the train was little damaged. [975] [976]

A passenger train bound for Malang collided with a minibus bound for the city of Solo on a level crossing in the Karanganyar district of Central Java, Indonesia, at 17:45 on Friday 3rd May 2002. Presumably due to a mistake on the part of a crossing keeper the barriers of the crossing had not been lowered, while the minibus driver failed to note the warnings of other drivers. Five people on the minibus were killed and another seven were injured. [977]

Five people died and two were hurt when train No. 3020, the Down *Kathgodam Express*, collided with a Jeep on a level crossing about a kilometre from Barauni station, in the Begusarai district of Bihar state, India, on Wednesday 22nd May 2002. [978]

A train from Guantanamo to Habana collided with a bus on a level crossing in the town of Jicotea, Cuba, in Villa Clara province, on or about Monday 3rd June 2002. Eight bus passengers, two of them boys aged 9 and 14, lost their lives and one man on the train was badly hurt. A preliminary investigation indicated that the bus driver had ignored a warning signal at the crossing. [979]

The Northeastern Railway Super Express from Kanpur to Kasganj collided with a bus at Chandi village in the Etah district of Uttar Pradesh, India, between Sahawar Town and Badhari Kalan stations, on Tuesday 4th June 2002. The collision carried the bus about 250 metres and threw it into a canal, killing a reported 34 people and injuring 26 more.

A mistake by a crossing keeper was apparently the cause of the accident. Police officer Satyendra Pal Singh stated, "The railway crossing gate was open. The gatekeeper has run away after the accident." [980] [981]

Travelling south from Bradford, the 6:15 a.m. Government of Ontario Transit (GO) train from there bound for Toronto collided with a 2000 Dodge Caravan minivan on a level crossing at Concord, Ontario, Canada, at about 7:25 a.m. on Friday 7th June 2002. Following the impact, the minivan was carried along the line for 30 metres before it fell into a ditch. Its five occupants were killed. None of the approximately 800 people on the train, which had

975 *Ananova*, 16th March 2002
976 www2.uol.com.br/tododia/ano2002/marco/170302/brasil.htm
977 *Channelnewsasia.com*, 5th May 2002
978 www.tribuneindia.com/2002/0524/nation.htm
979 www.cubanet.org/CNews/y02/jun02/03e4.htm
980 *Ananova* report, 5th June 2002
981 *The Hindustan Times*, Delhi, India, 4th June 2002

been travelling at 130 km/h, was hurt. The crossing was provided with barriers, warning lights and bells, which were found to be working correctly after the accident. There were rumours after the accident that the barriers had not come down before the accident, but a video camera on the train showed the van skidding beneath the lowered crossing barrier and stopping on the line in front of the train. It was stated to be not uncommon for drivers to ignore the crossing signals. A local, Darlene Shearer, stated, "I've noticed people shooting through here at quite a few times. (They) speed through the barriers because they don't want to wait for the trains to pass." [982] [983] [984]

The Southern Railway *Netravati Express* from Kurla to Thiruvananthapuram collided with a school van on a Crossing No. 224 at Punnol, between Tellicherry (now Thalassery) and Mahe stations, Kerala, India, at 08:50 on Wednesday 26th June 2002. A major tragedy was avoided as the van had already dropped off the pupils at Sree Narayana English Medium School but the bus driver, named variously as Ratheesh and Satheesan, was fatally injured and succumbed at the Indira Gandhi Hospital, while a child, Muhammad Ferin, was seriously hurt. The gates of the crossing had reportedly not been closed for the passage of the express. [985]

Nine jeepney passengers died and another nine were seriously injured when it was struck by a Philippine National Railways freight train on a level crossing at Bagong Pook, Laguna province, the Philippines, on Saturday 29th June 2002. [986]

The driver of a light truck overtook other vehicles stopped waiting for the train, ignored the warning lights at a level crossing near the town of Kobrin, Belarus, and drove onto the line, on the afternoon of Saturday 29th June 2002. A Prague – Moscow express travelling at almost 100 km/h collided with it, killing six people in it. The reckless driver was a local man, aged about 23 and his passengers were aged between 10 and 42. As a result of the accident the line was blocked for about two and a half hours. [987]

Five people lost their lives in a level crossing accident near Marlow, Illinois, United States, at a level crossing with a rural blacktop road, protected only by crossbucks, on Tuesday 9th July 2002. The accident happened when the van in which they were riding was struck by a Norfolk Southern Railway (NS) train. The victims were Connie Lorance, 39, from Opdyke, Illinois (the driver); her children Adam, 8, and Alyssa, 6; her nephew, Blake Trout, 5; and her niece Julie Trout, 11. Husband Robert "Mike" Lorance, 43, survived with serious injuries.

The Illinois Commerce Commission subsequently cited NS for its failure to remove vegetation obstructing the view of motorists at the crossing, quoting an Illinois law that stated, "Every railroad shall keep its right-of-way adjacent to its tracks reasonably clear of brush, shrubbery, trees, weeds, and crops and all unnecessary permanent obstructions such as unauthorized signs and billboards, for a distance of at least 500 feet each way from every grade crossing, where such things would materially obscure the view of approaching trains to travellers on the highway."

---

982 www.ctvnews.ca/content/sites.../870716.as
983 www.nationalp...493320.htm
984 www.ble.org/pr/news/headline.asp?id=3589
985 www.kaumudi.com/NEWS/june 28/kerala.stm
986 Internet report on Resolution No. 632 to Senate of the Philippines
987 www.brestonline.com/cgi-bin/Dailynews/Local/printnews.cgi?newsid1025622856,30675,English

The NS subsequently agreed to cut back the vegetation "as a sign of good faith," while disputing liability for the accident.

As for the provision of gates and lights at the crossing, John Blair, an engineer for the commission, pointed out that these cost upwards of $100,000 per site, and that busier crossings must have priority over rural ones such as that near Marlow, which were used by less than 100 road vehicles daily, stating, "If we had the funding, they'd all be signalized. But we have to prioritize." [988]

A bus carrying pilgrims from the Varishali district of Jharkhand and travelling from Tarapith to Kolkata was struck by an Eastern Railway goods train on the unmanned Crossing No. 8 (also known as Abdarpur Crossing) near the town of Suri, West Bengal, India, at about 03:40 on Friday 6th September 2002. Four bus passengers were killed outright and another four succumbed in a Suri hospital. The bus driver had failed to observe the approaching train.[989] According to some reports, the bus had been driven through the bars of the crossing. [990]

A 42-car CSX Transportation freight train from St Louis, Missouri, to Boston, Massachusetts, collided with a pickup truck on an unprotected level crossing near DeGraff, Ohio, United States, just after 12:30 p.m. on Sunday 8th September 2002. The Dodge Dakota was carried almost a mile by the train and burst into flames, burning the bodies of four people beyond recognition. Witness Tracy Reynolds told a reporter, "The front engine of the train was on fire and the truck was engulfed." Other witnesses pointed out that the road crossed the tracks at an acute angle, which made it difficult for motorists to observe eastbound trains, plus trees partially obscured the view. Witness, Lori Westlake, added, "These kids didn't have to die. If there was a bell or there was a light…" [991]

Don Dillon, 56 (the driver), Marti Hicks and her sons Natoris, 11, and Datoris, 10, were all killed more or less outright. A third son, Artoris, 12, was thrown from the bed of the truck, initially survived the accident and was taken to the Children's Hospital in Columbus, but he succumbed to his injuries the following day.

It was believed that Mr Dillon had failed to see the approaching train.

It was reported that there were 6,100 public road level crossings in the state of Ohio, 40 per cent of which had gates and lights, 15 per cent of which had just lights and the remaining 45 per cent of which had only warning signs. [992]

Santos station, Lisboa, Portugal, was the scene of a serious level crossing accident early on Wednesday 2nd October 2002. At 06:48 the 18-year-old male driver of a Peugeot 205 car carrying six young people – four men and two women – ignoring the flashing warning lights and ringing bells, went around the lowered half barriers of the crossing and was struck by a train. The train driver was also sounding his horn and a railwayman at the station gesticulated to the driver of the advancing car to stop, all to no avail. The car was dragged some 80 metres and reduced to the usual heap of scrap metal; its occupants were all killed. A taxi driver, his passenger and a woman whose house adjoined the crossing also witnessed the accident. By chance the woman had just been opening her window at that moment. She told a reporter, "I

---

[988] www.ble.org/pr/news/headline.asp?id=4267
[989] http://in.news.yahoo.com/020906/43/1uxzd.html
[990] www.hinduonnet.com/thehindu/holnus/04061405.htm
[991] NBC columbus.com News
[992] Ibid

was just opening the window. I saw the train coming, the gate closing. My God, I didn't want to believe it..."

When Instituto Nacional de Emergências Médicas (INEM) staff first arrived at the scene three of the victims were still alive but Dr João Madeira Lopes said that they had been trapped and gravely injured and it had been impossible to save them. It took a pair of cranes two and a half hours to extricate the car from beneath the train, by which time all its occupants were dead. The line was closed until 10:00, when single-line working was opened up between Cascais and Alcântara. The crossing equipment had to be checked and damaged parts of it replaced, a process likely to take two days, during which time the road over it would be closed.

Rede Ferroviária Nacional (Refer, the company for the infrastructure of Portuguese railways since privatisation in 1997) spokesman Rui Reis stated that Refer had a national policy of removing level crossings and that 200 had been removed already that year. Regarding the Santos crossing, there had already been discussions with the local authorities about its abolition. [993]

South Central Railway train No. 252 from Guntakal to Chikkajajur collided with an Andhra Pradesh State Road Transport Corporation bus on an unmanned level crossing between Obulapuram and Samalapuram, Andhra Pradesh, India, at about 08:45 on Sunday 17th November 2002. The train hit the rear of the bus and dragged it almost 500 metres. Five people were killed outright and 36 of the possibly 54 people on the bus were badly injured, one of whom subsequently succumbed on the way in Bellary hospital in Karnataka and another later. The bus had been on its way from Raidurg to Gavisiddeswaram. [994]

Minister of State for Railways Bandaru Dattatreya emphasised that it was mandatory for drivers to check before crossing the line, while conceding that many failed to do so. He stated that there was no provision for the payment of *ex gratia* sums to the victims of accidents at unmanned crossings, adding that it was proposed to convert all such crossings into manned ones in a phased manner. [995]

Andhra Pradesh Transport Minister M. Damodar Reddy, however, stated that compensation of 50,000 rupees would be paid to the families of child victims who had died and of one lakh (100,000) rupees to the families of the other deceased. [996]

The low sun getting into the eyes of a lorry driver was given as the reason for his failure to see an oncoming train at Ferreira, Free State province, South Africa, on the outskirts of Bloemfontein. A freight train collided with the lorry, which was dragged about 500 metres, during which a tractor fell off the back of the lorry. More seriously, four people riding on the lorry were killed more or less instantly and the lorry driver himself died from his injuries in a short time. The driver and assistant driver on the train, which was transporting new road vehicles, were both injured in the accident, which occurred at 17:45 on Thursday 21st November 2002. The line was blocked until the following morning. Two people had died in other accidents at the same crossing in the previous months. [997]

---

993 https://www.publico.pt/2002/10/03/jornal/seis-mortos-em-colisao-de-automovel-com-comboio-em-lisboa-175043
994 www.hinduonnet.com/thehindu/holnus/01172004.htm
995 *www.outlookindia.com/pti_print.asp?id=97978*
996 *Deepikaglobal.com*, 27th November 2002
997 www.news24.com/News24/South_Africa/0,1113,2-7_1288553,00.html

A tempo carrying a party of 24 people belonging to a wedding party was struck by the *Sutlej Express*, bound for Hissar, on an unmanned level crossing on the Ferozepur-Ludhiana line at Rajguru Nagar, near the village of village of Tharike, Punjab, India. The accident happened at about 19:45 on Thursday 28th November 2002. Despite seeing the approaching train, the tempo driver tried to cross. Six people were killed outright, including a one-year-old child, and 13 were injured, one of whom later died. The injured were treated in different hospitals in Ludhiana. [998]

Three adults, one of them the driver, and two schoolchildren, lost their lives when a train on the Bialystok – Warsaw line collided with a bus near Małkinia, Poland, on the afternoon of Monday 2nd December 2002. None of the train passengers were hurt but its driver suffered from shock. The children who lost their lives were aged 10 and 11 and lived in Zawisty Podleśne. Five other children aged between nine and 17 were injured, one seriously.

The bus, from Ostrowia, was driven directly in front of the train, and was carried several dozen metres by the impact. There were 25 people on it altogether, a number of them adults, as it was not a school bus *per se* but an ordinary service bus, some of the passengers on which were pupils returning home from school.

The crossing was unprotected and had been the scene of previous accidents. Repeated requests had been made for the provision of barriers or some other form of protection at it but nothing had been done.

The dead driver had been with the bus company for only three months and had been driving the bus involved in the accident for the first time, standing in for a colleague. He had never travelled over that route before. According to a surviving pupil he had neither stopped at the crossing nor even slowed down, perhaps because being unfamiliar with the route he had not realised that it was there. Witnesses also stated that the train had not whistled for the crossing, though this might have meant simply that they had not heard the whistle. At the time of the report on the accident the police were still attempting to determine who was to blame for it. [999]

Gross overcrowding of a bus was a factor in a level crossing accident that occurred early on Saturday 4th January 2003 in the Kambazuma suburb of Harare, Zimbabwe. The 75-seater Keystone company vehicle was carrying 115 people when it was struck by a freight train on a level crossing. The bus was carried 50 metres by the train and a number of passengers were trapped in its wreckage and had to be freed by firemen before being ferried to hospital in a fleet of ambulances.[1000] Four passengers were killed outright and the other 111 people on the bus were injured and taken to Harare Central and Parirenyatwa hospitals, where another passenger succumbed. Two days after the accident, 12 other passengers were still in a critical condition. The state-run Blood Transfusion Service announced that it had run out of emergency blood supplies, while basic drugs were also in short supply. Zimbabwe was in an economic crisis at the time, with acute shortages of food, hard currency, drugs and fuel following a drought and economic mismanagement. [1001]

A bus passenger told a reporter, "We even shouted at the driver to stop as the train was coming, but he just proceeded." The bus driver, who was little hurt, was subsequently arrested

[998] *Deepikaglobal.com*, 30th November and 1st December 2002
[999] https://wiadomosci.wp.pl/zderzenie-pociagu-z-autobusem-szkolnym-sa-zabici-i-ranni-6031943752459393a?proxied
[1000] *News24*, Cape Town, South Africa, 4th January 2003
[1001] *The Star online*, Petaling Jaya, Malaysia, 5th January 2003

and faced five counts of culpable homicide. [1002]

Five children aged between eight and 15 lost their lives and four other children were injured when the tractor trolley on which they had hitched a lift was hit by a train on the night of Tuesday 7th January 2003 on the unmanned Nizampur Crossing, Uttar Pradesh, India. A dozen children from Nizampur village were riding on the tractor trolley when the accident happened near Khurja station. After the accident, angry villagers attempted to set the train on fire and protested at the failure of the authorities to provide gates at the crossing, despite previous accidents there. [1003]

Nine people travelling on an autorickshaw lost their lives when it got stuck on an unmanned level crossing at Devarkodava, Andhra Pradesh, India, on Friday 10th January 2003. The vehicle was struck by a goods train proceeding from Kacheguda to Kurnool. [1004]

At about 23:20 on Friday 10th January 2003 a bakkie (pick-up truck) failed to give way to a freight train at a level crossing at Kanana, North West province, South Africa, near Klerksdorp. Seven people died in the accident, including the 43-year-old driver of the bakkie, and five were injured and taken to the Tshepong Hospital in Klerksdorp. The engine crew were unhurt. [1005]

A St Bernard Parish school bus, which was in collision with a train on the evening of Monday 20th January 2003 at Chalmette, Louisiana, United States, on the outskirts of New Orleans, was fortunately only occupied by its driver at the time of the accident. It was struck about three feet behind the driver's seat by the Norfolk Southern Railway train and pushed 50 feet along the line, before it collided with a fence and a utility pole. Driver Charles J Baggese III suffered moderate but not life threatening injuries in the 6.05 p.m. accident and was taken to hospital. He said that he had started from his home only moments before the accident to pick up children for a sporting event, and said he had not seen the train. Police issued him with a citation for failing to yield to the train. [1006]

At about 19:30 on Sunday 2nd February 2003 a Calais – Lille passenger train travelling at 96 km/h collided with a Renault 19 car carrying five young people on a level crossing at Hesdin, France. The car was crushed, carried 350 metres and its occupants were all killed. The driver of the train was slightly hurt but its 140 passengers all escaped injury. About 30 firemen from three fire stations and a large number of police attended the accident. It took the firemen a considerable time to cut the bodies from the wreckage, working under floodlights, and at 23:30 one body remained trapped in the wreckage. The crossing, with a narrow country road, lacked barriers but had stop signs, while to warn motorists of its presence it was preceded by three countdown markers, spaced at 50-metre intervals. [1007]

1002 *AllAfrica.com News*, 6th and 8th January 2003

1003 *The Tribune online*, Amritsar, India, 8th January 2003

1004 *Deepikaglobal.com*, 8th January 2003

1005 http://iafrica.com/news/sa/132263.htm

1006 www.nola.com/news/t-p/index.ssf?/base/news-0/1043220357325942.xml

1007 *L'Interaute*, 3rd February 2003

The Maurya Express of the Northeastern Railway collided with a Jeep on an unmanned level crossing near Nonapaar Halt, Uttar Pradesh, India, on the evening of Sunday 30th March 2003. The Jeep had come from Bharhe Chaura and was bound for Darauli in Bihar for a barat (a celebratory wedding procession). Five people were killed outright, six died in hospital and eight were injured. [1008]

Western Railway train No. 9006, the Up *Saurashtra Mail*, which had left Okha at 11:30 for Mumbai, was in collision with an oil tanker lorry on an unmanned level crossing at Lunsariya station, Gujarat, India, at about 19:00 on Monday 31st March 2003. Fire broke out following the collision and enveloped the last three bogies (coaches) of the train as well as the lorry. According to some reports at least five people died, including the lorry driver, and another 15 suffered serious burns.[1009] The tanker, travelling at speed and unable to stop, had rammed the side of the train.

The injured were taken to Rajkot hospital, where some were interviewed by a reporter. 'Little did 54-year-old Vasta Krishnaof Wadhwani village in Kheda district know that his attempts to save fellow passengers could leave him with 60 per cent burns.

'A wager commuting daily from Rajkot to Memdabad, Vasta was travelling in the third coach from the engine. "I was resting at the seat next to the door when I heard a huge blast. Before I could understand what the sound was, flames started entering my coach. People started pushing each other. There was an old lady. I helped her get out of the coach along with some aged people sitting near the windows. Meanwhile someone pushed me near the flames. I continued helping people get out of the train when I suddenly started feeling the flames on my skin. That's when I realised that my clothes had caught fire. Somehow I managed to remove the clothes and got out of the coach," he said.

'Ironically, while the rooms of other injured patients are full of friends and relatives, Vasta's room is empty. Vasta clarified, "I have told my family members about the accident, but they won't come as we can't afford the fare. My family is dependent on what little I earn."' [1010]

A major disaster occurred at Siófok, Hungary, just after 08:30 on Thursday 8th May 2003, when an express from Budapest to Nagykanizsa collided with a double-decker bus carrying 38 German tourists, most of them pensioners, on a level crossing. As a result of the collision the bus was cut in two, half of it being dragged about 150 metres along the line. Television footage showed part of the bodywork of the coach wrapped around the front of the locomotive.

Captain Tamás Hekker, the local police chief, stated that 30 of the bus passengers had died at the scene and four more had succumbed later. In the early afternoon several bodies, covered with blankets, were still lying by the side of the track. Seven other people were hurt. It was not stated if the bus driver had survived the accident. The reported number of dead was later corrected to 33, it being stated that a report that a passenger had died in hospital in the afternoon had been mistaken. It was confirmed that the bus driver was amongst the dead.

István Gálos, a 39-year-old engineer who was stopped in his car on the opposite side of the crossing when the accident happened, spoke of seeing the bus moving onto the crossing in front of the train. "The express whistled twice in an attempt to warn the bus driver but it (the train) could not pull up in time to prevent the accident."

[1008] *Deepikaglobal.com*, 4th April 2003

[1009] Ibid

[1010] http://cities.expressindia.com/fullstory.php?newsid=48042

Railway officials confirmed that the red warning lights at the ungated crossing had been flashing when the bus had started across the line. [1011]

Prime Minister Péter Medgyessy visited the scene and later called for the installation of barriers at all level crossings in his country; probably an option that would be very expensive. Meanwhile, German Chancellor Gerhard Schröder offered his condolences to the relatives of the victims and the German Embassy set up a crisis centre close by.

Siófok was a popular tourist centre on Lake Balaton, a freshwater lake whose waters were rich in minerals – especially calcium and magnesium – reputed to have a therapeutic quality.[1012] The coach passengers had booked their trip with travel operator Maxim Reisen in Cloppenburg, northern Germany and the travel firm had chartered the coach from Ursel Reisen in Loehne, near Bielefeld. [1013]

According to Javanese custom after a wedding, the parents of the bride, if they can afford it, hold a feast at their home to welcome their son-in-law. Thus, early on the morning of Monday 16th June 2003, a minibus containing a wedding party was travelling from Tegal to the town of Sragen in Central Java, Indonesia. It had almost reached its destination when it was struck by a train on a level crossing, killing the bride and groom and 13 wedding guests. The other seven guests were critically injured. A police officer stated that negligence on the part of the crossing keeper was believed responsible, saying, "He is being questioned as a suspect because the barriers at the crossing were not down when the train passed." [1014]

A Northeastern Railway Gonda – Bareilly – Mailani express collided with a tractor trolley carrying members of a marriage party on an unmanned level crossing near Holasia, between Kahraha and Mihinpurwa, Uttar Pradesh, India, on Wednesday 18th June 2003. Nine people died and 20 were injured, five of them seriously. [1015]

Five people died and four were injured when an express train from Timişoara to Iaşi collided with a van on a level crossing near Stroieşti, Romania, on Tuesday 24th June 2003. The van driver had reportedly ignored the red warning lights protecting the crossing. [1016]

On Sunday 6th July 2003 a freight train collided with a lorry on a near the town of Abu el-Matamer, Egypt, about 130 kilometres north of Cairo. There was no authorised level crossing at the point of the accident. The lorry driver and nine passengers on the vehicle were killed and six other passengers were hurt. [1017] [1018]

Another level crossing accident involving a school van occurred in India on Tuesday 22nd July 2003 on the Southern Railway, at the village of Yemur, Tamil Nadu. The special train

1011 *Daily Sitka Sentinel*, Sitka, Alaska, United States, 8th May 2003
1012 *Winnipeg Free Press*, Winnipeg, Manitoba, Canada, 9th May 2003
1013 www.swissinfo.org/sen/Swissinfo.html?siteSect=143&sid=1843741
1014 www.utusan.com.my/utusan/content.asp?y=2003&dt=0617&pub=Utusan_Express&sec=World&pg=wo_07.htm
1015 http://sify.com/printer_friendly.php?id=13175693&ctid=2&lid=1
1016 www.bt.no/utskrift.jhtml?articleID=166652
1017 http://edition.cnn.com/2003/WORLD/meast/07/06egypt.accident.reut
1018 http://www.dagsavisen.no/utenriks/apor/2003/07/740366.shtml

involved in the accident chanced to be carrying the Southern Railway general manager and other officials from Dindigul to Karur after a wayside inspection, while the van belonged to a private matriculation school at Thanthonimalai. As a result of the collision the van was dragged some distance and badly damaged. Four children were killed and six were seriously hurt. The van driver pleaded that because of a curve, he had not been able to observe the approaching train. [1019]

At 06:30 on Monday 4th August 2003 express train No. S1 from Hanoi to Ho Chi Minh City collided with a minibus carrying 22 Viet Cong war veterans aged between 70 and 85 and their families on a level crossing in the Cam Lô district of Quảng Trị province, Vietnam. Twelve people on the minibus were killed outright and the other twelve were seriously hurt, two of whom succumbed to their injuries in Quảng Trị Hospital. One of those who died was a 12-year-old girl. The veterans had been on a day trip to the beach organised by the Board of Representatives for Former Prisoners of War.

The Quảng Trị authorities granted a sum of 2,000,000 Vietnamese dong (VND) to the families of the deceased and 1,000,000 VND to the injured. Cam Lô district authorities had given initial allowances of 500,000 and 300,000 VND respectively to the two categories of victims.

The accident was blamed on the bus driver, who had attempted to negotiate the crossing despite warning signals. [1020]

Down passenger train No. 354 from Sargodha to Lalamusa collided with a packed 44-seater bus, reportedly carrying over 60 passengers, on a level crossing at Malikwal, Pakistan, at about 08:25 on Saturday 20th September 2003. According to the Edhi Foundation relief organisation, 34 people lost their lives in the accident, and a police official spoke of at least 40 dead, but Pakistan Railways and doctors quoted a somewhat lower figure of at least 27 dead and six injured; those who died comprising 24 men and three children. All the dead had been on the bus, though two of the injured were on the train, one of them being Driver Malik Abid Hussain.[1021] The injured were taken to hospital in nearby Mandi Bhawaldin.

Local resident Javed Navaz told a reporter, "I saw blood and human parts scattered everywhere. The bus looked like a skeleton." He added that dozens of residents had rushed to the scene to help, while later the police had arrived and taken away the dead and injured.

The residents were reportedly very angry that no measures had been taken to improve safety at the crossing, despite four earlier accidents in which at least four people had died.[1022] Protection at the crossing was completely lacking apart, perhaps, from warning signs. Mohammed Sabir, one of the residents, complained that about a year earlier, a train had collided with a lorry at the same place, killing two people, and that there had been a further accident after that. "Then we again contacted railway officials six months ago when another train hit a lorry, injuring one person. We wrote letters to the Minister for Railways about the danger we were facing but they took no notice." Nasrullah Ghalzai, a spokesman for Pakistan Railways, denied knowledge of any previous crashes and insisted that his department had not received any letters of complaint from local residents. [1023]

1019 www.hinduonnet.com/thehindu/holnus/02222205.htm

1020 *Vnagency.com.vn*, 4th and 7th August 2003 and 15th February 2004

1021 *The Daily Star*, Dhaka, Bangladesh, 21st September 2003

1022 www.hipakistan.com/en/detail.php?newsId=en3931&F_catID=&f_type=source

1023 *Bakersfield Californian*, Bakersfield, California, United States, 21st September 2003

Doctor Pervez Ahmad in the Mandi Bhawaldin Hospital said that hundreds of people had gathered outside to identify bodies and pray for the injured. [1024]

Railways Minister Ghaus Bakhsh Mehr refused to resign even if Pakistan Railways were found guilty of negligence in the accident. He admitted to reporters that unmanned level crossings not only affected train speeds but caused accidents, however, "We don't have funds to man all the crossings, nor can we close them down, keeping in view the public need." He said that he regretted the general public lack of awareness of road safety measures. "They would insist that a gatekeeper allow them to pass through the crossing even when the train is hardly a hundred yards away." He added that the railway had been paid 1,900,000 roubles for the introduction of a manned crossing at Malikwal, 1,100,000 roubles short of the required cost and despite repeated reminders it had not received the balance outstanding. Compensation would be paid to the victims only if the railway was found to be negligent. There was no proposal to reduce the number of level crossings in the country. [1025]

The initial inquiry into the accident fixed the number of dead at 27. It blamed the accident on the bus driver, saying that it should have been easy to judge the speed of the train even when it was a kilometre away, but the bus driver had misjudged it. Because he had been injured in the accident the train driver had been unable to apply the emergency brake and the fireman had done so. [1026]

Three pupils, two boys and a girl, aged 14 or 15, were killed and 38 pupils and teachers were injured, one of whom later died, when a southbound train travelling at 73 km/h collided with the Lung Mao Co school bus in which they were riding on a level crossing in Yingko, Taiwan, during the morning rush hour, at 07:38 on Tuesday 14th October 2003. The bus had just left Yulin Secondary School in the northern town of Shulin on a three-day graduation trip to the southern holiday resort of Kenting.

The road over the crossing was narrow and as the bus entered the crossing from one side, a lorry did so from the other. They could not pass one another on the narrow crossing, while 11 other buses were following the school bus, so its driver could not reverse off the crossing. Instead, he asked the lorry driver to set back but before the lorry could do so, the train had arrived.

Rescuers had to use ladders to climb into the overturned bus and rescue the injured while police restrained crying relatives who had come to the scene.

There had been a similar accident at the crossing on 20th July in which 16 people had been injured.

Thirty-four trains carrying 20,000 passengers were delayed by the accident. The line was reopened at 13:16.

The crossing was equipped with automatic barriers, warning bells and also, since June 2001, an emergency button, which if pressed could alert the driver of a train when it was 800 metres from the crossing. As the button was not pressed on the morning of the accident, the train driver had no warning of the danger until he was about 50 metres from the crossing, while it would take 270 metres for the train to stop.

Taiwan Railway Administration (TRA) Deputy Director and spokesman General Hsu Ta-wen stated that the railway had requested improvements to the crossing after the July accident and the Ministry of Transport and Communications had set aside 15,000,000 Taiwan New

---

[1024] *The Daily News*, Jacksonville, North Carolina, United States, 21st September 2003
[1025] *Dawn*, Karachi, Pakistan, 24th September 2003
[1026] www.dailytimes.com.pk/default.asp?page=story_29-9-2003_pg7_19

Dollars (NT$) for the task but the work had not yet been begun.

TRA official Chen Shih-cheng stated that there were in total 40 such crossings in the country, which were too narrow for rapidly growing cities and hence needed similar work, adding that the government had spent NT$ 1,300,000,000 on such improvements in the last decade.

The TRA announced that should the police investigation determine that the railway was at fault it would pay compensation of NT$ 20,000 to the families of those killed and NT$ 6,000 to those injured. [1027] [1028]

There were complaints after the accident that many road users were not aware of the emergency buttons and that the warning bells announcing the approach of a train only sounded 30 seconds before it arrived at the crossing, which did not allow sufficient time for the emergency button to be pressed. TRA Deputy Director Ta-wen stressed that in an emergency the button should be pressed at once, before the approach of a train was announced. [1029]

Seven members of the same family were killed when the autorickshaw in which they were travelling was struck by a Bhuj – Viramgaam passenger train on an unmanned level crossing near the town of Gandhidham, Gujarat, India on Sunday 26th October 2003. It appeared that the driver of the autorickshaw had failed to notice the train. [1030]

Nine pre-school children and two adults were riding in a school minibus when it was struck by a shunting train on a level crossing protected by warning bells and lights at Wanganui, New Zealand, at 08:10 on Wednesday 5th November 2003. The bus had been on its way to a local kohanga reo (Māori language and culture revival programme). The train clipped the back of the bus, throwing it into a set of traffic lights. All those in the bus were injured and four of the children were initially taken to the intensive care unit of Wanganui Base Hospital and the other five to the children's ward. Two of the badly hurt children were later flown to the Starship Hospital in Auckland. The crossing lights and bells were found to be working correctly.[1031] The 74-year-old bus driver was subsequently charged with six counts of careless driving causing injury. [1032]

A train travelling from Yoshkar-Ola, the capital of the Mari-El region, to Kazan, collided with a bus carrying 30 people from a state farm on a level crossing near Shelanger, Russia, 50 kilometres south of Yoshkar-Ola, on the night of Saturday 15th November 2003. Five people, presumably bus passengers, died in the accident and another 13 were hurt. A preliminary report blamed the bus driver for the accident. [1033]

On the evening of Tuesday 2nd December 2003 a pickup truck belonging to the Companhia Hidrelétrica do São Francisco was struck by a freight train of the Companhia Ferroviária Nordeste on an open level crossing in Itapiúna, Brazil. The train consisted of three locomotives and 29 waggons. As a result of the accident the leading engine of the train was

---

[1027] *TaiwanNews.com*, 15th October 2004
[1028] www.taipeitimes.com/News/taiwan/archives/2003/10/15/2003071699
[1029] www.taipeitimes.com/News/taiwan/archives/2003/10/16/2003071853
[1030] www.expressindia.com/print.php?newsid=25631
[1031] *New Zealand Press Association*, 5th and 7th November 2003
[1032] http://xtramsn.co.nz/news/0,,3762-3000279,00.html
[1033] *Ananova*, 16th November 2003

derailed and the pickup truck was wrecked, seven people in it being killed. They were later named as José Evandro Garcia Filho, Adriana Maciel de Freitas, Cícero Monteiro dos Santos, Maria Mariana Ribeiro Silva, her children Jaqueline Janine and Dario José, and José Cláudio da Silva Vieira, the driver of the truck. Cícero dos Santos and José Vieira were company employees, the others were being given a lift. One passenger in the vehicle, Cristina Pereira da Silva, 29, survived the accident with injuries and was taken to the José Frota Institute hospital in Fortaleza. It took rescuers five hours to remove the bodies of the victims.

There was no protection at the crossing other than warning signs. Police were hoping to interview Driver Pedro Paulo de Abreu of the Quixadá-to-Fortaleza train, surviving pickup passenger Cristina Silva and a motorcyclist witness, in order to determine how the accident had happened. [1034] [1035]

An accident similar in its cause to that at Fox River Grove, Illinois, over eight years earlier, occurred on the afternoon of Monday 5th January 2004 in Slidell, Louisiana, United States. Again, a school bus stopped a set of traffic lights at red with its rear foul of the railway tracks and before it could move forward, it was struck by a train. This time there were no pupils on the bus, all having been dropped off at the time of the 4:30 p.m. accident. Engineer Charles Mosby, driving the southbound 144-car Norfolk Southern Railway train, made an emergency brake application but was unable to avoid striking the bus. Travelling at about 25 mph, the train hit the back left corner of the vehicle, spinning it around. One of the back seats was crushed and the emergency door was forced open.

'Juanita Jenny, who was driving one of the cars stopped in front of the bus, said she was alarmed when the bus started over the tracks after the traffic light had turned red. "We kept inching forward to give her more room but we had a red light and couldn't go anywhere. The next thing I knew, the railroad crossing rams came down, and the train crashed into the bus."'

No one was hurt in the accident. Substitute bus driver Jamie Rancatore, 27, who had been filling in for a sick colleague, pleaded that the drivers of the two cars ahead of her had failed to pull forward enough to let her get off the crossing. It was pointed out that state law required drivers to have at least 50 feet clear beyond a crossing before entering it and she was given a ticket to appear in court for failing to comply with this requirement. [1036]

At least ten labourers lost their lives and 30 were injured when the lorry in which they were riding was struck by the *Noakhali Express*, bound for Dhaka, at about 06:30 on Monday 12th January 2004 and fell into a ditch. The accident happened at Nimtali Crossing, Bangladesh, not far from Tongi. There had been about 70 labourers on the lorry, mostly from Mymensingh and Jamalpur, who had been on their way to shovel earth at a construction site at Pubail. The injured were taken to Dhaka Medical College Hospital and Tongi Hospital. No one on the train was hurt. After some delay, it continued its journey and rail traffic was not greatly affected.

Tongi police blamed the accident on Crossing Keeper Mohammad Hanif, whom they believed might have fallen asleep prior to the accident. He was suspended pending an inquiry. Expressing his condolences to the families of the victims, Communications Minister Nazmul

[1034] https://diariodonordeste.verdesmares.com.br/seguranca/colisao-entre-toyota-e-trem-mata-sete-1.185757
[1035] https://ne10.uol.com.br/canal/noticias//noticia/2003/12/03/sete-pessoas-morrem-em-colisao-no-ceara-53008.php
[1036] www.nola.com/news/t-p/index.ssf?/base/news-1/1073372248319420.xml

Huda promised exemplary punishment to those whose negligence had caused the accident. [1037]

Another level crossing collision involving a school bus in the United States occurred on Tuesday 13th January 2004, this time at Bloomfield Hills, Michigan. Again, there were no children on the bus at the time of the accident; it was just on its way to pick them up for Lahset High School at 10:30 a.m. The gates, bells and lights of the crossing were working normally but for some unexplained reason the bus failed to stop and was driven past them, through the crossing barrier and into the path of a National Railroad Passenger Corporation (Amtrak) passenger train running on Canadian National Railway tracks. The bus was badly damaged and driver Clarence Bates, 68, was thrown from it and critically hurt. There were only two crew and seven passengers on the train, which was bound for Detroit and Chicago. None of them were hurt. After the accident the train returned to Pontiac and its passengers were invited to continue their journey by bus.

Factors such as the weather, road conditions, drugs and alcohol were ruled out as explanations for the accident. Not excluded at the time of the initial report into the accident were the mechanical condition of the bus and a medical episode affecting the bus driver. Police Chief Jeffrey Werner stated, "We have no idea why he drove through the gates. We're at a loss to explain why this happened." The driver had been operating buses for 25 years and had a clean record. In view of his age, a medical condition such as a heart attack seems a distinct possibility. [1038]

Three Chinese tourists, a Thai tour guide and the Thai bus driver were killed and 21 tourists were hurt when the coach in which they were riding was struck by a train at 16:30 on Monday 2nd February 2004, on a level crossing at Soi Nong Jab Rao, Thailand. The accident happened near the town of Sattahip and not too far from the beach resort of Pattaya. The injured were admitted to Krungthep Pattaya, Sattahip and Wat Yannasangwararam hospitals. Two members of the tourist party were not involved in the accident as they had been taken ill and remained in their hotel rooms rather than going on the trip. The accident was blamed on the bus driver, who it was stated had smashed through the crossing barrier into the path of a freight train carrying diesel oil. [1039] [1040]

Multiple vehicles were involved in a level crossing collision at Phagwara, Punjab, India, on the Northern Railway, at 11:25 on Monday 2nd February 2004. The *Swaraj Express*, bound for Jammu, had just left Phagwara station when it collided with a motorcycle, a scooter and a cycle rickshaw, killing five people and injuring a sixth. It was theorised that having seen the *Dadar Express* pass the crossing, the road users had assumed that all was clear. [1041]

Nine people lost their lives and four were injured in an accident on the unguarded Nazyvayevo-Dragunskaya level crossing involving a freight train and a minibus, west of Omsk, in Siberia, Russia, on Friday 20th February 2004. The injured, who were badly hurt, were taken to hospital in Cheremnovskaya. Rail traffic was disrupted for several hours and two passenger trains were delayed. Trains were diverted via an alternative route. The damaged

---

1037 *The Daily Star*, Dhaka, Bangladesh, 13th January 2004
1038 www.clickondetroit.com/print/2765243/detail.html
1039 www.bangkokpost.com/030204_News/03Feb2004_news19.html
1040 www1.chinadaily.com/cn/en/doc/2004-02/04/content_302859.htm
1041 *The Hindustan Times*, Delhi, India, 3rd February 2004

engine was sent for repair while the waggons of the train with damaged brakes were temporarily stabled in the nearest available siding. [1042] [1043]

Initial reports were that at least 30 people belonging to two marriage parties had been killed when train No. 5657, the Up *Kanchanjungha Express*, bound for Guwahati, collided with the crowded minibus in which they were riding on an unmanned Crossing No. SK-333 between Suraj Kamal and Kanke stations, West Bengal, India. The accident, on the Northeast Frontier Railway, occurred at 17:13 on Friday 27th February 2004. The Commission of Railway Safety report on the accident corrected the number of dead to 10, with 37 injured, 31 of them seriously, and one more person apparently died later, to increase the number of dead to 11.[1044] Those who lost their lives or were injured were all on the bus, with the exception of Driver N.K. Singh, who was slightly hurt. [1045]

The bus, carrying about 50 people, had been returning to Deogan from a wedding at Purnea in Bihar and had followed a Tata Sumo sports utility vehicle (SUV) over the crossing. The SUV had made it over the line; the minibus had not.

The accident occurred in a remote location and it had been some time before the police became aware of it from the guard of the train, who had stated that there were "a number of dead and injured." [1046]

Surviving bus passengers stated that the bus driver had been drunk and travelling at breakneck speed. Tiku Rishi, 25, related, "We had repeatedly requested the driver to slow down, but he did not heed our plea. He must have overlooked the approaching railway crossing altogether."

The accident was investigated by the Commission of Railway Safety and attributed to the failure of persons other than railway staff. The Commissioner made ten recommendations:

(1) The train vehicle units (TVUs) at Crossing No. SK-333 according to a 1998 census were 3642. However, a repeat traffic census in 2003 gave a considerably higher figure of 9054 units. A further post-accident census should be done.

(2) Local residents were demanding immediate manning of the crossing as there were a number of schools in the area. A cyclist had reportedly been run down at the crossing previously. The Railway could not produce any details of previous accidents. It should upgrade the crossing based on the traffic census and arrange for it to be manned.

(3) As a long-term measure, the Railway should replace the crossing with an underbridge.

(4) The railway should ensure the road surface and gradients within the gates at level crossings in terms of Annexure 9/1 para. 904 Item 15 of the Indian Railways Permanent Way Manual.

(5) Whenever feasible, grade separations should be provided at level crossings by the provision of underbridges while projects such as doubling or gauge conversion, involving considerable expenditure, were being pursued in an area, as this would enhance safety. Such safety enhancing measures should be clearly spelt out in the terms of reference for such projects so that the project executing authority would incorporate them and safety would be enhanced and ensured.

[1042] www.hindustantimes.com/onlineCDA/PFVersion.jsp?articl.../181_580536,00050003.ht
[1043] www.gateway2russia.com/st/art_214236.php
[1044] http://pib.nic.in/release/release.asp?relid=1381
[1045] http://economictimes.indiatimes.com/articleshow/msid-525304,prtpage-1.cms
[1046] *The Times of India online*, 28th February 2004

(6) Although the 2003 traffic census had given a TVU figure of 70.857 x 127.78 = 9054 units, the Railway had made no proposals to man the crossing as the vehicle units figure was <180, so it was not considered to qualify for manning in terms of the revised policy instructions No. 2003CE-I/LX/59 dated 11/6/2003. Amendment clarification should be issued to the Railways in this context.

(7) The mishap revealed a poor state of enforcement of provisions of Sections 123 and 131 of the Motor Vehicles Act, 1988. The State Civil Authorities should be pursued for the effective enforcement of the provisions of the Act so as to ensure safety at level crossings.

(8) Crew and station staff should be trained in handling/managing accidents/disasters.

(9) The Railway should execute a Memorandum of Understanding with hospitals/nursing homes available in the area for mobilisation and the provision of fast relief and medical attention to those injured in accidents.

(10) The Railways should be delegated suitable powers for the reimbursement of medical aid/services provided by private hospitals/nursing homes to the persons grievously hurt in similar cases. [1047]

In a strange sequel to the accident the two brides who had been married in the weddings preceding it, Basanti Devi and Lako Devi, were branded as witches by locals who threatened to burn them alive if they returned to their village. With their grooms – Dilip and Ranjit Rishi – they had gone over the crossing in a car minutes before the accident, something the villagers interpreted as an evil omen. The brides had gone into hiding. [1048]

At least ten and possibly 12 people died in a horrific level crossing accident at General Pinto, Argentina, at 05:08 on Wednesday 3rd March 2004 when a Ferroexpreso Pampeano freight train carrying foodstuffs from Bahia Blanco to Rosario collided with a Renault Trafic commercial van on a level crossing protected only by warning signs. Following the collision the vehicle caught fire and an attempt at rescue by a lorry driver who witnessed the accident was frustrated when the fuel tank of the van exploded. A fierce fire followed, all the occupants of the van died and by the time that firemen had extinguished the blaze their bodies had been largely consumed, leaving at least ten skulls. There had only been seats in the van for its driver and a companion, the other passengers sitting on a mattress in the rear of the vehicle and it was believed that the mattress might have intensified the fire.

The victims were believed to have been Bolivian retailers from the town of Luján de Cuyo in Mendoza province, who had been on their way to Buenos Aires to purchase clothes.

A thick fog prevailing at the time of the accident was doubtless a contributory factor. [1049]

At 06:20 on Tuesday 23rd March 2004, school bus driver Liu Changlong made the fatal mistake of entering an unprotected level crossing at Tieling, in Liaoning province, China, without looking carefully enough to check if a train was coming. As a result the bus, carrying 22 students of Zhangxiang Middle School in the Qingha district, was struck by a freight train. One student was killed and three students and the bus driver were badly hurt and taken to hospital. [1050]

[1047] http://civilaviation.gov.in/CRSS/Annual%20report03-04/Brief%20details%20of%20s...
[1048] www.iol.co.za, 2nd March 2004
[1049] www.eldia.com.ar/ediciones/20040304/policiales7.asp
[1050] http://news.xinhuanet.com/english/2004-03/23/content_1380591.htm

Eleven or twelve people died and 17 were injured when a Sinar Dempo intercity bus was struck by the PT Kereta Api Indonesia *Limek Sriwijaya* express on an unmanned level crossing between Martapura and Gilas stations, South Sumatra, Indonesia, before dawn on Thursday 1st April 2004. Those who lost their lives included Rusli bin Jali, the bus driver. Local residents took the injured victims to hospital, while the train, on its way from Lampung to Kertapati, was delayed for over four hours. The bus had been on its way from Jakarta to the town of Pagaralam. Police were investigating how the accident had occurred. [1051]

At about 8:30 a.m. on the morning of Wednesday 31st March 2004 a school bus stopped near a level crossing on Midvale Road in Big Beaver, Pennsylvania, United States, to pick up a pupil. Having done so, its driver attempted to turn around to continue the journey, at which point the rear wheels of the bus got caught in an embankment and it was firmly stuck, foul of the crossing. Keeping her cool when she was unable to free the bus, driver Patricia Simecak, 46, was able to usher her 14 pupils off the vehicle via the rear emergency door before the vehicle was struck by a 36-car Norfolk Southern Railway train from Chicago, Illinois, to Conway, Pennsylvania. No one was hurt and damage to the bus was minor. [1052]

The overloading of a car was a likely factor in a level crossing accident that occurred at 07:21 on Friday 9th April 2004 in the town of Martin de Yeltes, Spain. The Seat Ibiza car was carrying eight young people at the time of the accident and it was believed that its excessive weight had caused it to ground on the line. Only stop signs were provided at the crossing but the line was straight and visibility was good at the time of the collision. A light engine travelling from Fuentes de Oñoro to Salamanca collided with the vehicle and wrecked it. The driver of the diesel engine had sounded his horn several times when he saw the car. Four young men and a woman were killed outright and a 19-year-old woman succumbed to her injuries after admission to hospital. Another man and a woman were seriously injured and admitted to Salamanca hospital. The road on which the accident happened linked Martin de Yeltes with Fuente de San Esteban and tended to be used at night by young people to avoid the likelihood of breath testing for alcohol on the main N-620 highway. [1053] [1054]

The *Izmir Blue Train* overnight passenger service, which had left Polatli at 08:05 on Friday 16th April 2004 for Ankara, collided with a school minibus on an unprotected level crossing near Temelli station, Turkey. The minibus was dragged 200 metres by the train and seven pupils were killed. Six more pupils and the bus driver, Izzet Güler, 43, were injured and taken to Polatli State Hospital, the driver later being transferred to a hospital in Ankara. The bus had been carrying pupils from Fare and Yenihisar villages to the secondary school in Temelli. Television coverage of the accident showed distraught families who had rushed to the scene of the accident. The Polatli – Ankara line was closed for some time. [1055]

Train No. 625, the Up Siliguri – Alipurduar passenger, collided with a school bus on an unmanned level crossing at Radebank Tea Estate, West Bengal, India, at about 10:45 on

---

[1051] *The Jakarta Post*, Jakarta, Indonesia, 2nd April 2004

[1052] www.ble.org/pr/news/pf_headline.asp?id=9979

[1053] *El Pais*, Madrid, Spain, 10th April 2004

[1054] https://www.lavozdegalicia.es/noticia/galicia/2004/04/09/mueren-seis-jovenes-arrollado-coche-tren-paso-barreras/0003_100000047928.htm

[1055] https://www-hurriyet-com-tr.translate.goog/gundem/mavi-tren-ogrenci-minibusune-carpti-7-olu-218162?_x_tr_sl=tr&_x_tr_tl=en&_x_tr_hl=en&_x_tr_pto=sc

Tuesday 27th April 2004. The location of the accident was also given as having been near Cerron, an alternative spelling of Keron. Three children and the bus helper were killed and 68 were injured, 11 critically. The injured were taken to the North Bengal Medical College and Hospital, Jalpaiguri Hospital, Birpara Hospital and the Army Hospital at Binnaghuri. Ajar Oraon, the bus driver, was stated by the police to have absconded. It was said that he should have noticed the train in broad daylight. [1056] [1057]

A train collided with a minibus on a level crossing at Regar, Tajikistan, 60 kilometres west of the capital, Dushanbe, on Wednesday 28th April 2004. Three people on the vehicle, including its driver, were killed outright, six more succumbed to their injuries in hospital and another 15 were injured. [1058]

Barely a week after the West Bengal accident, on Wednesday 5th May 2004, another four children were killed and 14 people were injured when a train collided with a school van at Bhai Himmat Singh Nagar in the Punjab, India. The Maruti van had been taking home pupils of the Sri Guru Harkrishan Public School. It had left the school less than ten minutes before it was struck by a Hisar-bound passenger train on the unmanned crossing.

Locals rushed the injured to Sri Guru Tegh Bahadur Hospital, Bagga Hospital and Dayanand Medical College. Anxious parents, upon hearing of the accident, descended on the hospitals, together with Mrs Hardyal Kaur Chhina, the school principal. She told a reporter, "The children were excited to return home and narrate their experiences about Hardy Wood Fun Park to their parents. This batch of students had left for home, while we were still waiting for other groups to return when a parent called us to inform {us} about the tragedy."

Mr Sukhjwan Singh, the school gatekeeper, added, "The school closed at 2.30 p.m. Barely ten minutes later, the school received information about the tragedy. Within minutes we were at the accident site, where two children, Rupali and Simranjit, both from the same class, lay dead with their bodies badly mutilated. They were identified much later."

Gurminder Singh, the van driver, was critically hurt. His two children had been passengers in the van and his daughter Gurpreet suffered a head injury.

Mrs Ramandeep Kaur, school accountant, who remained at the scene of the crash until all the children had been removed to hospital, said that the children had been going home to Shimla Puri, Nirmal Nagar, Himmat Nagar and places nearby.

Most of the injured had suffered head injuries and were unconscious, so their bags were scanned to establish their identities. It transpired that there had been 18 pupils on the bus plus, of course, its driver, and it was believed that the large number of passengers might have prevented the van driver from seeing the train.[1059] School principal Hardial Haur stated that the van had been hired by the parents to take the children to a picnic at Hardy's World and its use had been nothing to do with the school. [1060]

La Villarreala Crossing, Spain, not far from Hellin station on the Madrid – Cartagena line, was the location of a serious level crossing accident at 07:15 on Friday 7th May 2004, when a Talgo train from Cartagena collided with a 19-year-old Opal Kadett car. The victims were five

---

[1056] www.thestatesman.net/page.print.php?usrsess=1&id=4180&c1id=1

[1057] www.ndtv.com/addons/printpage.asp?id=52992&ca11id=1&caption=undefined

[1058] www.thecouriermail.news.com.au/printpage/0,5942,9428120,00.html

[1059] www.tribuneindia.com/2004/20040506/main1.htm

[1060] *The Times of India online*, 6th May 2004

Bolivian migrant workers who had been on their way to a farm in Jumilla, about 30 kilometres away from the crossing, probably to work in the apricot fields. They had apparently been taking a short cut to avoid the attentions of the Civil Guard, as they lacked the proper papers. They were named as Alberto Escvalante Pardo, 30; Maria Adriana Pérez, 35; Mariela Salvatierra, 35; her husband Jesús Subirales Durán, 29; and Eldi Guzmán Teco, 29. All were killed. Maria had left five children in Bolivia and Eldi, who was five months pregnant, had two children also in Bolivia. There should have been a sixth person in the car, Eldi's cousin Luis. He explained afterwards, "I did not accompany them because I had not the money to pay for my share of the petrol. I thus was able to save my life but I am sad rather than happy, because my cousin has lost her life."

The car was carried about 400 metres by the collision. The train did not escape either, being well derailed. Two of its 150 passengers suffered slight injuries.

A Red Nacional de los Ferrocarriles Españoles source stated that the crossing was 'fully protected,' being provided with automatic half barriers, warning lights and bells. Unfortunately the car driven had chosen to ignore the warnings and zig-zag around the lowered barriers. On the other hand, union member Gonzalo Aparicio pointed out that the accident location was a black spot for railwaymen, with a level crossing following a curve, and called for its closure, as promised by the previous Minister of Public Works, Francisco Álvarez-Cascos. [1061]

Another serious railway accident occurred in India at about 10:15 on Wednesday 12th May 2004 when Northeastern Railway train No. 4015, the Up Sadbhavna Express collided with a three-wheeler (tempo) on unmanned Crossing No. 30 between Sagarpali and Phebhna, in the Ballia district of Uttar Pradesh. Five women, a man and two children died at the scene and five more people died later in hospital, while four were injured. Two others were seriously hurt. The victims, members of a marriage party, were all from the same family. The Chief Minister of the state announced an *ex gratia* payment of one lakh (100,000) rupees to the families of the deceased and of 50,000 rupees to the injured. [1062] [1063]

At 07:36 on Sunday 16th May 2004 a freight train collided with a bus near Tashbunar station in the Odessa region of Ukraine, carrying the vehicle 500 metres. Shortly after the accident the number of dead had increased to 16, while another 29 seriously injured people were taken to two hospitals 50 kilometres away. Neither the train nor the track was much damaged and rail traffic was not affected. The accident was blamed on an infringement of the traffic regulations by the bus driver, who survived the accident and at the time of the report was being questioned by police. [1064]

The *Mutiara Timur* passenger train, travelling from Surabaya to Banyuwangi, collided with a Akas Asri company bus, on its way from Jember via Banyuwangi to Bali, on a level crossing in the village of Garahan, East Java, Indonesia, at about 03:15 on Wednesday 26th May 2004. Dragged some metres, the bus finally ended up in a rice paddy. At least seven people, all on the bus, lost their lives and 21 were injured. Killed outright when the train hit the front of the bus were the bus driver, named Munarso, and his assistants, M. Syair and Suheri. Four bus passengers later succumbed in hospital. A PT Kereta Api Indonesia

---

1061 *El Pais*, Madrid, Spain, 8th May 2004
1062 *Deepikaglobal.com*, 12th May 2004
1063 http://news.xinhuanet.com/english/2004-05/12/content_1465756.htm
1064 www.itar-tass.com/eng/prnt.html?NewsID=819869

spokesman stated that traffic on the Surabaya – Banyuwangi line had not been affected. [1065]

The accident was blamed on the failure of Crossing Keeper Muhairi to lower the barriers of the crossing. He was arrested and was being questioned by police about his failure to do so. [1066]

At 04:50 on Tuesday 15th June 2004 a freight train collided with a minivan on a level crossing at Wiradesa, near the city of Pekalongan, Central Java, Indonesia, killing seven people, all on the minibus, and injuring three. Again, the failure of the crossing keeper to close the barriers before the arrival of the train was blamed for the accident.[1067] Police stated that because he had fallen asleep, Crossing Keeper Sarmidi, 55, had neither lowered the barriers nor sounded the warning bell until both the Jakarta – Surabaya train and the minivan were almost at the crossing, when the minivan driver had accelerated to get across as the barriers were belatedly being closed. The railwayman was charged under Article 356 of the Criminal Code with neglect of duty resulting in death and faced a maximum sentence of five years' imprisonment.

PT Kereta Api Indonesia officials were able to quickly clear the line and there were no delays to other trains. [1068]

Probably late on Wednesday 17th June 2004 an Izmir – Ankara train collided with a car on a level crossing near Balikesir Organized Industry Zone, Balikesir, Turkey. Five people in the car were killed. [1069]

A light engine collided with an autorickshaw on manned Crossing No. 5, between Latur and Latur Road stations, Maharashtra, India, on the Solapur division of the Central Railway on Saturday 19th June 2004. Six people died in the accident and two were injured. [1070]

Superintendent Guillermo Enrique Cortez Pinto, 47, of the Paine fire brigade; his wife, Fresia Retamal Fuentes, 45; their daughters Yasmín, 17, Valeria, 15, and Catalina; together with Fernando Hernández, 23; and Héctor González, 19, all lost their lives when the Nissan V16 car in which they were riding was struck by an Empresa de los Ferrocarriles del Estado (EFE) Terra Sur passenger train on a level crossing at Bascuñán, near Paine, Chile, at 00:52 on Saturday 26th June 2004. The two young men were the boyfriends of Yasmin and Valeria. [1071]

The crossing at which the accident occurred was fully protected with barriers and a crossing keeper. The exact circumstances of the accident are not clear from the newspaper reports on it, but it was stated that when it occurred, Crossing Keeper Juan Vidal had not been at his post, while the crossing barriers had presumably been open for road traffic to pass. There had been a dense fog and the Santiago to Temuco train had been running somewhat late, though it had not been travelling faster than the permitted maximum speed of 140 km/h.

EFE President Luisa Ajenjo stated that the railway company would accept responsibility

---

[1065] *The Jakarta Post*, Jakarta, Indonesia, 27th May 2004
[1066] www.antara.co.id/e_berita.asp?id=156009&th=2004
[1067] www.iol.co.za, 15th June 2004
[1068] *The Jakarta Post*, Jakarta, Indonesia, 16th June 2004
[1069] www.turkishpress.com/turkishpress/news.asp?ID=21284
[1070] Government of India Ministry of Civil Aviation and Tourism Commission of Railway Safety Annual Report for 2004-2005
[1071] https://www.emol.com/noticias/nacional/2004/06/26/151436/siete-personas-mueren-en-accidente-ferroviario.html

for the accident if a judicial investigation found that it had been the fault of the crossing keeper, even though the keeper was an employee of an external company. [1072]

Only three days after the disastrous derailment at Mekece (see above) a train collided with a minibus on a level crossing near the Provincial Directorate of Public Works in Aydin, Turkey, killing 15 people and injuring several others. The accident happened at about 18:00 on Sunday 25th July 2004 and involved regional train No. 31703 from Izmir to Denizli. The minibus was carrying members of the Inan, Demir and Ece families from Esentepe village and of the Seyrek family from Yeniköy. They were returning from a circumcision party in Aydin, where the operation had been performed on sons Kaan and Faruk of their relative Yener Buluk.

The driver of the minibus, Mustafa Inun, drove around the lowered half barriers of the crossing, clipping one of them, and into the path of the train. The minibus was dragged 300 metres and destroyed. Mustafa, his wife Fatma, their son Bahtiyar, their daughter Meral; Kezban Ece, the wife of the minibus's owner Mehmet Ece, who himself had stayed in the village; their daughters Burcu and Kezban and their son Onur; Dilaver Demir, his wife Ayşe, their daughter Seçil; Çetin Seyrek, his wife Zehra, and their daughters Merve and Fatma were all killed. Fatma, Zehra, and Ayşe Demir (the latter a daughter with the same first name as her mother?) and Sevim Ece, who survived the accident with serious injuries, were taken to Aydın State Hospital. When rescuers arrived at the scene seven victims had still been alive but three of them succumbed shortly afterwards.

The level crossing at the Public Works Directorate was busy on Sunday, as it was every day. While several vehicles were stopped in front of the barriers waiting for the train to pass, Mustafa Inan had overtaken them and entered the crossing. Why he had done so remained unknown. A post-mortem determined that he had not been under the influence of alcohol, while tests carried out after the accident showed that the crossing equipment was working correctly.

İbrahim Akıncı, the train driver, aged 50 with 24 years' operating trains, told investigators that he had whistled for the crossing and applied his brakes in emergency when he had seen the minibus entering it. He testified, "I did my best, but the minibus driver just ignored me." He was released after questioning. [1073]

A train collided with a bus carrying workmen or farmers on an unauthorised level crossing near the town of Badr, in Buheira province, Egypt, north-east of Cairo in the Nile delta, on Tuesday 3rd August 2004. Ten people died and another 11 were injured. [1074]

Although it only had a nominal capacity of eight people, a Land Rover was carrying 20 persons when it was in collision with a Tanzania Zambia Railway Authority (Tazara) freight train on a level crossing at Kiyogozi village, Tanzania, on Sunday 8th August 2004. Five people in the Land Rover were killed outright and the other 15 were taken to St Francis Hospital at Ifaka, where another five soon succumbed to their injuries. [1075]

Omari Mahimbo, the driver of the Land Rover, was among the fatalities. In the aftermath of the accident the hospital, which was obviously a small one, was reportedly short of drugs,

1072 https://www.emol.com/noticias/nacional/2004/06/29/151703/efe-respondera-por-accidente-de-paine-si-guardavias-es-culpable.html
1073 https://www.hurriyet.com.tr/gundem/olume-boyle-daldi-244281
1074 *The Daily Gleaner*, Kingston, Jamaica, 4th August 2004
1075 www.ippmedia.com/cgi-bin/ipp/print.pl?id=18081

so many of them having been used in treating the injured from the accident. [1076]

Train No. 2662, the Down Swatantrata Senani Express, collided with a van on an unmanned level crossing between Handia and Janghiganj stations, Uttar Pradesh, India, on the Varanasi – Allahabad line of the Northeastern Railway on Wednesday 1st September 2004. The van was dragged over 2½ kilometres and 'totally damaged.' Five men, two women and two children in it were killed. A four-year-old boy was taken from the wreckage alive, though critically injured, and was rushed to a local hospital. [1077]

Near the city of Blitar, Indonesia, a passenger train from Surabaya to Blitar collided with a pick-up truck packed with people bound for the Muslim Isra Miraj festival, on a level crossing on Sunday 5th September 2004. It was theorised that the large number of passengers had prevented the driver of the truck from seeing the train. Also, instead of using the main level crossing in the town, which was gated, the driver had taken a popular short cut, intended only for the use of pedestrians and ungated. Eleven people on the truck, including a three-year-old boy, were killed and two other passengers were seriously hurt. Both pick-up and train drivers were detained for questioning. [1078]

Suicide by train is unfortunately not uncommon. Often the would-be suicide is a pedestrian, but now and then the suicide attempt involves a motorist parking his or her car on a level crossing and waiting for a train to collide with it. In most such collisions the car is reduced to a heap of scrap metal and the train is little damaged, but as in most things, there are always exceptions. Thus, on the evening of Saturday 6th November 2004 such a collision resulted in the serious derailment of the train and the death of five passengers and the train driver, as well as of the intended suicide. The accident occurred at Ufton Crossing, England. The location of the accident was referred to in both the press at the time and the report on it as Ufton Nervet, the name of the nearby village, but the place, which previously boasted a signal box, has always historically been known by the railway as Ufton Crossing, so that name will be used here.

Involved in the accident was the 17:35 London Paddington to Plymouth high speed train (HST) (1C92), in charge of two train managers (newspeak for guards) and Driver Stanley Martin, 54. The train left Reading one minute late at 18:03 and was travelling at just under 160 km/h when it arrived at Ufton Crossing at 18:11.

The signal box at Ufton Crossing had been abolished at the end of 1977 and automatic half barriers had been provided at the crossing. On the evening of the accident a Mazda 323 car driven by a Mr Brian Drysdale, 48, arrived at the crossing just after 18:00. He drove onto the line, manoeuvred his car so that it was directly in the path of an approaching train, stopped and waited. He had apparently decided to kill himself because he feared that he had contracted Acquired Immunodeficiency Syndrome as a result of a previous sexual relationship. His actions were observed by an off-duty police officer who used the emergency telephone at the crossing to contact the signalman at Reading but by then the train was too close at hand for it to be warned. Upon striking the vehicle the leading wheelset of the front power car of the train was derailed to the left. As chance would have it, 95 metres beyond the crossing was a set of facing points leading to a Down Goods Loop, and it was the presence of this set of points that caused the general derailment of the train and the serious consequences of the accident. In addition to the fatalities 71 passengers were badly enough hurt to need to be taken to hospital,

1076 www.ippmedia.com/cgi-bin/ipp/print.pl?id=18336
1077 www.deepikaglobal.com/search_newsdetail.asp?newscode=69460
1078 www.etaiwannews.com/Asia/2004/09/06/1094437717.htm

18 of whom were seriously injured. [1079]

Having witnessed the accident, the off-duty police officer was immediately able to inform the signalman and the emergency services were thus alerted without delay.

Various surviving passengers gave their accounts of the disaster. Mario Iotti, on a business visit to the United Kingdom, related, "At first it went all black. Some other passengers were trying to calm other people down in the carriage. We also had a priest in there, he was saying 'please remain calm.' Two guys in my car managed to get the hammers and break out of the top windows (as the carriage was on its side). I managed to get out through the door at the back. It was our car that had the fatality in it. One person was close to the windows and went through and got trapped underneath the train. It was very unpleasant.

"Some girls in front of me had their faces covered in blood and I remember someone saying they had a broken arm. I guess that 20 to 30 people were trapped in that carriage. Inside the carriage there was glass everywhere. The really dramatic thing was people flashing their mobiles to see (in the dark). For some, this kind of thing happens very quickly, but for me it happened very slowly. At first it felt like we were going over a bump, like hitting something. A couple of seconds later the train tilted. You felt like something was definitely not right. I heard the noise of the wheels screeching very loud. There was broken glass all over the track. The whole thing tilted to the left of the tracks. My carriage remained on the tracks but on its side, sliding along. I could see some people at the side of the tracks. I saw one carriage smashed in completely. It was a big pile of wreckage. As far as I can recall some people were struggling, with broken legs and arms, but 95% were without major injuries, which is great. It is unbelievable to think to you could go through that kind of thing and even survive. To see from the inside a train derailing is really unbelievable. Now you just worry that a lot more people were injured as severely."

Another passenger was Jonny Saunders, a BBC Radio Five live reporter, who recalled, "There was suddenly this extraordinary stopping sensation as if someone had pulled the emergency cord, but it carried on and carried on. We came to a juddering halt and suddenly the lights went off, screaming, shouting, and it went pitch black, then total chaos in the carriage for a few moments. I was incredibly lucky because the carriage I was in didn't actually go over on its side, but the carriage in front of me did go over on its side and the carriage behind me also went over on its side. I tried to get the hammer to break the glass. We managed to eventually to get out. We were lucky in our carriage, but certainly there would be people less lucky. We managed to get out and were taken to a nearby pub and people were being treated by the paramedics who were at the pub. It was pretty scary stuff.

"I was in the back. What they had done on this train, was not normal and I get this train most days, is that they put the first class section towards the front of the train. It was pitch black so it was difficult to say what was going on, but it derailed between Reading and Newbury. There was lots of screaming and shouting and people obviously started panicking. All I can tell you is what happened in our carriage, but there were lots of people getting up onto their seats and trying to get off the train. There was a real smell as well of the fumes everywhere. One person tried to light a cigarette and someone said don't light up a cigarette. It was very, very scary indeed and people all around here, myself included, are in complete shock." [1080]

The accident inquiry noted that the track at the crossing had been in good condition and there had been no problems with the signalling, which by this date comprised multiple-aspect colourlights with full track circuiting, nor with the crossing equipment. Nor had there been any problems with the vehicles of the train or their maintenance.

[1079] Rail Safety and Standards Board report

[1080] *BBC News*, London, 7th November 2004

The accident report continued with a general discussion on level crossing safety, noting that since the earliest days in Great Britain it had been the responsibility of the railway to fence off lines using reasonable means, usually with the provision of manually operated swinging gates at level crossings. With the great increase in road traffic and with the dual objectives of preventing delays to road users and reducing labour costs, automatic half barrier (AHB) crossings had been substituted for gated crossings from the 1960s. Such crossings were in use worldwide.

Unlike manually operated crossings, AHB crossings were not connected to the lineside signalling. They were only installed on crossings where the line speed did not exceed 100 mph (160 km/h) and to reduce delay to road users the barriers did not descend very long before the arrival of a train (the minimum warning time at line speed was 27 seconds).

Previous accidents at automatic level crossings involving passenger fatalities had occurred at Hixon in 1968 and Lockington in 1986 (covered in previous volumes of this series). Both accidents had been the subject of investigations and recommendations had been made. As in other countries, crossings were being abolished whenever possible.

The report went on to describe Ufton Crossing in detail. The subject of obstacle detection at automatic crossings was then touched upon and it was concluded that such a detection system would not be able to warn a train in time for it to stop, though the train might be able to reduce speed.

The influence of design of the train on its becoming derailed and the effectiveness of the vehicle and coach structures in protecting the passengers and crew were examined in detail. The report then went on to describe the rescue operation, detail the experiences of passengers, describe the recovery operation and so on.

The report concluded that the initial derailment of the train had resulted from a collision with a stationary motor car, which had caused the flange of the left-hand wheel of the leading bogie of the front power car to climb over the left-hand rail. The subsequent derailment of all the vehicles of the train had occurred shortly afterwards, when it had reached a set of facing points, which had caused the derailed left-hand wheel to follow the left-hand stock rail of the points, pulling the trailing axle of the leading bogie into derailment, followed by all the other bogies of the train. The underlying causes relating to the behaviour of the car driver were outside the remit of the inquiry.

The inquiry made 23 recommendations:

(1) Consideration should be given to the establishment of a team dedicated to the implementation of Network Rail's stated level crossing strategy of giving priority to crossing closures. The team should bring vigour to a joint approach by the railway and road authorities to negotiating the extinguishing of rights-of-way to allow level crossings to be closed. Particular emphasis should be placed on locations where the line speed was high.

(2) Where considerable expenditure was justified for safety reasons, vigorous consideration should be given to the provision of a bridge, rather than an upgrade to the crossings, thus allowing crossing closure whilst maintaining the right of way.

(3) Given that a suitable alternative road route existed via Tyle Mill, means of achieving the closure of Ufton Crossing should be investigated.

(4) The Network Rail level crossing risk assessment process should be revised so that the proximity of track and lineside features that could exacerbate the effect of a derailment were included as a factor.

(5) Research should be undertaken to establish whether a practical system could be developed to detect and warn drivers of an obstruction on an AHB level crossing.

(6) The operation and design of the AHB level crossing emergency telephone system should be reviewed from a human factors, environmental and technical viewpoint and modified to promote successful contact with the signalman.

(7) The work of the National Level Crossing Safety Group should be vigorously supported and given high priority for funding and resources.

(8) Where there existed a random mixture of steel-framed concrete road units and polymer concrete (non-framed) road units in the four-foot of a track on a level crossing, there should be a policy that allowed for the re-configuration of the units into a grouping of the former type of unit at the running-on end of the crossing followed by a grouping of the latter type of unit at the running-off end. The policy should be applied whenever the opportunity arose for such a reconfiguration of the units in conjunction with their temporary removal from the track for other planned work on the crossing.

(9) The exemption for axle loads greater than 17 tonnes from the general requirement in Railway Group Standard GM/RT 2100 to fit obstacle deflectors to new-build leading vehicles should be reviewed, taking into account the mechanism of derailment of the leading power car at Ufton Crossing.

(10) The feasibility of modifying the front end shape of HST power cars and/or fitting an obstacle deflector in order to reduce the derailment risk following a collision with a large obstacle such as a motor vehicle should be investigated and implemented, if practicable, on power cars undergoing a life extension programme.

(11) A programme of research should be pursued in order to assess:

(a) Whether there could be a net safety benefit in fitting seat belts in passenger vehicles.

(b) The percentage usage of the belts that would be required in order to realise a degree of safety benefit commensurate with the costs of fitting and maintaining the belts.

(c) Whether that percentage usage was achievable in long-term service operation.

This information should form part the input to a study to determine whether the fitting of seat belts was reasonably practicable.

(12) The requirements in Railway Group Standard GM/RT 4506 and Association of Train Operating Companies Standard AV/ST 9001 relating to the provision of laminated windows should be extended to cover vehicles undergoing major internal refurbishment. The Railway Safety and Standards Board (RSSB) research work on the provision of windows for emergency egress should be accelerated so that implementation of the requirements on vehicles undergoing major refurbishment was not delayed by lack of information on the optimum disposition of toughened and laminated windows.

(13) The requirement for RSSB research project T118 should be reviewed so as to identify opportunities to bring forward the completion date of the work on the role of couplers and bogie retention in whole-train dynamic behaviour in accidents.

(14) There should be a requirement to apply photo-luminescent markings to emergency equipment so as to enable passengers to locate it in poor lighting conditions.

(15) Safety information should be provided in the vicinity of passenger seats where it was practicable to do so.

(16) Light sticks should be provided for the use of passengers in all trains where emergency lighting was not available.

(17) The provision of high-integrity emergency lighting, which would function in a situation where the main lighting supply was lost, should be mandated for all new passenger

vehicles and for those undergoing major internal refurbishment.

(18) Where passengers were required to use a hammer to break toughened glass windows as part of a vehicle egress procedure, the hammer should be of a sufficient strength to avoid it being broken by the action of the passenger. RSSB research project T424 should be extended to include an investigation into a suitable standard of design or designs of hammer. Where windows were double-glazed, the signing should clearly indicate this fact together with the appropriate action to be taken by the passenger in order to break and remove both panes of glass.

(19) The responses of on-train staff when confronted with an accident should be reviewed, with particular emphasis on the evacuation of passengers from a derailed coach and the protection of the line. The lessons learned should be used as appropriate to reinforce the training programme together with ongoing exercises and briefings.

(20) Consideration should be given to the inclusion, in the emergency training module for on-train staff, of the opportunity for them to visit a facility providing first-hand experience of the conditions inside a rolled coach.

(21) Consideration should be given to the emergency training of on-train staff, as currently practised by First Great Western, to determine if there were elements in the training that could usefully be transferred to other train operators.

(22) A review should be undertaken of the procedural arrangements with specialist investigation organisations in respect of the attendance and crash worthiness specialists at the site of a train derailment.

(23) The present emergency command structure in Network Rail, at the three levels of Gold, Silver, and Bronze, should be reviewed in respect of the way in which it was applied to major rail accidents. The review should consider all aspects of the requirements, especially the seniority, competence and experience of the appointees to the executive positions within the structure together with the communications between the different levels. [1081]

An inquest on the victims of the accident concluded that it had resulted from the suicide of Brian Drysdale and that the train driver and passengers had been "unlawfully killed." The coroner stated that the accident was "a unique set of circumstances that had resulted in catastrophic consequences."

The crossing was replaced by an overbridge in mid-December 2016. [1082]

Five employees of Avondale Mills in Graniteville, South Carolina, United States, had just completed a two-hour shift and were on their way home from work at about 8:15 a.m. on Wednesday 10th November 2004 when the 1994 Buick car in which they were riding was struck by Norfolk Southern Railway (NS) train on Ascauga Lake Road Crossing in the town. The car was dragged about a quarter of a mile down the line and all five of its occupants were killed outright. They were named as Connie Bodie, 26, of Williston, about 30 miles away (the driver); Nathaniel Hay, 21, of Williston; Michael Mack, 39, of Williston; Monica Myers, 41, of Springfield; and Tammy Wilson, 36, of Williston.

Darleen Stewart, of North Augusta, who witnessed the accident, said, "She drove up and stopped. She had her tires on the track, but it was almost as if her car just froze. The train had no choice but to hit her." The crossing was equipped with flashing lights and warning bells but had

[1081] Rail Safety and Standards Board report
[1082] Wikipedia – Ufton Nervet rail crash

no barriers. Witnesses stated that the lights and bells had been operating, despite which vehicles had continued to cross. The first two had managed to get across in front of the train, the car with the workers in it had been next. It was common practice, according to the same witnesses, for motorists to chance their luck at the crossing in order to avoid possible long delays.

NS spokesman Robin Chapman explained that the train involved was not an ordinary freight or passenger service but a track measurements working, consisting of a locomotive and two measurement cars. It had been going from Augusta, Georgia, to Columbia, South Carolina, and had been travelling within the 49 mph speed limit at that point.

A local newspaper wrote, 'In light of the multiple fatality crash, a spokesman for Avondale Mills, Stephen Felker, said he hopes the issue of cross bars will be revisited.

'Ronnie and Dawn Smith, who live directly across from the crash site, said the road and intersection are dangerous. "I think a lot of people take chances out there and try to outrun the train," Ronnie Smith said. "This wasn't the first time."

'Late Wednesday evening the only reminder of the deadly crash was five newly painted crosses at the intersection, eulogizing the lives of those killed. Cars continued to travel through the intersection and work crews began to clear up their initial investigations. Highway Patrol, Norfolk Southern and the Coroner's Office will continue the investigations. Autopsies are scheduled for today.' [1083]

Following the accident there were calls for the speed of trains running through the town to be reduced, which demands were resisted by the NS, on the grounds of loss of efficiency and productivity. [1084]

The members of an extended family, on their way home from a graduation ceremony at a nursing collage in Tegal for Triana Sari, aged 22, were riding in a Toyota Kijang minivan that was struck by the *Kamandanu* executive-class passenger train from Jakarta to Semarang, on an unprotected level crossing near Padaarja village, Central Java, Indonesia, on Saturday 27th November 2004. A local resident said afterwards that he had screamed at the minivan driver, named Sobirin, to try and warn him of the fast-approaching train, but Sobirin had driven on. The van was carried 'dozens of metres' before the train could pull up. Twelve people were killed and the only survivor, Triana herself, was critically injured and taken to Texin Hospital in Tegal. The train was delayed 45 minutes. [1085]

Only four days after the Padaarja accident, at about 20:30 on Wednesday 1st December 2004, the *Gaya Baru Malam* economy train from Surabaya to Jakarta collided with a lorry packed with passengers on an unprotected level crossing in the hamlet of Kauman, Central Java, Indonesia, about 12 kilometres east of Purworejo. Six of the people on the lorry were killed outright and five were injured, one of whom passed away after admission to Purworejo Hospital. Another thirteen passengers and the driver, named Budiyono, returned home.

An eyewitness named Karyadi, who lived near the crossing, said that the people on the truck had been on their way to an Islamic service being held in the Sunan Geseng Mosque in Kauman. All were from the village of Kemiri, about 20 kilometres west of Purworejo.

Another witness, named Tuyem, said, "I was surprised by a loud noise followed by screaming, just as I was about to go to bed."

[1083] *Aiken Standard*, Aiken, South Carolina, United States, 11th November 2004
[1084] Ibid, 3rd February 2005
[1085] *The Jakarta Post*, Jakarta, Indonesia, 1st December 2004

Lorry Driver Budiyono had apparently failed to see the train. [1086]

Five Czech soldiers who had returned from Afghanistan not long before were killed when the T-815 army lorry in which they were riding was struck by a passenger train on a level crossing near Prostějov, Czech Republic, about noon on Thursday 9th December 2004. Six other soldiers were injured, three of them seriously, and two people on the train were slightly hurt.[1087] The troops had been returning from a shooting range.[1088] The line between Prostějov and Olomouc was expected to be out of use until Saturday. [1089]

At about 16:35 on Thursday 3rd February 2005 a tractor trolley carrying labourers from Ramtek was in collision with a local Ramtek – Nagpur passenger train on the unmanned Crossing No. KR 5, Maharashtra, India (newspaper reports refer to the location as Borda Crossing and Kanhan Crossing). The marriage party riding on the trailer, which included a large number of women and children, were returning from a wedding in Ramtek taluka to Kachurwadi village, the bride and groom having left earlier on another vehicle. Initial reports were that at least 35 people had died and another 15 were injured, some seriously. The injured were taken to the Jawaharlal Nehru Hospital in Kanhan village and the bodies of the dead to a Kamptee government hospital.[1090] There had been about 70 people on the trailer in all.

South East Central Railway Deputy General Manager Jagdish Kumar stated that the train had been travelling at 45 mph and the driver had applied his brakes in emergency when he had seen the tractor trailer on the crossing. No one on the train was hurt and it was not derailed. He added, "We had not installed a gate at the crossing because very few trains pass this way. But after this tragedy we will consider posting a guard." [1091]

Sudarshan Shende, who worked at a nearby school, witnessed the accident. He told a reporter that at least three vehicles carrying a large number of wedding guests had climbed the slope leading to the crossing and the others had passed over it when the last vehicle, the tractor trailer, had pulled up to the crossing and stopped. He said, "They stopped for a while, and I could hear them arguing. Some wanted to cross, others wanted to wait. Then the driver decided to cross. The tractor got over, but the train rammed into the trailer." [1092]

The final report on the accident stated that the number of known dead had risen to 56, while of 14 injured 11 were still in a critical condition.[1093] Most of the victims were from the same family.

Five young people aged between 18 and 22, three boys and two girls, lost their lives when a train collided with their car on an automatic half barrier level crossing on Nierstraat, in Diepenbeek, Belgium, at about 01:00 on Saturday 12th February 2005. A freight train from Bilzenhad had broken down and stopped a short distance from the crossing. Seeing the stationary train and assuming that this was the only reason the barriers were lowered, car

---

1086 Ibid, 3rd December 2004

1087 https://www.mo.gov.cz/scripts/detail.php?id=3961#:~:text=Ve%20%C4%8Dtvrtek%209.,s%20vojensko u%20Tatrou%20T%2D815

1088 www.guardian.co.uk/worldlatest/story/0,1280,-4663459,00.html

1089 http://news.xinhuanet.com/english/2004-12/09/content=2315128.htm

1090 www.deepikaglobal.com/search_newsdetail.asp?newscode=91593

1091 *BBC News*, London, 4th February 2005

1092 www.guardian.co.uk/worldlatest/story/0,1280,-4776391,00.html

1093 www.reuters.com/printerFriendlyPopup.jhtml?type=topNews&storyID=7532919

driver Johan Surinx drove his car around the lowered barriers, directly into the path of a fast-moving train approaching on the other line. At least one other driver had done the same thing, while others had made U-turns and taken alternative routes. [1094] [1095]

At about 16:00 on Sunday 13th February 2005 Northeast Frontier Railway (NFR) train No. 626 from Alipurduar to New Jalpaiguri Junction collided with a pick-up van carrying members of a wedding party on an unmanned crossing at Radebank Tea Estate, West Bengal, India, between Banarhat and Karan stations. The van had been on its way from Sukhani Tea Estate to Ambari Tea Estate and had been carrying friends and relatives of the groom, Sanjib Khalko. The groom himself had been riding ahead of the pick-up in a Maruti van. Possibly unaware of the tragedy behind him, he had continued his journey and his marriage to Ruma Bhagat went ahead that evening regardless.

'Witnesses said the train hit the van with an ear-splitting crash, the impact crushing the vehicle into a mangled mess. The van was carried a good 200 yards by the train. Two bodies were found smashed against the cow-catcher.' At least 11 people died in the accident, seven at the scene and four on the way to hospital, while another seven were injured. Those who lost their lives included Sudama Prasad, 32, the owner and driver of the van; and five members of the Bharat Band Party from Malbazar, hired to play brass instruments and drums at the ceremony. The injured were taken to the North Bengal Medical College and Hospital in Siliguri and Sulkapara Tea Estate and Malbazar State General Hospitals in Jalpaiguri. The line was re-opened at about 19:30.

The same crossing had been the scene of a collision with a school bus in April of the previous year (see above). NFR Alipurduar manager Pampa Babbar stated that, "The crossing is unmanned because we have categories based on train and road traffic data." Also, "The van driver did not possess a valid license. We are probing the accident." Likewise, NFR assistant commercial manager Tarak Nath Bhattacharjee, asked why the crossing had not been manned despite previous accidents there, stated, "The Traffic Vehicle Unit Act does not permit us to man all the level crossings. However, at all the unmanned level crossings, the department has introduced certain safety measures and the drivers should follow those." 'The locals, however, have refused to accept the railway's logic, accusing them of shying away from their responsibility of passenger safety.' [1096] [1097]

An express from Szczecin Główny to Lublin collided with a school bus on an unprotected level crossing in the town of Władzin, Poland, about 15 kilometres from Mińsk Mazowiecki, at about 15:00 on Tuesday 15th March, 2005. The impact was powerful enough to overturn the vehicle several times and two pupils aged 14 and 15 lost their lives, as did the bus driver. Eleven other children were injured and were taken to hospitals in Mińsk Mazowiecki and Otwock. There had been about 30 pupils on the bus from a school in Kołbiel. The accident was being investigated, though insufficient caution on behalf of the bus driver was thought to be the likely cause. [1098]

1094 https://www.lavenir.net/archives-journal/2010/02/18/accident-mortel-a-un-passage-a-niveau-a-diepenbeek-le-tribunal-ne-statue-pas-UECU7Y2WNRB77GS5I3T3QUUGCI/

1095 https://www.standaard.be/binnenland/vijf-doden-bij-ongeval-in-diepenbeek-update/43665163.html

1096 www.telegraphindia.com/1050214/asp/others/print/html

1097 www.thestatesman.net/page.print.php?usrsess=1&id=68683&c1id=6

1098 https://pl.wikinews.org/wiki/2005-03-15:_Zderzenie_autobusu_z_poci%C4%85giem_pod_Mi%C5%84skiem_Mazowieckim

A Gwalior – Guna passenger train collided with a tractor-trolley loaded with stone at about 11:00 on Thursday 17th March 2005. The accident happened at the unmanned Crossing No. 57, in the Ratahara area of Madhya Pradesh, India. Five men riding on the trolley were killed. [1099]

At 02:24 on Monday 18th April 2005 a freight train collided with a scheduled bus on an unprotected level crossing near Vozhayevka station, in the far-eastern Amur region of Russia. Seven bus passengers were killed and 18 others of the 27 on it were injured. According to one report the bus driver had ignored the red warning lights at the crossing. [1100] [1101] [1102]

At about 07:20 on Monday 25th April 2005 five men and three women were killed more or less outright when a freight train collided with a Phabziri company bus on a level crossing at Muraleni, Limpopo province, South Africa. The crossing was near Schoemansdal and about 20 kilometres from Makhado. Fifty-six other bus passengers were injured. The most seriously injured were airlifted to the provincial hospital at Polokwane while the others were taken by road to Elim, Louis Trichardt Memorial and Tshilidzini Hospitals. Three of the injured died later. No one was hurt on the train, which was travelling from Musina to Makhado; the bus had been taking workers and villagers to Makhado. [1103] [1104]

Bus passenger Mr Alex Mavhetha, who was injured in the accident, said: "The bus was overcrowded, with some people standing and, as I was facing the direction where the train was coming from, it was visible, but the driver didn't stop at the railway crossing."

Likewise Captain Ailwei Mushavhanamadi of the police stated of the bus driver, "He should have stopped, but he just drove on." The accident was therefore clearly the bus driver's fault, despite which charges of culpable homicide were laid against both the train driver and the bus driver. Asked how the police could open a case against the train driver while it was the bus driver who failed to stop at the railway crossing, Mushavhanamadi said that, as part of police investigations into an accident like this, it was the police's duty to open a culpable homicide for both drivers, as that would make the investigation easier. He also confirmed that the bus driver had been arrested by the police at Ha-Mashamba on Tuesday, and that he would appear in the Makhado Magistrate's Court soon.[1105] The case against the bus driver was postponed for the collection of further evidence and it has not been possible to determine whether he eventually stood trial or if so, what the outcome was.

A passenger train from Colombo to Kandy collided with a bus carrying 99 people on a level crossing in the town of Polgahawela, Sri Lanka, on Wednesday 27th April 2005. Railwayman E.R. Jayaratna, who witnessed the accident, stated, "There were other vehicles waiting, but this bus overtook them to get to the line. Its driver thought that he could accelerate and get across in front of the train but it did not work out."

The train driver, H.A. Sirisena, stated, "The signal was Green and the level crossing gates were closed. Then I saw the bus trying to cross…and the next thing I knew the engine was hitting the rear of the bus. After the first impact, the bus spun around and hit the train again. I

[1099] www.deepikaglobal.com/search_newsdetail.asp?newscode=96183

[1100] www.theaustralian.news.com.au/printpage/0,5942,15005779,00.html

[1101] www.tass.ru/eng/prnt.html?NewsID=1944643

[1102] www.mosnews.com/news/2005/04/18/amurbus.shtml

[1103] www.news24.com/*News24*/South_Africa/News/0,,2-7-1442_1696446,00.html

[1104] www.news24.com/*News24*/South_Africa/News/0,,2-7-1442_1695480,00.html

[1105] https://zoutpansberger.co.za/articles/news/3163/2005-04-29/the-worse-bus-crash-in-57-years-have been in this job for 41 years and this is the first time that something like this has happened."

Bus passenger Priyalatha Imbulgoda related, "I was sitting at the very back of the bus. I saw flames coming from the front. There was little time to do anything. Some neighbours from my village were sitting next to me. I don't know what happened to them." [1106]

Operating Superintendent G.T.P Chandratilleke indicated that the crossing was one of over 900 in the country provided with automatic half barriers and that accidents due to vehicles going around the lowered barriers were common, though they were not usually of such a magnitude.

Police spokesman Rienzie Perera added that, "Our initial investigation suggests that two buses were racing one another in order to reach Colombo first."

Initially it was reported that 50 people had lost their lives in the accident. Later the number of dead was corrected to 36. The higher figure arose from a number of the victims rendered unconscious in the accident having been counted as being dead. Sixty people were injured, 30 of them seriously, of whom nine were said to be in a critical condition. Most of them were taken to a local hospital, where Dr Ananda Goonesekara confirmed the number of fatalities as being 35. Nine of the worst hurt, including a 4-year-old girl and a 10-year-old with head and chest injuries, were taken to the National Hospital in the capital, Colombo, 80 kilometres away. Among those treated in the local hospital was the bus driver, who was only slightly hurt and was arrested there by the police. The bus conductor was also arrested there, perhaps partly for his own safety, as the police had to guard the two from the attentions of angry survivors. The number of fatalities eventually reach 41

The private bus was on its way from Dambulla to Colombo. As a result of the collision it was dragged 100 metres along the line, then 'burst into flames and ended up a mangled heap of metal. Open suitcases and passengers' clothes were strewn about.' Witnesses reported that people, some of them on fire, had jumped through the windows of the bus after the collision. All the victims had been on the bus. About 50 soldiers were deployed to assist in the rescue operation. [1107]

After visiting the scene Transport Minister Felix Perera commented, "We can clearly say that the fault lies with the driver and conductor of the bus."

President Chandrika Kumaratunga professed in a statement that she had been, "shocked and saddened" by the accident and agreed that, "Initial reports...(indicate) that this was the fault of an errant private bus crew who defied traffic rules." She added that she had directed the transportation minister to prepare a detailed report on the accident. [1108]

The bus driver and conductor were subsequently arrested and there were suggestions that the driver, Abdul Wahid Hafrool, might face murder charges and even be liable for the death penalty.[1109] Such proved, indeed, to be the case, for in April 2013 the driver and his conductor, M.G. Buddhika Ruwan Kumara, were sentenced to death at Kurunagala High Court for causing 41 deaths, as well as receiving 35 concurrent sentences of 10 years' rigorous imprisonment for causing injury and ordered to pay compensation of 5,000 rupees to each of the injured, with an alternative of another year's imprisonment.[1110] The handing out of a death sentence for causing death in a railway accident through negligence is, as far as the author is aware, almost unprecedented outside of Soviet Russia during the Stalin era, Communist Yugoslavia, and possibly East Germany; presumably its imposition was an indication of the

---

[1106] *BBC News*, London, 27th April 2005
[1107] *The Wilson Daily Times*, Wilson, North Carolina, United States, 27th April 2005
[1108] *The Daily Times*, Farmington, New Mexico, United States, 28th April 2005
[1109] http://icwales.icnetwork.co.uk/printable_version.cfm?objectid=15453411&siteid=50082
[1110] https://www.onlanka.com/news/driver-conductor-sentenced-to-death-for-causing-deaths-of-41-passengers.html

shock caused by the magnitude of the accident. There has, in fact, been a moratorium on executions in Sri Lanka since 1976, at least up to July 2025.

The driver of a van failed to give way to a train on an open level crossing near the town of Guercif, Morocco, early on Saturday 30th April 2005. The train hit the van, killing seven people and injuring four others. [1111]

Returning in a bullock cart from Bioara, where they had been shopping for the grandson of Baldev Singh Sutar, 60, a family consisting of the aforenamed, his wife Nabada Bai, 50; their daughter Kalabai, 35; and her daughters Rekha, 18; Manju, 17; Manisha, 7; Deepak, 5; and an unnamed baby girl, were all killed when the cart was struck by the Okha – Gorakhpur express train No. 5054 at an unmanned Crossing No. 122, near Pachor-road, Madhya Pradesh, India, at about 18:30 on Monday 9th May 2005. (The ages and relationships of the victims vary in different accounts.) Baldev's sons fainted when they reached the scene and saw what had happened.

One of the two bullocks pulling the cart was also killed, the other survived. The bodies of the victims were scattered over a distance of 200 metres and parts of the cart were carried 500 metres. One of the victims, a woman, was still alive when the engine driver reached her but passed away within a few minutes. No one on the West Central Railway train was hurt. Relatives of the deceased complained that there was no trace in the wreckage of the jewellery or other items the family had been carrying, raising the suspicion of possible looting. [1112] [1113]

At about 1:30 p.m. on Thursday 12th May 2005 a Wentworth School District school bus carrying fifth grade pupils (10 and 11 year olds) arrived at a level crossing with the Burlington Northern Santa Fe Railway close to the bank of the Mississippi River, in Hannibal, Missouri, United States.

The bus driver had passed over the crossing, then had found that the parking area on the opposite side, which he wished to go to, was occupied, so reversed his vehicle in order to take an alternative route. As he had been doing so, the rear of his vehicle fouled the railway and had then been stuck by a freight train moving at 25 mph. Thirty-three pupils from the Heritage Intermediate School in Wentzville were subsequently taken to hospital, though only five were found to have slight injuries, while a sixth, who was more seriously hurt, was kept in overnight for observation.

There had been 37 pupils in total on the bus, plus a few parents, on a field trip from Wentzville, 70 miles further south. After the accident a Hannibal School District bus was provided to take the uninjured pupils back to Wentzville. [1114] [1115]

East Central Railway train No. 3226, an Up Patna – Darbhanga intercity express, collided with a Tata Sumo Jeep on the unmanned Crossing No. 26, between Bachhwara and Satha Jagat stations, Bihar, India, on the Sonepur Division of the railway on Tuesday 31st May 2005. Warning lights were provided at the crossing, which the driver of the Jeep either failed to notice or ignored. Five people on the Jeep were killed outright and two were fatally hurt.

[1111] www.moroccotimes.com/news/printArticle.asp?id=6161
[1112] www.hindu.com/thehindu/holnus/001200505092101.htm
[1113] www.centralchronicle.com/20050510/1005002.htm
[1114] *The Joplin Globe*, Joplin, Missouri, United States, 13th May and 28th June 2005
[1115] *The Sedalia Democrat*, Sedalia, Missouri, United States, 13th May 2005

Traffic on the Barauni – Samastipur line was disrupted for several hours. [1116]

Another bad level crossing accident happened in Ukraine at about 15:00 on Thursday 2nd June 2005, near the village of Novoselovka in the Odessa region. A local bus was struck by a 39-waggon coal train on an unprotected crossing, killing 13 or 14 people according to different accounts, two of them children, and injuring eight more, half of them seriously. The accident was believed to have been the fault of the bus driver, who survived, failing to obey the warning lights at the crossing. [1117]

A second accident occurred when the breakdown train going to the site of the first collided with a taxi on a level crossing in the Poltava region, killing three taxi passengers and seriously injuring a fourth. [1118]

A serious level crossing collision occurred in Israel on Tuesday 21st June 2005 when a passenger train from Tel Aviv to Be'er Sheva, carrying at least 300 and possibly 400 people, collided with a lorry loaded with coal dust near Kibbutz Revadim, not far from the town of Kiryat Gat, at about 18:00. The lorry proved to be too substantial an obstruction for the train, which was travelling at 140 km/h, to merely push it aside and instead it was badly derailed, while the lorry itself was split into three pieces. A reporter stated, 'Witnesses spoke of horrific scenes, with passengers thrown from carriages and somersaulting rail cars. Rescue teams found mangled wreckage strewn around a nearby sunflower field after at least two rail carriages overturned on impact. Witnesses said many people who had been standing, because the train was full, were flung from the carriages.'

A slightly injured passenger quoted in the *Haaretz* newspaper stated, "People flew into the (first) car, began bleeding; pieces of glass from the windows flew on them, there was a lot of chaos."

Army Radio reporter Daphna Arad, who had been a passenger on the train, said, "It took about 15 or 20 minutes for the rescue services to arrive. We felt really helpless. All we could see around us was fields. We had no idea where we were. Soldiers took out their bandages and began to treat the injured as much as possible. I approached a woman who looked alright but had slipped onto the floor. She said that she was pregnant and was very worried about her baby. I looked for an old man who had been sitting next to me and I saw that half his head was coated in blood."

Medic Dudi Greenwald related, "It's a horrible sight. It looks like a terror attack. One of the carriages is upside down and it's impossible to tell what's inside. It's the worst accident I've ever seen." [1119]

Over 60 ambulances went to the scene, which they had difficulty in reaching, while Israeli Defence Force helicopters were also drafted in to assist in the evacuation of the injured. Ten hospitals took about 195 injured in all. A total of 248 people in all were hurt. Five people, including both the train driver, Leonid Turk, 46, and the lorry driver, Leonid Galinski, 51, were declared dead at the scene and three of the injured succumbed within the next few hours.[1120] A chance event that had increased the severity of the accident was the faulty air-conditioning system in the rear coaches of the train, which had caused many passengers to

[1116] www.patnasdaily.com/news05/june/060105/seven_dead_in_train+jeep_accident.html

[1117] www.itar-tass.com/eng/prnt.html?NewsID=2098667

[1118] http://en.for-ua.com/news/?id=944

[1119] *The Chronicle Telegram*, Elyria, Ohio, United States, 22nd June 2005

[1120] www.israelnewsagency.com/israeltraincrash8890622.html

move to the front of it, the part most affected by the collision. [1121]

Scheduled radio and television programmes were interrupted with reports from the scene, while officials stressed repeatedly that it had been an accident and not a terrorist attack. Media reports indicated that it was believed that the lorry driver had been working continuously for 20 hours, and there was speculation that due to tiredness he might have dozed off at the wheel.

The accident happened at a crossing protected only by stop signs.

Transportation Minister Meir Sheetrit promised that a report on the causes of the accident would be made within a week. [1122]

It emerged after the accident that the lorry driver had accumulated no less than 24 traffic violations in the previous four years, including ones for driving at excessive speed and failing to obey a stop sign.[1123] His driving record was therefore anything but good. Rafi Eliyahu, the company whose lorry he had been driving, pleaded, however, that most of the violations had been of a technical nature, such as the lorry not having a fire extinguisher. The company itself had been the subject of a police investigation three months earlier due to reports that its drivers were working for 30 or 40 hours continuously, well in excess of the maximum legal limit of 12 hours.

The investigation into the accident reportedly found that the lorry driver had contributed significantly to the accident because he had not obeyed a stop sign at the crossing and had been driving for more than 12 hours. However, an accident could have occurred in certain circumstances even if he had obeyed all the regulations. The train driver, on the other hand, had acted correctly and could not have prevented the accident. His train had been travelling at high speed and he had sounded his horn when the lorry was about 20 metres from the crossing. A barrier at the crossing would have significantly reduced the likelihood of an accident. A crossing keeper had been stationed there for part of the time but on the day of the accident his hours of duty had ended at 15:30. [1124]

Three weeks after the accident a train collided with a car that had stalled on a level crossing near Ahuzam and one person, probably the car driver, was killed and 63 were injured. Again, the train was derailed. After this second accident there was a suggestion from the police that trains should no longer operate in the south of the country, where both accidents had occurred. The transportation minister rejected the suggestion, stating that, “We won’t stop the trains, and there’s no need to react with hysteria and panic and get the public hysterical.” Measures would, however, be taken immediately to close all unauthorised agricultural crossings. [1125]

Partly in response to the accidents Israel Railways General Manager Yosef Mor resigned on 19th July. [1126]

In a sequel to the accident, in September 2008 three Israeli Railway officials – Deputy Head of Maintenance Surin Seuvitch, Assistant Director of Research and Development Ilia Volkov, and Transportation Department Director Harel Even – were indicted for criminal negligence and negligent homicide. It was alleged that they had failed to take the necessary measures to prevent the Revadim accident, despite it being known to be a dangerous level crossing and the fact that a near-fatal accident had occurred at the same crossing a month earlier. They could, it was maintained, have restricted the speed of trains over the crossing or alternatively could have stationed a crossing keeper there. They had also failed to implement

---

[1121] www.israelnn.com/print.php3?what=news&id=84340
[1122] *BBC News*, London, England, 22nd June 2005
[1123] www.israelnn.com/print.php3?what=news&id=84413
[1124] www.haaretz.com/hasen/spages/594097.html
[1125] www.haaretz.com/hasen/spages/598206.html
[1126] www.haaretz.com/hasen/spages/602099.html

the findings of an internal report, which had investigated previous accidents.

With regard to the Azuham accident, Israel Railways had done nothing to correct a dangerous problem with regard to it, nor had the crossing been closed pending attention to its deficiencies, as ordered by a chief executive officer. [1127]

A Jeep carrying eight people was struck by the Palanpur – Samdadi express on an unmanned level crossing at Ramsan – Jenal, Banaskantha, Gujarat, India, on Monday 27th June 2005. Five of those riding in the were killed and the other three were seriously hurt. [1128]

Sunday 24th July 2005 saw another bad level crossing accident in Indonesia, this time in the Kisaran Timur district of Asahan regency, North Sumatra. A 45-seater PO Pinem company bus from Ujung Batu to Medan was struck by the *Sri Bilah* train from Medan to Rantau Prapat, killing at least nine people and injuring 23. The dead comprised eight bus passengers and one of its drivers, named Budi, who was sleeping at the time, having been relieved by his co-driver. All the casualties had been riding in the rear of the bus, the part struck by the train. The accident occurred at about 03:00. The dead and injured were taken to Kisaran general hospital. Little damage was caused to the train and rail services were barely affected.

The crossing was a manned one and the accident had resulted from the crossing barriers not having been lowered for the train. The crossing keeper disappeared after the accident and police were searching for him. [1129]

A fourth bad level crossing accident occurred in Indonesia on Wednesday 10th August 2005, this time on an unmanned crossing in the town of Baturaja, South Sumatra. At about 04:00 a bus travelling from Medan to Jakarta and carrying about 40 people was struck by a freight train carrying coal. Ten people were killed and 27 injured, of whom at least 12 were badly hurt and admitted to Palembang general hospital. All had been on the bus. [1130]

At about 3:30 p.m. om Thursday 11th August 2005 a Ferrocarriles Nacionales de México freight train was in collision with a Termogas lorry hauling two tankers of butane gas on a level crossing in the town of Lucio Blanco, Mexico. Following the collision one of the tankers burst into flames and the other exploded half an hour later, while part of the train was derailed.

A newspaper related, ‘The resulting explosion reduced cement homes near the tracks to little more than piles of charred rubble. Corrugated metal roofing collapsed into some of the structures. “It was terrible,” said Jorge Batres. “It was like an earthquake.” He described the scene south of the crash site as looking like a battlefield. “All is destroyed.”’

The accident and its results were described in great detail in a Texas newspaper. ‘According to police and witnesses a truck hauling the two tankers crossed the tracks as the train was passing, hitting both tankers. The tankers landed on their backs on both sides of the train. The tanker on the north side of the tracks spun around as it was spitting fire from a big hole. The fire reached the stalls and the houses nearby, burning some of them to the ground. The second tanker was on its back on the south side of the tracks.

‘Federico Covarrubias, owner of the filling station near the accident scene, said they

---

[1127] https://www.ynetnews.com/articles/1,7340,L-3598018,00.html
[1128] www.ahmedabad.com/index/printpage/article/16888/section/14
[1129] *The Jakarta Post*, Jakarta, Indonesia, 25th July 2005
[1130] *The Jakarta Post*, Jakarta, Indonesia, 11th August 2005

watched the tanker as its tires burned for at least 30 minutes. "We then saw the whole tanker engulfed in flames and started running down the street," he said. "We heard one explosion, then a second one." Covarrubias said if the fire department had arrived at the scene sooner, he believed they would have extinguished the fire from the burning tank.

'The explosion was so powerful that a large chunk of metal from the tanker flew more than 100 yards into the air and smashed into the wall of a house, bounced twice on a street and landed on a pickup truck. Covarrubias said four of his workers suffered minor burns in the blast.

'Francisco Serrata, who operates a tire repair shop about 200 yards from the tracks, said he saw the truck heading toward the train. "The train kept blowing its horn," he said. "But the truck didn't stop." Serrata said he heard a big explosion and started running away from the shop. "The impact knocked me down," Serrata said, showing scrapes on his left elbow, "I have never seen anything like it."'

'For others there was tragedy. '"Where is he? Where is he?" yelled an older man on Friday after he was told one of his grandchildren was found dead near the wall of an empty shed. "Oh my God." The boy, identified as Victor Izaguirre de Leon, was last seen running away from Thursday's collision of a Termogas tandem-tanker truck carrying butane gas and a train at 3.30 p.m. that erupted in a violent explosion and fire that destroyed or damaged nearly everything in the six-block area that comprises the small town. Witnesses said Izaguirre ran toward a shed near the side of the train tracks and apparently sought protection in a makeshift shelter. He was found covered in long stems of grass and debris. It appeared that Izaguirre, 12, had been struck by a piece of metal. His uncle, José Luis Vallejo, had searched Thursday night and all morning Friday for the boy. Vallejo ran toward a grassy area and told Victor de Leon about his grandson. De Leon started weeping and nearly fainted after he was told that Izaguirre de Leon was dead. The boy used to help his family sell scrap metal on the side of the road near the railroad tracks. The boy's father and a younger brother were also injured by the explosion and taken to a hospital in Monterrey, an unidentified woman said.'

On Friday the only other known fatality was a 17-year-old girl, Basilia Santos Perez. Forty-four other people were taken to hospital in Monterrey and across the border to Harlingen, Texas, several in a critical condition. Others survived but lost all their possessions and their homes. '"I was inside, but we all ran – me, my mother, my wife. I was carrying my one-year-old baby girl," said Felipe Amaya, 28, standing outside what had once been his house. "Clothes, furniture, the kitchen – all was lost."

'Inside the Mercado Chavez, one of the larger stores, a computer monitor used with the cash register was destroyed. Everything, from boxed milk to an ice-cream freezer, was covered with black soot.

'Manuel Gonzáles Torres on Friday walked across the rubble that a day before had been crafts and pottery from all over Mexico, to show where his store had ended and living quarters began. A blackened and warped stove still stood, but all else was lost in a tangle of broken walls and roofing. "There's nothing left," he said.

'Like other residents, Felipe Amaya, 28, and Gonzáles Torres said officials told them the propane company would pay their damages. "We're waiting for someone to help us," Amaya said. A man who answered the phone at Termogas' offices in Monterrey said the company had sent people to Lucio Blanco to assess the situation but wouldn't comment further.'

By the following day the number of fatalities had risen to four, with the death in Monterrey hospital of Eric Bautista, 29, and Reyna Briceño, while the newspaper that reported their deaths also carried another story of Lucio Blanco residents who had suffered greatly. 'Sitting in the middle of her ruined house, 66-year-old Maria Petra Sanchez de Araiza said she has lost everything. "I have been living in this house for the last 43 years," she said. "This was our livelihood." Sanchez says she was watching television Thursday when an explosion shook her home, the largest in a strip of small shops and single family dwellings on the north side of the railroad tracks, where a truck hauling butane gas and a train collided. "I saw fire and smoke heading toward my house," she said. "The fire came inside and I immediately ran toward the back." She took shelter behind her running dishwasher and a tiny space in the wall and waited for maybe a minute before emerging. "I heard things coming down from the roof and everything else collapsing," she said. "I ran out of the back door and started running down the street, where I found my husband."

'The Lucio Blanco resident plans {to} stay with neighbours while her home is inspected for damage. "We have to get back on our feet," she said.

'Yolanda Silva, whose roadside shop was blown away by the blast, considers herself lucky that the changaritto {street vending cart} was closed at the time of the accident. "We could have been wiped out," she said. "We were about three blocks from here when we heard a large boom."

'It's unclear if the Mexican government will offer help to the people of Lucio Blanco who are looking to rebuild and return to their lives before the accident that killed four people and injured several more. "Things are different here in Mexico," Silva said. "I hope we get help to put up another changaritto, because this has affected a lot of people."

'Scores of homes, roadside shops, and as many as a dozen vehicles were destroyed by the blast. To rebuild and replace what was lost won't be easy, a government official said. Lupita Flores de Suarez, a member of the state of Tamaulipas House of Deputies, represents District 7, which includes Lucio Blanco. "We are going to set up a temporary office so these people can address their problems," Flores said. "Right now we are telling them to bring the title of their properties so we can file them and keep them in one place."

'For Sanchez, losing everything includes the documents that proved her home ever existed. "What property title?" she said. "Everything that we have is right here."'

Two more victims of the accident had succumbed by Monday 15th. One was Daniel Arriaga, 33; the other was Rodolfo Briceño, 38, brother of Reyna. Juan Briceño, another brother, related that he had helped to extricate his siblings and had seen then that they had been badly hurt and were not likely to survive. "I helped them to get out of the taqueria {food stall} and took them to Matamoros. My brother's last words were 'Don't leave me. Don't leave me.' They worked at the taqueria less than a block from the railroad tracks. They were making an honest living."

'The Arriagas, Briceños and others affected by the explosion said they have no idea if they would ever get help either from the government or from or from the company that owns the truck.'

Initially it had been believed that the lorry driver, Ramón Calderón, and the unnamed crew of the train had both died in the accident. Later, it transpired that the two of them had survived, though injured, and had been taken to Monterrey hospital. Calderón had been driving along a narrow highway, not authorised for navigation by heavy lorries, while rather than stopping at the unprotected crossing where a view of trains was obscured by lineside buildings, he had attempted to negotiate the crossing at a speed of 40 mph.

Two more victims of the accident succumbed to their injuries later, one of them Faustino Izaguirre, father of Victor. His wife, Alejandra de Leon, was left to nurse their other son, Angel, 6, who had suffered both second- and third-degree burns over 30% of his body and a cardiac arrest, which had severely affected his brain. He was also in constant pain. In March 2006 she was said to be still struggling for financial assistance and hoping to send him to the United States for further treatment.

'Imelda Buenrostro and Felipe Santos are also trying to get back on their feet. The two operated a roadside curio shop within a block from the train crossing and lost everything. Buenrostro still has scars caused by the burning butane on her shoulder, arms and legs. "The man working for the insurance company came by and offered us $7,000 to settle the matter," she said. "I don't think that is enough, but I am now afraid that if I don't take that money I am going to get nothing. I don't know what to do right now."

Others were a little more fortunate. 'But Maria Petra Sanchez de Araiza considers herself one of the lucky ones. Her house was destroyed by the blast, but within days, someone gave her a canopy to use as a shelter. A few months later, the administration of Matamoros Mayor Baltazar Hinojosa Ochoa donated building materials to the victims who lived near the tracks and who suffered losses from the blast. Sanchez de Araiza said they built a small wooden room where she and her husband live. And recently they were the beneficiaries of a church sponsored programme that has new homes built by volunteers from All Saints Episcopal Church of San Benito {Texas}.'

In August 2006, on the first anniversary of the accident, a Texas newspaper carried another report on the lives of those affected by the accident. 'It was a year ago that a tandem-truck hauling butane gas collided with a passing train, leaving a trail of destruction. Since the accident that claimed the lives of eight people, things have not changed much here. Residents said the train still cruises by at the same speed, while trucks and passenger cars have not slowed down at the rail crossing.

'"We are planning to open a store in there," Catarina Santiago Rosa said, pointing at a building still under construction. "It has been hard for us, because a year later we are still working on this project." She said the old shop she had was across the street from the new one. "We had a lot of stuff before," said Santiago. "We used to sell tin, ceramic, and we had a food booth. Right now we are barely getting on our feet." Nevertheless Santiago and her daughter-in-law, Carolina Garza Lopez, said they consider themselves fortunate. The two of them are now living in brick homes built for them by a San Benito church congregation. "We are grateful for this but we have no money today," she said. Garza said another school year is about to start, but she hasn't bought clothes or school supplies and has not paid the fee to register her young daughter to attend elementary school. "We got money from the insurance company, but it's all gone today," Garza said, referring to the settlement she received from the death of her husband, Eric Bautista.

'A couple of miles up the road, Alejandra de Leon tends to her six-year-old son Daniel Izaguirre. The boy suffered major injuries in the blast that killed his father and older brother. "He started showing a smile about four months ago," de Leon said. "He is getting a little better but doesn't talk." She said Daniel will turn 7 on Thursday and would have attended second grade later this month. De Leon said she has been receiving financial help from a Monterrey-based company that provided insurance for the truck company. "I have also been getting assistance from some people on the U.S. side of the border, and I feel grateful for that. However, de Leon said she repeatedly sought help to send her son to a hospital in Texas, but she was told it wasn't possible. In the meantime, she said she will be taking care of the boy for as long as she can. "He loves to be outside in that cart," she said, pointing toward a stroller under the carport. "Right now, I bring him here early in the morning and late in the evening."

Another accident occurred at the Lucio Blanco crossing on Wednesday 11th October 2006, when the driver of an articulated lorry attempted to beat a train over the crossing. This time two people were hurt and were taken to hospital, one of whom was Alejandro Briceño, another brother of siblings Reyna and Rodolfo, who had lost their lives in the accident the previous year. The lorry driver fled after the accident. No measures had apparently been taken by this time to improve the protection at the crossing, though a petition had been sent to the Department of Transportation for the provision of crossing barriers plus speed bumps on the road approaches to the crossing.

There was mention after the earlier accident of the likelihood of lorry driver Calderón being charged with negligent homicide and causing criminal damage in connection with it, but it has not proved possible to find any reports of a trial. [1131]

On Sunday 14th August 2005 a bus carrying 24 passengers and bound for Ho Che Minh City was struck by a train on a level crossing in the Hoài Tân district of Binh Dinh province, Vietnam. The bus was thrown 30 metres through a high-tension power line and into a rice paddy. Six people on it were killed outright and 19 were critically hurt, four of whom died on the way to Bồng Sơn Hospital and at least one more died later. Bus driver Hoàng tức, 40, had reportedly failed to observe the approaching train; he was among the badly injured. The bus passengers had been residents of Hoài Nhơn district, bound to Ho Che Minh City for work or to visit relatives. The railway made an immediate offer of 1,000,000 Vietnamese dong to the victims and their families. [1132]

At about 11:30 on Wednesday 31st August 2005 The *Lokmanya Tilak Express* collided with a maxicab on an unmanned level crossing at Iglur, Karnataka, India, on the outskirts of Bangalore. Five people who had been riding in the vehicle were killed and two were injured, one critically. The maxicab was dragged about 15 metres. Its badly damaged remains were quickly cleared from the line but the engine of the train was too badly damaged to proceed with the train and another had to be procured to take it forward. [1133]

A pick-up truck serving as a school bus and carrying Muslim pupils aged between 14 and 19 home from the Thamma-withya Islamic School, a private secondary school for children of the local elite, was struck by a passenger train bound for Bangkok on an unprotected level crossing in the village of Yupoe, Yala province, Thailand, at about 16:00 on Monday 12th September 2005. Nine students and the pick-up driver were killed and the other ten students were critically hurt. A police spokesman stated that the pick-up driver had entered the crossing without looking carefully enough to determine if a train was coming. The truck was dragged more than 500 metres and pinned beneath the train, until another train was used to pull the first one back. The release of the victims therefore took several hours. A large number of distraught relatives, friends and classmates maintained a vigil outside of Yala Hospital, where the victims were taken. [1134]

Seven occupants of a pick-up truck were killed and nine were hurt, four seriously, when it was hit by a train on an unmanned crossing near Shah Maqsood, Pakistan, four kilometres from Haripur, on Wednesday 28th September 2005. There had been 20 people on the truck, all related to one another, who had been returning from a marriage ceremony in Donali Shah Maqsood village. In a hurry for some reason, the pickup driver had attempted to cross in front of train No. 11, the Up *Chenab Express*, bound for Havelian. Instead of his getting across, the train caught the vehicle and dragged it about 200 metres. [1135]

[1131] *The Brownsville Herald*, Brownsville, Texas, United States, 12th, 13th, 14th, 16th, 17th and 18th August 2005 and 5th May, 11th August and 12th October 2006

[1132] http://thanhniennews.com/print.php?catid=3&newsid=8542

[1133] www.deccanherald.com/deccanherald/sep12005/index2034262005831.asp

[1134] www.newpaper.asia1.com.sg/printfriendly/0,4139,94462,00.html

[1135] www.dawn.com/2005/09/29/top13.htm

Ramadhan Mututu, the driver of a matatu (privately owned minibus) on its way from Sio-Port to Kisumu failed to reduce speed, overtook another matatu that was already stopped at Kisian Crossing, Kenya, seven kilometres from Kisumu, and having also ignored the blasts of its warning horn, collided with a passenger train travelling from Kisumu to Butere. Bus conductor Patrick Kusimba said that they had seen the train but his driver seemed determined to cross in front of it. He related, "Everything went so fast I could not figure out what happened. I thought we had passed the railway crossing but found myself in a pool of blood. It is my first accident since I started plying this route four years ago."

The accident happened at about 09:15 on Sunday 16th October 2005. The minibus was 'reduced to twisted metal' by the impact with the train, the engine of which was detached from its coaches and derailed. Of the 24 passengers on the minibus, four were killed outright and the others were all badly hurt and taken to the nearby New Nyanza General Hospital. With the subsequent death of several of the injured the final number of dead was at least 10. According to different reports driver Mututu was either killed at the scene or seriously injured. Driver Gilbert Obala and possibly Assistant Driver Arnold Mango on the train were also hurt and taken to the Aga Khan Hospital in Kisumu. None of the 36 train passengers was hurt. Assistant Driver Mango stated that they had seen the speeding minibus and hooted to no effect, then they had applied the brakes. He related, "We tried all that we could, but the driver of the matatu was speeding and ignored us. We hooted several times and that is why some of the vehicles stopped to give way."

Rachel Buoga was a passenger in another matatu stopped at the crossing and witnessed the accident. "In a flash of a second we saw a huge mound of dust and {heard} cries coming from the vehicle and the train. The train derailed as it swept the vehicle to the bridge," she said.

Ms Irene Adhiambo, 29, was a passenger on the minibus involved, travelling with her five-year-old niece, Caren Andati, who was critically hurt. She recounted she had been asleep when the crash happened. "We boarded the minibus at Sio-Port going to Kisumu from where we were to take a bus to Nairobi. I was woken up by a big bang."

Local resident Philip Orao described the crossing as a black spot and urged the government to install speed bumps on its road approaches to slow down traffic. [1136]

The busy Kisumu – Busia – Kampala highway was blocked for hours by the wreckage and passenger trains between Kisumu and Nairobi were suspended for at least two days. [1137] [1138]

A faulty clutch resulted in a school bus stalling on a level crossing near Tournai, Belgium, on Monday 17th October 2005. Its driver was unable to move it off the crossing but was able to evacuate all of his young passengers before a local train arrived at the crossing and collided with it. No one was hurt but the train was derailed and suffered significant damage, as presumably did the bus. [1139]

As it was crossing the line at the unmanned Crossing No. 23 near Uppal Khalsa village, Punjab, India, at about 11:55 on Saturday 5th November 2005, a school van developed a fault and stopped on the line. Its driver, Harvinder Kumar, 25, unable to move the van off the crossing and seeing a train approaching, attempted to flag it down, then, seeing that it could not stop, he fled the scene. The van belonged to the S.D. Model School in Mandi Gobindgarh

[1136] www.eastandard.net/print/news.php?articleid=30656

[1137] www.nationalmedia.com/dailynation/nmgcontententrty.asp?category_id=1&newsid=59493

[1138] www.nationalmedia.com/dailynation/printpage.asp?newsid=59565

[1139] http://72.14.207.104/search?q=cache:JWrgCg_OOyIJ:www.brftv.be/nachrichtenex/sh...

and was carrying seven pupils at the time. Two of them managed to jump clear and escaped but the other five were still in the van when it was struck by a Ludhiana – Lohian diesel multiple unit (DMU) passenger train. Simaranjit Kaur, 6, and her brother Amrit Pal Singh, 4, were killed, and the other three pupils were seriously hurt and were rushed to the Christian Medical College in Ludhiana.

Jalandhar Deputy Commissioner Ashok Gupta stated that 50,000 roubles would be granted to the next of kin of the dead students and 10,000 roubles for the treatment of those injured.

'Meanwhile, heart-rending scenes were witnessed at the house of the deceased brother-sister duo. Hundreds of people, including residents of the village, descended on the place to console the bereaved family. Harnek Singh, father of the deceased, fell unconscious as soon as he received news about the death of his two kids.' [1140]

Passenger train No. 62201 from Adana to Mersin collided with a lorry carrying farmworkers on an uncontrolled level crossing in the village of Arıklı, Turkey, near the town of Yenice, in the Tarsus district of Mersin, at about 06:30 on Wednesday 23rd November 2005. There were no gates at the crossing and the lorry driver, Mahfuz Özalp, 36, either failed to notice the approaching train or thought that he could get across in front of it. The lorry was dragged 100 metres by the train and cut in two. Ten of its passengers, eight of them women and three of them from the same family, were killed or fatally hurt and the lorry driver and 31 workers were injured, 16 of them seriously. Many ambulances were sent to the scene from Tarsus, Mersin and Adana. Sixteen workers and the lorry driver were taken to hospitals in Mersin and Tarsus and the other 15 to hospitals in Adana. The workers had migrated from the eastern and southeastern parts of the country to the Tozkoparan neighbourhood of Tarsus and had been on their way to harvest lemons near the scene of the crash.

Selim Caglia, 23, who was injured, told a reporter, "We were going to a lemon orchard. I didn't notice the train coming, I was half asleep, like many of the others. The scene was awful after the accident. I saw the bodies of some of my relatives and had to help some other relatives who were injured." [1141]

After an investigation of the scene by the Public Prosecutor, the bodies of the nine workers who had died at the scene were taken to Tarsus State Hospital. Upon hearing about the accident, many citizens rushed to both the hospitals and the accident scene. The workers' relatives were described as having suffered nervous breakdowns. The train drivers, named as Ramazan Ayar and Ahmet Mithat Gülsoy, were detained while the investigation was in process; one of them was possibly an assistant driver. The Adana-Mersin line was reopened at noon.

A news report noted that while the issues of transporting people in lorries and of uncontrolled level crossings were once again on the agenda, Türkiye Cumhuriyeti Devleyt Demiryollari (TCDD), the state railway company, had announced that it is not their responsibility to provide protection at all level crossings. TCDD Regional Director Hasan Tashin Köprülü commented, "If under and overbridges are not constructed and the authorities does not take the necessary precautions, it is impossible to prevent such accidents."[1142] The news report pointed put that the Adana-Mersin railway had passed through uninhabited areas when it was built but due to population growth, the areas were now much more populated and level crossing accidents were thus more frequent. [1143]

[1140] www.tribuneindia.com/2005/20051106/main5.htm
[1141] www.mg.co.za/printPage.aspx?area=/breaking_news/breaking_news_international_news/html
[1142] www.turkishpress.com/news.asp?id=80748
[1143] https://www.hurriyet.com.tr/gundem/tren-kazasi-10-olu-3555104

CHAPTER 13

RAIL/RAIL CROSSING ACCIDENTS

At locations where railways and roads intersected, the idea of having them cross on the level had obvious advantages in terms of economics over the cost of building an over or underbridge, especially if the level crossing were unmanned and the construction of a bridge would require the building of substantial approach embankments or possibly approach cuttings leading to an underbridge. With road users being limited in the early days to pedestrians, cyclists, and horse-drawn vehicles the dangers to a train from a collision on such a level crossing were also likely to be minimal. For railways crossing other railways on the level, though, the situation was very different. A collision with another train on a crossing was likely to be far more serious than a collision with, say, a horse and cart, and furthermore a train driver approaching a level crossing and seeing another train nearing or on it could not pull up in anything like the distance that a road user could. It is no surprise therefore that in most countries, where the railway system was either built directly by the state or under state supervision, rail/rail crossings were not provided. It was only in countries where the railways were built by competing private companies without state interference that such crossings existed, most especially in the United States, where lines were often built as cheaply as possible and the economy of not needing to construct a bridge was doubtless an important consideration. Thus, the United States had many rail/rail level crossings, Great Britain at one time had about 20 or so and its colonies, such as Ireland and Canada, had a handful. There was at least one in Cuba and there were also a number in Argentina, where the railways were constructed by British companies, such as Circunvalacion Crossing, Argentina, where the broad-gauge Buenos Aires Great Southern and the metre-gauge Province of Buenos Aires crossed, and there were one or two in other countries, such as at Istravoz (later Hilal), Turkey, where the Oriental Railway crossed the Smyrna – Cassaba Railway, and at Warthausen, Germany, where a major disaster had occurred in 1944. As has been pointed out, the American companies also seemed to have a perverse proclivity for building rail/rail crossings where they could easily have been avoided, perhaps because in the early days train speeds were low, trains were infrequent and night operation was unusual, so that the dangers were not fully appreciated. By 1991 many rail/ rail crossings had disappeared, even in the United States, either by replacing them with bridges or with the closure of one or both of the lines involved, but some remained and there would be a handful of collisions at them in the next 15 years. Only one crossing of any significance remained in Great Britain, at Newark, which was fully signalled, and perhaps a handful still existed in other countries. If so, no accidents have come to light at them.

At 2:30 p.m. on the afternoon of Saturday 7th March 1992 eastbound Gateway Western Railway (GWW) freight train No. 13227, travelling at 26 mph, collided with the side of southbound Kansas City Southern Railway freight train No. 714, travelling at 9 mph, at the crossing of the two lines at Sheffield Junction, Kansas City, Missouri, United States. Four locomotives and six cars of the trains were derailed and four crew members were treated at the scene by paramedics and declined further treatment. 15,000 gallons of diesel fuel from damaged tanks on the locomotives was also spilt and caught fire; the blaze was extinguished by Kansas City firefighters. The derailed cars were reportedly carrying grain, paper products and new automobiles. [1144]

An investigation by the Federal Railroad Administration determined that the GWW train had been unable to stop short of the crossing of the two lines because of a brake failure, the angle cock being closed at the rear of the first car. Contributing to the accident had been inadequate test procedures in the performance of the initial air brake test, which had failed to reveal the problem. [1145]

Southbound Norfolk Southern Railway (NS) freight train No. 255L5 from Detroit, Michigan, to Fort Wayne, Indiana, and eastbound Conrail freight train No. TV 220 from Chicago, Illinois, to Columbus, Ohio, collided on the rail/rail crossing at Butler, Indiana, United States, at about 4:48 a.m. on Wednesday 25th March 1998. The leading diesel unit of the two pulling the NS train hit the fifth car of the Conrail train, resulting in the derailment of both diesel units and five cars of the NS train and three cars of the Conrail train. Engineer Jim Gilbert and Student Engineer Anthony Louros of the NS train jumped immediately before the collision and escaped with slight injuries. Conductor Howard Rose of the same train either decided not to jump or did not have time to and was killed. The crew of the Conrail train was unhurt. Damage to the fuel tanks of the NS locomotives led to the loss of about 7,000 gallons of fuel oil. Initially there was also some concern regarding the spillage of a possibly hazardous white powdery substance but Conrail officials stated that this was only whey, a harmless dairy product. The crossing was protected by colourlight signals controlled remotely from Dearborn, Michigan, and the accident was determined to have been due to the student engineer operating the NS train overrunning the stop signal protecting the crossing, which should have been displaying a red aspect but was possibly unlit when the train reached it. [1146] [1147] [1148]

The National Transportation Safety Board (NTSB) investigated the accident. Its report reached 22 conclusions:

(1) The following factors had not contributed to the accident: the mechanical condition of the trains and track, the weather, the operation of the Conrail signalling system and the interlocking signal system, alcohol or drug use and the qualifications of the train crews.

(2) Norfolk Southern had lacked adequate safeguards to prevent student engineers from being placed in untenable situations, in which rules and procedures were disregarded.

(3) Allowing local variations in training requirements promoted operational inconsistencies and hindered uniform compliance with the student engineer training programme.

(4) Unless all Norfolk Southern operating personnel were informed of restrictions on the operation of trains by student engineers, a situation in which a student engineer operated a train without proper supervision could occur again.

[1144] *The Lawrence Daily Journal World*, Lawrence, Kansas, United States, 8th March 1992
[1145] Summary of Accidents Investigated by the Federal Railroad Administration Calendar Year 1992
[1146] *UTU Daily News Digest*, 25th and 26th March 1998
[1147] *Pharos-Tribune*, Logansport, Indiana, United States, 26th March 1998
[1148] *Kokomo Tribune*, Kokomo, Indiana, United States, 26th March 1998

(5) Norfolk Southern's engineer training programme was not adequate to prepare student engineers to cope with all known or anticipated operational requirements system-wide, such as operating trains with the long hood forward.

(6) Not providing feedback, because the student engineer had not reached the point at which he was scheduled to be formally evaluated by a road foreman (that is, he had not worked in the Lake Division for 1 month) was inconsistent with the goals of effective training.

(7) Norfolk Southern's oversight of student engineers during on-the-job training was inadequate.

(8) Norfolk Southern crew co-ordination, communication, and cab-discipline had been inadequate in the events leading up to the accident.

(9) The engineer, as the individual responsible for train handling and care, and the conductor, as the individual responsible for ensuring that rules and instructions were followed, had disregarded their responsibilities during the accident trip.

(10) The accident could have been avoided had the conductor and engineer complied with Norfolk Southern rules and instructions, which included observing and confirming all signal aspects; actively supervising the student engineer, particularly one who was unfamiliar with the territory; and not reading or engaging in other activities.

(11) Because of the engineer's and conductor's lack of vigilance and decision not to call clear signals, the crew of Norfolk Southern train 255L5 had failed to react to either an approach or a possible dark signal at milepost 111.

(12) Had there been procedures to actively engage crew members in observing and confirming all signal aspects, such as recording the aspects, train crew members would have been more likely to call signals in compliance with the operating rules.

(13) The engineer and conductor had failed to observe and call signal 111 because they had been distracted, inattentive or possibly asleep.

(14) A fully implemented positive train separation system would have prevented the accident.

(15) This and other accidents investigated by the National Transportation Safety Board demonstrated that railroad safety would be enhanced if crew members received train crew resource management training.

(16) The Norfolk Southern Lake Division dispatching centre had lacked an effective procedure for identifying reported signal malfunctions of undetermined causes for further monitoring.

(17) Had Norfolk Southern's maintenance programme responded to and corrected the twice-reported signal deficiencies at signal 111, the signal would not have continued to go dark intermittently.

(18) Had the train been equipped with an in-cab voice recorder, recorded crew communications would have been available, which could have provided valuable clues in reconstructing the accident.

(19) Because of the absence of data relative to the accident time frame and the randomness of the dark signal occurrences, the aspect displayed by intermediate signal 111 to the Norfolk Southern train just before the accident could not be determined.

(20) Although the product improvement announcement issued by Harmon Industries addressed the capacitator problem, replacement of the capacitators was not just an improvement but needed to be made a requirement for the safe operation of Electro

Code 4 units.

(21) The reduced visibility on the student's side of the cab as the train approached milepost 111, compounded by a possible dark signal and by the student engineer's lack of experience, might have caused the student to miss the signal and the reflectorised milepost marker.

(22) Better co-ordination by Norfolk Southern with DeKalb County might have prevented the delay in identifying the possibly hazardous material transported on the Norfolk Southern train.

The NTSB determined that the probable cause of the accident had been the failure of the engineer and conductor of train 255L5 to comply with operating rules (specifically, their failure to observe and confirm signal aspects and their failure to continuously and directly supervise the student engineer) and the failure of Norfolk Southern to ensure employees' compliance with operating rules. Contributing to the accident had been Norfolk Southern Corporation's failure to ensure that its locomotive engineering training programme provided effective, timely training, oversight and feedback to ensure that students were adequately prepared for operational situations. Also contributing to the probability of the accident occurring had been the failure of Norfolk Southern Corporation's signal maintenance programme to respond to a reported signal deficiency.

The NTSB report ended with a good number of recommendations.

To the Federal Railroad Administration:

(1) It should review Norfolk Southern's 49 CFR 240 submission, Certification of Locomotive Engineers, specifically "Section 5: Training, Testing and Evaluating Persons Not Previously Certified," to determine whether the company's training programme was adequate for training new engineers and any deficiencies found should be required to be corrected.

(2) In co-operation with the Class 1 railroads, the American Short Line and Regional Railroad Association, the Brotherhood of Locomotive Engineers and the United Transportation Union, it should be develop crew resource management training required for all crew members, which addressed at a minimum:

(a) Crew member proficiency

(b) Situational awareness

(c) Effective communication and teamwork

(d) Strategies for appropriately challenging and questioning authority

(3) It should direct Harmon Industries and the railroad carriers to identify and replace all faulty Electro Code 4 capacitators and make follow-up inspections to ensure that corrective actions had been taken.

To Norfolk Southern Corporation:

(1) It should develop and implement methods to improve employee compliance with company rules and instructions.

(2) It should develop and implement procedures that actively engaged crew members in observing and confirming all signal aspects.

(3) It should inform all operating personnel of their responsibilities regarding student engineers.

(4) It should assign supervisors dedicated exclusively to student engineers who would, at a minimum:

(a) Meet with student engineers at the start of the on-the-job training phase to ensure that the student engineers were aware of the conditions under which they could operate a train and that they knew what to do if these conditions were not met.

(b) Track the student engineer's daily train assignments, daily crew assignments, and performance evaluations.

(c) Provide timely feedback and advice to student engineers on a continuing basis.

(5) It should provide student engineers with formal training in all known or anticipated operational requirements system-wide, including operating trains with the long hood forward.

(6) It should conduct a comprehensive assessment of the locomotive engineer training programme and revise it, as necessary, to ensure that student engineers consistently operated with and were mentored by coach-trained engineers and that engineer training reflected actual operating conditions.

(7) It should provide employees, especially trainees, with effective strategies for dealing with crew members who knowingly disregarded the operating rules.

(8) In co-operation with the Federal Railroad Administration, the American Short Line and Regional Railroad Association, the Brotherhood of Locomotive Engineers and the United Transportation Union, it should be developing crew resource management training required for all crew members that addressed at a minimum:

(d) Crew member proficiency

(e) Situational awareness

(f) Effective communication and teamwork

(g) Strategies for appropriately challenging and questioning authority.

(9) It should designate dedicated personnel to record and track all signal malfunctions and repairs in order to identify recurring, unresolved failures.

(10) It should conduct an audit to identify all communities through which it transported hazardous materials and, in co-ordination with these communities, develop, implement, and keep current written emergency response plans and procedures for handling hazardous material releases. The procedures should address, at a minimum, key railroad personnel and means of contact, procedures to identify the hazardous materials being transported, identification of resources for technical assistance that might be needed during the response effort, procedures for the co-ordination of activities between railroad and emergency response personnel, and the conduct of disaster drills or other methods to test emergency response plans.

To the Class 1 Railroads and Amtrak:

(1) In co-operation with the Federal Railroad Administration, the American Short Line and Regional Railroad Association, the Brotherhood of Locomotive Engineers and the United Transportation Union, they should be developing crew resource management training required for all crew members that addressed at a minimum:

(a) Crew member proficiency

(b) Situational awareness

(c) Effective communication and teamwork

(d) Strategies for appropriately challenging and questioning authority.

To the American Short Line and Regional Railroad Association:

(1) In co-operation with the Federal Railroad Administration, the Class 1 Railroads, the Brotherhood of Locomotive Engineers and the United Transportation Union, it should be developing crew resource management training required for all crew members that addressed at a minimum:

(e) Crew member proficiency

(f) Situational awareness

(g) Effective communication and teamwork

(h) Strategies for appropriately challenging and questioning authority.

To the Brotherhood of Locomotive Engineers:

(1) In co-operation with the Federal Railroad Administration, the Class 1 Railroads, the American Short Line and Regional Railroad Association and the United Transportation Union, it should be developing crew resource management training required for all crew members that addressed at a minimum:

(i) Crew member proficiency

(j) Situational awareness

(k) Effective communication and teamwork

(l) Strategies for appropriately challenging and questioning authority.

To the United Transportation Union:

(1) In co-operation with the Federal Railroad Administration, the Class 1 Railroads, the American Short Line and Regional Railroad Association and the Brotherhood of Locomotive Engineers, it should be developing crew resource management training required for all crew members that addressed at a minimum:

(m) Crew member proficiency

(n) Situational awareness

(o) Effective communication and teamwork

(p) Strategies for appropriately challenging and questioning authority. 1149

Some weeks after the accident, the Brotherhood of Locomotive Engineers petitioned the Federal Railroad Administration to introduce a rule banning the running of trains with the leading locomotive in the long nose forward position (as the leading locomotive had been in the Butler accident) whereby the body of the locomotive made it difficult for the engineer to observe signals. [1150]

At about 7:02 a.m. on Tuesday 23rd March 1999 eastbound Conrail freight train KAEL-3 from Kankakee, Illinois, to Elkhart, Indiana, collided with southbound Union Pacific Railroad (UP) freight train ZYCMX-23 at the crossing of the two lines at Momence, Illinois, United States. The Conrail train struck the side of the leading unit of the UP train, derailing both locomotives and the first 16 cars of the UP train and the first locomotive of the Conrail train. Leaking fuel from the second locomotive of the UP train then started a fire in the wreckage, though this was soon extinguished by firemen. Both conductors and both engineers were slightly injured; the Conrail conductor declined medical treatment at the scene and the other three crew were released after treatment at the Riverside Medical Center in Kankakee. The UP estimated

[1149] National Transportation Safety Board report

[1150] www.ble.org/pr/news/longnose.html

the damage to its train at $1,654,000. Conrail put the damage to its train at rather less – $137,000.[1151] The UP train had been carrying auto parts from Chicago, bound for Mexico.

A local newspaper wrote, 'About 20 departments responded to the crash, including one from the Joliet Mobile Oil refinery, which provided a tanker containing foam to clean up diesel fuel.

'"As the situation developed, we were very thankful no one was seriously injured," said Ken McCabe, chief of investigations for Kankakee County Sheriff's Department. That sentiment was echoed by others in the neighborhood, included a neighbor who pulled the Conrail engineer from the engine for fear a fire would start. We don't want another scene like Bourbonnais" {which had happened but a week earlier and only 10 miles away – see level crossing collisions chapter} said James Coulter, 36, Momence. "They were more or less in shock, didn't know what happened," he said of the victims.

'David Hoeksema called 911 after hearing a crash he likened to the crinkling sound of aluminium foil being crushed, a sound that continued as the cars piled into each other. Other neighbors said they often hear thuds as the trains change cars at the intersection. "But this noise was louder," said Lela Roberson. "I looked out of the window and saw black smoke," she said. "I was hoping nobody was hurt. At least it wasn't like Bourbonnais." [1152]

The National Transportation Safety Board (NTSB) published a brief report on the accident. It noted that the Conrail train had left Kankakee, Illinois, about an hour earlier. It had been moving at 44 mph when Engineer Tim Norred observed the Distant signal for Momence Junction displaying Approach-Restricting. He had applied the automatic brakes, reducing the speed to 27 mph, but about a minute later he had released the brakes again. He told the NTSB investigators that he had realised that the Approach-Restricting indication might mean that the next signal at Momence was at Stop and that a UP train could be passing over the crossing.

As the train had continued around a curve tall trees obscured the view of Home signal at the crossing, and as the train had exited the curve the low rising sun meant that the signal still could not be distinguished with the naked eye, despite which the engineer had taken no further action to slow the train. The conductor, riding with the engineer in the cab of the locomotive, was becoming concerned but did not say so. Instead he had taken a pair of binoculars from his bag, by means of which he could see that the Home signal was indeed Red. He had confirmed that it was, told the engineer and the latter had applied the brakes, though he did not at once make an emergency application and he did not activate the two-way end-of-train device at all. Had he done so, it would have increased the force of the brake application. The train had slowed following this further brake application but did not stop before striking the UP train.

Post-accident tests showed no problems with either the train's brakes or the signalling. However, safety depended entirely on the engineer obeying signal indications.

The NTSB determined that the probable cause of the collision had been the failure of the Conrail engineer, despite the crew's difficulty in seeing the Home signal, to slow the train enough to be able to stop it before it entered the crossing. Contributing to the accident had been the conductor's delayed action in warning the engineer to slow the train even though the conductor had realised that the train might be travelling too fast to stop at the crossing. Also contributing to the accident had been the lack of any safety redundancy system capable of preventing a collision in the event of human failure. [1153]

[1151] National Transportation Safety Board report

[1152] *The Star*, South Holland, Illinois, United States, 25th March 1999

[1153] National Transportation Safety Board report

Momence, Illinois, United States, was again the scene of a side collision at about 7:45 a.m. on Thursday 24th November 2005, this time involving a southbound Union Pacific Railroad (UP) freight train bound for Little Rock, Arkansas, and an eastbound Norfolk Southern Railway (NS) train heading for Pittsburgh, Pennsylvania. As a result of the collision two locomotives and 14 cars of the UP train and three locomotives and 31 cars of the NS train were derailed. Three UP crew members were slightly hurt. There was also a leakage of potentially flammable grain alcohol from a damaged car and a few nearby houses had to be evacuated.[1154] The Kankakee County Emergency Services and Disaster Agency was soon able to announce that the spill was contained. The NS train had been going over the crossing when the UP train had run into its side. [1155]

[1154] www.suburbanchicagonews.com/send_story/printstory.asp?HTMLpath=/couriernews/to...
[1155] www.nbc5.com/print/5397788/detail.html

CHAPTER 14

FIRES

Fires often breake out on trains following derailments or collisions. While wooden coaches were still in use in passenger trains, an overturned stove used for heating or spilt engine coals from a steam locomotive might set them ablaze. Later on, the substitution of gas for oil for train lighting proved to be another hazard, and a number of accidents were made much more severe by fires started by leaking or damaged gas cylinders until its dangers led to its replacement by electricity. This chapter will confine itself, however, to what might be termed primary fires, that is, accidents where there was no precipitating derailment or collision but merely a fire. Serious primary fires were considerably rarer than what might be termed secondary fires, but they did occur from time to time from a variety of causes, while more minor fires were fairly common.

A fire broke out on the Bangladesh Railway *Simanta Express* after it had left Poradah Junction, Bangladesh, on Tuesday 14th January 1992. Three carriages were affected and according to a government source 27 people died, while another source spoke of 36 deaths. Other witnesses and a stationmaster spoke of a possible 150 deaths, as many people had jumped from the train.

A passenger named Abukaer Khan, whose 14-year-old son was one of many passengers reported missing, said that the fire had burnt for half an hour before the train was stopped.

An electrical short circuit was blamed by officials for the fire. They stated that the train driver and four other railwaymen had been suspended and an investigation into the accident had been opened. [1156]

The S-5 sleeper coach of train No. 8001, the South Eastern Railway Down Bombay – Howrah mail, was full to capacity early on the morning of Wednesday 26th October 1994 and passengers who had been unable to get reservations were sleeping on the coach floor. Most of those in the coach were returning to Kolkata after having been on a package tour to Goa. When fire broke out in the coach, the passengers were asleep and according to a newspaper report the alarm was raised by the guard of a train passing in the opposite direction, who had alerted the staff at Chakradharpur station. The report of the Commission of Railway Safety states that the fire was observed by the assistant stationmaster at Lotapahar as the train passed through there at 02:12. As the train was running on the broad-gauge Rourkela – Tata electrified section of the railway and was being hauled by an electric locomotive, he was able to get the current switched off and the train duly came to a halt between Lotapahar and

---

1156 *New York Times*, New York, New York, United States, 14th January 1992

Chakradharpur stations, Bihar, India, at about 02:15. When the accumulation of smoke had begun to cause the passengers breathing problems, they had become aware of the fire and attempted to pull the alarm chain but this had been damaged by the fire itself and did not work. The train crew, oblivious of the fire, had thus made no attempt to stop the train until the failure of the current had stopped it for them, when they had realised what was wrong. Meanwhile, a number of passengers had jumped from the moving train to escape the fire. [1157]

'The ill-fated train with the remaining unaffected coaches arrived at Howrah station in the afternoon, while a relief train which was to bring the injured and the stranded passengers from Chakradharpur was held up for several hours because of a rail roko agitation {a form of protest involving the blocking of train tracks} by the alleged Jharkhand supporters. The *Gitanjali Express,* in which some of the passengers of the 8001 down Bombay were accommodated, got stranded and was running several hours behind schedule.

'A pall of gloom loomed large at the Howrah station special inquiry office with people jostling with each other for having a glimpse of the casualty list. With tears rolling down from the eyes of the unfortunate relatives losing their near and dear ones, the station turned into a wailing ground.' [1158]

An early report indicated 27 fatalities in the fire, though relatives complained of others who were missing but whose names did not appear on the passenger list. The final number of dead according to the Commission of Railway Safety was 46, with another 17 injured, four of them seriously. [1159]

An accident due to, perhaps, the upsetting of the kerosene stove of a tea vendor, was quoted as the cause of the fire. The Commission of Railway Safety report theorised that the fire had resulted from 'the accidental ignition of some combustible material, which was present under the lowest shelf of the pantry rack of a coach, possibly by a burning match-stick thrown carelessly by someone, and its continued burning unchecked and unnoticed resulting in an uncontrollable fire.'

The Commission made a number of recommendations in the aftermath of the fire:

(a) For unreserved compartments the Railway should formulate rules and policies to check unauthorised occupation and means for tackling fire including checks at stations.

(b) Adequate numbers of fire extinguishers should be provided in both front and rear brake vans and in each AC and non-AC coach to fight train fires. Their number and type should be specified by the Railway Administration.

(c) The guards in both the front and rear brake vans should be provided with adequate and suitable fireproof clothes and gear to fight fires (the items to be specified and laid down by the Railway Administration).

(d) All the running staff including commercial and pantry staff must be trained to fight fire.

(e) The Railway Administration should specify the type and number of the firefighting equipment that should be provided at stations to fight fires of a serious nature.

(f) Firefighting equipment should form part of the Accident Relief Medical Equipment van. The type and number of such equipment should be adequate to fight serious fire cases. The Railway Administration should specify the type and number of these equipment and lay down specific guidelines for their use.

[1157] *The Times of India*, Mumbai, Maharashtra, India, 22nd November 2011

[1158] *Indian Express*, Madras, Tamil Nadu, India, 27th October 1994

[1159] Government of India Ministry of Civil Aviation and Tourism Commission of Railway Safety Annual Report for 1994-1995

(g) Additional coach doors were recommended so that passengers had more escape routes in cases of emergencies. The Railway Administration should undertake a timely review of the design of the coaches.

(h) The alarm chain apparatus should be more efficient/effective so that the train stopped within the shortest possible time. The Railway Administration should undertake a review of the present system to identify deficiencies and suggest and implement corrective measures. [1160]

Fire broke out in a three-tier S-8 second-class compartment of train No. 2137, the Down *Punjab Mail* from Mumbai to Firozpur, at about 03:40 on Saturday 19th February 2000, between Duskheda and Savda stations, Maharashtra, India. The accident occurred in the Bhuswal-Khandwa division of the Central Railway, 456 kilometres from Mumbai, where the train had left the Chhatrapati Shivaji Terminus at 19:10 the previous evening.

The fire spread very rapidly to the adjacent S-9 coach before the train had come to a stand and according to eyewitnesses, engulfing six of the 18 carriages of the train within seven minutes, coaches S-4 to S-7 being the others involved. The gatekeeper at Redgaon Crossing reportedly saw that the train was on fire and exhibited a Red flag to it, while passengers were pulling the communication chain. First reports stated that 18 passengers had died in the fire, most of them succumbing while they were asleep, and another 15 were injured. The second-class coaches each had a seating capacity of 74 passengers.

Various surviving passengers spoke of their experiences. D.S. Aswal, an instructor with a marine academy in Noida, said, "The fire moved like a sea wall and there was a strong chemical smell, like that of burning leather."

Anees Khan declared that the fire had begun in berth one and had very quickly engulfed the whole of coach S-8, giving its passengers little chance to escape.

Mohammad Afzal spoke of his being traumatised by the sight of an elderly Sikh couple who he had witnessed being burnt to death. He said that most of those who had died had done so after becoming unconscious due to smoke inhalation.

Mohammed Quereshi, who had been on his way from Mumbai to Agra, related that passengers travelling in the other bogies had tried their best to rescue their fellow travellers but the train had not stopped for five or seven minutes after the communication chain had been pulled.

Gurjit Kaur, who had been riding in coach S-8, stated that the passage between it and coach S-9 had been open up until 02:30 and she had used it earlier. Shortly after the fire had begun she had rushed towards the passage, which had then been closed. She agreed that the train had taken six or seven minutes to stop, after, she thought, the engine driver had heard the cries of the passengers.

After a long battle firemen finally extinguished the blaze at 08:30. The six affected bogies (coaches) were then detached at Savda station and six fresh coaches having been added to the train, it resumed its journey at 11:20.

The injured were taken to Bhusawal Railway Hospital. Most of them were suffering from smoke inhalation but one passenger who had jumped from the train had sustained a serious head injury.

Various theories circulated as to the cause of the fire. The possibility of an electrical short circuit was ruled out as the coach was not equipped with air conditioning. Some witnesses

---

[1160] Ibid

spoke of passengers who had been drinking alcohol getting into a scuffle, suggesting that in fracas a cigarette butt might have been discarded. Railway officials denied a rumour that the fire had originated in the pantry car, while one official commented that "marriage parties have a tendency to light small portable stoves in trains, the fire could have started from one of those." Carelessness on the part of a cigarette smoker could not be ruled out.

Railway Minister Mamta Banerjee expressed her shock and grief over the fire and offered her condolences to the families of the bereaved, while Central Railway General Manager K.B. Sankaran and senior officials left by train for the accident site.

An Indian newspaper described the anguish of the relatives of possible victims of the fire. 'Meanwhile, anxiety is writ large on the face of 78-year-old Phulchand Singh, a resident of Mumbai, as he hesitantly inquires for the nth time about the arrival of a list of casualties. Phulchand, who paces up and down restlessly despite his advanced years, is one among the numerous relatives gathered at Chhatrapati Shivaji Terminus, to inquire about the exact fate of their kith and kin. "My daughter-in-law and my granddaughter are aboard the train. They had left for Agra for the wedding of a relative," he says, even as he runs his wrinkled fingers through the list of information provided by the railways.

'Gulchand Agerwal, whose wife and daughter are on the train, slouches into a rickety chair, placed beside the information officer, with a blank expression on his face even as the police personnel stationed to provide information move about briskly performing their job. Gulchand, who first heard the news after his relative called up in the afternoon, has been glued to the information centre, hoping for some news to trickle in. However, his brother's assurance that his wife and daughter could well be safe as they were travelling in S-7 and not S-8, which witnessed maximum casualties does little to ease the furrows on his forehead.' [1161]

A further account of the accident pointed out that the train would likely have been moving at 60 or 70 km/h by Rangaon Crossing, 12 kilometres after leaving Bhusawal station, and that it would take at least three minutes for it to pull up from that speed.

The account continued, 'The train came to a halt even before it had completely crossed the Rangaon level crossing. Burnt pieces of wood from the window, frozen lumps of melted glass were scattered all around.

'For the passengers trapped in the gas chamber that the S-8 compartment had become even one minute would have seemed like eternity. Survivors say that the smoke spread rapidly in the compartment, blinding and choking them. No one heard a blast, which rules out a bomb explosion. Since most passengers were disorientated, having woken up from deep sleep, the thought of opening the windows to let in fresh air did not occur to them. All the windows were closed to keep out the winter air. Even when the train stopped, it would have taken a couple of minutes for the passengers to jump out, given the commotion and panic in the dark smoke-filled and overcrowded compartments.'

The article continued, detailing the fate of two passengers who did not survive, one a child and the other a would-be rescuer. 'Jaswinder Gujral, who lost his eight-year-old daughter Sneha in the fire, said that although he had held his daughter's hand they were separated in the stampede. "I was pushed out of the train by the force of those behind me but Sneha was left behind," he said. The survivors could not re-enter the compartment to save those left behind because of the dense and toxic fumes resulting from the burning of resin and foam seats. Eyewitnesses said that Praveen Kumar Chanchada of Delhi who went into the compartment to rescue those rendered unconscious by the fumes never came out. His relatives identified him by the 'kada' {a steel or cast iron bangle worn by Indians, especially Sikhs} he was wearing and the brass button of his jeans. Meanwhile the railways have confirmed that the vestibule door

[1161] www.indian-express.com/ie/daily/20000220/ina20054.html

between two compartments was shut, as is the practice, after 10 p.m. to prevent theft.' [1162]

Two days after the fire, Minister of State for Railways Digvijay Singh, ruling out the possibility that the fire had starred accidentally, for example from a discarded cigarette butt, declared that, "This is a terrorist act. I am not ruling out the involvement of some ISI agents." (ISI was an abbreviation for Inter-Services Intelligence, the Pakistani intelligence agency, suspected of supporting terrorist attacks in India.) Meanwhile, the Commissioner of Railway Safety was interviewing the staff at Savda and Duskheda stations, Guard R. Bari of the train and Crossing Keepers Ashok Onkan and Rajaram Oneha. [1163]

Egypt suffered a catastrophic train fire at about 02:00 on Wednesday 20th February 2002 involving an old slow-moving passenger service from Cairo southbound to Luxor. The train stopped at almost every station along the route and also carried daily newspapers to towns and villages along the River Nile. It was the eve of the Islamic Eid al-Adha holiday, the most important of the year, and as a result the train was grossly overcrowded with people going to spend the holiday with their families, while animals were also carried on the train and the compartments were stuffed with luggage.

Police stated that the fire had broken out in one of the 11 carriages of the train when it was some kilometres short of the town of El Ayyat. Each carriage, they added, had a nominal capacity of 150 passengers but had been carrying twice that number, meaning that there would have been over 3,000 people on the train. Another account stated that the train had been composed of 16 carriages and carrying 4,000 passengers. Before it was finally stopped the train travelled some distance and the flames had spread to six other carriages.

Fuad Amin, 22, a construction worker, told a reporter that he had jumped from the moving train, breaking his hand in the process. He explained that the train had been so crowded that he and six friends had been forced to sit on the floor. The first sign of trouble had been his hearing shouts and screams, which he had initially attributed to a fight on board. Then he had seen flames and people running, one of them a woman whose clothes were on fire. "People were running like crazy," he related in a weak voice.[1164] He ran too, until he found a window that had been smashed open. At first he hesitated to jump because of the speed at which the train was running, but then, "I thought that I was going to die anyway, so I jumped." He remembered nothing more until he woke up in hospital and did not know what had become of his friends. [1165]

From his hospital bed another survivor told a Reuters reporter, "We pushed each other and we were suffocating from the smoke. We threw each other out of the windows."

According to witnesses, the escape of some of the passengers had been frustrated by the barred windows of the carriages. A charred body of one of the victims was found between the bars.

At the time of the above report (about 10:00 Greenwich Mean Time on the day of the accident) the number of known dead, initially put at 20, then 50, and then 100, had already risen to 215 and it would soon rise a good deal further as the remains of more victims were discovered. Thirty ambulances and 25 fire engines had reportedly been dispatched from Cairo to the scene, which a police officer described as "hell." About 75 injured passengers were taken to hospital. It took firefighters several hours to extinguish the blaze.

1162 *The Times of India*, Mumbai, Maharashtra, India, 23rd February 2000

1163 www.tribuneindia.com/2000/20000222/narion.htm

1164 *Yahoo! News* Headlines, 20th February 2002, 10:06

1165 *The Guardian*, London, England, 20th February 2002

The official Middle East News Agency (MENA) stated that an initial investigation indicated that the explosion of a gas cylinder in the dining car of the train had caused the fire. [1166]

By the time that Prime Minister Atef Obeid reached the scene of the fire, at the village of Reqa al-Gharbiya, a little later in the morning, the number of known dead had risen to 254 as rescuers continued to search the wreckage. He admitted that he could not be certain as to the cause of the fire, though he speculated that the blame lay with "the materials that the passengers carry with them, such as gas cylinders and small stoves." He added that, "Until now, there is no proof of any laxity in safety precautions." [1167]

Shortly afterwards the death toll had climbed sharply to 342 and Ahmed al-Sherif, director of the state-owned Egyptian Railway Authority, stated that there had been nothing like it in the recent or distant past. Commenting on the MENA report as to the origin of the fire, he explained that the third-class train had no dining car, though passengers often brought gas cylinders and small stoves onto the train for cooking, despite regulations forbidding the practice. [1168]

A little later, Health Ministry undersecretary Ahmed Abdulaziz stated that the death toll had reached 350. Meanwhile, Social Affairs Minister Amina el-Gindy promised that his government would pay LE (Egyptian pounds) 3,000 to the families of the dead and LE 1,000 to those injured.

A reporter described the accident scene. 'One Egyptian rescuer in a bloodied galabiya robe, plastic gloves and sandals on his feet worked his way through one of the compartments in search of survivors but found no one alive. Other rescue workers wearing paper masks fought through the stench of burnt flesh to lift bodies from the train. Scattered on the ground outside the train were the scorched remains of clothes and shoes, singed papers and notebooks.' [1169]

Mosques nearby were opened to accommodate the victims and villagers supplied blankets, food and hot drinks to uninjured but stranded passengers. Meanwhile, the Egyptian Red Crescent mobilised several hundred volunteers from eight branches to provide first aid, transport to hospitals, food, blankets and financial assistance and psychological support to the victims and their families.[1170] The line was closed until further notice.

A report published at about 18:00 Greenwich Mean Time on the day of the catastrophe stated that Giza medical officials were at that time putting the death toll at 366, revising it down from 373 because some amputated limbs had been counted as separate bodies. Only 36 bodies had been identified. The recovery work had, Undersecretary Abdulaziz announced, ceased just before sunset, and it was not expected that any more bodies would be discovered.

'Mohammed Mokhtar, tears streaming down his cheeks, told MENA he had saved himself and a woman from death by pushing her out of a window and jumping out behind her. But he said he had been too late to save her four-year-old son, whose little body had already been consumed by the flames. "I don't know how I can live with all this death," he said. [1171]

Another report added, 'Survivors reported difficulties escaping from the moving and crowded train. Many of the windows were barred and some of the dead were found trapped under seats and huddled by closed doors between carriages. Egyptian trains are frequently very crowded, more so on this occasion as Egyptians are about to observe Aid al-Adha, a

---

[1166] *Yahoo! News* Headlines, 20th February 2002, 10:27

[1167] *Ananova*: Egyptian premier visits 250-death train fire scene

[1168] *Ananova*: Train fire death toll rises to 342

[1169] *Yahoo! News* Headlines, 20th February 2002, 13:56

[1170] Egyptian Red Crescent Information Bulletin, 21st February 2002

[1171] *Yahoo! News* Headlines, 20th February 2002, 17:54

three-day holiday marking 40 days after the end of the fasting month of Ramadan. Many are deterred from travelling by road where the death rates are even higher.

'Police praised the driver for separating the burning carriages from the rest of the train, saving this saved hundreds more lives. Those who were trapped died in what Gen. Abdel Aal, head of the rescue operations, likened to a "burning prison."

'{Prime Minister} Obeid ruled out a technical fault for the tragedy or an error by the train crew, despite the long distance travelled by the burning carriages before coming to a halt. "The fire was started by cookers used by passengers during the journey," Obeid told reporters on the scene. But several survivors interviewed in hospital said they saw no stoves used. "There was no room to put out the stoves," said Sami Zoheir Hanna, who managed to escape through a door and suffered only bruises.

'The driver was not immediately aware of the fire because it erupted in the rear carriages but reacted well once he had realised, the top rescue official said. "The driver had an excellent reaction when he quickly separated the burning carriages from the others before the flames could reach them," Abdel Aal said.

'The accident comes at a difficult time for the Egyptian government, which is battling an economic crisis caused by a drastic decline in tourism.

'Accidents on the 150-year-old railway network in Egypt are all too common, a legacy of decades of under-investment and general overcrowding, according to critics.' [1172]

A report in an Australian newspaper carried accounts from other surviving passengers. 'Ali Thabet climbed out of the window of a speeding train and held on for his life on Wednesday as flames rushed through his carriage and leapt into others. Mr Thabet, 20, and his father, Ali Moustafa, jumped and survived. "I just tried to save my life," Mr Thabet said as he sat at his father's hospital bed while rescue workers pulled bodies from the wreckage. "I held on until the fire made by feet and legs burn, then I let go."

'"The fire started to get really close," said Yassir Fouad Mohammed, 28, who also escaped. "But there was no place for us to get out. We couldn't get through to the doors because of the crowds."

'Witnesses described panicked women, their dresses on fire, rushing forward and slamming into a wall of frantic passengers. They surged forward and died together, one on top of the other. "The dead people were all piled up in one place in each car, like they were trying to escape," said Mamdouh Salem, one of the first ambulance officers at the scene. Outside, the carriages were burnt down to the metal. Inside, rescue workers struggled to pull out the remains. The heat had been so intense that everything – luggage, seats and people – had melted.' [1173]

The day after the fire, Prosecutor General Abdel Wahid promised that there would be "severe punishments" were there any proof of negligence in connection with the catastrophe, while, 'Train driver Mansour Youssef el-Qams said on seeing flames he stopped the train and ordered his crew to separate the burning carriages from the rest of the train. They tried to extinguish the blaze with four extinguishers but strong winds fuelled the fire.' [1174]

On Friday, President Mubarak proclaimed, "I will not allow any effort to hide the truth or cover up any aspect of what happened, because the calamity is great, the accident grave. We have ordered competent authorities to conduct a complete investigation to clearly determine responsibility and to hold accountable anyone proven to have fallen short in their duties or who were careless in providing safety." The same day, Transport Minister Ibrahim Demiri and

[1172] *AllAfrica.com*, 21st February 2002

[1173] *The Age*, Melbourne, Victoria, Australia, 22nd February 2002

[1174] *Ananova*: Egypt train disaster investigators vow 'severe punishments.'

Ahmed al-Sherif, director of the state-owned Egyptian Railway Authority, both tendered their resignations. [1175]

Meanwhile, the banned opposition Muslim Brotherhood blamed the catastrophe on "gross negligence" and the opposition newspaper *al-Wafd* demanded that those responsible should be brought to trial, writing in an editorial, 'This is more than gross negligence. We need to know who was responsible and hang them in public squares and curse them for what they have done to the helpless Egyptian people.' [1176]

Speculation as to the cause of the fire continued, with an alternative theory that it might have been caused by an electrical short circuit, but this theory was apparently rebutted when investigators found several small stoves amongst the charred remains of the worst damaged coaches, such stoves being commonly used by passengers to heat food and water for tea or coffee. [1177]

Towards the end of March 11th, medium and low-ranking railway officials were charged with various offences, such as not ensuring that the train had sufficient functioning fire extinguishers, allowing too many passengers to board the train so that passengers could not reach the communication cord to warn the driver, allowing the train to run for 15 minutes after the fire had first broken out and falsifying papers stating that the train was in a fit state to carry passengers. [1178]

Defence lawyers argued that the accused were being made scapegoats for years of negligence and mismanagement by the upper echelons of government and the railway authority. At the end of September, the accused were all acquitted. The court ruled that the cause of the fire was unknown. [1179]

The final death toll in the catastrophe was put at the time of the trial at 361, though another source states that it was officially 383 and that this figure, too, was a gross underestimate. The true figure will doubtless never be known. [1180]

A disaster reminiscent of the July 1978 Taunton train fire in England (see previous volume of this series) occurred on the Société nationale des chemins de fer français (SNCF) at about 02:15 on Wednesday 6th November 2002, near Nancy in eastern France. Again, a sleeper train was involved, this time train No. D261, which had left Paris l'Est at 22:58 for Strasbourg and München, Germany, with through carriages for Frankfurt am Main, Germany, and Wien, Austria, and was booked to run non-stop between Paris and the German border town of Kehl.

The alarm was reportedly raised by the Nancy station staff who saw smoke billowing from the train as it was passing through, upon which they raised the alarm and cut off the electric current, bringing the train to a halt about a kilometre further on. Firemen reached the scene at 02:22. About 130 passengers escaped unharmed but 12 passengers – six men, four women and two children – who had been riding in the leading two sleeping cars lost their lives, most or all it was believed from smoke inhalation, and nine others were injured. Chief firefighter Jean-Louis Modere stated, "We quickly realised the seriousness of the situation. The catastrophe was amplified by the fact that it was in a confined space. The fire was limited, and the amount of smoke quickly became catastrophic." [1181] [1182]

[1175] *Yahoo! News* Headlines, 22nd February 2002

[1176] *BBC News*, London, England, 22nd February 2002

[1177] *The Globe and Mail*, Toronto, Ontario, Canada

[1178] *Ananova*: Eleven railway employees charged over Egypt's worst train disaster

[1179] South African Broadcasting Corporation, 29th September 2002

[1180] Wikipedia – 2002 El Ayyat railway accident

[1181] www.guardian.co.uk/worldlatest/story/0,1280,-2150279,00.html

Regional official Jean-Francois Cordet commented that as well as French passengers there had been various other nationalities amongst the victims, who had comprised a mixture of tourists and business travellers. The injured – four Germans, three Britons, an American and a French national – had been taken to the university hospital in Nancy. The 12 who died were American Salvatore Michael Amore, 43; his wife Jeanne, 43; their children Emily, 12, and Michael Bernhardt, 8; and Salvatore's mother Susanne, 72; three German men, a Russian man and woman, a Hungarian man and a Greek woman.

Surviving passengers spoke of panic within the train, with screaming passengers making their escape by breaking windows and climbing out through them, while those in other coaches had seized their possessions and exited via the doors. French passenger Marc Giraud said that smoke had filled his compartment as soon as the door was opened. He had grabbed an emergency ladder, smashed the window of the compartment and he and his fellow passengers had climbed out. He told a TF1 French television reporter, "We got out through that mouse hole. I don't know how we got out through there." German(?) passenger Susan Stroembum related that she had been sleeping in another car and when she had awoken flames had been visible through the window. "Someone screamed, 'The train is burning,'" she related at Stuttgart station, where 55 survivors were given food and drink by Deutsche Bahn.

The affected sleeping cars were owned by Deutsche Bahn and were provided with fire extinguishers but did not have fire or smoke detectors. The sleeping car where the fire had originated dated from 1964. It had been extensively renovated in 1999, overhauled in 2001, and subjected to a technical examination two days before the accident.

Initially it was theorised that an electrical fault had caused the fire. Later, it was determined that this was not the case. Instead, the fire had originated in a kitchenette hot-plate in the attendant's compartment of the leading sleeping car, which had set fire to a jacket left hanging too close to it. The German Train Attendant Volker Janz, 53, had been asleep in his bunk when the fire had begun. Upon waking and discovering it, he had first attempted to extinguish it with water. When this had failed he panicked and instead of using a fire extinguisher or operating the emergency alarm to stop the train, he had then gone in search of his conductor, thus allowing the fire to spread unchecked. He subsequently refused to turn himself in to the authorities until a Nancy judge had issued an international arrest warrant for him, whereupon he did surrender himself. He faced possible multiple charges of involuntary homicide through negligence. [1183]

The accident was investigated by the Bureau d'enquêtes sur les accidents de transport terrestre (BEA-TT), the French accident investigation board, which found that numerous causes and elements had exacerbated the situation during the fire, specifically the following:

(1) The faulty use of the hot-plate and working space, the initial cause of the fire.

(2) The low level of preparedness of the sleeping-car attendant when faced with an emergency situation.

(3) The haphazard communication between the members of the train crew in the emergency.

(4) The locking of the corridor door, which had prevented any passengers escaping through that corridor, and the blocking of access doors with hooks, which had delayed the rescuers.

(5) The susceptibility of the sleeping car to fire because of the very nature of certain materials found in it (which had contributed to the quick development of the blaze and the production of smoke and toxic gases) and because of the limited anti-fire devices (the lack of fire detectors and of extinguishers near the service area).

[1182] www.thisislondon.co.uk/til/jsp/modules/Article/print.jsp?itemId=1912502

[1183] www.stanfordadvocate.com/news/local/state/hc-18191013.apds.m0251.bc-ct--fran...

(6) The window-breaking hammers had not been visible enough to enable use of the compartment emergency exits.

The BEA-TT made a number of recommendations:

(1) The responses of sleeping-car staff when faced with an emergency situation should be strengthened:

(a) By holding a staff meeting at each train departure so that they were aware of specific communication and alerting arrangements, as was the case for all SNCF trains with sleeping-cars.

(b) By including emergency situations more specifically in the instructions and training for this category of staff, particularly for service providers, and by requiring an acceptable foreign language fluency on international trains.

(2) Sleeping-car characteristics that worked against safety, and were due mostly to the outdated design of the cars, dating back some 40 years, should be addressed before this type of car was put back into use on the French national railway network. Safety improvements required included the following:

(a) Locking devices that did not comply with the Union Internationale des Chemins de fer (UIC) regulations, and had been installed on access and corridor doors for security reasons, should be removed.

(b) The emergency exits and associated devices (window-breaking hammers, windows) as well as extinguishers should be upgraded.

(c) An audit of fire susceptibility, based on the UIC recommendations, should be carried out, with upgrading in line with its findings.

(d) The safety and ergonomics of the electrical equipment used for catering needed improvement.

(3) In addition, and in cooperation with the other partner railway undertakings, the SNCF should:

(a) Check that all the measures listed here above had been applied to all sleeping-cars used on the French network, since the age of the fleet could make it prone to similar defects.

(b) Plan to equip all these sleeper-cars with smoke detectors and a public-address system, by a given date which would be set after studying implementation arrangements.

(c) Study the introduction of a secure ventilation system in case of fire, to avoid toxic smoke propagation.

(d) Improve means of communication to be used by on-board staff, with at the very least an intercom system.

(4) Finally, at the instigation of SNCF and the French Ministry of Transport (through its Land Transport Directorate), all these measures should lead to the updating, clarification and reinforcement of the existing construction standards as well as of the interoperability rules being prepared at an international level. [1184]

Indian Prime Minister Atal Behari Vajpayee was forced to cancel his planned trip to a Commonwealth summit in Australia following a horrific fire on a train as it pulled out of the station at Godhra Junction, Gujarat, India, on Wednesday 27th February 2002. The train

[1184] Bureau d'enquĕtes sur les accidents de transport terreste report involved in the fire was the Western Railway *Sabarmati Express* from Varanasi City, Uttar Pradesh, to Ahmedabad, Gujarat, had begun its journey on Monday 25th February.

In February, thousands of devotees of the Hindu deity Rama, known as Karsevaks, had travelled from Gujarat to the town of Ayodhya in Uttar Pradesh at the behest of the Vishva Hindu Parisad, a right-wing Hindu nationalist organisation, to participate in the Purnahuti Maha Yagna ceremony. On Monday 25th, 1,700 Karsevaks and Hindu pilgrims had boarded the express at Ayodhya for the return journey. The express arrived at Godhra Junction at 07:43 on the Wednesday morning, about four hours late. As the train was leaving the station the emergency chain was operated, bringing it to a halt again; according to the driver there were multiple chain pullings. At this point, the train was reportedly attacked by a mob of about 2,000 Muslims, who first stoned it and then set it alight. [1185]

The attack was the latest in a whole series of Hindu-Muslim conflicts. Ayodhya, according to Hindu tradition, was the birthplace of Rama, while since the 16th century it had been the location of a mosque. In December 1992, the mosque had been destroyed by Hindu nationalists, who subsequently planned to erect a temple to Rama there. Construction of the Rama temple had been due to begin on 15th March, in defiance of a court order forbidding it. The conflicts between Christians and Muslims over holy sites in Jerusalem come to mind.

Early reports stated that the bodies of 15 children, 25 women and 17 men had been recovered from the four coaches involved in the fire and that dozens more passengers were injured, including 38 who were in a critical condition. Passenger Sanjay Kumar, who later reached Ahmedabad in the undamaged portion of the train, told a reporter, "We have been through hell. We don't know the fate of many passengers who were with us."

Rakesh Kimani, 18, who lived near the scene, related, "I heard screams as I came out of the house. I saw a huge ball of fire. I saw...people putting out their hands and heads through the windows trying to escape. It was a horrible sight." [1186]

The Additional Director General of Gujarat Police, J. Mahapatra, declared that "miscreants had kept petrol-soaked rags ready for use much before the train had arrived at Godhra."

Meanwhile, Prime Minster Vajpayee was appealing for calm, stating, "I appeal to the people to be patient, exercise restraint, and be calm. Do not let emotions sway you. The unity and communal harmony of India will have to be protected at all costs. The government is very worried about this incident."

The day after the fire, it was announced that 51 people, all of them Muslims, had been arrested on charges of arson, rioting and looting. More arrests followed, while the number of fatalities rose to 59. The number of deaths in the train fire was, however, completely eclipsed by the possibly 2,000 people who died in three days of communal violence in Gujarat in the aftermath of the blaze.

Whether or not the fire was arson or accidental remains disputed, though an account from a Hindu passenger named Rakeshbhai Kantibhai Patel, related a year later, if true, would appear to confirm arson. He stated that he had been in the bathroom of the train when he had heard rocks raining on its roof. '"I heard noises of people shouting outside: 'Kill them! Cut them up!" Patel said. He looked out and saw men with sticks attacking the train car. The train erupted in panic. Patel used his handkerchief to bandage a woman wounded by a rock. Then a burning rag was thrown into the train. The passengers doused that fire with water, he said. Then someone poured a flammable liquid into the car, and a burning torch. Within minutes, the coach was full of smoke. Passengers were choking. Outside, the mob was trying to break open the doors, he said. Patel and two dozen others leaped out and began pulling others out.

[1185] Wikipedia – Godhra train burning
[1186] *BBC News*, London, England, 27th and 28th February 2002

But dozens were already burned. Many died screaming. "Even now, after one year, I cannot forget that day. I won't forget for the rest of my life," Patel said.' [1187]

A study by the Gujarat Forensic Science Laboratory concluded that the fire had been started by the ignition of an inflammable liquid, which the attackers had poured into the coach, and that the fire had definitely begun within the train.

At the beginning of March, the Gujarat government established the Nanavali-Meta Commission to investigate the disaster. The first part of the report, published in 2008, blamed the fire on arson, naming a Muslim cleric and a dismissed police officer as the principal conspirators.

A separate report, compiled by Supreme Court Justice Umesh Chandra Banerjee, and first published in January 2005, concluded, however, that the fire had been accidental. In October 2006, the findings of the Banerjee Report were dismissed by the Gujarat High Court.

In February 2011, 94 people stood trial in connection with the 2002 incident, charged with murder and conspiracy. At the end of the trial, 31 were convicted, of whom 10 were sentenced to death and 20 to life imprisonment. The 63 acquitted due to lack of evidence included Maulvi Saeed Umarji, supposedly the chief conspirator. The convictions were confirmed on appeal, though the death sentences were later commuted to life imprisonment. [1188]

A few minutes after it had left the station at Ludhiana, Punjab, India, at 03:35 on Thursday 15th May 2003, fire broke out on the Northern Railway train No. 2903, the *Golden Temple Mail* (previously known as the *Frontier Mail*). It apparently began in the S-4 coach and spread rapidly. Twelve fire tenders took almost two hours to bring the blaze under control, and after they had done so, the charred and unrecognisable bodies of 35 passengers were recovered from the worst-affected S-4 coach and three more were taken from the S-5 coach. At the scene, the firemen had difficulties obtaining enough water as there was no suitable source nearby. The fire had broken out when the train was near Phaguwal village and villagers from there were the first on the scene. The fire was reportedly visible from four kilometres away. 'Sanjiv Sud and Mammohan Singh, residents of Phaguwal village, said they rushed to the spot after seeing leaping flames. Soon, 300 people from the small village rushed to the burning train with whatever relief material they could arrange. "Initially we took water in buckets but it was inadequate," Sud said, adding the villages immediately switched on a tubewell for drawing water. They also arranged food and water for the passengers.' [1189]

'Anguished relatives searched for loved ones in a charred pile of bodies littering the tracks where a crowded passenger train caught fire in Punjab on Thursday, leaving at least 38 people dead… "I have lost my wife and both my sons in this ghastly accident," said soldier M.K. Shah, as he sat weeping by the side of their now shrouded bodies.

'Relatives of Anil Bhatia, a private sector employee, were searching for his wife. "Anil is in hospital with burns but please help us to trace Anita," his wife, said one of the couple's relatives, pointing at the badly-charred bodies.

'Three economy-class carriages at the rear of the express were burned, officials at the site told a news correspondent, adding that the bodies of 35 of those killed were brought out from a single coach. The remaining other three bodies were retrieved from the two other smouldering carriages, an official said, adding that 11 of those killed were children. The train, travelling from Mumbai to the holy Sikh city of Amritsar, had caught fire just 10 minutes after

[1187] www.guardian.co.uk/worldlatest/story/0,1280,-2437749,00.html
[1188] Wikipedia – Godhra train burning
[1189] http://timesofindia.indiatimes.com/cms.dl1/html/uncomp/articleshow?msid=46454342

it steamed out from Amritsar. Rail authorities had rushed relief and medical teams from several cities to the site of the accident, the spokesman added.' [1190]

One surviving passenger related that she had tried in vain to raise the alarm upon becoming aware of the fire: 'In the wee hours of the morning, thirty-year-old Manju Sharma's screams seemed to fall on deaf ears. It was 3.45 a.m. Most of the passengers on the train were fast asleep; others who were awake took her to be insane. Manju was on-board the ill-fated S-4 compartment of the Amritsar-bound Frontier Mail in which 38 people were killed on May 15 as the bogey burst into flames. "I was shouting, running about telling people to pull the chain but nobody took me seriously. They thought I had gone mad," Manju told HindustanTimes.com in a telephonic conversation from Ludhiana. It was a reality no passenger was ready to believe. She was the first to have noticed fire emanating from the upper berth of the adjacent compartment. A student of BIT {Bhagwati Institute of Technology}, Ghaziabad, she was travelling on berth no 32 to Jalandhar.

'"At around 3.45 a.m., when I was just close to sleep, I saw a small spark on the upper berth of the next compartment. Perhaps, it was a short-circuit fire," she says. "The spark grew into a blaze and engulfed the man sleeping on the berth. Before anything could be done, the fire spread, fast – from one seat to another and soon to the floor of the berth. I ran towards the S-1 compartment, frantically asking people to stop the train, but no one heeded my screams," she recalls. "It was only in the S-1 compartment that a chartered accountant from Jalandhar, alarmed by my screams and seeing the fire, pulled the chain, bringing the train to a halt. But, by then, the fire had already spread to the adjacent compartments," says Manju. As soon as the train screeched to a halt, Manju and her two co-passengers were the first to jump out of the train, shouting out to people, urging them to detrain. With regret in her voice Manju says: "Had someone pulled the chain earlier, heeding my alarms, many lives would have been saved."

'Manju is now safe in her home in Jalandhar but the ghastly scenes with not let her sleep. She says: "I see people on fire; babies being thrown out of the train. It was a journey of traumas…" [1191]

Another woman passenger saved herself and her two children by her own initiative: 'In a rare act of courage, 30-year-old Neelam Sharma, along with her two minor kids, jumped to safety from the burning Golden Temple Mail. "My wife and two kids, Jashan (4) and Harry (5) are safe and have reached Amritsar," Neelam's husband Subhash Sharma told UNI {United News of India} on the phone from his residence at Kalyan in neighbouring Thane district.

'Western Railway's chief spokesman Shailendra Kumar said Neelam was inside the S-5 coach of the ill-fated Mumbai – Amritsar Golden Temple Mail which caught fire near Ludhiana early on Thursday morning. A relieved Sharma said Neelam took the courageous step of jumping out of the train, barefoot in darkness, and consequently saved her children. "There was chaos after the accident. Smoke filled inside the compartment following the fire in the S-3 and S-4 coaches. She was not able to see anything," he said. "But in the darkness she went towards the door and then jumped. Some other passengers followed suit," he added. "It was a big fire and people were running helter-skelter," said Sharma, who runs a spare parts business at Kalyan.' [1192]

'Indian infantry sergeant AD Singh and six of his troopers on board the train risked their lives to drag out smoke-blinded passengers in pitch darkness from a carriage where the blaze originated. "We were in a different coach when we heard screaming from the carriage just behind us and when we entered that we found piles of baggage blocking its exits," Singh said.

[1190] www.keralanext.com/news/index.asp?id=10633

[1191] www.hindustantimes.com/onlineCDA/PFVersion.jsp?a.../181_253967,00090001000l.ht

[1192] http://sify.com/printer_friendly.php?id=13151803&ctid=2

"I pulled the chain and applied the emergency brakes and we brought out as many (passengers) as we could. The heat was so much that the roof above our heads was melting and the smoke was choking." A visibly exhausted Singh said. "People were screaming, weeping, and they were in panic," the soldier, covered in soot, added. [1193]

Conductor Sudhir Prakash and Travelling Ticket Examiners Ajit Kumar and Jasbir Singh were reportedly suspended a couple of days after the fire, following allegations from passengers that they had not been in the coaches that they should have been at the time the fire broke out. [1194]

Regarding the fire, a Railway Ministry spokesman in Delhi stated that there was no indication of sabotage, saying: "Prima facie, it is a case of electric short circuiting or sparking or leakage of some inflammable objects being carried by passengers." [1195]

Several theories circulated as to the origin of the fire. H.C. Gupta, a member of the Railway Safety Board, ruled out an electrical cause, stating that there was no evidence of any short circuit. He also ruled out arson and was of the opinion that it might have begun with the bursting of a stove in the S-4 coach. [1196]

Passengers suggested different causes: 'Army Havildar A.D. Singh, who was travelling in the S-4 coach, said he saw a co-passenger throwing a cigarette butt inside the bogie after which the entire compartment caught fire. He added the fire spread to other coaches and would have engulfed more if the passengers had not pulled the emergency lever and stopped the train.

'Another passenger, Safi Pitoliwali, travelling with three other families from Daund, said when he got up early in the morning he saw electrical wires near a toilet in his coach, S-5, catching fire. Pitoliwali said he immediately raised an alarm, alerting other passengers and pulled the chain. The train stopped after 3-4 minutes. A police officer travelling on the train also passed on a message to his colleagues.' [1197]

The preliminary report of the Commission of Railway Safety stated that they had been unable to reach any definite conclusions as to the cause of the fire and were awaiting forensic reports. The final report of the commission does not appear to be available; probably it could not reach a definitive conclusion either. [1198]

A power failure between two underground stations in Pyongyang, North Korea, on about Sunday 22nd May 2005, was the precipitating event in a fire that reportedly cost the lives of three to five people. When the power was restored an electrical cable caught fire, generating toxic gases from burning plastic, which rendered train passengers unconscious as they attempted to escape from the station. [1199]

A fire broke out in the penultimate coach of 22 of the Southern Railway *Cheran Express* from Chennai to Coimbatore between Sulur Road and Irgur stations, Tamil Nadu, India, on Tuesday 19th July 2005. A lawyer became aware of thick smoke and fire coming from a toilet as the train was approaching Irgur at 05:40. Another passenger, one of the staff of a courier

[1193] www.news24.com/*News24*/World/News/0,6119,2-10-1462_1359902,00.html
[1194] http://timesofindia.indiatimes.com/cms.d11/html/comp/articlesshow?msid=46774538
[1195] www.hindustantimes.com/onlineCDA/PFVersion.jsp?a.../181_253417,000900010001.ht
[1196] www.deepikaglobal.com/latestnews.asp?ncode=1800
[1197] www.Rediff.com/cns/print.jsp?docpath=/news/2003/May/15fire.htm
[1198] http://sify.com/printer_friendly.php?id=13171948&ctid=2&lid=1
[1199] http://quote.bloomberg.com/apps/news?pid=71000001&refer=japan&sid=ar3ZfULLI9S4

company, then made his way towards the spot to save his luggage, which he had left close to and inside the toilet. He found that the toilet door, which was already alight, was locked from the inside, and the two men then pulled the alarm chain. At the same time the guard had also seen smoke coming from a toilet and alerted his driver, who stopped the train. Once the train had come to a stand the guard and driver detached the burning coach and the rear brake van from the remainder of the train and almost 100 passengers were able to escape from the affected coach with their luggage. Firemen were able to extinguish the fire by about 06:45.

After the fire had been extinguished the body of a man aged about 35 or 40, charred beyond recognition, was discovered in the toilet. Forensic experts also recovered a lungi (a men's skirt tied around the lower waist) and a towel soaked in kerosene from the toilet. One theory was that the victim had accidentally started the fire while lighting a cigarette or beedi in the locked toilet and that the kerosene had been concealed there out of the eyes of the ticket examiner. The alternative theory was that the victim had deliberately set himself alight in an act of self-immolation. [1200] [1201]

[1200] http://spiritofchennai.com/news/chennai-news/7699.html

[1201] www.hinduonnet.com/thehindu/thscrip/print.pl?file=2005072017690100.htm...

CHAPTER 15

CHEMICAL EXPLOSIONS

Railway accidents involving explosions are principally of two types; those which might be defined as chemical explosions and locomotive boiler explosions. Accidental chemical explosions, such as those due to mishaps with blasting powder or munitions, are fairly rare, while by 1991, with very few steam locomotives remaining in service other than on preserved (so-called heritage) railways, locomotive boiler explosions were becoming a rare class of accident. This chapter deals with chemical explosions.

Explosions involved several waggons of a Soviet military freight train on its way from Artemovsk in Ukraine to Marneuli in Georgia on Wednesday 2nd October 1991, They occurred as the train was passing through a densely populated area of the Georgian capital of Tbilisi and according to a report on Georgian radio 19 people were injured. An anonymous spokesman for the Georgian Interior Ministry reported that some people had been killed, though he had no details of casualties. He added that the waggons, which had presumably been loaded with artillery shells, had caught fire and then begun to explode. News agencies stated that there had been a failure of the overhead line equipment and the cables had showered sparks onto waggons carrying explosives, possibly gunpowder or munitions. They added that the series of explosions had begun near Didube station at about 09:15.

Some doubt existed as to the cause of the explosions, as the independent Interfax news agency reported that they had been set off by gunmen firing on the train. They did not identify the gunmen or provide any other details. Georgian President Zviad Gamsakhurdia had been locked in a political standoff with armed opponents, but Tengiz Kitovani, commander of the rebel National Guard, denied that his troops had been involved in the explosions. [1202]

An explosion in the motor of train E 3040, a railcar operating between Hamm and Aachen, occurred between Kamen and Dortmund, Germany, on Tuesday 3rd March 1992. Pieces of metal were thrown 40 metres and one person was injured. [1203]

On Sunday 3rd July 1994 an explosion occurred on a passenger train near the city of Sunchang, southern China, killing 18 people and injuring 49 others, 10 of them critically. Investigators theorised that explosives brought onto the train by a passenger had been set off

---

[1202] *The Telegraph*, Nashua, New Hampshire, United States, 2nd October 1991
[1203] *Katastrophen der deutschen Bahnen Teil II* by H.J. Ritzau, p. 202

by the jarring motion of the train. [1204]

An apparently fresh explosion occurred on a passenger train bound from Zhanjiang to the tourist centre of Guilin in Guanxi province, China,[1205] on Saturday 13th August 1994, this time killing 12 and again injuring about 50. Explosives or flammable materials carried by passenger were again blamed for the accident, reports of which were barred from the Chinese media due to fears that the tourist industry of the region might be harmed. [1206]

Like the July accident, brief reports of that which reportedly occurred in August only appeared some days later in Hong Kong newspapers. It is possible that there was in fact only one accident, in July, though given the interval of time between the reports this seems unlikely.

At least 16 people were hurt in an explosion that occurred on a train at location in southern India on Monday 4th August 1997. The cause of the blast, which occurred while most of the passengers on the Bangalore-bound train were asleep, was unclear. [1207]

An explosion occurred in a coach of South Eastern Railway No. 454 Down passenger train at Km.318/2-10 at Bhowra station, West Bengal, India, on the Adra Division of the railway, on Tuesday 15th September 1998. No further details are known. [1208]

On Thursday 22nd April 2004 South Korean media sources reported that two fuel trains had collided at Ryongchŏn in North Korea, killing or injuring 3,000 people (the accident was reported in some American newspapers as '3,000 die.'). The North Korean government declared a state of emergency and shortly afterwards cut telephone lines to the West in a vain attempt to prevent news of the accident reaching abroad. Residents of the Chinese city of Dandong, close to the North Korean border, who talked to relatives living in Ryongchŏn, nevertheless learnt of the disaster before telephone lines to China were also cut and the news was quickly leaked to the South Korean media. The Yonhap news agency in South Korea quoted a witness who stated, "The area around Ryongchŏn station has been turned into ruins as if it had been bombarded. Debris from the explosion soared high into the sky and drifted to Sinuiju."[1209] (Sinuiju is situated only about 2 kilometres from the Chinese border, Ryongchŏn being about 16 kilometres further south.)

Attempts at concealing the disaster having failed, the next day North Korea appealed to the United Nations for international aid and the Red Cross was allowed into the area, though no other outside agencies were let in. The Red Cross reported that 160 people had died in the explosion and 1,300 had been injured; official casualty figures were 54 killed and 1,249 cases of injury. A Chinese author stated in June 2004 that, 'At least 161 men and women were killed in the explosion, including 76 primary school students and one Chinese citizen. More than 1,300 were injured. Within a few kilometres of Ryongchŏn station, 1,850 homes were completely wrecked, 6,350 were partially collapsed. Twelve public facilities were completely destroyed, and ten were partially damaged.'

[1204] *The Bakersfield Californian*, Bakersfield, California, United States, 6th July 1994
[1205] *The Straits Times*, Singapore, 19th August 1994
[1206] *The Lethbridge Herald*, Lethbridge, Alberta, Canada, 17th August 1994
[1207] *The Guardian*, London, England, 5th August 1997
[1208] Government of India Ministry of Civil Aviation and Tourism Commission of Railway Safety Annual Report for 1998-99
[1209] *Daily Sitka Standard*, Sitka, Alaska, United States, 22nd April 2005

Locals were stated to have believed at first that the United States had carried out a nuclear attack on them, but listening to their radios could get no information regarding what had happened from the state media for several days. On Sunday 25th April, the anniversary of the Korean People's Revolutionary Army, news of the accident was made public in a roundabout fashion. The next day the *Rodong Sinmun*, the North Korean government newspaper, assured locals that government institutions would assist them. Meanwhile, the Korean Central News Agency (KCNA), the North Korean equivalent of Yonhap, published a story implying that portraits of North Korean leaders were considered more valuable than the lives of ordinary children. The story, possibly true, ran: "As a massive explosion struck the building of the Ryongchŏn elementary school and fire broke out in class, a teacher named Han Un Suk, 32 years old, evacuated portraits of President Kim Il Sung and Comrade Kim Jong Il to a safe place and then went to save seven children – and perished in the process." [1210]

Various theories were suggested to account for the explosion:

(1) The initial suggestion was that it had followed a collision between trains carrying petrol and liquefied petroleum gas, perhaps supplies sent from China to alleviate an ongoing fuel shortage.

(2) Diplomats and aid workers proposed that contact with a live electric cable had triggered the explosion of materials being carried in waggons during shunting operations, the materials being intended for use in canal construction. North Korean officials had apparently corroborated this theory to the Russian Itar-Tass news agency and the Japanese Kyodo news service.

(3) The Xinhua Chinese news agency reported that there had been a leak of ammonium nitrate, used both as a component of fertilizer and as a rocket fuel. The KCNA appeared to agree with this version, stating that the explosion had resulted, "because of an electrical contact due to carelessness, while waggons loaded with ammonium nitrate fertilizer were being shunted."

(4) A notable aspect of the explosion was that it had occurred at 12:10 on the very day that leader Kim Jong-il had passed through the station in an armoured train just a few hours earlier, returning from an official visit to China. Was this just a coincidence or had it been a failed assassination attempt? Had his train been involved in the explosion, it might have been obliterated. One objection to this theory was that his train had already passed through Ryongchŏn several hours before the explosion occurred, though this could have been because his would-be assassins had mistaken his train for an ordinary passenger working and so had left it alone. Kim Jong-il himself apparently believed that it had been an assassination attempt, triggered by the use of a mobile phone (cell phone), and the possession of such phones were subsequently banned in North Korea. The Supreme Leader also ordered the execution of several transport officials in the aftermath of the explosion, according to a defector. South Korean intelligence officials, on the other hand, believed that the explosion had been an accident. [1211]

[1210] https://www.nknews.org/2019/01/the-ryongchon-blast-why-one-of-north-koreas-worst-disasters-remains-a-mystery/

[1211] Wikipedia – Ryongchon disaster

CHAPTER 16

BOILER EXPLOSIONS

By 1991, steam locomotives had given way to diesel or electric traction just about everywhere and only three boiler explosions occurred in the next 15 years, one on a heritage railway in the United States and the others on sugar works railways in Cuba.

At about 7:20 p.m. on Friday 16th June 1995, the crownsheet of locomotive No. 1278 failed at Gardners, Pennsylvania, United States, as it was pulling a six-coach Gettysburg Passenger Services Inc. excursion train on the Gettysburg Railroad, a short-line heritage railway. The instantaneous release of steam into the engine cab seriously injured the engineer and two fireman, especially the former, who suffered third-degree burns over 65% of his body and had to be taken by air ambulance to a burns centre in Philadelphia. The firemen were taken to hospital by a road ambulance. None of the 310 train passengers or other train crew was injured. The engine sustained some damage, mostly to its firebox, grates and crownsheet. The accident was the subject of a full investigation by the National Transportation Safety Board (NTSB).

The NTSB report noted that on the day of the accident engine No. 1278 had already made two round trips between Gettysburg and Biglersville. The third and final trip of the day, known as the "dinner train," was to be from Gettysburg to Mount Holley Springs, where the passengers were to have two hours to enjoy meals at local restaurants before returning on the train to Gettysburg. From a point near Aspers (the Wolf Pit) the train had been assisted in the rear by a freight train.

A check valve between the feed water hear pump (feed pump) and the boiler had been leaking all day, despite the check valve having been recently repaired, and First Fireman Christopher Miller stated that as soon as the train had left Wolf Pit the feed pump had been turned off because the check valve had been throwing water onto the track, causing the driving wheels of the engine to slip. Then, as soon as the train was moving, he maintained that the feed pump had been turned back on. Second Fireman Eddie Sharrah said, however, that the feed pump had still been turned off when he had relieved Fireman Miller at Gardners and that he had then turned it on "all the way." Fireman Sharrah said that the changeover had taken place at Pond Road, a mile short of Gardners. The firemen agreed that it was possible to tell by sight and sound that the feed pump was working, though the NTSB investigators pointed out that the feed pump could continue to move with little or no water flowing. Both firemen testified that they had frequently checked the water glass (water gauge)(sight glass) and that the level of water in the boiler had appeared to be normal. As the train was moving at 15 mph at Gardners its crownsheet had collapsed.

First Fireman Miller told investigators: "We got to the top of the grade and levelled off and...we had a normal water reading, had plenty of steam. I got up about 30 seconds before the crownsheet failed. I got up to put coal in the corners, because it's an automatic stoker, and it fans [the coal] so it won't hit the corners, and I decided to wait until we got across the crossing and up around the bend before we started up the next grade. Fireman Sharrah took the fireman's seat, and I decided against...putting coal in. So, I went to the door and waved to all the people at the crossing. And about 5 seconds later is when we had the accident. Well, [when] the crownsheet failed...it just sounded like a muffled .22 [rifle] pop. I instinctively turned towards the noise....I remember getting hit with—it just got dark because of all the soot and the smoke in the cab. And I remember feeling intense heat and thinking...I've got to get out of here. I jumped out and...yelled at [the engineer's wife] to call 911. And then I thought about Engineer James Cornell and I started back up just past Fireman Sharrah. I knew Fireman Sharrah was all right then, because I saw him. He was limping, but I knew he was all right. Then I went up to find Engineer Cornell and found him on the other side of the train, lying there. And he asked for his wife. So I ran back and got her. And she handed me the phone, and I finished the 911 call."

Second Fireman Sharrah testified: "Fireman Miller stepped down to fire the back corners of the firebox, and I told him to take a break and I'd take over firing, which, I guess, gave a time span of about 5 minutes from that point until the explosion. There was like an initial poof sound, but then there was like a second explosion, which is what jarred the fire doors open and dumped everything back into the cab. All the steam and a lot of the coal just blew back into the cab. We had the feed pump on and had the stoker on. I shut the stoker and the feed pump down. And at first, just not knowing where everything was coming from, I jumped forward in the cab between the boiler and the outside wall of the cab to try to get away from it. After about 10 or 15 seconds, I realized it wasn't getting any better, and that's when I came back to the seat. And you couldn't see anything as far as the doorway or anything. So, that's when I climbed up on the seat and jumped out the window. Almost immediately after the explosion, Fireman Miller went out the doorway. And to my knowledge...Engineer Cornell apparently stayed on until it stopped."

The helper engineer stated that Engineer Cornell had applied the brakes and the conductor had called: "Emergency! Stop! Stop! Stop!" over his radio.

Engineer Cornell was in hospital for six months and needed further treatment later. Fireman Miller suffered second- and third-degree burns over 10% of his body and took about a month and a half to recover. Fireman Sharrah sustained second- and third-degree burns and leg fractures and was in hospital for several weeks.

Analysing the accident, the NTSB stated that it had been immediately clear that there had been a failure of the crownsheet. This failure had resulted from a combination of factors related to boiler maintenance and steam locomotive operation, due to lack of training, knowledge and application.

The events leading to the failure had begun when Fireman Miller had turned off the feed pump at Wolf Pit, preventing water from entering the boiler. The feed pump had remained off for about 10 minutes until Fireman Sharrah had taken over, while the engine was working hard climbing the gradient and thus using more water than otherwise. Neither fireman had been concerned that the boiler pressure had fallen from 230 pounds per square inch (psi) to 175 psi, because they had assumed that the water glass would alert them if the level fell dangerously.

Engineer Cornell stated that he normally kept the water level in the boiler lowish to minimise the chance of water entering the cylinders and to get "good steaming," though this reduced the level of water above the crownsheet and hence the safety margin. As the engine began to ascend the gradient, the water would have run to the back of the boiler so the water glass should have shown a higher level than when the engine was on the level. Instead, it had remained showing only half full, unchanged from the start at Wolf Pit, but the engine crew had failed to notice the danger. With the engine working hard the water level in the boiler had fallen further and at some point the highest point of the crownsheet had been uncovered. The turning back on of the feed pump at Gardners had come too late to prevent the failure.

A progressive drop in the water glass level had not been seen by the firemen because of the restricted condition of the spindles leading to the water glass, due to a higher than normal water level (false head) and/or free-floating scale. Hard scale had blocked 75% to 85% of the water glass spindles. The lowest gauge cock had been totally blocked by deposits, the middle gauge cock had been half blocked, the upper gauge cock had been clear. Steam locomotive experts did not believe that such a large amount of scale could have accumulated between monthly boiler washings. Engineer Cornell and Fireman Miller acknowledged that their method of boiler washouts had not been thorough. The water glass had also not been properly illuminated, though both firemen had stated that they had had no problem in reading it.

The attempts at water treatment to prevent water with a high mineral content being used in the boiler appeared, in the words of the NTSB report, 'to be irregular, rather than part of a planned and researched policy.' No test results could be found to determine the effectiveness of the treatment.

In describing the boiler washout, Fireman Miller had stated that only four of the 29 washout plugs had been removed, making even the most conscientious effort to wash out the boiler of doubtful effectiveness. Due to a lack of knowledge and training, he had also failed to use the proper equipment.

Gauges showing the stoker-jet pressure, steam-heat pressure and feed pump pressure had been removed from the engine, making it impossible to determine if the feed pump pressure had been high enough to force water into the boiler.

The leaking check valve had also decreased the water supply to the boiler.

The engine crew had stated that they had not used the injector. Had they attempted to do, so it would not have worked properly anyway because it had the wrong type of steam valve disc.

None of the various railwaymen except the owner of the Gettysburg Railroad (Engineer Cornell's father) knew how to correctly blow down and verify the gauge glass.

There was no clear division of responsibility between the members of the engine crew, especially between the two firemen.

The NTSB reached 12 conclusions regarding the accident:

1. The explosion in the locomotive had resulted from a crownsheet failure caused by having too little water in the boiler.
2. Because the water-glass spindles had been restricted, the water glass had not been able to represent the water level in the boiler accurately.
3. Although the engineer had signed the Federal Railroad Administration's forms No. 1, certifying that the work had been done, the water-glass spindles and gauge cocks had not been cleaned on a monthly basis.
4. The water glass had not been illuminated as required.
5. Gettysburg Passenger Services, Inc., had not had a comprehensive water treatment programme.
6. The boiler washing procedure described by the fireman had been inadequate to ensure that the boiler was properly and thoroughly cleaned as required by Federal Railroad Administration regulations.
7. Because the feed-pump gauge had been missing, the train crew had had no reliable way of

determining whether feed-pump pressure was overcoming boiler pressure and delivering water to the boiler.

8. Because the wrong type of disc had been installed in the injector, it would have been difficult to use the injector to add water to the boiler.
9. The firemen had not known, because they had not been properly taught, how to blow down the water glass or test the gauge cocks.
10. There had been no clear division of responsibility among the members of the crew in the accident, particularly between the two firemen.
11. Gettysburg Passenger Services, Inc., management had not been aware of the Hours of Service Act.
12. Gettysburg Passenger Services, Inc., had had no effective formal training or certification programme, and its on-the-job training had been based on second- and third-hand expertise.

The NTSB determined that the probable cause of the firebox explosion on steam locomotive 1278 had been the failure of Gettysburg Passenger Services, Inc., management to ensure that the boiler and its appurtenances had been properly maintained and that the crew had been properly trained.

The NTSB report concluded with a series of recommendations.

To the Federal Railroad Administration:

(a) It should require that each operating steam locomotive have either a water column or a water glass in addition to the water glass and three gauge cocks already required.

(b) It should require steam-locomotive operators to have a documented water-treatment program.

(c) It should describe basic responsibilities and procedures for functions required by regulation, such as blowing down the water glass and washing the boiler.

(d) In cooperation with the Tourist Railway Association, Inc., it should promote awareness of and compliance with the Hours of Service Act.

(e) In cooperation with the National Board of Boiler and Pressure Vessel Inspectors and the Tourist Railway Association, Inc., it should explore the feasibility of requiring a progressive crown-stay feature in steam locomotives.

(f) In cooperation with the National Board of Boiler and Pressure Vessel Inspectors and the Tourist Railway Association, Inc., it should develop certification criteria and require that steam-locomotive operators and maintenance personnel be periodically certified to operate and/or maintain a steam locomotive.

(g) In cooperation with the National Board of Boiler and Pressure Vessel Inspectors and the Tourist Railway Association, Inc., it should update 49 Code of Federal Regulations Part 230 to take advantage of accepted practical modern boiler-inspection techniques and technologies, to minimise interpretation based on empirical experience, and to maximise the use of objective measurable standards.

To the National Board of Boiler and Pressure Vessel Inspectors:

(a) It should cooperate with the Federal Railroad Administration and the Tourist Railway Association, Inc., in exploring the feasibility of Federal regulations requiring progressive crown-stay failure features in steam locomotives.

(b) It should participate with the Federal Railroad Administration and the Tourist

Railway Association, Inc., in developing criteria to be used in the periodic certification of steam-locomotive operators and maintenance personnel.

(c) It should participate with the Federal Railroad Administration and the Tourist Railway Association, Inc., in updating 49 Code of Federal Regulations Part 230 to take advantage of accepted practical modern boiler-inspection techniques and technologies, to minimise interpretation based on empirical experience, and to maximise the use of objective measurable standards.

To the Tourist Railway Association, Inc.:

(a) In cooperation with the Federal Railroad Administration, it should promote awareness of and compliance with the Hours of Service Act.

(b) It should encourage its members who operate steam locomotives to cooperate with the Federal Railroad Administration and the National Board of Boiler and Pressure Vessel Inspectors in exploring the feasibility of Federal regulations requiring progressive crown-stay failure features in steam locomotives.

(c) It should encourage its members who operated steam locomotives to participate with the Federal Railroad Administration and the National Board of Boiler and Pressure Vessel Inspectors in developing criteria to be used in periodically certifying steam-locomotive operators and maintenance personnel.

(d) It should encourage its members who operated steam locomotives to participate with the Federal Railroad Administration and the National Board of Boiler and Pressure Vessel Inspectors in updating 49 Code of Federal Regulations Part 230 to take advantage of accepted practical modern boiler-inspection techniques and technologies, to minimise interpretation based on empirical experience, and to maximise the use of objective measurable standards. [1212] [1213]

Baldwin 2-8-0 locomotive No. 1382 exploded at the Augusto Cesar Sandino sugar mill (No. 105), Cuba, on Monday 25th February 2000. It was wrecked and its two engine crew were seriously and very possibly fatally injured.

A detailed report on the state of the locomotive after the accident stated that its entire boiler had been pushed back about a metre on the frame and tipped about 20 degrees to the right. All the frame attachments and fittings had sheared off and the rear of the frame was bent down and distorted. Parts of the valve gear and piping had been badly bent from the force of the explosion. A section of the boiler shell from the rear of the smokebox to the front of the steam dome was missing. It had broken into pieces during the explosion, and several of them scattered around the area. It appeared that the reinforced area around the steam dome might have stopped the rupture of the boiler toward the rear. Examination of the boiler shell near the steam dome revealed what appeared to be sheared edges and a clean-looking break. There was no evidence of wasting or corrosion. It was difficult to tell where the initial failure had occurred, but one likely place seemed to be at the joint with the front tube sheet. The other might have been at the bottom of the boiler, where years of untreated water and the resulting corrosion might have thinned the shell. The front tube sheet was in one piece, although badly bent and distorted. It had come to rest still attached to a piece of the boiler shell. A few rivets in a section of approximately 1/6th of the boiler circumference held the two together. The other 5/6ths of the rivets had pulled out around the circumference of the tube sheet. Some had remained in the boiler shell and some had left with the tube sheet. A few "safe-ends" were

[1212] *Gettysburg Times*, Gettysburg, Pennsylvania, United States, 19th June 1995
[1213] National Transport Safety Board report

visible in the tube sheet, revealing where tubes had been repaired using this method. The tubes were still in the remains of the boiler, hanging out and bent in all directions. Some were wrapped around the shell and pointing toward the rear of the loco. All had pulled cleanly out of the front tube sheet. The smokebox shell was still in place, due no doubt to the strength of the cylinder saddle attachment, and two large frame braces on the exterior. The force of the explosion had shattered the smokebox door. Two small pieces remained attached to the hinges. The rest was missing. The stack was also sheared off and broken approximately in half. There was no sign of the headlight, turbogenerator, or the front sand dome. All that remained of the rear sand dome was a section of the base.

A jammed safety valve was blamed for the explosion. [1214]

A further boiler explosion occurred in Cuba in mid-March 2002 at the Australia sugar mill (No. 303) involving Baldwin 2-8-0 locomotive No 1513. Five railwaymen were badly hurt, one of whom later died in hospital. Lack of water was given as the cause. [1215] [1216]

[1214] https://www.internationalsteam.co.uk/trains/cuba0007.htm
[1215] https://david-longman.com/Cuba_Matanzas.html
[1216] https://www.internationalsteam.co.uk/trains/cuba0201.htm

CHAPTER 17

VANDALISM

'Accidents' due to vandalism and sabotage have always been a problem for the railways. They are obviously very closely related, inasmuch as both represent deliberate attempts to cause trains to crash, as opposed to accidents caused through non-deliberate carelessness, negligence or purely accidentally. Here they will, however, be treated separately, as they are slightly different. This separation is perhaps a little artificial, and if the perpetrator of a given act is not caught or his or her motives not sufficiently obvious, it may be difficult to decide which it was, but it has nevertheless been felt that the distinction is worth making.

In an accident that perhaps came closer to sabotage than vandalism the *Colonial,* a National Railroad Passenger Corporation (Amtrak) passenger train from Boston, Massachusetts, to Newport News, Virginia, United States, had almost completed its journey when it was derailed on a set of facing points at 9 p.m. on Wednesday 12th August 1992. About 70 of its passengers were injured, none seriously. A CSX Transportation spokesman stated that the lock of a switch had been broken and the switch opened. The engine was derailed and overturned against the side of a cutting and the five leading coaches of the train also left the track but remained upright.

Passenger David Yancey, 20, from Newport News, said that the lights had flashed and then gone out. He and another passenger had held on to one another until the train had stopped rocking. "It was pretty damned scary," he told a reporter. "It was done, it was over, just like that," he added, snapping his fingers, "but it felt as if it went on for two years." [1217]

Two Coast Guardsmen who had been stationed in the area, Raymond Gary Bornman Jr., of Baltimore, Maryland, and Joseph Lee Loomis, of St Charles, Illinois, both aged 20, were subsequently charged and convicted of causing the derailment. They had been seen in the area immediately after the derailment. Bornman pleaded guilty and was sentenced to 16 years and eight months' imprisonment; Loomis was found guilty after a trial and received 17 years and six months. Loomis told the judge that he was extremely sorry, saying: "The last thing I ever wanted to do was hurt anyone. The whole time I thought I was incapable of a derailment. I guess I was wrong." [1218]

On Sunday 16th May 1993 local train N 6181 bound for Dortmund collided with three blocks of wood that had been placed on the track by vandals at Dahl, Germany, on the line between Lüdenscheid and Hagen. The control car of the train was derailed, fell down an embankment and collided with a parked car, the driver of which had just left it. Fifteen people

---

[1217] *Times-News*, Burlington, North Carolina, United States, 13th August 1992

[1218] *The Indiana Gazette*, Indiana, Pennsylvania, United States, 21st December 1993 were injured. [1219]

A piece of concrete five centimetres thick was placed on the track at Burgdorf, Germany, by an unknown person or persons and was struck by an express on Tuesday 1st June 1993, fortunately without any serious result. [1220]

An act of vandalism at Saint-Leu-d'Esserent, France, on Wednesday 1st December 1993 resulted in the deaths of a train driver and three passengers and injury to 23 other people, four being seriously hurt. As a Persan – Beaumont – Creil railcar reached a point about 400 metres south of Saint-Leu-d'Esserent station, travelling at just over 70 km/h, it collided with a cast iron expansion plate weighing 32 kg, was derailed, ran off the track for 170 metres and ended up foul of the opposite running line, where it was struck by a suburban empty stock train, which tore out the side of its leading car. The three passengers who lost their lives were secondary school students. The driver of the empty stock was slightly hurt.

When the cause of the derailment became apparent the police carried out an extensive investigation in the neighbourhood, which involved over 200 people. A boy aged 16½, who lived near the spot, was eventually identified as the culprit. He confessed that he had been walking along the track between the station and his school at Creil when he had come across the expansion plate blocking his path. He had then picked it up and put it on the track "to see what would happen when a train struck it." He could not really explain why he had done so and maintained that he had not expected his act to have serious consequences. His companions, who had known nothing of what he had done, said that he seemed very upset afterwards at what had happened. He had never been in trouble with the law before.

The teenage vandal subsequently found himself on trial at the Oise Assizes, charged with 'voluntarily causing the derailment of a train resulting in deaths and injuries'. Even though he had been a minor at the time of the act, he faced a possible sentence of 20 years' imprisonment. In the end he was sentenced to a term of seven years. [1221] [1222]

Subsequently, the Société nationale des chemins de fer français took the decision to fence off the area around the tracks in the Île-de-France, though it was another 10 years before this was done on the line in question. [1223]

A leaflet-scattering device exploded just before 07:00 on Monday 2nd May 1994 inside a carriage of a Metro train stopped at Parque del Este station in Caracas, Venezuela. Five passengers were slightly injured by the blast, which also damaged the carriage roof and blew out several windows. Leaflets found at the scene claimed responsibility for the apparent terrorist attack on behalf of the hitherto unknown "Urban Commando Union of the Popular Resistance Front." Later, the apparent terrorist bomb turned out to have been a dangerous prank by teenagers who had used high school laboratory equipment. Two middle-class high school students were soon arrested by the police and reportedly confessed to the 'bombing.' The teenagers had taken their "joke" to the length of scattering leaflets at the scene claiming the attack to have been in the name of the non-existent "Urban Commando Union of the

---

1219 *Deutsche Eisenbahn-Katastrophen* by Hans Joachim Ritzau, p. 165
1220 Ibid
1221 https://www.leparisien.fr/oise-60/un-precedent-tragique-06-03-2011-1344938.php
1222 https://www.humanite.fr/-/-/accident-de-saint-leu-un-acte-de-malveillance
1223 Wikipedia – Liste des accidents ferroviaires en France dans les années 1990

Popular Resistance Front." [1224]

A serious case of vandalism occurred near Branchton, Scotland, late in the evening of Saturday 25th June 1994. In an area prone to such anti-social behaviour, the 22:45 electric multiple unit (EMU) passenger train from Wemyss Bay to Glasgow Central collided with some concrete cable trough covers that had been placed on the single track. The train was derailed and its front collided with the pier of a road overbridge, killing Motorman Arthur McKee, 35, and passenger Alan Nichol, 21. Ironically, Nichol had chosen a seat immediately behind the motorman to reduce the risk of injury due to broken glass from windows smashed by stones thrown by vandals. There had only been the motorman, the guard and four passengers on the train.

ScotRail and Railtrack offered a reward of £10,000 for information leading to the arrest of those responsible, while a newspaper described the effect of the vandalism on those close to the two men who had died. 'Mr Arthur McKee, Snr., father of the dead driver, said: "I could not have had a better son and it hurts to think how needless his death was. We can't take in what has happened."

'Mr Nichol, a student, was to have been married later this summer to Miss Tricia Darrah, 22, who said: "I don't know how I'm going to face the future without him. I hope to God they catch the people who caused this before they can wreck anybody else's lives." [1225]

Another newspaper quoted Mr Steve Mannion, British Transport Police assistant chief constable. 'Mr Mannion said rail vandalism in Scotland had been decreasing, but the fatal crash was a setback to a long-running anti-vandalism campaign, which included visits to schools. "We will be redoubling our efforts and have started putting extra police patrols on lines," he said. As school holidays approach, police and rail officials appealed to the public to be vigilant and to report all trespassers on the railway.'

Teenagers Gary Dougan and Craig Houston, both aged 17 and both from nearby Greenock, were subsequently found guilty of culpable homicide and sentenced to 15 years' imprisonment. [1226] [1227]

South Central Railway train No. 2430, the *Rajdhani Express* from Hazrat Nizamuddin to Bangalore City, consisting of a diesel locomotive and 11 coaches, was slightly derailed at about 15:59 on Sunday 14th August 1994 between Garladinne and Anatapur stations, Andhra Pradesh, India. Only the leading pair of wheels of the leading bogie of the engine left the rails. No one was hurt and damage amounted only to an estimated 6,000 rupees but the Commission of Railway Safety nevertheless carried out an investigation. It determined that the derailment had resulted from the placement of a key on the outer rail 'by some urchins.' [1228]

Via Rail train No. 66 from Toronto, Ontario, to Montreal, Quebec, was passing through Brighton, Ontario, Canada, at about 6:20 p.m. on Sunday 20th November 1994 when it struck a metre-long length of rail, which ruptured a fuel tank and set several of its cars on fire. The leading first-class car was directly sprayed with burning fuel. The blaze spread to the second

1224 https://scholarworks.sjsu.edu/cgi/viewcontent.cgi?article=1144&context=mti_publications
1225 *The Daily Telegraph*, London, England, 28th June 1994
1226 *The Times*, London, England, 27th, 28th and 30th June, 2nd July and 23rd December 1994
1227 Wikipedia – Greenock rail crash
1228 Government of India Ministry of Civil Aviation and Tourism Commission of Railway Safety Annual Report for 1994-95

coach and passengers as far back as the sixth car – which was occupied by a wheelchair basketball team – saw flames outside of their windows.

Bob Bosada, 57, a Toronto lawyer who had been a passenger in the first-class car, told a reporter, "Heat was cracking the windows while we were still moving. When we stopped, the heat and smoke became overpowering…so we began breaking windows to get out." He had broken his hand in doing so.

A witness to the accident from outside was Richard Cook, 19, who had been leaving his house near the line when he heard a scraping sound and saw something speed through a level crossing. "You couldn't recognise it as a train because it was all on fire," he related. "It was going very fast. It kind of looked like a comet, just like on TV with a lot of sparks and flame."

Firemen extinguished the blaze at 9:15 p.m. By that time, six other trains bound for Ottawa, Toronto and Montral had been delayed.

About 44 of the 407 passengers on the train were taken to hospital and six were detained overnight. Some had suffered broken bones or burns, though none of the injuries was life-threatening. A number had hurt themselves jumping from the train before it stopped.

Police seized the length of rail and police dogs scoured the area, while a police helicopter hovered overhead. Brighton Provincial Police Sergeant Ron LeClair declared, "Chunks of rail don't just jump on the tracks by themselves," indicating that an accidental cause of the mishap had all been excluded.

At about 2:30 a.m. on the day of the incident, another Via Rail train had been partially derailed at Rimouski, Quebec, Canada, large concrete blocks having been jammed in a set of points. Police blamed young vandals for this incident, in which no one was hurt. The previous weekend a driver had managed to stop a Via Rail train before striking concrete blocks, which had been left on the track at Rimouski. Via spokeswoman Dianne Graham in Toronto stated, "Vandalism is something we have to deal with from time to time, though we're still speculating."

Meanwhile, Via spokesman Paul Raynor ruled out any connection between the Rimouski incidents and the Brighton fire, saying, "We do run 400 trains across the country 24 hours a day, seven days a week. It's not entirely out of the realm of possibility that you could have two incidents like this, however rare they might be, on the same day." [1229]

A Hamburg – Kiel railcar ran into 30 pieces of concrete placed on the line by unknown perpetrators at Elmshorn, Germany, on the evening of Wednesday 23rd November 1994 and its control car was derailed. No one was hurt in the accident on the line between Hamburg and Neumünster. [1230]

Albertslund, Denmark, a suburb of København, was the location of a derailment due to vandalism at an unspecified date in March 1995. An S-train suburban passenger working was thrown off the rails after striking some concrete slabs, which two boys had placed on the track. No one was apparently hurt. [1231]

On Sunday 11th June vandals placed a piece of reinforced concrete on the overhead line equipment at Göttingen, Germany. The windscreen of the cab of an oncoming train was

[1229] *The Lethbridge Herald*, Lethbridge, Alberta, Canada, 21st November 1994
[1230] *Deutsche Eisenbahn-Katastrophen* by Hans Joachim Ritzau, p. 167
[1231] https://www.toglyde.dk/filer/ulykker.pdf

shattered when it collided with part of the slab but the motorman was unhurt. [1232]

Two teenagers, aged 16 and 18, placed slabs of concrete on the track near Rathenow, Germany, on Thursday 12th November 1998. Caught a few days after their prank, which caused considerable damage to a passenger train, they reportedly stated that they had done so "for fun," a common motive for vandalism. [1233]

Lecce – Roma Eurostar train No. 9358 ran into concrete blocks placed on the line near Trinitapoli, Italy, between Foggia and Bari, on Thursday 8th April 1999, and was derailed. Its driver was slightly hurt. [1234]

Just after 21:30 on the evening of Thursday 29th April 1999, Eurostar No. 9424, a Salerno – Roma – Milano Pendolino, ran into some concrete blocks placed on the track near Castel Guelfo, Italy, three kilometres beyond Parma and between there and Fidenza. No one was hurt but the train was stopped in open country for some time before it proceeded slowly to Fidenza. After stopping there for several minutes, it proceeded to Milano but due to damage to the front of the power car it could not travel at over 50 km/h and thus reached Milano Central station 37 minutes late at 23:00. The perforated blocks had been placed alongside the line in connection with engineering work. Ferrovie dello Stato sources stated that it was unlikely that the blocks would have caused a derailment. It was presumed that the obstructions had been placed on the line by vandals. [1235]

At 10:40 on Tuesday 18th April 2000 passenger train L 4581 from Namur to Charleroi collided with some concrete blocks placed in a set of points at Aiseau-Presles, Belgium. It was thereby derailed, fouled the opposite running line and was struck by passenger train IR 3331 travelling in the opposite direction. Driver Antonio Mastronardi, operating one of the trains, was killed, and 21 people were injured, three seriously. [1236]

Three boys aged eight, nine and 13 subsequently admitted jamming small pieces of concrete in the points and then placing a metre-long block of concrete across the tracks. Chief Prosecutor Cedric Visart de Bocame said that the boys would have to appear in a special juvenile court but that as they were all minors, their parents were legally and financially responsible for their actions. He added that: "It was a game to start with…to derail a train…They probably didn't realise what the effect of such a thing would be." [1237]

Three boys, two aged 14 and one aged 13, had been fishing at a pond near Rockwood, Ontario, Canada, on Sunday 14th May 2000, when after about half an hour they became bored and decided that it would be fun to pile some debris onto nearby railway tracks and see it 'explode' when a train hit it. Shortly afterwards, a westbound Via Rail passenger train with 116 passengers and six crew arrived and despite an emergency brake application by its engineers could not avoid striking the metre-high obstruction. Fortunately, it was not derailed. About two hours later the teens built a second pile of debris and were joined by three more

---

[1232] *Deutsche Eisenbahn-Katastrophen* by Hans Joachim Ritzau, p. 168

[1233] https://taz.de/!1313762/

[1234] *Today's Railways* Europe No. 43

[1235] *La Stampa*, Torino, Italy, 30th April 1999

[1236] Internet web page: News on the SNCB/NMBS April 2000

[1237] http://www.ble.org/pr/news/headline041900b.html

teens. A westbound freight train then arrived but its engineer, having been warned of the earlier incident, was able to pull up his train in time.

In July, the original three teens appeared in court and pleaded guilty to mischief endangering life. Sentencing was reserved to September. The second group of teens was due to appear in court later. [1238]

A Dakota Minnesota and Eastern Railroad train carrying corn was diverted into a siding at Brookings, South Dakota, United States, and there collided with eight parked cars loaded with soybeans. Conductor Brad Davis, who had a wife and five children, was killed and Engineer Dennis Baum was seriously injured. It was clear that someone had deliberately reset the points for the siding and railroad president Kevin Schieffer called it a "cold blooded action," adding that a composite sketch of someone seen leaving the scene had been made. [1239]

Andrew Golz, aged about 21 at the time of the accident, from California, was later convicted of the "prank." Before sentencing at Sioux Falls he turned to the families of the two railwaymen and told them: "I'm very sorry for the pain I've caused you." He was sentenced to 25 years' imprisonment and ordered to pay over $2,800,000 to the families of the victims and the railroad. (Where was he expected to get such a sum from?) His family told the court that he had bipolar disorder and suffered from bouts of depression, which made him impulsive. He had been sent to live with an aunt in Brookings in July 2000 in the hope that getting him away from his friends would help, but a month later he had broken the lock of the switch (points) with a hammer, set them for the siding and placed at least one plastic garbage bag over the points indicator, so that the engineer of an approaching train could not see that it was showing Red. [1240]

Via Rail passenger train No. 15, from Halifax, Nova Scotia, to Montreal, Quebec, was derailed at about 14:24 on Thursday 12th April 2001 on Canadian National Railway tracks at a set of facing points at Stewiacke, Nova Scotia, Canada. The train consisted of two locomotives and 14 coaches and was carrying 132 passengers. As a result of the accident, nine of the coaches of the train were derailed when the third and following cars were diverted from the main line onto a siding and demolished a farm supply building. Twenty-two people on the train were injured and taken to hospitals in Truro and Halifax; nine of them were seriously hurt, five of whom were train staff. The four occupants of the building escaped unhurt. [1241]

A Stewiacke resident called the emergency services at 14:26. Nine fire engines, eight road ambulances, two support vehicles and an air ambulance attended the accident.

The accident was investigated by the Transportation Safety Board of Canada (TSB).

It was immediately clear after the accident that it had been caused by vandalism as the switch lock and chain of the points in question had been damaged. The siding had not been used for industrial purposes for some time and had been last used six months previously by maintenance of way personnel.[1242] Three youths were seen by residents to leave the scene shortly after the derailment and they were arrested the next day, though only one, a 13-year-old boy, was subsequently charged with committing mischief to endanger life and criminal negligence causing bodily harm.[1243] The case against the teen dragged on for some time but at the beginning of November 2002, to the great distress of his mother, he was sentenced to six

---

1238 http://www.toronto.globaltv.com/ont/news/stories/news-20000711-112938.html

1239 *The Herald Bulletin*, Anderson, Indiana, United States, 27th August 2000

1240 https://www.postbulletin.com/news/23-year-old-sentenced-for-derailment-that-killed-conductor

1241 *The Lethbridge Herald*, Lethbridge, Alberta, Canada, 13th April 2001

1242 Transportation Safety Board of Canada report

1243 *The Lethbridge Herald*, Lethbridge, Alberta, Canada, 14th and 16th April 2001 months in jail plus two years' probation.[1244] Unfortunately, this does not seem to have reformed him, for his family subsequently moved from Stewiacke to Cole Creek, Nova Scotia, where in July 2003 he and his half-brother were in court charged with torturing a cat and hanging it from a tree. [1245]

The TSB report noted that it was not their function to determine civil or criminal liability and they would concentrate on other aspects of the event related to safety. There had been no deficiencies in the condition or operation of the train or in the track or signalling. Dynamic forces acting on the unsecured facing points had caused them to move under the train, diverting part of it into the siding and leading to its derailment.

With regard to trespassing, the town of Stewiacke had grown up on either side of the railway, which provided a convenient short cut between its two parts. There was limited fencing and signage to deter people crossing the line and no comprehensive effort to deter trespassing.

The standard lock provided at the switch (points) was significantly easier for a vandal to break than a high-security lock, thus increasing the risk of tampering.

With the points unsecured but not having moved more than ¼ inch the protecting signal would remain showing clear, while if the points were moved after a train had passed the protecting signal, the signal would revert to Danger too late to warn the train crew. Equally, the switch target provided at the points themselves would remain showing a Green indication to a train if the points were merely unlocked and only slightly open.

The location of the wheelchair boxes in some of the coaches of the train had obscured the emergency hammers for breaking the windows, also the beds in some of the sleeper cars had not been properly secured.

There had been no problems with the emergency response to the accident.

The TSB noted that since the accident, measures had been taken to reduce trespass in the Stewiacke area such as school presentations and additional signage, while CN had removed the TU-29 switch involved and the associated redundant siding. At other locations, more high-security locks were being installed at switches.

On trains, the emergency hammers were being relocated and the locking of sleeper car beds improved. Other measures taken were an improvement in the emergency egress from the kitchen area, the storage of glassware and cutlery in that area was to be improved and employees were to be provided with portable radios and flashlights. [1246]

A Canadian National Railway freight train was derailed as it was entering a siding between Melville and Saskatoon, Saskatchewan, Canada, on Wednesday 8th August 2001. Two engines, four empty cars and one loaded with salt left the rails and the line was blocked for some hours. No one was hurt. Two children were found to have tampered with a set of points and were charged with mischief. [1247]

A Madrid – Cartagena Talgo passenger train was derailed near Tobarra, Spain, on Saturday 4th January 2003. The engine, generator car and leading carriage were derailed and overturned and two women passengers in that carriage were killed, while 28 people were injured, three

---

1244 *The Brandon Sun*, Brandon, Manitoba, Canada, 17th April and 7th December 2002

1245 Ibid, 12th July 2003

1246 Transportation Safety Board of Canada report

1247 http://www.trainscan.com/news/scan/s0109

seriously. It was found that vandals had placed concrete blocks on the line. [1248]

A three-car electric multiple unit (EMU) train from Beach station, Chennai, to Tirumala on the Mass Rapid Transit Section was derailed near Fort station, Tamil Nadu, India, at about 15:40 on Friday 23rd May 2003. The motorman saw that there was an obstruction on the line and applied the brakes, despite which two of the cars of the EMU left the rails after encountering a second obstruction 50 metres beyond the first. No one was reported hurt. Two pieces of broken rail had been jammed in the gap between the running rail and the guard rail. Divisional Manager M.S. Jayanath ruled out sabotage as a cause of the derailment, which he attributed instead to what he described as the actions of "miscreants," by which he presumably meant vandals. [1249]

Stones placed on the rails were suspected to have been the cause of the derailment of two coaches of the *Malopolska* express from Gdynia to Przemysl at Pruszków, Poland, in the western suburbs of Warszawa on Monday 26th May 2003. The train was travelling at 120 km/h at the time. Five passengers were injured, 700 metres of track were damaged and the line was blocked for several days. [1250]

On Saturday 29th November 2003 a Tranz Rail freight train, consisting of an engine and four waggons, which was bound for Tauranga, had not progressed very far from Hamilton, New Zealand, when it was derailed by a piece of steel jammed into some points, in what the railway company described as a successful and sophisticated act of vandalism. No one was hurt but Police Detective Sergeant Neville Ross commented that: "If the passenger train had come down before the goods train then that is pretty serious stuff; it has gone beyond a practical joke." Workmen were busy on the Sunday re-railing the train. [1251]

An overnight international train from Oslo Sentrum in Norway, which was due in Stockholm, Sweden, at 06:50 on Monday 5th September 2005, was stopped for almost five hours outside of Katrineholm, Sweden, after it had collided with some manhole covers the police believed had been intentionally placed on the tracks. The collision damaged the brakes of the train, making it unsafe to continue. The approximately 230 passengers were eventually evacuated at about 10:00. No-one was hurt. [1252]

On Saturday 17th September 2005 a Companhia Paulista de Trens Metropolitanos passenger train on line D, travelling from Luz in central São Paulo to Rio Grande da Serra collided with a paving stone, which vandals had placed on the track, in the town of Mauá, Brazil, was derailed and collided with two poles supporting the overhead line equipment, bringing the wires down. Three people were slightly hurt and buses were laid on in place of trains until the line could be re-opened. [1253]

[1248] *Today's Railways* Europe 87
[1249] http://hinduonnet.com/stories/2003052407900300.htm
[1250] *Today's Railways* Europe 91
[1251] *Nzoom.com* web page
[1252] http://www.aftenposten.no/english/local/article1108936.ece?service=print
[1253] http://www1.folha.uol.com.br/folha/cotidiano/ult95u113178.shtml

CHAPTER 18

SABOTAGE

Four distinct motives for sabotaging trains may be distinguished (after Robert B. Shaw)

(a) As an aid to intended robbery, either of the train passengers or to steal, for example bullion being carried by the train.

(b) To satisfy a personal grudge against the railway company, such as by an ex-employee who feels that he has been wrongly dismissed.

(c) To hinder the operation of the railway, as for example during a strike.

(d) As an act of guerrilla warfare or terrorism, often against an occupying enemy power. Sabotage against the railways has been a feature of many conflicts from the American Civil War onwards.

The period covered by this volume saw numerous of sabotage, including the planting of bombs in train carriages, in a good number of countries such as Russia, Turkey, Cambodia, India, Pakistan, and so on, often in connection with independence movements, most of the acts sabotage during this period, therefore, belonging to the fourth of the categories outlined above.

Early on the morning of Wednesday 6th February 1991 South Central Railway train No. 721 Up, the *Dakshin Express* from Hyderabad Deccan to Hazrat Nizamuddin was picking up speed after leaving Kagaznagar station, Andhra Pradesh, India, between Vempalli and Sirpur Town, when it was derailed at 03:30. A bomb blast either as the train was passing or sometime earlier had created a pit almost a metre wide and damaged half a kilometre of track in total. Fourteen coaches were derailed and A. Raghunath, a salesman for Nagarjuna Fertilizers Limited; and a man named Rehman, described as a railway employee (travelling as a passenger?) were killed outright. Of the 900 passengers on the train another 18 were wounded.

Police found detonators and other explosive materials, bags and the wires of a bomb near the site of the derailment. Dogs from Hyderabad traced the presumed perpetrators to a nearby forest. It was thought that they might have employed a time bomb or a remote-controlled device. Left-wing Naxalite activists of the People's War Group were suspected of having been behind the attack. [1254]

A bomb ripped through the rear carriage of a passenger train from Moscow to Baku, the capital of Azerbaijan, on Thursday 30th May 1991. Eleven passengers were killed and 22 were injured. The bombing occurred near Khasavyurt station, USSR, in the Soviet autonomous

---

[1254] *Indian Express*, Madras, Tamil Nadu, India, 7th February 1991 region of Checheno-Ingus, situated north of Azerbaijan.[1255] There had been previous attacks on trains in the region. Azerbaijan was locked in a conflict with neighbouring Armenia over the region of Nagorno-Karabakh, a region of Azerbaijan mainly populated by ethnic Armenians. [1256]

An official stated, "As it was a passenger train and not a freight train with an explosive or dangerous cargo, there is no doubt that it was a terrorist attack." Soviet television announced that the KGB was investigating the attack. [1257]

Two months after the above attack, a bomb apparently exploded on another Soviet passenger train near Temirgau station, Dagestan, USSR,[1258] early on Wednesday 31st July 1991. This second attack happened about 15 kilometres from the Caspian Sea community of Makhachkala. One coach of the train was derailed and 16 people died and 20 were injured. Spokesman Oleg Seyenkovsky stated that the explosion had been "an act of subversion," while at the same time admitting that the cause of the explosion was not immediately known. The national and Azerbaijani KGB were investigating. [1259]

South Eastern Railway train No. 2815, the Down *Neelachal Express*, was derailed between Haridaspur and Jenapur stations, Orissa, India, on Monday 24th June 1991. There was no loss of life but two people were wounded, one badly. Damage caused was estimated at 450,000 rupees. The Commission of Railway Safety determined that two pairs of fishplates had been completely opened by 'miscreants.' [1260]

Northeast Frontier Railway train No. 2456, the Up *Brahmaputra Mail*, was derailed between Ghagrapur and Rangiya stations, Assam, India, on the broad-gauge section of the Alipurduar Division of the railway at about 14:40 on Tuesday 6th August 1991. Two people died and four were injured. The Commission of Railway Safety reported that the derailment had resulted from: 'Removal of fishplate of the left-hand rail joint by the miscreants.' [1261]

Shortly after it had started from Chaman, Pakistan, for Quetta, the capital of Baluchistan province, on Monday 2nd September 1991, a bomb exploded on a passenger train, killing at least eight people and injuring 27. No one claimed responsibility but previous similar attacks had been blamed by the Pakistani authorities on Soviet agents in neighbouring Afghanistan, which was occupied by the Russians at the time. [1262]

Sikh militants, fighting for an independent homeland, derailed a train consisting of an engine and two coaches, which was on a routine patrol to search for bombs on the track on Sunday 6th October 1991. The attack took place as the train was crossing a bridge at Kallichan, Punjab, India, 35 kilometres east of Amritsar. The explosion of the remotely

[1255] https://www.azerbaycan24.com/en/armenia-s-terrorist-acts-against-azerbaijan-facts/

[1256] *The Orange County Register*, Irvine, California, United States, 31st May 1991

[1257] *The News Herald*, Panama City, Florida, United States, 1st June 1991

[1258] https://www.azerbaycan24.com/en/armenia-s-terrorist-acts-against-azerbaijan-facts/

[1259] *The Transcript*, North Adams, Massachusetts, United States, 31st July 1991

[1260] Government of India Ministry of Civil Aviation and Tourism Commission of Railway Safety Annual Report for 1991-92

[1261] Ibid

[1262] *The Lethbridge Herald*, Lethbridge, Alberta, Canada, 3rd September 1991 controlled bomb separated the engine from its coaches and although the engine remained on the track, the coaches fell 10 metres to a dry river bed. The attackers then opened fire on them. Of the ten security people on the train, four Punjab state policemen, two railway policemen and two home guard militiamen were killed and two home guard militiamen were wounded. [1263]

A bomb exploded in a coach of a Bombay Victoria Terminus (now Mumbai CST) to Ambarnath local train in the station at Kalyan, Maharashtra, India, on the evening of Friday 8th November 1991. Eight passengers died in the explosion and another 65 were wounded. A man named Ravinder Singh, alias Bittu, was later convicted of planting the bomb and sentenced to life imprisonment. He was released on parole in 2007 and then absconded, being recaptured in 2016. [1264]

Separatist rebels from the Karen minority blew up the Moulmein – Yangon railway line at an unstated location in Myanmar on about Saturday 12th October 1991, derailing a train, killing two passengers and wounding five others. [1265]

An explosion tore up a portion of the track of the main line between Belfast and Londonderry near Eglinton, Northern Ireland, on the evening of Thursday 23rd January 1992, shortly before the arrival of a passenger train. The driver of the train was able to reduce speed but could not pull up in time to prevent the derailment of its leading carriage. None of the 32 train passengers or its crew was hurt. [1266]

Two time bombs exploded inside a freight train stopped in the suburban Athens train station of Ath Rendis, Greece, on Wednesday 25th March 1992. The blasts, which destroyed two automobiles in a shipment of German-made BMWs on the train, had been preceded by an anonymous warning call to the station. The Greek ultra-leftist terrorist group November Seventeen and the affiliated "Popular Uprising" cell had claimed responsibility for several attacks against German corporate targets in the previous 12 months. [1267]

At least seven people were killed and 26 were wounded when bandits attacked an express train near Hyderabad in southern Pakistan on about Friday 17th April 1992, according to police reports. [1268]

Central Railway train No. 8033 Down, the Ahnedabad – Howrah express, was derailed at 04:05 on Thursday 9th July 1992 between Dhamangaon and Talni stations, Maharashtra, India, on the Badnera – Wardha section of the Nagpur division. Eleven coaches were derailed, five of which capsized; 15 people lost their lives and 28 were injured, ten seriously. The injured were taken to Sevagram Medical hospital in Wardha and Yavatmal Government hospital.

[1263] *The Southern Illinoisian*, Carbondale, Illinois, United States, 7th October 1991
[1264] https://www.mid-day.com/news/india-news/article/1991-Kalyan-blast-convict--who-jumped-parole-in-2007--held-in-Punjab-17536190
[1265] *The Daily Telegraph*, London, England, 13th November 1991
[1266] *The Daily Gleaner*, Kingston, Jamaica, 25th January 1992
[1267] https://scholarworks.sjsu.edu/cgi/viewcontent.cgi?article=1144&context=mti_publications
[1268] *The Independent*, London, England, 28th April 1991

Doctors were taken to the scene on a light engine from Dhamangaon and Army jawans (soldiers) from Pulgaon also rendered assistance. [1269]

The accident was investigated by the Commission of Railway Safety, whose report included the testimony of a good number of surviving passengers.

Shri Amit Nandi, 19, and his sister Kum. Jamma Kumari, 21, said they had been travelling in the S2 coach. They had been awake from 03:30 onwards to the time of the accident, had heard a rattling noise and felt the coach riding unsteadily until it had capsized. They had also seen water coming from the tank in the coach, which had been pierced by a rail passing through it. Amit had escaped with bruises; his sister had suffered head and leg injuries. Both had been brought to Wardha, where they had been admitted to Sevagram hospital.

Shri Anant Soni, 34, related that he was from Raipur in Madhya Pradesh (since late 2000 in Chhattisgarh) and had been travelling with his wife and three children. Up to the derailment the journey had been uneventful. The coach they had been in had capsized, after which it had become dark inside. He had sustained a liver injury and fractures to his right hand and leg. His wife had escaped with slight injuries and his children had been unhurt. He praised the help given by a Dhamangaon voluntary organisation.

The stories of the other passengers were similar.

Driver P.D. Bendale and Assistant Driver Prabkaker Chabildar of train No. 7339 Down, a Dadar – Nagpur express, stated that they had left Dhamangaon at 02:54 (about an hour before train No. 8033) and reached Pulgaon at 03:12. The journey had been normal and they had seen no one near the line.

Driver A.G. Bezalwar and Assistant Driver B. Mangaroo testified that had been operating an "N" Box down special goods train from Badnera to Wardha, this having been the train immediately preceding train No. 8033. They had passed through Dhamangaon at 03:07 and Talni at 03:20. The train had ridden smoothly and they had observed no trespassers.

Driver K. Kaloo and Assistant Driver S.E. Hedaoo of an Ajni – Bhusaval Up Special goods train said that they had passed through Pulgaon at 02:15, Talni at 02:30 and Dhamangaon at 02:45. The journey had been uneventful and they had seen no one near the line.

Driver R. Gammai and Assistant Driver P.K. Vatts stated that they had operated Up passenger train No. 1398, which had passed Talni at 02:54 and Dhamangaon at 03:12. They had felt no jerks and seen no trespassers.

Guards D.S. Mohare, D.M. Shrivas and A.X. Sayyad, of trains 1398 Up, the Ajni goods train and the "N" Box special respectively, corroborated the accounts of their enginemen.

Driver Waman Ramraoji of train No. 8033 stated that 5 or 6 minutes after leaving Dhamangaon he had felt a severe jerk at kilometre 714/22. Sensing danger, he had applied the brakes in emergency and the train had stopped at kilometre 715/2. Instructing Assistant Driver Bhalchandra Shyamlal to protect the Up line, which had been fouled by derailed coaches, he then had tried to use his portable telephone but two emergency sockets had been damaged. He and Guard D.N. Vaidya had then sent spare Guard Jafarali Sahebali to the nearest level crossing to raise the alarm. He had found two pairs of dismantled fishplates at Km 715/0 on the Up line, plus an old tie-bar and some loose keys of steel sleepers. He and the guard had subsequently found fishplates lying on the track and several elastic rail clips at Km 714/22, the point of the initial derailment. He put the speed of his train at the time of the accident at 85 km/h (the Hasler track recorder indicated 100). He had not seen anyone about. He attributed the derailment to sabotage.

[1269] www.parliamentofindia.nic.in/1sdeb/1s10/ses4/12140792.htm

Assistant Driver Shyamlal spoke of protecting the line with detonators and stopping a light engine bringing doctors to the scene at about 05:55. He had not seen any trespassers.

Guard Vaidya described a severe jerk and the vacuum dropping to zero when the train left the rails. Finding that the emergency phone sockets were damaged, he had sent the spare guard to the nearest level crossing. With his driver he had found missing track fittings, suggesting sabotage. He had seen no outsiders near the track.

Assistant Guard P.M. Patil stated that he had protected the Down line. He had seen no one.

Guard Sahebali stated that a violent jerk had announced the derailment. Finding that it was not possible to use the portable telephone, he had run towards Dhamangaon to engineering crossing No. 76/C at Km 713/2, about two kilometres distant, getting there in 12 minutes. From the crossing he had informed the stationmaster at Dhamangaon at 04:35, before returning to the accident site with the crossing keeper and some gangmen. Like the driver and Guard Vaidya, he had observed missing fishplates.

Station Superintendent V.Z. Alone at Pulgaon stated that upon learning of the derailment he had proceeded to the scene on a light engine with two doctors, some gangmen and a pointsman.

Assistant Stationmaster S.N. Sharma spoke of being informed of the derailment at 04:40 from the engineering crossing, whereupon he had informed Control and organised the relief arrangements from his end. He had had no previous reports of interference with the permanent way or theft of fittings.

Switchmen Chotelal Dhanna and Kaitlash Yadavrao, on duty at Talni on the 8th and 9th of July and Pointsman Bhimrao Kisan at Talni were questioned as to the interfering with the siphon pipes in Talni yard, which had delayed the departure of NGP Up special goods train from Talni for 40 minutes. Driver B.M. Deshmuk, Assistant Driver B. Fakhira and Guard M. Alexander of this train denied seeing any outsiders but Pointsman Kisan stated that he had asked Control for security protection after two miscreants had threatened him with a knife.

Telecommunications Inspector S.K. Deb at Wardha stated that the damage to the emergency telephone sockets had been fresh.

Carriage Foreman E.R. Rajhans at Bhusaval stated that the rolling stock of the accident train had been examined and was in good condition.

Carriage Depot Officer B.P. Sharma at Ahmedabad, Chief Train Examiners Fakir Chand and M.P. Sharma, and Head Train Examiners Prakash Manlik and Antilkumar Nigam detailed their examinations of the rolling stock of the accident train.

Gateman Adinath Shivram at engineering crossing No. 76 described being informed of the accident by Guard Jafarali (here spelt Zafarali in the report, in which both spellings are used at different points) and proceeding to the site with gangmen and water for the passengers.

Gangmates Nama Uahav and Samrao Babya stated that there had been no problems with the track since its renewal in 1987-88, nor any theft of track fastenings. Their tool box had been locked and intact after the accident. Two experienced gangmen could, they thought, dismantle the fishplates and fastenings in about 15 to 20 minutes.

Permanent Way Mistry (foreman/supervisor) Hari Vithoba described being assaulted by a former gangmate named Roopchand Sitharam. He was not aware of any criminal activity relating to the track in the previous three years.

Permanent Way Inspector (III) S.B. Gautam at Pulgaon agreed that there had been no problems with track maintenance since 1987/88. He considered that tampering with the elastic track fastenings by the saboteurs could have caused the long welded rail to penetrate several coaches of the accident train, as it had done. He estimated that two experienced gangmen could have sabotaged the track in 7 to 8 minutes.

Permanent Way Inspector V.K. Verma at Pulgaon stated that he had arrived at the accident scene on the light engine and related what he found. He described the track maintenance and stated that there had been no problems with it. He considered that the rail must have been freed by the removal of fastenings to cause it to enter the coaches. He put the time required for two experienced gangmen to commit the sabotage at 10 to 15 minutes.

Railway Protection Force Sub-Inspector Madan Gorakhnah detailed the thefts of permanent way material, which had occurred in the area under his jurisdiction during 1992, most of which had not been in the Dhamangaon – Pulgaon section.

Assistant Engineer S.K. Garg testified that he had been the first "officer" to arrive at the accident scene, with the breakdown train, at 06:54, when he had found the damage much as described by the driver and guard. He agreed that a loosened rail could have penetrated the coaches. The track maintenance had not posed any problem. He did not think that the assault on the permanent way mistry was evidence of that gangman's involvement in sabotage.

Civil Engineer (South) S.C. Sharma concurred with the comments made by the permanent way inspectors. He considered that 4 to 6 men could have carried out the sabotage in 12 to 15 minutes.

Assistant Divisional Medical Officer Dr Maniratan Chakraborty stated that he had been a passenger on the train. He had given what medical aid he could with the limited resources available, pending the arrival of the first doctors from Pulgaon. He suggested possible improvements in the first aid arrangements.

Chief Engineer S.S. Ramteke stated that the piercing of five coaches of the train by a rail demonstrated that all the elastic rail clips over a length of at least 130 metres must have been removed.

Divisional Railway Manager R.K. Palham described the damage to the track as related by previous witnesses. He concluded that the derailment must have resulted from tampering with the track by miscreants.

Commissioner of Railway Safety Dr M. Mani detailed the tests conducted and observations made after the accident.

Discussing the accident the commissioner fixed the time of the derailment at 03:59.

The speed of the train, as per the Hasler recorder, had been 100 km/h.

At least seven possible causes could have led to the derailment:

(1) Track subsidence due to rain, washaways and so on (a 'convulsion of nature', as the report termed it). The weather had been fine and no flooding or rain had preceded the accident.

(2) An obstruction on the line. None had been seen by the engine crew and there was no evidence of any.

(3) Excessive speed and/or traction shock due to sudden braking. The speed of the train had been within the permitted range and the vigilance device on the engine had been operative. The effect even of sudden braking on the 1 in 330 rising gradient at the accident site would have been minimal.

(4) An engine defect. There was no evidence of anything having been wrong with the engine.

(5) A defective vehicle. None of the undamaged coaches had shown any defects. Pre-accident examinations of the train had revealed no problems and no passengers had reported rough riding prior to the derailment.

(6) Track defects. The track leading up to the accident point was in a satisfactory condition.

(7) Sabotage. The evidence of various witnesses indicated that there was overwhelming evidence of serious tampering with both the Up and Down lines as well as with the emergency phone sockets.

The saboteurs would have had a clear 40 minutes to damage the Down line and almost 59 minutes to interfere with the Up line; ample time for them to carry out the task.

No specific motive for the sabotage was evident. There were no valuables being carried on the train; it was not carrying any VIPs and there had been no attempt at theft after the accident. No terrorist group had claimed responsibility. As both tracks had been damaged it was fortunate that train No. 2042, the Up *Najivan Express* from Madras to Ahmedabad, had been running 45 minutes late that night.

The Government Railway Police agreed that it was a case of sabotage and were investigating the affair.

Trials on developing an anti-pilferage device for elastic rail fastenings had so far proved unsuccessful. They should be vigorously continued.

Two travelling ticket examiners and a conductor riding on the train had failed to provide assistance after the accident (a third ticket examiner had been injured).

The public information system at Nagpur had failed to keep the public informed about the accident.

The guard of the train had lacked a portable telephone.

Railway Medical Doctor Chakraborty suggested the equipping of trains with two powerful searchlights, the provision of first-aid boxes to both the driver and guard and the training of coach attendants in first-aid techniques.

Commissioner Mani concluded that the derailment had been due to discontinuity of the track caused by the removal of fishplated joint fixtures on the buffer rails as well as the elastic fastenings on a portion of the long welded track, due to sabotage.

Primary responsibility for the accident rested with the unknown saboteurs. Travelling Ticket Examiners D.D. Lange and P.R. Patil and Conductor D.N. Sole, who had failed to render assistance after the derailment, and Deputy Coaching Controller M.V. Ghodeshwar, responsible for the failure to update the public about the accident, should be suitably disciplined by the Railway Administration.

Regarding remarks and recommendations, the commissioner noted that the replacement of all buffer rails on the Central Railway system by converting shorter into longer welded rails was, in his words, a 'stupendous task' but the following course of action needed to be taken on a war footing:

(a) Vulnerable areas should be identified and their buffer rails eliminated as a priority.

(b) After oiling and greasing, two fish bolts should be burred on all buffer rail joints.

(c) Joints should be eliminated from long welded rails, including machined joints behind switch expansion joints.

(d) Security patrolling should be introduced at vulnerable points, in consultation with the railway police.

(e) An effective system for monitoring the maintenance of buffer rails should be evolved.

The commissioner made several other recommendations:

(1) The guards of long-distance trains should be provided with portable telephones by all means.

(2) The work of doctors at accident sites, especially at nights, would be improved by the use of portable searchlights, the provision of first-aid boxes to the train crew and coach attendants, the training of coach attendants in first-aid, and ticket examiners and coach attendants obtaining the names of railway doctors and officials travelling on the train at the start of the journey.

(3) Given the large number of elastic rail clips on the main lines, research efforts into anti-sabotage devices should be seriously revived.

(4) Long-distance trains should have nominated coaches, so that the rake integrity was maintained, while an integrated system of carriage maintenance should be speedily implemented at Ahmedabad.

(5) The large-scale absenteeism of permanent way noticed during the inquiry should be set right. [1270]

Partiya Karkerên Kurdistanê (Kurdistan Workers Party, PKK) guerrillas derailed a passenger train travelling from Mus to the eastern city of Elazig, Turkey, on Tuesday 1st September 1992. Four passengers were seriously hurt. [1271]

Shi'ite rebels fighting the Iraqi government of Saddam Hussein reportedly attacked and destroyed a train carrying Republican Guards at an unstated location south of Baghdad, Iraq, killing a number of the guards, according to the state-run Tehran radio. The report added that the train had been on a special mission taking troops from the Iraqi capital to combat rebels in the south. The date of the alleged attack was not given.

The official Iraqi news agency stated on 4th October 1992 that a high-ranking government official had denied reports of any "disturbances" in the south of the country, while making no mention of the Iranian radio report. [1272]

As a train was standing in the station at Dairut, Egypt, on Tuesday 6th October 1992 a suspected Islamic extremist prepared to throw a bomb from one of its bathrooms at people standing on the platform. His plans were frustrated when the device went off prematurely, killing him, a father and his three-year-old son. Ten other people were wounded, two of them seriously, including the train conductor, who suffered a fractured skull. [1273] [1274]

Two time bombs exploded within ten minutes of one another on a passenger train as it was travelling between Patiladaha and Sarbhog, Assam, India, on Tuesday 13th October 1992, killing 25 passengers and wounding 50 or more. Police Chief S.V. Subramanian would not speculate as to the identity of the bombers, beyond pointing out that the train had been passing through an area where militants from the ethnic Bodo tribe were known to be active. [1275]

1270 Government of India Ministry of Civil Aviation and Tourism (Commission of Railway Safety) Railway Accident Investigation Report on Derailment of 8033 Down Ahmedabad – Howrah Express at Km 714/20-22 between Dhamangaon and Talni Stations on the Badnera – Wardha Electrified Double Line Section of Nagpur Division of Central Railway at 03-59 hours on 09-07-1992

1271 https://scholarworks.sjsu.edu/cgi/viewcontent.cgi?article=1144&context=mti_publications

1272 *The Guardian*, London, England, 5th October 1992

1273 *The Lethbridge Herald*, Lethbridge, Alberta, Canada, 6th October 1992

1274 *Daily Herald*, Arlington Heights, Illinois, United States, 8th October 1992

1275 *The Janesville Gazette*, Janesville, Wisconsin, United States, 13th October 1992

Several warning messages were sent by the Irish Republican Army (IRA) to different radio stations in London, England, stating that a bomb had been planted on an East London railway line on Wednesday 21st October 1992 but: 'Forty minutes after the IRA's Dublin statement, the British Transport Police said no security alerts were under way. Twenty minutes later the bomb exploded, hitting the passing 4.22 Liverpool Street to Cheshunt train near Silver Street station.'

The IRA then accused the London police of ignoring the bomb warnings: 'British police have made no effort to act on authenticated warnings that our volunteers have left a bomb on the railway line. This latest incident underlines the reckless policy being pursued by British officials.'

Several London radio stations denied having received any warning. Scotland Yard refused to comment.

Later, Scotland Yard stated that they believed the bomb had been planted at night and that many trains had passed over it before it exploded.

There were two other bomb attacks later the same day, in one of which a bomb was planted on a railway line in North London. [1276]

Two railwaymen were killed and 27 people were injured on Sunday 25th October 1992 when a train struck a mine planted by Kurdish separatist rebels and fell into the Murat River between Tatvan and Elazig, Turkey. Security forces later clashed with rebels of the Kurdistan Workers Party (PKK) at the scene of the derailment and one soldier was killed in the engagement, according to the semi-official Anatolia news agency. [1277] [1278]

Four passengers were injured on Friday 18th December 1992 when a bomb wrapped in a leather bag exploded in a suburban train in Bombay, near the northern suburb of Borivali, Maharashtra, India. No one claimed responsibility for the blast but the explosion followed riots in the city, triggered by the destruction of a Muslim shrine in Ayodhya on 6th December by Hindu militants. [1279]

The New Mon State Party rebel group and anti-government students known as the All Burma Students' Democratic Front blew up a freight train travelling between Ye and Moulmein in southeastern Myanmar on Monday 8th February 1993, killing nine people and injuring 18 others. Some or all of the casualties were presumably passengers. [1280]

Georgian Interior Ministry spokesman Valiko Gogolashvili announced that at least three carriages of an international train from Yerevan, Armenia, to Tbilisi, Georgia, had been derailed in the Marneuli region of Georgia, on Friday 26th February 1993. He stated that according to initial reports no one had died but he could not give the number of injured. Railway and Interior Ministry officials stated that a remotely controlled device had been set off, damaging the track, just before the train had reached the spot. The sabotage cut Armenia's only rail link with the outside world. It occurred in an area where there had previously been

---

[1276] *The Independent*, London, England, 22nd October 1992
[1277] Ibid, 26th October 1992
[1278] https://scholarworks.sjsu.edu/cgi/viewcontent.cgi?article=1144&context=mti_publications
[1279] Ibid
[1280] Ibid

repeated attacks on a strategic gas pipeline. [1281]

Thirteen passengers died and another 12 were wounded as a bomb exploded on train No. 260, travelling from the resort of Kislovodsk, Russia, to Baku, Azerbaijan, on Sunday 28th February 1993. The explosion occurred as the train was leaving Gudermes station, 30 kilometres from Grozny, in the self-proclaimed republic of Chechnya, Russia.[1282] It was thought that the bombing might have been in retaliation for the Georgia sabotage two days earlier.

All the bogies (coaches) of a South Eastern Railway passenger train bound for Lohardagga were derailed at about 12:30 on Tuesday 20th April 1993 at Km 452/14, near Birgora Halt, between Tangarbansli and Nagjua stations, Bihar, India, on the narrow-gauge, non-electrified, single-line Ranchi – Lohardagga section of the Adra Division of the railway, due to the explosion of a charge of dynamite on the track. Fifteen people died in the attack and 18 were injured, nine seriously. The injured were taken to the nearby Holy Family Hospital and the Rajendera Hospital in Ranchi. [1283]

'Railway Minister C.K. Jaffer Sharief expressed grief at the mishap and asked the officials concerned to speed up relief measures and ex gratia payments to the persons affected.' [1284]

Suspected Khmer Rouge guerrillas attacked a train in the Mung Russei district of Battambang province, Cambodia, on the evening of Wednesday 5th May 1993. The attack was the latest of a series of terrorist attacks aimed at disrupting the first free elections to be held in the country for some decades. According to a United Nations service, at least 13 civilians were killed in the attack and dozens more were wounded. Senior state railway official Kim Hin in Phnom Penh gave different figures of 19 dead and 42 wounded. It was unclear if the deaths had occurred as a direct result of the detonation of the two mines, which had derailed the third and fourth carriages of the train, or in the guerrilla attack with B-40 rocket grenades and automatic rifles that followed. Kim Hin stated that the locomotive and seven coaches of the train had burnt out as a result of the attack, in which about 100 guerrillas had also robbed passengers of their valuables. [1285] [1286]

Shortly after 14:30 on Monday 2nd August 1993 an unidentified group, believed to belong to the Khmer Rouge, attacked a passenger train at a location about 20 kilometres east of the town of Kampot, in the province of the same name, Cambodia. At least 10 people were killed and about 30 were wounded, while the guerrillas stole about 100 new motorcycles, electronic equipment and passengers' luggage.[1287] The attack occurred in an area that was a known sanctuary for Khmer Rouge guerrillas.[1288] As in the May attack, the train had run over mines and then been raked by gunfire and shoulder-held rockets.[1289] The attackers reportedly included about 20 black-clad women comrades who waved pistols and AK-47 rifles and screamed death threats. [1290]

[1281] *Sun Journal*, New Bern, North Carolina, United States, 26th February 1993
[1282] *Pacific Stars and Stripes*, Tokyo, Japan, 2nd March 1993
[1283] http://alfa.nic.in/1sdeb/1s10/ses6/0421049301.htm
[1284] *Indian Express*, Madras, Tamil Nadu, India, 21st April 1993
[1285] *The Janesville Gazette*, Janesville, Wisconsin, United States, 6th May 1993
[1286] *The Daily Gleaner*, Kingston, Jamaica, 7th May 1993
[1287] *The Times*, London, England, 4th August 1993
[1288] *The Orange County Register*, Irvine, California, United States, 3rd August 1993
[1289] *Daily News-Record*, Harrisonburg, Virginia, United States, 3rd August 1993
[1290] *The Times*, London, England, 5th August 1993

Khmer Rouge guerrillas attacked a train travelling from Phnom Penh to the north-western city of Battambang, Cambodia, on the morning of Sunday 15th August 1993. Two people were killed and four were seriously wounded.[1291] About 30 attackers derailed the engine and four waggons with a mine at a point about 100 kilometres out, then attacked the train with hand grenades, B-40 shoulder-fired rockets, and AK-47 assault rifles. The attack lasted about 25 minutes before the guerrillas retreated upon the arrival of government soldiers. [1292]

During the evening rush hour, bombs exploded within minutes of each other in the carriages of commuter trains in Manila and in Pasay City, the Philippines, on Tuesday 11th May 1993, wounding 23 people. Police said the devices, which were grenades attached to timers, had damaged three train coaches and disrupted one of the capital's most popular modes of transport. Muslim extremists were suspected of having planted the bombs. [1293]

First reports stated that up to 100 people had been killed and a similar number wounded in an attack on a train near the town of Quipungo in Huila province, Angola, on Thursday 27th May 1993. The attack was believed to have been carried out by União Nacional para a Independência Total de Angola (UNITA) guerrillas. Two United Nations (UN) helicopters had been dispatched to the scene and had airlifted 40 survivors to the capital, Luanda. Peace talks between the Angolan government and UNITA, monitored by the UN, had collapsed two weeks previously. [1294]

More information on the attack emerged at the beginning of the following month in a statement from General João de Matos, chief of staff of the Angolan armed forces, who stated that, "It was a barbarous attack on a purely civilian train," adding that the death toll had been considerably higher than originally announced, about 300. He explained that the train in question had run daily between Matala and the port of Namibe (renamed Moçâmedes in 2015) since the end of the civil war in 1975. There had been 3,000 people on the train. An explosion had been detonated beneath its rear carriages, causing them to derail and overturned, crushing many passengers, and the guerrillas had then attacked the survivors. The engine and front carriages had remained on the rails and continued the journey.

UNITA countered that there had been many soldiers on the train, most of them dressed in civilian clothes, together with arms and ammunition to launch a new offensive in the south of the country. [1295]

A further attack on a train was made by UNITA exactly a month later on Sunday 27th June 1993, between the towns of Matala and Olivenca, Angola, again in Huila province. At least 19 people died in this attack and two others were seriously wounded. The victims had been railwaymen of the Caminho de Ferro de Moçâmedes company. [1296]

Late on Tuesday 13th July 1993 an explosion on a railway line 45 kilometres from Vladikavkaz, Russia, the capital of the republic of North Ossetia, derailed a freight train bound for Prokhladnoy. No one was injured, but authorities suspended rail traffic to clear the wreckage. Ossetian separatists were suspected. The Caucasus mountain region had erupted in fighting in 1991, when secessionists in Georgia's South Ossetia had tried to unify with

---

[1291] Author's personal notes, source mislaid.

[1292] https://scholarworks.sjsu.edu/cgi/viewcontent.cgi?article=1144&context=mti_publications

[1293] Ibid

[1294] *The Medicine Hat News*, Medicine Hat, Alberta, Canada, 31st May 1993

[1295] Ibid, 7th June 1993

[1296] https://scholarworks.sjsu.edu/cgi/viewcontent.cgi?article=1144&context=mti_publications

Russia's North Ossetia. A peacekeeping force led by Russia essentially ended the fighting in 1992, but the insurgents had not given up the war entirely. [1297]

Train No. 5609, a North Eastern Railway Up express, was derailed on Friday 13th August 1993 between Mankapur and Motiganj stations, Uttar Pradesh, India. Twelve people were wounded, four of them seriously, and damage estimated at 1,147,730 rupees was caused. The Commission of Railway Safety attributed the derailment to tampering of the track by 'miscreants.' [1298]

On Sunday 22nd August 1993 Muslim separatists campaigning for a separate Islamic state in Southern Thailand planted two mines on the track, which failed to detonate, beneath a passenger train from Bangkok to Sungai Kolok, at an unstated location in Narathiwat province. They then instead attacked the train with assault rifles, killing a female student and wounding eight others, three critically. A note left at the scene by them urged non-Muslims to stop abusing Muslims and demanded an independent Patani province. The Patani United Liberation Organisation claimed responsibility for the attack. [1299] [1300]

An act of sabotage at Osmangarh, Bangladesh, on Saturday 16th October 1993, was apparently one of those carried out with the intention of robbery. A gang of 50 bandits interfered with the track, with the result that five of its 11 carriages were derailed and about 120 metres of track were damaged. The bandits then robbed the passengers at gunpoint but most of them fled after police appeared, though one was captured. According to Bangladesh Railway officials none of the 400 passengers was hurt, though several newspapers, on the other hand, alleged that up to 200 had been injured. [1301] [1302]

A bomb exploded on a commuter train stopped in the station at Matunga, Maharashtra, India, in the centre of Bombay, at about 13:40 on Friday 29th October 1993. At least 15 people were wounded but no one was killed. No one claimed responsibility for the act. [1303]

Following the fall of the Soviet Union, a civil war had broken out in Tajikistan in May 1992, which would last until 1997. This conflict was the background to the derailment of a freight train between Kurgan Tyube in Tajikistan and Termez in neighbouring Uzbekistan on Thursday 25th November 1993. No one was hurt in the derailment, which had been carried out by anti-government forces who had infiltrated from Afghanistan and planted four mines to disrupt traffic on the important main line. [1304]

Bombs exploded on five trains in India on the night of Sunday 5th December (the first) and the morning of Monday 6th (the other four) 1993. Three of the explosions occurred on

---

[1297] Ibid

[1298] Government of India Ministry of Civil Aviation and Tourism Commission of Railway Safety Annual Report for 1993-94

[1299] *The Times*, London, England, 23rd August 1993

[1300] *Pacific Stars and Stripes*, Tokyo, Japan, 24th August 1993

[1301] *Herald Bulletin*, Anderson, Indiana, United States, 18th October 1993

[1302] *European Stars and Stripes*, Darmstadt, Germany, 18th October 1993

[1303] https://scholarworks.sjsu.edu/cgi/viewcontent.cgi?article=1144&context=mti_publications

[1304] *Syracuse Herald-Journal*, Syracuse, New York, United States, 26th November 1993 different *Rajdhani Express* services, a fourth on the *Andhra Pradesh Express* and the fifth on the *Flying Ranee Express*. The Sunday bombing occurred on train No. 2306, the Down *Rajdhani Express* bound for Howrah, at Karbigwan station, injuring two waiters in a pantry car. The second explosion occurred at 05:04 on train No. 2301, the Up *Rajdhani Express* from Howrah to New Delhi and the three other explosions followed in the next couple of hours. The most serious attack in terms of casualties was that on the *Andhra Pradesh Express* near Secunderabad, Andhra Pradesh, in which one or two people died. [1305]

A passenger from one of the trains described his experiences. 'Speaking to ENS {Etesh News Service} before he was whisked by a team of railway officials and doctors from the platform, Dileep Mehta, the back of his shirt stained with dried blood, recalled with a shudder how he "missed death by sheer luck." At exactly 6.12 a.m. the ill-fated C-7 coach of the train was enveloped in smoke and darkness with shrapnel from the explosive flying all around. The bomb, of medium intensity, was placed on a chair near the front vestibular section. "I was fast asleep," Mehta recounted. "The loud boom woke me up but within seconds I was losing consciousness. When I regained consciousness I found my shirt was soaking with blood. The first thing I did was change my shirt."

'For the next 30 minutes, panic gripped the railway staff on board the train. Frantic announcements were made for any doctor on the train to render aid to the injured. This is when Matilda Fernandes, a young nurse, sitting in coach marked H-A rushed towards C-7 and the immediate assistance provided by her probably saved the life of Raja Ram, who was bleeding profusely from head and neck. "I did whatever I could in that crucial moment. The man was in a bad shape," Matilda told ENS afterwards. A group of doctors which was on board but in another compartment also attended to the four injured persons.

'M.S. Bhardwaj, a retired ship captain, who was in the ill-fated coach with his wife and two small children said that immediately after the explosion the coach was filled with a pungent smelling gas. The blast created a small hole on the metallic floor of the coach. Bhardwaj opined that in all probability the explosive had been planted somewhere between Surat and Baroda. "Lots of people entered the coach and started piling their luggage on the chair cars."

'Bhardwaj as well as many other passengers of the C-7 coach complained that even after the blast the railway staff did not provide them with alternative accommodation in other coaches. "We shifted on our own," they said. The official relief team caught up with the victims at Amli station. At Mughal-Sarai the passengers were served tea and biscuits and provided {with a} telephone facility to call their relatives and family members. At New Delhi station officials set up an assistance booth in the morning. A team of four doctors was also stationed there to attend to the needs of passengers who travelled by these trains.'

No group immediately claimed responsibility for the explosions, though the timing, on the first anniversary of the destruction by Hindus of the Muslim mosque at Ayodhya, suggested religious extremists. Railway Minister Sharief stated, "The planting of these devices on selected prestigious trains seems to indicate a design to create a scare among the public." [1306]

A Red Nacional de los Ferrocarriles Españoles (RENFE) commuter train was derailed by sabotage at 06:20 on the morning of Wednesday 15th December 1993 at Sant Pol de Mar, Spain, near Barcelona. It was believed at the time that more than one saboteur must have been involved in the act. Between midnight and the time of the accident, in a quite sophisticated attempt at sabotage, the rail nearer the sea had been sawn through, about 50 nails securing the rail to the wooden sleepers had been removed and the rail had been pushed inwards. The break in the rail

[1305] *The Chronicle Telegram*, Elyria, Ohio, United States, 6th December 1993
[1306] *Indian Express*, Madras, Tamil Nadu, India, 7th December 1993

had been concealed with adhesive tape and a cable had been connected to the ends of the rail so as to maintain the electrical circuit and prevent a break in the track circuit, putting the protecting signals to Danger. Fortunately, the train was only travelling at low speed and despite the saboteur's attempts at concealment the train driver had realised at the last moment that something was not right and applied his brakes. None of the 160 passengers were injured in the actual derailment, though a man hurt himself after alighting when he fell into a stream in the darkness. Five of the six cars of the train left the track. The line was blocked all day, with passengers being bussed between Arenys de Mar and Pineda stations. [1307]

Less than three weeks after the Sant Pol de Mar derailment, another Red Nacional de los Ferrocarriles Españoles passenger train was the victim of an act of sabotage between Blanes and Margrat de Mar, Spain, at a bridge over the Tordera River. Again, the derailment occurred in the early morning, at 06:10 on Tuesday 4th January 1994, at a location only 15 kilometres from the site of the earlier attack. This time there were only 10 passengers and crew on the train, none of whom was hurt.

A man who had also murdered his father with a crossbow was found guilty of having caused the second derailment (and the first?) and was still in prison 20 years later. [1308] [1309]

Two bombs exploded on an express train near Hardwar, Uttar Pradesh, India, in the northern Himalayan foothills, early on Friday 21st January 1994. The train was en route from Allahabad to Dehra Dun in Uttar Pradesh state. Six people were injured. Police were reportedly searching for two youths who had got off the train and left luggage on board. [1310]

At about 01:30 on Saturday 19th February 1994 shots were fired at the sleeping car of a luxury Cairo – Luxor express as it was passing through Baqur, Egypt, just north of Abu Tig, the attackers, concealed in the darkness, firing from an adjoining cane field. The side of the car was riddled with bullets, some windows were shattered and four people were injured, none of them seriously. Eleven people were quickly arrested by the police. [1311]

On Wednesday 23rd February 1994 the shooting was followed by the explosion of a bomb on another passenger train, also near Abu Tig, Egypt. This time, five foreign tourists and five Egyptians were wounded. Responsibility for the attack was claimed by al-Gamaa al-Islamiya, an Islamic group that had been fighting for two years to overthrow President Mubarak's secular government and had warned foreign tourists and investors to leave the country or else be caught up in the violence. [1312]

20 Yanvar (20th January) station on the Baku underground, Azerbaijan, was the location of a suicide bombing at 13:00 on Wednesday 19th March 1994. A time bomb planted beneath a seat in the leading coach of a train exploded as it stopped at the station, killing 14 people, including the bomber, named as Oklay Gurbanov, and wounding 49 others. The leading car of

---

[1307] https://www.saul.cat/article/5345/reporter-gener-de-1994-els-autors-del-descarrilament-del-tren-a-sant-pol-eren-professional

[1308] https://www.servimedia.es/noticias/acto-sabotaje-posible-causa-descarrilamiento-tren-barcelona/1410846931

[1309] https://elpais.com/diario/1994/01/05/espana/757724418_850215.html

[1310] https://scholarworks.sjsu.edu/cgi/viewcontent.cgi?article=1144&context=mti_publications

[1311] *European Stars and Stripes*, Darmstadt, Germany, 20th February 1994

[1312] *The Telegraph*, Nashua, New Hampshire, United States, 24th February 1994 the train was destroyed and part of the station roof collapsed. [1313] [1314]

At 08:30 on Sunday 3rd July 1994 there was a second bomb attack on the Baku underground, this time with the use of a remotely controlled device, when a bomb exploded between 28 May and Ganjlik stations and about 500 metres short of the latter. Thirteen people died and 42 were wounded in this second attack. Hundreds of passengers were evacuated after the explosion and the line was shut for some hours. [1315]

Eleven people who were judged indirectly responsible for the first attack were sentenced to terms of imprisonment: two of them to life terms and the other nine to 15 years. The one person responsible for the second attack was also sentenced to life imprisonment. [1316]

A powerful bomb exploded beneath the rail tracks early on 30th March 1994, derailing a 15-carriage passenger train from Bangkok to the Malaysian border town of Sungai Kolo near Chana station, Songkhla province, Thailand, and injuring some passengers. The bomb appeared to have been homemade, but was sufficiently powerful to severely damage the tracks, the locomotive and two carriages. No one claimed responsibility, but the authorities blame "separatists." [1317]

A bomb killed three passengers on a train at an unstated location in southern Georgia on Saturday 9th April 1994. A 14-year-old girl was also seriously injured by the explosion. The train was travelling from the Georgian capital, Tbilisi, to Yerevan, the capital of Armenia. 'The area of Georgia where the explosion occurred housed many Azeri refugees,' the report on the attack added. [1318]

An explosion early on Wednesday 13th April 1994 ripped apart a passenger train carriage in the south Russian province of Dagestan, killing four persons. The blast occurred at Dagestanskiye Ogni station, 30 miles from the border with Azerbaijan. [1319]

A bomb exploded on a Delhi-bound express train just after it had left the station at Rajnandgaon, Madhya Pradesh, India, on Sunday 5th June 1994. Two people were killed and 25 injured. Maoist Naxalite guerrillas were believed responsible for the attack. [1320]

A passenger train that had left the capital Colombo on Friday evening for the north of the country had stopped at Anuradhapura station, Sri Lanka, at 04:00 on Saturday 9th July 1994, when a bomb exploded on it, killing a Tamil civilian and injuring 11 others. No group claimed responsibility for the attack but there were fears of a Tamil terrorist campaign beginning before the parliamentary elections on 16th August. [1321]

[1313] *Delaware County Sunday Times*, Upper Darby, Pennsylvania, United States, 20th March 1994
[1314] Wikipedia – 1994 Baku Metro bombings
[1315] *The Cedar Rapids Gazette*, Cedar Rapids, Iowa, United States, 4th July 1994
[1316] Wikipedia – 1994 Baku Metro bombings
[1317] https://scholarworks.sjsu.edu/cgi/viewcontent.cgi?article=1144&context=mti_publications
[1318] Author's personal notes – source mislaid.
[1319] https://scholarworks.sjsu.edu/cgi/viewcontent.cgi?article=1144&context=mti_publications
[1320] Ibid
[1321] *The Daily News*, Fort Walton Beach, Florida, United States, 10th July 1994

The derailment of a passenger train on the Trans-Siberian railway had probably been caused by a mine laid on the tracks, according to officials. Eight passenger carriages and a dining car of a train left the rails near Chita, Russia, about 5,600 kilometres south-east of Moscow, on the night of Thursday 21st July 1994, the *Itar-Tass* news agency stated. There was no mention of casualties. 'Gangsters' were blamed for the attack.[1322] [1323]

Nine people were killed and scores were injured when Khmer Rouge guerrillas attacked a train in southern Cambodia, taking more than 200 passengers captive, including three foreigners. The rebels detonated two mines, one under the engine of the train and the other under the rear carriage on Tuesday 26th July 1994, as the train passed through the Kompong Trach district of Kampot province, about 130 kilometres south of the capital, Phnom Penh.

Pich Kimsreag, director of the state railway, stated: "After the mine explosion, the Khmer Rouge started firing heavily at the train. They want to show that they are still strong enough, even though the government has declared them outlaws."

Five civilians, two soldiers and two train militia were killed in the ambush and many others were wounded. The guerrillas then robbed the passengers before marching more than 200 of them into the nearby jungle near Phnom Voar, the director added. Passengers from the Kampot area were later released but scores of others from Phnom Penh were still being held, as were the three foreigners. [1324]

Another report stated that 50 guerrillas had attacked the train, which had been heading for Sihanoukville, and that 13 people had been killed, five of them soldiers and three of them Vietnamese passengers, whom it was thought had been executed later.[1325] The attack was said to have been the most serious since Cambodia's rebirth 14 months earlier. [1326]

The three foreign hostages, Frenchman Jean-Michel Braquet, 27; Briton Mark Slater, 28; and Australian David Wilson, 29, were later also executed by the guerrillas. Their bodies were exhumed at the beginning of November. [1327]

South Central Railway Bijapur – Guntkal fast passenger train No. 290, consisting of a steam locomotive and six coaches, was derailed at kilometre 36/4-1 between Hole-Alur and Mallapur stations, Karnataka, India, at 10:40 on Saturday 13th August 1994. The third and fourth coaches left the rails and the fifth and sixth capsized. Six passengers were injured and the damage was estimated at 117,600 rupees. The Commission of Railway Safety determined that the derailment had been caused by a boundary rail post, which had been uprooted and placed inside the tracks by some 'miscreants.' [1328]

A small bomb exploded under the seat of a train that had arrived in Dublin Connolly, Eire, from Belfast on Monday 12th September 1994, slightly injuring two passengers. The Ulster Volunteer Force (UVF) claimed the attack and said that it had planted six other bombs, though a search turned up nothing. Northern Ireland Police had warned the authorities in the south the

---

1322 *The Guardian*, London, England, 23rd July 1994
1323 *The Daily Telegraph*, London, England, 23rd July 1994
1324 *The Daily Telegraph*, London, England, 28th and 29th July 1994
1325 *The Times*, London, England, 28th July 1994
1326 *The Independent*, London, England, 29th July 1994
1327 *Brandon Sun*, Brandon, Manitoba, Canada, 3rd November 1994
1328 Government of India Ministry of Civil Aviation and Tourism Commission of Railway Safety Annual Report for 1994-95

previous week to expect attacks by Protestant extremists in retaliation for the decision by Irish Prime Minister Albert Reynolds to meet with Sinn Féin, the political arm of the Irish Republican Army (IRA). [1329]

Northeast Frontier Railway train No. 4055, the Down Brahmaputra Mail, consisting of a WDM-2 Class diesel locomotive and 18 coaches, was derailed at 17:10 on Wednesday 12th October 1994 between Patiladaha and Bijni stations, Assam, India. The rear of the engine and 10 coaches left the rails and four of the coaches capsized. One person died and 12 were hurt, four seriously. Damage was estimated at 7,051,000 rupees. Unauthorised tampering with the track by persons unknown was given by the Commission of Railway Safety as the cause of the derailment. [1330]

At least six people were killed and 15 injured early on Friday 28th October 1994, when a time bomb exploded on a crowded train in the northwestern province of Battambang. The device, which detonated after the train had left for Sisophon, had been hidden in a vegetable basket. Khmer Rouge guerrillas, who had warned that people travelled on trains at their own risk, were blamed for the attack. [1331]

About 20 Khmer Rouge guerrillas attacked a passenger train on Monday 2nd January 1995 as it was travelling from Battambang province to Phnom Penh, the capital of Cambodia. Eight passengers were killed and 40 others were wounded in the attack in Kampong Chhang province, about 90 kilometres from Phnom Penh. Five mines planted on the track stopped and presumably derailed the train, which the guerrillas then ran towards, firing AK-47 rifles and grenade launchers. [1332]

Northeast Frontier Railway train No. 5908, the Down *Brahmaputra Mail*, was close to the Down Home signal of Nailalung station, Assam, India, at about 04:45 on Saturday 25th February 1995, when two bombs exploded in the second coach behind the engine. In total the train consisted of a YDM-2 diesel locomotive and 16 coaches. Twenty-six passengers died in the blast and 19 were injured, 10 of them seriously. [1333]

The target for the bombing had been soldiers returning from Manipur state, where they had been monitoring local elections. Assam State Police Chief K. Rudra told the press that, "There are definite clues that the Naga militants are involved in the incident. Some militants boarded the train and planted time bombs in the coach before escaping."

A newspaper report added, 'Rebels of the National Socialist Council of Nagaland, a Maoist group, have been fighting the Indian Army for more than 40 years.' [1334]

The Commission of Railway Safety made a number of recommendations in relation to the bombing:

[1329] https://scholarworks.sjsu.edu/cgi/viewcontent.cgi?article=1144&context=mti_publications
[1330] Government of India Ministry of Civil Aviation and Tourism Commission of Railway Safety Annual Report for 1994-1995
[1331] https://scholarworks.sjsu.edu/cgi/viewcontent.cgi?article=1144&context=mti_publications
[1332] *The Bakersfield Californian*, Bakersfield, California, United States, 4th January 1995
[1333] Government of India Ministry of Civil Aviation and Tourism Commission of Railway Safety Annual Report for 1994-1995
[1334] *The Sunday Review*, Towanda, Pennsylvania, United States, 26th February 1995

(a) The intelligence system of the Railway Protection Force needed improvement.

(b) A system of obtaining intelligence reports from the State Government Intelligence Wing and other intelligence wings such as that of the defence forces, the Union Home Ministry, the Border Security Force and so on needed to be evolved.

(c) There should be a review and augmentation of trained personnel and equipment with the Railway Protection Force to prevent damage to railway property.

(d) Ticket and luggage checking in unreserved as well as reserved compartments should be ensured.

(e) The carriage of explosive materials in railway compartments in violation of the Railway Rules should be prevented.

(f) Army/Central Reserve Police Force and other defence personnel and paramilitary forces should not be permitted to carry inflammable materials, including liquor and ammunition, in railway compartments without proper authorisation by the Railway.

(g) The Railway Protection Force should carry out an investigation in cases of suspected sabotage.

(h) Care should be taken that important clues were not destroyed or materials taken away by outsiders from the site of an accident.

(i) The recording of joint observations and the preservation of clues should be done as per the Railway's accident manual. [1335]

A bomb exploded on a Réseau Express Régional (RER) train as it was pulling into the Saint-Michel station, Paris, France, at 17:30 on Tuesday 25th July 1995. A British newspaper wrote: 'A witness on the train reported an ear-splitting explosion before the tunnel filled with smoke. People were lying on the ground and not moving, he said. "We arrived on the platform at Saint-Michel when I heard a loud explosion. Everything went black and the window of the carriage exploded. There was a smell of gunpowder. On the platform of Saint-Michel I saw a lot of smoke. There were some people lying on the ground; some were seriously injured." [1336]

A bomb weighing 6.5 pounds (0.5 kilograms) and equipped with a timing device had been hidden beneath a seat.

The Paris fire brigade brought its so-called Red Plan into effect after the explosion, the station was evacuated, and a helicopter landing pad was installed for air ambulances to take the most critically injured to hospital. A nearby café was turned into a rescue centre for the treatment of some of the less badly hurt. 'They straggled out of the subway station, some with tears streaming down their smoke-blackened faces, their clothes tattered and bloodstained,' a newspaper correspondent wrote. Four people were killed outright and 87 were taken to hospital, three of whom later succumbed to their wounds.

Two anti-terrorist investigators, François Battut and Patrick Lalande, hurried to the station. No group immediately claimed responsibility for the bombing and possible suspects included Algerian terrorists, Bosnian Serbs and groups linked to Carlos the Jackal. France supported the government in Algeria, its former colony, which was opposed by militant Islamic groups, while French aircraft had attacked Bosnian Serbian positions in retaliation for the death of two

[1335] Government of India Ministry of Civil Aviation and Tourism Commission of Railway Safety Annual Report for 1994-1995

[1336] *The Times*, London, England, 26th July 1995

French peacekeeper soldiers. [1337]

Another bomb exploded on a Réseau Express Régional train between Saint-Michel and Museum d'Orsay stations, Paris, France, at 07:05 on Tuesday 17th October 1995. This time, no one was killed but 29 people were wounded, five very seriously, some losing legs. The bomb consisted of hex nuts loaded into a gas cylinder, a trademark of Islamic Algerian terrorists, and this time the Armed Islamic Group claimed responsibility for the bombing and for seven previous attacks or attempted attacks in Paris, including that at Saint-Michel station in July.

Parisians were naturally frightened: "We're all a little bit traumatised right now. It's happened too many times. There's a psychosis among the population. No one knows what to do," said passenger Anne Guescoux. [1338]

Islamic extremists were blamed for the derailment of a train near Ksar el-Boukhari, Algeria, on Tuesday 8th August 1995.[1339] Eight people reportedly died and 40 were wounded. [1340]

On Tuesday 26th September 1995 a train was stopped by terrorists between Algiers and Blida, its passengers were ordered off and its coaches burned. Two days later, on 28th September, a bomb exploded near Algiers, Algeria, beneath an engine sent to clear the line, killing five people and injuring 11, presumably railwaymen. [1341]

Northern Railway train No. 3308, the Down Ganga Sutlej Express, had passed the Down Distant signal of Sirhind station, Punjab, India, on the evening of Sunday 27th August 1995, when at 22:58 its driver felt a sudden jerk and the vacuum dropped. It was found that the sixth to the 13th coaches had derailed. Three passengers died, eight were injured, two seriously, and damage amounted to 2,690,000 rupees. The Commission of Railway Safety found that unidentified persons had removed elastic rail clips from both sides of a rail over a length of about 40 metres. [1342]

At about 23:34 on Saturday 7th October 1995 the engine and six coaches of South Central Railway train No. 288 Down from Guntakal to Tirupati were derailed and capsized at Kilometre 86/8-9 between Nallacheruvu and Tanakallu stations, Andhra Pradesh, India. Four passengers were hurt, two seriously, and damage amounted to 348,000 rupees. Unauthorised persons had tampered with the track, the Commission of Railway Safety reported. [1343]

At about 1:30 a.m. on Monday 9th October 1995 an Amtrak passenger train from Miami, Florida, to Los Angeles, the *Sunset Limited*, was derailed as it crossed a bridge over a gully in the desert near Palo Verde, Arizona, United States, north-east of Hyder, on the Union Pacific Railroad line between Phoenix and Yuma. The whole train of two locomotives and 12 coaches left the rails and three sleeping cars and a dining car fell 30 ft into the dry bed of the creek. There were 248 passengers and 20 crew on the train, of whom one person was killed and about 78 were wounded, some 20 seriously. Those hurt included a 31-year-old woman on her honeymoon, who was reported to be in good condition in St Joseph's Hospital in Phoenix,

---

[1337] *European Stars and Stripes*, Darmstadt, Germany, 26th and 27th July 1995
[1338] Ibid, 19th and 20th October 1995
[1339] *Herald-Journal*, Syracuse, New York, United States, 9th August 1995
[1340] https://scholarworks.sjsu.edu/cgi/viewcontent.cgi?article=1144&context=mti_publications
[1341] Ibid
[1342] Government of India Ministry of Civil Aviation and Tourism Commission of Railway Safety Annual Report for 1995-1996
[1343] Ibid

Arizona. The person who died was named as Mitchell Bates, 41, a sleeping car attendant.

A local newspaper wrote: 'Les Johnson had been sound asleep in his coach seat of the Amtrak train when he woke up lying on the floor. "It was a sudden stop. No one realized we were in a wreck," said Johnson, a Yuma resident. Johnson, 41, was in a car towards the rear of the ill-fated Amtrak train that was sabotaged in a remote stretch of desert about 100 miles east of Yuma early on Monday. Five people were scheduled to get off the train at the next stop, which is Yuma, said Dawn Soper, Amtrak spokeswoman in Washington, D.C. Amtrak has refused to release a passenger list or reveal the names of those getting off in Yuma. Johnson was returning from Houston, where he had gone to be with his father in his final days after being diagnosed with terminal cancer. The car in which Johnson slept didn't tumble off the trestle 30 feet into the ravine.

'Passengers from the next car came through, prompting panic as they urged people to get off the train. Johnson said he calmed them down and told them to wait for a conductor to appear. A devoutly religious man, Johnson asked the throng to pray that no one was hurt. Silence reigned for perhaps 15 seconds. "No one said anything, not even the babies," Johnson said. The prayer finished, a conductor arrived. He asked for all able-bodied men to disembark and help rescue people from the sleeper cars that were pitched on their sides in the wash. By the light of a full moon, Johnson clambered into a ravine. He and others stood on the top of the cars and pulled people out of the wreckage. They gave them blankets and comforted them. Johnson helped remove a window of a sleeping car to free a woman trapped inside. Most had bumps and bruises. But others were more seriously injured. "I stayed with a woman who had a big gash on her head. She had chest pain and had a pacemaker. I stayed with her till she was airlifted," Johnson said.

'Among those dispatched to the scene was a medical evacuation team from Yuma Proving Ground and a Search and Rescue helicopter from the Marine Corps Air Station. Fifteen Yuma based Rural Metro workers with six ambulances also were dispatched to the scene. YPG received a call for help about three hours after the wreck, said Chuck Wullenjohn, base spokesman. The four-man crew arrived at about 6.30 a.m. "They shuttled two elderly people who had head and leg injuries to a make-shift triage center about six miles away." They were in total shock. I don't think they even knew we transported them," said Maj. Keith Handy, helicopter pilot. The four man SAR crew also was contacted about the same time as YPG.

'Meanwhile Johnson's wife Lynn drove to the Phoenix train station, where uninjured passengers were to arrive in the afternoon. Until she arrived at the station, she did not know if her husband was safe. An FBI agent who had a passenger list told her that Johnson was OK. Johnson and the other passengers were herded onto buses and taken to a Holiday Inn before being reunited with loved ones. The FBI didn't question passengers individually at the hotel, Johnson said. He signed papers that summarized what had happened and had to state if there was anyone he knew of who would want to harm him. Only then was Johnson reunited with his bride of six months. "I thank God. It could have been worse. I could have been dead. We were blessed," Johnson said. Johnson, who was making his first train trip, said this would also be his last. "I'll take Greyhound or fly," he said.

Examination of the scene of the derailment showed that a 3-foot-long section of steel rail weighing 18 pounds had been removed, with the gap being bridged by a length of wire to prevent the protecting signals from being put to Danger by the break in the track circuit. In addition, 29 spikes holding the rails to the wooden crossties (sleepers) over a length of 19 feet had been removed. Thus, the derailment had clearly been caused by an act of sabotage and within hours, the Federal Bureau of Investigation (FBI) had been called in to assist in the investigation.

It was estimated that someone with a basic knowledge of railways could have carried out the sabotage in about 10 minutes.

Four typewritten notes found at the scene had mentioned recent Federal raids on right-wing extremists at Waco, Texas, and Ruby Ridge, Idaho, and had been signed "Sons of Gestapo." This was apparently a previously unknown terrorist group and it was unclear if indeed such an extremist group existed, or whether the notes had been merely a ploy to throw investigators off the track of the real perpetrators. A substantial reward was offered for information leading to the arrest of the saboteurs and an extensive investigation was undertaken, but 30 years on, the mystery as to who derailed the train remains. [1344] [1345]

The South Central Railway zone of India was again the target of an attack at 20:00 on Saturday 18th November 1995 when its train No. 7048, the Down *Goutami Express* from Secunderabad to Kakinada Port, was derailed at Kilometre 213/8-10 on the approach to Ghatkesar station, Andhra Pradesh. One passenger died, 68 were injured, 19 seriously, and damage was estimated at 9,646,000 rupees. There had been 'unauthorised tampering with the track by persons unknown.' [1346]

Protesters against the government of the Prime Minister removed a portion of track near the station at Tongi, Bangladesh, on Saturday 30th December 1995, derailing the locomotive and two carriages of the train. Reports on the accident did not mention any casualties. Over 30,000 activists had blocked railways and roads throughout the country. [1347]

Train No. 2301, the Up *Rajdhani Express* of the Eastern Railway, was derailed at 00:03 on Tuesday 20th February 1996 after it had passed the Up Home signal at Muthani station, Bihar, India. Two people died, two were injured, one seriously, and damage was estimated at 5,950,000 rupees. Sabotage was noted by the Commission of Railway Safety, who also noted that a switch joint rail had been interfered with by the removal of fishplates, bolts, nuts and Pandrol clips. [1348]

At least 12 people were killed and 20 were injured on Tuesday 12th March 1996 when a mine derailed an engine and two coaches of a train between the northwestern towns of Oran and Tlemcen, Algeria. [1349]

A fuel train was derailed by a homemade bomb laid by Muslim militants on Friday 9th August at Naciria, Algeria, about 30 kilometres north-east of Algiers. The blast caused massive damage but no casualties. [1350]

Islamic terrorists were suspected of being responsible for the explosion of a bomb beneath a passenger train from Algiers to Oran near Baba Ali, Algeria, on the evening of Thursday 24th October 1996. Eight people died and 30 were wounded, 10 seriously. An insurgency had

---

1344 *The Yuma Sun*, Yuma, Arizona, United States, 10th, 11th, 12th, 13th and 15th October 1995

1345 Wikipedia – 1995 Palo Verde, Arizona, derailment

1346 Government of India Ministry of Civil Aviation and Tourism Commission of Railway Safety Annual Report for 1995-1996

1347 *Sun Journal*, New Bern, North Carolina, 31st December 1995

1348 Government of India Ministry of Civil Aviation and Tourism Commission of Railway Safety Annual Report for 1995-1996

1349 https://scholarworks.sjsu.edu/cgi/viewcontent.cgi?article=1144&context=mti_publications

1350 Ibid

been in progress for 4¼ years between the militants and the secular Algerian government. [1351]

Friday 26th April 1996 saw the derailment of the engine and 10 coaches of a passenger train from Argun in Chechnya to Mozdok in North Ossetia, Russia. The train had not proceeded very far when the derailment occurred at about 15:00 local time, a bomb having damaged the track. The train was carrying a mixture of civilians and servicemen, Mosdok being the main military base for war operations against the Chechens. An army officer and the train conductor were killed in the derailment. Following the dissolution of the USSR at the end of 1991, a secessionist regime in Chechnya, led by Dzhokhar Dudayev, had been fighting for independence and Russia had been at war with the rebels since 1994. On the Sunday prior to the derailment, Dudayev had been killed in a Russian air strike, after which the rebels had vowed to avenge his death. [1352] [1353]

Four passengers were killed and another 16 were seriously hurt when a bomb exploded in the fourth carriage of a train on the Moscow underground at about 21:15 on Tuesday 11th June 1996. A small device had exploded beneath a seat between Tulskaya and Nagatinskaya stations. A newspaper correspondent wrote: 'Rescue workers struggled in thick billowing smoke to move out passengers while anti-terrorist experts sifted debris to assess the seriousness of the threat. Soon after the explosion a fireman at Tulskaya said: "There are three dead and lots of injured. It went off while the train was still in the tunnel. It's a complete mess."

'Irina Dovoshenk, a paramedic, said: "Dolls were strewn around inside the carriage." She added that about 250 passengers in three carriages had been affected. The blast came a day after authorities stepped up security in Moscow.'

No one claimed responsibility for the bombing. Moscow Mayor Yuri Luzhkov stated: "This terrorist attack is part of a chain of measures to complicate or make impossible or create chaos in the run-up to the presidential elections." The elections were to be held the following Sunday. [1354] [1355]

Other perpetrators suggested were Chechen rebels and even the Japanese Aum Shinri Kyo doomsday cult, responsible for a nerve gas attack on the Tokyo underground in 1995. [1356]

A bomb exploded in the coach of passenger train No. 606 from Astrakhan to Volgograd while it was stopped in the station at Trubnaya, Russia, about 15 kilometres south of Volgograd, on Monday 12th August 1996. A 44-year-old woman died of her wounds in hospital and eight other people were wounded, three of them seriously. A bomb containing about a kilogram of trinitrotoluene (TNT) had been planted under the floor of a carriage. A previous bomb had exploded on an empty train stabled on a siding and a third had been defused a few days earlier. Some officials had blamed the attacks on Chechen rebels fighting for the independence of the southern republic. The train had been en route from Astrakhan to Volgograd. [1357]

Eight minutes after midnight on Thursday 19th December 1996 a bomb exploded on an underground train in St Petersburg, Russia. Placed beneath a seat, the bomb blew out the doors and windows of the second car of a train between Lenin Square and Vyborgskaya stations. There had been only two passengers in the coach, one of whom, a young man, suffered from slight concussion but did not need to go to hospital. Aide Nikolai Savelev stated

---

[1351] *Times-News*, Twin Falls, Idaho, United States, 27th October 1996

[1352] *The Orange County Register*, Irvine, California, United States, 27th April 1996

[1353] *Clovis News Journal*, Clovis, New Mexico, United States, 28th April 1996

[1354] *The Times*, London, England, 12th June 1996

[1355] *The Orange County Register*, Irvine, California, United States, 12th June 1996

[1356] *The Medicine Hat News*, Medicine Hat, Alberta, Canada, 14th June 1996

[1357] *The Garden City Telegram*, Garden City, Kansas, United States, 12th August 1996 that: "It was obviously a terrorist act." The bomb resembled that used in the June attack in Moscow. [1358]

Nine people were killed and seven others injured on the night of Tuesday 21st May 1996 when a mine exploded under a train at an unspecified location on the main line between Yangon and Mandalay, Myanmar. Officials blamed the blast on Karen National Union (KNU) guerrillas, who had been fighting the central government for autonomy since 1949. The KNU was by this date the only major ethnic minority guerrilla army not to have signed a cease-fire with the military government. [1359]

On Friday 21st June 1996 one passenger was killed and five were wounded when a bomb exploded on a Djibouti-bound train about 20 kilometres from Dire Dawa, Ethiopia. No group claimed responsibility for the attack. [1360]

After a soldier had become suspicious of a bag on a commuter train that had left Colombo at 17:15 on Wednesday 24th July 1996, it was stopped in the suburban station of Dehiwala, Sri Lanka, and bomb disposal officers were called. Before they had arrived, railwaymen arranged for the train to move away station platform, upon which a number of passengers, wrongly believing that the all-clear had been announced, reboarded it. Two bombs, which had been planted in different carriages, then exploded at about 18:15, killing possibly 70 people and injuring over 450.

A British newspaper correspondent in Colombo wrote, 'Two carriages were blown apart by what are thought to have been separate devices, and part of the station was demolished. A survivor said: "The station was filled with smoke. When it cleared, I saw people fallen all over the station."

'The injured, who were found lying amid smouldering debris, were carried in cars, buses, and scooter rickshaws through traffic to nearby hospitals, which are ill-equipped for such a disaster.

'Last night's initial death tally of at least 67 is expected to rise substantially in coming days. One hospital treating 275 of the injured said last night that at least 70 were fighting for their lives.' [1361]

The train had been bound for Aluthgama, about 50 kilometres from Colombo.

Lakshman Sooriyawardene, 23, an office worker who had just left the train, related, "I was walking to the exit when I heard two explosions behind me. When I turned around, the whole station was filled with smoke. People were screaming and running."

No one claimed responsibility for the attack but it had occurred on the week marking the 13th anniversary of their campaign for an independent homeland in the largely Sinhalese country, the Tamils claiming discrimination in education and jobs. Over 43,000 people had died in the civil war between Tamils and Sinhalese since 1983. Policeman D.M. Tennakoon, one of the earliest on the scene, stated, "This is obviously the work of the Tamil Tigers. The bombs are similar to others they have exploded." [1362]

[1358] *The Lethbridge Herald*, Lethbridge, Alberta, Canada, 20th December 1996

[1359] https://scholarworks.sjsu.edu/cgi/viewcontent.cgi?article=1144&context=mti_publications

[1360] Ibid

[1361] *The Times*, London, England, 25th July 1996

[1362] *The Cedar Rapids Gazette*, Cedar Rapids, Iowa, United States, 25th July 1996

On state television, the office of President Chandrika Bandaranaike Kumaraninga accused the Liberation Tigers of Tamil Eelam (LTTE) of attempting to provoke an anti-Tamil backlash from the Sinhalese majority. "The president requested the people to remain calm and help the government maintain law and order at this juncture without playing into the hands of the LTTE, which has a hidden agenda of inflaming another Black July and alienating Sri Lanka in the international community," the office announced. (Black July was 24th July 1983, when rioting Sinhalese killed 400 Tamils after Tamil rebels had killed 13 soldiers, provoking a separatist war.) [1363]

Police arrested two Tamils shortly after the explosion, whom passengers stated had been acting suspiciously. [1364]

Militant anti-nuclear protesters opposed to the transport of radioactive waste by rail in CASTOR (an acronym for 'Cask for the storage and transport of radioactive material') trains cut through a 1.5-metre length of rail near Dannenberg, Germany, on Thursday 22nd August 1996. Unfortunately, a railcar carrying 12 passengers was the first train to appear and was partially derailed. It does not seem as if anyone was hurt. The protesters left "Stop Castor" leaflets at the scene. [1365]

Central Railway train No. 1078, the Up *Jhelum Express* from Jammu Tawi to Pune, was derailed in the early morning of Saturday 14th September 1996 shortly beyond the station at Raver, Maharashtra, India, on the Bhusaval Division of the railway. One person was killed and 19 were injured, four seriously.

The train had been coasting down a 1 in 135 falling gradient on a straight stretch of track at Km. 477 when it left the rails after passing the Raver Up Advanced Starting signal. The derailment was blamed on sabotage, fishplates having been removed from several joints. [1366]

On Thursday 31st October 1996 a freight train was derailed by sabotage at Thannamunai, Sri Lanka, just outside of Batticaloa. Following the derailment the drastic step was taken of closing the line, at first only between Valaichenai and Batticaloa and then as far back as Polonnaruwa,[1367] 100 kilometres short of Batticaloa. The rails and sleepers over the 50 kilometres between Punanai and Batticaloa were used by the army and security forces to build and reinforce security posts and checkpoints, while more sleepers were allegedly taken by locals and sold for profit.[1368] It was only after the signing of a peace accord between the Government and the Liberation Tigers of Tamil Elam in March 2002 that the line was finally re-laid and reopened in April 2003, then was closed again twice because of problems such as sleepers being removed, before finally being permanently brought back into use in December 2003. [1369]

A bomb planted in a toilet compartment of the second car of the *Tezgam Express* from Karachi to Rawalpindi exploded as it was passing Tandoi village, Pakistan, some 65

1363 *The Lethbridge Herald*, Lethbridge, Alberta, Canada, 25th July 1996

1364 *The Medicine Hat News*, Medicine Hat, Alberta, Canada, 26th July 1996

1365 https://www.gorleben-archiv.de/chronik/1996-2/

1366 Government of India Ministry of Civil Aviation and Tourism Commission of Railway Safety Annual Report for 1996-1997

1367 https://www.tamilnet.com/art.html?catid=13&artid=10773

1368 https://www.tamilnet.com/art.html?catid=13&artid=8762

1369 https://www.tamilnet.com/art.html?catid=13&artid=10773

kilometres south of Rawalpindi, on Sunday 3rd November 1996. Five passengers died and 20 were wounded. [1370]

Two men unscrewed bolts on the track near Novgorod, Russia, on Sunday 10th November 1996, but their efforts were frustrated when railwaymen noticed damage to the track and imposed a speed restriction. Thus, only the locomotive of what was described as a luxury express bound for St Petersburg was derailed and no one was hurt. Trains usually passed the spot travelling at 100 km/h. The culprits were subsequently arrested; according to the police, their object had been robbery. [1371]

Northern Railway train No. 1077, the Up *Jhelum Express*, consisting, behind its engine, of all-steel anti-telescoping Integral Coach Factory carriages, arrived at Platform No. 6 of Ambala Cantonment station, Haryana, India, at 01:55 on Monday 2nd December 1996. Here, its electric locomotive was detached and a diesel locomotive was substituted at 02:05. As the train was about to depart, at 02:20 a major bomb explosion occurred in the S-4 sleeper coach 12th from the engine. A fire followed the explosion and there was heavy damage to both the coach and the platform shelter, as well as the setting fire to a stack of cotton bales. Ten people died in the blast and 29 were injured, 17 of them seriously. [1372]

Jasjit Singh, who had just arrived at the station on another train, said, "There was smoke all around and the entire platform was strewn with limbs and body parts." No one claimed responsibility for the bombing, though Islamic Kashmiri separatists had previously tried to derail trains in the area with lineside bombs. [1373]

There was another bomb explosion on the Réseau Express Régional in Paris, France, at 18:05 on Tuesday 3rd December 1996. This time it happened as a train was pulling into the Port-Royal station. Two passengers were killed outright, 35 were seriously hurt, seven said to be in a critical condition, and about 50 more were slightly hurt.

A passenger who had been riding in the rear of the train told French radio that the explosion had blown off the doors of the second coach of the train, while the shock had jolted the entire train forward. This passenger was not wanting in courage for he had then seized a fire extinguisher and made his way through the panicking passengers to the seat of the blast. He related: "The car was still burning. There were no more seats. The doors were folded in half." He added that mutilated people lay on the ground, with one body already being covered by a blanket.

'Witnesses described a scene of panic, of thick black smoke, the chilling wail of ambulances and of paramedics frantically carrying away the wounded on stretchers. "I saw lots and lots of smoke and I heard a big boom," said a man who gave his name only as Jean-François. "People were crying and in a state of shock." The first ambulances were on the scene within two minutes, a witness said. The wounded were rushed to a nearby military hospital specializing in trauma victims.

'"It's starting again," said Estelle Campos, clutching her 6 and 12-year-old sons tight near the subway station. "It's always the same people who pay."

[1370] https://scholarworks.sjsu.edu/cgi/viewcontent.cgi?article=1144&context=mti_publications

[1371] Author's personal notes – source mislaid.

[1372] Government of India Ministry of Civil Aviation and Tourism Commission of Railway Safety Annual Report for 1996-1997

[1373] *The Medicine Hat News*, Medicine Hat, Alberta, Canada, 2nd December 1996

'The blast was chillingly close to the site of the worst of last year's bombs, at the St Michel station. The July 28 bomb killed eight people and wounded 84 others and launched a four-month bombing campaign that put Paris under siege. A sombre President Jacques Chirac condemned "these unacceptable acts, these barbaric acts that always attack innocent people. The government and I are determined to fight against terrorism in all its forms," he said. "No stone will be left unturned." [1374]

As in the attacks the previous year, the bomb had been fashioned from a 28-pound gas canister. The use of a battery-powered detonator also suggested Algerian Islamic terrorists. [1375]

As train No. 4055, the Down *Brahmaputra Mail* from Dimapur to New Delhi, was running between Km. 216/7 and 217/0, between Kokrajhar and Fakiragram stations, Assam, India, on the Bongaigaon – New Coochbehar broad-gauge, single-line, non-electrified section of the Alipurduar Division of the Northeast Frontier Railway, at about 19:15 on Monday 30th December 1996, it was wrecked by two explosions. Thirty-three people died as a result of the sabotage and 62 were injured, 25 seriously. [1376]

A newspaper report described the event and its aftermath. 'Hacking through tropical forest, rescuers today pulled out dozens of corpses from the wreckage of an express train blown off the tracks by a bomb in northeastern India. Rescuers cut away metal sheets of the train cars and sliced through steel bars to extricate 38 bodies, said S.P. Mehta, a railway official. Another 80 people were taken to hospitals in nearby Alipur Duar by a special train. "Many of the injured are in a critical condition and could die," Mehra said.

'No one claimed responsibility for Monday's explosion, but it came in the heart of an area that militants of the Bodo tribe want as their own autonomous region. Their campaign has been mounting in violence in recent months, as rival Bodo factions compete to demonstrate their strength.

'Seven minutes after the New Delhi bound Brahmaputra Express pulled out of Kokrajhar station a huge explosion threw the engine and the next four coaches off the track. The train originated in Gauhati, the state capital of Assam, about 125 miles to the east. Residents of Sesapani, a Bodo village, were shaken from their bamboo huts by the blast shortly after dark and rushed to the site to help the injured. But all organised rescue was hampered by the lack of paved roads to the railway line, a 15-minute walk through teak and bamboo forests. Trees were cut down to make way for rescue vehicles.

'By midday today, hundreds of railway workers and federal and local police had nearly completed the search for bodies, Mehta said. As they scoured the debris for clues, preliminary investigations indicated the bomb was placed on the track and detonated by remote control, said K.K. Gupta, a railway safety official.

'Police found strands of wire near the track that may have connected the explosive device to a battery-operated detonator hundreds of yards away. If that be the case, the bombing would be similar to other recent attacks blamed on Bodo rebels that have killed at least 25 people. The blast severed the only rail line through the area through the narrow land route between Bangladesh to the south and Bhutan to the north. Most residents of the remote and undeveloped area cannot afford plane fares and roads are poor.

[1374] *The News Herald*, Panama City, Florida, United States, 4th December 1996

[1375] *European Stars and Stripes*, Darmstadt, Germany, 4th and 5th December 1996

[1376] Government of India Ministry of Civil Aviation and Tourism Commission of Railway Safety Annual Report for 1996-1997

'The Bodos have been struggling for a separate homeland for 20 years, claiming their culture is threatened by Muslim emigres from Bangladesh. In 10 years at least 1,000 people have been killed in the dispute. Hoping to quell the revolt, India agreed to a Bodo Autonomous Council to represent Bodo interests but refused to create a separate state that would give the Bodos greater financial and political control. Some Bodo factions want complete independence from India.' [1377]

More than 1,640 feet of track between Cerrejón Norte and Puerto Bolivar, Colombia, was damaged by dynamite, causing the derailment of the engine and 27 cars of a freight train carrying 270 tonnes of coal and the spillage of 3,700 gallons of diesel on Wednesday 9th April 1997. The Intercor mining enterprise suffered a loss of between $3,000,000 and $5,000,000. The left-wing Ejército de Liberación Nacional (National Liberation Army) (ELN) group was thought responsible for the attack.

An attack on the same stretch of railway on Monday 15th September 1997 damaged 980 feet of track and caused the derailment of 26 cars carrying 270 tonnes of coal. This time the attack was blamed on the 59 Front of Fuerzas Armadas Revolucionaries de Colombia – Ejército del Pueblo (Revolutionary Armed Forces of Columbia – People's Army) (FARC).

On Saturday 20th September 1997 the ELN blew up 230 feet of track in Magdalena province, Colombia, derailing seven cars of coal. This time the train was on its way from a mine owned by the American Drummond corporation to the port of Santa Marta.

Finally, on Sunday 28th September the ELN were thought responsible for the detailing of 18 cars carrying 200 tonnes of coal, again on the Cerrejón Norte – Puerto Bolivar line in Colombia. [1378]

At least 21 people died and 20 more were wounded when a mine exploded beneath a train at Oued el Kerma, in the Algiers suburb of Guê de Constantine, Algeria, at 08:00 on Friday 25th April 1997. One side of the railcar was destroyed and it was partly derailed. The attack occurred on the line between Algiers and Bilda, about 50 kilometres to the south and the location of the main army garrison in the country. Security sources stated that a home-made device was responsible.

There was no immediate claim of responsibility but Muslim insurgents hoping to disrupt parliamentary elections on 5th June were supposed to have been responsible. There had been attempts to topple the government since the army had cancelled an election in January 1992, which might have given power to the Islamic Salvation Front, since banned. [1379]

Half of a carriage was torn apart when a bomb exploded on a train in the station at El Harrach, Algeria, on Wednesday 25th June 1997. A reporter wrote that the station was still full of smoke an hour after the mid-morning explosion. 'Cushioned seats, glass and even train doors were strewn across the platform and blood stains dotted the area.' Fifty people were wounded, most of whom needed hospital treatment, but perhaps surprisingly, no one was killed. A home-made bomb was blamed for the blast.

Anxious to play down the violence in the country, the government did not immediately confirm the explosion and no mention of it was made on state television or radio of the blast, which was due, it was assumed, to Islamic terrorists, though there was no immediate claim of

---

[1377] *The Orange Leader*, Orange, Texas, United States, 31st December 1996

[1378] https://2009-2017.state.gov/documents/organization/19808.pdf

[1379] *Saturday Free Press*, Winnipeg, Manitoba, Canada, 26th April 1997 responsibility. [1380]

Another home-made bomb exploded on a train early on Saturday 28th June at an unstated location in Algeria, killing one passenger and wounding three. [1381]

The Arab-language newspaper *Al Khabar* stated that two bombs had exploded on a passenger train between western Algeria and Algiers on Saturday 23rd August 1997, killing eight people and wounding 22. [1382]

On Tuesday 14th October 1997 a bomb exploded on a train shortly after it had left Algiers, Algeria, for the south, injuring four people and badly damaging a coach. [1383]

An explosive device was detonated in the toilet of passenger train No. 24 from Moscow to St Petersburg as it neared Torbino, Russia, 225 kilometres from its destination, at 18:30 on Friday 27th June 1997. Three people were killed outright, two were fatally hurt and another 11 were taken to hospital in St Petersburg and other towns. Those who lost their lives included 13-year-old Yevgeny Tsvetkov and Yuri Kusnetzov, a Second World War veteran. The blast had a force of 200 to 400 grams of TNT and destroyed part of the coach; it also damaged the brake pipe, bringing the train to a halt. After the coach involved had been detached, the remainder of the train was able to continue its journey.

It was believed that the explosion had been set off by the individual in the toilet, but not clear if it had been a terrorist attack or suicide. [1384] [1385]

A bomb exploded in a second-class general sitting coach of Northern Railway passenger train No. 336 from Ganganagar to Ambala at Laihra Khana Halt, Punjab, India, on the Bhatinda – Rajpura broad-gauge section of the railway, at 14:00 on Tuesday 8th July 1997. Thirty-five people were killed in the explosion and 63 were injured, 36 of them seriously. [1386]

The train had only shortly before left Bhatinda station and was carrying 500 passengers in its five coaches, some of them pilgrims bound for the Ganges River. One of the passengers was Beera Kaur, who had seen her husband, Lakshman Chari, killed in the blast. She said, "I heard a deafening explosion and within seconds the coach was engulfed in fire and nauseating smoke."

A police officer, speaking on condition of anonymity, stated that the bomb had apparently been placed beneath the floor of the coach and equipped with a timing device. The floor and roof of the coach had been blown off by the force of the blast.

There was no immediate claim of responsibility for the attack. The Punjab had been the scene of a 10-year secessionist campaign by Sikh militants, fighting for an independent homeland. The rebellion had been crushed in the early 1990s but sporadic violence continued. [1387]

Train No. 9734/4710, the *Shekhawati Express* of the Northern Railway, was derailed at 02:05 on Monday 4th August 1997 between Mahendragarh and Kanina Khas stations, Haryana, India. The Commission of Railway Safety determined that the accident had resulted from the

---

[1380] *The Lethbridge Herald*, Lethbridge, Alberta, Canada, 26th June 1997

[1381] *International Herald Tribune*, Paris, France, 26th June 1997

[1382] *The Guardian*, London, England, 25th August 1997

[1383] *European Stars and Stripes*, Darmstadt, Germany, 16th October 1997

[1384] *Saturday Free Press*, Winnipeg, Manitoba, Canada, 28th June 1997

[1385] *Odessa American*, Odessa, Texas, United States, 29th June 1997

[1386] Government of India Ministry of Civil Aviation and Tourism Commission of Railway Safety Annual Report for 1997-1998

[1387] *The Daily Inter Lake*, Kalispell, Montana, United States, 9th July 1997 'removal of fishplates, bolts and nuts at a rail joint by unidentified outside persons.' [1388]

Hundreds of passengers escaped unscathed on Monday 11th August 1997 when a bomb derailed a train in the Barpeta district of Assam, India, 150 kilometres north of Guwahati, the state. The bomb left a one-metre crater and destroyed a section of track. Seven coaches of the train were derailed.

None of the several militant separatist organisations fighting for a homeland in Assam claimed responsibility for the attack, though police suspected that the United Liberation Front of Assam had been the perpetrators. [1389]

An explosion 'blew off' a coach of No. 175 Up passenger train between Kathalpukhuri and Silarubari, Assam, India, on Thursday 14th August 1997. Eight people died and five were wounded. [1390]

On Tuesday 19th August 1997 a goods train was derailed by a bomb at Bridge No. 292, between Salekati and Basugaon, Assam, India, wounding the driver and assistant driver. [1391]

The Arab newspaper *Al Khabar* stated that two bombs exploded on Saturday 23rd August 1997 on a train running between western Algeria and the capital, Algiers. Eight people were killed and 22 were wounded. [1392]

Three bombs exploded on the *Golden Temple Mail* train from Mumbai to Amritsar in the Punjab at Ghaziabad, Uttar Pradesh, India, on Wednesday 1st October 1997. Two or three people were killed and about 35 were wounded. [1393]

In what were believed to be attacks carried out by Muslim extremists to mark the fifth anniversary of the destruction of the historic Babri Masjid mosque in the city of Ayodhya by Hindus, bombs exploded on trains at three locations in Tamil Nadu state, India, on Saturday 6th December 1997. The first explosion, in the rear coach of a train at Tiruchi, killed two women. A second explosion in the *Cheran Express* killed four people and wounded eight at Erode, while four people also died and 40 were wounded in the third explosion at Thrissur. [1394]

A powerful bomb exploded as a high-speed train was passing the spot near San Sebastian, Spain, on Sunday 22nd February 1998. The engine and one of the carriages of the train were slightly damaged. There were no reports of injury but rail traffic was disrupted for three hours. The train had only just left Irun station. Euskadi Ta Askatasuna (ETA), a Basque nationalist terrorist group, were suspected to have been responsible. [1395]

Eighteen people died on Monday 23rd February 1998 when a bomb exploded underneath a train near El Affroune, Algeria, about 15 kilometres south of Algiers. Another 25 people were

---

1388 Government of India Ministry of Civil Aviation and Tourism Commission of Railway Safety Annual Report for 1997-1998

1389 *International Herald Tribune*, Paris, France, 12th August 1997

1390 http://www.axom.faithweb.com/news/aug1.html

1391 Ibid

1392 *The Guardian*, London, England, 25th August 1997

1393 *The Times*, London, England, 2nd October 1997

1394 *CNN World News*, 6th December 1997

1395 http://home1.gte.net/tpnews/news-arch.htm

injured. The militant Armed Islamic Group were believed responsible for the bombing. The explosion followed an attack on Saturday on the nation's most important gas pipeline. [1396]

Two bombs exploded on an Algiers – Oran passenger train on Tuesday 5th May 1998 as it was entering a tunnel near Aïn Torki, Algeria. Fifteen people were wounded, four of whom were said to be in a serious condition. [1397]

Condeck, Algeria, in Ain Defla province, was the location of another train bombing at 08:50 on Thursday 11th June 1998. An initial report stated that 12 people had died and 21 had been wounded; the figures were later amended to 17 dead and 31 wounded. Again, a home-made bomb was stated to have been responsible. [1398] [1399]

Ain Defla province, Algeria, was the scene of another train bombing on Wednesday 12th August 1998, this time near Aïn Torki. Seven passengers on the Oran – Algiers train lost their lives and 11 were wounded. Security forces launched a hunt for the bombers. [1400]

On Thursday 26th February 1998 a bomb exploded in an economy coach of a passenger train bound for Karachi at an unstated location in Punjab province, Pakistan. Three people were killed and 19 were wounded. [1401]

On Monday 9th March 1998 a bomb planted under a seat exploded on an express train near the town of Pattoki, Pakistan, killing 11 people and wounding about 35. At 07:35 the next morning, Tuesday, another bomb exploded on a train as it reached Walton station, outside of Lahore. The train was packed with 800 passengers, variously described as labourers and factory workers, from Kasur, 80 kilometres to the south and close to the border with India. Ten people died and more than 80 were hurt in this second attack. [1402]

Ambulance driver Mohammed Hussain, speaking after the Walton bombing, said that body parts had been strewn across the station, making them difficult to identify, and that police had employed hacksaws to prise free trapped, weeping victims. "Smoke was pouring out of the carriages and many of the bodies were burned beyond recognition or ripped apart," he related.

Passenger Liaquat Ali said, "There was blood everywhere and people were screaming and weeping."

Following the second bombing paramilitary troops fanned out across Lahore to protect mosques and government buildings, as earlier on Monday there had been a third bombing at a courthouse in Sukkur. Meanwhile, a Foreign Ministry statement in Islamabad announced that there was conclusive evidence that the bombings and 10 previous ones had been masterminded and executed by India's Research and Analysis Wing, its foreign intelligence agency. [1403] [1404]

Another time bomb exploded on the *Khyber Mail* at 04:00 on Sunday 7th June 1998, killing 26 people and wounding at least 45. The explosion occurred near Khairpur, Pakistan, in Sindh province. Again, Pakistan blamed Indian intelligence forces for the attack. The bomb had

---

1396 *The Southern Illinoisian*, Carbondale, Illinois, United States, 24th February 1998
1397 *International Herald Tribune*, Paris, France, 6th May 1998
1398 *The Lethbridge Herald*, Lethbridge, Alberta, Canada, 12th June 1998
1399 Ibid, 14th June 1998
1400 *The Medicine Hat News*, Medicine Hat, Alberta, Canada, 13th August 1998
1401 *The Robesonian*, Lumberton, North Carolina, United States, 28th February 1998
1402 *The Times*, London, England, 11th March 1998
1403 *The Medicine Hat News*, Medicine Hat, Alberta, Canada, 11th March 1998
1404 *The Robesonian*, Lumberton, North Carolina, United States, 11th March 1998

reportedly been set to explode halfway between Karachi and Peshawar. [1405]

Twelve people died and 41 were injured when train No. 3231 of the Eastern Railway, a crowded Howrah – Danapur express, was derailed at Bakhtiyarpur station, Bihar, India, at about 07:17 on Tuesday 28th April 1998.[1406] The injured were taken to the Patna Medical College Hospital, Nalanda Medical College Hospital and another hospital in Patna, and the uninjured passengers were also taken to Patna. Railway Board chairman V.K. Agarwal and General Manager S. Ramanathan supervised the rescue and relief operation, while Railway Minister Nitish Kumar also went to the scene. The Railway authorities announced compensation of 15,000 rupees for the families of those who had lost their lives, 5,000 rupees for those seriously hurt and 500 rupees for those slightly hurt. Repair work was being carried out "on a war footing" and it was hoped to reopen the line by 10:00 the next day.

The derailment was apparently due to sabotage, as railway officials stated that "Prima facie, the cause of the accident is the removal of fishplates on the right side of the tracks."

Whatever the cause of the derailment, Minor Irrigation Minister Vijay Krishna demanded the resignation of his colleague, the railway minister, on moral grounds, on the basis that he had failed to provide for passenger safety and check track maintenance. [1407]

The South Central Railway *Narayanadri Express* from Secunderabad to Tirupati was derailed at an unstated location in Southern India at about 04:10 on Monday 29th June 1998, the engine and 10 carriages leaving the track. More serious results were probably averted because the train was quickly pulled up and the casualties were limited to slight injuries to 20 passengers. The derailment was blamed on the removal of fishplates. [1408]

South Central Railway train No. 7056, the Hyderabad – Narasapur express, was derailed on Tuesday 8th October 1998, probably due to an act of sabotage. Eleven passengers died and 37 were injured, 16 seriously.[1409] The accident occurred between Sattenapalle and Pedakurupadu stations, Andhra Pradesh, India, in the Vijayawada Division of the railway. The locomotive and nine coaches left the track. 'One of the coaches was completely jammed and gas cutters were used to remove passengers trapped inside…' [1410]

After receiving information from locals, police and Army personnel located and successfully defused two explosive devices placed on the track between Domohani and Falakata, India, on Sunday 20th June 1999. Early on the next day, Monday 21st, a mine planted on the track derailed a few waggons of a goods train between Domohani and New Jalpaiguri stations, West Bengal, India. Two detonators, a nine-volt battery and pieces of electromagnetic wire were recovered at the scene.

A considerably more serious case of sabotage was then to occur at noon on Tuesday 22nd June 1999 when a bomb hidden in a briefcase exploded on the platform at New Jalpaiguri station itself. This time, nine people were killed, three of them Army jawans (soldiers) bound

---

[1405] Independent Television broadcast, 7th June 1998

[1406] Government of India Ministry of Civil Aviation and Tourism Commission of Railway Safety Annual Report for 1998-1999

[1407] https://www.Rediff.com/news/1998/apr/28train.htm

[1408] *Danger Ahead! Historic Railway Disasters* – Sabotage derails Narayanadri express

[1409] http://pib.nic.in/ar.../r261098.htm

[1410] www.tribuneindia.com/1998/98oct09/spotlite.htm

for Kargil, and 70 others were wounded. Extensive damage was caused to one bogie (carriage) of the Sikkim – Mahanda express, which had been waiting at the station, and the *Darjeeling Mail*, which was also in the station, was damaged to a lesser extent. This time, a nine-volt battery, pieces of electromagnetic wire and 3.1 kg of explosive were recovered at the scene. [1411]

Rebels blew up a train as it entered Kibossi station, Republic of Congo, on about Sunday 25th July 1999. Dozens of passengers were wounded and were taken to hospital in the capital, Brazzaville, 50 kilometres away. No one was reported killed. The army blamed the attack on so-called Ninja rebels loyal to the former Prime Minister, Bernard Kolelas. The rebels had carried out frequent previous attacks on the railway. [1412]

A goods train was derailed by sabotage between Sarupeta and Barpeta, Assam, India, on Saturday 31st July 1999. [1413]

United Liberation Front of Asom (ULFA) militants were blamed for an attempt to derail Northeast Frontier Railway train No. 5622, the Up *North East Express*, near the station at Bijni, Assam, India, early on Saturday 7th August 1999. What was described as 'a powerful bomb' went off beneath a pilot engine that had preceded the express, which served its purpose and frustrated the attempt. No one was hurt, though traffic was disrupted. [1414]

At about 02:10 on Monday 9th August 1999 a remotely controlled device was planted by the ULFA under a bridge between Salekati and Kolrajhar stations, Assam, India. The Northeast Frontier Railway *Rajdhani Express* from New Delhi to Guwahati was shortly due but the 'victim' of the explosion that followed was instead a goods train, six bogies of which were derailed. Its driver, assistant driver and three security personnel were wounded. [1415]

Two days on, at about 20:45 on Wednesday 11th August 1999, the value of employing a pilot locomotive was again demonstrated when one exploded a mine planted under a bridge over the River Paschnoi between Majbat and Dhekaijuli Road stations, Assam, India, on the metre-gauge section of the Northeast Frontier Railway between Rangia and Tezpur. The engine was derailed and its driver injured but train No. 5617, the Up *Arunachal Express*, which had been following the pilot engine, escaped. [1416]

Thirteen people were injured, one fatally, when the Northern Railway Lucknow – Saharanpur express was derailed at 06:55 on Saturday 11th September 1999 between Haldaur and Bijnaur stations, Uttar Pradesh, India, on the Gajraula – Najibabad section of the line. Nine sleeper cars and an air-conditioned second-class coach were affected. A 2.19-metre length of rail had been removed. [1417]

At least 14 people probably lost their lives when a bomb exploded in a coach of the Poona Express at about 20:15 on Thursday 11th November 1999. The blast occurred in a third-class carriage crowded with pilgrims returning from the Vaishno Devi Hindu temple, shortly after the train had left Pathancot station, Punjab, India. The blast derailed the train, bringing it to a halt. Three passengers were killed outright, 10 more had already died of their wounds and 50 were hurt, some with severe burns, at the time of the report. No one claimed responsibility for

[1411] www.ganashakti.com/old/1999/990628/bengal.htm

[1412] https://www.thenewhumanitarian.org/news/1999/07/27/dozens-injured-railway-blast

[1413] www.axom.faithweb.com/news/aug1.html

[1414] *Calcutta Online*, Calcutta, West Bengal, India, 8th August 1999

[1415] www.swordoftruth.com/swordoftruth/archives/byauthor/shamolimitra/tndofitia.html

[1416] Ibid

[1417] www.Rediff.com/news/1999/sep/11nr.htm

the attack but militant groups seeking an independent Kashmir were thought to be likely to blame. [1418]

A landmine exploded under the engine of a Chemin de Fer Djibouto-Éthiopien passenger train from Djibouti to Addis Ababa at Melka Jebdu, Ethiopia, about 12 kilometres from Dire Dawa, on Saturday 13th November 1999. The driver and assistant driver were killed and a third railwayman, probably a conductor, was critically wounded. There was no claim of responsibility.

In a previous attack on an express bound for Djibouti in the Dire Dawa area, presumably also involving a landmine, at an unstated date in August, a driver and assistant driver had been wounded. [1419] [1420]

A crude bomb exploded on the Shamli – Saharanpur passenger train at Old Dehi station, India, at 16:05 on Thursday 6th January 2000. The train was not due to leave until 17:00, so there was hardly anyone on it and the 20 people who were wounded had all been standing on the platform and were struck by shards of glass. Only four were detained in hospital. [1421]

Suspected United Liberation Front of Asom (ULFA) militants derailed a coal train bound for Haryana at 01:00 on Wednesday 19th January 2000.[1422] The derailment occurred between Bokajan and Chungajan, Assam, India. [1423]

A bomb exploded on the *Sealdah Express* bound for Calcutta shortly after it left the station at Vijaypur, Jammu and Kashmir, India, on the late evening of Thursday 10th February 2000. Three people were killed outright, two died shortly afterwards, 10 were admitted to hospital and others were treated and discharged in the explosion, which wrecked the S3 and S4 coaches, between which it had been set off by remote control. At least three improvised explosive devices had been planted on the track but two had failed to go off due to a technical fault, preventing a much more serious accident in the packed train. Ten of the coaches of the train were derailed. Security was already tight, with passengers being frisked and luggage screened, and it was stated that it would be tightened further.[1424] Railway Minister Mamata Banerjee ruled out an inquiry into the explosion, stating that it was obviously a terrorist attack rather than an accident, though security would be further tightened. [1425]

Sixteen passengers riding on the Northern Railway *Gomti Express* were wounded when a bomb exploded in coach 10 of the train on the evening of Thursday 17th February 2000. The train was passing the East cabin at Jhinjhak station, Uttar Pradesh, India, at the time. Police blamed a crude bomb of the sort used by criminals in the area. [1426] [1427]

Only a couple of minutes after it had pulled out of the station at Hyderabad, Pakistan, on Saturday 5th February 2000 a bomb exploded on a passenger train. The train was passing

[1418] *Danger Ahead!* 26th April 2002

[1419] *Danger Ahead!* Currents: 2 Die in Ethiopian train explosion

[1420] http://www.../ND11-15.HT

[1421] https://m.Rediff.com/news/2000/jan/06bomb.htm

[1422] www.axom.faihweb.com/archives/jan2.html

[1423] www.axom.faithweb.com/news/aug1.html

[1424] https://m.Rediff.com/news/2000/feb/10blast1.htm

[1425] https://www.tribuneindia.com/2000/20000212/main2.htm

[1426] https://www.tribuneindia.com/2000/20000219/spotlite.htm

[1427] https://www.satp.org/satporgtp/countries/pakistan/database/railwayattack.htm

through a shantytown at the time, heading for Mirpur Khas. Four people were killed outright, another four died later and 40 were wounded. No one claimed responsibility for the act or a second bombing, not on the railway, on the same day. [1428]

A bomb exploded in the rear coach of a crowded passenger train, again as it was pulling out of a station in Hyderabad, Pakistan, on Sunday 16th July 2000, this time killing 10 people and injuring 30. Those who lost their lives included a five-year-old girl and a 10-year-old boy. The governor of Sindh province, Mohamadian Soomro, stated that foreign forces might be behind the bombing. Pakistani officials frequently blamed India for such incidents; India always denied any involvement. Police ordered a high-level investigation and security at railway stations was tightened up. [1429] [1430]

Two empty tank cars of a freight train were derailed by an explosion between Gudermes, Chechnya, and Khasavyurt, Dagestan, Russia, on Friday 10th March 2000. No one was hurt. Chechen rebel separatists were suspected of being responsible. [1431]

Northeast Frontier Railway train No. 5719, an Up intercity express, was derailed between Aluabari Road and Taiabpur stations, India, on the Kishanganj – Siliguri Junction metre-gauge section of the Katihar Division of the railway, on Tuesday 21st March 2000. One person was killed and six were injured, three of them seriously. Fishplates and fish bolts had been removed from track joints by a person or persons unknown. [1432]

Train No. 105, the Up *Shalimar Express* from Karachi bound for Lahore, was derailed at 13:05 on Tuesday 4th April 2000 between Buradabad and Jhumpir stations, Pakistan, and close to the latter. Nine coaches of the train left the rails and collided with the girders of a bridge. Four passengers were killed and 59 were injured and taken to hospitals in Hyderabad, Karachi, Makli and Thatta. Fifty of them were discharged after treatment.

Pakistan Railways General Manager Iqbal Samad Khan rushed to the scene to supervise the rescue operations and stated that an inquiry would immediately be opened into the derailment. He added that, "Sabotage cannot be ruled out as during initial inquiries, three bolts of the fishplates on each side of the Up tracks were found missing." [1433]

A goods train was derailed by sabotage between Rangiya and Ghograpar, Assam, India, on Sunday 30th July 2000. [1434]

Two coaches of passenger train No. 5716 were derailed by sabotage on the following day, Monday 31st, between Khandikar and Goreswar, Assam, India. This time the results were much more serious than the day before, with nine dead and six wounded. [1435]

[1428] *The Chronicle Telegram*, Elyria, Ohio, United States, 6th February 2000
[1429] *International Herald Tribune*, Paris, France, 17th July 2000
[1430] *European Stars and Stripes*, Darmstadt, Germany, 17th July 2000
[1431] https://www.rferl.org/a/1093486.html
[1432] Government of India Ministry of Civil Aviation and Tourism Commission of Railway Safety Annual Report for 1999-2000
[1433] *Dawn*, Karachi, Pakistan, 6th April 2000
[1434] www.axom.faithweb.com/news/aug1.html
[1435] http://www.axom.faithweb.com/news/aug1.html

Eleven people died and about 40 were wounded when a bomb, probably concealed in a piece of luggage, exploded in the rear coach of the *Sabarmati Express* from Muzaffarpur to Ahmedabad at about 21:50 on Monday 14th August 2000. The explosion occurred near Rauzagaon, Uttar Pradesh, India, about 30 kilometres from Faizabad, where the wounded were admitted to hospital. Railway spokesman Akshay Kumar said that because of the holiday there had only been about 70 passengers in the compartment (coach); had the train been full as normal the number of dead would have been much higher. It was theorised that the bomb might have been set off accidentally on the train. Police blamed Pakistan's Inter-Services Intelligence agency, involved in an armed struggle with India over border territories between the two countries. [1436]

An explosion occurred in the GS coach of a Northern Railway passenger train between Shergarh and Raman stations, Punjab, India, on Thursday 26th October 2000. Two passengers were killed, 17 were seriously hurt and two were slightly wounded. Damage was estimated at 50,000 rupees.

The Commission of Railway Safety made three recommendations in relation to the attack:

(1) Police escorts might be re-introduced on all passenger trains, wherever they had been withdrawn, particularly on trains passing through sensitive areas.

(2) Telephone communications between manned level crossings and adjoining stations should be provided with a minimum of delay wherever such works had been sanctioned.

(3) The reliability of overhead control communication should be of a very high order. Priority should be given to replacing them with underground cables. Also better co-ordination with the Centre for the Development of Telematics (a government-owned telecommunications development) was essential for the lines maintained by them. [1437]

The Oromo Liberation Front (OLF) in a press release claimed that it had derailed a train carrying troops and weapons near Kora, Ethiopia, on Tuesday 14th November 2000. It stated that a mine had wrecked the engine and six waggons, causing heavy casualties, and accused the Ethiopian government of moving troops from the northern part of the country to its southern and eastern parts in a "futile attempt to put out the liberation struggle of the Oromo people." It added that the previous week, another freight train taking vegetables and goods from Dire Dawa to Djibouti had been derailed near the town of Hada-gala, though the statement did not make it clear if this, too, had been due to sabotage or had been an accident. [1438]

A series of almost simultaneous bomb explosions occurred in Manila, the capital of the Philippines, on Saturday 30th December 2000, which was Rizal Day, a public holiday in memory of national hero José Rizal, executed by the Spanish in 1896 when the country was still a Spanish colony. Bombs exploded on a bus, at the airport and so on, but the most serious attack in terms of casualties was an explosion on a Manila Light Rail Transit System train close to Blumentritt station, which killed at least 11 people and injured about 50. [1439]

Ticket Clerk Mari Vicpaglan told a reporter: "The train was approaching when I heard the explosion in the front coach. It was so loud. I tried to help them. I felt dizzy because of the

[1436] https://www.news24.com/10-dead-in-indian-train-bomb-20000814

[1437] Government of India Ministry of Civil Aviation and Tourism Commission of Railway Safety Annual Report for 2000-2001

[1438] *Eritrea Profile*, Vol. 7, No. 37, 18th November 2000

[1439] *Barstow Desert Dispatch*, Barstow, California, United States, 31st December 2000 number of people pleading for help."

The government initially blamed the attacks on the Communists, the police being of the opinion that Islamic extremists were more likely to be to blame. [1440]

The police proved to be correct, members of the Jemaah Islamiyah terrorist group later being convicted of causing the bombings. [1441]

At least two people died and 29 were reportedly injured when the *Udayan Express* was derailed between Gunabati and Sadhradi stations, Bangladesh, at about 04:50 on Wednesday 7th February 2001. M.A. Rahim, Director General of Bangladesh Railway, blamed the derailment on miscreants who had removed a fishplate on Bhaichakhali bridge (No. 204). As a result, the engine and leading four coaches of the train had plunged from the structure into a canal. The accident toll would have been higher had not the muddy canal and shock-absorbing steel coaches lessened the impact.

Relief trains were sent to the spot from Akhaura and Pahartoli and work was described as going on around the clock to restore communications. Meanwhile, an investigation into the affair was opened and Communications Minister Anwar Hossain Manju ordered the railway to take extra measures to prevent a repetition of the attack, including running trains at reduced speed. [1442]

The sabotage was related to a countrywide protest against a High Court ruling the previous December banning existing fatwas, or Islamic edicts, which could subject women to torture for alleged adultery and prevent their mixing and working with men. [1443]

Farmers working during the night on a rice-irrigation project observed about 20 people doing something stealthily on the bridge and in response to their information, police arrested a local student, who apparently had links with international Islamic militants.

Traffic on the Dhaka – Chittagong main line was restored the next day, though the engine and coaches, which had fallen 25 feet into the canal, remained there pending the obtaining of a breakdown crane capable of retrieving them. Meanwhile, cases had been filed against a number of individuals under Sections 8.10 and 12 of the Public Safety Act and Section 126 of the Railway Act. [1444]

Driver Shariff of Nizamabad – Mudhked passenger train 582 was able to avert a serious accident when upon hearing a strange sound between Dharmabad and Karkheli stations, Maharashtra, India, he applied the brakes. Despite this, his engine and five coaches were derailed and tilted over but no one was hurt in the derailment at 00:45 on Wednesday 28th February 2001. Upon alighting from the engine he discovered that the rails and fishplates had been removed over a length of 30 metres. Locals aggrieved at the allocation of funds in the railway budget were suspected of having been responsible for the sabotage. [1445]

An explosion derailed a passenger train in the southern Songkhla province, Thailand, on Saturday 7th April 2001, slightly injuring seven people. Four people died in other attacks the

---

[1440] *Appeal-Democrat*, Marysville, California, United States, 31st December 2000

[1441] Wikipedia – Rizal Day bombings

[1442] *The Independent*, Dhaka, Bangladesh, 8th February 2001

[1443] www.expressindia.com/news/daily/20010207/00703001.htm

[1444] *The New Nation*, Internet Edition, 9th February 2001

[1445] *The Hindu*, Chennai, Tamil Nadu, India, 1st March 2001 same day on a station and a hotel. The Patani United Liberation Organisation, fighting for an independent Islamic state in the south of the country, was blamed for the attacks. [1446]

Northern Railway train No. 4007 Up, the *Sadbhwanna Express* from Muzaffarpur to New Delhi, was derailed between Koiripur and Maharani Paschim stations, Uttar Pradesh, India, on Saturday 19th May 2001, on the Zafarabad – Sultanpur broad-gauge section of the Lucknow Division of the railway. One passenger died, two were seriously injured and two were slightly injured. The Commission of Railway Safety determined that the derailment had been due to 'tampering with the track' by unknown persons. [1447]

A bomb exploded in a coach of Northern Railway Down passenger train No. 372 from Hardwar to New Delhi near Km. 39/12-11 between Modinagar and Muradnagar stations, Uttar Pradesh, India, at 15:40 on Tuesday 14th August 2001. Three people were killed and 30 were wounded. The wounded were taken to the civil hospital in Muradnagar. [1448] [1449]

Seven of the 12 coaches of the *North East Express* were derailed by a bomb at about 19:00 on Tuesday 25th September 2001 between Chaprakata and Bijni, Assam, India, near the village of Baghmara on the Northeastern Railway. About 100 people were hurt but no one was killed, despite the damage done to the coaches, one of which burst into flames. The explosives had been concealed beneath a bridge and set off by a radio-controlled device. The outlawed National Democratic Front of Bodoland was thought responsible. [1450]

The rear coach of a Riga – Vilnius passenger train was derailed near Kaunas, Lithuania, after the engine and four coaches had passed the spot safely. The derailment occurred on Friday 6th July 2001, which also happened to be the country's State Day. No one was hurt. The derailment was clearly the result of sabotage, rail bolts having been unscrewed over a three-metre length of track. [1451]

A passenger train heading from Luanda to Dondo, Angola, was derailed by União Nacional para a Independência Total de Angola (UNITA) guerrillas who had planted an anti-tank mine in the track on Sunday 12th August 2001. The fuel tank of the engine was punctured in the derailment, starting a fire, and the guerrillas opened fire on the passengers who tried to flee the burning train. An initial government statement indicated that 16 people had died but the final death toll among the probably 800 people on the train would be much higher than this. In addition, over 50 victims were early on stated to have reached hospitals in Luanda and Ndalatando, some in a serious condition.[1452] The attack occurred at a location variously described as Maria Teresa and Zenza do Itombe (the places are within a few kilometres of one another.)

The train consisted of four passenger coaches, two general freight waggons and two

---

[1446] www.dawn.com/2001/04/08/intl.htm
[1447] Government of India Ministry of Civil Aviation and Tourism Commission of Railway Safety Annual Report for 2001-2002
[1448] https://m.Rediff.com/news/2001/aug/14blast.htm
[1449] Government of India Ministry of Civil Aviation and Tourism Commission of Railway Safety Annual Report for 2001-2002
[1450] *Dawn*, Karachi, Pakistan, 26th September 2001
[1451] ...909/06/2002Pravda.RU Lithuania: Train Crash Near Kaunas Tracked Down To Sabotage
[1452] www.africast.com/article.php?newsID=21529

waggons loaded with drums of petrol. A survivor stated that the driver had attempted to keep the train moving following the explosion, until the attackers had shot him dead. Some of the guerrillas had been in uniform, others were in civilian clothes. [1453]

UNITA subsequently admitted responsibility for the attack, stating that the train had been a legitimate target because it "was being escorted by a battalion of the government army with munitions." The attack coincided with a United Nations visit intended to determine whether conditions were right for elections to go ahead the following year. By the time of this later report on Tuesday 14th the bodies of 152 victims had been buried in a common grave near the scene of the attack and 146 people had been reported wounded. [1454]

By the following day, the number of known dead had risen to 252.

The final number of reported dead was no less than 422, with another 229 people stated to be missing and unaccounted for. [1455]

At about 09:00 on Thursday 27th December 2001 about 50 members of the Karen National Liberation Army, the armed wing of the Karen National Union, derailed a Mawlamyinr – Ye passenger train between Htinshu and Kyaungywa stations, Myanmar. Two passengers were killed and six were wounded, while another 16 were taken hostage. [1456]

Maoist Communist Centre (MCC) ultras blew up a section of track at a level crossing near the station at Parasnath, Jharkhand, India, in the early hours of Wednesday 20th February 2002, derailing 12 compartments (coaches) of the Hatia – Parna express. No casualties were reported. The action was part of a 24-hour bandh (general strike) in protest at a Prevention of Terrorism Ordinance brought against Naxalite groups. [1457]

Two men and a woman died and 28 people were wounded, 10 seriously, when a bomb exploded at midnight on Wednesday 13th March 2002 in a coach of the Northern Railway *Dhanbad Express*, near Doraha station, Punjab, India. The coach caught fire after the explosion, causing most of the injuries. Some of the passengers also jumped off the train in panic. Most of the victims were migrant labourers.

Chief Minister Amarinder Singh stated that the bomb had been of a crude nature with an explosive of low intensity. He stated that the government would award 100,000 rupees to the families of the deceased and 5,000 rupees to the wounded.

There was a suggestion that Bhushan Ram, one of those who had died, had been the person carrying the bomb and that it had exploded prematurely. [1458]

Ninja militiamen who had refused to abide by a nationwide disarmament programme derailed a passenger train in the Pool department of the Republic of Congo at an unspecified date in March 2002, killing two passengers.

In a further attack in Loulombo, Pool, Republic of Congo, on Saturday 24th August 2002

1453 www.africast.com/article.php?newsID=21627
1454 www.africast.com/article.php?newsID=21636
1455 www.actsa.org/Angola/apm/apm0801.htm
1456 http://www.myanmar-information.net/press/Article2002.../jan19.ht
1457 *Hindustan Times*, New Delhi, India, 20th February 2002
1458 https://www.tribuneindia.com/2002/20020315/main7.htm

Ninja guerrillas derailed a freight train from the port of Pointe-Noire to the capital Brazzaville with a mine. An initial report stated that at least 30 people were injured, though there was no report of fatalities. However, a further report stated that the bodies of 12 unauthorised passengers who had secretly boarded the train had been recovered from the wreckage. After the derailment, the guerrillas looted the derailed waggons of food and medicines, with the assistance of several hostages, and set fire to fuel containers. [1459] [1460]

What was described as a 'powerful land mine' derailed another freight train between Pointe-Noire and Brazzaville at an unstated location in the Pool department of the Republic of Congo on about Monday 30th September 2002. A waggon was blown off the line, an engine was damaged and a number of people were injured. [1461]

On Thursday 2nd May 2002 leftist guerrillas derailed an empty coal train on the Cerrejón Norte – Puerto Bolivar line in Colombia. At 05:00 on Thursday 6th June another empty coal train was likewise derailed by sabotage on the same line. [1462]

Eleven people died and 44 were injured, 14 seriously, when train No. 2402 Down, the New Delhi – Patna *Shramjeevi Express* of the Northern Railway, was derailed at about 03:35 on Sunday 12th May 2002 between Khetasarai and Mehrawan stations, Uttar Pradesh, India, on the Faizabad – Zafrabad section of the line. The second of the two engines of the train and the following 13 of the 24 coaches of the train left the track. The train had been diverted from its normal route due to the derailment of a goods train. Railway officials attributed the derailment to sabotage, though government officials blamed the failure of a bridge. [1463]

A newspaper reported on the reaction of some of the survivors of the accident in arriving at Patna. 'Reaching the familiar environs of Parna, Vinod Paswan, Rajni and Ashutosh Sinha felt a kind of relief that they had never experienced before. They and other passengers were brought to Patna by a special train… "It was the happiest moment in my life when I smelt the air of Patna (on Sunday night). I was lucky to reach my destination unscathed," said Rajni, a Delhi based student.

'However, Paswan was not so fortunate. He lost an ear in the mishap but was still glad that he is alive to tell the tale. "When the train derailed it felt like an earthquake had hit us. The bogie rolled thrice over after the mishap," he recalled while getting first-aid from a government doctor.

'Their faces pale from the experience of having escaped death and grievous injury, the survivors recalled the traumatic event. "I found myself in a nearby field after the accident as I was sitting at the door of a coach," said Nalindrer Kumar. He praised the Muslim students of a local madarsa (seminary) near the accident site. "They were the first to help us and provided much relief," he said. Similar praise came from a grateful Ganga Prasad Jha.

'However, for 60-year-old Ayesha Khatoon, it is not yet over as there is no trace of her only young son. "I was sleeping when the accident occurred. I survived the accident but could not locate my son Javed as it was dark. I requested the police and other people to search for him," a worried Khatoon said…However, the arrival of the special train carrying survivors did not bring joy to everyone, as some anxious relatives got upset on not finding their dear or near

[1459] www…/0.1113,2-11-39_1247462_00.htm
[1460] *CNN.com*, 27th August 2002
[1461] www.new.../0,1113,2-11_1265383,00.htm
[1462] http://www.petroleumworld.com/story8247.htm
[1463] http://alfa.nic.in/1sdeb/1s13/ses9/130502.html

ones. Sisters Pallavi and Sangita could not find their brother-in-law who was travelling in the ill-fated train along with his wife. The only ray of hope was the fact that their names were not in the list of those injured or deceased, they said.' [1464]

A goods train carrying grain from Zimbabwe was derailed at an unstated location in Mozambique on Wednesday 5th June 2002. Then on Friday 14th June, a goods train from South Africa carrying cement and citrus fruit was derailed 67 kilometres from Maputo, Mozambique, though this time only two wheels of one waggon left the rails. Both derailments were due to someone having loosened the rails.

Two different motives were suggested for the sabotage; one was robbery and the other was attacks by people who were unhappy with the recent restructuring of the railways and ports, which had resulted in job losses. [1465]

A bomb exploded on the track as a train was passing Lod, Israel, during the early-morning rush hour of Sunday 30th June 2002. Three people were slightly wounded and train services were disrupted. Although there had previously been frequent suicide attacks on buses, this was the first reported attack on a train. [1466]

A bomb exploded in a sleeping compartment of a Northern Railway Bareilly – New Delhi passenger train as it was stopped in Chandausi station, Uttar Pradesh, India, at 20:05 on Wednesday 3rd July 2002, injuring 10 people. [1467]

About a fortnight later it was reported that the police had arrested two local criminals named Badaun and Veermal, who had planted the bomb to implicate a gram pradhan (president of a village council) with whom they were involved in a long-standing feud. [1468]

The force of the explosion of a remotely controlled bomb weighing about 11 pounds beneath a passenger train near the city of Yavneh, Israel, was sufficient to injure its driver in the abdomen, though the train was not derailed. The attack, on Sunday 21st July 2002, was blamed on Palestinian militants. [1469]

Three crude bombs exploded on a Rampurhat – Barharwa passenger train on Thursday 15th August 2002 as the train was nearing Kotalpukur station, Jharkhand, India. Fifteen people were injured. Police believed that the bombs were being ferried by miscreants for use elsewhere and had gone off accidentally. [1470]

Eastern Railway train No. 2301, the Up *Rajdhani Express* from Howrah to New Delhi, was derailed at about 22:39 on Monday 9th September 2002 at Km 508/19-21 on the approach to the Dhawa bridge (No. 445) between Rafiganj and Deo Road stations, Bihar, India, on the

1464 *Redcliff.com*, 19th August 2002
1465 http://news.xinhuanet.com/english/2002-06/15/content_442545.htm
1466 *BBC World News*, 30th June 2002
1467 https://www.tribuneindia.com/2002/20020704/spotlite.htm
1468 https://timesofindia.indiatimes.com/city/lucknow/two-held-in-chandausi-train-blast-case/articleshow/16195905.cms
1469 *The Daily Times*, Farmington, New Mexico, United States, 22nd July 2002
1470 https://www.satp.org/other-data/india-maoistinsurgency-jharkhand-pakur/explosions_2002

double-line electrified section of the Mughalsarai Division of the railway. The engine remained on the line after the derailment, the leading four coaches fell down the embankment at the Delhi end of the bridge, the following coaches fell with the bridge, which collapsed, and were piled up in the river, and only the 16th coach (derailed) and the 17th and 18th coaches at the rear of the train remained on the line at the Howrah end of the bridge. The results of the accident were very serious – 107 passengers lost their lives, 60 were seriously hurt and 117 were slightly hurt. (This is a corrected official death toll. The initial official figure was somewhat higher, and a Wikipedia account speaks of 130 bodies recovered, with some being recovered from the river some weeks after the accident.)

Rescuers, who including local military personnel, were hampered by the region's poor roads, which had become muddy due to recent rain. Locals attempted to give what aid they could, and 125 people were pulled to safety by morning, but nothing could be done for those trapped in the carriages that had fallen into the river. There had been over 1,000 passengers on the train. [1471]

No less than 143 people gave evidence at the Commission of Railway Safety inquiry into the accident, beginning with several witnesses who spoke of fishplates and nuts being found lying around on the approach to the bridge after the accident.

Driver Ramdeo Shaw told the inquiry that his train had been running at 126/127 km/h as it arrived at the bridge and there had been no indication of any problem until he had felt a jerk near the bridge as if his train had hit something, his engine had begun shaking and the brake pressure dropped, when he had applied the brakes in emergency and stopped at Km 509/7. Going back, he had found that the rear axle of his engine was derailed and nearly all of the train behind was wrecked, as stated above.

Co-driver A.K. Sinha corroborated the driver's account.

Locomotive Inspector Umesh Singh, who had also been on the locomotive, testified that the driver and co-driver had been alert and had properly exchanged signals at different points.

Guard D.N. Singh Yadav spoke of having felt a number of severe jerks at the bridge and seeing, after the train had stopped, the extent of the accident with the collapsed bridge and coaches in the river or dangling from the structure. He had then set about protecting the line and had advised Radiganj station using his walkie-talkie radio. He had also personally seen a fishplate, nuts and bolts removed from the track on the approach to the bridge.

Driver Arjun Yadav and Assistant Driver N.K. Singh of train No. BOXNE, which had passed the bridge about 22:18 on the Down line; Driver Baldeo Mahat and Assistant Driver H.K. Pramanik of train No. 3010, the *Doon Express*, which had passed the bridge at about 22:34 on the Down line; Driver Gulam Sah and Assistant Driver Birbal Kunar Sinha, of train No. BCNE/7303, which had passed the bridge on the Up line about 22:00; and Driver Zafar Ahmed and Assistant Driver B.K. Pandey of express train No. 3151, which had passed the bridge on the Up line at about 21:51 on the Up line, all denied having seen anyone near the bridge, or having heard of any sabotage in the area.

Rafiganj Stationmaster J.K. Singh described the traction power going off about 22:40 and then having been informed of the accident, first by the driver and then by the guard of train No. 2301, whereupon he had informed Control and various other officials.

Assistant Stationmaster M. Siddique in Rafiganj East cabin, Assistant Stationmaster H. Rehman in Rafiganj West cabin, Assistant Stationmaster Ram Briksh Yadav in Guraru East cabin and Stationmaster Shakil Ahmed at Guraru all testified that there had been nothing wrong with train No. 2301 when it passed them.

[1471] Wikipedia – Rafiganj train wreck

Junior Permanent Way Engineer Y.M. Singh described finding fishplates and bolts lying beside the line after the accident, also a piece of rail, which he thought had been thrown out during the derailment by the lateral thrust of the engine.

Section Engineer (Permanent Way) B. Naskar gave similar evidence.

Junior Permanent Way Engineer A.K. Vatshyan described work done on the track at the accident site in the previous weeks.

Gangmen Devi Dayal, Ram Dhani and 18 other members of Gang No. 6 gave evidence that the line was only patrolled two times at night.

Chief Controller R. Chaturvedi at Mughalsarai stated that the accident had been reported to him at 22:40.

Divisional Engineer II Abhay Kumar at Mughalsarai described in detail the damage to the track on the approach to the bridge.

The statements of various other witnesses, detailed in a report that extends to 324 pages, will not be reproduced here for the sake of space. The accident report then deals with various tests done and observations made in connection with them.

Discussing the accident, Commissioner Mahesh Chand noted that it could be assumed to have occurred at 22:39. The speed of the train he fixed at 126 km/h.

The commissioner dealt with the various possible causes of the accident in turn:

(a) There was no evidence of any abnormal weather resulting in any damage to the bridge piers; the water level at the bridge had been over 4 metres below the danger level.

(b) The crew of train No. 2301 had not seen any obstruction on the line and there was no evidence of anything having fallen from a preceding train.

(c) There had been no speed limit at the bridge and the train had been travelling within the maximum permitted speed of 130 km/h. The tachograph indicated that the train had been travelling at a uniform speed up to the point of derailment, excluding poor enginemanship.

(d) There was no evidence of any engine defect.

(e) There was likewise no evidence of a defect in the rolling stock, while as the derailment had occurred in mid-section on plain track, a signalling problem could be excluded.

(f) The permanent way in the rear of the derailed coaches had been in good condition.

(g) The masonry of the bridge was in good condition and its steel girders showed no signs of corrosion.

(h) Various witnesses, both railwaymen and members of the public, had reported finding fishplates, bolts and nuts on the approach to the bridge and the three railwaymen on the engine had felt a severe jerk at that point. The commissioner himself stated that he was 'convinced that there was deliberate removal of fishplates and rail sleeper fastenings (elastic rail clips) of the 9.01 m long cess side rail on the Up line, before the passage of the ill-fated train.' There had been two time gaps of 22 and 16 minutes between the passage of trains, which would have given miscreants long enough to interfere with the track, while the area was known to be sabotage-prone. In view of these considerations the commissioner stated that, 'I am convinced that the above derailment took place due to tampering of track by miscreants.'

The commissioner thus concluded that the derailment had been due to sabotage.

The primary responsibility for the accident rested with the saboteurs.

Secondary responsibility rested upon the law and order authorities for not taking adequate pre-emptive measures in an area known for sabotage and upon the railway administration for the same reason.

Blameworthy were Gangmen Sarvashri Kuldeep and Ganpat, working as patrolmen, Section Engineer Naskar, Assistant Engineer Mukesh Gupta, Divisional Engineer Kumar, the Mughalsarai Finance Branch and the Track Directorate of Research Designs and Standards Organisation at Lucknow.

Commissioner Chand made a large number of recommendations:

(1) Stretches of track on the Eastern Railway that were affected by miscreant activity should be identified and speeds should be reduced appropriately with immediate effect.

(2) The following anti-sabotage measures might be considered in the affected area:

   (a) A pilot locomotive should be run ahead of the *Rajdhani Express* and other superfast trains.

   (b) There should be security patrolling of tracks.

   (c) The fish bolts and bolts in switch expansion joints should be burred.

   (d) The elastic rail clips on every third sleeper should be jammed with epoxy or any other suitable material.

   (e) The Railway might consider the safety involved in the running of superfast trains at night in the affected area.

(3) The Railways might take any other appropriate measures that they deemed fit to eliminate accidents on account of tampering with the track by miscreants and also to minimise the consequences of such accidents.

The report noted that the Chief Safety Officer, Eastern Railway, had issued an instruction that: '*With immediate effect and until further advice the speed of all trains running in those divisions* {Danapur, Mughalsarai and Dhanabad} *between 20:00 hours and 05:00 hours should be restricted to 100 Km/h. A light engine should run ahead of Rajhdani Express in those divisions between 20:00 hours and 05:00 hours as a pilot.*'

(4) Detailed and clear-cut instructions should be issued over security patrolling.

(5) There was a need for a further reduction in speed in the Gaya – Dehri - on- Sone section to 75 Km/h at night. A similar review should be made for other sections affected by miscreant activities.

(6) Security patrolling should be done in sabotage-prone areas.

(7) An anti-sabotage design of fittings on concrete sleepers should be evolved and put in place in sabotage-prone areas within a year.

(8) No surplus/released/maintenance reserve track material should be left unguarded near the track, particularly in sabotage-prone areas.

(9) Single cut rails and buffer rails at vulnerable locations such as high embankments, sharp curves, bridge approaches and on major bridges should be eliminated.

(10) Hammers should be provided in air-conditioned coaches so that passengers could break open the glass in case of accidents/emergencies.

(11) Emergency lights should be provided at different places in coaches, which would automatically switch on in case of emergency.

(12) Emergency exits should be provided in all coaches to enable passengers to escape in case of fire, accidents and so on. Although the Railway Board had issued instructions

on this point way back in 1999, the same were yet to be implemented.

(13) Mouldable and shock-absorbing materials with round edges should be used in the internal furnishing of coaches. The seats/berths should be made such that passengers were not thrown out of their positions in case of an accident.

(14) The riding qualities of the coaches of high-speed trains should be checked periodically by an objective system of evaluation by instrumentation.

(15) The dash pot oil should be checked after every trip for the *Rajhdani Express* coaches. A friction-type snubber or any better alternative should be used in place of the present system of dash pot oil arrangement to have maintenance-free service.

(16) Compliance with R.P.C. 4 {a document relating to rolling stock maintenance) should be certified and sent to the Commissioner of Railway Safety for information before the introduction of new trains.

(17) A facility for retrieving data from Memotel-type speedometers should be provided in trip sheds.

(18) Safety precautions for passengers travelling in coaches should be prominently displayed both in coaches and on the back of tickets.

(19) Apart from the existing insurance system available for rail passengers, they should be allowed to be additionally insured by the insurance companies.

(20) Doctors travelling as passengers should be encouraged to divulge their identity by issuing them with concessional tickets on condition that they attended to passengers in case of emergencies/accidents.

(21) The track structure on open-deck bridges should be modified and strengthened.

(22) The renewal of bridge timbers and the wooden layout under points and crossings and special layouts should be completed as soon as possible. Until then, speeds should be reduced as required.

(23) Open-deck plate girders and underslung girders should be discouraged in new constructions.

(24) At least six to 10 satellite phones should be kept at a central place and moved to the site of an accident whenever an emergency arose.

(25) Some 10 coffins (after anti-termite treatment) should be kept in each Accident Relief Medical Equipment van and some more at a central place. These could be moved when required.

(26) *Prima facie* causes of accidents where the Commissioner of Railway Safety was likely to conduct an inquiry should not be announced by the railway authorities.

(27) All technical circulars, letters and policy guidelines pertaining to safety should be made available to the Commission of Railway Safety by the railway board as well as by the zonal railways.

(28) Mobile telephones with roaming facilities should be made available by the railway administration to all the technical officers of the Commission, i.e. the Chief Commissioner of Railway Safety, the Commissioners of Railway Safety and the Deputy Commissioners of Railway Safety. [1472]

[1472] Ministry of Civil Aviation (Commission of Railway Safety) Railway Accident Investigation Report on Derailment of 2301 Up Howrah – New Delhi Rajdhani Express train at Km. 508/19-21 between

Eight bogies (coaches) of the *Quetta Express* from Rawalpindi to Quetta met with an accident at 10:05 on Thursday 26th September 2002 at Bridge No. 130-B, a 200-ft-long structure between Dambuli and Dinghra stations, Pakistan, on the Sibi-Jacobabad section of the railway in Baluchistan province. An early report stated that seven people had died and 57 had been injured, of whom 47 had been taken to hospital in Sibi, with the seriously hurt being transferred to Quetta and elsewhere. Other reports spoke of 14 dead and 60 injured and of the recovery of 12 bodies. The railway favoured the lower figure and Chief Controller Jalil ur Rahman in Quetta added that many of the survivors of the accident had simply made their way to a parallel main road and there obtained transport further. A later report stated that in addition to the 12 bodies recovered from the wreckage, four of the injured had succumbed to their injuries, raising the death toll to 16. [1473]

Four relief trains went to the scene from Quetta and Sukkur together with four road ambulances, while assistance was also provided by local villagers, railwaymen, Pakistani Army soldiers and Army helicopters.

The accident cost Pakistan Railways a total of 25,000,000 rupees, not including compensation paid to victims.

According to eye-witnesses the diesel locomotive and one bogie had crossed the bridge when it collapsed and the leading bogie had fallen into the nullah, followed by the seven following ones. Another source said that only one bogie had actually fallen 4.5 metres from the structure, though two were left hanging from it. A senior Sibi official stated, "Six pillars of the railway bridge collapsed, causing the disaster." The structure had been built by the British over a century earlier, across the bed of a stream that was normally dry but was prone to flooding after heavy rain. Railway officials expressed the opinion that the bridge, which was described as 'creaky', had been weak and had simply collapsed. [1474]

Guard Muhammad Rafiq described feeling a jerk after the engine and leading coach had crossed the bridge. Looking in his mirror, he had seen coaches falling and a cloud of dust. He said that an air-conditioned coach, two sleeper cars and five economy-class coaches had left the track, the engine and leading a rear three carriages remaining on the line. [1475]

Railwayman Mohammed Afzal, who had been a passenger on the train, said that he had been shaken from his sleep when the train had suddenly come to a stop in a remote area. "I looked out of a window and saw two carriages hanging over the bridge," he related. Six of the dust-covered coaches had been lying on their sides. The engine had remained on the line and proceeded to a station 20 kilometres away for help.

Although the cause of the accident was apparently obvious and due to the failure of an old bridge, Transport and Communications Minister Javed Ashraf spoke of wooden sleepers apparently having been removed and the possibility of sabotage, adding that all but the leading coach had fallen into the dried-up river bed.

On 2nd October General Manager (Operations) Iqbal Samad Khan made a statement that seemed to strengthen the sabotage theory, saying that there were signs of a derailment a rail length short of the bridge, where a fishplate had apparently been removed. Three bridge pillars had been damaged but he attributed this to the results of the derailment and not to any inherent weakness of the bridge. The final verdict on the cause of the accident, though, would have, he

---

Rafiganj and Deo Road stations on the Gaya – Mughalsarai Grand Chord Double Line Broad Gauge Electrified Section of Mughalsarai Division of Eastern Railway on 09-09-2002.

1473 www.balochistanpost.com/item.asp?ID=2845

1474 *Dawn*, Karachi, Pakistan, 27th September 2002

1475 http://w.../default.asp?page=story_27-9-2002_pg1_

said, to await the results of an inquiry. [1476]

South Central Railway (SCR) train No. 7685, otherwise the *Bangalore Express*, which had left Kachiguda, near Hyderabad, at 19:00 the previous evening, was derailed at about 00:35 on Saturday 21st December 2002 between Pagidirai and Pendekallu stations, Andhra Pradesh, India. The train was travelling at 90 km/h when the engine and nine of its coaches – two general compartments, a luggage van and six sleeper compartments – left the track, seven of the coaches overturning. Nineteen people were killed outright or fatally injured and about 87 were hurt, some seriously.

Driver Koteshwar Rao told reporters at Pendekallu station that the derailment had occurred near a culvert, which two of the bogies had fallen into. SCR General Manager S.N. Singla stated that a preliminary investigation had shown that there was a break in the track and that the Commissioner for Railway Safety, South Central Region, had opened an investigation. [1477]

'Most of the passengers of the Bangalore Express, which derailed in Kurnool district in Andhra Pradesh in the early hours today were fast asleep, not realising what fate had in store for them. "It was around one in the night when I tumbled down from my middle berth and the next thing I could feel was blood streaking down my face," 52-year-old Surendranath Modi, almost on the verge of tears, said. Modi, working as Principal of a Junior College in Mahbubnagar in Andhra Pradesh, was coming to Bangalore to visit his daughter working in software major Infosys here. "I was administered first aid and now I am being taken to the railway hospital here. I just want to see my daughter," a visibly shaken Modi, who has a bandage around his forehead, told PTI.

'For the couple, Perves and Ayesha Banu, it was a horrifying experience. "The train took a sharp turn in the night and we fell down with great force, not knowing what had happened." Perves, a car driver in Bangalore, had gone to Kurnool to visit his in-laws and was on his way back with his wife when the mishap occurred. The couple were sent home after being administered first-aid.

'"I just cannot get over the experience. I just want to get out of this place (railway station)," 47-year-old Indrani Bhattacharya from Kolkata, who had a narrow escape, said. She had arrived here on a special train from Kurnool. Indrani, along with her niece Swati Mukherjee on a tour, was coming to Bangalore to visit her married daughter who is settled here. "Late in the night, I fell down from the middle berth followed by my niece sleeping in the upper one. We knew that something had gone radically wrong. And once we struggled out, we could hear wails coming from the tilted bogies. I am thankful to God that I am alive," Indrani, in the middle of a tearful reunion with her daughter who had come to the station to receive her, said.

'R.V. Haranur, and his wife, one of the few composed unhurt passengers, who were coming to Bangalore from Hyderabad to attend a marriage function, said, "The train was going at breakneck speed when the driver applied sudden brakes bringing the train to a grinding halt. When we came out, we could see some derailed bogies in the moonlight and hear groans. Since the accident came in the wee hours, help came only at 4 a.m., when we were put into a relief train to Guntakal. I just cannot forget the shell-shocked expression of a pregnant woman in her late twenties who was in our compartment. The poor lady started crying her heart out and it took a while for fellow passengers to pacify her," Haranur said.' [1478]

The cause of the derailment was disputed. General Manager Singla stated, "It was all a planned act of a Naxalite group." He said that the rail had been cut and iron filings found at

[1476] https://www.emerald.com/insight/content/doi/10.1108/dpm.2003.07312cac.004/full/html
[1477] www.outlookindia.com/pti_print.asp?id=106398
[1478] www.outlookindia.com/pti_print.asp?id=106385

the accident site below the tracks also indicated the use of hacksaw blades, four or five of which had been recovered nearby. In addition, the texture of the filings powder had been smooth, while had it been a normal (accidental) break the texture would have been granular. “A rail has been cut. Who has cut it, why he has cut it, it is for the police to find out,” Singla added. He explained that the track at the broken joint had been covered with a piece of cloth so that the break would not be noticed by the train driver. Meanwhile, the track had been ready to bring back into use but the reopening had been delayed by the police, pending examination of the site by forensic experts. [1479]

Some state government officials and police, however, believed that the excessive speed of the train or poor track maintenance could have caused the derailment. T. Devender Goud, Andhra Pradesh home minister, stated that the speed of the train at the derailment point, over a line opened only the previous year, should have been limited to 40 km/h, but had instead been 90 km/h and, “This could have caused the accident.” [1480]

Allegations that substandard track had caused the derailment were dismissed as “politically motivated” and “without any rhyme or reason” by a top railway official. [1481]

The Commission of Railway Safety concluded that the derailment had occurred: ‘due to tampering with the track by way of cutting the left-hand rail with a hacksaw and thereafter pushing its Pagidirai end inside by an unauthorised person other than railway staff.’ [1482]

The Northeast Frontier Railway *Kamrup Express*, from Dibrugarh to Howrah, was derailed at about 01:30 on Monday 20th January 2003 just after it had left the station at Diphu, Assam, India, the engine, an empty general compartment, a parcels van and a brake van leaving the rails. No one was hurt. A spokesman stated that a bomb planted on the track had caused the derailment, the explosion leaving a crater almost a metre wide and about half a metre deep. The single broad-gauge section of line between Diphu and Lumding was blocked and about 60 replacement buses were brought into use. Police were unsure who was responsible for the attack, possibly the United People’s Democratic Solidarity group or the Nagas. [1483]

Ninja rebels derailed a military train from Pointe Noire to Brazzaville, near Loulombo, Pool department, Republic of Congo, on Wednesday 5th February 2003. Ordinary passenger trains having been suspended since the previous April due to guerrilla attacks, a number of civilians were also riding on the train and one soldier and one civilian were killed, while several other people were injured. [1484]

A bomb exploded in a first-class ladies’ compartment of a Central Railway local train, which had left Mumbai CST station at about 20:00 on 13th March 2003 for Karjat. The explosion occurred at about 20:45 just as the train had reached Mulund station, Maharashtra, India. It was powerful enough to rip off the roof of the compartment and first reports said that seven people had been killed outright and about 30 were wounded.[1485] The figures were later increased to 11 dead and 65 wounded.

1479 www.outlookindia.com/pti_print.asp?id=106512
1480 www.guardian.co.uk/worldlatest/story/0,1280,-2265753,00.html
1481 www.outlookindia.com/pti_print.asp?id=106513
1482 https://eparlib.nic.in/bitstream/123456789/759683/1/lsd_13_12_20-02-2003.pdf
1483 www.outlookindia.com/pti_print.asp?id=113324
1484 www.guardian.co.uk/worldlatest/story/0,1280,-2386356,00.html
1485 www.gunnah.com, 13th March 2003

'The deafening explosion as the train was pulling into Mulund suburban station panicked commuters, most of them office workers on their way home. Briefcases, handbags and paper files fluttered down on platform 3 at Mulund as office workers, reinforced by the fire department rescuers and rail staff, rushed the injured to hospitals. The impact was so powerful it blew a gaping hole in the metal structure and left seats a heap of twisted metal.'

The banned Students' Islamic Movement of India, which wanted India to become an Islamic state, was suspected of being behind the attack. [1486]

National Democratic Front of Bodoland militants were suspected of being responsible for a bomb that derailed several waggons carrying cement belonging to a goods train at Sesapani, between Fakiragram and Kokrajar, Assam, India, at about 21:30 on Friday 20th June 2003. No one was hurt. Their target had been the Northeast Frontier Railway *Avadh Assam Express*, which usually passed at about the hour of the attack. [1487] [1488]

At about 02:00 on Sunday 6th July 2003 the engine and two coaches of a metre-gauge South Central Railway Pakala – Bellary passenger train were derailed between Kurapalakota and Thummalakunta stations, Andhra Pradesh, India. Assistant Driver D. Mani jumped from the engine and was killed, Driver N.S. Gupta suffered head injuries and 22 passengers were slightly hurt. Minister of State for Railways Bandaru Dattatreya blamed sabotage for the accident, stating, "According to information I received, 13 boulders were found on the track resulting in the engine getting capsized. Some sleepers and a fishplate were also missing, which clearly points to the hand of some anti-social elements in the accident." [1489]

Maoist Communist Centre activists blew up tracks with dynamite and derailed three or four trains in Bihar state, India, on Monday 14th July 2003. Three coaches of East Central Railway Muzzafrapur – Gorakhpur train No. 519 were derailed between Bhairoganj and Kharpokhra; train No. 285, the Darbhanga-Narkatiaganj Express, was thrown off the line by the removal of fishplates near Nadda; three coaches of train No. 5219, the Khula Dharbhanga express, were derailed at Randhadrapur; and there was possibly a fourth derailment. All occurred at low speed as trains were entering or leaving stations and there were no casualties. The MCC had organised a bandh (general strike) against increasing police brutalities. [1490]

At least five people died when two apparently homemade explosive devices were set off beneath a passenger train near Kislovodsk, Russia, in the Stavropol region of Chechnya, at about 07:00 on Wednesday 3rd September 2003. Several coaches of the train, a local service from Kislovodsk to Minaralniye Vody, caught fire after the explosion, and at least another 29 of its approximately 50 passengers were wounded. [1491]

Rebels reportedly blew up an oil train at an unstated location in the Chechnya region of Russia on about Monday 27th October 2003. Two waggons of the train were derailed but there was no fire. A serviceman was killed in the attack. [1492]

A powerful bomb tore in half a carriage of passenger train No. 6309, Kislovodsk to

[1486] www.news24.com/News24/World/News/0,6119,2-10-1462_1333159,00.html
[1487] www.deepikaglobal.com/latestnews.asp?ncode=3219
[1488] http://sify.com/printer_friendly.php?id=13177631&ctid=2&lid=1
[1489] www.Rediff.com/news/2003/jul/06ap.htm
[1490] *Mid Day Mumbai*, Mumbai, Maharashtra, India, 15th July 2003
[1491] https://www.aljazeera.com/news/2003/9/3/russian-commuter-train-blown-up
[1492] http://www.guardian.co.uk/worldlatest/story/0,1280,-3320547,00.html

Mineralnye Vody, in an attack on the train a few kilometres from Yessentuki (also spelt Essentuki), Chechnya, Russia, on Friday 5th December 2003. The wreckage then caught fire. An early report stated that at least 28 people had died on the spot, four after they had been taken to a local hospital and four elsewhere. The number of dead later rose to 46, while about 100 passengers were wounded and taken to hospital and another 50 with minor wounds were treated on the spot. Among the dead was the apparent suicide bomber, whose body parts investigators found next to a bag, which had contained plastic explosives equivalent to 10 kilograms of trinitrotoluene (TNT).

'Television images showed one of the carriages as a mangled wreck of metal with shreds of bloodied clothing hanging from rods poking out. The roof was collapsed and the windows blown out. The other train cars looked intact.'

Russian Interior Minister Boris Gryzlov promised: "We will find those who have committed this terrorist act. The earth will burn under their feet. These animals will never feel safe anywhere. We will find them and punish them according to the law."

Alexander Gurov, chief of the State Duma defence committee, stated to Moscow Echo radio that, "Since practically all our attacks are linked to Chechnya, it is clear that Chechnya rebels are behind the blast." [1493] [1494]

Fuerzas Armadas Revolucionaries de Colombia (FARC) guerillas derailed 17 cars of a coal train between Cerrejón Norte and Puerto Bolivar, Colombia, on Friday 19th September 2003. The carriage of 70,000 tonnes of coal a day was interrupted. [1495]

The locomotive and five bogies of the Northern Railway *Shalimar Express*, bound for New Delhi, were derailed when a bomb exploded in the toilet of its leading coach at about 22:06 on Sunday 26th October 2003 between Samba and Gagwal stations, India. No one was hurt. 'Intelligence officials said the blast may have been the handiwork of militants from across the border {i.e. in Pakistan} who wanted to divert the entire police force of the area had it been a major blast and secure the release of one of their associates held in jail there.' [1496]

United Nations (principally American) forces had invaded Iraq in March 2003 and after less than a month, had toppled the regime of Saddam Hussein, but sporadic opposition to the invading forces continued. On Thursday 30th October 2003 an improvised bomb set fire to four shipping containers being carried on a freight train near Fallujah, Iraq. No one was hurt but locals carried off computers, tents, bottled water and other goods. [1497]

At 09:00 on Saturday 15th November 2003 an explosion derailed a military train between Samara and Tikrit, Iraq. Again, no one was hurt. [1498]

A bomb derailed eight of the 20 carriages of a freight from Baghdad to Monsul train near Samara, Iraq, on Saturday 6th December 2003. There were no casualties. [1499]

A train bound for Habbaniyah was attacked at an unstated location in western Iraq on

---

[1493] https://www.aljazeera.com/news/2003/12/5/forty-killed-in-russian-train-attack

[1494] http://newsfromrussia.com/accidents/2003/12/05/51672.html

[1495] http://reuters.com/printerFriendlyPopup.jhtml?type=worldNews&storyID=3479226

[1496] http://sify.com/printer_friendly.php?id=13292314&ctid=2&lid=1

[1497] *The Guardian*, London, England, 30th October 2003

[1498] www.news24.com/News24/World/Iraq/0,,2-10-1460_1446028,00.html

[1499] www.seacoastonline.com/news/12082003/world/b.htm

Wednesday 31st December 2003 and was derailed by a rocket-propelled grenade. The attackers then stripped the engine of the train of its batteries but took nothing else. [1500]

Militants hurled a powerful bomb at the engine of a goods train carrying onions and Amul (Anand Milk Union Limited) dairy products at about 22:55 on Friday 31st October 2003 when the train was near the station at Dangtol, Assam, India. Driver Niranjan Charkarborty was seriously wounded and Assistant Driver Jyotish Sharma was less badly hurt. [1501]

Not for the first time, an attempt to wreck a passenger train was frustrated on Wednesday 14th January 2004 when by chance it was running late and was preceded instead by a goods train. United Liberation Front of Asom militants had intended the Northeast Frontier Railway *Brahmaputra Mail* to be their target but that day it was running 30 minutes behind time and thus, a goods train was derailed instead, between Namati and Amguri stations, Assam, India. No one was hurt. Fishplates had been removed from the track. [1502]

A suspected suicide bomb tore open a car of a train on the Moscow Metro in Russia at about 08:40 on Friday 6th February 2004, killing at least 39 passengers and wounding another 129. The explosion occurred between Avtozavodskaya and Paveletskaya stations.

Vice Mayor Valery told reporters: "It (the train) had moved 300 metres (from Avtozavodskaya station) when there was an explosion by the first door of the second coach. Although the bomb was not packed with bullets, the blast damaged the metal parts of the coach and blew out its windows. The third coach was also damaged."

Russian Deputy Interior Minister Alexander Chekalin stated that the bomb might have been concealed in a briefcase or rucksack, while police surveillance video reportedly showed two suspicious-looking people, one a man and the other a woman, boarding the train carrying briefcases.

President Vladimir Putin condemned the blast as an act of terrorism, perhaps aimed at pressuring the Russian leadership ahead of the March elections, and repeated that the government would never negotiate with terrorists. He added that: "Only with the united efforts of the world community can we deal with this plague of the 21st century."

Chechen separatist leader Aslan Maskhadov in Grozny denied that Chechen rebels had been responsible for the blast. [1503]

Four almost co-ordinated suicide bombings occurred on the Cercanias commuter train system of Madrid, Spain, on Thursday 11th March 2004, four days before the country's general elections. In the worst terrorist attack in the history of the country, 200 people eventually died and about 2,500 were wounded. The last victim to succumb was in a coma for 10 years before finally passing away in 2014.

The trains involved had all left Alcalá de Henares station between 07:01 and 07:14.

At 07:37 a bomb exploded on train No. 21431 at Atocha station, followed by two more at 07:38 in the fourth to sixth coaches. A fourth bomb failed to explode and was later dealt with

[1500] www.albawabba.com/news/printArticle.php3?sid=267047&lang=e
[1501] www.ndtv.com, 1st November 2003
[1502] www.hinduonnet.com/thehindu/holnus/00414213059.htm
[1503] https://edition.cnn.com/2004/WORLD/europe/02/06/moscow.blast/

by bomb-disposal experts.

At about 07:38, two bombs exploded in the fourth and fifth coaches of train No. 21435 as it was leaving El Pozo del Tio Raimundo station. Two other bombs were later dealt with by a bomb-disposal team.

At about 07:38, a bomb exploded in one coach of train No. 21713 at Santa Eugenia station.

At 07:40, bombs exploded in the first, fourth, fifth and sixth coaches of train No. 17305, 800 metres from Atocha station. [1504]

Of the victims who died, 149 were Spanish, 16 were from Romania, six were from Ecuador, four each were from Bulgaria, Peru and Poland, two each were from Colombia, Dominican Republic, Honduras, Morocco and Ukraine and one each were from Brazil, Chile, Cuba, France, Senegal and the Philippines. At least 34 people were believed to have died in the first bombings, at least 70 at El Pozo, at least 17 at Santa Eugenia and least 59 in the train approaching Atocha. [1505]

Initially, the attacks were blamed on the Basque separatist group Euskadi Ta Askatasuna (Basque Homeland and Liberty) (ETA), although ETA normally gave a telephone warning of its attacks and had never carried out such a serious one. ETA denied responsibility and later the focus switched to Islamic terrorists, protesting at Spain's involvement in the war in Iraq against Saddam Hussein, which the ruling conservative Partido Popular party of President José Maria Aznar had entered into without the approval of parliament. The Moroccan Islamic Combatant Group then became the focus of the investigations into the bombings.

In February 2007 the trial of 29 defendants linked to the attacks began. All were described as local Islamic extremists inspired by the internet, not directly linked to any foreign Islamic group and with no connection to ETA. At the end of October, 21 of them were found guilty and were sentenced to various terms of imprisonment, two of them to more than 40,000 years, though they were not likely to serve more than 40. [1506]

South East Central Railway train No. 406B, an Up Bishrampur – Bilaspur passenger, was derailed at 22:40 on Friday 5th March 2004 near Bridge No. 85 over the Halfali River between Darritola and Udalkachhar stations, Chhattisgarh, India, on the Annapur – Bishrampur broad-gauge, single-line section of the Bilaspur division. Two passengers died in the accident and 24 were injured, 13 seriously. An investigation by the Commission of Railway Safety determined that unknown person had loosened fishplates and elastic clips on the right-hand side of the line on the Darritola approach to the bridge. [1507]

At about 20:43 on Tuesday 23rd March 2004 East Central Railway train No. 2303, the Up *Poovra Express* from Howrah to New Delhi, was travelling at 40 to 45 km/h between Fatuha and Banka Ghat stations, Bihar, India, when a bomb exploded in its S-10 coach. Three passengers were injured, one seriously, and damage estimated at 15,000 rupees was caused.

The Commission of Railway Safety carried out an investigation and made one recommendation:

(a) The Government Railway Police and the State Police might be approached by the Railway Administration to intensify random checks by searches and deploying dog squads etc. to

[1504] Wikipedia – 2004 Madrid train bombings

[1505] http://news.bbc.co.uk/1/hi/world/europe/3504912.stm

[1506] Wikipedia – 2004 Madrid train bombings

[1507] http://pib.nic.in/release/release.asp?relid=1366

deter the anti-social elements from carrying explosive material in the trains. [1508]

Seven of the 18 coaches of express train No. 34 from Moscow to Vladikavkaz, the capital of North Ossetia, were derailed by an explosive device at 07:15 on Saturday 29th May 2004 near the village of Elkhotovo, Russia. It was believed that the device had been planted on the track, though also believed possible that it had been on the train itself. Six people were slightly hurt. [1509] [1510]

A pressure-triggered landmine exploded beneath a train near the station at Alpatovo, Russia, at about 16:00 on Thursday 10th June 2004. One coach was derailed, a serviceman was injured and the track was slightly damaged. [1511]

Late on Saturday 19th June 2004 United Liberation Front of Asom (ULFA) militants derailed eight waggons of a coal train by setting off a powerful explosion. The train driver was injured.[1512] The attack occurred between Tingram and Makum, Assam, India.

ULFA militants derailed a goods train by a bomb placed on the track on Thursday 26th August 2004 between Pratapkhata and Choutara stations, Assam, India, three waggons leaving the line. In other bomb attacks the same day on two buses, six persons were killed. [1513]

A freight train exploded a mine at Genç, Bingöl province, Turkey, on Saturday 28th August 2004, wounding a train driver. Kurdish separatists were suspected of being responsible.[1514] Five waggons of the Tatvan – Elazig train were derailed. [1515]

A train carrying military equipment was derailed by a bomb planted on the tracks between Kale and Kurt stations, Turkey, on Tuesday 14th September 2004. Six waggons were derailed but no one was hurt. [1516]

An act of sabotage in Thailand on about Monday 15th November 2004 was apparently one of those resulting from a grievance held by a former railway employee. The Ching Mail bound for Bangkok ran over three signal posts laid on the track at Tha Ma Fueng village and was derailed. Several passengers were slightly injured.[1517] Worse was an explosion in the fuel tank of the engine, which had ended up in a ditch, which seriously injured two workmen who were attempting to remove it from the underside of the diesel locomotive using a blowtorch.

The perpetrator of the derailment was allegedly one Sanguan Kitsuan, 39, a repairman who had been sacked by the railway three months previously and had carried out the sabotage with the help of three accomplices. [1518]

[1508] http://civilaviation.nic.in/ccrs/accidents/2003-2004/DETAILS%20OF%20%20ACCIDEN...
[1509] http://www.itar-tass.com/eng/prnt.html?NewsID-881547
[1510] http://www.sanluisobispo.com/mld/sanluisobispo/news/world/8790614.htm?template=c...
[1511] http://www.itar-tass.com/eng.prnt.html?NewsID=932072
[1512] www.keralanext.com/news/index.asp?id=39567
[1513] www.webindia123.com/news/m_details.asp?newscode=68668&catcode=ENG3...
[1514] http://www.kurdmedia.com/news.asp?id=5414
[1515] http://www.turkishpress.com/turkishpress/news.asp?ID=25902
[1516] http://www.turkishpress.com/turkishpress/news.asp?ID=27144
[1517] *MCOT: TNA* English news, 21st November 2005
[1518] *The Nation*, Bangkok, Thailand, 21st and 24th November 2004

Ten bogies of the *Meghna Express*, which had left Chandpur at 05:00 on Sunday 16th January 2005 for Chittagong, left the track at 05:45 near the station at Hajiganj, Bangladesh. According to different accounts 40, 100 or even 200 passengers were injured, none apparently that badly, while no one was killed. The railway authorities maintained that only 13 people had been hurt. A preliminary investigation found that fishplates and fish-bolts had been removed from the track. [1519]

Bombs planted by the Balochistan Liberation Army (BLA) damaged the main line between Karachi and Quetta, near Sibi, Pakistan, for the second time in three days on Monday 24th January 2005 and at about 22:00, a trolley carrying five railway engineers who were examining the line was derailed at the point where about a metre of track had been ripped up, injuring them all. [1520]

On Friday 4th February 2005 train No. 21, the Down *Chiltan Express* from Quetta to Faisalabad escaped derailment when the crew of a preceding pilot engine discerned a large crater caused by a bomb at Dera Ghazi Khan. The report did not indicate if the engine was itself derailed or if it was stopped in time. [1521]

On Friday 18th March 2005 both train No. 22, the Up *Chiltan Express*, and train No. 21, the Down *Chiltan Express*, were attacked, this time by bombs placed on the trains themselves. An explosion in the fifth coach of the Up train as it was about to leave Mach station killed one passenger and wounded five more at about 15:15. The second explosion occurred in the Down train in the Mushkaf area; it killed another passenger and again wounded five. [1522]

Train No. 21, the Down *Chiltan Express*, was again the target for a terrorist attack on Tuesday 12th April 2005. A Pakistan Railways official stated that: "A powerful bomb exploded in the toilet of an economy class bogie." This happened at about 16:10 between Aab-i-Gum and Peshi stations. A large hole was blown in the coach but the train did not derail and continued to Peshi station, from where seven injured passengers were taken to Sibi Civil hospital, two of them being admitted. [1523]

A security forces special train, 'combing the track for explosives', was derailed by two home-made bombs, each weighing about 20 kg, which had been planted on the track at kilometre 1031, between Sungai Padi and Koke Saya stations, Thailand, at about 06:15 on Sunday 27th March 2005. As a result the engine was separated from the single coach it was pulling and both 'rolled down the side of the track.' The saboteurs had set off the bombs by detonation cords attached to them and following the derailment, sprayed the train with bullets. Police on the train replied and after a ten-minute gun battle the insurgents retreated. Nineteen of the 11 police and 11 railwaymen on the train were injured by shrapnel and gunshot wounds. Following the attack the railway unions called for stronger security measures in the area to protect their members from further attacks. [1524]

A day after militants had staged concerted attacks at 63 locations to steal arms and probably explosives, they derailed a northbound passenger train, which was just picking up speed after leaving the station at Sungai Kolok, Thailand, in Yala province, on Thursday 27th October 2005. Four bombs had been planted on the track, two of which failed to explode. It was believed that

1519 http://nation.ittefaq.com/artman/publish/printer_15544.shtml

1520 www.dailytimes.com.pk/print.asp?page=story_25-1-2005_pg1_4&ndate=1/25/2005

1521 www.theadvertiser.news.com.au/printpage/0,56492,12145226,00.html

1522 www.hipakistan.com/en/detail.php?newsId=en77198&F_catID=&f_type=source

1523 www.dawn.com/2005/04/13/top11.htm

1524 www.bangkokpost.com/News/28Mar2005_news01.php

they had been set off by remote control. Only one of the approximately 100 passengers on the train was hurt; others jumped from the train in panic after the explosions. [1525]

The locomotive and six bogies of train No. 20, the Down *Khushhal Khan Khattak Express*, from Karachi to Peshawar, were derailed near Bijarani station, Pakistan, on Wednesday 1st June 2005. Driver Baba Rais Ahmed stated that his train had been moving at considerable speed when he had seen that the track ahead was unbolted. He had applied his brakes but the train had left the rails on the approach to the station. No loss of life was reported. [1526]

A pilot engine preceding the East Central Railway *Mahananda Express* from New Delhi to Katihar was derailed between Fatuha and Banka Ghat stations, Bihar, India, on the Danapur division of the railway, on Thursday 9th June 2005. Fishplates and rail clips had been removed from the track. No one was hurt but trains Nos. 3039, the Up *Janata Express*; 3050, the *Amritsar Express*; 3111, the *Lal Quila Express*; and 2391, the *Shramjeevi Express*, were all either delayed or diverted. Railway Minister Lalu Prasad Yadav instructed railway officials to increase track patrolling and hire more gangmen to do track inspections. 'The railway is conducting a probe into the incident. So far no arrest has been made.' [1527]

The driver of train No. 381 from Grozny to Moscow saw an explosion occur on the tracks immediately in front of him at Uzunovo station, Russia, at 07:10 on Sunday 12th June 2005. Four coaches of the train were derailed and 42 people were hurt, though less than 10 were admitted to hospital.[1528] No one was killed. The line was blocked until 05:00 the next morning. Investigators determined that an explosive device with a power equivalent to three kilograms of trinitrotoluene (TNT) had been used and there were wires leading from the crater caused by the explosion to the spot where the perpetrators had hidden. [1529]

A pressure mine exploded beneath a track maintenance train at about 02:00 on Saturday 25th June 2005 between Kurush and Karlan-Yurt, Dagestan, Russia, injuring the driver and assistant driver and derailing the engine and four flat cars. Traffic was then suspended for some time, affecting five passenger and two freight trains. [1530]

On Thursday 7th July 2005 an explosive device went off under a locomotive between Makhachkala and Tarki, Dagestan, Russia, slightly damaging it and the track. There was no derailment and no one was hurt. [1531]

A bomb exploded beneath the leading car of a passenger train at 05:30 on Sunday 24th July 2005, killing a 20-year-old woman passenger and wounding the driver, assistant driver and three passengers. The attack occurred between Khasavyurt and Kizilyurt, Dagestan, Russia. There had only been 16 passengers on the train, which was derailed by the blast. The bomb was stated to have been equivalent to 11 kilograms of TNT. [1532]

A report added that, 'Dagestan, which borders Chechnya, has seen increasing violence believed to be connected with insurgents and criminal gangs. The number of terrorist attacks

---

[1525] http://natiomultimedia.com/2005/10/28/headlines/index.php?news=headlines_18998915.html
[1526] www.dawn.com/2005/06/02/top12.htm
[1527] http://www.patnadaily.com/news05/june/061005/train_accident_averted.html
[1528] http://www.tass.ru/eng/prnt.html?NewsID=2126284
[1529] http://en.rian.ru/russia/20050614/40517155-print.html
[1530] http://www.mosnews.com/news/2005/06/25/dagtrain.shtml
[1531] http://www.interfax.ru/e/B/politics/28.html?id_issue=11332698
[1532] http://www.interfax.ru/e/B/politics/28.html?id_issue=11353550

in the region has more than doubled this year to 70, a human rights group said last week.' [1533]

A Makhachkala – Astrakhan train which had left Kizilyurt at 20:00 on Saturday 27th August 2005 was derailed by a makeshift bomb at about 23:00 at Kilometre 63 of the Kizilyurt – Kurush line, Dagestan, Russia. The engine and two carriages were derailed but preliminary reports indicated that there had been no casualties. Explosives equivalent to 5 kilograms of TNT left a crater two metres in diameter and 70 centimetres deep. [1534]

The engine and two waggons of a freight train from Besian to Nazran were derailed, probably by a remotely controlled device, on Friday 16th September 2005 between Kantyshevo and Maiskoye, Russia, on the border between Ingushetia and North Ossetia. The line was closed after the explosion amid fears that a second explosive device might be present. [1535]

A small explosive device derailed one vehicle of a train on the outskirts of Khasavyurt, Dagestan, Russia, early on Thursday 8th December 2005. No one was hurt and only a metre of track was damaged. [1536]

A remotely controlled mine exploded beneath what was described by Kurdish sources as a military train and by other sources as mail or passenger train No. 52862 close to the station at Süveren, Bingol province, Turkey, at 10:20 on Friday 1st July 2005. Five soldiers or security guards were killed outright when at least four coaches of the train were derailed and 12 people were wounded, one of whom died later. There were about 45 people on the train. A relief train sent to the scene was derailed in turn by a second mine without stated further casualties, while a third mine was discovered and rendered safe. Hêzên Parastina Gel (People's Defence Forces) (HPG), the military wing of the Partiya Karkerên Kurdistanê (Kurdish Worker's Party) (PKK) claimed responsibility for the attacks, which it stated were dedicated to the recently martyred Erzerum Province Commander Serxwebum (Ahmet Okur), who had been killed 'during an operation.' [1537] [1538]

The PKK derailed a freight train near Altunkent, Erzincan province, Turkey, on Thursday 7th July 2005. Seven waggons were derailed and overturned but this time there were no casualties. [1539]

A mail train from Elazig to Muş was derailed by a mine near Kale-Kurt station, Muş province, Turkey, on Tuesday 26th July 2005. No casualties were reported and there was no immediate claim of responsibility. [1540]

A remotely controlled mine exploded beneath a passenger train near the southern city of Gaziantep, Turkey, on Friday 19th August 2005. Both track and train were damaged but it was not derailed and none of its 83 passengers or crew were hurt. [1541]

On the morning of Thursday 7th July 2005 the author was at his tea break at work when he logged on to the computer to find a BBC News report headed 'London Attacks,' which began, 'A series of explosions has ripped across central London, killing at least 37 people and

[1533] http://www.times.spb.ru/index.php?action_id=2&story_id=257
[1534] http://www.tass.tu/eng/prnt.html?NewsID=2356655
[1535] http://www.tass.tu/eng/prnt.html?NewsID=2420424
[1536] http://www.hindu.com/thehindu/holnus/003200512081223.htm
[1537] http://www.kurdishinfo.com/modules.php?name=News&file=print&sid=2860
[1538] http://www.arabnews.com/services/print/print.asp?artid=66311&d=3&m=7&y=2005&h1=Bo...
[1539] http://icwales.icnetwork.co.uk/printable_version.cfm?objectid=15714363&siteid=50082
[1540] http://www.rednova.com/modules/news/tools.php?tool=print&id=188085
[1541] http://www.jpost.com/servlet/Satellite?pagename=JPost/JPArticle/Printer&cid=11244...

injuring many hundreds more.' [1542]

That morning, Islamic terrorists had carried out a series of co-ordinated suicide attacks on public transport in London, England, targeting three underground trains and a bus. The first three bombs exploded on trains within 50 seconds of one another at 08:49. The initial attack was on London Underground eastbound Circle Line six-car train No. 204 between Liverpool Street and Aldgate stations. An explosion on westbound Circle Line six-car train No. 216 quickly followed. This train had just left Edgware Road station for Paddington. The third attack took place on the Piccadilly line, on southbound six-car train No. 311 between King's Cross St Pancras and Russell Square stations. Finally, at 09:47 a fourth explosion occurred on a double-decker bus at Tavistock Square.

The final number of fatalities in the attacks (including the four suicide bombers) was 56. Of the innocent victims, 32 had been British nationals, three were from Poland and one each was from Afghanistan, Australia, France, Ghana, Grenada, India, Iran, Israel, Italy, Kenya, Mauritius, New Zealand, Nigeria, Romania, Sri Lanka, Turkey and the United States (joint American/Vietnamese citizenship). Six of the underground victims had died at Edgware Road, seven at Aldgate and 26 at King's Cross. The confined space of the deep-level Piccadilly line as opposed to the relatively spacious 'cut-and-cover' Circle Line would have increased the effect of the blast there. Almost 800 other people were wounded.

The suicide bombers were later identified as Hassib Hussain, 18; Germaine Lindsey, 19; Shehzad Tanweer, 22; and Mohammad Sidique Khan, 30. Three of them were immigrants from Pakistan, the fourth, Lindsey, was Jamaican and a convert to Islam. None of them had previously been known to the authorities. Their stated motives, recorded on a pre-attack videotape, were to induce the British government to withdraw its forces from Afghanistan and Iraq and to punish British civilians for supporting a government they believed continued to oppress Muslims in Palestine, Iraq, Afghanistan and Chechnya. [1543]

On Thursday 21st July 2005 four further bomb explosions occurred in London, England. Again, there were three on London Underground trains and one on a bus and again, they were carried out simultaneously. The train attacks occurred at Warren Street, The Oval and Shepherd's Bush stations, the bus attack on the Hackney Road. None of the explosions was very powerful, no one died and only a few people were hurt. [1544]

Train No. 2381, the *Shramjeevi Express* from Patna to New Delhi, was between Hariharpur and Koiripur, Uttar Pradesh, India, at about 17:15 on Thursday 28th July 2005 when a powerful bomb exploded in the general compartment third from the engine. An initial report spoke of nine dead and 57 wounded, with 20 of the wounded being in a critical condition. Alerted by the blast and fire, local villagers rushed to the spot and the wounded were taken to Badlapur hospital, the most seriously hurt being transferred to the Banaras Hindu University Hospital in Varanasi. The Northern Railway Lucknow Division accident relief train was also dispatched to the scene, arriving there at about 19:15, while medical vans were rushed from Lucknow, Faizabad, Varanasi and Mughalsarai. Initially it was not clear if the explosion had been due to a bomb or was the result of the accidental explosion of a gas cylinder.

Uttar Pradesh Chief Minister Mulayam Singh Yadav announced an *ex gratia* payment of 1 lakh (100,000) rupees to the families of the dead, 50,000 rupees to the seriously injured and 25,000 rupees to those who were slightly hurt. [1545] [1546]

[1542] http://news.bbc.co.uk/1/shared/spl/hi/uk/05/london_blasts/html/default.stm
[1543] Wikipedia – 7 July 2005 London bombings
[1544] http://www.guradian.co.uk/worldlatest/story/0,1280,-5157507,00.html
[1545] www.telegraphindia.com/1050729/asp/others/print.html

Bomb-disposal experts also went to the accident scene and soon confirmed that the explosion had been due to a bomb, apparently concealed in a suitcase and set off by a timing device. Suspicion fell on two young men who had appeared nervous and had twice tried to stop the train by pulling the communication cord. When other passengers had prevented them doing this, they had jumped from the moving train just before the blast. [1547]

The interim report of the Commission of Railway Safety into the explosion concluded that it had been due to the setting off by a timer of a mixture of RDX (Royal Demolition Explosive – hexahydro-1,3,5-trinitro-1,3,5-triazine), aluminium nitrate and fuel oil. The materials and timer had been concealed in a suitcase placed beneath a seat in the coach. [1548]

On about Sunday 18th September 2005 insurgents blew up a train carrying petroleum products in Dora district of Baghdad, Iraq, and set it on fire. [1549]

A train carrying soldiers and military trucks, which had left Bangkok on the morning of Tuesday 18th October 2005 for the deep south, exploded two of three bombs, which had been planted on the track three kilometres beyond Yala station, Thailand, at about 09:30. No one was hurt but damage was caused to the locomotive and a waggon. Fearing an ambush, Driver Nulmamadkhan Osatan first set back 500 metres, then reversed further to Yala station. The third bomb was later defused. Each had weighed about 15 kg and had been constructed from fire extinguishers packed with explosives, which had been buried under the track. [1550]

Only one passenger of the 100 on the train was injured and rushed to a nearby hospital after two bombs derailed it between Sungai Kolok and Sungai Padi, Thailand, on Thursday 27th October 2005. Insurgents had planted explosives on the track and detonated them from nearby bushes. The blast caused a crater a metre deep and 3.2 metres wide. Hours earlier, the militants had launched 63 co-ordinated attacks on non-railway targets in Narathiwat, Pattani and Yala provinces. Rail traffic to the deep south of the country was suspended after the attack. [1551]

1546 www.indianexpress.com/print.php?content_id=75292
1547 www.indiadaily.com/breaking_news/42107.asp
1548 www.newkerala.com/2.php?action=fillnews&id=19945
1549 www.arabmonitor.info/printnews.php?idnews=11348
1550 www.bangkokpost.com/News/20Oct2005_news07.php
1551 *The Nation*, Bangkok, Thailand, 27th and 28th October 2005

CHAPTER 19

MISCELLANEOUS ACCIDENTS

The vast majority of railway accidents fall into one of a fairly small number of categories such as derailments, collisions, bridge failures and so on. A few, though, do not easily fall into any obvious group, hence this chapter on miscellaneous accidents. Also included in this chapter are a large number of accidents involving the release of hazardous or toxic chemicals, a type of accident that has only become common in modern times.

A newspaper reported that a disaster had been narrowly averted shortly after midnight on Friday 4th January 1991 at Stein, Switzerland. A freight train from Basel, consisting of an engine and 14 tank waggons, had been approaching the station there, travelling at 80 to 100 km/h according to Chemins de fer fédéraux suisses spokesman Max Kappeler. For reasons not immediately clear, eight waggons had become separated from the remainder of the train near a set of points. The detached waggons had become derailed, and five of them had overturned. Three of these, which were seriously damaged, had leaked their contents and caught fire. Each waggon had been carrying 85,000 litres of petrol.

The Stein fire brigade arrived on the scene 20 minutes after the accident. At first glance, the scale of the disaster had not been fully understood, stated Stein Fire Chief Hansjörg Huber, and so the surrounding fire departments and the cantonal fire service had not immediately been alerted. When the magnitude of the accident became clear, they had been, and eventually more than 900 men (764 firefighters from Basel, Rheinfelden, Wettingen, Brugg, and Aarau, among others) battled the flames. Around 100 vehicles, two CFF firefighting and rescue trains from Zurich and Olten, and three fireboats from the cities of Basel, Aarau, and Wettigen were made available. The operations were led by the Aargau cantonal police, which responded with 160 personnel. After extinguishing one of the burning cars during the night, the firefighters focused on preventing the risk of explosion by cooling the undamaged cars with water and decided to let the other two cars burn themselves out.

Two hours after the accident the cantonal police sounded the general alarm and advised the population over loudspeakers to stay at home and close their doors and windows. In addition, 200 people living close to the danger zone were evacuated and taken to municipal civil defence shelters, though they were allowed to return home from 07:00 after specialists had checked that there were no petrol fumes in the basements of their homes. Local schools remained closed for the day. According to Walter Frick, spokesman for the cantonal fire service, an as yet undetermined quantity of petrol had escaped into the River Rhine. However, environmental pollution was limited, he added. The village of Stein was closed to traffic throughout the morning, while the N3 national road between Basel and Frick and the Frick-Mump regional railway line were also closed. Intercity and express trains were diverted via Olten-Hauenstein.

The cantonal authorities reportedly praised the prompt response of the police and fire brigade, stating that their rapid intervention had prevented a real catastrophe.[1552] However, a few days after the accident there were criticisms in the press that the rescue operation could have been carried out more effectively. [1553]

Seven cars of a Buffalo and Pittsburgh Railroad freight train bound for Buffalo, New York, were derailed at Butler, Pennsylvania, United States, at 3:45 a.m. on Wednesday 17th July 1991. About 200 gallons of irritant, toxic and flammable methyl methacrylate, a chemical used in the manufacture of plastics, leaked from one of the derailed cars before it could be righted. Railwaymen were able to pump several thousand more gallons of the chemical from the car. About 100 workers from the nearby Armco Advanced Materials steel mill had to be evacuated for a short time. [1554]

Fourteen cars of a 42-car Southern Pacific Transportation freight train from Los Angeles to Oakland were derailed just after 12 noon on Sunday 28th July 1991 at Seacliff, California, United States. The accident happened at a point where the line ran along the Pacific coast, the final derailment occurring after the cars involved had been dragged along for several miles in a shower of sparks. Four of the derailed cars contained aqueous hydrazine and naphthalene, both reasonably toxic. As it was a Sunday, there were a large number of people on the beach adjoining the railway and 250 residents of oceanfront houses, surfers from the Mussel Shoals and Rincon beaches and oil facility workers had to be evacuated. A 10-mile stretch of the Ventura Freeway was also closed by the California Highway Patrol because it was unclear if fumes emanating from the wreckage were due to smoke from burning grass or were toxic substances from the spilt chemicals. Serious traffic jams then resulted, while the Amtrak passenger service between Los Angeles and San Francisco was also disrupted, with replacement buses provided between Los Angeles and Santa Barbara. A local resident and a television cameraman who had ventured too close to the fumes suffered nausea and were treated at hospital and released.

The finding of a broken axle by a fire investigator at a point three miles before the final point of derailment made it obvious that an overheated journal ('hot-box') had been the cause of the accident. According to witnesses there had been sparks coming from the train already at Camarillo, 20 miles short of the accident point.

Conductor Bob Nagle stated that he had been unaware that there was anything wrong until the final derailment and separation of the train had caused its brakes to come on in emergency. He said his train had been travelling at 56 mph at the time, within the permitted line speed of 60 mph, and added: "I looked back and saw clouds of dust and I knew we had a problem." [1555]

On Monday 2nd September 1991 two diesel units of a Consolidated Rail Corporation (Conrail) freight train were struck by an empty steel cable spool that was projecting from the side of a gondola car of a freight train travelling in the opposite direction and a brakeman was killed. The accident happened near Silver Creek, New York, United States. The National Transportation Safety Board found that the loading of the gondola car had not complied with the Association of American Railroads rules. The steel bands, which had secured the spools, had been installed at the facilities of a private shipper and the crimps on the bands had been

---

[1552] *Journal de Jura*, Bienne, Switzerland, 5th January 1991

[1553] *Feuille d'Avis de Neuchâtel*, Switzerland, 5th January 1991

[1554] *Leaser-Times*, Kittanning, Pennsylvania, United States, 17th July 1991

[1555] *Los Angeles Times*, Los Angeles, California, United States, 29th July 1991 improperly executed and hence had come loose. [1556]

A passenger train and a freight train collided on the morning of Friday 25th September 1992 in the station at Næstved, Denmark. Although there was considerable damage to both trains, no one was apparently badly hurt and it was not at first realised that there was a leakage of flammable, toxic and possibly carcinogenic acrylonitrile from one of the derailed freight waggons. Later, it became evident that between 2,000 and 4,000 litres of the chemical had been lost. The spillage was covered with foam; the leak was plugged and the remaining 67,500 litres of acrylonitrile in the tank were pumped out. The initial clearance operation lasted 11 days and contamination of soil was not entirely dealt with until 1996. A total of 52 people were admitted to hospital, mostly just for observation. [1557]

At about 23:30 on Wednesday 13th January 1993 seven tank waggons of a freight train, each containing 80,000 litres of petrol, left the rails as the train exited the portal of a tunnel south of La Voulte-sur-Rhône, France. Four of the derailed waggons suffered damage in the derailment sufficient to release their contents, which ignited. Fifteen or 20 minutes later there was an explosion and a ball of fire. Streams of burning petrol spread to houses situated 20 metres from the line and burning fuel getting into sewers generated a series of explosions.

A protective cordon with a perimeter of 600 metres was set up and 1,000 inhabitants had to be evacuated in the middle of the night. By midday on Thursday, 250 firemen had brought the fire under control but about 100 residents were re-housed for over four months while decontamination work was undertaken. The driver of the train was able to pull the leading portion of it clear of the blaze. In doing so he suffered superficial burns, as did two residents, while three other residents sustained various injuries in their rush to escape. Fifteen houses in all were destroyed, the same number of vehicles were destroyed and a rainwater pumping station was put out of action, damaged by explosions of burning petrol.

An estimated 300,000 litres of petrol had escaped from the derailed waggons, of which 200,000 was burnt off or vapourised and 100,000 litres penetrated the subsoil. Of the latter quantity 8,500 litres entered the aquifer. Only the remaining 20,000 litres of petrol could be recovered from the four damaged tank waggons. There was pollution of agricultural wells and contamination of 2.6 hectares (26,000 square metres) of land. The total cost of the derailment was estimated at 70,000,000 francs.

Two investigations into the derailment determined that it had been due to an overheated axle ('hot-box') on one of the waggons. Marks on the sleepers showed that the faulty waggon had been dragged along the line for 300 metres before it had reached a set of points that had initiated the final derailment. [1558]

Norfolk Southern Railway freight train No. 16713, consisting of three locomotives and 90 cars, was travelling at 28 mph when it became derailed near Harriman, Tennessee, United States, at 5:26 a.m. on Monday 5th July 1993. Twenty-nine cars left the rails and one containing volatile and highly flammable methyl ethyl ketone caught fire and exploded. A total of 150 people had to be evacuated while firemen dealt with the fire. No one was hurt.

[1556] National Transportation Safety Board report

[1557] European Commission report July 2000 Lessons learnt from train accidents recently occurred in the EU web page

[1558] https://www.aria.developpement-durable.gouv.fr/wp-content/files_mf/A4225_ips04225_002.pdf

Despite an investigation, the cause of the derailment could not be determined. [1559] [1560]

Wikipedia states that a serious train accident occurred in the Xiangfan section of China on Saturday 15th January 1994 in which seven passengers and a worker (railwayman?) died and 12 passengers and 11 workers (railwaymen?) were injured. No other details, such as the nature of the accident, are given and it has not been possible to trace any other information regarding it. [1561]

Situated on the Pacific Ring of Fire tectonic belt, the American state of California is highly vulnerable to earthquakes. While hardly comparable to the San Francisco catastrophe of 1906, a major quake occurred in the San Fernando Valley north-west of Los Angeles at 4:31 a.m. on Monday 17th January 1994. In total, 57 people died in the quake. The event is relevant to this book because it also resulted in the derailment of a Southern Pacific Transportation freight train from Houston, Texas, to Roseville, California, at Northridge, California, United States. All four locomotives and 29 cars of the train left the rails. Sixteen of them were carrying sulphuric acid but there was no spillage and no one was hurt. [1562]

A Quebec Northshore and Labrador Railway freight train from Sept Iles, Quebec, to Labrador, Newfoundland and Labrador, was derailed at mile 159.75, between Seahorse and Dolliver, Newfoundland and Labrador, Canada, on the evening of Tuesday 8th February 1994. Six tanker cars of Bunker C (heavy fuel) oil and two of bentonite (aluminium silicate) left the track and an estimated 87,000 gallons of Bunker C oil was spilt. Much of the oil collected in a drainage ditch on the eastern side of the track but some reached an ice-covered river. The very low temperatures (-70 Celsius at the time of the derailment) prevented the oil from migrating further but also made the clean-up operation very difficult. Measures to remove the oil dragged on for months. No one was hurt in the accident and the cause of the derailment was not stated. [1563]

On Tuesday 22nd February 1994 a seven-car train on the Nemuro line of the Hokkaido Railway in Japan was derailed by cyclonic winds, the two leading cars overturning and a third car being derailed. The accident happened on an embankment and a curve. No casualties were recorded. Following the accident research was carried out by the Railway Technical Research Institute, as a result of which wind barriers were installed and regulations were set out whereby train services were restricted with regard not only to the absolute wind velocity but also with regard to the direction from which the wind was blowing. [1564]

The most serious railway accident to occur in Switzerland for 10 years happened at 14:25 on Monday 21st March 1994. As a direct Brig – Lausanne – Biel – Romanshorn passenger train had passed Däniken station and was moving at 125 km/h, it came into collision with the counterweight of a crane, weighing 100 tonnes, which cut through the windows of one first-class and two second-class carriages. The leading van of the train was derailed but it only

---

[1559] https://railroads.dot.gov/sites/fra.dot.gov/files/fra_net/15541/Accidents_Sum_1993.pdf

[1560] *Altoona Mirror*, Altoona, Pennsylvania, United States, 6th July 1993

[1561] Wikipedia – List of rail accidents in China

[1562] https://www.upi.com/Archives/1994/01/17/Quake-blamed-for-train-derailment/9674758782800/

[1563] Environment Canada – Bunker C Fuel Rail Spill Case Study web page.

[1564] https://www.sciencedirect.com/science/article/abs/pii/S0167610502002738

came to a halt a kilometre further on.

The line was only closed for a few hours, during which period Intercity and Eurocity trains were diverted via Sufr-Zofingen or Pratteln, while local passengers were carried on bus replacement services. The line was reopened at 17:25.

The material damage caused by the accident was estimated at a million francs and compensation for the victims was expected to total nine million francs.

By 30th March nine people had died as a result of the accident and 23 were injured, one of whom was still critically ill. Criminal proceedings for negligent homicide, negligent bodily harm and disruption to rail traffic had been opened against three Chemins de fer fédéraux suisses employees. The crane operator stated that a railway employee had given him permission to swing the crane foul of the line and that he had not seen the train. [1565] [1566]

At the end of June 1997 a court at Olten awarded the head of safety and the crane safety manager at Däniken suspended sentences of one and two months' imprisonment on charges of negligent homicide and causing negligent bodily harm. The third accused employee was acquitted. [1567]

Union Pacific Railroad freight train Extra 6187 West from Fort Worth, Texas, to Big Spring, Texas, on Saturday 11th June 1994 was a big train, consisting of four diesel units and 124 cars. It was negotiating a curve at a speed of 45 to 50 mph when three units and the leading 31 cars left the rails at 13:00 at Eastland, Texas, United States. One of two derailed cars, which contained hydrochloric acid, was punctured and a total of 18,000 gallons of acid escaped. Some of the cars fell into the north fork of the Leon River, and a tributary stream had to be dammed to prevent the acid from entering the river. The accident occurred in a sparsely populated area but 50 people were evacuated overnight. Of the other derailed cars 22 had been empty, one had been loaded with plastic pellets and the remainder had been carrying pipes.[1568] One person was slightly hurt.

The National Transportation Safety Board determined that various track defects had combined to cause the derailment. It reported that the track was wide to gauge, there was improper support of rail joints, improper rail anchorage and improper placement of crossties (sleepers). [1569]

At 02:56 on Wednesday 29th June 1994 seven of the 50 waggons of a freight train from Basel to Geneva were derailed in the station at Lausanne, Switzerland. The train had entered the station on track no. 3 and when it reached points No. 76a, at the eastern side of the station, the 25th waggon, which was empty, rose, then fell adjacent to the tracks, but running on the sleepers, supported by the edge of the platform edge, it remained upright and in line. The train then made a service stop at the station in order to change drivers. Upon its restarting, the wheels of the derailed waggon were trapped between the rail and the platform edge. At the end of the platform, at another set of points, the two waggons preceding the empty one, together with the 11 subsequent ones, were then also derailed. Of the 14 waggons that had left the track, either totally or partially, five of them, including three tank waggons containing hazardous freight, were lying sideways on the line. One was carrying thionyl chloride and the

---

[1565] *Le Journal du Jura*, Biel, Switzerland, 22n and 30th March 1994
[1566] *Feuille d'Avis de Neuchâtel*, Switzerland, 23rd March 1994
[1567] *Le Journal du Jura*, Biel, Switzerland, 15th July 1997
[1568] *Texas City Sun*, Texas City, Texas, United States, 12th June 1994
[1569] https://www.ntsb.gov/investigations/AccidentReports/Reports/RAB9603.pdf

other two epichlorohydrin. The three tank waggons had had capacities of 47,000 litres of epichlorohydrin and 24,000 litres of thionyl chloride respectively. The first waggon containing epichlorohydrin was slightly punctured, and 400 litres of the toxic, slightly volatile and flammable chemical were released.

The alarm was sounded at 03:14 a.m. and over 500 firefighters and police officers eventually attended the scene to plug the car openings, recover the product that had spread, and ventilate the sewers. The entire district, accounting for 1,500 residents, was evacuated as a safety measure during the emergency response, 300 residents at once and the remainder in the early morning. Over the ensuing four days, the tanks of both the epichlorohydrin and thionyl chloride were emptied, the overturned cars were placed upright, discharged or returned to their recipients. The local population had to be evacuated a second time when hazardous operations such as righting the overturned waggons were carried out. The entire neighbourhood was cordoned off, and the station was completely closed, as were three schools. No explosion occurred and no casualties were reported.

The initial derailment was blamed on a 'switching error,' presumably meaning that the points had been moved under the train. [1570] [1571]

Westbound Canadian Pacific Railway freight train No. CP 971-14 was travelling at about 30 mph when it became derailed at about 06:30 on Monday 17th October 1994 at Mile 180.05, near Lethbridge, Alberta, Canada. The train consisted of three engines and 35 waggons. Six tank waggons containing methanol left the rails and four of them leaked their contents, losing about 230,700 litres of the flammable and toxic liquid. A lorry driver witnessed the derailment and notified the emergency services. A commercial area of Lethbridge had to be evacuated until the spillage had been contained and then removed. No one was hurt.

The accident was investigated by the Transportation Safety Board of Canada, who determined that the derailment had resulted from a broken rail, initiated by the propagation of undetected fatigue cracks in a rail which had worn beyond condemnable limits. The Class 111A tank cars involved in the derailment were also more susceptible to release of their contents upon derailment and impact than pressure tank cars, and yet there were a number of toxic and volatile liquids which were still permitted to be carried in these minimum standard Class 111A tank cars.

A joint project to develop rail inspection technology was being conducted and the present technology was being examined to see if it could be improved. Meanwhile, immediate action to reduce the likelihood of contents leakages from Class 111A tank cars, for example by design changes, was recommended. [1572]

Torrential rain led to flooding and damage to the track of a railway line, causing the derailment of a train carrying fuel to a depot near the town of Durunka, Egypt, at about 06:30 on Wednesday 2nd November 1994. Fuel escaping from the derailed waggons was ignited by some electrical wires and then carried into the town of 22,000 people, where it set fire to many houses.

Mohamed Abdel-Rahman escaped the fire with his wife and seven children but his home was destroyed by the fire. He said: "It was like winds of fire coming down the mountain."

Butcher Ahmed Mohamed, who announced that he had lost 11 family members in the fire, said: "There was fire everywhere. There were children run over by people and killed in the

---

[1570] *La Gruyère*, Bulle, Switzerland

[1571] https://www.aria.developpement-durable.gouv.fr/accident/5515_en/?lang=en

[1572] Transportation Safety Board of Canada report

flight out. About 40 army conscripts manning the depot died as well."

Ahmed Sharif Eddin, a taxi driver, spoke of looking out of his house and seeing: "A wave of people running to the mosque screaming, 'There is only one God.' They thought it was the day of judgement," he added.[1573]

In a complex series of events two of the fuel tanks at the depot itself had also been ignited by being struck by lightning, but Assiut Province Governor Samih al-Saeed explained that the fire at the depot was not linked to the fire in the town. He told reporters that at least 167 people had died in the fire, while survivors said that they expected the number of dead to rise dramatically, because after the accident many houses remained on fire, while others were submerged by the floodwaters. Another, probably slightly later report, stated that 258 bodies had been recovered by midday, while by the following day the number of fatalities was stated to exceed 475.[1574] Twenty hours after the derailment, firefighters were still trying to extinguish the blaze.

A reporter wrote: 'The town was almost deserted after its occupants were evacuated to schools and government buildings in the nearby city of Assiut. Charred animals littered the streets and there were blackened tree stumps. A mosque was turned into a blackened hulk. Large parts of the town were levelled. Bodies littered the streets before being picked up by ambulances...The government sent in emergency provisions to help the thousands left homeless by the fire and floods, Governor el-Saeed said. But survivors interviewed in an Assiut mosque where they had taken refuge said they had yet to receive help.'

Meanwhile, there was criticism of the government for erecting the fuel depot near the town. Its tanks were intended to hold 15,000 tons of petroleum as strategic reserves for the armed forces. [1575] [1576]

Two trains carrying crude oil and petrol collided at Kumagaya, Japan, on Thursday 18th May 1995. As a result, there was a spillage of some 60,000 litres of oil, which via the irrigation system spread over a 10-kilometre area. [1577]

A Southern Pacific Transportation freight train from Houston, Texas, to West Colton, California, was derailed at 4:45 a.m. on Wednesday 10th January 1996 near Sentinel, Arizona, United States. The train consisted of five diesel units and 110 cars. One locomotive and 23 cars left the rails, the derailed cars including one containing naphtha distillate and a second carrying butyl acrylate. Originally it was maintained that there had been no leakage of chemicals, other than diesel fuel from one of the locomotives, but later it was stated that 20,000 gallons of distillate had been lost, though none of the butyl acrylate. The derailment had occurred in a sparsely populated area and there was no need for an evacuation but two crew members were taken to the Good Samaritan Hospital in Phoenix for a check-up, examined and released.

A hairline fracture in a rail was blamed for the derailment. Arizona Corporation Commission safety chief Gary Smith said that the rail had split apart beneath the train, adding: "It's a defect that goes undetected mechanically or visually before it fails." An inspection car had been directed into a siding two miles short of the derailment site by the dispatcher to allow the freight train to overtake it. Smith did not consider that the inspection car would have

---

[1573] *The Indiana Gazette*, Indiana, Pennsylvania, United States, 3rd November 1994
[1574] *The Daily Interlake*, Kalispell, Montana, United States, 4th November 1994
[1575] *The Daily News*, Huntingdon, Pennsylvania, United States, 3rd November 1994
[1576] *The News Herald*, Panama City, Florida, United States, 3rd November 1994
[1577] https://www.ejrcf.or.jp/jrtr/jrtr06/pdf/topics.pdf

detected the fracture anyway, saying: "This is a very rapid failure and it happens very quickly. You just don't really know it's there until, unfortunately, it fails." [1578]

At about 5:55 a.m. on Wednesday 21st February 1996 Southern Pacific Transportation freight train 1ASRVM-18 ran away and became derailed in the Tennessee Pass, Colorado, United States. There were actually three separate derailments as the out-of-control train had broken into three parts. Damage to different waggons in the derailment resulted in the loss of 51,706 gallons of sulphuric acid and 19,733 gallons of triethylene glycol.

The train consisted of two diesel units and 82 cars and was on its way from East St Louis, Illinois, to California. It was in charge of Conductor Steven Hudson, 45, with Engineer Jimmy Roybal, 41, and Student Engineer Cecil Wayne Reagan, 34. The engineer and student engineer were both killed in the derailment. The injured conductor made his way to a road and was subsequently admitted to hospital with a broken collarbone. About 20 other people were treated at the Vail Valley Medical Center after complaining of burning throats and skin and eye irritation, doubtless due to exposure to the acid. Four members of a family living on a nearby farm had to be evacuated.

Clearance of the wreck and the spilt acid and glycol was complicated by the ground being covered by thick snow, so it was several days before the line could be reopened. Fortunately, the acid contained in four other tank cars was not spilt. There had also been fears initially that the acid might flow into the Eagle River or its tributaries but instead it had collected in low spots near the line. [1579]

The National Transportation Safety Board (NTSB) noted that post-accident investigations had shown no problems with the train's brakes. The event recorder indicated that small, incremental brake pipe reductions had been made in an attempt to control the increasing speed of the train, until it had increased to the point where the brake system was unable to control it. Such a technique would have been typical of an inexperienced engineer. Finally, an emergency brake application had been made but by then the train had been going too fast for it to be effective. Although the body of Engineer Roybal had been found at the controls, Conductor Hudson stated that Student Engineer Reagan had been driving the train and it seemed likely that Engineer Roybal had taken over the controls at the last moment.

The NTSB concluded that the probable cause of the accident had been the mismanagement of the air brake system by the student engineer, allowing the speed of the train to increase to the point where the system no longer had the ability to stop the train, and the failure of the locomotive engineer to ensure proper train control. [1580]

Seventeen homes had to be briefly evacuated around Cushing, Minnesota, United States, on Wednesday 28th February 1996 after a Burlington Northern Santa Fe Railway freight train was derailed. U.S. Route 10, which ran alongside the railway, was also closed. Of the 32 cars that left the track, four were carrying anhydrous ammonia, some of which escaped, causing nine people, three of them members of the train crew, to need hospital treatment after exposure to the fumes. The evacuees were allowed to return in the evening but were told that they would have to leave again while the cars were re-railed. [1581]

[1578] *The Yuma Daily Sun*, Yuma, Arizona, United States, 10th and 11th January 1996

[1579] *Gazette Telegraph*, Colorado Springs, Colorado, United States, 22nd, 23rd, 24th and 25th February 1996

[1580] National Transportation Safety Board report

[1581] *Altoona Mirror*, Altoona, Pennsylvania, United States, 29th February 1996

On Monday 4th March 1996 Wisconsin Central freight train LO-22-4 from Stevens Point to Neenah was approaching Weyauwega, Wisconsin, United States, travelling at 48 mph, when it became derailed at 5:49 a.m. The train consisted of two locomotives and 81 cars. Both engines and the leading 16 cars passed a switch (set of points) without incident but the next 34 cars, back to the 50th, were derailed. Seven of the derailed cars were loaded with liquefied petroleum gas (LPG), another seven were loaded with propane and two were carrying sodium hydroxide. The flammable contents ignited and the resultant fire consumed seven of the cars, destroyed a lineside feed mill building, brought down high-tension lines and disrupted the water and natural gas supplies of the town. About 3,155 residents of Weyauwega were evacuated and Highways 10 and 110 and all local roads were closed off. [1582]

A local newspaper detailed the stories of some of those affected. 'Laurie Mapes and her three young children had no money and a big need for diapers for her 2-year-old son when they arrived at a Red Cross center in nearby Waupaca after hastily leaving their Weyauwaga home. "The kids are very confused. My 9-year-old is angry because she is worried about the kittens," Mayes said. She referred to the family cat with two 5-week-old kittens – all left behind at the home. "It probably wouldn't be so bad if I could take care of our animals and get what I need."

'Robert Dunbar, 61, and his mother Dorothy, 86, went to the Hendrickson Center, a community center that the Red Cross set up as a shelter for displaced people. "I heard the first (railroad) car go up – ba-wooom. It didn't shake the house but you could hear a loud noise like a sonic boom," Dunbar said.' [1583]

The decision to evacuate the town had been made an hour after the derailment, when it became known that cars containing LPG were involved in the fire, while at the same time the firefighters were withdrawn because of fears of a disastrous boiling liquid expanding vapour explosion (BLEVE), the fire being allowed to burn itself out.

The residents were not allowed to return home until 20th March,[1584] though those whose homes were not too close to the derailment site were allowed to return briefly on Friday 8th to collect pets. [1585]

Another resident was relieved to be reunited with her pet after a longer wait. 'A week ago, Susan Weiss sat in a church sobbing. Kynda, her cat, was trapped in a home deemed too close to a pile of burning, derailed propane tankers to be visited by an armored car pet rescue. But any tears shed Friday were tears of joy for Weiss. Kynda was back at her side – thanks to a clandestine rescue that even she knows nothing about. "I don't know who did it. I don't know first names or anything. It was like a cloak-and-dagger operation. But I got her with me," a jubilant Weiss said in a telephone interview from the Kaukauna motel where she is staying while this community of 1,790 remains evacuated by the train wreck. After nearly two weeks of being alone in the city empty of people for fear of a massive explosion, Kynda was "very, very, thirsty but otherwise OK," Weiss said. "She is really quite spunky here," the car lover said, chuckling. "Now she is curling around my feet, wondering why I am talking on the phone and not petting her." [1586]

The accident was investigated by the National Transportation Safety Board, which commented that pieces of broken rail from the "heel" area of a switch point rail had been recovered in the wreckage. The broken rail had displayed failure characteristics indicating that

---

[1582] National Transportation Safety Board report

[1583] *The Janesville Gazette*, Janesville, Wisconsin, United States, 5th March 1996

[1584] Ibid, 22nd March 1996

[1585] Ibid, 9th March 1996

[1586] Ibid, 16th March 1996

the fractures had originated from a bolt-hole crack. Their examination and analysis of the broken rails indicated that the bolt-hole crack had been present for some time. The examination had also revealed that the rails and joint bars displayed many characteristics indicative of problems in the joint and bolt-hole area. These characteristics had, the NTSB commented, been tell-tale signs of a problem, which should have been observed and acted upon by well-trained, vigilant track inspectors and their supervisors. [1587]

At about 4:10 a.m. on Thursday 11th April 1996 19 cars of Montana Rail Link freight train 01-196-10 left the rails near Alberton, Montana, United States. Among the derailed cars were a tank car containing chlorine, which ruptured, releasing 130,000 pounds of the poisonous gas; another car lost 17,000 gallons of potassium cresylate and a third 85 dry gallons of sodium chlorate.

About 1,000 residents had to be evacuated, 350 were treated for chlorine inhalation, of whom 123 were stated to have sustained injury, and nine, including both of the train crew, were admitted to hospital. Worse, a 'hobo' who had been riding on the train died of chlorine poisoning. Chlorine drifting across Interstate 90 caused a number of accidents before the road was closed. The governor of Montana declared a local state of emergency. [1588]

A local newspaper reported, 'As soon as trucker Layne Atwood caught a whiff of the foul mist enveloping Interstate 90 in western Montana he knew he and his son were in trouble. "It was burning our lungs and eyes," Atwood said from his hospital bed. "It takes your breath away...You feel like your lungs are on fire." The two Cardston, Alberta, truckers were carrying a load of tulips from Vancouver, British Columbia, to Chicago when their 18-wheeler entered a fog of chlorine gas from derailed, ruptured, railroad tank cars. They stopped to help some motorists who had pulled over to the side of the road, and then proceeded 150 miles to their morning's destination, a truck stop in Butte. They were later taken to a hospital.' [1589]

The evacuation area was reduced on 14th April and residents were allowed closer to the derailment site temporarily to feed and water livestock and collect possessions and pets. The evacuation was not finally lifted until 28th April, after all the chlorine from other tank cars had been removed.

The accident was investigated by the National Transportation Safety Board (NTSB). It noted that the train had consisted of three locomotives, 36 loaded and 35 empty cars: 25 of the loaded cars had contained hazardous substances. It had been moving at 40 mph when its engineer and conductor had felt a patch of rough track and it had derailed.

The NTSB concluded that the probable cause of this accident was a rail section that had fractured in vertical split head mode under the passing train. The rail had reduced load-bearing capacity because of head and gauge face wear. Contributing to the cause of the accident had been the inability of the ultrasonic rail defect detection equipment to readily identify internal defects in rails with surface defects. [1590]

At about 17:30 on Saturday 1st June 1996 a freight train consisting of an engine and 18 waggons of vinyl chloride monomer, travelling from Tessenderlo in Belgium to Schkopau in Germany, was negotiating a facing crossover from track 3 to track 1 at Schönebeck, Germany, when the rear bogie of its sixth waggon became derailed. The train then separated between its

[1587] National Transportation Safety Board report
[1588] National Transportation Safety Board report
[1589] *The Daily Inter Lake*, Kalispell, Montana, United States, 12th April 1996
[1590] National Transportation Safety Board report

sixth and seventh waggons and the engine and leading vehicles came to a stand 400 metres further on due to an automatic brake application when the brake pipe was pulled apart. The waggons following the sixth also left the rails and piled up upon one another. The third waggon from the rear fouled the overhead line equipment, there was an explosion and a fireball swept the allotments that adjoined the railway. Four other waggons also caught fire and a column of smoke rose perhaps 800 metres into the air.

The local volunteer fire brigade concentrated initially on rescuing people from the adjoining allotments and for some time remained unaware of how many waggons were derailed and of their contents. Only when a helicopter had been called in to survey the site did they discover that the front of the train was stopped some distance down the line, thus it was 19:12 before they were able to obtain a consist of the train's contents from the driver and became aware that it was carrying vinyl chloride.

By chance, the BASF chemical factory at Ludwigshafen learnt of the accident from radio broadcasts, whereupon they had contacted the Schönebeck firemen and then proceeded to the scene with their own BASF fire engines, and also sent specialists to the site from their branch plant at Schwarzheide. The fires were finally extinguished by 07:30 the next morning but it was over a fortnight before all the remaining vinyl chloride could be pumped from the tank waggons.

The cause of the derailment was later determined by the Eisenbahn-Bundesamt (Federal railway authority) to have been a worn set of points and wide track combined with the hunting of the tank waggons as they had negotiated the crossover, exacerbated by the design of the waggons, which had lacked baffles to disrupt the swirling of their liquid contents. Significant defects were also noted by the Bundesverkehrsministerium (Ministry of Transport) in 11 of the 18 waggons of the train. No one was found to be to blame for the accident. [1591]

Probably as a result of driver error, two freight trains were involved in a sidelong collision on Wednesday 19th June 1995, at Hagen-Vorhalle, Germany. Five waggons were derailed, three of them overturned and 25 tonnes of highly inflammable isopropanol was spilt from one of them, a tank waggon. There was no fire. Damage was estimated at 1,000,000 deutschmarks. [1592]

Just before 8 a.m. on Saturday 10th August 1996 a Canadian National Railway freight train from Toronto to Sarnia was derailed at Kerwood, Ontario, Canada. Thirty-five of the 112 cars of the train left the rails and a tank car loaded with sulphuric acid was ruptured, losing about 4,000 litres.

'Eileen Woods, 77, watched from her general store as "cars jack-knifed right before my eyes. Men arrived in my door in outfits to handle toxic materials, and we were told to leave." Like most people, she left to stay with family outside town as police blocked off roads.

'Brian Campbell, 38, was getting into his truck to go to work shortly before 8 a.m. when he heard the crash. "I heard a lot of clatter and I looked up and saw cars flying off the tracks. They were flipping over and piling up," he said later, standing near an industrial yard strewn with paper and debris from the wreckage.'

About 200 people were evacuated but were allowed to return at 7 p.m. By late Sunday most of the acid had been cleared up; it had not escaped into any waterways. Some sort of equipment failure was blamed for the derailment. [1593]

[1591] *Eisenbahnunfälle bei der Deutschen Bahn* by Erich Preuß, pp. 132-5
[1592] *Deutsche Eisenbahn-Katastrophen* by Hans Joachim Ritzau, p. 153
[1593] *The Lethbridge Herald*, Lethbridge, Alberta, Canada, 11th and 12th August 1996

At about 11:20 on Thursday 29th August 1996 St Lawrence and Hudson River Railway eastbound freight train No. 902-29 suffered a derailment at Dalhousie Mills, Quebec, Canada. The train was travelling at nearly 65 mph when 36 of its waggons left the rails. The train consisted of two locomotives, 41 loaded and 28 empty waggons. About 1,900 litres of hydrogen peroxide were lost from one waggon and a small fire broke out. The residents of 40 homes were evacuated for five hours as a precaution. No one was hurt.

The Transportation Safety Board of Canada determined that the derailment had been caused by the hunting (oscillation) of car MSDR 81026 as a result of the excessive wear of its bolsters and side frames, which had meant that its suspension damping components were unable to sufficiently resist the rolling of the vehicle. The high speed of the train had probably increased the severity of the hunting. The worn parts of the car had not been recognised as safety defects.

The speed of freight trains was reduced after the accident to lessen the likelihood of excessive hunting. Measures were also taken to make the consists of trains more quickly available to emergency workers after an accident, to improve the placing of emergency calls by train crews and to research waggon component wear. [1594]

Kleine Scheidegg station, in the Swiss Alps, is at 2,061 metres the highest junction station in Europe and the meeting point of the Wengernalpbahn lines to Wengen and Grindelwald and the nine-kilometre-long Jungfraubahn to Jungfraujoch. Both the Wengernalpbahn and the Jungfraubahn are rack railways but they have different gauges and are not connected.

On the morning of Monday 11th November 1996 the föhn wind was blowing exceptionally strongly, forcing the cancellation of Wengernalpbahn trains between Kleine Scheidegg and both Wengen and Gridelwald stations. However, this left about 100 people, many of them foreign tourists, stranded at Jungfraujoch. These people were able to return to Kleine Scheidegg over the Jungfraubahn and were then transferred to a 50-tonne Wengernalpbahn power car for the journey to Grindelwald. However, on this second part of their journey the föhn wind, estimated by the Bern cantonal police to have been blowing at 150 km//h, proved too strong for the train and at an exposed stretch of the line on the way to Grindelwald, at 15:05, a particularly fierce gust of wind overturned the train. As a result four passengers were injured, one suffering a broken arm. The injured were evacuated by a snowplough and taken to hospital in Interlaken, while the remaining passengers returned, partly on foot, to Kleine Scheidegg, from where they were evacuated in the evening using tracked vehicles. [1595] [1596]

On Thursday 12th December 1996, a tank waggon carrying corrosive and toxic hydrosilicofluoric acid became derailed in Jämsä, Finland, and its contents began to leak out. The leakage increased when the fire department sought to block the leak, since the tank waggon was in poor condition. The fire department then transferred the contents of the waggon into another brought to the scene of the derailment. Some of the acid leaked onto the ground and the soil was subsequently neutralised with lime.

The sequence of events had begun when another train was unable to climb a hill south of the Jämsä railyard. The train pulling the waggon that subsequently derailed came to assist in pulling this second train. Towards the end of this procedure, when the first train began to brake, the second train, which was being pulled behind, pushed the waggons together. When

[1594] Transportation Safety Board of Canada report
[1595] *Le Nouvelliste*, Sion, Switzerland, 12th November 1996
[1596] *La Liberté*, Fribourg, Switzerland

this occurred, the elastic limits of the buffers of the waggon in front of the one carrying the acid were reached and its buffers were broken. The tank waggon carrying the acid then mounted the buffers of the waggon in front, and as a result, the waggon derailed at the next set of points. The contents of the tank waggon, which had been constructed in 1971 and lined with rubber in 1990, then began to leak, although it had not been struck.

The firemen climbed to a dangerous height to pull slings over the leaking tank waggon, even though the current of the overhead line equipment had not been isolated. The delay in doing this was due to the fact that another waggon was being brought next to the leaking waggon for the transfer of the load. The firemen were, however, aware that the electricity had not been cut off.

The economic loss from the accident was less than 500,000 Finnish marks (equivalent to about US$ 100,000 or 84,000 euros).

According to the regulations, the leaking tank waggon should have been lined with neoprene rubber, which was known to be able to contain the hydrosilicofluoric acid. However, an analysis carried out by the VTT (State Technical Research Centre) showed that the lining was polyurethane, and there was no information available on the chemical reaction between this and hydrosilicofluoric acid. [1597]

North Eastern Railway train No. 5323, the Up *Gandak Express*, was blown over by severe cyclonic winds between Chauradano and Adapur stations, Bihar, India, on the Bairangnia – Narkatiaganj Junction metre-gauge, single-track section of the Samastipur Division, at 20:28 on Saturday 24th May 1997. No one was killed but four people were injured, all but one of them seriously. Damage was estimated at 20,400 rupees.

The Commission of Railway Safety made 10 recommendations in relation to the accident:

(1) On the north side of the track at the accident site, plantations should be done within the railway boundary as per the standard section for embankments, so as to retard the velocity of the storm. Similar action should be taken at all other places in the vulnerable sections.

(2) To safeguard against the shifting of the track due to wind forces, minimum metre gauge (MG) track of 75 lb rails on (M+6) sleeper density, CST-9 plates with LWR ballast section should be used.

(3) The highest priority should be given for metre gauge to broad gauge conversion between Raxaul and Darbhanga stations, which were in close proximity to the Nepal hill ranges from where severe storms affect train operations.

(4) In the vulnerable sections, all the stations should be provided with Anemometers and not limited to those on either side of major bridges only as specified in the Railway Board's letter No. 58/W-3/SG/172 dated 7/3/1964.

(5) The despatch of weather forecasting messages from the Patna Meteorological Centre should be arranged by the railway, through telephone or fax to avoid delay, which now occurred for two days, making the forecast irrelevant.

(6) In the face of severe storms, directions to open the doors and windows of the coaches should be displayed at prominent places at all the stations in the vulnerable sections and also written inside the coaches.

1597 https://www.turvallisuustutkinta.fi/en/index/tutkintaselostukset/railroad/tutkintaselostuksetvuosittain/raideliikenne1996/c231996rpiifluorivetyhappovaunusuistumi.html

(7) Telephonic communications between guards and drivers with speakers in the coaches provided so they could warn the passengers to open the doors and windows in the face of storms.

(8) Air-conditioned coaches should not be run in the vulnerable section, viz. Raxaul – Darbhanga, during the three months when severe storms generally occur.

(9) The Railway should examine the design of the MG coaches so as to lower their centre of gravity, making them more stable against overturning.

(10) The communication system between Gorakhpur and Darbhanga stations should be improved. As an immediate solution, VHF communication should be provided at stations that were more than eight kilometres apart. [1598]

On Saturday 5th July 1997, southbound Deutsche Bahn freight train No. 54755 included some Rs 669 flat cars with Niesky bogies loaded with steel pipes, each of which pipe was about seven metres long and weighed between 6 and 8 tonnes. The pipes had not been properly secured with steel reinforcements when they were loaded onto the waggons, unnoticed during the transfer of the waggons to the railway, and during a subsequent inspection of the train at Seelze marshalling yard.

Travelling in the opposite direction along the Main – Weser double-track main line was regional express No. 3602 from Frankfurt (Main) Hbf to Kassel Kbf, composed of double-decker rolling stock. As the trains were passing one another near the station at Neustadt, Germany, several of the pipes being carried on one of the waggons became loose and one of them struck the coaches of the express, damaged its leading three carriages and then penetrated the fourth. As a result, six passengers lost their lives, two were seriously injured and 14 were slightly injured.

The public prosecutor subsequently began criminal proceedings against seven accused but later the case was dropped, because it was unclear to what degree the improper loading of the pipes and to what degree the hunting (oscillation) of the waggons had been responsible for the accident.

No previous accidents involving waggons Rs 669 type had occurred and upon examination after the accident they were found to be in order. Even so, the Eisenbahn-Bundesamt (Federal railway authority) ordered that their Niseky bogies must be reprofiled before they were permitted to again be run at a speed of 100 km/h. [1599]

A freight train was derailed at Ixtlahuaca, Mexico, on Friday 25th July 1997 and there was a spillage of flammable, toxic and carcinogenic propylene oxide. Although there were no human casualties there was considerable contamination of the ground. The exact location of the accident is unclear as there are three localities with this name in Mexico in different states, and the brief report of the accident does not indicate which was involved. [1600]

At about 14:00 on Saturday 23rd August 1997 a Minsk – Orsha freight train that included

[1598] Government of India Ministry of Civil Aviation and Tourism Commission of Railway Safety Annual Report for 2001-2002

[1599] *Eisenbahnunfälle bei der Deutschen Bahn* by Erich Preuß, pp. 139-41

[1600] .../search%3Fq%3DESCARRILAMENTO%2BDE%2BR%25C3%258DO%26start%3D30%26ie%3DUTF04-05-03

28 tanker waggons was passing through the station at Borisov, Belarus, when eight of the waggons became derailed and two of them, loaded with 120 tonnes of propane butane gas, were punctured and caught fire. The fire spread to involve other waggons in the train carrying rubber and Bavarian beer. The station and surrounding area were hastily evacuated. When the firemen arrived they faced problems because of the low water pressure and were forced to try and obtain supplies from the adjoining Berezina River, negotiating its steep banks to do so. Later, a fire train arrived from Minsk with powerful pumping equipment and foam, the latter being employed from 16:00. At 20:15 the fire was under control and the station could be re-opened. Fortunately, there was no explosion. The derailment was thought to have probably been due to a defective wheel on one of the waggons. [1601]

A major catastrophe in Yaoundé, the capital of Cameroon, on Saturday 14th February 1998, had its origins in a collision, apparently head-on, between a train of oil tankers coming to the capital from the oil fields bordering Nigeria and a second train, possibly of empty tank waggons. As a result of the collision one or both of the trains was derailed and there was a considerable spillage of oil, or possibly petrol. As has happened elsewhere at different times, usually in poor countries, local people, including taxi drivers, women and children, then began collecting the spilt fuel in whatever containers they could find, in the hope of selling it for a small sum. The fuel was then ignited, perhaps from a lighted cigarette, and a massive fireball enveloped the people who were helping themselves to the fuel and anyone else who chanced to be nearby. The accident happened in a densely populated part of the city, close to the bus and taxi terminals.

The emergency services were quickly on the scene but were unable to do much because of the intensity of the blaze, other than contain it to the extent that it did not spread to a nearby petrol depot at Nsiam. It was at least a day before the fire burnt itself out and a pall of black smoke hung over the scene for a good deal longer. Early estimates of the number of dead put the figure at 120, with another 150 seriously injured, and the final death toll was over 200.

The cause of the initial collision was not known. [1602] [1603] [1604]

A Norfolk Southern Railway freight train was being switched at Montview Yard on Tuesday 31st March 1998 and its engine had been detached from its cars, whereupon the latter began to run away down an incline. The 65 cars continued downhill about three miles before they collided, at 9:37 a.m., with a second stabled train of 61 cars near the centre of Lynchburg, Virginia, United States. Three locomotives and 10 cars were derailed in the collision and a fire broke out involving acetone from one of the derailed cars and diesel fuel from the locomotives. A 50-block area around the accident site was evacuated and 260 pupils of Dearington Elementary School were moved to another school.

A newspaper wrote, 'Ruth Stonell, 91, said she could feel the collision about a block from her home. "It knocked me against the window sill, and I thought for sure it was going to knock me to the floor. I didn't know which way to run," she said.

'Mary Reed, 60, a neighbor of Ms Stonell, said her house shook and her windows rattled. "I saw all the black smoke and yellow flames coming straight up," Ms Reed said. "I was starting to choke, and I had to get away from it. It gives you a bad taste in your mouth and makes your throat burn."'

[1601] .../14/07/2002WWW, Belarus, The "*Vecherny Minsk*" newspaper, Tank-trucks were derailed
[1602] *Danger Ahead*: Train Explosion Kills 120 in Cameroon
[1603] http://news.bbc.co.uk/1/hi/despatches/56771.stm
[1604] Wikipedia – Yaoundé train explosion

People were allowed to return home at about 3:30 p.m., firemen having by then extinguished the fire, though not before it had caused a considerable amount of damage. 'The fire spread from the wreckage and destroyed a wood frame storage building nearby where the city stored snowplows, salt, calcium chloride and other equipment used to clear snow from the roads. Damage to the structure and road clearing equipment was estimated at $1 million.'

A National Transportation Safety Board (NTSB) investigator stated that initial findings indicated that there had been no brakes on the cars, except a couple of hand brakes, which had been applied but were insufficient to hold them. The air brakes of the cars would have worked had they been applied. It has not proved possible to access the NTSB report, if one was published. [1605]

CSX Transportation freight train Q316 from Huntington to Parkersburg had only travelled a few miles when it became partly derailed at Cox Landing, West Virginia, United States, at 12:37 p.m. on the afternoon of Saturday 20th June 1998. Emogene Vanatter, who witnessed the accident from the hill overlooking the line where he lived, said: "The train was going slowly and looked as if it was about to stop, as it often does, then all of a sudden there was a big bang and we saw a cloud of dust and smoke and knew there had been a train wreck."

Thirty of the 148 cars of the train left the rails and landed in a zig-zag position near the bank of the Ohio River. One was upside down and others ended up across the tracks. About 21,500 gallons of formaldehyde was lost from a tank car and about 500 residents were forced to evacuate, while others, including Emogene Vanatter, were told by the sheriff to shelter in place. The conductor of the train, six local residents and nine firemen were taken to St Mary's Hospital in Huntington with breathing difficulties, treated and discharged. [1606]

The National Transportation Safety Board made a full investigation of the accident. It listed six findings:

(1) The operation of the train had not caused or contributed to the severity of the derailment.

(2) The derailment had not been caused by a mechanical defect in the empty covered hopper, which was the first car to derail.

(3) Drainage in the accident area had been inadequate and, as a result, the roadbed in the derailment area had likely become water-saturated, rendering the track structure unable to maintain track integrity under the load of train Q316.

(4) The full potential of the Cabell County Operation Respond Emergency Information System software could not be realised, or even evaluated by emergency responders, because of a lack of information about the system in Cabell and Wayne Counties' emergency response plan and because exercises involving the system had not been included in periodic disaster drills.

(5) Because the Ohio River Road Volunteer Fire Department had not participated in Cabell and Wayne Counties' disaster drills and because its officers had been unfamiliar with the counties' emergency response plan, the incident commanders had not used all available resources to assist in the emergency.

(6) Railroad wreckage-clearing operations and pipeline operations had not been effectively coordinated and unified under an effective command structure, which had placed excavation personnel at risk while they worked in the vicinity of a natural gas line.

The NTSB determined that the probable cause of the derailment had been an unstable

[1605] *The Winchester Star*, Winchester, Virginia, United States, 1st and 2nd April 1998
[1606] *Cumberland Times-News*, Cumberland, Maryland, United States, 21st June 1998

roadbed, which had resulted from the inadequate or ineffective measures taken by CSX Transportation, Inc., to permanently correct known drainage problems in the accident area.

The NTSB made a series of recommendations relating to the accident.

To the Federal Railroad Administration:

(1) It should review both the implementation and the management oversight of CSX Transportation's track inspection and maintenance programs for the Ohio River Subdivision and take the actions necessary to ensure the safe passage of trains and the safe shipment of hazardous materials through the area.

To CSX Transportation, Inc.:

(1) It should perform a comprehensive engineering analysis and evaluation of track and roadbed conditions on the Ohio River Subdivision and develop a plan and a timetable for correcting existing or potential deficiencies, including inadequate drainage, which might affect the safe passage of trains, and the safe shipment of hazardous materials through the area. It should provide to the NTSB a schedule to correct the deficiencies found during the evaluation.

(2) It should develop and implement incident coordination procedures that would ensure that safety-critical operations during wreckage-clearing activities were coordinated with all parties involved in those activities.

(3) It should review and revise, as necessary, in light of this accident, its community outreach and training assistance programs to ensure that all emergency response groups that might be called upon to respond to an incident or accident involving its railroad, received the necessary training on a timely and recurring basis.

To Cabell and Wayne Counties' Local Emergency Planning Committee:

(1) It should revise its emergency response plan to incorporate information about the capabilities and use of Operation Respond Emergency Information System software.

(2) It should include, in its periodic disaster drills, exercises designed to familiarise emergency responders with the capabilities and use of Operation Respond Emergency Information System software.

(3) It should include in its periodic disaster drills all emergency response agencies within its jurisdiction, including the Ohio River Road Volunteer Fire Department, and ensure that those agencies were aware of Cabell and Wayne Counties' emergency response plan and its implementation.

(4) In cooperation with CSX Transportation, Inc., it should develop and implement incident coordination procedures which would ensure that safety-critical operations during wreckage-clearing activities were coordinated with all parties involved in those activities.

To the Mountaineer Gas Company:

(1) In cooperation with CSX Transportation, Inc., it should develop and implement incident coordination procedures that would ensure safety-critical operations during wreckage-clearing activities were coordinated with all parties involved in those activities. [1607]

At about 3 p.m. on Monday 29th June 1998 an Iowa Interstate Railroad freight train was stopped on a bridge over the Iowa River in Iowa City, Iowa, United States, due to a tornado

[1607] National Transportation Safety Board report

warning, when the wind was powerful enough to throw some of its cars off the structure and into the river. No one was hurt.[1608] The same day, the same storm, with winds moving at an estimated 95 mph, derailed some cars of an Illinois Central Railroad freight train south of Tolono, Illinois, United States. [1609]

Just before 6:30 p.m. on Monday 5th October 1998 six or seven cars of a Buffalo and Pittsburgh Railroad freight train were derailed at Ridgway, Pennsylvania, United States, and there was a spillage of sulphuric acid from a damaged tank car, which reportedly left a pungent haze over the adjoining Clarion River and led to the evacuation of about 100 people from the village of Portland Mills. Most of them went to stay with friends or relatives, though the Red Cross set up a shelter a few miles away in Red City. Large quantities of lime were brought in to neutralise the acid and by 3 a.m. the following morning the spill had been contained, though the tank continued to leak. The acid had come from Valleyfield, Quebec, Canada, and was destined for a fertiliser company in Florida. The cause of the derailment was not immediately known. [1610] [1611]

Seven of the 120 cars of a CSX Transportation freight train left the track at 6 a.m. on Friday 6th November 1998 just south of Point Pleasant, West Virginia, United States, and a cloud of dilute hydrochloric acid then formed when a tank car was breached. There was no evacuation but residents living within a mile of the site were advised to remain indoors until the cloud had dissipated, two roads were closed, and the Ohio River was likewise closed to boat traffic. No one was hurt. [1612]

At 16:28 on Saturday 6th February 1999 eastbound Canadian National Railway freight train No. M 304-41-05 from Hornepayne, Ontario, to Toronto, Ontario, became derailed at Mile 248.5 of the Ruel Subdivision near Neswabin, Ontario, Canada. The train consisted of two locomotives, 57 loaded and 14 empty waggons. The loaded waggons had been carrying a variety of hazardous chemicals including vinyl chloride, benzene, anhydrous ammonia, methanol and ethylene glycol.

At about 16:15 the train had passed a Wayside Inspection System (WIS) monitor near Oba, Ontario, which had alarmed and warned the train crew of overheated bearings, hot wheels or dragging equipment. The alarms did not differentiate between these faults and the dragging equipment detector had given false alerts for three previous trains and was therefore considered to be faulty itself. Slowing their train, its crew had contacted the rail traffic controller in Toronto for advice. The latter had in turn spoken to his colleague in Edmonton, it was decided that the Oba WIS was giving false alarms again, and the train crew were instructed to proceed at unrestricted speed. Shortly afterwards the train derailed. Twenty cars, from the 21st to the 40th, left the rails, the total contents of a tank waggon loaded with benzene were lost, and a fire involving the benzene and some lumber from an adjoining waggon burnt for several days. The derailment occurred in an uninhabited area and no one was hurt.

The Transportation Safety Board of Canada (TSB) issued a comprehensive report on the accident which reached 13 conclusions:

1608 *Cedar Rapids Gazette*, Cedar Rapids, Iowa, United States, 30th June 1998
1609 *The Telegraph*, Alton, Illinois, United States, 1st July 1998
1610 *The Daily News*, Huntingdon, Pennsylvania, United States, 6th October 1998
1611 *The Progress*, Clearfield, Pennsylvania, United States, 6th October 1998
1612 *The Robesonian*, Lumberton, North Carolina, United States, 6th October 1998

(1) The roller bearing at the L-3 location on the south side of car CN 604697 had overheated and seized, resulting in a burnt-off axle journal and the derailment of 21 cars.

(2) The definition of "faulty tape" had allowed rail traffic control mechanical service representatives (RTC Mechs) and hot box operators (HBOs) to disregard scans which met or exceeded TP-105 criteria. They had believed that the information on the tape did not represent actual train conditions. This had resulted in a train with a defect being allowed to proceed.

(3) Unclear lines of authority and communication difficulties had precluded a highly cooperative analytical environment. This had resulted in the train not being stopped once the HBO had realised it was possible that the Wayside Inspection System (WIS) indications reflected a safety problem.

(4) The train crew members had not performed a walking inspection of their stationary train after having been provided with multiple audible alarms of possible defective conditions; therefore, they had not detected the single derailed car at the Oba location, thus missing an opportunity to greatly reduce the consequences of the overheated roller bearing.

(5) Company procedures (TP-105) regarding the monitoring of the bearings identified as "warm" subsequent to the scan at MacDuff, located 13.6 miles before the WIS at Oba, had not been followed. This had resulted in the information from the Oba site being interpreted in isolation. A train crew member performing an inspection for possible bearing distress, after having been stopped by a WIS, might not reliably identify a developing problem due to the tools, ambient temperature and time factors involved in performing the inspection. Therefore, a defective bearing might go undetected and remain in service.

(6) Other types of physical inspections performed by train crews, either from the locomotive or on the ground observing a passing train or pull-by inspections by wayside employees and car inspectors in train yards, could not reliably detect a bearing failure or identify a developing bearing failure.

(7) Safety inspections of stationary cars by certified car inspectors in train yards could not ensure the future condition of the bearing from origin to destination, and railways could operate trains for significant distances after performing such inspections. There were subdivisions not equipped with WISs where railways relied on physical inspections (i.e. safety inspections, wayside inspections, passing trains, and train crews) to ensure the safe mechanical condition of the bearings on their trains.

(8) Because journal bearings on some types of equipment were not accurately scanned by wayside hot box detectors (HBDs), and since most equipment, including some used on high-speed passenger trains, did not have on-board journal bearing heat detection, axle failure and subsequent derailment might occur when developing problems on bearings went undetected.

(9) Monthly testing of the HBDs and hot wheel detectors (HWDs) at Canadian National (CN) and some other Canadian railways was not performed at the alarm threshold levels; therefore, it was not known if the alarm system would activate at the minimum threshold levels.

(10) The CN had one TEST car which was capable of dynamically testing the HBDs and it had not travelled in eastern Canada for a period of over three years. The car did not dynamically test hot wheel detectors (HWDs) or dragging equipment detectors (DEDs). No Canadian railway other than CN was known to be dynamically testing its WIS sites.

(11) The CN dynamically tested its HBDs at speeds typically no greater than 60 mph; therefore, it is not known if the alarm system would activate as intended at maximum

passenger train speeds.

(12) The number of WIS sites to be monitored, combined with the task-intensive and memory-reliant nature of the HBO's duties, made it difficult for the HBO to follow all TP-105 tracking requirements, reducing the potential for early detection of distressed bearings.

(13) Communication failures between key users of the system, combined with differences in procedures, practices and systems knowledge, indicated that there was no consistent systems approach to quality management for the WIS safety devices.

(14) There was currently no regulatory requirement for a railway company to have safety devices such as the WIS in place despite the risks that might exist in its railway operations (e.g. high-speed passenger travel and dangerous goods).

(15) There were no minimum regulatory guidelines for the direction of railways pertaining to installation, inspection, calibration, alarm levels, and quality management of the WIS.

(16) Transport Canada had no national program to audit and monitor WISs, to ensure the safety of operating trains as it related to failing bearings, wheels, and dragging equipment while travelling on territories equipped with WISs.

Following the accident a large number of safety initiatives were undertaken by Canadian National to address the safety deficiencies identified in the areas of reference manuals, work procedures, job aids, and supervision relating to the activities of the rail traffic control mechanical service representatives (RTC Mechs) and hot box operators (HBOs):

(1) The CN issued a document titled Wayside Inspection Systems Chart Handling Procedures, which was considered to be a supplement to TP-105 to provide definite processes for chart reading and understanding roles and responsibilities of the RTC Mech and HBO, and training was provided to HBOs and RTC Mechs on the changes. The document contained descriptions of faulty tapes, including signs of possible derailment, and the actions which should be taken by specific parties.

(2) The CN stated that it had now ensured that, when the Wayside Inspection System (WIS) issued an alarm directly to a train, this alarm took precedence. The CN further explained that no amount of experience, judgment or feeling on the part of the HBO or the RTC Mech could override such an alarm.

(3) The RTC Mech was no longer allowed to override the HBO's decision regarding TP-105 decisions. The CN stated that TP-105 was now applied strictly and was covered in paragraph 3(c) of the above-mentioned document.

(4) The faulty tape criterion was tightened up. The CN is considering making it an absolute stop of the train when a chart is deemed to be "faulty." The CN stated that some charts, which could be deemed faulty, were now being read as indicating a potential derailment due to the Neswabin accident.

(5) The CN put system maps on the HBOs' desks to help them conceptualise the various subdivisions.

(6) A series of "team building" sessions was provided to all RTC Mechs and HBOs, including the spares who normally worked the HBO job. The training was provided by Grant MacEwen College in Edmonton and was comprised of 2 two-day sessions. The curriculum consisted of: - communications - conflict management - stress management - focus on performance.

(7) The CN had now ensured that its supervisory program would include the HBOs. It had

implemented audits of the journal advance warning signal (JAWS) and other reports, which were completed by the HBO; more direct contact with the supervisors; and operator training that stressed the point of more communication between HBOs and their supervisors.

(8) The CN had dedicated considerable capital and operating funds to the further expansion of its hot box detector (HBD) network. Main route HBD spacings continued to be reduced. Further, in April 2000, CN had lowered the alarm threshold for TP-105 hot alarms on 20 branch line subdivisions that have HBDs spaced 40 miles apart or more.

The TSB pointed out that there were several other issues, which, if addressed, might enhance both the efficiency and safety afforded by the existing system, and which the CN might wish to consider:

(1) When operating in an exploratory manner, the systems knowledge of all employees involved was important. Without a solid understanding of the complete system, employees might take action unaware of the detrimental effect it might have on other aspects of the system. Specifically, slight oversensitisation of HBDs might seem like a very reasonable option to Signals and Communications (S&C) technicians since, from their perspective, they were increasing the safety margin of the system. However, this action could result in unwarranted alarms to HBOs and train crews, increasing their workload and reducing their trust in the system. Also contributing to the workload of the HBO was the interface for the Devtronics system, which currently did not take into consideration all the task requirements of the HBO (i.e., the number of WIS sites to be monitored, tracking of trains over previous and subsequent WIS sites, identification of car numbers). Further automation of this process in the future could assist in making even better use of WIS data, while reducing employee workload.

(2) The WIS system was designed to detect distressed bearings and wheels by measuring their temperature as they passed by the scanner at train speed and to identify dragging equipment. If the heat level exceeded a pre-determined threshold, an alarm was activated. The CN tested its systems monthly in a static state and at a temperature well below the alarm threshold level, with the understanding that, if the system was set up at the testing temperature, it would detect an overheated bearing at the alarm threshold level under the dynamic condition of a passing train. Static testing of the dragging equipment detector (DED) involved a pass/fail functionality test. Dynamic testing was typically conducted semi-annually; however, it did not test DEDs nor did it test at maximum train speeds. In the absence of dynamic testing procedures which incorporated maximum train speeds, the threshold heat levels, and the simulation of dragging equipment, there was a risk that the system would not function accurately under all conditions.

(3) The Association of American Railroads (AAR)-approved tempilstik normally provided to train crews to aid them in their determination of an overheat condition had significant limitations. This was further compounded by such things as cold ambient conditions (if it was extremely cold or raining) and the time it might take before the task could be performed (if the car was near the end of the train). There presently existed other tools that could assist train crews in this task and provide them meaningful information about bearing temperature. For example, in cases where the suspect bearing was not readily apparent (e.g. the temperature was insufficient to melt the tempilstik), the train crew using a remote (hand-held) infrared temperature gun could perform a comparative analysis in the field (i.e. a train-side or car-side average) even if the bearing had cooled down below the melting threshold of the tempilstik. In this way, relative temperatures might indicate which bearing was giving a problem.

In concluding its report, the TSB wrote: 'It is recognized that the majority of railway operations in Canada are monitored by advanced technological means, such as WISs, and that the railways continue to make significant monetary investments to improve their existing systems. However, there still are subdivisions where WISs are installed at intervals in excess of 25 miles, thereby reducing the probability that distressed axle journal bearings will be detected before failure. There are also locations where axle bearing temperatures on passenger trains and/or shipments of dangerous goods are not monitored by WISs at all. Notwithstanding the foregoing, some equipment by its design is ill-suited for bearing temperature detection by WISs, irrespective of the detector spacing. Although axle failures are rare, there can be severe adverse consequences associated with such an event resulting in a risk to persons, property, and the environment. The Board is concerned that, without an industry standard that will detail the frequency and adequacy of roller bearing inspections, there may be less-than-adequate protection against the catastrophic failure of railway axles in some operational settings.' [1613]

In April 1999 a North Atlantic Treaty Organisation (NATO) operation known as Operation Allied Force had been launched against the Federal Government of Yugoslavia, aimed at ending its oppression of Albanians in Kosovo. The operation had begun with attacks on purely military targets, then had been expanded to include strategic and economic targets. Thus on Monday 10th April 1999 a NATO fighter targeted a bridge over the Grdelica gorge, a guided missile striking the centre of the bridge. The strike took place at the moment passenger train No. 393 from Belgrade to Ristovac was at the bridge. Despite this, the pilot of the aircraft made a repeat pass and fired a second missile, both striking the train.

At least 20 and possibly 55 civilian passengers died in the attack, only eight of whom could be positively identified. The attack caused immediate controversy and was condemned by various human rights organisations, including Amnesty International.

Supreme Allied Commander Europe General Wesley Clark stated that the train had been travelling too fast and that the missile had been too close to the target for it to be diverted in time. The first missile had been fired from a significant distance from the target, and the pilot had allegedly not been able to recognise the train visually. Realising that the train had been hit but believing that he could still complete the mission by striking the end of the bridge where the train had already passed, the pilot had then made another pass and fired a second missile. This one, too, had hit the train. Clark described the second hit as an "uncanny accident" in which the train had continued moving into the target. [1614] [1615]

Twenty-nine cars of the 76 forming a westbound Montana Rail Link freight train were derailed two miles west of Paradise, Montana, United States, at about 4 p.m. on Sunday 11th June 1999. Four of the cars ended up in the Clark Fork River, including one that spilt liquid asphalt and another that lost cans of beer. A second car of asphalt went into the river but was not breached, while a car containing propane was stuck in a sandbar and was also not breached. Other cars in the train had contained corn syrup, corn, soybean meal and processed grain. The asphalt was not considered a major hazard as in contact with cold water, it rapidly solidified. At the time of the derailment the temperature had been 91 degrees Fahrenheit (33 Celsius) and it was believed that heat distortion of the rails (a "sun kink") had caused the accident. [1616]

[1613] Transportation Safety Board of Canada report
[1614] Wikipedia – Grdelica train bombing
[1615] http://www.ble.org/pr/news/headline0413a.html
[1616] *The Daily Inter Lake*, Kalispell, Montana, United States, 12th July 1999

Shortly before 11 p.m. on Saturday 24th July 1999 a Burlington Northern Santa Fe Railway freight train consisting of three diesel units and 112 cars was passing through Moyie Springs, Idaho, United States, when 12 of its cars jumped the tracks. Six of these were tank cars carrying anhydrous ammonia, one of which had its outer and inner shells pierced by a rail, and 33,500 of the gas escaped and formed a 'potentially deadly cloud,' as a newspaper report described it. Fifty-four people were evacuated from their homes, the Hemlocks Motel and the Twin Rivers resort in Moyie Springs, and spent the night in a Red Cross shelter in Bonners Ferry High School, while Highway 2 was closed off. The line was of course blocked and throughout Sunday and early Monday Amtrak passengers were taken by bus between Whitefish, Montana, and Seattle, Washington, or Portland, Oregon, while about 15 freight trains were diverted over the Montana Rail Link line between Billings, Montana, and Sandpoint, Idaho.

No one was hurt in the derailment. Heavy overnight rain and strong winds helped to dissipate the cloud and the evacuation order was rescinded at 8:35 a.m. on Sunday. Fortunately, the ammonia did not enter the Kootenai River, 45 feet from the track. The cause of the derailment was unclear. [1617] [1618]

Two locomotives and ten tank cars of an Alaska Railroad freight train bound for Anchorage from the Williams Company oil refinery in the North Pole became derailed at 1 a.m. on Sunday 31st October 1999, at a location variously given as near Hurricane, near Gold Creek and Canyon, Alaska, United States. About 8,500 gallons of jet fuel was lost from one of the derailed cars. Some of the spillage accumulated in a frozen beaver pond and workmen used a boom and skimmers to prevent any fuel from reaching the Indian River, though their efforts were hampered by the cold weather and deep snow. [1619]

The derailment was later attributed to the spreading of the track by the forces exerted by the wheels of two switcher (shunting) engines, which had formed part of the train. These usually had a wider side-to-side swing than other engines, which should have been limited by the installation of coupler blocks before the engines had begun their journey at Fairbanks, but the blocks had not been fitted. [1620]

Freight train No. 70 from Ballsh to Tirane was derailed in the village of Kraps, Albania, at 21:30 on Thursday 11th November 1999. Six waggons left the rails, five of them containing gasoil, of which 332,900 gallons, worth 17,316,000 Leks, was lost. No one was hurt. The derailment was thought to have been due to the poor condition of the track. [1621]

The power of the forces of nature was once again demonstrated in Denmark on Friday 3rd December 1999. Northern Europe was affected by a storm with very strong winds, with speeds up to 200 km/h, which were sufficient to derail a freight train between Lunderskov and Kolding. About 250 metres of track were torn up and the line was blocked for several days. [1622]

1617 Ibid, 26th July 1999

1618 *The Times-News*, Twin Falls, Idaho, United States, 26th July 1999

1619 *Fairbanks Daily News-Miner*, Fairbanks, Alaska, United States, 2nd November 1999

1620 https://www.alaskarails.org/news/ARR-news99.html

1621 Albanian Telegraphic Agency report, 12th November 1999

1622 https://www.railorama.dk/nyheder/1998-1999/

A southbound Alaska Railroad freight train heading from the Williams Company oil refinery in North Pole to Anchorage was derailed at 2 a.m. on Wednesday 22nd December 1999 at a location known as Gold Creek, Alaska, United States. The train consisted of four locomotives and 53 cars, of which three locomotives and 15 cars were derailed and five cars leaked an estimated 120,000 gallons of jet fuel. None of the three train crew was hurt and they were able to continue south on the leading engine, which had remained on the line.

At 9:30 a.m. a train carrying railwaymen and officials of the Department of Environmental Conservation and the Environmental Protection Agency set out for the site from Talkeetna, 36 miles to the south, followed by about 50 workmen and state officials on a diesel railcar. The derailment site was accessible only by rail. There had been fears that the oil might reach the Susitna River, 1,000 feet west of the line, but this did not happen, large quantities of it being absorbed by the deep snow and more collecting in a drainage ditch. Oil remaining in the derailed cars was transferred by workmen to other cars.

Following this second derailment and spillage in two months, the chief of Alaska Railroad and former Alaska governor Bill Sheffield ordered senior management to review all operation of the fuel trains, which formed 45% of the company's business, and within 10 days to develop a list of changes that could be made immediately, followed by longer-term recommendations.

The derailment was later attributed to an obstruction on the line. [1623] [1624]

On the evening of Thursday 30th December 1999 Canadian National Railway (CNR) No. U-783-21-30 was travelling westward from Saint-Romuald, when it suffered a derailment at Mile 50.84, near Mont-Saint-Hilaire, Quebec, at about 19:00. Some of its cars fouled the opposite running line and were immediately struck by eastbound train No. M-306-31-30, which was passing at the time. The brakes of train No. 783 had come on in emergency as the engines had passed one another and looking back, its engineer had seen a ball of fire towards its rear and its conductor broadcast an emergency on the train radio.

Train 783, commonly known as an Ultratrain, was a unit train, one of those worked in a continuous cycle carrying hydrocarbons between Saint-Romuald and Montréal, Quebec. It consisted of two engines and 64 cars.

Train 306 consisted of two locomotives, 17 loaded and 33 empty cars.

As a result of the derailment the cars were scattered across both tracks and in the ditches on either side of the tracks. The locomotives from train 306 were engulfed in flames and lay on the south side of the track, in the ditch between the railway tracks and Laurier Boulevard (Highway 116).

Thirty-five tank cars from train 783 derailed (the 11th to 45th cars). Thirty of these sustained considerable damage and were demolished at the site. The tanks of these cars were punctured and torn open, and the safety valves and isolation valves were ripped off. The other five cars sustained minor damage. Eleven of the derailed cars had been carrying about 1 million litres of petrol and 24 contained about 2.3 million litres of heating oil. Some 790,000 litres of petrol and 1.9 million litres of heating oil burned in the fire or were not recovered. Approximately 255,000 litres of petrol and 330,000 litres of heating oil were transferred to tanker trucks and tank cars.

Both locomotives from train 306 were destroyed as a result of the impact and fire. The front of the lead locomotive was lodged in the ditch between the railway tracks and Highway

---

1623 *Fairbanks Daily News-Miner*, Fairbanks, Alaska, United States, 24th December 1999

1624 https://www.alaskarails.org/news/ARR-news99.html

116, at an underground gas pipeline. The first 26 cars of this train were derailed; 24 of them sustained considerable damage and were demolished at the site; the other two sustained minor damage. Two cars that had not derailed sustained major damage in the fire. The derailed cars included a hopper car containing 79,900 kg of sodium chlorate, which released about half of its contents in the accident; however, the product did not come into contact with the hydrocarbons. The sodium chlorate and the contaminated soil were transferred to specially prepared containers when the site was cleaned up.

The hydrocarbons from train 783 and cargo from some of the cars from train 306 burned for four days, creating a smoke plume about 500 m high. The smoke affected air quality in the immediate area of the accident site, necessitating the evacuation of residents of the Domaine des Hurons in Sainte-Madeleine. Approximately 350 families living within a two-kilometre radius of the accident site had to be temporarily evacuated. Two locomotives and 61 cars were damaged in the accident. The engineer and conductor of train No. M-306-31-30 were fatally injured in the accident.

The fire was isolated and brought under control during the night of 30th to 31st December but continued to burn for four days. In addition, the accident resulted in damage to private and public property and to the environment.

As might be expected, the Transportation Safety Board of Canada (TSB) investigated the accident. Its report reached 21 conclusions:

(1) The two crew members on train 306 had been fatally injured in the deflagration, which had occurred when the lead locomotive had struck the 15th car of train 783.

(2) In all probability, the 15th car of train 783 had derailed when the rail head section had separated.

(3) The direction of the chevron markings observed on the fracture surfaces indicated that the break had originated in the pre-crack area located in the lower corner of the rail base.

(4) The mould seam of the weld resting against the tie plate, the mismatch between the rail ends and the use of a straight base plate had created conditions conducive to the development and propagation of cracks.

(5) The existing pre-crack had been sufficient to cause the rail to break under the effect of the stresses caused by the low temperatures and wheel impact loads lower than the wheel impact load detector (WILD) limits set by Canadian National (CN) for car set-off.

(6) Only specific inspections of the weld and measurements of the mismatch between the rail ends would have shown that the weld did not meet the existing standards and that it was resting against the tie plate.

(7) Like the other railway companies, the CN had discontinued ultrasonic inspection of thermite welds. Inspection of flash butt welds was still required, even though their quality was more uniform and their rate of defects is lower.

(8) Manual ultrasonic testing, as practised by CN, would not have detected the pre-crack in the base, but nevertheless represented an essential safety tool, especially on tracks where rail flaw inspection frequency was low.

(9) Without an ultrasonic inspection of the entire periphery of the weld, defects in the base could not be detected and might cause rail failure.

(10) Since welds were critical elements of track infrastructure, quality control by independent personnel was imperative for immediate identification and rectification of dangerous conditions or inadequate work methods.

(11) Although the industry recognised that wheels generating high-impact loads might

damage railway equipment and infrastructure, and although the CN had clearly established a causal link between high-impact loads and rail breaks, there were still no standards or regulations in this regard.

(12) Since the WILD communication system was not fail-safe, a communication breakdown might hamper its operation and allow defective wheels to continue in service at the risk of breaking a rail or damaging equipment.

(13) Since the weighing of cars was left to the shippers' discretion and the CN no longer checked car weights, an incorrect weight entered by a client might go unnoticed, increasing the risk to employees and the public.

(14) Because the electronic data interchange (EDI) software did not freeze erroneous information when new data were "suspended," the system could continue to generate inaccurate train consists even though the correct information was available.

(15) The inspection frequency recommended by Transport Canada (TC) – once every six years – was not conducive to ensuring that the transportation and handling of dangerous goods was always consistent with the *Transportation of Dangerous Goods Regulations.*

(16) An emergency response plan based on TC's document TP 9285 would have facilitated and accelerated the response by defining roles, establishing appropriate response methods and providing an inventory of the foam and equipment available in the region.

(17) The transportation of large quantities of liquid hydrocarbons on unit trains through urban areas created a risk.

(18) The increased use of WILDs and more frequent testing for internal rail defects had probably contributed to the overall decline in in-service rail breaks and related derailments.

(19) The absence of design and construction standards for the locomotive event recorders had impeded efforts to understand rail accidents and improve the safety of this mode of transportation.

(20) The firefighters had performed their role in an effective manner, even though they had been facing a major disaster involving a mode of transportation not covered by their practical training.

(21) Since the contaminated soil of the railway roadbed, the highway and the ditches had been dug up and removed for treatment, the long-term environmental impact had been minimal.

The TSB noted that since the accident the CN had instructed track supervisors to check the position of existing thermite welds and to adjust any that were resting against tie plates. It had initiated a visual random inspection of 4,000 newly installed field welds; 10 per cent of these welds had been examined and only one defective had been found. It had also increased the frequency of rail testing on main tracks in eastern Canada to four times per year.

In 2000, new transmission devices had been installed by CN on each wheel impact load detector (WILD) to improve protocol compatibility, while a new monitoring system and new procedures had been implemented at the rail traffic control centre so that transmission errors could be detected and timely corrective action taken.

A second WILD communication alarm monitor had been added in the office of the rail traffic control mechanical service representative (RTC Mech) in Edmonton to ensure that any communication failure with a field site was acted on quickly. The CN had also initiated weekly WILD system performance review meetings between RTC Mechs and Signals and

Communications technicians to ensure that any maintenance or operational problems were identified and corrected in a timely manner.

The TSB commented that the introduction of unit trains carrying liquid hydrocarbons through urban areas created unusual operating conditions, which were not adequately addressed by the existing safety regulations. Although the accident had occurred in a sparsely populated area, the Ultratrain's route took it through many urban areas where the risk was much greater. It therefore recommended that Transport Canada should review the requirements for emergency response plans to ensure that the transportation of liquid hydrocarbons was consistent with the risks posed to the public.

Further recommendations in the report were that Transport Canada should ensure that the design specifications for locomotive event recorders included provisions regarding the survivability of data and that Transport Canada should review the requirements for the inspection and quality control of thermite field welds, to ensure that an adequate level of safety was maintained on all types of tracks.

Finally, the TSB voiced concerns over the possible risks associated with the electronic exchange of shipping documents and the possibility of human errors leading to an incorrect description of the contents being transported. It made no recommendations regarding this matter.

On the evening of Sunday 9th January 2000 a Ferrocarriles Chiapas-Mayab freight train was stopped on a slight incline about 11 kilometres south-west of Tapachula, Chiapas state, Mexico, when nine waggons became uncoupled and began to run back down the bank. They increased speed until finally they left the rails on a curve. Several tank waggons were punctured in the derailment and 130,000 gallons of petrol and diesel oil were spilt. Local officials and troops sealed off the area while workmen laboured to recover the spilt fuel or deactivate it with chemicals. Road links to several local towns and to neighbouring Guatemala were shut off.

José Herrera, regional director of civil defence for the state, commented: "We think criminal hands disconnected the brakes, causing the waggons to run away and reach a speed of about 98 km/h before they overturned on the Paso del Toro curve." [1625]

It was hoped that the spillage might be cleared in a couple of days. [1626]

A Wisconsin Central freight train was derailed on the evening of Sunday 16th January 2000 at Kiel, Wisconsin, United States, and the leakage of a small amount of methanol led to the evacuation of about 200 residents. They were allowed to return home after five hours. [1627]

Faulty loading caused a train derailment in Columbus, Ohio, United States, on or about Wednesday 19th January 2000. Containers were stacked up too high on a car to pass beneath the bridge carrying Nationwide and Front Streets, struck the bridge and three cars were thrown off the rails. [1628]

Southbound Ontario Northland Railway (ONR) freight train No. 402 left Englehart, Ontario, for North Bay, Ontario, at about 13:40 on Tuesday 14th March 2000. It consisted of two

---

[1625] *The New Mexican*, Santa Fe, New Mexico, United States, 11th January 2000
[1626] *The Yuma Daily Sun*, Yuma, Arizona, United States, 11th January 2000
[1627] http://hv.greenspun.com/bboard/q-and-a-fetch-msg.tcl?msg_id=002LSm
[1628] Seven days… archive of rail accidents January 2000 web page

locomotives, 38 tank cars loaded with sulphuric acid, 11 boxcars loaded with lumber products, one empty boxcar and one car loaded with steel ingots and was actually a combination of two separate trains from Kidd Creek, Ontario, and Noranda, Quebec. Pre-departure inspections and brake tests had been carried out at Kidd Creek and Noranda and another brake test was conducted after the two trains had been combined at Englehart. A wayside hot bearing/wheel and dragging equipment detection system located at Mile 80.6 had not registered any defects. At about 16:10, while the train was rounding a right-hand curve at Mile 63.4, near Temagami, Ontario, Canada, a train-initiated emergency brake application had occurred. The event recorder transcript indicated that the train speed had been 39 mph at that time.

Before the emergency brake application, the brakes had been fully released and the throttle had been in the No. 1 position. After conducting the necessary emergency procedures, the crew had determined that 29 cars had derailed, beginning with the 13th car behind the locomotives and extending to and including the 41st car. Twenty-five tank cars of sulphuric acid were involved in the derailment, 12 of which lost varying amounts of acid. Four cars had had their tank shells breached as a result of impact with wheels, bolsters and/or couplers from preceding derailed cars. Ten cars had lost product as a result of damage to their top fittings; two of these cars were also punctured. Six of the ten cars sustained damage to their rupture disc assemblies. Five cars located at the south end of the derailment site rolled down an embankment when their forward speed had been near zero. Despite this rather gentle roll, the top fittings (mainly the rupture disc/safety vent assembly) on four of the five cars had sheared off. Some of the cars had sustained damage to their air lines and/or the loading hole covers. The loss of sulphuric acid was estimated to be 386,000 litres. About 144,000 litres of acid flowed into Martin Creek, which ran parallel to the rail track at the south end of the derailment site. An estimated 35,000 litres of sulphuric acid that spilled into the creek migrated to Hornet Lake. Twenty-three of the derailed cars were extensively damaged. No one was hurt but some fish and other wildlife were killed.

The Transportation Safety Board of Canada (TSB) investigated the derailment. Its report reached 11 conclusions:

(1) The deteriorated track ties (sleepers), combined with the cracked and broken joint bars, had resulted in a weakened track structure that could no longer support the applied stresses of rail traffic.

(2) The use of a temporary measure (gauge rods) for an extended period of time had masked deteriorating track conditions.

(3) The rapid turnover of track maintenance personnel had resulted in a loss of continuity in the monitoring of track conditions, and the use of less than fully trained supervisors and track inspectors had created a situation where deteriorating track conditions would not be noticed.

(4) The ONR's quality assurance practices for track inspection and maintenance had not led to the necessary corrective measures to ensure that the previously identified weakened track at the derailment location was repaired.

(5) The failure of the top fitting on Class 111 tank cars had caused sulphuric acid to leak, resulting in an environmental impact.

(6) Postponement of the needed maintenance (tie installation and gauge correction), without the implementation of a temporary slow order to reduce the lateral forces applied to the track in the derailment area, had increased the risk of derailment.

(7) The industry-wide practice of visually inspecting ties could not adequately assess individual tie strength.

(8) The inability of the current industry methods of inspection to detect small fatigue cracks

in joint bars had increased the vulnerability of the track structure at joint locations.

(9) In the absence of formal track standards for the use of gauge rods, track maintenance staff were expected to limit the use of these devices based upon their knowledge and understanding of generally accepted track maintenance principles.

(10) The absence of a standard requiring that Class 111 tank cars used in the transportation of sulphuric acid be equipped with top fitting protection designed to minimise the loss of product during rollover presented a continuing risk to the public and the environment.

(11) The ONR's emergency plan had been implemented in an effective and timely manner and had minimised the effect of the spill on the environment.

The TSB noted that since the accident the ONR had undertaken various initiatives:

(1) A gauge restraint measurement test had been conducted. This measured the track strength by applying non-destructive vertical and lateral loads. The ONR had used the test results to identify critical weak sections of track and had taken corrective action to replace defective ties where required. The ONR intended to perform such tests annually. Corrective actions to replace or repair defective track structure would be completed wherever weak conditions were found.

(2) The ONR was completing training for track maintainers and track maintenance foremen. Track supervisor training would be provided to all existing supervisors and other personnel who might be required to work as relieving track supervisors. By December 2001, a formal procedure for certification of track supervisors and inspectors would be developed and implemented.

(3) The use of gauge rods was being reviewed and might result in the development of a standard.

(4) The ONR had implemented two track inspection units that measured and detected wide gauge. These inspection units were being utilised on a regular schedule throughout the ONR rail system. A track geometry test car would continue to be used twice annually.

(5) In 2000, the ONR had installed 50,000 track ties concentrated on curves between North Bay and Temagami. The 2001 tie program consisted of installing 100,000 track ties on the ONR rail system.

(6) The transient nature of the track maintenance workforce would be discussed with the unions in upcoming negotiations.

(7) A consultant engaged by ONR to review rail infrastructure and maintenance standards and practices had reported that the overall condition of track, bridges and facilities compares favourably with any viable rail operation in North America, including that of Class 1 carriers.

In addition, the TSB noted that an Association of American Railroads (AAR) Tank Car Committee Task Force (including a Transport Canada representative) had been formed and it had been agreed and subsequently adopted by the AAR that the specifications for fittings for acid cars should be revised. The specifications would be applicable to all sulphuric acid cars, new and existing, and would provide requirements to be met to achieve an adequate level of top fitting protection. For example, safety vents and threaded assemblies that had proven to be vulnerable during the Temagami derailment would be prohibited. An implementation period of 10 years was proposed. [1629]

[1629] Transportation Safety Board of Canada report

A little before midnight on Saturday 4th November 2000 Barbara Meinecke was awoken by an acrid odour, which she thought resembled that of burnt plastic. "I kept smelling this terrible smell," she said. "It was so bad it made my throat raw. It was bad enough to get me out of bed." When she got up to investigate she saw that there were police cars turning into the street outside and shortly afterwards she heard warnings to leave, shouted over police bullhorns. She lived just a few streets away from the site of a derailment in Scottsbluff, Nebraska, United States. With her husband and neighbours she left her home to take shelter in Gering High School, which accommodated over 1,000 people that night, though later Mr and Mrs Meinecke went on to her brother's house instead.

At about 10:30 p.m. that Saturday 18 of the 79 cars of a Burlington Northern Santa Fe Railroad (BNSF) freight train had derailed. Three of the cars had been torn open and released about 80,000 gallons of toxic and carcinogenic benzene. In all, about a third of the 15,000 inhabitants of Scottsbluff were evacuated. From 2 a.m. they were allowed to return home at their own risk, only for some to then be evacuated a second time when it was feared that they had been allowed home too soon. [1630]

A few days after the derailment it was theorised that a coupling falling from a car and getting beneath the wheels of another car had probably caused it. [1631]

Eastbound Canadian National Railway freight train M336-41-10, from Winnipeg, Manitoba, to Thunder Bay, Ontario, was travelling at 34 mph when 17 of its cars became derailed at 07:45 on Monday 11th December 2000 at Mile 50.3 of the Kashabowie Subdivision, at Shabaqua, Ontario, Canada. The derailed cars included two loaded tank cars containing methanol, one of which was breached, spilling about 100,000 litres of the toxic chemical, most of which entered the Shebandowan River. The derailment occurred in an unpopulated, forested area. Persons living close to the river, downstream, were cautioned not to consume water drawn from wells near the river. There were no injuries and the environmental damage was minimal.

The Transportation Safety Board of Canada report on the derailment reached three conclusions:

(1) The forces generated by an undesired emergency brake application (due to a car with a defective brake valve) on a piece of rail exhibiting marginal ductility, high internal tensile stresses and a pre-existing crack had led to the rail failing and the derailment.

(2) The rapid propagation of cracks in the rail during times of low temperatures called into question the ability of widely spaced (twice-yearly) ultrasonic testing regimes to ensure a safe operating environment in areas that experienced severe winter conditions.

(3) The railway response, and that of agencies charged with public and environmental safety, had been timely, effective and comprehensive.

The TSB noted that Transport Canada (TC) would be contacting CN representatives in the near future to determine if problems related to chromium rail had been experienced before and to obtain information as to the extent of its use. Based on the response, TC would decide upon any safety action to be taken, if required. The CN had increased and continued to increase its rail testing frequency on core and secondary lines where it was warranted, especially in the

[1630] *The Hutchinson News*, Hutchinson, Kansas, United States, 6th November 2000
[1631] *The Hays Daily News*, Hays, Kansas, United States, 12th November 2000

winter months. [1632]

The day after the Shabaqua derailment, Tuesday 12th December 2000, Canadian Pacific Railway (CPR) freight train LDRS-12, consisting of a locomotive and 55 loaded cars, left Lloydminster, Saskatchewan, Canada, at about 21:50, proceeding southward on the Lloydminster Subdivision, en route to Wilkie, Saskatchewan. At about 22:25, while travelling at approximately 27 mph, a train-initiated emergency brake application occurred. After emergency procedures had been executed, inspection of the train revealed that 13 cars, the 26th to the 38th car in the train, had derailed near Lone Rock, Saskatchewan. Twelve tank cars and approximately 600 feet of track were destroyed.

Seven of the derailed cars had contained kerosene and six had contained liquid asphalt. About 84,000 litres of kerosene and 150,000 litres of liquid asphalt were lost as a result of the derailment. About 48,000 litres of kerosene were lost from the 30th car (NCTX 20736) through four holes torn open in the bottom of the tank. The 34th car (CP400335), which lacked a head shield, had received a head shell puncture and lost about 34,000 litres of kerosene. The 29th (CP400281), 32nd (CP400295), and 33rd (CP400305) cars, which did not have manway housings, had lost about 2,000 litres of kerosene from leaks through damaged safety vents. The safety vent on car CP400281 had broken off at the nipple connection to the manway cover, allowing the valve and stem to drop into the tank, leaving a 1½-inch hole. As the car was leaning at a 45-degree angle, kerosene had leaked through this opening. Tank cars CP400295 and CP400305 had minor leaks through the safety vent due to deteriorated O-rings. The 27th, 35th, and 36th cars, carrying liquid asphalt, had leaked approximately 150,000 litres of their loads as a result of damage to the tank shells. The 27th car received punctures to the head shell at both ends of the tank car. The 35th car (GATX 89182), not having a head shield, had received an impact resulting in a one-foot hole in the "A" end head shell. The 36th car (GATX 72480) had a small leak at a tank shell puncture. Five of the 13 cars had sustained damage to the bottom unloading valves. There had been no protection for the bottom outlets or top fittings on these cars.

There were no reported injuries. The hamlet of Lone-Rock, the closest habitation, was about a mile south of the accident site. The weather at the time was -30° Celsius, clear and calm with unlimited visibility.

The Transportation Safety Board of Canada investigated the accident. Its report contained four findings:

(1) The derailment had occurred in cold weather due to the brittle fracture of the rail at a joint where there were existing pre-cracks at an elongated bolt hole.

(2) Although the track inspection program had included electronic testing and regular inspection by track forces using a Hi-rail vehicle, the inspections had not been adequate to detect damaged bolt holes or the loose joint assembly.

(3) The inadequate diameter of the brake support bracket pad of tank shell steel of car NCTX 20736 had resulted in the stress loads being transferred through the pad to the tank shell, rather than the bracket breaking away from the pad, resulting in the breach of containment.

(4) The risk of a breach of containment during derailments was increased when dangerous goods were shipped in tank cars without head shields and top fitting protection.

Following the derailment the TSB noted that:

[1632] Transportation Safety Board of Canada report

(1) Transport Canada had advised that the issue of the existence of loose rail joints and bolts at the location of the derailment would be examined by its regional office.

(2) Tank car NCTX 20736 had been one of a fleet of cars owned by PLM International Inc. (PLM) and identified by the Association of American Railroads (AAR) as having a potential problem with the dimension and welding of the pad to the tank shell. The AAR and the car owner were working together to ensure that all cars in the fleet identified with similar brake support pad size were inspected at the next shopping of the car. At the beginning of 2003, there were 14 out of a total of 1,208 identified cars left to be inspected. Of these, 7 cars were in active service. It was planned to have them inspected by the end of March 2003. Another 4 cars were in storage and would have the necessary changes completed before they were returned to service. The remaining 3 cars had been assigned "to be scrapped" status. Car owners other than PLM had been notified of the problem and requested to initiate a modification program similar to PLM's to conclude by 10th October 2003.

(3) The CPR had contracted an engineering firm to reclaim the soil contaminated from the tank car containment breach. The accidental release of kerosene and liquid asphalt had required extensive remedial action to protect the environment. About 7,125 cubic metres of kerosene-contaminated soil had been excavated from the derailment site and transported to a land farm treatment area. The soil was to be reclaimed in accordance with provincial and federal regulatory requirements until it had been returned to its natural state. The top nine inches of soil had been tilled, allowing hydrocarbons to vent to the atmosphere. The soils were sampled two to three times a year. When the soil met the provincial criteria, it was returned to the excavation site. To date, three lifts had been returned to the excavation site. This left two lifts for 2003, which should complete the mediation of the contaminated soils. Earth dykes had been constructed to contain the liquid asphalt, which had solidified and had not infiltrated the soil. About 481 tonnes of solidified asphalt had been excavated and disposed of at a resource recovery facility. [1633]

At 7:30 p.m. on the evening of Tuesday 2nd January 2001 a Union Pacific Railroad freight train from Nogales to Tucson, travelling at 41 mph, was approaching Sahuarita, Arizona, United States, when 19 of its 107 cars became derailed and about 20,000 gallons of sulphuric acid were lost from two of them. Some 96 local residents were evacuated and a two-mile stretch of the Old Nogales Highway, which paralleled the railway, had to be closed for more than a day while the spillage was dealt with. The two-man crew was unhurt but a fireman had to be treated in hospital after inhaling the fumes.

Gene Olivas, 35, who was evacuated along with his wife and son, told a reporter: "My eyes were watering, burning. It was scary, especially for my wife, because she's pregnant."

The acid had been destined for a metal-extraction plant at Ray Junction, Arizona. [1634]

Early on Sunday 5th August 2001 a Burlington Northern Santa Fe Railway freight train en route from Teague to Houston suffered the derailment of 21 of its 84 cars on a curve between Magnolia and Pinehurst, Texas, United States. Many of the derailed cars were tank cars, three of which spilt about 11,000 gallons of acetic anhydride and 23,000 gallons of 2-ethylhexyl acrylate, both chemicals being irritant to skin, eyes and respiratory system. As a result, 230 residents were evacuated. "'It's a very strong, noxious odor inside the house," said Darlene Mims, who was one of those evacuated. "It's nauseous down there. I was really sick."

1633 Transportation Safety Board of Canada report

1634 *Yuma Daily Sun*, Yuma, Arizona, United States, 3rd and 4th January 2001

There were fears that if it rained, the spilt chemical might get into drainage ditches and creeks.

It was hoped that the residents might be able to return home on Tuesday. [1635]

Three waggons of a freight train that had come from the works of the national chemical company were derailed in the Thiaroye-sur-Mer suburb of Dakar, Senegal, on Sunday 30th September 2001. A quantity of phosphoric acid was spilt from the derailed waggons and affected 26 homes situated close to the railway line. Some 250 people were temporarily made homeless and 80 needed hospital treatment. No one was killed. According to some witnesses the train had been travelling at high speed prior to the derailment. Measures were being considered to prevent a repetition of the accident. [1636]

At about 3:30 p.m. on Sunday 23rd December 2001 part of a CSX Transportation freight train was derailed in Charlotte, New York, United States. At least seven of the 42 cars of the train left the track, two tank cars containing methylene chloride and acetone spilt their contents and a fire broke out. The engineer was injured when he jumped from the train and two empty lineside houses were damaged by being struck by the derailed cars and by the fire. [1637]

At about 1:37 a.m. on Friday 18th January 2002 eastbound Canadian Pacific Railway freight train 292-16 was travelling at 41 mph when it became partly derailed near Minot, North Dakota, United States, 31 of its 112 cars leaving the track. Five tank cars loaded with anhydrous ammonia suffered what was termed catastrophic ruptures and a plume of ammonia vapour covered an area occupied by 11,600 people. One resident was fatally poisoned, 11 were seriously injured and 301 residents, 19 emergency responders, the engineer and the conductor were slightly hurt. Early on, the residents were advised to shelter in place rather than attempt to evacuate the area. The man who died and some of the seriously injured had either not heard or had disregarded this advice and had attempted to flee. Damage from the accident exceeded $2,000,000 and 'environmental remediation' $8,000.000.[1638] The National Transportation Safety Board conducted a comprehensive investigation of the accident.

'Roger Bredahl needed a machine to help him breathe, and his face was still red from burns a day after he was caught in a deadly cloud of gas from a fertilizer spill after a train derailment. "Everything was burning," the Onalaska, Wis. Man said from his hospital bed Saturday. I have trouble breathing and swallowing.

'Also hospitalized was Andrew Hinton, who had rushed out of his house with his parents after the derailment, only to be forced back by the anhydrous ammonia fog. "I thought we were about ready to die," Hinton said.' [1639]

The NTSB noted that the ammonia had originated at Medicine Hat, Alberta, Canada. The train had come from South Edmonton, Alberta, and was destined for St Paul, Minnesota. It had consisted of two locomotives, 86 loaded and 26 empty cars. Fifteen of the cars had been loaded with anhydrous ammonia, 11 with styrene monomer and 10 with liquefied petroleum gas (LPG).

[1635] *Port Arthur News*, Port Arthur, Texas, United States, 7th August 2001
[1636] http:…./200110010507.htm
[1637] *The Post Standard*, Syracuse, New York, United States, 24th December 2001
[1638] National Transportation Safety Board report
[1639] *The Janesville Gazette*, Janesville, Wisconsin, United States, 20th January 2002

The NTSB report reached 15 conclusions:

(1) Train crew qualifications and train operations had not been factors in the accident, and there was no evidence found that crew fatigue or alcohol or drug use had been causal or contributory to it.

(2) The derailment had not been caused by a mechanical or component failure of any of the train's rolling stock.

(3) The joint bars at the east end of the plug rail had fractured under the previous train or as the accident train had passed over the joint, and after the joint bars had fractured, the rail itself had also fractured and broken away, causing train 292-16 to derail.

(4) Canadian Pacific Railway inspection procedures before the accident had been inadequate to properly inspect and maintain joints within continuous welded rail, and these inadequate procedures had allowed undetected cracking in the joint bars at the accident location to grow to a critical size.

(5) Federal Railroad Administration requirements regarding rail joint bars in continuous welded rail were ineffective because they did not require on-the-ground visual inspections or non-destructive testing adequate to identify cracks before they grew to critical size and resulted in joint bar failure.

(6) Canadian Pacific Railway's track procedure manual had been confusing and thus had not provided employees with clear guidance on the practices to be followed in installing and maintaining continuous welded rail.

(7) The Federal Railroad Administration's oversight of the Canadian Pacific Railway's continuous welded rail program had been ineffective because the agency had neither reviewed the program nor ensured that its track inspectors had copies of the program to determine if the railroad was in compliance with it.

(8) The catastrophic fracture of five tank cars had increased the severity of the accident by exposing residents to high concentrations of toxic vapours from the instantaneous release of 146,700 gallons of anhydrous ammonia and to the rocketing of portions of tank cars.

(9) The low fracture toughness of the non-normalised steels used for the tank shells of the five tank cars, which had catastrophically failed in this accident, had contributed to the cars' complete fracture and separation.

(10) The use of tank cars built before 1989 and fabricated from non-normalised steel to transport US Department of Transportation class 2 hazardous materials under current operating practices had posed an unquantified but real risk to the public.

(11) The research program proposed by the Federal Railroad Administration to model the dynamic forces and evaluate the crashworthiness of tank cars in accident conditions was incomplete without a plan to validate the predictive model.

(12) A materials standard to define the minimum level of dynamic fracture toughness for the material in all tank cars that transported class 2 hazardous materials over the entire range of operating temperatures would provide greater assurance that the tank car materials would perform in a safe manner in accident conditions.

(13) Before the accident, the Minot emergency responders, the city of Minot and the Canadian Pacific Railway had conducted a disaster preparedness exercise, which had enhanced the effectiveness of the emergency response to the anhydrous ammonia release on January 18th, 2002.

(14) The Ward County 911 dispatchers had provided accurate and timely information to the residents of Minot even though the Ward County 911 system had received more than

2,800 calls immediately following the accident.

(15) The decision by the chief of the Minot Rural Fire Department to shelter the residents of Minot in their homes during the anhydrous ammonia release had represented an effective response to the emergency.

The NTSB determined that the probable cause of the derailment of Canadian Pacific Railway train 292-16 had been an ineffective Canadian Pacific Railway inspection and maintenance program, which had not identified and replaced cracked joint bars before they had completely fractured and had led to the breaking of the rail at the joint. Contributing to the severity of the accident had been the catastrophic failure of five tank cars and the instantaneous release of about 146,700 gallons of anhydrous ammonia.

The NTSB made several recommendations.

To the Federal Railroad Administration:

(1) It should require all railroads with continuous welded rail track to include procedures (in the programs filed with the Federal Railroad Administration) that prescribed on-the-ground visual inspections and non-destructive testing techniques for identifying cracks in rail joint bars before they grew to critical size.

(2) It should establish a program to periodically review continuous welded rail joint bar inspection data from railroads and Federal Railroad Administration track inspectors and, when determined necessary, it should require railroads to increase the frequency or improve the methods of inspection of joint bars in continuous welded rail.

(3) It should instruct Federal Railroad Administration track inspectors to obtain copies of the most recent continuous welded rail programs of the railroads that fell within the inspectors' areas of responsibility and require that inspectors used those programs when conducting track inspections.

(4) It should conduct a comprehensive analysis to determine the impact resistance of the steels in the shells of pressure tank cars constructed before 1989. At a minimum, the safety analysis should include the results of dynamic fracture toughness tests and/or the results of non-destructive testing techniques that provided information on material ductility and fracture toughness. The data should come from samples of steel from the tank shells from original manufacturing or from a statistically representative sampling of the shells of the pre-1989 pressure tank car fleet.

(5) Based on the results of the Federal Railroad Administration's comprehensive analysis to determine the impact resistance of the steels in the shells of pressure tank cars constructed before 1989, as addressed above, it should establish a program to rank those cars according to their risk of catastrophic fracture and separation and implement measures to eliminate or mitigate this risk. This ranking should take into consideration operating temperatures, pressures, and maximum train speeds.

(6) It should validate the predictive model the Federal Railroad Administration was developing to quantify the maximum dynamic forces acting on railroad tank cars under accident conditions.

(7) It should develop and implement tank car design-specific fracture toughness standards, such as a minimum average Charpy value, for steels and other materials of construction for pressure tank cars used for the transportation of US Department of Transportation class 2 hazardous materials, including those in low-temperature service. The performance criteria must apply to the material orientation with the minimum impact resistance and take into account the entire range of operating temperatures of the tank car.

To the Canadian Pacific Railway:

(1) It should finalise and submit to the Federal Railroad Administration its revised continuous welded rail maintenance program and ensure that all maintenance employees were trained in the requirements of the new program. [1640]

Twenty-two of the 65 waggons of a Pipri – Muzaffargarh special oil train left the rails between Daur and Bandhi stations, Pakistan, on Saturday 9th February 2002 and there was a large spillage of furnace oil. The cost of the lost oil and damage caused by the derailment was estimated at between 3,000,000 and 5,000,000 rupees.[1641] A couple of weeks later, two men were arrested for stealing some of the spilt oil with the connivance of railway staff. [1642]

On Saturday 16th February 2002 the engine and nine waggons of a freight train were derailed at Osnabrück, Germany. One of the waggons, containing acetonitrile, was punctured and a fire broke out that took firemen several hours to bring under control. An unknown quantity of the flammable and harmful chemical also seeped into the soil. No one was hurt but the accident was likely to prove expensive in terms of damages. A broken rail was later determined to have caused the derailment. [1643]

Also on Saturday 16th February 2002 near Zell am See in Austria, a single-car local passenger train was derailed and overturned by an exceptionally strong gust of wind. A woman passenger was killed and several others were seriously hurt. [1644]

Eight of the 79 cars of a CSX Transportation freight train from Hamlet, North Carolina, to New Orleans, Louisiana, were derailed in the northwestern part of Atlanta, Georgia, United States, on Saturday 15th March 2002. Five of the cars contained hexamethylenediamine, a corrosive chemical used in the production of nylon, which can cause skin and eye irritation and respiratory difficulties. The chemical leaked from one car and 100 people had to be evacuated from nearby streets, while a woman and three children riding on a MARTA bus nearby were taken to Grady Memorial Hospital after possible exposure to the chemical. [1645] [1646]

A Canadian National Railway freight train from Toronto, Canada, to Chicago had 35 of its 58 cars derail at Potterville, Michigan, United States, at 12:30 p.m. on Monday 27th May 2002. Nine of the derailed cars were carrying liquefied petroleum gas (LPG), specifically propane, which leaked from at least two of them. Two other cars were carrying sulphuric acid but these were not punctured. Because of the potential danger of fire and explosion, 2,200 inhabitants of Potterville had to be evacuated.

A newspaper reported, '"It derailed right in front of our house," said Kelli Barcomb, who lives with her husband and three children in the Independence Commons mobile home park. "It was kind of scary because we're usually out there and we watch it go by."

'Dozens of police and fire officials used bull horns and went door to door trying to clear

---

[1640] National Transportation Safety Board report

[1641] http://www.jang.com.pk/thenews/feb2002-daily/11-02-2002/national/n4.htm

[1642] *Dawn* internet edition, 21st February 2002

[1643] *Neue OZ online*: Ursache schienbruch

[1644] http://www.*Ananova*.com/news/story/sm_710787.html

[1645] *The Bakersfield Californian*, Bakersfield, California, United States, 17th March 2002

[1646] *Sunday Telegraph*, Bluefield, West Virginia, United States, 17th March 2002 the town and shut down all entrances to the city. The Michigan Department of Transportation shut down both exits to Potterville off Interstate 69 as well as sections of other roads. Consumer Energy cut off the electrical service to most of the town at the request of fire officials, said utility spokesman Doug Crawford. “Nothing like this has ever happened here,” said Benton Township – Potterville Fire Capt. T.J. Richardson.

‘Residents with nowhere else to go were given food and water at the sheriff's office. More than 90 people and 17 volunteers were at the shelter, but several left after learning Canadian National would reimburse residents for hotel rooms. Julie Potter arrived at the shelter about 4 p.m. with her three cats. She had been mowing her lawn when the train derailed and immediately fled. “I'm a nervous wreck,” Potter said. The Potterville schools have closed for the day Tuesday, the sheriff's department said.’ [1647]

The line through Potterville, used by more than 30 trains daily, was of course blocked. Freight trains were diverted to Norfolk Southern and CSX Transportation routes and Amtrak services between Toronto and Chicago were cancelled. The situation was being monitored by the National Transportation Safety Board, but their spokeswoman Lauren Peduzzi said that although they considered the accident as having been serious, they did not have the resources to carry out an investigation into it. [1648]

The Potterville residents were allowed to return home on Friday 31st after the LPG had been pumped out or burnt off. [1649]

Two of the three locomotives and the leading 24 cars of a 141-car Norfolk Southern Railway freight train 15T from Knoxville, Tennessee, to Birmingham, Alabama, were derailed at about 11:20 a.m. on Sunday 15th September 2002 at Farragut, Tennessee, United States, not far from Knoxville. The train was carrying a mixed freight including aluminium, bricks, limestone and military vehicles but it was the rupture of a tank waggon carrying sulphuric acid that was a cause for concern. Almost 11,000 gallons of the acid were lost and vaporised to form a toxic cloud, leading to the evacuation of at least 3,000 people.

‘With a billowing cloud of fuming sulfuric acid hovering over her neighbourhood, Lisa Perkins knew that it was time to get her four children out of the area. The family, along with their pet Chihuahua, was among 42 people sleeping on cots Sunday night at Bearden High School after a train carrying 10,600 gallons of the hazardous chemical derailed in the Farragut community near Knoxville. No one was injured, but at least 3,000 people were evacuated. Perkins' husband John stayed behind. “I kept calling my husband, trying to get him to come up here, and he said, ‘If I'm going to go, I'm going to go,’” she said. “My main concern was the kids. I told my husband, ‘If you're going to be stubborn, stay here. But I have to make sure these kids get to safety.’”’ [1650]

The evacuation order was lifted early on Tuesday. [1651]

The National Transportation Safety Board (NTSB) investigated the accident. It found that about two hours before the derailment the crew of eastbound train No. 703 had received an unexpected signal check and had discovered that the spring switch at Boyd Siding, near Farragut, was not properly closed. The switch (points) was trailing for their train, which was therefore able to negotiate it without any problem. The dispatcher was informed and advised a signal maintainer to investigate, adding that there was no immediate hurry, as no westbound

[1647] *The Daily Press*, Escanaba, Michigan, United States, 28th May 2002
[1648] Ibid, 29th and 31st May 2002
[1649] *Clovis News Journal*, Clovis, New Mexico, United States, 2nd June 2002
[1650] *Laurel Leader-Call*, Laurel, Mississippi, United States, 17th September 2002
[1651] Ibid, 18th September 2002

trains were due for a while, so the maintainer had his breakfast before finally arriving at the switch at about 11 a.m. The maintainer could see nothing obviously wrong with the switch but knew that he would need a track warrant to block the line before carrying out a more detailed inspection. He contacted the dispatcher and they agreed to wait until train 15T and a coal train had passed before the granting of the track warrant. Train 15T then arrived and the maintainer became a witness to the derailment. Post-accident inspection showed that a bolt had broken off at the bottom of the switch, in a position where it would have been invisible to the maintainer upon his arrival, and the detached bolt had prevented the switch from closing properly.

The NTSB determined that the probable cause of the derailment was:

(1) The decision by the train dispatcher and signal maintainer to allow the train to pass over the spring switch at maximum authorised speed before the switch had been adequately inspected or clamped closed.

(2) The lack of company procedures requiring that train dispatchers, after receiving a report of a problem involving a main track switch, should immediately stop trains or implement an appropriate speed restriction in the affected area.

The NTSB made two recommendations.

To the Federal Railroad Administration:

(1) It should require that train dispatchers, upon receiving reports of potentially hazardous conditions involving a main track segment or switch, stop train movements or immediately implement an appropriate speed restriction for the affected area and remove the restrictions only after the completion of those inspections and/or repairs that were essential for the safe movement of trains.

To the Norfolk Southern Railway:

(1) It should require that its train dispatchers, upon receiving reports of potentially hazardous conditions involving a main track switch, immediately stop train movements or implement an appropriate speed restriction for the affected area and remove the restrictions only after the completion of those inspections and/or repairs that were essential for the safe movement of trains. [1652]

A Canadian National Railway (CN) freight train derailed near Amite, Louisiana, United States, on Saturday 12th October 2002. The 19th to the 41st cars of the 116-car train left the track and there was a spillage of 15,000 gallons of hydrochloric acid and 14,000 gallons of styrene monomer. An evacuation of part of Tangipahoa Parish was necessary for three days while the spill was being dealt with, 784 homes and 40 businesses being affected. The derailment was blamed on a broken rail. A CN spokesman stated that the track was visually checked twice a week and also subject to an annual ultrasonic examination. [1653]

At about 9:04 a.m. on Sunday 9th February 2003 northbound Canadian National Railway freight train M33371, travelling at about 40 mph, had 22 of its 108 cars derail at Tamaroa, Illinois, United States. Toxic and flammable methanol was lost from four of the derailed cars and a fire broke out following the derailment. In addition, hydrochloric acid was lost from two derailed cars and formaldehyde and vinyl chloride monomer from one car each. About 850 residents were evacuated from a three-mile radius of the accident site, including the entire village of Tamaroa, and those who lived closest to the accident site were not allowed home

[1652] National Transportation Safety Board report
[1653] https://ble-t.org/news/broken-rail-cited-in-amite-derailment/

until 8 a.m. on Friday 15th. One worker injured in the clean-up operation was the only casualty. Total damage caused by the derailment was estimated at $1,900,000.

The National Transportation Safety Board carried out an in-depth investigation of the accident. It noted that the train was a regular working from Memphis, Tennessee, to Chicago, Illinois, and because it included hazardous materials it was designated a "key train." Nineteen cars had been taken on at Carbondale Yard, from where the train had consisted of 76 loaded and 32 empty cars. Passing Tamaroa, 26 miles north of Carbondale, the engineer had felt a "tug," then a surge, the brakes had gone into emergency and he had notified the dispatcher that his train had derailed. Five minutes after the derailment, the engineer and conductor had seen that it was on fire.

The NTSB report listed nine findings regarding the derailment:

(1) The response to the accident by emergency personnel and the train crew had been timely and effective and had enhanced the safety of the community.

(2) Fatigue cracks had developed in the rail from areas of untempered martensite {a very hard form of steel crystalline structure} at points where bond wires had been exothermically welded to the rail head.

(3) The placement of the bond wire welds at Tamaroa, which had been authorised by a Canadian National signals supervisor, had not been in accordance with the company's exothermic bond wire welding policy, which had been promulgated two months before.

(4) The known soft ballast condition in the area of the insulated joint had increased the amount of rail flexing in that area, which, in turn, had significantly increased stresses in the rail.

(5) The increased stresses due to the flexing of the rail in the area of the insulated joint had caused the propagation of the cracks, which had originated in areas of untempered martensite at the rail head, causing the rail to fail only 17 days after installation.

(6) ERICO Products, Inc. exothermic bond wire welding application literature had not been entirely consistent with recommended industry practices.

(7) Laboratory tests and examination of field welds indicated that little or no tempering of the affected steel occurred during the exothermic welding process.

(8) The American Railway Engineering and Maintenance-of-Way Association Manual for Railway Engineering had not adequately addressed the proper placement of bond wire welds.

(9) Because most current rail failure-tracking methods did not record the cause of rail failures that might have occurred at locations where bond wires had been welded or brazed, no adequate evaluation could be made of the effects of these welds on rail integrity in the US rail system.

The NTSB determined that the probable cause of the derailment was Canadian National's placement of bond wire welds on the head of the rail just outside the joint bars, where untempered martensite associated with the welds had led to fatigue and subsequent cracking, which, because of increased stresses associated with known soft ballast conditions, had rapidly progressed to rail failure.

The NTSB made four recommendations.

To the Federal Railroad Administration:

(1) It should require in 49 Code of Federal Regulations Part 213, "Track Safety Standards," that rail cracks originating from bond wire attachments should be identified as rail defects and that information should be collected on the methods and locations of those attachments.

(2) It should require in 49 Code of Federal Regulations Part 225, "Guide for Preparing Accident/Incident Reports," that derailments caused by rail cracks originating from bond wire attachments should be reported with a specific cause code and that information on the methods and locations of those attachments should be provided in the accident narrative.

To ERICO Products, Inc.:

(1) It should revise the instructions for its Cadweld welding systems to address the proper placement of exothermic bond wire welds, especially in the vicinity of joint bars, and to make users aware that these welds created untempered martensite that could, under certain conditions, lead to fatigue cracking and rail failure.

To the American Railway Engineering and Maintenance-of-Way Association:

(1) It should modify its Railroad Engineering Manual and/or its Communications and Signals Manual so both addressed the proper placement of exothermic bond wire welds and high-temperature pin-brazings and to include information that these welds and brazings created untempered martensite that could, under certain conditions, lead to fatigue cracking and rail failure. [1654]

The 126 inhabitants of the hamlet of Villa El Marco, just outside of Melipilla, Chile, were woken from their sleep by the sound of sirens at 01:40 on Tuesday 1st April 2003 and told by the police that they must leave their homes as quickly as possible. Nine waggons of a Transportes Andrés Pirazzoli (Transap) freight train had derailed close to their homes and five of them, each containing 47,000 litres of sulphuric acid, had been breached and were losing their contents.

"I woke up and it was difficult to breathe because of the acid fumes which had come into my house," related Alicia Aguilera. "We had to leave with what we were wearing."

Firemen from Melipilla, Santiago, San Miguel and Ñuñoa arrived at the scene, assisted with the evacuation and closed the road to Melipilla. Meanwhile, a Servicio Agricola y Ganadero (Agricultural and Livestock Service) spokesperson told a reporter that because of the clay soil at the point of the derailment the ground was likely to absorb the acid and it would not therefore travel very far or get into the water supply.

Fernando Pérez, the mayor of Melipilla, was not happy, pointing out that two years earlier, when the sulphuric acid was being transported by road, a lorry had overturned, and they had been assured that transport by rail would be safer. [1655]

Empresa de los Ferrocarriles del Estado (EFE), the state railway company, confirmed that they owned the line on which the accident had happened but stated that the Transap company were responsible for the maintenance of the locomotives and rolling stock used on the freight trains. [1656]

The derailment was attributed to an undetected broken rail. [1657]

In accordance with normal practice, a Burlington Northern Santa Fe Railway freight train from Dayton, Texas, to Galesburg, Illinois, was preparing to stop after its crew had received a tornado warning when 39 of its 118 cars, situated about the middle of it, were derailed, possibly because they had been struck by a severe burst of wind. The derailment occurred five

---

1654 National Transportation Safety Board report
1655 http://members.fortunecity.es/treneschilefotos/noticias/noticia17.html
1656 http://members.fortunecity.es/treneschilefotos/noticias/noticia18.html
1657 http://members.fortunecity.es/treneschilefotos/noticias/noticia41.html

miles east of Matfield Green, Kansas, United States, at 5:55 p.m. on Thursday 8th May 2003. One of the derailed cars leaked sulphur dioxide. The accident happened in a rural area but 250 residents within a radius of five miles were evacuated. One resident who suffered breathing problems needed hospital treatment and six firefighters and law enforcement officers were treated at the scene. Many residents were allowed home after four hours and all were back home by Saturday. Highway 117 was closed until Friday evening. Six cars of liquefied petroleum gas were not damaged. [1658] [1659]

Southbound Canadian National Railway (CN) freight train No. A-450-31-21 left North Bay, Ontario, Canada, on Wednesday 21st May 2003 for Toronto, Ontario. It consisted of three locomotives, 103 loaded and nine empty cars. The weather was 20°C, sunny and clear, with a slight north-east wind. At about 11:44, while travelling at 59 mph with the throttle in the No. 5 position, an undesired emergency brake application (UDE) occurred. After conducting the necessary emergency procedures, the crew inspected the train and found that 49 cars, the 47th to the 95th from the engine, had derailed and were piled up, immediately south of the Highway 12 crossing at Mile 68.9, in the village of Gamebridge, Ontario. Some 1,700 feet of track was destroyed, beginning at the south end of the Highway 12 crossing and extending southward through the crossing at Concession Road A. The derailed cars comprised 21 Class 111A tank cars loaded with sulphuric acid, two empty box cars, and 26 box cars loaded with paper. Approximately 250 tons of acid was spilt. Emergency responders contained the leaking product and extinguished a small fire. An underground fibre-optic cable on the west side of the track was severed, disrupting service for 16 hours. Highway 12 was closed in the vicinity of the derailment area until the morning of 26th May 2003. Some 50 people had to be evacuated due to the toxic nature of the acid and to ease clean-up operations. Two firefighters suffered minor fume inhalation, and a local citizen suffered minor acid burns to his feet.

The Transportation Safety Board of Canada (TSB) report on the accident contained six findings:

(1) The derailment had probably occurred due to wheel climb and/or wheel lift when the train was travelling over the cross-level variations near the Highway 12 crossing.

(2) Due to the improper calibration of Canadian National Railway's TEST car, the severity of the track defects immediately north of the Highway 12 crossing had been incorrectly identified as priority defects, resulting in inadequate corrective or protection measures being taken for near-urgent or urgent defects.

(3) Although the track inspection program had included regular inspection by track forces, the inspections had not detected and corrected the near-urgent and combination track surface defects leading up to the point of derailment.

(4) The current practice of not recording emerging defects on regulatory inspection reports might prevent this information from being shared with other railway personnel involved in track inspections on the territory.

(5) With the increased workload for maintenance personnel and the high volume of traffic on the Bala Subdivision, the time available for routine maintenance and inspection was reduced, increasing the risk of emerging track defects going unattended and growing more serious.

(6) The marshalling of empty cars between blocks of loaded cars was not considered causal in this incident, but it had left the train particularly susceptible to in-train forces,

---

[1658] *The Salina Journal*, Salina, Kansas, United States, 9th May 2003

[1659] *Lawrence Journal-World*, Lawrence, Kansas, United States, 11th May 2003 especially during an emergency brake application, and had increased the risk of derailment and extensive track and equipment damage.

The TSB report noted that as a result of the potential failure to protect or repair improperly identified track geometry defects, Transport Canada had issued a notice pursuant to Section 31 of the *Railway Safety Act* to Canadian National Railway. The company had responded that the previously incorrectly identified defects had been protected or corrected, and that it had initiated the following additional actions:

(1) All defect settings on the TEST car were being audited to ensure compliance with *Railway Track Safety Rules* (TSR) standards.

(2) A daily procedure had been developed and implemented that required TEST car operators to review and validate defect parameter settings and track class before testing operations.

(3) Since the derailment, two additional TEST car runs were being scheduled over the Bala Subdivision. All defects identified during these tests were being properly protected and corrected.

(4) Two additional inspections using contracted track geometry vehicles with gauge restraint technology were scheduled on the Bala Subdivision in 2003.

In addition, Transport Canada continued to follow up on the issue of train marshalling with the industry. [1660]

A Ferrocarril Centro-Atlántico freight train was involved in a major derailment in the early hours of Tuesday 10th June 2003 near the town of Uberaba in Minas Gerais state, Brazil. Eighteen of the 33 waggons of the train left the track and there was a loss of over 700,000 litres of toxic and inflammable methanol, inflammable octanol and iso-butanol and potassium bifluorate. An unknown quantity of the chemicals escaped into the Congonhas stream, a tributary of the Uberaba River, so that the water supply of the town of 270,000 people had to be cut off, with supplies of drinking water brought in by road. In addition, an explosion and fire caused further problems and in the evening, firemen were still struggling to quell the blaze. No one was apparently hurt, though a fair amount of wildlife was killed by the fire and the contaminated water. A television report stated that the driver had detached his engine from the train and fled the scene. The train had been on its way from a petrochemical centre at Camacari in Bahia state to Paulinia in São Paulo state. [1661] [1662]

In the aftermath of the accident Cia Vale do Rio Doce, the mining company that owned the Ferrocarril Centro-Atlántico, was fined 10,000,000 reais (equivalent to $3,000,000) for causing the pollution. It stated that it would appeal. [1663]

Three earthquakes affected the Miyagi prefecture of Japan at 00:13, 07:13 and 16:56 on Saturday 26th July 2003. A four-car passenger train was derailed on the Ishinomaki line at Kanan but none of its 10 passengers were hurt. [1664]

Exactly two months later, the Japanese island of Hokkaido was struck by two earthquakes at 04:50 and 06:08 on Friday 26th September 2003. An express train was derailed as a result at

---

[1660] Transportation Safety Board of Canada report

[1661] http://www.planetark.org/avantgo/dailynewsstory.cfm?newsid=21149

[1662] http://quote.bloomberg.co.../news?pid=71000001&refer=latin_america&sid=aioGoupM2sQ

[1663] http://www.planetark.org/avantgo/dailynewsstory.cfm?newsid=21177

[1664] http://www.japantimes.co.jp/cgi-bin/getarticle.p15?nn20030727a2.htm

Onbetsu, injuring one of its 39 passengers. [1665]

The engine and 18 waggons of a 35-waggon freight train from Maschen to Hagen were derailed near the station at Osnabrück, Germany, at about 22:30 on Wednesday 17th March 2004. A tank waggon of propane-butane mixture was ruptured and a fire broke out, with flames reaching a height of about 80 metres. Some 86 residents had to be evacuated while about 200 firemen fought the blaze and were able to prevent any further explosions from a second tank waggon of propane-butane and a third waggon of propylene. The remainder of the waggons were not carrying any hazardous materials. The main line was blocked for hours, with long-distance trains being diverted via Minden and Verden. The cause of the derailment was not immediately known.[1666] The two train drivers suffered from shock.

At about 5:03 a.m. on Monday 28th June 2003 westbound Union Pacific Railroad (UP) freight train MHOTU-23 came into sidelong collision with eastbound Burlington Northern Santa Fe (BNSF) freight train MEAP-TUL-126-D at the west end of the siding (crossing place) at Macedon, Texas, United States, a few miles west of San Antonio. The trains were on a UP line over which the BNSF had trackage rights (running powers). As a result of the collision four diesel units and the leading 19 cars of the UP train and 17 cars of the BNSF train were derailed. The 16th car of the UP train, containing liquefied chlorine under pressure, was punctured and a cloud of the toxic gas enveloped the area. Conductor Heath Pape, 23, of the UP train and two local residents, an elderly woman and her daughter, died of chlorine poisoning.[1667] The UP engineer, 23 residents and six emergency responders needed hospital treatment, mostly for respiratory problems associated with the chlorine. Damage was estimated at $5,700,000.

The National Transportation Safety Board investigated the accident. It noted that the BNSF train had originated in Eagle Pass, Texas, was bound for Tulsa, Oklahoma, and consisted of a locomotive(s) and 123 empty cars. Its crew had been expecting to take it to San Antonio but en route had been instructed that they were to tie down (stable their train) at Macedon.

The UP train had started from Houston, Texas, to Tucson, Arizona, and from San Antonio consisted of four locomotives and 74 cars. At 4:54 a.m. it had passed a clear colourlight signal at Alamo Junction, travelling at 46 mph, and speed had then been reduced to 22 mph at 4.57½ as it had passed a signal displaying Advance Approach (flashing Yellow, prepare to stop at second signal, speed not to exceed 40 mph) because the BNSF train was nearing the west end of the siding at Macedon. The speed of the train had then increased and it had passed the signal at the east end of Macedon siding at 5:01, travelling at 46 mph. This signal had been displaying Approach (steady Yellow, prepare to stop at next signal, speed not to exceed 30 mph). The BNSF train had been entering the west end of Macedon siding. The speed of the UP train had then diminished marginally to 44 mph but there had been no further attempt to stop it. According to the crew of the BNSF train, they had dimmed the headlight of their engine, without any reaction from the crew of the UP train, which they could see was going too fast. The UP train then ran past the Stop signal at the end of the siding and collided with the 63rd car of the BNSF one.

A 911 emergency call was made by a resident of Nelson Road, near the derailment site, at 5:06 a.m. He stated that he was having difficulty breathing and that there had been a train derailment. However, the part of his statement regarding the derailment, made in a weak

---

[1665] http://www.japantimes.co.jp/cgi-bin/getarticle.p15?nn20030926c1.htm

[1666] https://www.spiegel.de/panorama/zugunglueck-beinahe-katastrophe-in-osnabrueck-a-291164.html

[1667] *Del Rio News-Herald*, Del Rio, Texas, United States, 29th June 2004 voice, was not picked up, so the first rescuers did not therefore realise that there had been a train accident or that the trains were blocking their access to victims via a level crossing. Three residents trapped in their home could not be rescued until 11:46 a.m. and were badly affected by the chlorine. The engineer and conductor of the UP train had attempted to make their escape on foot until the conductor had been unable to go any further and the engineer had been too weak to help him. The engineer himself was critically injured. The BNSF crew were not affected by the chlorine.

The NTSB came to 13 findings related to the accident:

(1) On the day of the accident, the crew of the Union Pacific Railroad train had failed to operate their train in accordance with operating rules and in compliance with wayside signal indications, including failing to take any action in response to the Stop signal at the west end of the Macdona siding.

(2) The following had been neither causal nor contributory to the accident: weather conditions, signal operation, equipment performance, track condition, drug use, or crew training and qualifications.

(3) The Union Pacific Railroad engineer's combination of sleep debt, disrupted circadian processes, limited sleep through the weekend, and long duty tours in the days before the accident had likely caused him to start the accident trip with a reduced capacity to resist involuntary sleep.

(4) The Union Pacific Railroad conductor's lack of sufficient rest before reporting to work, the disruption to his previous work/rest pattern resulting from his change in work schedule, and his alcohol consumption on the evening before the accident had likely combined to reduce his capacity to remain awake and alert during the accident trip.

(5) Neither the engineer nor the conductor of the Union Pacific Railroad train had made effective use of the time available to them, between the time they had been released from their previous assignments and the time they had been called for the accident trip, to obtain rest.

(6) The engineer of the Union Pacific Railroad train had likely experienced one or more periods of microsleep early in the accident trip, and these had probably been followed by a deeper descent into sleep as the train had travelled past the signal at the east end of the Macdona siding.

(7) The conductor of the Union Pacific Railroad train had most likely been asleep during much of the accident trip.

(8) Even though the initial dispatch of emergency response resources to the accident had been timely, the overall execution of the incident command process during the response effort had not been timely, effective, or appropriate.

(9) The chlorine tank car, which had released a portion of its load in this accident, had been punctured during the derailment by impact with the left side frame member of the flatcar loaded with steel plates, which had been four cars ahead of it in the train.

(10) The shell of the punctured chlorine tank car in this accident would have been susceptible to catastrophic fracture if it had experienced a large penetration of several feet or more.

(11) The unpredictability of their work schedules might have encouraged the Union Pacific Railroad engineer and conductor to delay obtaining rest in the hope that they would not be called to work until later on the day of the accident.

(12) Limbo time, which was limited neither by Federal regulation nor railroad operating

rules, could be a factor in crewmember fatigue in that required rest periods did not take into account the extended hours of wakefulness before the rest period began.

(13) The Macdona, Texas, accident was another in a long series of railroad accidents that could have been prevented had there been a positive train control system in place at the accident location.

The NTSB determined that the probable cause of the collision had been the Union Pacific Railroad train crew's fatigue, which had resulted in the failure of the engineer and conductor to appropriately respond to wayside signals governing the movement of their train. Contributing to the crewmembers' fatigue had been their failure to obtain sufficient restorative rest prior to reporting for duty because of their ineffective use of off-duty time and the Union Pacific Railroad train crew scheduling practices, which had inverted the crewmembers' work/rest periods. Contributing to the accident had been the lack of a positive train control system in the accident location. Contributing to the severity of the accident had been the puncture of a tank car and the subsequent release of poisonous liquefied chlorine gas.

The NTSB made a number of new recommendations and reiterated some recommendations from earlier reports on other accidents.

New Recommendations:

To the Federal Railroad Administration:

(1) It should require railroads to use scientifically based principles when assigning work schedules for train crewmembers, which considered factors that impacted sleep needs, to reduce the effects of fatigue.

(2) It should establish requirements that limited train crewmember limbo time to address fatigue.

To the Union Pacific Railroad:

(1) It should use the Macdona collision as a case study in fatigue awareness training to illustrate the shared responsibilities of the carrier to provide an employee the opportunity for adequate sleep and of the employee to acquire sleep sufficient to work at a safe level of alertness, and the options available if adequate sleep was not obtained.

To the Brotherhood of Locomotive Engineers and Trainmen and the United Transportation Union:

(1) It should use this accident as a fatigue case study to illustrate the responsibility of the carrier to provide an employee the opportunity for adequate sleep and the responsibility of the employee to acquire sleep sufficient to work at a safe level of alertness, and the options available if adequate sleep was not obtained. It should present this case study to its members at meetings, through written materials, and via other appropriate methods.

Reiterated Recommendations:

To the Federal Railroad Administration:

(1) It should conduct a comprehensive analysis to determine the impact resistance of the steels in the shells of pressure tank cars constructed before 1989. At a minimum, the safety analysis should include the results of dynamic fracture toughness tests and/or the results of non-destructive testing techniques that provided information on material ductility and fracture toughness. The data should come from samples of steel from the tank shells from original manufacturing or from a statistically representative sampling of the shells of the pre-1989 pressure tank car fleet.

(2) Based on the results of the Federal Railroad Administration's comprehensive analysis to determine the impact resistance of the steels in the shells of pressure tank cars

constructed before 1989, as addressed above, it should establish a program to rank those cars according to their risk of catastrophic fracture and separation and implement measures to eliminate or mitigate this risk. This ranking should take into consideration operating temperatures, pressures, and maximum train speeds.

(3) It should validate the predictive model the Federal Railroad Administration was developing to quantify the maximum dynamic forces acting on railroad tank cars under accident conditions.

(4) It should develop and implement tank car design-specific fracture toughness standards, such as a minimum average Charpy value, for steels and other materials of construction for pressure tank cars used for the transportation of US Department of Transportation class 2 hazardous materials, including those in "low-temperature" service. The performance criteria must apply to the material orientation with the minimum impact resistance and take into account the entire range of operating temperatures of the tank car.

(5) It should require railroads to implement operating measures, such as positioning tank cars toward the rear of trains and reducing speeds through populated areas, to minimise impact forces from accidents and reduce the vulnerability of tank cars transporting chlorine, anhydrous ammonia, and other liquefied gases designated as poisonous by inhalation.

(6) It should determine the most effective methods of providing emergency escape breathing apparatus for all crewmembers on freight trains carrying hazardous materials that would pose an inhalation hazard in the event of unintentional release and then require railroads to provide these breathing apparatus to their crewmembers along with appropriate training. [1668]

A Railway Systems of Zambia freight train bound for the Konkola Copper Mines in Chingola, Zambia, was derailed close to its destination on Wednesday 21st July 2004. As a result of the derailment three tanker waggons carrying sulphuric acid lost their contents into the Mindolo stream. The latter was a tributary of the Kafue River, from which the Nkana Water and Sewage Company obtained its water supply, and the company was forced to cut off water to users in nearby Kitwe and Kalulushi until the spilt acid had been neutralised by lime. [1669]

Ten or 12 cars of a 75-car eastbound Dakota Minnesota and Eastern Railroad freight train were derailed at about 11 p.m. on Tuesday 27th July 2004 at Balaton, Minnesota, United States. About 30,000 gallons of ethanol escaped from one tank car and about 15,000 gallons from a second. Elienore Myhre told a reporter: "I heard it all. It was a terrible, terrible noise. Like cars clanging when they get dropped off. It was something. All of a sudden it was quiet."

Because of the danger of a possible fire, 75 nearby residents were evacuated overnight and either stayed with relatives and friends or were accommodated in the library of Balaton Public School. Dykes were quickly set up to prevent the ethanol from getting into Lake Yankton. No one was hurt. [1670] [1671]

On the evening of Monday 9th August 2004 a 17-waggon freight train bound for Duiwelskloof was derailed at Solomondale, Limpopo province, South Africa, and a

[1668] National Transportation Safety Board report

[1669] http://allafricacom/stories/printable/200407220197.html

[1670] https://www.firehouse.com/rescue/news/10509186/dozens-evacuated-after-minnesota-train-derailment

[1671] https://ble-t.org/news/train-derailment-forces-evacuation-in-minn/

considerable quantity of sulphuric acid was spilt, the spill covering an area of 50 square metres. The main road between Polokwane and Duiwelskloof was closed as residents were warned of possible dangers from the acid but there was apparently no evacuation. No one was hurt. [1672] [1673]

Canadian National Railway train No. U-781-21-17, a petroleum product unit train, consisting of two engines and 68 loaded cars, en route from the Ultramar Canada Inc. refinery in Lévis, Quebec, Canada, to Montréal, Quebec, had travelled just over 11 miles when a train-initiated emergency brake application occurred at about 14:40 on Tuesday 17th August 2004. Its crew found that it had become derailed at Mile 3.87 of the Lévis Subdivision, in a marshy area of the Grande plée Bleue, near Saint-Henri-de-Lévis, Quebec. Eighteen cars, the 22nd to 39th, had left the rails and about 200,000 litres of petrol and diesel fuel had spilt into the marshy area, but the spilt product was recovered. There were no injuries. The crew had not observed any irregularity in the operation of the train or the condition of the track before the emergency brake application.

The City of Lévis Public Safety Department was immediately notified of the accident and authorities implemented the city's emergency plan. In addition to the Lévis Fire Department and CN staff, representatives of various levels of government, the petroleum industry, and clean-up companies arrived at the site at different intervals. Hydrocarbon fumes had affected the air quality in the immediate area of the accident site, which required the area within a one-kilometre radius of the derailment site to be cordoned off and evacuated for a period of four days. Hydrocarbon cleaning and recovery operations were started once the liquid and vapour leaks were controlled. The hydrocarbons that had spilled into the ditches and marshy areas were contained by booms and dykes to limit the spread of the contamination. Two clean-up companies subsequently recovered the hydrocarbons and decontaminated the water and soil.

The Transportation Safety Board of Canada listed nine findings in relation to the accident:

(1) Track subgrade settlements accumulating under the repeated effect of loads had gradually resulted in distortion and realignment of peat fibres, which most likely had led to the sudden punching failure {a sudden major settlement of the peat layer beneath the track}.

(2) The cars had been unable to negotiate the sudden collapse of the track brought on by subgrade failure and had thus derailed.

(3) Areas where the track subgrade was built on saturated peat and where surfacing work was required on a regular basis to maintain track profile were susceptible to punching.

(4) Axle weight, tonnage, and train speed were factors that could contribute to the punching phenomenon.

(5) Railway inspection procedures and technologies, based on surface observations, were not able to detect the impending risk of collapse.

(6) Despite measures taken by the regulator and the industry, most non-pressurised tank cars used in the transport of hydrocarbons and other dangerous goods remained vulnerable to puncture and continued to present risks, even following impacts at moderate operating speeds.

(7) Further research was required to better understand shear punching failure.

(8) A conservative evaluation of the risk of slope failure revealed that the railway fill placed

---

[1672] http://www.news24.com/News24/South_Africa/News/0,,2-7-1442 1570318,00.html

[1673] http://www.sabcnews.com/south_africa/general/0,2172,85390,00.html

over peat deposits was stable in the event of slope failure under the load of the cars and locomotives.

(9) The response of the City of Lévis fire department had been effective and successful in protecting the public and others on the site.

On 19th August 2004, Transport Canada (TC) imposed a speed restriction of 25 mph between Mile 1.0 and Mile 6.0 of the line, while in the immediate vicinity of the derailment, the speed was reduced to 10 mph.

In June 2006, after several accidents in the United States involving hazardous material spills, the Federal Railroad Administration (FRA) had begun an in-depth assessment of the construction standards of tank cars used to transport hazardous goods, with a view to minimising the risk of spills occurring during derailments. TC was working closely with the FRA to address the same issues and ensure harmonisation. In addition, TC, the FRA, and the industry had signed a memorandum of cooperation in April 2007 to develop a better understanding of the factors contributing to high-pressure tank car safety and to enhance the effectiveness of railway-specific hazardous material bulk packaging under the project "Next Generation Tank Car."

The subgrade failure had occurred suddenly and without warning when all indications were that the track was in good condition and that there were no early signs of distress. The railway inspection technologies and procedures, mainly based on evaluations of track conditions at surface level, had not been effective in assessing the condition and the behaviour of the subgrade and detecting the impending risk of collapse.

Many miles of Canadian railway tracks were built along rivers or across marshy areas where peat was present. Therefore, these locations were exposed to the conditions encountered in this occurrence. Given these circumstances, a better understanding of the failure mechanism and the effect of railway loading on peat subgrade was needed to mitigate the risks inherent to these types of organic soils.

The TSB recognised the efforts by the railway and the regulator to address the issues related to soil and rock slope stability. Industry had taken important initiatives such as the Railway Ground Hazard Research Program to better understand, identify, and mitigate risks relating to geology and the behaviour of railway track subgrade. These initiatives had been focused mainly on slope stability and on the behaviour of soils of glaciolacustrine origin, even though the recent development of ground-penetrating radar technology could potentially apply to peat areas.

The study commissioned by the TSB and carried out by the Civil Engineering Department of Université Laval had determined punching shear as a failure mechanism. It had also indicated that the areas where the tracks were built on saturated peat, and where surfacing work was required regularly to maintain the profile of the track, were susceptible to punching. The shear punching mechanism was consistent with site observations and integrated results from tests performed in the laboratory and in the field. However, additional research efforts were required to enhance understanding of their phenomena and reduce risk. Therefore, the TSB recommended that the Department of Transport and the railway industry conduct in-depth studies on the behaviour of saturated organic materials under cyclic loading.

There had been a significant spill of hydrocarbons when the tank shells and heads of the Class 111A tank cars involved in the derailment were breached, even though the derailment had happened in a marshy area where the surrounding terrain was particularly soft. The Class 111A tank cars' weaknesses had been acknowledged by the regulator and industry, resulting in measures to mitigate risk in the event of a derailment. The number of products these cars were allowed to transport had been reduced when the *Transportation of Dangerous Goods Regulations* were amended and new tank car construction standards were established by TC.

However, the safety enhancements included in the new standards did not apply to Class 111A tank cars with a maximum gross weight of 263,000 pounds or less, or to other non-pressurised tank cars. Consequently, a large number of the existing tank cars carrying dangerous goods would be vulnerable to puncture, even during derailments at moderate operating speeds. The TSB therefore recommended that the Department of Transport should extend the safety provisions of the construction standards applicable to 286,000-pound cars to all new non-pressurised tank cars carrying dangerous goods. [1674]

Two waggons of a freight train were derailed with a spillage of sulphuric acid on Tuesday 31st August 2004 at Huachipa, Peru. Police and firemen evacuated residents who lived within 100 metres of the spillage. The local mayor blamed the derailment of the train, which was bound for Callao, on the poor condition of the track, and stated that he had notified the Ferroviaria Andina company, so that it could take action regarding the matter. [1675]

An oil train from Chittagong to Sylhet, carrying a mixture of diesel, kerosene, octane and petrol, became derailed at Mallickpur Crossing, Bangladesh, at about 19:30 on the evening of Thursday 16th September 2004. The contents of some of the derailed tank waggons were spilt and at about 21:00 a massive fire erupted, with flames reportedly reaching a height of 100 feet. Burning oil covered an area of 600 square metres and gutted one house as well as destroying a large tree. Two women in the house were injured and were the only casualties in the accident.

After much effort, firefighters were able to bring the blaze under control at 03:00 the following morning and the line was reopened on Friday afternoon. Six tanker waggons were destroyed and there was significant damage to the local ecosystem. Railway sources put the cost of the accident at 1,000,000 taka (8,100 US dollars at late July 2025 exchange rates) for the damaged waggons, 500,000 taka for the damaged track and 5,000,000 taka for the loss of oil, though oil traders estimated the cost of the lost oil as being considerably higher. The driver of the train was suspended pending the results of an investigation into the cause of the derailment. [1676] [1677]

East Japan Railway passenger train Toki No. 325 from Tokyo to Niigata, a Joetsu Shinkansen Line 'bullet train,' was derailed near Nagaoka station, Japan, at 17:56 on Saturday 23rd October 2004. Eight of the 10 cars of the train were derailed, the rear three tilting over by as much as 40 degrees. None of the 151 train passengers were hurt but it was the first derailment of a 'bullet train' since they had been introduced in 1964.

Passenger Eiichi Nakamura, 59, related, "The train was running normally when suddenly it swayed badly and I almost hit my face on the back of the seat in front of me."

The train had been derailed by a powerful earthquake. Seismometers located every 20 kilometres along the line were designed to automatically cut the power to trains and apply their brakes should an earthquake occur but even so, a train travelling at 210 km/h would take 2.5 kilometres and 1 minute 45 seconds to come to a halt. The accident train had travelled 1,637 metres in a derailed state. A possible major disaster was averted because train No. 406, due to leave Nagaoka station in the opposite direction at 17:58, was stopped in the station by

[1674] Transportation Safety Board of Canada report

[1675] http://wwww.24horaslibre.com/imprimir.php?id=1094062228

[1676] http://nation.ittefaq.com/artman/publish/printer_12434.shtml

[1677] http://www.thedailystar.net/2004/09/18/d4091801099.htm

the earthquake. [1678] [1679]

Fourteen of the 55 cars of a CSX Transportation freight train from Decatur to Indianapolis were derailed about a mile north of Paris, Illinois, United States, at about 8 a.m. on Wednesday 22nd December 2004. Hydrochloric acid was lost from a tank car as well as diesel fuel from two train engines, forcing the evacuation of about 250 people from nearby homes. By the afternoon, some of the residents were already being allowed home. No one was hurt. [1680]

What up to the present time (February 2025) is probably the world's worst railway accident in terms of fatalities, occurred on Boxing Day, Sunday 26th December 2004, in Sri Lanka. It had a very unusual, if not unique cause, in that a train was struck by a tsunami following the Sumatra-Andaman earthquake in the Indian Ocean. The death toll from the accident, perhaps 1,500 or 1,700 people, was also only a fraction of the total number of lives lost in various Asian countries due to the earthquake and tsunami, which was estimated at about 230,000.

On the morning of the accident train No. 50, the *Matara Express*, left Colombo Fort station at about 06:50 on its regular run northwards along the Coastal Line to the town in its name (the train is often wrongly referred to as the *Samudra Devi (Queen of the Sea)*, which was another service). It was in charge of Guard Wannigaranta Karunatilake with Assistant Guard R.U.A. Gunaratna,[1681] Driver Janaka Fernando and Assistant Driver Sivaloganathan. As well as being the Christmas holiday weekend it was the Buddhist full moon holiday, so the train was very well filled, with more than 1,500 passengers with ordinary tickets plus others with season tickets or government passes, perhaps 1,750 in all. The eight-coach train was being pulled by Class M2a diesel engine No. 591 Manitoba. As its name implies, the Coastal Line runs close to the sea, which at Telawatta, a few kilometres south of the accident site, is only 200 metres distant.

The earthquake occurred at 07:59 and was registered by the Sri Lankan seismic monitoring station within minutes but it was not at first thought likely that the tsunami it caused would reach Sri Lanka. When Control at Maradana (Colombo) were notified of the tsunami they were able to stop eight other trains on the Coastal Line but were unable to reach the *Matara Express*, which was at Ambalangoda station, as the staff there were all out on the platform attending the train and only answered the telephone after it had left. The staff at stations further south had reportedly fled from the impending tsunami.

At a point near Hikkaduwa station, about 12 kilometres north of Ambalangoda, the train came to a halt as the first giant wave of the tsunami reached the island and water surged around it. Many local villagers, believing that the train was solid enough to resist the force of the tsunami, then climbed onto the roofs of its coaches to avoid being swept away, while others took shelter on the landward side of the train. Entering the carriages, the water caused a panic amongst the train passengers. After 10 minutes, a second, even more massive wave struck the train and was powerful enough to pick it up and throw it against the trees and houses alongside the track, in the process crushing those who were stood behind it. Most of the passengers in the carriages, unable to escape, drowned when they were filled with water, while those on the coach roofs were thrown off and either drowned or were crushed. The

---

[1678] http://mdn.mainichi.co.jp/news/20041025p2a00m0dm007000c.html

[1679] www.japantimes.co.jp/cgi-bin/getarticle.pl5?nn20041025a3.html

[1680] *Southern Illinoisian*, Carbondale, Illinois, United States, 23rd December 2004

[1681] https://www.hindustantimes.com/india/same-train-same-staff-different-run/story-sjYu9TkOQ0ptgt9YldZxoO.html locomotive was carried 100 metres to end up in a swamp, its driver and assistant driver being among the dead, though the guard and assistant guard survived. [1682]

Guard Karunatilake had taken off his uniform and dived into the water to rescue a girl after the first wave had struck the train. His own home was struck and badly damaged by the tsunami; his wife and children had only survived by climbing onto its roof.[1683] The water reached an estimated height of between 7.5 and 9 metres above sea level or 2 to 3 metres higher than the roof of the train.

As might be expected the massive scale of the tsunami disaster, of which the accident to the train was only a small part, overwhelmed the authorities and it was many hours before any rescuers came to the aid of the survivors. The fact that a train had been involved at all only became certain at about 16:00 when its wreckage was spotted by the crew of an Army helicopter. Even then, it was a long time before help arrived and in the meantime many badly injured victims died. The recovery of the bodies of the dead went on for over a week and eventually 900 corpses were recovered, while others had been swept out to sea or in some cases removed by relatives for burial. A forensic team from Colombo photographed and fingerprinted unclaimed bodies at Batapola Hospital so that the records could be stored and looked at after the bodies had been buried. A Buddhist monk, Baddegama Samitha, helped perform funeral rites along with his students on 204 of the retrieved bodies that had not been identified or claimed, and these were buried three days later in a Buddhist ceremony near the damaged railway line. He said, "This was the only thing we could do. It was a desperate situation. The bodies were rotting. We gave them a decent burial." In addition to the normal Buddhist rituals, in recognition that some of the anonymous victims had probably been Muslims, Christians, or even atheists, a moment's silence was held in their honour. [1684]

About 150 people who had been on the train survived.[1685] One British survivor related his story. 'Shenth Ravindra, 25, from Crawley, Sussex, believes he may be one of only five Western survivors from the train which was caught up in the Boxing Day disaster. After meeting his mother, Niru Ratman, 54, and his brother Guy, 19, a student at Cardiff University, at Heathrow airport today, Mr. Ravindra told how he had clambered through waters filled with dead bodies after taking refuge on the roof of a house. Mr. Ravindra, who six years ago escaped with his life from a serious car accident, explained: "I was on the train from Colombo to Hikkabuwa when it stopped outside of Hikkabuwa. I heard lots of screaming outside and saw lots of women running towards the train. The next thing I knew, this huge wave comes and hits the train, the train buckles and tilts at 45 degrees. I climbed to the roof; there were lots of children everywhere grabbing hold of me and then I saw the second wave come and push the carriage towards a house. I managed to jump on to the house by which time there were lots of dead bodies everywhere, lots of children. I thought it was just a freak tide; I thought the water level was going to go out and any minute now there would be help. The most frightening aspect was when I saw the second wave coming, there was lots of screaming, the dynamics of the horizon had changed. I just saw this wall of water coming towards us and then the screaming and the shouting."

'He added, "There was a little girl who was with me on top of the roof, then I decided to leave the house. I tried to bring her with me, but she wasn't coming with me, there were a lot of Sri Lankan people there as well. I stated my intention that I wanted to go off the house. I said, 'I think another wave might come or the house might collapse,' but they were saying, 'How do you know there's another wave?'" He said a Swedish woman called Karina as well

---

[1682] Wikipedia – 2004 Sri Lanka tsunami train wreck

[1683] https://www.hindustantimes.com/india/same-train-same-staff-different-run/story-sjYu9TkOQ0ptgt9YldZxoO.html

[1684] http://seattletimes.nwsource.com/cgi-bin/PrintStory.pl?document_id=2002133726&zse...

[1685] Wikipedia – 2004 Sri Lanka tsunami train wreck

as two other Swedish tourists had survived, as well as a Western man. With Karina, Mr. Ravindra managed to climb down from the house. "I went down the house. There were a lot of dead bodies at that time, so I was making my way through the dead bodies trying not to make eye contact. Within a couple of steps, I lacerated my leg quite badly. We walked about two kilometres to a Buddhist pagoda." After receiving treatment from Buddhist monks, Mr. Ravindra then met an American Sri Lankan who had lost her mother in the train but had an uncle, who owned a farm nearby. "She said, 'Come with me.' She had lost a mother in the train accident as well. We walked a further three kilometres."' [1686]

Four tank waggons containing sulphuric acid belonging to Konkola Copper Mines and destined for Copperbelt province, Zambia, became derailed in Kabwe, Zambia, on Wednesday 5th January 2005. Three of them overturned and the acid was spilt over the surrounding countryside. A breakdown train sent to the accident site was derailed in turn and Railway Systems of Zambia (RSZ) suspended passenger train services between Livingstone and Chiliabombwe until the mess had been cleared up. RSZ blamed the frequent derailments on the line on the poor condition it had been in, with many rotten sleepers, when it had been inherited from the state Zambia Railways, with the rainy season as a contributing factor. [1687]

At about 2:39 a.m. on Thursday 6th January 2005 northbound Norfolk Southern Railway (NS) freight train No. 192, travelling at about 47 mph, arrived at Graniteville, South Carolina, United States, where it was diverted via a wrongly set switch (facing points) onto a siding. Standing on the siding was NS freight train No. P22, and a collision followed, in which both locomotives and 16 of the 42 cars of train No. 192 and the locomotive and two cars of train No. P22 were derailed. Amongst the derailed cars of train No. 192 were three tank cars of chlorine, one of which was breached, and a large quantity of the toxic gas was released. Engineer Christopher Seeling, 28, of train No. 192, seven residents of Graniteville and a Canadian lorry driver died of chlorine poisoning and no less than 554 people were taken to hospital, of which 75 were admitted. A total of 5,400 people living within a mile radius of the accident site were evacuated for several days, while damage was estimated at over $6,900,000. The National Transportation Safety Board (NTSB) carried out a full investigation of the accident. [1688]

As might be expected the accident was given extensive coverage in local newspapers, of which one extract must suffice. 'Rodney Johnson's eyes were still swollen and burning when he arrived at the American Red Cross Shelter at First Baptist Church in Aiken early Thursday afternoon, hoping to find out if several of his co-workers on the night shift at Avondale Mills made it to safety after the long day of chaos following the train derailment. Johnson, a machine operator, rushed as many of his fellow workers to safety at Aiken Regional Medical Centers in the back of his truck, but some he had to leave behind.

'What was the most frightening experience of his life began when he left his co-workers to use the rest room. He returned to complete terror. "I went to the bathroom at about 2.35 a.m. and while I was in there I started smelling a Chlorox smell," said Johnson, who did not hear the wreck just a short distance from the mill. "When I opened the door I saw a green mist coming towards me. I stepped up to see what it was and ran to my supervisor. He said to get them out."

'In seconds he was in pain and coughing uncontrollably. It tore me all up," Johnson said. My eyes burned and lungs hurt. I couldn't breathe at all. All I could think about was breathing

[1686] http://news.scotsman.com/print.cfm?id=3944128&referringtemplate=http%3A%2F%2Fn...

[1687] http://allafrica.com/stories/printable/200501060029.html

[1688] National Transportation Safety Board report

and getting to rescue. The parking lot where some people were gathered was even worse, so Johnson ran to the plant's mixing room, where he got a respirator. "I put it on, ran through the green fog to get my truck with my eyes watering," Johnson said. "I pulled my truck up as close as I could and told them to jump in. I blew my horn and told them to jump in." Johnson called his wife on his cell phone and had her call 911, while he raced his pickup truck, packed with as many people as he could possibly fit, to the hospital. He remembers taking around 15 people to safety. "I didn't see the wreck – all I thought about was getting to the hospital," Johnson said. "I was just trying to get to the hospital as quickly as I could. They say I saved a couple of lives."

'Johnson said the chlorine gas poured into the building, and he was very concerned about his friends still there when he left. "It came all inside the place," Johnson said. "I don't think I would have survived without my cell phone and a respirator.' [1689]

The NTSB noted that the line involved was non-signalled (so-called dark territory), with a dispatcher at Greenville, South Carolina, giving authority for train movements by issuing track warrants. At 8:11 a.m. the dispatcher had issued a track warrant to Conductor Mike Ford of train No. P22 to work between Milepost SA 51.0 on the Aiken line and Warrenville, the junction of the Aiken line with the Augusta – Columbia main line. Graniteville lay a short distance north-east of Warrenville on the line to Columbia. During the morning sidings were shunted on the Aiken line, then at 2:10 p.m. another track warrant authorised them to work sidings on the Augusta – Columbia main line. The day's work should have finished with the train being stabled in the siding at Warrenville but the train crew were behind time and so decided to stable the train in the sidings at Graniteville instead. Their train was duly left there, a taxi picking up them up afterwards, but none of the train crew thought to reset the switch leading into the sidings for the main line, or to check that it had been so reset.

Brakeman James Thornton told the NTSB investigators: "In my mind, when I left, everything was properly lined back to the main line. I had no doubts in my mind when I left there." But he admitted: "I am not 100 per cent sure that I did {reset the points}. I would say I might have made a mistake."

Engineer Benjamin Aiken pleaded that he had not been able to see the position of the main line points from his engine when in the siding and he had not known: "how the brakeman handled it."

Conductor Ford stated: "While we were working in the plant I was off in the field and I never got near the switch...We usually have job briefings, but I never told my brakeman to make sure that the switch was lined and locked for the main line movement. I never told him and I never touched the switch myself. It was in my mind when I arrived at the mill. I should have done a briefing to ensure that the switch was lined back."

The track warrant for train No. P22 was cleared at 7:54 p.m. and there were then no further movements over the line until a track warrant was issued at 2 a.m. for train No. 192 from Macon, Georgia, to Columbia, South Carolina, the warrant extending between Macon and Summit, South Carolina. As the train reached Graniteville the conductor recalled the engineer applying the brakes in emergency and crying, "The target {switch banner} is wrong!" At 2:40 a.m. the Greenville dispatcher was advised, probably by Engineer Seeling: "Dispatcher. This is train...192...We need emergency assistance in Graniteville. We just...hit a switch...at Graniteville...It was lined off the main line and we went around there and...hit an engine that was in the track...We went through the switch at 45 mph. We need help."

The NTSB came to 11 findings regarding the accident:

[1689] *Aiken Standard*, Aiken, South Carolina, United States, 7th January 2005

(1) Neither train equipment defects nor the track condition had been causal or contributory to the accident.

(2) In regard to the crew of train 192, fatigue, crew training and qualifications, and drugs and alcohol were not factors in the accident.

(3) In regard to the crew of train P22, neither crew qualifications and training nor fatigue was causal or contributory to this accident, and no evidence was found to suggest drug or alcohol use.

(4) The execution of the emergency response to this accident had been timely, appropriate, and effective.

(5) The crew of train P22 had failed to reline a main-line switch after using it, leading to the subsequent and unexpected diversion of train 192 into an industry track where it had struck train P22 and derailed.

(6) The crew of train P22 had failed to reline the main line switch for one or more of the following reasons:

(a) The task of relining the switch had been functionally isolated from other tasks the crew had been performing.

(b) The crewmembers had been rushing to complete their work and secure their train before reaching their hours-of-service limits.

(c) The crew had achieved their main objective of switching cars and had been focused on the next task of securing their equipment and going off duty.

(d) The switch had not been visible to the crew as they had worked, leaving them without a visual reminder to reline it.

(7) Had the conductor of train P22 held a comprehensive job briefing at the Avondale Mills industry track, as required by Norfolk Southern operating rules, the crew might have attended to the main-line switch, and the accident might not have occurred.

(8) At the speed train 192 had been travelling as it entered Graniteville, the distance required for the train crew to perceive the banner of the misaligned switch, react to it, and bring the train to a safe stop had been greater than the distance available.

(9) The chlorine gas release in this accident had resulted when the shell of the 9th car on the train had been punctured by the coupler of the 11th car.

(10) As shown in the Graniteville accident, even the strongest tank cars in service could be punctured in accidents involving trains operating at moderate speeds.

(11) Had the engineer of train 192 been wearing appropriate, fully functioning emergency escape breathing apparatus when he had walked away from the collision site, he might not have succumbed to the effects of chlorine gas inhalation.

The NTSB determined that the probable cause of the accident was the failure of the crew of Norfolk Southern train P22 to return a main-line switch to the normal position after the crew had completed work at an industry track. Contributing to the failure was the absence of any feature or mechanism that would have reminded crewmembers of the switch position and thus would have prompted them to complete this final critical task before departing the work site. Contributing to the severity of the accident had been the puncture of the ninth car in the train, a tank car containing chlorine, which had resulted in the release of poisonous chlorine gas.

The NTSB made four recommendations regarding the accident.

To the Federal Railroad Administration:

(1) It should require that, along main lines in non-signalled territory, railroads installed an automatically activated device, independent of the switch banner, which would, visually or electronically, compellingly capture the attention of employees involved with switch operations and clearly convey the status of the switch both in daylight and in darkness.

(2) It should require railroads, in non-signalled territory and in the absence of switch position indicator lights or other automated systems that provided train crews with advance notice of switch positions, to operate those trains at speeds that would allow them to be safely stopped in advance of misaligned switches.

(3) It should require railroads to implement operating measures, such as positioning tank cars toward the rear of trains and reducing speeds through populated areas, to minimise impact forces from accidents and reduce the vulnerability of tank cars transporting chlorine, anhydrous ammonia, and other liquefied gases designated as poisonous by inhalation.

(4) It should determine the most effective methods of providing emergency escape breathing apparatus for all crewmembers on freight trains carrying hazardous materials that would pose an inhalation hazard in the event of unintentional release and then require railroads to provide these breathing apparatus to their crewmembers along with appropriate training. [1690]

The derailment of a freight train near the town of Gwanda, Zimbabwe, at about 08:30 on Wednesday 26th January 2005, led to fears of an environmental disaster. The 36-waggon Beitbridge-Bulawayo Railway train was carrying diesel oil and fertiliser but the part of its load, which caused the problem, was sulphuric acid, at least 40,000 gallons of which was feared spilt when 14 of its waggons left the track. The acid escaped into a stream, which was a tributary of the Mtshabezi River, from which the town derived its water supply. Police cordoned off the area and residents were warned not to drink water from the river or the lake formed by its dam, while efforts were being made using barricades to limit the extent of the spillage and minimise its effects. No one was hurt. [1691]

A freight train from Cordoba to Tierra Blanca had six of its waggons leave the track in the community of Presidio, Veracruz state, Mexico, near the city of Tezonapa, on the afternoon of Sunday 20th March 2005. Each waggon was carrying about 61,000 litres of corrosive sodium hydroxide, used in soap manufacture, but only about 500 or 600 litres of the chemical was lost, some of it seeping into the ground. About 20 families living nearby were evacuated after some complained of nausea. No one was hurt. [1692]

Ten waggons of a northbound Nigerian Railways Corporation freight train carrying petroleum were derailed in the Bodija district of Ibadan, Oyo state, Nigeria, on the evening of Saturday 7th May 2005. Three waggons containing premium motor spirit lost their contents, contaminating a large residential area and causing a good many people to flee their homes, for fear of a possible disastrous fire and/or explosions. Security agents and members of the state fire service were put on red alert and a large detachment of police was employed to prevent people from attempting to collect the spilt petrol. In the end there was no fire and no one was hurt.

Three differing theories were suggested as causes for the derailment. Two of them placed the

---

[1690] National Transportation Safety Board report

[1691] http://allafrica.com/stories/printable/200501280269.html

[1692] http://www.billingsgazette.com/printer.php?id=1&display=rednews/2005/03/22/stories/world/...

blame on the heavy rainfall, which it was said had either washed away a portion of the track or washed sand onto it. The third theory was that illegal lineside dwellers had dumped rubbish onto the line, which had got under the wheels of the train. [1693]

At about 20:00 on Wednesday 15th June 2005 a freight train was derailed between Zubtsov and Aristovo, Russia. Twenty-four of 60 tank waggons of fuel oil left the rails, there was a major spill of oil and the derailment occurred only 150 metres from the Vazuza River, a tributary of the Volga.[1694] Measures were taken to contain the spill, despite which an estimated 200 tonnes of oil reached the river, forming an oil slick between 40 and 60 metres wide. There were concerns about groundwater contamination but fears of a threat to Moscow's water supply were played down.[1695] By 19th June a total of 126 tons of the spilt oil had been recovered[1696] and the clean-up operation was reported to have been completed on Tuesday 21st.[1697] However, on 24th June, pollution was reported to have increased, either, it was thought, due to higher temperatures or seepage into rivers from contaminated groundwater. [1698]

Environmental groups stressed that the number of train derailments in Russia due to technical faults was two or three times greater than in Western countries but a Rossiyskie zheleznye dorogi (Russian Railways) (RZD) spokesman stated that present derailment was probably due to track subsidence following heavy rain rather than a technical problem. The driver had apparently felt his train give a sideways kick and had first reduced speed to 20 mph and then applied the brakes in emergency. [1699]

At about 06:00 on Thursday 14th July 2005 a freight train consisting of perhaps 40 tanker waggons of benzene became derailed in a suburb of Fushun, Liaoning province, China. One of the derailed waggons ruptured, and leaking benzene caught fire.

Crossing Keeper Yang Guowen related: "The engine had just passed through and I saw thick smoke billowing up from a spot about 100 metres west of my crossing. Then there was this furious fire."

Resident Chen Yafeng, 45, had been in her courtyard at the time. She heard the initial explosion as the benzene caught fire and was knocked down by the force of the blast. Opening her eyes, she was confronted with thick black smoke and metre-high flames close to her. All nearby residents were quickly evacuated by the police for fear of the blaze setting off explosions in other tanker waggons, while the local fire brigade rushed to the scene. The firemen were divided into three groups, one group being employed keeping the remaining tanks cool to forestall further explosions, while the other two teams were tasked with dousing the fire, which had spread to nearby houses.

After three hours, fires that had begun in the houses had been extinguished but the intense heat had deformed other tank waggons. Deputy Fire Chief Qian Junkai said: "It was a very dangerous situation. There could have been further explosions at any moment."

By about 11:30 the estimated 40 tons of benzene that had escaped had burnt itself out and the fire had been extinguished. Only three residents were slightly hurt. The cause of the initial

---

[1693] http://allafrica.com/stories/printable/200505090926.html
[1694] http://www.mosnews.com/news/2005/06/16/trainaccid.shtml
[1695] http://en.rian.ru/russia/20050616/40535705-print.html
[1696] http://www.mosnews.com/news/2005/06/19/tvercleanup.shtml
[1697] http://news.monstersandcritics.com/mediamonitor/printer_1017963.php
[1698] http://www.tass.ru/eng/prnt.html?NewsID=2170629
[1699] http://www.themoscowtimes.com/stories/2005/06/17/042.html

derailment was being investigated. [1700]

At 05:09 on Wednesday 3rd August 2005 Canadian National Railway (CN) freight train M30351-03 from Edmonton, Alberta, to Vancouver, British Columbia, became derailed near Wabamun, Alberta, Canada. Forty-three of its cars left the rails, including one loaded with pole-treating oil, one loaded with toluene and 25 loaded with Bunker C (heavy fuel oil). About 700,000 litres of heavy fuel oil and 88,000 litres of pole treating oil were spilt, causing extensive biological, environmental and property damage. In addition, about 20 residents living in the immediate area had to be evacuated. No one was hurt.

The train consisted of three locomotives, 117 loaded cars, 13 of which contained residue and 10 of which were empty. Nothing unusual had occurred on the journey from Edmonton until, as it was travelling at 46 mph, there was a train-initiated emergency brake application. The cars in the middle of the train derailed in three separate groups; the front group contained all the waggons of fuel oil.

The Transportation Safety Board of Canada (TSB) investigated the accident. Its report reached 13 conclusions:

(1) The train had derailed when one or more rail breaks had occurred under it.

(2) The fragmentation of the maintenance rail had resulted from the formation of transverse defects along the full 40-foot length of it. The transverse defects had initiated in areas of shelling, which had resulted from high traffic density and loading.

(3) The rail breaks had occurred when pre-existing detail fractures had grown, undetected, to a critical size.

(4) Had the maintenance rail's mechanical properties, quality, and strength all matched those of the parent rail in the curve, the defects would have been less likely to develop.

(5) Although it was a common industry practice, the use of maintenance rails of different mechanical properties, quality, and strength to existing rails increased the risk of defect development, growth, and premature failure.

(6) Despite the requirements of Appendix D of Canadian National's Standard Practice Circular 3200, rails with internal defects could sometimes be misclassified as being free of defects due to the limitations of current ultrasonic inspections and the detectability of certain defects.

(7) Despite advances in ultrasonic testing technology, the detection of all internal rail defects remained problematic.

(8) In the absence of industry standards for rail fatigue life, rails could remain in track beyond their fatigue limit, thereby increasing the risk of sudden rail failure and derailment.

(9) Given the limitations of ultrasonic inspection technology and methods, the continued development of effective complementary defect management strategies was necessary to reduce the risk that defects would remain undetected, grow to critical size, and cause rail failure.

(10) The hazardous properties of Bunker C and pole-treating oil had not been understood and effectively communicated to enable preventive mitigation of the associated risks to residents, workers, and the environment.

(11) Had Canadian National's Dangerous Goods Emergency Response Plan (DGERP) and

[1700] http://www2.chinadaily.com.cn/english/doc/2005-07/16/content_460709.htm

its Environmental Emergency Response Plan (EMP 3) been activated under a unified command (UC) system, the management of the spill would have been more efficient.

(12) The Bunker C was not classified as a dangerous good because it was never at a temperature greater than or equal to its flashpoint at any time while in transport.

(13) An effective provincially developed emergency plan to respond to spills of this magnitude could have mitigated the impact of the accident.

The TSB noted that since the accident, Alberta Environment (AENV) and the CN were continuing to check water and soil samples for possible contamination. AENV had established an Environmental Protection Commission to deal with future contamination events and the commission in turn had created an environmental support team, the Alberta Environment Support and Emergency Response Team (ASERT) and an agency to respond to emergencies, the Alberta Emergency Management Agency (AEMA).

The CN had established additional response equipment capability, strengthened its dangerous goods programme, reviewed its emergency response plans, was working with shippers to identify environmentally sensitive non-regulated commodities and had developed a computer-based reporting and monitoring system for its track personnel.

Transport Canada was working with the companies to modernise the Railway Track Safety Rules.

Turning to actions still required, the TSB made two recommendations:

(1) The Department of Transport should establish minimum standards for the quality and strength of maintenance rails.

(2) The Department of Transport should establish standards requiring that rails approaching their fatigue limit be replaced.

Finally, the TSB expressed concern about the emergency response to the derailment, such as confusion between first responders, the lack of booms to contain the spilt oil and the failure of Environment Canada to establish environmental response protocols with its provincial counterparts, so as to ensure an adequate and comprehensive early response to environmental damage as a result of rail transportation accidents. [1701]

At about 07:20 on Friday 5th August 2005 Canadian National Railway (CN) freight train A47151-05 became derailed at a bridge over the Cheakamus River near Garibaldi, British Columbia, Canada. No one was hurt but about 40,000 litres of corrosive sodium hydroxide were lost into the river, causing extensive environmental damage.

The train involved in the accident had left Squamish, British Columbia, at 06:24, for Prince George, also in British Columbia. It was a very large train, consisting of five locomotives, an empty box car, three tank cars of sodium hydroxide, 103 empty cars, two locomotives and another 37 empty cars. All the locomotives were capable of being brought into operation, it being a so-called distributed power train, in which additional operating locomotives positioned, for example, at the midway point, two-thirds of the way back or at the rear of the train, enabled the operation of longer, heavier trains while providing a means to control or minimise in-train forces.

Approaching Cheakamus, the speed of the train had dropped and the engineer concluded that the locomotives in the remote consist (those marshalled towards the rear of the train) were not providing power. When speed decreased to 8.5 mph he brought another unit at the front of

[1701] Transportation Safety Board of Canada report

the train into operation. Speed then increased to 34 mph before the throttle was closed for a 20 mph speed restriction. Somewhat further on, several wheel slips occurred, then as the train was negotiating a set of reverse curves approaching the Cheakamus River Bridge, travelling at 14 mph, the brakes went into emergency and the train crew found that nine cars had left the track. Four of the derailed cars had fallen from the bridge, one of them being one of the tank cars carrying sodium hydroxide.

The Transportation Safety Board of Canada (TSB) investigated the accident. Its report reached 12 conclusions:

(1) The train had derailed on a 1.97 per cent ascending grade when it had string-lined to the inside of the 12-degree, 20-minute left-hand curve over the Cheakamus River Bridge.

(2) The combination of excessive locomotive tractive effort and trailing tonnage, along with long-short car coupling, had produced high lateral forces and a correspondingly high lateral/vertical (L/V) ratio and wheel lift, causing the train to string-line on the curve.

(3) The distributed power (DP) locomotives had been set up incorrectly in North Vancouver, resulting in significant electrical faults when the remote locomotive consist was brought online, and the activation of protection circuits had prevented them from developing horsepower.

(4) Although Train Line 2 (TL2) alarms had been generated by the faults in the remote consist, they had not resulted in the first crew (North Vancouver to Squamish) identifying the problem or initiating any action.

(5) Because the train had been marshalled with earlier-model locomotives (GEF-40 Dash 8s) leading the head end and also controlling the remote consist, no direct indication of the inoperative state of the remote locomotives had been provided to either crew. Deployment of two of the four available GEF-44 Dash 9s in these positions would have provided this indication.

(6) With the DP locomotives unavailable, train A471 had become a conventional train and subject to conventional train restrictions, which had not been applied.

(7) Bringing the lead locomotive online had resulted in additional horsepower and driving axles exceeding powered axle and horsepower restrictions for conventional trains.

(8) Canadian National (CN) had prepared a Safety Integration Plan and provided informal training in DP setup and operations. However, in the absence of a formal risk assessment, CN had resumed long DP train operations without adequate consideration of the value of retaining and using local knowledge and experience in the operation of long DP trains. This had resulted in a lack of training and proper supervision, which had contributed to the derailment.

(9) The General Operating Instructions (GOIs) restrictions in effect at the time had been inadequate to capture all empty cars close to 80 feet in length, increasing the risk of derailment due to excessive trailing tonnage behind these types of cars.

(10) In the absence of specific details on locomotive faults identified by TL2 alarms, and their effect on locomotive operation, it had been difficult for locomotive engineers to determine the appropriate action when alarms occurred.

(11) This occurrence was the first of four similar derailments involving long, empty DP trains that had occurred on the Squamish Subdivision between 5th August and 5th December 2005. While circumstances differed, the common feature of all four derailments was that derailed trains had string-lined to the low rail, or inside of sharp curves. The string-lining had occurred due to high lateral forces creating a high L/V ratio, wheel lift and derailment.

(12) The multi-agency response to the incident had been well coordinated and effective.

The TSB noted that following the accident and three subsequent derailments, various notices and orders had been issued to the CN regarding train handling, length and tonnage over the severe curvature and gradients of the line in question. After a two-month period free of further derailments after that of 5th December, a new notice and order issued on 6th March 2006 increased the length permitted of northbound DP trains to 114 cars in the Squamish area. Conventional trains were restricted to a length of 80 cars.

The TSB also expressed safety concerns that the railways were not marshalling their trains so that the safest technology was available to locomotive engineers. In addition, locomotive engineers might not be able to differentiate between the many alarms in the cab and as a consequence, might not always effectively prioritise the alarms and take corrective action. It considered that the railways would benefit from working with human factors specialists and locomotive manufacturers to design alarms that would clearly identify the nature of the fault. [1702]

An estimated 24,000 gallons of sulphuric acid spilt from two train cars and contaminated the Santa Cruz River following a derailment of a Ferromex freight train at about 15:00 on Thursday 11th August 2005, near Santa Cruz, Sonora state, Mexico. Private well-users in the South and North River Road areas and along State Route 82 were asked not to use water from their wells until the situation had been mitigated. A team of emergency-management officials monitored the water flow at the border with the United States, about 65 river kilometres to the north, throughout the night. Public works crews from Nogales and Santa Cruz County built a number of berms in the river north of the border about 1,000 feet apart. Firefighters from the Nogales Fire Department continuously tested the pH levels in the river. By midnight on August 11th, the pH had decreased to four. After being treated with lime, the pH levels had risen to a neutral seven by 11:00 on Friday 12th. [1703]

A 17-car Kansas and Oklahoma Railroad freight train from Hutchinson to Wichita suffered a partial derailment at the start of its journey in Hutchinson, Kansas, United States, just before 11 p.m. on Tuesday 30th August 2005. Only three cars left the track but two of them were tank cars containing denatured alcohol, 29,000 gallons of which was lost from one. The derailment occurred in a mostly industrial area but between 40 and 50 homes had to be evacuated while the spillage was dealt with. No one was hurt. [1704]

At 4:56 a.m. on Saturday 15th October 2005 westbound Union Pacific Railroad (UP) freight train ZYCLD 13 ran into the rear of freight train MPBHG 15, which was stationary in the yard at Texarkana, Arkansas, United States. As a result of the collision a tank car containing flammable propylene gas was punctured and the gas, being heavier than air, flowed along the ground. When it reached a nearby house it was ignited from an unknown source and exploded, killing a woman inside. A second adjoining house was destroyed by fire. As a precaution, 3,000 people had to be evacuated. [1705]

The accident received considerable press coverage. 'A plume of smoke covered the south end of the city, and at least seven people went to hospital emergency rooms with complaints

---

[1702] Ibid

[1703] http://www.sulphuric-acid.com/techmanual/plant_safety/safety_accidents-Archive.htm

[1704] *The Hutchinson News*, Hutchinson, Kansas, United States, 1st September 2005

[1705] National Transportation Safety Board report

of respiratory problems. At least two homes were destroyed – including one where the victim died – and several vehicles were totalled in the quarter mile area surrounding the accident, police spokesman Chris Rankin said. A 209-foot-long railroad bridge also caught fire and was destroyed, a Union Pacific spokesman said.

'The propylene tank was still burning Saturday evening, but the fire was under control and the smoke had thinned out, Rankin said. Union Pacific officials planned to let the tank burn out and estimated that it would take until early on Sunday morning. Police cancelled the evacuation order Saturday afternoon, about 10 hours after the derailment.

'Initially the police thought the chemical involved was vinyl acetate, which releases poisonous fumes. Officers went door to door, urging thousands of people in a 2-by-5 mile area to move to the north side of town. Propylene is less dangerous. "The smoke was so thick it blocked out daylight," Rankin said.

'A Union Pacific train coming from Chicago hit the back of another train in the rail yard, causing the eight cars to derail, said Union Pacific spokesman Mark Davis. None of the railroad crew members was injured, he said.

'Rankin said the switching yard is behind the police station, about a quarter mile from the Texas border and also serves Kansas City Southern and Amtrak. Amtrak spokesman Marc Magliari said a southbound passenger train was delayed two hours by the derailment but was eventually rerouted. He said the damage would likely mean delays for Amtrak's Texas Eagle until the site was cleaned up.

'Wadley Regional Medical Centre was in the evacuation area, and several patients and nurses also complained of nose and eye irritation, spokeswoman Shelby Brown said.' [1706]

The National Transportation Safety Board (NTSB) carried out an investigation into the accident. It noted that train ZYCLD 13 had left Pine Bluff, Arkansas, at 11:34 p.m. the previous day for Big Sandy, Texas, and had consisted of two locomotives and 57 loaded cars. At McNeil, Arkansas, the crew of train ZYCLD 13 had encountered a Red signal, which had then changed to a Flashing Yellow. At Stamps, 18 miles further on, they had had Flashing Yellow signals and at Lewisville, 4½ miles beyond Stamps, a Red, which had then changed to a Yellow. They had thus been aware that they were following train MPBHG 15. At East Texarkana the train had been stopped at a Red signal, which had cleared to Yellow at 4:13 a.m. and Flashing Yellow at 4:24 a.m., but the locomotive event recorder indicated that the train had not set off again until 4:45 a.m.. The engineer and conductor could not explain why they had not restarted their train before, and would not admit that both of them had been asleep. The next signal, close to the entrance to Texarkana yard, had been Yellow and was passed with the train travelling at 19 mph. There were no further signals in the yard to indicate that there was a train ahead and all trains were meant to travel at restricted speed. The speed of the train had then been maintained at about 17 mph until the engineer had made an emergency brake application at the last moment.

In the darkness and with restricted visibility, it was not at first realised exactly what had happened. The propylene did not explode at once but only at 5:08 a.m., 12 minutes after the first 911 emergency call reporting a strange odour and four minutes after the yardmaster and the 911 dispatcher had first discussed the possibility of propylene being involved. The NTSB identified various failures in the communications and co-operation between the railway and the city authorities:

(1) The yardmaster had failed to notify the 911 dispatcher about the accident.

(2) The yardmaster had failed to provide the 911 dispatcher or the incident commander with

---

[1706] *The News*, Port Arthur, Texas, United States, 16th October 2005 a complete list of the hazardous materials in train MPBHG 15.

(3) After the evacuation of the yard no UP employee had remained there to answer telephone enquiries and none had checked in at the command post to co-ordinate with the incident commander.

(4) There had been a lack of co-ordination between the manager of rail operations and the fire department had resulted in an unnecessary duplication of effort.

The NTSB determined that the probable cause of the accident was the failure of the crew of train ZYCLD 13 to remain attentive and alert and thereby able to stop short of an observable standing train. Contributing to the severity of the accident was the puncture of a tank car during the collision, which had resulted in the release of propylene and a fire.

The NTSB made four recommendations.

To the city of Texarkana:

(1) It should coordinate with all the regional and local transporters of hazardous materials, such as railroads and trucking companies, to establish effective communications and response plans and to conduct periodic joint emergency response drills and exercises.

To the Union Pacific Railroad:

(1) It should implement measures to ensure that all of its field personnel understood and complied with its procedures for responding to hazardous materials incidents, with particular emphasis on timely notifications and appropriate coordination with local emergency responders.

To the International Association of Fire Chiefs:

(1) It should notify its members about the circumstances of the accident in Texarkana, Arkansas, on October 15, 2005, and urge them to coordinate with all regional and local transporters of hazardous materials, such as railroads and trucking companies, to establish effective communications and coordination through joint emergency response drills and exercises.

To the Association of American Railroads and the American Short Line and Regional Railroad Association:

(1) It should notify its members about the circumstances of the accident in Texarkana, Arkansas, on October 15, 2005, and urge them to coordinate with all communities adjacent to their railroad yards and along their hazardous materials routes to establish effective communications and coordination through joint emergency response drills and exercises. [1707]

Five of the six coaches of East Japan Railway Inaho express train No. 14 from Akita to Niigata were derailed at about 19:15 on Christmas Day, Sunday 25th December 2005, shortly after the train had crossed a 600-metre-long bridge over the Mogami River between Sagoshi and Kita-Amarume stations, Japan. Four passengers were quickly known to have lost their lives and about 30 of the estimated 44 people on the train were taken to local hospitals, including Driver Takashi Suzuki. Three other passengers were missing for some time and the body of one was later found in the wreckage; the other two had survived and left the scene. [1708]

It soon became evident that a strong gust of wind had blown the train over. Conductor Takahiro Nakabachi stated that he had felt a jolt as if the train was being pushed from behind

[1707] National Transportation Safety Board report

[1708] www.japantimes.co.jp/cgi-bin/makeprfy.p15?nn20051226a3.htm

soon after it had crossed the bridge, then the lights had gone out and it had stopped suddenly.[1709] Prior to the accident train services were suspended when wind speeds reached 108 km/h and train speeds were to be reduced to 25 km/h when wind speeds exceeded 90 km/h, however, an anemometer near the scene had only recorded a wind speed of 72 km/h at the time of the accident. The train had been travelling at 100 km/h when it was derailed, within the speed limit of 120 km/h.[1710] An investigation theorised that a sudden microburst of intense wind had probably derailed it.[1711] Following the accident, the decision was taken to suspend train operation if the wind speed exceeded 90 km/h. [1712]

[1709] www.japantimes.co.jp/cgi-bin/getarticle.p15?nn20051227a4.htm
[1710] www.asahi.com/english/Herald-asahi/TKY200512270140.html
[1711] www.yomiuri.co.jp/dy/national/20060104TDY02006.htm
[1712] www.yomiuri.co.jp/dy/national/20060108TDY02001.htm

INDEX

TABLE OF MAJOR RAILWAY ACCIDENTS 1991-2005

Date	Location	Railway Company	Description	Passenger Fatalities
6 Jan 1991	Between Budge Budge and Ballygunj, West Bengal, India	Eastern	Rear collision in fog	8
15 Feb 1991	Karchana, Uttar Pradesh, India	Northern	Derailment on points	15
26 February 1991	Near Ciudad Serdán, Puebla, Mexico	Ferrocarriles Nacionales de México	Head-on collision, single line?	5
4 Mar 1991	Nacala, Nampula, Mozambique	Portos e Caminhos de Ferro de Moçambique	Runaway train, derailment	96
6 Apr 1991	Near Santo Domingo, Villa Clara, Cuba	Ferrocarriles Nacionales de Cuba	Derailment, track defects?	56
16 Apr 1991	Near Chittagong, Bangladesh	Bangladesh	Head-on collision?	5
15 May 1991	Between Onodani signal station and Shigarakikyu, Japan	Shigaraki-Kogen	Head-on collision, single line	37
30 May 1991	Near Khasayvurt, Checheno-Ingush, USSR	Soviet	Explosion, sabotage	12
22 Jun 1991	Dire Dawa, Ethiopia	Chemin de Fer Djibouto-Éthiopien	Derailment	50
31 Jul 1991	Temirgau, Dagestan, USSR	Soviet	Explosion, sabotage	15
31 Jul 1991	Lugoff, South Carolina, USA	CSX Transportation	Derailment, points moved under train	8
3 Aug 1991	Between Kaliningrad and Svetlogotsk, USSR	Soviet	Level crossing collision with lorry, derailment	5
4 Aug 1991	Sonik, Uttar Pradesh, India	Northern	Head-on collision	11
18 Aug 1991	Dayaoshan Tunnel, China	China	Fire in tunnel	15
28 Aug 1991	14th Street-Union Square, New York, USA	Metropolitan Transportation Authority	Derailment in tunnel, excessive speed, motorman intoxicated	5
29 Aug 1991	Piran Ghaib, Pakistan	Pakistan	Collision with shunting goods	6
2 Sep 1991	Near Chaman, Pakistan	Pakistan	Bomb exploded on train	8
6 Sep 1991	Mvoungouti, Republic of Congo	Chemin de fer Congo-Océan	Head-on collision	133
5 Oct 1991	Dalhem, Belgium	Société nationale des chemins de fer belges	Runaway train, derailment	7

Date	Location	Railway Company	Description	Passenger Fatalities
6 Oct 1991	Kallichan, Punjab, India	Northeast Frontier	Derailment, sabotage, part of train fell from bridge	8
17 Oct 1991	Melun, France	Société national des chemins de fer français	Head-on collision	23
19 Oct 1991	Pondugula, Andhra Pradesh, India	South Central	Derailment on points, collision	13
30 Oct 1991	Between Makalidurga and Thondebhavi, Karnataka, India	Southern	Derailment, train ran into rockfall	29
8 Nov 1991	Kalyan, Near Bombay, Maharashtra, India	Central	Bomb exploded on train	12
15 Nov 1991	Zaoqiao, Taiwan	Taiwan	Sidelong collision	30
7 Dec 1991	Between Jawanwala Shehr and Harsar Dehri, Himachal Pradesh, India	Northern	Derailment, 'faulty loading,' high speed	27
14 Jan 1992	Near Poradah Junction, Bangladesh	Bangladesh	Fire	27
8 Feb 1992	Near Narwana, Haryana, India	Northern	Bomb exploded on train	6
12 Feb 1992	Badrshein, Egypt	Egyptian State	Rear collision	11
25 Feb 1992	Pichkurir Dhal, West Bengal, India	Eastern	Train diverted onto wrong line, collision	5
3 Mar 1992	Podsosenka, Russia	Rossiyskie zheleznye dorogi	Head-on collision, fire	35
21 Mar 1992	Between Tettu and Kavali, Andhra Pradesh, India	South Central	Derailment, broken rail	5
23 May 1992	Dorud, Iran	Islamic Republic of Iran	Head-on collision, single line?	26
5 Apr 1992	Tsunduru Junction, Andhra Pradesh, India	South Central	Rear collision	7
17 April 1992	Central Tanzania	Tanzania	Derailment	16
27 Apr 1992	Near Hyderabad, Pakistan	Pakistan	Train attacked by bandits	7
10 Jun 1992	Caluso, Italy	Ferrovie dello Stato	Head-on collision, single line	5
9 Jul 1992	Between Dhamangaon and Talni, Maharashtra, India	Central	Derailment, sabotage	15
3 Aug 1992	Between Budkovce and Drahňov, Slovakia	Železnice Slovenskej republiky	Collision	6
7 Aug 1992	Near Lazaro Cárdenas, Michoácan, Mexico	Ferrocarriles Nacionales de México	Head-on collision, single line?	10
17 Aug 1992	Kazichene, Bulgaria	Balgarski darzhavni zheleznitsi	Rear collision?	8

Date	Location	Railway Company	Description	Passenger Fatalities
18 Aug 1992	Grünbach, Austria	Österreichische Bundesbahnen	Head-on collision, single line	5
5 Sep 1992	Kirodimalnagar, Madhya Pradesh, India	South East Central	Rear collision	37
Sep 1992	Between Baghdad and Basra, Iraq	Iraqi Republic	Train attacked by rebels	?
13 Oct 1992	Between Patiladaha and Sarbhog, Assam, India	Northeast Frontier	Bombs exploded on train	25
1 Nov 1992	Reti, Pakistan	Pakistan	Rear collision	15
15 Nov 1992	Northeim, Germany	Deutsche Reichsbahn	Derailment, defective vehicle, collision	10
30 Nov 1992	Hoofddorp, the Netherlands	Nederlandse Spoorwegen	Derailment, excessive speed	5
18 Jan 1993	Near Semnan, Iran	Islamic Republic of Iran	Collision	24
18 Jan 1993	Gary, Indiana, USA	Chicago South Shore and South Bend	Sidelong collision	7
30 Jan 1993	Ndethia Geithia River Bridge, Kenya	Kenya	Bridge collapsed under train	117
8 Feb 1993	Between Ye and Moulmein, Myanmar	Myanma	Derailment, sabotage	?
28 Feb 1993	Gudermes, Chechnya, Russia	Rossiyskie zheleznye dorogi	Bomb exploded on train	13
28 Mar 1993	Gupo, South Korea	Korean National	Derailment, track subsidence	78
20 Apr 1993	Birgora Halt, between Tangarbansli and Nagjua, Bihar, India	South Eastern	Derailment, sabotage	15
5 May 1993	Mung Russei, Battambang, Cambodia	Chemins de fer du Royaume du Cambodge	Train derailed and fired on by Kher Rouge guerrillas	13
27 May 1993	Between Gyumri and Nalband, Armenia	Haykakan yerkat'ughi	Runaway waggons, collision	30
27 May 1993	Quipungo, Huila, Angola	Caminho de Ferro de Moçâmedes	Derailment, sabotage by UNITA guerillas	225
1 Jun 1993	Samalpatti, Tamil Nadu, India	Central	Derailment, broken axle, collision	21
27 Jun 1993	Between Matala and Olivenca, Angola	Caminho de Ferro de Moçâmedes	Derailment, sabotage	19
10 Jul 1993	China	China	Rear collision	40
15 Jul 1993	Between Haya Ghat and Thalwara, Bihar, India	North Eastern	Head-on collision, single line	22

Date	Location	Railway Company	Description	Passenger Fatalities
31 Jul 1993	Visoi, Kenya	Kenya	Head-on collision?	5
2 Aug 1993	Near Kampot, Cambodia	Chemins de fer du Royaume du Cambodge	Train attacked by Khmer Rouge guerrillas	10
2 Aug 1993	Near Vega de Anzo, Spain	Ferrocarriles Españoles de Via Estrecha	Head-on collision, single line, in tunnel, fire	10
3 Aug 1993	Mairwa, Bihar, India	North Eastern	Derailment, defective vehicle	23
13 Sep 1993	Malema, Mozambique	Portos e Caminhos de Ferro de Moçambique	Derailment	77
21 Sep 1993	Between Chhabra Gugar and Bhulon, Rajasthan, India	Western	Head-on collision, single line?	78
22 Sep 1993	Big Bayou Canot Bridge, Alabama, USA	CSX Transportation	Bridge collapsed under train, fire	42
27 Sep 1993	Temara, Morocco	Office National des Chemins de Fer du Maroc	Rear collision, explosion, fire	?
3 Oct 1993	Nordstrand, Norway	Norges Statsbaner	Runaway engine, collision	5
2 Nov 1993	Between Depok and Citayam, Indonesia	Kereta Api Indonesia	Head-on collision, single line	18
24 Nov 1993	Birlanagar Junction, Madhya Pradesh, India	Central	Collision with buffalo, derailment, collision	13
12 Dec 1993	Zifta, Egypt	Egyptian State	Train diverted onto wrong line, head-on collision	12
2 Jan 1994	Near Bukedea, Uganda	Uganda	Derailment, excessive speed? 'Faulty loading'? Track defects	65
15 Jan 1994	Xiangfan, China	China	?	7
16 Feb 1994	Bangrod, Madhya Pradesh, India	Western	Rear collision	8
8 Mar 1994	Marianhill, Near Pinetown, Natal, South Africa	Transnet	Derailment, excessive speed?	67
19 Mar 1994	20 Yanvar, Baku, Azerbaijan	Azərbaycan Dəmir Yolları	Explosion, sabotage	14
20 Mar 1994	Central Taiwan	Taiwan	Level crossing collision with lorry, derailment	7

Date	Location	Railway Company	Description	Passenger Fatalities
21 Mar1994	Shaba, Zaire	Société Nationale des Chemins de Fer Zaïrois	Derailment	43
21 Mar 1994	Däniken, Switzerland	Chemins de fer fédéraux suisses	Collision with jib of crane fouling line	9
3 Apr 1994	Near Sohawa, Pakistan	Pakistan	Derailment	13
15 May 1994	Pinlaung, Myanmar	Myanma	Derailment	25
20 May 1994	Arjawinangun, Java, Indonesia	Perusahaan Umum Kereta Api	Head-on collision	9
c2 Jul 1994	Near Sunchang, China	China	Explosion	18
3 Jul 1994	Between 28 May and Ganjlik, Baku, Azerbaijan	Azərbaycan Dəmir Yolları	Bomb exploded on train in tunnel	13
20 Jul 1994	Between Mewa Nawada and Seohara, Uttar Pradesh, India	Northern	Derailment, broken rail	7
26 Jul 1994	Phnom Voar, Cambodia	Chemins de fer du Royaume du Cambodge	Train mined and fired upon by Khmer Rouge guerrillas	9
11 Aug 1994	Near Dvuluchnoye, Russia	Rossiyskie zheleznye dorogi	Derailment, collision	21
13 Aug 1994	Zhanjiang and Guilin, China	China	Explosion	12
13 Aug 1994	Gurjaani, Georgia	sakartvelos rk'inigza	Rear collision	24
20 Aug 1994	Kalla-Seghira, Tunisia	Société Nationale des Chemins de Fer Tunisiens	Head-on collision, fire	21
4 Sep 1994	Mora, Mexico	Ferrocarriles Nacionales de México	Derailment, track subsidence? Excessive speed?	12
11 Sep 1994	Tithorea, Greece	Organismos Sidirodromon Ellados	Derailment, excessive speed	5
22 Sep 1994	Near Tolunda, Angola	Empresa do Caminho de Ferro de Benguela	Derailment, defective vehicle	300
26 Oct 1994	Between Lotapahar and Chakradharpur, Bihar, India	Southeastern	Fire	46
28 Oct 1994	Battambang, Cambodia	Chemins de fer du Royaume du Cambodge	Bomb exploded on train	6

Date	Location	Railway Company	Description	Passenger Fatalities
2 Dec 1994	Szajol, Hungary	Magyar Államvasutak	Train diverted onto wrong line, derailment	29
30 Dec 1994	Near Wuntho, Myanmar	Myanma	Runaway train, derailment	102
31 Dec 1994	Between Yelahanka Junction and Chennasandra, Karnataka, India	Southern	Derailment	8
2 Jan 1995	Kampong Chnnang, Cambodia	Chemins de fer du Royaume du Cambodge	Train derailed and fired upon by Khmer Rouge guerrillas	8
13 Jan 1995	Hili, Bangladesh	Bangladesh	Head-on collision	100
25 Feb 1995	Nailalung, Assam, India	North East Frontier	Bomb exploded on train	26
27 Feb 1995	Éibar, Spain	Eusko Trenbideak	Derailment, excessive speed	5
3 Apr 1995	Near Prachuap Khiri Khan, Thailand	State Railway of Thailand	Collision with lorry on open level crossing, derailment, fire	11
14 May 1995	Between Lokur and Danishpet, Tamil Nadu, India	Southern	Head-on collision, single line	49
29 May 1995	Tukh Crossing, Cairo, Egypt	Egyptian State	Derailment, engine defect	15
1 Jun 1995	Near Kalubathan, Bihar, India	Eastern	Train diverted onto wrong line, collision	45
24 Jun 1995	Between Krouná and Předhradi, Czech Republic	České Dráhy	Runaway waggons, collision	18
25 July 1995	Saint Michel, France	Réseau Express Régional	Bomb exploded on train	7
8 August 1995	Near Ksar el Boukhari, Algeria	Société Nationale des Transports Ferroviaires	Derailment, sabotage	5
20 Aug 1995	Firozabad, Uttar Pradesh, India	Northern	(Collision with buffalo), rear collision	298
24 Oct 1995	Trowek Bridge, near Tasikmalaya Indonesia	Perusahaan Negara Kereta Api	Runaway train, derailment	17
28 Oct 1995	Ulduz and Nariman Narimanov, Baku, Azerbaijan	Azərbaycan Dəmir Yolları	Fire in tunnel	289
21 Dec 1995	Badrashin, Egypt	Egyptian State	Rear collision in fog	75
8 Jan 1996	Kongaish, Bangladesh	Bangladesh	Head-on collision, single line?	16
17 Jan 1996	Mwekera, Zambia	Zambia	Head-on collision, single line?	30

Date	Location	Railway Company	Description	Passenger Fatalities
8 Feb 1996	Near El Rahmaniya, Egypt	Egyptian State	Collision with lorry on open level crossing, derailment	8
16 Feb 1996	Georgetown Junction, Maryland, USA	CSX Transportation	Sidelong collision, fire	8
13 Mar 1996	Between Oran and Tiemcen, Algeria	Société Nationale des Transports Ferroviaires	Derailment, sabotage	12
9 Apr 1996	Kiazi, Zaire	Sizarail	Runaway train, derailment	100
18 Apr 1996	Domingarh, Uttar Pradesh, India	Northeastern	Head-on collision	54
23 May 1996	Between Yangon and Mandalay, Myanmar	Myanma	Derailment, sabotage	9
31 May 1996	Near Litvinovo, Siberia, Russia	Rossiyskie zheleznye dorogi	Runaway waggons, collision	18
7 Jun 1996	Lukanda Crossing, near Kapiri Mposhi, Zambia	Zambia	Collision with lorry on open? level crossing, derailment?, fire	?
14 Jul 1996	Near Thénia, Algeria	Société Nationale des Transports Ferroviaires	Collision	7
24 Jul 1996	Dehiwala, Sri Lanka	Sri Lanka	Bomb exploded on train	69
18 Sep 1996	Japeri, Brazil	Rede Ferroviária Federal	Runaway train, collision	15
24 Oct 1996	Near Baba Ali, Algeria	Société Nationale des Transports Ferroviaires	Bomb exploded under train	8
3 Nov 1996	Tandoi, Pakistan	Pakistan	Bomb exploded on train	5
2 Dec 1996	Ambala Cantt, Punjab, India	Northern	Bomb exploded on train, fire	10
30 Dec 1996	Kokrajhar and Fakiragram, Assam, India	Northeast Frontier	Derailment, sabotage	33
3 Feb 1997	Al-Radisiyah, Egypt	Egyptian State	Rear collision	11
10 Feb 1997	Caguasal, Near Ciego de Avila, Cuba	Ferrocarriles Nacionales de Cuba	Collision with light engine? Head-on collision, single line?	13
3 Mar 1997	Mehr Shah, Near Khanewal, Punjab, Pakistan	Pakistan	Runaway train, derailment in sand drag at end of trap siding	136
31 Mar 1997	Uharte Arakil, Spain	Red Nacional de los Ferrocarriles Españoles,	Derailment, excessive speed	18
15 Apr	Lagos, Nigeria	Nigerian	Derailment	5

Date	Location	Railway Company	Description	Passenger Fatalities
1996				
25 Apr 1997	Oued el Kerma, Algeria	Société Nationale des Transports Ferroviaires	Derailment, sabotage	21
29 Apr 1997	Rongjiawan, Hunan, China	China	Rear collision	90
5 May 1997	Reptowo, Poland	Polskie Koleje Państwowe	Derailment, defective points?	12
27 Jun 1997	Near Torbino, Russia	Rossiyskie zheleznye dorogi	Bomb exploded on train	5
5 Jul 1997	Neustadt, Germany	Deutsche Bahn	Collision with load projecting from freight train on adjoining line	6
8 Jul 1997	Laihra Khana Halt, Punjab, India	Northern	Bomb exploded on train	35
27 Jul 1997	Faridabad, Haryana, India	Central	Rear collision	13
8 Aug 1997	'300 Km from Baghdad,' Iraq	Iraqi Republic	Collision	50
14 Aug 1997	Between Kathalpukhuri and Silarubari, Assam, India	Northeast Frontier	Derailment, sabotage	8
23 Aug 1997	Algeria	Société Nationale des Transports Ferroviaires	Bombs exploded on train	8
Summer 1997	Between Baekam and Hyesan, North Korea	Chosŏn Minjujuŭi Inmin Konghwaguk Ch'ŏldosŏng	Runaway train, derailments	?
8 Sep 1997	Crossing No. 365, between Saint-Antoine-de-Breuilh and Saint-Foy-le-Grande, France	Société national des chemins de fer français	Collision with petrol tanker on automatic half barrier level crossing, fire	12
14 Sep 1997	Between Naila and Champa, Madhya Pradesh, India	Southeastern	Derailment, track defects, train fell from bridge	88
19 Sep 1997	Southall, England	Railtrack Great Western	Sidelong collision, fire	7
22 Sep 1997	Sucat and Alabang Muntinlupa, the Philippines	Philippines National	Runaway coaches, collision	7
20 Oct 1997	Between Kinshasa and Matadi, Democratic Republic of Congo	Société Nationale des chemins de fer du Congo	Head-on collision, single line?	12
28 Oct 1997	Near Guangzhou, China	China	Level crossing collision with lorry	6
24 Dec 1997	Rustam Sargana, Pakistan	Pakistan	Head-on collision in fog	35

Date	Location	Railway Company	Description	Passenger Fatalities
5 Jan 1998	Near Karachkum, Tajikistan	Rohi ohani Toçikiston	Collision with shunting goods, fire	15
5 Jan 1998	Between Karna and Masit, Punjab, India	Northern	(Collision with cow), rear collision in fog	51
31 Jan 1998	Sihala, Pakistan	Pakistan	Runaway train, derailment in sand drag at end of trap siding	11
6 Feb 1998	Na Bon, Thailand	State Railway of Thailand	Derailment	9
23 Feb 1998	Near El Affroune, Algeria	Société Nationale des Transports Ferroviaires	Bomb exploded on train	18
6 Mar 1998	Jyväskylä, Finland	VR-Yhtymä Oyj	Derailment, excessive speed	9
9 Mar 1998	Pattoki. Pakistan	Pakistan	Bomb exploded on train	11
10 Mar 1998	Walton, Lahore, Pakistan	Pakistan	Bomb exploded on train	10
24 Apr 1998	Parli Vaijnath, Maharashtra, India	South Central	Runaway waggons, collision	24
28 Apr 1998	Bakhtiyarpur, India	Eastern	Derailment	12
10 May 1998	Naoti, Bangladesh	Bangladesh	Rear collision	17
22 May 1998	Malabvi, Malawi	Malawi	Runaway train, derailment	18
3 Jun 1998	Eschede, Germany	Deutsche Bahn	Derailment, broken tyre, bridge fell onto train	97
7 Jun 1998	Near Khairpur, Pakistan	Pakistan	Bomb exploded on train	26
11 Jun 1998	Near Condeck, Algeria	Société Nationale des Transports Ferroviaires	Bomb exploded on train	17
18 Jul 1998	Near Goudiri, Senegal	Société nationale des chemins de fer du Sénégal	Derailment	6
12 Aug 1998	Ain Torki, Algeria	Société Nationale des Transports Ferroviaires	Bomb exploded on train	7
8 Oct 1998	Between Sattenpalle and Pedakurapadu, Andhra Pradesh, India	South Central	Derailment, sabotage?	11
18 Oct 1998	Kafr-el-Dawar, Egypt	Egyptian State	Runaway train, derailment	?
30 Oct 1998	Near Goudiri, Senegal	Société nationale des chemins de fer du Sénégal	Derailment	7

Date	Location	Railway Company	Description	Passenger Fatalities
20 Nov 1998	Between Epa and Dancheon, North Korea	Chosŏn Minjujuŭi Inmin Konghwaguk Ch'ŏldosŏng	Runaway train, derailment	1,500
26 Nov 1998	Between Khanna and Chawapalli, Punjab, India	Northern	Derailment, broken rail, collision	212
3 Dec 1998	Between Mhasawad and Shiroli, Maharashtra, India	Central	Derailment	8
3 Jan 1999	Near Tiefa, Liaoning, China	China	Collision	24
13 Jan 1999	Darus, Sudan	Sudan	Rear collision	7
15 Mar 1999	McKnight Road Crossing, Bourbonnais, Illinois, United States	Illinois Central	Collision with articulated lorry on automatic half barrier level crossing, derailment, fire	11
24 Mar 1999	Near Man Eaters Junction, Kenya	Kenya	Derailment, excessive speed	32
10 Apr 1999	Between Shirbin and Bilqas, Egypt	Egyptian State	Head-on collision, single line, fire	7
12 April 1999	Grdelica, Yugoslavia	Jugoslavenske željeznice	Train bombed by Allied aircraft	20
4 June 1999	Between Ghanapur and Ippaguda, Andhra Pradesh, India	South Central	Derailment, track defects	9
22 Jun 1999	New Jalpaiguri, West Bengal, India	Northeast Frontier	Bomb exploded on train	9
9 Jul 1999	Near Hengyang, Hunan, China	China	Derailment, excessive speed	9
16 Jul 1999	Between Farah and Baad, Uttar Pradesh, India	Central	Derailment, defective vehicle, collision	21
2 Aug 1999	Gaisal, West Bengal, India	Northeast Frontier	Train diverted onto wrong line, head-on collision, fire	282
18 Sep 1999	Between Wah and Taxila, Pakistan	Pakistan	Runaway engine, collision	20
5 Oct 1999	Ladbroke Grove, England	Railtrack Great Western	Head-on collision, fire	29
11 Nov 1999	Near Pathankot, Punjab, India	Northern	Bomb exploded on train	14
2 Dec 1999	Near Glenbrook, New South Wales, Australia	State Rail Authority	Rear collision	7

Date	Location	Railway Company	Description	Passenger Fatalities
17 Dec 1999	Near Ywathagyi, Myanmar	Myanma	Head-on collision?	70
4 Jan 2000	Åsta, Norway	Norges Statsbaner	Head-on collision, single line, fire	16
5 Feb 2000	Hyderabad, Pakistan	Pakistan	Bomb exploded on train	8
6 Feb 2000	Brühl, Germany	Deutsche Bahn	Derailment, excessive speed	9
10 Feb 2000	Near Vijaypur, Jammu and Kasmir, India	Northern	Bomb exploded on train	5
19 Feb 2000	Between Duskheda and Savda, Maharashtra, India	Central	Fire	18
8 Mar 2000	Nakameguro, Japan	Teito Rapid Transit Authority	Derailment, track and vehicle defects	5
12 Apr 2000	Hinda, Republic of Congo	Chemin de fer Congo – Océan	Train diverted onto wrong line? collision	11
16 Apr 2000	Near Thaton, Myanmar	Myanma	Derailment	7
1 May 2000	Mekar Sari, Indonesia	PT Kereta Api Indonesia	Head-on collision	7
22 May 2000	Near Diamniadio, Senegal	Société nationale des chemins de fer du Sénégal	Derailment	12
28 Jun 2000	Batang Anai, West Sumatra, Indonesia	PT Kereta Api Indonesia	Runaway train, derailment	32
16 Jul 2000	Near Hyderabad, Pakistan	Pakistan	Bomb exploded on train	10
28 Jul 2000	Between Jaraguá and Perus, Brazil	Companhia Paulista de Trens Metropolitanos,	Runaway train, collision	9
31 Jul 2000	Between Khandikar and Goreswar, Assam, India	Northeast Frontier	Derailment, sabotage	9
14 Aug 2000	Near Rauzagaon, Uttar Pradesh, India	Northern	Bomb exploded on train	11
15 Aug 2000	Near Lela, Kenya	Kenya	Derailment, excessive speed	13
10 Sep 2000	Pelabuhan and Krengseng, Batang, Indonesia	PT Kereta Api Indonesia	Head-on collision, single line	13
11 Oct 2000	Between Intundla Siding and Ingwe Siding, Zimbabwe	National Railways of Zimbabwe	Head-on collision, single line	14

Date	Location	Railway Company	Description	Passenger Fatalities
11 Nov 2000	Kitztsteinhorn Tunnel, Austria	Kapruner Gletscherbahnen	Fire in tunnel (funicular railway)	150
14 Nov 2000	Near Kora, Ethiopia	Chemin de Fer Djibouto-Éthiopien	Derailment, sabotage	?
2 Dec 2000	Between Sarai Banjara and Sadhoogarh, Punjab, India	Northern	Derailment, broken rail? Defective vehicle?	45
28 Dec 2000	Near Mieso, Ethiopia	Chemin de Fer Djibouto-Éthiopien	Derailment	13
30 Dec 2000	Blumentritt, The Philippines	Manila Light Rail Transit System	Bomb exploded on train	11
9 Jan 2001	Near Kotri, Pakistan	Pakistan	Derailment, defective vehicle	15
10 Jan 2001	Mvoungouti, Republic of Congo	Chemin de fer Congo – Océan	Runaway train, collision, fire	45
28 Feb 2001	Heck, England	Railtrack London North Eastern	Derailment, obstruction on line, collision	6
2 Mar 2001	Nacavala, Mozambique	Portos e Caminhos de Ferro de Moçambique	Train diverted onto wrong line, head-on collision	8
14 Mar 2001	Guayacanes, Cuba	Ferrocarriles Nacionales de Cuba	Rear collision	5
27 Mar 2001	Pécrot, Belgium	Société nationale des chemins de fer belges	Head-on collision	5
1 Jun 2001	Sindaw Bridge, Myanmar	Myanma	Bridge collapsed under train	?
17 Jun 2001	Near Muzaffargarh, Pakistan	Pakistan	Derailment	5
22 Jun 2001	Bridge No. 924, between Parappanangadi and Kozhikode s, Kerala, India	Southern	Bridge collapsed under train	52
8 Jul 2001	Between Hathras City and Hathras Road, Uttar Pradesh, India	North Eastern Railway	Collision with mini truck on open level crossing	6
10 Aug 2001	Near Maria Teresa, Angola	Caminho de Ferro de Luanda	Train derailed by UNITA guerrillas, fire	422
16 Aug 2001	Near Girdhaur, Bihar, India	Northern	Runaway waggons, collision	10
18 Aug 2001	Between Bujjamuwa and Yattalgoda, Sri Lanka	Sri Lanka	Derailment, excessive speed over new track, 'faulty loading'?	14
2 Sep 2001	Cirebon Kejaksan, West Java, Indonesia	PT Kereta Api Indonesia	Head-on collision	39

Date	Location	Railway Company	Description	Passenger Fatalities
16 Dec 2001	Greenlands, Free State, South Africa	Transnet	Derailment, excessive speed through crossover	7
25 Dec 2001	Ketanggungan Barat, West Java, Indonesia	PT Kereta Api Indonesia	Head-on collision	45
13 Jan 2002	Near Rambukkana, Sri Lanka	Sri Lanka	Runway train, derailment	15
5 Feb 2002	Near Charlottedale, South Africa	Transnet	Rear collision	24
20 Feb 2002	Near Reqa-al-Gharbiya, Egypt	Egyptian State	Fire	361
26 Feb 2002	Wampersdorf, Austria	Österreichische Bundesbahnen	Runaway train, head-on collision	7
27 Feb 2002	Near Godhra Junction, Gujarat, India	Western	Fire, arson	56
16 Mar 2002	Near Qila Sattar Shah, Pakistan	Pakistan	'Buffer stop collision' at end of trap siding	10
10 May 2002	Potters Bar, England	Railtrack London North Eastern	Derailment, defective points	6
12 May 2002	Between Khetasarai and Mehrawan, Uttar Pradesh, India	Northern	Derailment, sabotage	11
25 May 2002	Tenga, Mozambique	Portos e Caminhos de Ferro de Moçambique	Runaway coaches, collision	205
24 Jun 2002	Between Igandu and Msagali, Tanzania	Tanzania	Runaway train, collision	288
16 Jul 2002	Hyderabad, Pakistan	Pakistan	Bomb exploded on train	10
20 Jul 2002	Rometta Marea, Sicily, Italy	Ferrovie dello Stato	Derailment, track defects	7
24 Aug 2002	Near Loulombo, Pool, Republic of Congo	Chemin de fer Congo – Océan	Derailment, sabotage, fire	12
9 Sep 2002	Dhawa Bridge (No. 445), between Rafiganj and Deo Road, Bihar, India	Eastern	Derailment, sabotage, bridge collapsed under train	107
26 Sep 2002	Bridge No. 130-B, between Damboli and Dinghra, Pakistan	Pakistan	Train fell from bridge? Bridge collapsed under train? Sabotage?	8
27 Oct 2002	Azampur, Bangladesh	Bangladesh	Head-on collision	8
6 Nov 2002	Near Nancy, France	Société national des chemins de fer français	Fire	12

Date	Location	Railway Company	Description	Passenger Fatalities
9 Dec 2002	Tersobo, Java, Indonesia	PT Kereta Api Indonesia	Derailment, underbridge struck and damaged by high road vehicle	6
10 Dec 2002	Near Coliseo, Cuba	Ferrocarriles Nacionales de Cuba	Derailment, excessive speed?, defective points?	14
21 Dec 2002	Between Pagidirai and Pendekallu, Andhra Pradesh, India	South Central	Derailment, sabotage	19
3 Jan 2003	Ghatnandur, Maharashtra, India	South Central	Rear collision	18
7 Jan 2003	Between Kraaifontein and Muldersvlei, Western Cape, South Africa	Transnet	Rear collision	10
31 Jan 2003	Near Waterfall, New South Wales, Australia	State Rail Authority	Derailment, excessive speed	6
1 Feb 2003	Between Dete and Mambanje Siding, Zimbabwe	National Railways of Zimbabwe	Head-on collision, single line, fire	45
18 Feb 2003	Jungangno, South Korea	Daegu Metropolitan Transport Corporation	Fire in tunnel, arson	198
1 Mar 2003	Near Alishan, Taiwan	Alishan Forest	Runaway train, derailment, part of train fell from bridge	17
13 Mar 2003	Mulund, Maharashtra, India	Central	Bomb exploded on train	11
15 May 2003	Ludhiana, Punjab, India	Northern	Fire	36
3 Jun 2003	Near Chinchilla, Spain	Red Nacional de los Ferrocarriles Españoles,	Head-on collision, single line, fire	14
22 Jun 2003	Rajapur Road and Vaibhavradi, Maharashtra, India	Konkan	Derailment, train ran into landslide	52
2 Jul 2003	Warangal, Andhra Pradesh, India	South Central	Derailment in sand drag, train fell from bridge	14
3 Sep 2003	Near Kislovodsk, Russia	Rossiyskie zheleznye dorogi	Bomb exploded under train, fire	5
26 Nov 2003	Near Matadi, Democratic Republic of Congo	Société nationale des chemins de fer du Congo	Derailment, train ran into landslide	10
5 Dec 2003	Near Yessentuki, Chechnya, Russia	Rossiyskie zheleznye dorogi	Bomb exploded on train	46
27 Dec 2003	Near Ye, Myanmar	Myanma	Derailment, track defects?	5

Date	Location	Railway Company	Description	Passenger Fatalities
6 Feb 2004	Between Avtozavodskaya and Paveletskaya, Moscow, Russia	Rossiyskie zheleznye dorogi	Bomb exploded on train in tunnel	39
11 Mar 2004	Atocha, Spain	Red Nacional de los Ferrocarriles Españoles	Bomb exploded on train	34
11 Mar 2004	Near Atocha, Spain	Red Nacional de los Ferrocarriles Españoles	Bomb exploded on train	59
11 Mar 2004	El Pozo del Tio Raimundo, Spain	Red Nacional de los Ferrocarriles Españoles	Bomb exploded on train	70
11 Mar 2004	Santa Eugenia, Spain	Red Nacional de los Ferrocarriles Españoles	Bomb exploded on train	17
22 Mar 2004	Between Mindouli and Kingoyi, Republic of Congo	Chemin de fer Congo – Océan	Derailment	31
16 Jun 2004	Between Veer and Karanjadi, Maharashtra, India	Konkan	Derailment, train ran into rockfall, part of train fell from bridge	15
22 Jul 2004	Near Mekece, Turkey	Türkiye Cumhuriyeti Devleyt Demiryollari	Derailment, excessive speed	39
1 Aug 2004	Near Aunglan, Myanmar	Myanma	Derailment, track defects?	7
11 Aug 2004	Tavsancil, Kocaeli, Turkey	Türkiye Cumhuriyeti Devleyt Demiryollari	Head-on collision during temporary single line working?	7
24 Aug 2004	Between Kanbalu and Chatthin, Myanmar	Myanma	Head-on collision, single line? Derailment?	?
6 Nov 2004	Ufton Crossing, England	Network Rail Wales and Western	Collision with car on automatic half barrier level crossing, derailment	5
12 Nov 2004	Near Padre Burgos, the Philippines	Philippines National Railways	Derailment, track defects, slightly excessive speed	13
14 Dec 2004	Between Bhangala and Mirthal, Punjab, India	Northern	Head-on collision, single line, fire	36
26 Dec 2004	Hikkaduwa, Sri Lanka	Sri Lanka	Train derailed by tsunami	1,500
7 Jan 2005	Near Bolognina di Crevalcore, Italy	Ferrovie dello Stato	Head-on collision, single line in fog	13
26 Jan 2005	Glendale, California, United States	Union Pacific	Collision with obstruction on line, collisions, fire	10

Date	Location	Railway Company	Description	Passenger Fatalities
5 Mar 2005	Near Babakwal, Pakistan	Pakistan	Derailment, track defects	6
12 Mar 2005	Near Lăng Cô, Vietnam	Đường sắt Việt Nam	Derailment, excessive speed	13
25 Mar 2005	Near M'Foumbou, Republic of Congo	Chemin de fer Congo – Océan	Collision with tree, derailment	6
21 Apr 2005	Samlaya Junction, Gujarat, India	Western	Train diverted onto wrong line, collision	14
25 Apr 2005	Between Tsukaguchi and Amagasaki, Japan	West Japan	Derailment, excessive speed	106
19 May 2005	Near Bandar Lampung, Indonesia	PT Kereta Api Indonesia	Rear collision	6
21 Jun 2005	Near Kibbutz Revadim, Israel	Israel	Collision with lorry on open level crossing, derailment	6
1 Jul 2005	Near Süveren, Bingol, Turkey	Türkiye Cumhuriyeti Devleyt Demiryollari	Derailment, sabotage	6
7 Jul 2005	Between Liverpool Street and Aldgate, England	Transport for London	Bomb exploded on train	7
7 Jul 2005	Edgware Road, England	Transport for London	Bomb exploded on train	8
7 Jul 2005	Between Russell Kings Cross St Pancras and Russell Square, England	Transport for London	Bomb exploded on train	27
13 Jul 2005	Sarhad, Sindh, Pakistan	Pakistan	Double collision	160
28 Jul 2005	Between Hariharpur and Koiripur, Uttar Ptadesh, India	Northern	Bomb exploded on train	13
31 Jul 2005	Near Shenyang, Liaoning, China	China	Rear collision	5
3 Oct 2005	Datia, Madhya Pradesh, India	North Central	Runaway train, derailment in sand drag	12
29 Oct 2005	Bridge No.61, between Valigonda and Ramannapet, Andhra Pradesh, India	South Central	Bridge washed away, train fell into river	115
9 Nov 2005	Barwadih, Jharkhand, India	East Central	Runaway train, sidelong collision	9
5 Dec 2005	Near Bimlagarh, Orissa, India	Southeastern	Runaway train, derailment	12
25 Dec 2005	Between Sagoshi and Kita-Amarume, Japan	East Japan	Train blown over by the wind	5

ABOUT THE AUTHOR

The author has been interested in railways for many years, ever since he was invited up a signal box at the age of 14 back in 1966. Not long afterwards he 'cabbed' the engine of a goods train that was shunted out of the way of a faster train and its driver showed him a copy of *Red for Danger*, the late L.T.C. Rolt's book about British railway accidents. This was the first railway book that he ever bought, several months' saved-up pocket money going towards a paperback copy of it. Later he came across a copy of *A History of Railroad Accidents, Safety Precautions and Operating Practices* by Robert B. Shaw, dealing with American railroad accidents, and his interest in the subject grew.

When the local signal boxes were all done away with in a modernisation scheme in the mid-1980s he began a more detailed study of world railway accidents. There was no internet back then, of course, and many hours were spent in the local reference library slogging through bound volumes of *The Times*, writing out longhand in large notebooks details of every accident found. Later it was possible to view the paper on microfilm and print pages, which made things a little easier, and with the coming of the internet the amount of information available vastly increased. Some years ago, the author contributed to *Railway Disasters of the World* by P.W.B. Semmens (which deals only with 20th-century accidents) and more recently the idea came of writing his own book or books on world accidents, starting at the beginning.

Made in the USA
Monee, IL
05 November 2026

67988021R10345